Smart Card Handbook

Second Edition

Smart Card Handbook

Second Edition

W. Rankl
W. Effing
Giesecke & Devrient GmbH, Munich, Germany

Translated by
Kenneth Cox
Wassenaar, The Netherlands

First edition translated by
Chanterelle Translations
London, UK

JOHN WILEY & SONS, LTD
Chichester • New York • Weinheim • Brisbane • Singapore • Toronto

First published under the title *Handbuch der Chipkarten* by Carl Hanser Verlag
© Carl Hanser Verlag, Munich/FRG, 1999
All Rights reserved
Authorized translation from the 3rd edition in the original German language
published by Carl Hanser Verlag, Munich/FRG

Copyright © 2000 by John Wiley & Sons, Ltd
               Baffins Lane, Chichester,
               West Sussex, PO 19 1UD, England

               National 01243 779777
               International (+44) 1243 779777
               e-mail (for orders and customer service enquiries): cs-books@wiley.co.uk
               Visit our Home Page on http://www.wiley.co.uk or http://www.wiley.com

Other Wiley Editorial Offices

John Wiley & Sons, Inc., 605 Third Avenue,
New York, NY 10158-0012, USA

Wiley-VCH Verlag GmbH
Pappelallee 3, D-69469 Weinheim, Germany

Jacaranda Wiley Ltd, 33 Park Road, Milton,
Queensland 4064, Australia

John Wiley & Sons (Canada) Ltd, 22 Worcester Road
Rexdale, Ontario, M9W 1L1, Canada

John Wiley & Sons (Asia) Pte Ltd, 2 Clementi Loop #02-01,
Jin Xing Distripark, Singapore 129809

Library of Congress Cataloguing-in-Publication Data
Rankl, W. (Wolfgang)
  [Handbuch der Chipkarten. English]
  Smart Card Handbook/W. Rankl, W. Effing: transalted by Kenneth Cox. – 3rd ed
    p. cm.
  ISBN 0 471 98875 8
  1. Smart cards – Handbooks, manuals, etc. I. Effing, W. (Wolfgang) II. Title.
TK7895.S62 R3613.2000
006 – dc21

                                                 00-043346
British Library Cataloguing in Publication Data
A catalogue record for this book is available from the British Library
ISBN 0 471 98875 8

Produced from Word files supplied by the translator
Printed and bound in Great Britain by Bookcraft (Bath) Ltd
This book is printed on acid-free paper responsibly manufactured from sustainable
forestry, in which at least two trees are planted for each one used for paper production

Contents

Foreword

This foreword is somewhat unusual, as it has not been written for just *any* book.

This book is a compilation of current knowledge – the 'state of the art' – relating to the new technology known as the Smart Card, and in particular to Smart Cards containing microprocessors.

The functions and effects of these new integrated circuit cards will reach far beyond any of the tasks and applications that are currently known, planned or even envisaged. They will affect our behavior as citizens.

I use the term 'citizens' rather than 'people' deliberately, as the future impact of the Smart Card will affect us as members of society and as members of organizations. It will affect us as members of nations and as members of economic groups or communities, and thus as citizens.

Smart Cards should however not become instruments of more or less subtle coercion. They should instead serve us, as citizens, with the richness of their full potential.

If we remain alert, Smart Cards will not become the agents of oppressors, to be used to bring about desired forms of behavior. On the contrary, they are capable of stimulating our spontaneity. In my opinion, the ultimate attraction of the Smart Card lies more in this future role than in its functions as a medium for financial transactions, authorization and identification according to our current ideas – no matter how economically and politically important these functions may be in the future.

You may thus ask along with me, 'Smart Card – *quo vadis*?'

- Will it lead to a not so brave new world in which citizens can be more easily monitored, and thus more easily ruled?
- Or will it perhaps, with the help of well-conceived exploitation of its technical potential, lead individuals to recapture their self-determination and their privacy – or at least to maintain the status quo?

The authors have astutely limited the chapter describing applications to examples, as neither they, nor in all likelihood any one of us, can presently even approximately delimit the full application potential of the Smart Card.

In 1978, the then director of the Department of Patents at the French computer firm Honeywell Bull SA said to me, 'The Smart Card business will one day be as important as the computer business is today.' We can already see how right he was.

The development of a new technology or its techniques always results from steps taken by very many interested parties with widely differing interests. These men and women include basic and applied scientific researchers, engineers, businessmen, 'decision-makers' and many others.

What motivates these people?

Curiosity, the basic drive to find out, is of course one of the incentives: 'Is it possible, can it be done?' Then there is certainly the desire for prestige and – even further away from the research laboratory, the drawing board and the computer – there are ultimately money and power, frequently acting in combination.

However, is not progress, as a form of continuous development, a natural part of life? We can also call it evolution – the trial and error of life. How could we play this game if we were not driven by our curiosity, our desire to find out whether the next trial may prove *not* to be an error?

Considering the complexity and the ever-new, ever-changing interdependencies that progress creates, it is easy to see how the assessment of the consequences of a given technology can become mired in the purely speculative. Such an assessment seeks to chart an unknown future. It starts with the beginnings of a particular development and attempts to plot the manifold possibilities of its ultimate growth into a broad technical discipline. We may well ask whether this is not too difficult a task.

I do not believe that we can overestimate the future importance of the Smart Card, and I thus believe that we probably still cannot fully anticipate its future importance. It is a powerful tool, and it is much more important as a semi-anonymous tool than the PC is or ever will be.

Given that the Smart Card is developing into a new technology in its own right, should we not be asking ourselves whether we are opening a Pandora's box? What might it hold in store – increased monitoring of the individual and greater control by the state?

'The file card is the subtlest tool of terror'. – E. v. Salomon

How much more, then, would the *Smart* Card, in combination with existing mainframe computers and database networks (with their potential for data comparison), be able to contribute to the 'file card' of the present and the future, if – yes, if it were to become the instrument of an impersonal organization, such as the bureaucracy?

If it were however subject to the authority of its owner, it would become a vehicle for the journey to increased independence of the individual.

I believe we are at an important crossroads, one that has already confronted earlier generations of engineers, businessmen and politicians faced with emerging technologies. I would thus like to remind my younger readers in particular of some wise Latin advice: *et respice finem* ('consider where it ends').

Ask yourselves where *your* path will lead you.

The Smart Card, with the help of software, can form the basis of the most different imaginable systems. If it is, in a manner of speaking, intimately matched to its individual possessor, it can – and I repeat – create an enormous potential.

Let us choose the *right* path, the path of independence! Let us meet on the royal road of self-determination (if it is still possible to breach the barricades...).

Do not read this book cynically or indifferently, but rather use it with skeptical optimism. Let us, as citizens, prove ourselves neither incapable nor unworthy of the proffered emancipation.

This is, indeed, an unusual foreword for a technical text. I have enjoyed writing it, because I believe this to be an important book. It contains information that is both accurate and important at this time:

- *Accurate,* because its authors are members of a very select group of people, the core group of pioneers in the development of Smart Card technology. They are familiar with the most intimate secrets of this new technology, and they share them in this text. In earlier times, they might have been regarded as high priests.
- *Important,* because it effectively provides us with the building blocks for the creation of new combinations and configurations.

I wish this book success in establishing itself as the standard work on the subject of Smart Cards, which serves:

- as a textbook for the beginner,
- as an encyclopaedia for the curious layman,
- as a reference book for the professional,
- as a state-of-the-art description for the inventor,
- as a foundation for those seeking to create new systems and applications.

The book deserves such success – just a glance at the index shows the amount of information it contains.

Even the late Helmut Gröttrup, who accompanied me on the early steps of my journey, could have nothing more to add.

We owe the authors and publishers a debt of gratitude.

Jürgen Dethloff
Spring, 1999

Preface to the Second Edition

This is the second English edition of The Smart Card Handbook. It has been considerably extended, compared to the first edition, and it has been updated in many areas to represent the current state of technology.

As with the previous editions, we have described and explained this ever-growing technology in as much detail as possible, according to the motto 'better one sentence too many than one word too few'. There are also more examples, explanatory drawings and photographs, since these are the simplest way to help the reader to understand increasingly complicated interrelationships. A new feature in this edition is hierarchical classification charts, which are intended to make systematic aspects understandable to the reader in graphic form. These additions, extensions and improvements have resulted in a book that is nearly three times as long as the first edition in 1995.

At the start of the 1990s, it was sometimes necessary for people working with Smart Cards to explain to new acquaintances exactly what a Smart Card was. In the meantime, this concept has extended outside of the domain of the specialist, and almost everyone knows what it means. The small, colorful plastic cards with their semiconductor chips are spreading out from their homeland – France and Germany – over the entire world. No other technology will be able to match this triumphal advance in the near future, especially considering that the technology is still in the initial developmental stage and there is no end or consolidation in sight.

While Smart Card technology progresses in leaps and bounds, it is naturally not possible to update *The Smart Card Handbook* at the same rate. It represents the present state of technical knowledge, and if it should happen that certain things come to be seen in a different light at a later date, we can only remark that no-one knows what the future will bring. In spite of this, or perhaps just for this reason, we welcome every comment, suggestion and proposed improvement, so that this book can continue to cover the subject of Smart Cards as completely as possible. In any case, an errata document will be made available on the web server of John Wiley & Sons [Wiley], along with any necessary addenda to the book.

Munich, February 1999

Wolfgang Rankl
[Rankl@gmx.net], [Rankl]

Wolfgang Effing
[Effing_Wolfgang@compuserve.com]

Symbols and Notation

General

- Commands used with Smart Cards are printed in upper-case characters (for example: SELECT FILE).
- The least-significant bit is designated '1' rather than '0', according to the ISO standards.
- Length specifications for data, objects and all countable quantities are shown in decimal form, in agreement with the usual practice in Smart Card standards. All other values are shown as hexadecimal numbers and identified as such.
- The prefixes 'kilo' and 'mega' have the values of 1024 (2^{10}) and 1,048,576 (2^{20}), respectively, as is customary in the field of information technology.

Representation of symbols and numbers

"ABC"	ASCII value
'00'	hexadecimal value
°0°, °1°	binary values
42	decimal value
Bn	byte number n (for example: B1)
bn	bit number n (for example: b2)
Dn	digit number n (for example: D3)

Logical functions

$\|$	chaining (of data elements or objects)
$\oplus$	logical XOR operation
$\wedge$	logical AND operation
$\vee$	logical OR operation
$a \in M$	a is an element of the set M
$a \notin M$	a is not an element of the set M
$\{a, b, c\}$	the set of elements a, b, c

Cryptographic functions

enc $_{X\,n}$ (K; D)   encryption using the algorithm X and an n-bit key,
                                      with the key K and the data D
                                      (for example: **enc** $_{DES\,56}$ ('1 ... 0'; 42)).

dec $_{X\,n}$ (K; D)   decryption using the algorithm X and an n-bit key,
                                      with the key K and the data D
                                      (for example: **dec** $_{IDEA\,128}$ ('1 ... 0'; 42).

S := **sign** $_{X\,n}$ (K; D)   generating the signature S using the algorithm X and
                                      an n-bit key, with the key K and the data D
                                      (for example: **sign** $_{RSA\,512}$ ('1 ... 0'; "Wolf")).

R := **verify** $_{X\,n}$ (K; S)   verifying the signature S using the algorithm X and
                                      an n-bit key, with the key K
                                      (for example: **verify** $_{RSA\,512}$ ('1 ... 9'; 42)).

Result = OK/NOK

References

See: '...'   This is a cross-reference to another location in the book.

[...]   This is a reference to a World Wide Web site listed in the
                                     Appendix.

[Xxx NN]   This is a cross reference to additional literature or standards
                                     listed in the Appendix. The format is:
                                     Xxx ∈ {surname of the first-named author} and
                                     NN ∈ {last two digits of the year of publication}.

Program Code Conventions

The syntax and semantics of the program code used in this book are based on those of the standard BASIC dialects. However, the use of explanations in natural language within a code listing is allowed, in order to promote the reader's comprehension of the code. Naturally, although this makes the programs easier to understand, it means that it is not possible to automatically convert the code into machine code. This compromise is justified by the significant improvement in readability that it provides.

:=	assignment operator
=, <, >, <>, <=, >=	comparison operators
+, -, *, /	arithmetic operators
NOT	logical not
AND	logical and
OR	logical or
||	chaining operator (for example: coupling two byte strings)
_	end-of-line marker for multiline instructions
// ...	comment
IO_Buffer	variable (printed in italics)
Label:	jump or call location (printed in bold)
GOTO ...	jump
CALL ...	function call (subroutine call)
RETURN	return from a function (subroutine)
IF ... THEN ...	decision, type 1
IF ... THEN ... ELSE ...	decision, type 2
EXIST	test for presence (for example: an object or data element)
LENGTH (...)	calculate the length
SEARCH (...)	search in a list; search string in parentheses
STATUS	query the result of a previously executed function call
WITH ...	starts the definition of a variable or object as a reference
END WITH	ends the definition of a variable or object as a reference

Acronyms

3DES	triple DES (see glossary)
A3, A5, A8	GSM algorithms 3, 5 and 8
ABA	American Bankers Association
ABS	acrylonitrile butadiene styrene
ACK	acknowledge
ACD	access control descriptor
ADN	abbreviated dialing number
AFNOR	Association Française de Normalisation (see glossary)
AGE	Autobahngebührenerfassung [motorway toll collection]
AGE	automatische Gebührenerfassung [automatic toll collection]
AID	application identifier (see glossary)
Amd	amendment
AND	logical AND operation
ANSI	American National Standards Institute (see glossary)
APACS	Association for Payment Clearing Services
APDU	application protocol data unit (see glossary)
A-PET	amorphous polyethylene terephthalate
API	application programming interface (see glossary)
ARM	advanced RISC machine
ASC	application-specific command
ASCII	American Standard Code for Information Interchange
ASIC	application-specific integrated circuit
ASK	amplitude-shift keying
ASN.1	abstract syntax notation one (see glossary)
ATM	asynchronous transfer mode
ATM	automated teller machine
ATR	answer to reset (see glossary)
BASIC	Beginners All-purpose Symbolic Instruction Code
BCD	binary coded digit
Bellcore	Bell Communications Research Laboratories
BER	basic encoding rules
BER-TLV	basic encoding rules – tag, length, value
BEZ	Börsenevidenzzentrale [Geldkarte electronic purse clearing center]
BGT	block guard time
BIN	bank identification number
bit	binary digit
BS	base station
BWT	block waiting time

CA	certification authority
CAD	chip accepting device
CAFE	Conditional Access for Europe (EU project)
CAP	card application
C-APDU	command APDU (see glossary)
CAPI	crypto-API (application programming interface)
CASCADE	Chip Architecture for Smart Card and portable intelligent Devices
CASE	computer-aided software engineering
CBC	cipher block chaining
CC	Common Criteria
CCD	card coupling device
CCD	charge-coupled device
CCITT	Comité Consultatif International Télégraphique et Téléphonique (now ITU) (see glossary)
CCR	chip card reader
CCS	cryptographic checksum (see glossary)
CD	committee draft
CDM	card dispensing machine
CEN	Comité Européen de Normalisation (see glossary)
CENELEC	Comité Européen de Normalisation Eléctrotechnique [European Committee for Electronics Standardization]
CEPT	Conférence Européenne des Postes et Télécommunications (see glossary)
CFB	cipher feedback
CHV	card holder verification
CICC	contactless integrated-circuit card
CISC	complex instruction set computer
CLA	class
CLK	clock
CMOS	complementary metal oxide semiconductor
COS	chip operating system (see glossary)
CRC	cyclic redundancy check (see glossary)
CRCF	clock rate conversion factor
CRT	Chinese remainder theorem
Cryptoki	cryptographic token interface
C-SET	chip SET (secure electronic transaction)
CT	card terminal
CVM	cardholder verification method
CWT	character waiting time
DAD	destination address
DAM	draft amendment
DB	database
DBF	database file
DCS	digital cellular system
DEA	data encryption algorithm

DECT	digital enhanced cordless telecommunications
	(formerly: digital European cordless telecommunications)
DES	data encryption standard
DER	distinguished encoding rules
DF	dedicated file (frequently also: directory file) (see glossary)
DFA	differential fault analysis (see glossary)
DFÜ	Datenfernübertragung [long-distance data communication]
DIL	dual in-line
DIN	Deutsche Industrienorm [German industrial standard]
DIS	draft international standard
DO	data object
DoD	US Department of Defense
DOV	data over voice
DPA	differential power analysis
dpi	dots per inch
DRAM	dynamic random-access memory (see glossary)
DSA	digital signature algorithm
DTAUS	Datenträgeraustausch [data medium exchange]
DVD	digital versatile disc
DVS	Dateiverwaltungssystem [file management system]
EBCDIC	extended binary-coded decimal interchange code
EC	elliptic curve
ec	Eurocheque
ECB	electronic code book
ECC	elliptic curve cryptosystem
ECC	error correction code (see glossary)
ECTEL	European Telecom Equipment and Systems Industry
EDC	error detection code (see glossary)
EDI	electronic data interchange
EDIFACT	Electronic Data Interchange for Administration, Commerce
	and Transport
EEPROM, E^{2}PROM	electrically erasable programmable read-only memory
	(see glossary)
EF	elementary file (see glossary)
EFF	Electronic Frontier Foundation
EFTPOS	electronic fund transfer at point of sale
EMV	Europay, MasterCard, Visa (see glossary)
EPROM	erasable programmable read-only memory (see glossary)
ESD	electrostatic discharge
ESPRIT	European Strategic Programme of Research and Development
	in Information Technology (EU project)
ETS	European Telecommunication Standard (see glossary)
ETSI	European Telecommunications Standards Institute
	(see glossary)
etu	elementary time unit (see glossary)

f	following page
FAQ	frequently asked questions
FAR	false-acceptance rate
FBZ	Fehlbedienungszähler
	[key fault presentation counter, retry counter]
FCB	file control block
FCFS	first come, first serve
FDN	fixed dialing number
FEAL	fast data encipherment algorithm
FET	field effect transistor
ff	following pages
FID	file identifier (see glossary)
FIFO	first in, first out
FIPS	Federal Information Processing Standard (see glossary)
FPGA	field programmable gate array
FRAM	ferroelectric random-access memory
FRR	false-rejection rate
FSK	frequency shift keying
FTAM	file transfer, access and management
GND	ground (earth)
GPS	global positioning system
GSM	Global System for Mobile Communications,
	formerly: Groupe Special Mobile (see glossary)
GUI	graphical user interface
GTS	GSM Technical Specification
HiCo	high-coercivity
HSM	hardware security module
HSM	high-security module
HSM	host security module
HTML	hypertext markup language
HTTP	hypertext transfer protocol
HV	Härte Vickers
HW	hardware
I/O	input/output
I^2C	inter-integrated circuit
IATA	International Air Transport Association
IBAN	international bank account number
ICC	integrated circuit card
ID	identifier
IDEA	international data encryption algorithm
IEC	International Electrotechnical Commission
IEEE	Institute of Electrical and Electronics Engineers
IEP	inter-sector electronic purse
IFD	interface device
IFS	information field size

IFSC	information field size for the card
IFSD	information field size for the interface device
IIC	institution identification codes
IMEI	international mobile equipment identity
IMSI	international mobile subscriber identity
INS	instruction
INTAMIC	International Association of Microcircuit Cards
IP	Internet Protocol
IPES	improved proposed encryption standard
IrDA	Infrared Data Association
ISDN	integrated services digital network
ISF	internal secret file
ISO	International Organization for Standardization (see glossary)
IT	information technology
ITSEC	Information Technology Security Evaluation Criteria (see glossary)
ITU	International Telecommunications Union (see glossary)
IuKDG	Informations- und Kommunikations-Gesetz [German information and communications law]
IV	initialization vector
IVU	in-vehicle unit
JCF	Java Card Forum
JECF	Java electronic commerce framework
JIT	just in time
JOS	Java operating system
JTC1	Joint Technical Committee One
JVM	Java virtual machine
K	key
KD	derived key
KFPC	key fault presentation counter
KID	key identifier
KM	key master
KS	key session
KVK	Krankenversichertenkarte [medical insurance card]
LAN	local-area network
Lc	command length
lcd	largest common denominator
Le	expected length
LEN	length
LIFO	last in, first out
LND	last number dialed
LOC	lines of code
LoCo	low-coercivity
LRC	longitudinal redundancy check
LSAM	load secure application module

lsb	least-significant bit
LSB	least-significant byte
LFSR	linear feedback shift register
M	month
MAC	message authentication code (see glossary)
ME	mobile equipment
MF	master file (see glossary)
MFC	multifunctional smart card
MIME	multipurpose Internet mail extensions
MIPS	million instructions per second
MCT	multifunctional card terminal
MLI	multiple laser image
MM	moduliertes Merkmal [modulated marker]
MMI	man machine interface
MMU	memory management unit
MOO	mode of operation
MOSAIC	microchip on surface and in card
MOSFET	metal oxide semiconductor field effect transistor
MoU	memorandum of understanding (see glossary)
MS	mobile station
msb	most-significant bit
MSB	most-significant byte
MTBF	mean time between failures
NAD	node address
NBS	US National Bureau of Standards (see glossary)
NCSC	National Computer Security Center (see glossary)
NDA	non-disclosure agreement
NIST	US National Institute of Standards and Technology (see glossary)
nok	not OK
NPU	numeric processing unit
NRZ	non return to zero
NSA	US National Security Agency (see glossary)
NU	not used
OBU	on-board unit
OFB	output feedback
OR	logical OR operation
OS	operating system
OSI	Open Systems Interconnections
OTA	Open Terminal Architecture
OTA	over the air
OTASS	over the air SIM services
OTP	one-time password; one-time programmable; open trading protocol
OVI	optically variable ink

P1, P2, P3	parameters 1, 2 and 3
PA	power analysis
PB	procedure byte
PC	personal computer
PC	polycarbonate
PC/SC	personal computer/Smart Card
PCB	protocol control byte
PCD	proximity coupling device
PCMCIA	Personal Computer Memory Card International Association
PCN	personal communication networks
PCS	personal communication system
PDA	personal digital assistant
PES	proposed encryption standard
PET	polyethylene terephthalate
PETP	partially crystalline polyethylene terephthalate
PGP	Pretty Good Privacy
PICC	proximity ICC
PIN	personal identification number
PIX	proprietary application identifier extension
PKCS	public key cryptography standards (see glossary)
PKI	public key infrastructure
PLL	phase-locked loop
POS	point of sale
POZ	POS ohne Zahlungsgarantie [POS without payment guarantee]
PPC	production planning and control
PPM	pulse position modulation
PPS	protocol parameter selection
prEN	pre Norme Européenne
prETS	pre European Telecommunication Standard
PROM	programmable read-only memory
PSAM	purchase-secure application module
PSK	phase-shift keying
PTS	protocol type selection
PTT	Postes Télégraphes et Téléphones [post, telegraph and telephone]
Pub	publication
PUK	personal unblocking key
PVC	polyvinyl chloride
PWM	pulse width modulation
R-APDU	response APDU (see glossary)
RAM	random-access memory (see glossary)
REJ	reject
RES	resynchronization
RF	radio frequency
RF-ID	radio-frequency identification
RFC	request for comment

RFID	radio-frequency identification
RFU	reserved for future use
RID	record identifier
RID	registered application provider identifier
RIPE	RACE Integrity Primitives Evaluation (EU project)
RIPE-MD	RACE Integrity Primitives Evaluation message digest
RISC	reduced instruction set computer
RND	random number
ROM	read-only memory (see glossary)
RSA	Rivest, Shamir and Adleman encryption algorithm
SAD	source address
SAM	secure application module
SC	Smart Card
SCC	Smart Card controller
SCQL	structured card query language
SDL	specification and description language
SEMPER	Secure Electronic Marketplace for Europe (EU project)
SEPP	secure electronic payment protocol
SET	secure electronic transaction
SFI	short file identifier (see glossary)
S-HTTP	secure hypertext transfer protocol
SigG	Signaturgesetz [signature law] (see glossary)
SigV	Signaturverordnung [signature decree] (see glossary)
SIM	subscriber identity module (see glossary)
SIMEG	subscriber identity module expert group (see glossary)
SM	secure messaging
SMD	surface-mounted device (see glossary)
SMG9	Special Mobile Group 9 (see glossary)
SMIME	secure multipurpose Internet mail extensions
SMS	short messages service
SPA	simple power analysis
SQL	Structured Query Language
SQUID	superconducting quantum interference device
SRAM	static random-access memory (see glossary)
SSC	send sequence counter
SSL	secure socket layer
STARCOS	Smart Card Chip Operating System (product of G&D)
STC	sub-technical committee
STT	secure transaction technology
SVC	stored value card (product of Visa International)
SW	software
SW1, SW2	status word 1, 2
SWIFT	Society for Worldwide Interbank Financial Telecommunications
T	tag
TAB	tape-automated bonding

TAL	terminal application layer
TAN	transaction number (see glossary)
tbd	to be defined
TC	technical committee
TC	thermochrome
TCOS	Telesec Card Operating System
TCP	transport control protocol
TCSEC	Trusted Computer System Evaluation Criteria (see glossary)
TDES	triple DES (see glossary)
TLV	tag, length, value (see glossary)
TMSI	temporary mobile subscriber identity
TOE	target of evaluation
TPDU	transmission protocol data unit (see glossary)
TS	technical specification
TTCN	the tree and tabular combined notation
TTL	terminal transport layer
TTL	transistor transistor logic
UART	universal asynchronous receiver transmitter
UIM	user identity module (see glossary)
UML	unified modeling language
UMTS	Universal Mobile Telecommunication System
URL	uniform resource locator
USB	Universal Serial Bus
USIM	universal subscriber identity module
VAS	value-added services
Vcc	power supply voltage
VCD	vicinity coupling device
VEE	Visa Easy Entry (see glossary)
VKNR	Versichertenkartennummer [insurance card number]
VLSI	very large-scale integration
VM	virtual machine
Vpp	programming voltage
W3C	World Wide Web Consortium
WAN	wide-area network
WG	working group
WORM	write once, read multiple
WWW	World Wide Web (see glossary)
XOR	logical exclusive-OR operation
Y	year
ZKA	Zentraler Kreditausschuß [coordinating committee of the major German banks]

1

Introduction

This book has been written for students, engineers and the technically minded who would like to find out more about Smart Cards. It attempts to cover as much of this wide-ranging topic as possible, in order to provide the reader with an overview of the fundamentals of the field and the current state of its technology.

We have put great emphasis on the practical approach. The wealth of pictures, tables and references to real applications is designed to help the reader become familiar with the subject rather faster than would be possible with a strictly scientific presentation. Hence, the book makes no claims to be scientifically complete, but rather to be useful in practice. This is also the reason why the explanations have been kept as concrete as possible. At many points, we had to choose between scientific accuracy and ease of understanding, and we have tried to strike a happy medium between the two. Where this proved impossible, we have always preferred comprehensibility.

The book is designed so that it can be read in the usual way, from front to back. We have tried as much as possible to avoid forward references. The designs of the individual chapters, in terms of structure and content, allow them to be read separately without any loss of understanding. The comprehensive index and the glossary allow this book to be used as a reference work. If you want to know more about a specific topic, the cross-references in the text and the annotated index of standards will help you find the relevant text.

At this point, we must mention one small proviso. This book deals mainly with microprocessor cards in the credit card format. As far as space permits, we have also described memory cards and card terminals.

A large number of acronyms have unfortunately become entrenched in the field of Smart Card technology, as they have in so many other areas of technology and in everyday life. This makes it particularly hard for the newcomer to gain familiarity with the subject. We have tried to minimize the use of these cryptic and often illogical abbreviations. Nevertheless, in many cases we have had to choose a middle way between the internationally accepted Smart Card jargon used by specialists and the vernacular terms that are more easily understood by the layman. If we have not always been successful, at least the very comprehensive list of acronyms at the front of the book should help to overcome any barriers to comprehension, which we hope will be short-lived. An extensive glossary at the end of the book explains the most important technical concepts and supplements the list of acronyms.

Active learning is indisputably much more effective and interesting than passive learning. Thus, for example, you can learn a new language more quickly and easily by spending some time in a country in which the language is spoken. Our Smart Card simulation program for PCs, which has become well known since its introduction with the second edition, will be available shortly in a new and improved version. It is now 'freeware' and can be found on the Internet site of the publisher [Wiley]. Good luck and good learning!

1.1 The History of Smart Cards

The proliferation of plastic cards started in the USA in the early 1950s. The low price of the synthetic material PVC allowed robust, long-lasting cards to be produced. These were much more suitable for use in everyday life than the paper and cardboard cards previously used, which could not adequately withstand mechanical stress and climatic effects.

The first all-plastic payment card for general use was issued by the Diners Club in 1950. It was intended for an exclusive class of individual, and thus also served as a status symbol, allowing the holder to pay with his or her 'good name' rather than with cash. Acceptance of these cards was initially restricted to the more select restaurants and hotels, which led to this type of card being referred to as a 'travel and entertainment' card.

The entry of Visa and Mastercard into the field led to a very rapid proliferation of plastic money, at first in the USA, with Europe and the rest of the world following a few years later.

Today, credit cards allow the traveler to shop without cash anywhere in the world. The cardholder is never without means of payment, yet he or she avoids exposure to the risk of loss through theft or other hazards that are difficult to anticipate, particularly whilst travelling. Using a credit card does away with the tedious task of exchanging currency when travelling abroad. These unique advantages have helped credit cards to become rapidly established worldwide. Many hundreds of millions of cards are produced and issued annually.

At first, the cards' functions were quite simple. They initially served as data carriers that were secure against forgery and tampering. General information, such as the card issuer's name, was printed on the surface, while personal data elements, such as the cardholder's name and the card number, were embossed. Furthermore, many cards had a signature field, in which the cardholder could sign his or her name for reference. In these first-generation cards, protection against forgery was provided by visual features, such as security printing and the signature field. Consequently, the system's security depended quite fundamentally on the quality and conscientiousness of the retail staff accepting the cards. However, this was not an overwhelming problem, due to the card's initial exclusivity. With increasing proliferation in card use, these rather basic features no longer proved sufficient, all the more so since threats from organized crime were growing apace.

Increasing handling costs for the merchants and banks made a machine-readable card necessary, while at the same time the card issuers' losses due to customer insolvency and fraud grew from year to year. It became apparent that both the security measures against fraud and tampering, as well as the basic functions of the cards, had to be extended and improved.

The first improvement consisted of a magnetic strip on the back of the card. This allowed digital data to be stored on the card in machine-readable form, as a supplement to the visual data. This made it possible to minimize the use of paper receipts, which were previously essential. However, the customer's signature on a paper receipt, as a form of personal identification, still remains a requirement in classical credit card applications.

New applications can however be devised in which paper receipts are unnecessary. This would allow a long-standing goal to finally be achieved, which is the replacement of paper-based transactions by electronic data processing. Identification of the card user, which up to now has been accomplished using signatures, must be accomplished in a different manner to make this possible. The use of a secret personal identification number (PIN) that is compared to a reference number has become generally accepted. The reader is surely familiar with this technique, which is used for bank machines (automated teller machines).

The embossed card with a magnetic strip is still the most commonly used type of payment card. Magnetic strip technology suffers from a crucial weakness, however, in that the data stored on the strip can be read, deleted and rewritten at will by anyone with access to the appropriate equipment. It is thus unsuitable for the storage of confidential data. Additional techniques must be used to ensure confidentiality and to protect against tampering. For example, the reference value for the PIN can be stored either in the terminal or in the host system in a secure environment, instead of on the magnetic strip. Most systems that employ magnetic-strip cards thus have on-line connections to the system's host computer for security reasons. However, this generates considerable data transmission costs. In order to reduce costs, solutions must be sought that allow card transactions to be executed off-line without putting the system's security at risk.

The development of the Smart Card, combined with the expansion of electronic data processing, has created completely new possibilities for solving this problem. Enormous progress in microelectronics in the 1970s made it possible to integrate data storage and arithmetic logic on a single silicon chip measuring a few square millimeters. The idea of incorporating such an integrated circuit into an identification card was contained in a patent application filed by the German inventors Jürgen Dethloff and Helmut Grötrupp as early as 1968. This was followed in 1970 by a similar patent application, made by Kunitaka Arimura in Japan. However, the first real progress in the development of Smart Cards came when Roland Moreno registered his Smart Card patents in France in 1974. It was only then that the semiconductor industry was able to supply the required integrated circuits at acceptable prices. Nevertheless, many technical problems still had to be solved before the first prototypes, some of which contained several integrated circuit chips, could be transformed into reliable products that could be manufactured in large numbers with adequate quality and reasonable prices. Since the basic inventions in Smart Card technology come out of Germany and France, it is not surprising that these countries played the leading role in the development and marketing of Smart Cards.

The great breakthrough was achieved in 1984, when the French PTT (postal and telecommunications services) successfully carried out a field trial with telephone cards. In this field trial, the Smart Cards immediately proved to meet all expectations with regard to protection against tampering and high reliability. Significantly, the breakthrough for Smart Cards did not come in an area where traditional cards were already used, but in a new application. Introducing a new technology in a new application has the great advantage that compatibility with existing systems does not have to be taken into account, so that the capabilities of the new technology can be fully exploited.

A pilot project was conducted in Germany in 1984–85, using telephone cards based on a variety of technologies. Magnetic-strip cards, optical-storage (holographic) cards and Smart Cards were used in comparative tests. The Smart Card proved to be the winner in this pilot study. In addition to a high degree of reliability and security against tampering, Smart Card technology promised the greatest flexibility in future applications. While the older but less expensive EPROM technology was used in the French telephone card chips, newer EEPROM chips were used from the start in the German telephone cards. The latter type of chip does not need a programming voltage supply. However, the unfortunate consequence is that the French and German telephone cards are mutually incompatible. It appears that even after the introduction of the Euro, French and German telephone cards will for a short time still not be usable in each other's countries of origin.

Further developments followed the successful trials of telephone cards, first in France and then in Germany, with breathtaking speed. By 1986, several million 'smart' telephone cards were in circulation in France alone. The total number reached nearly 60 million in 1990 and several hundred million worldwide in 1997. Germany experienced a similar development, with a time lag of about three years. These systems were marketed throughout the world after the successful introduction of the public Smart Card in France and Germany. Telephone cards incorporating chips are currently used in over 50 countries.

Progress was significantly slower in the field of bank cards, which is partly due to their greater complexity in comparison to telephone cards. These differences are described in detail in the following chapters. Here we would just like to remark that the development of modern cryptography has been just as critical for the proliferation of bank cards as the developments in semiconductor technology.

With the general expansion of electronic data processing in the 1960s, the field of cryptography experienced a sort of quantum leap. Modern hardware and software permitted the implementation of complex and demanding mathematical algorithms, which made possible a degree of security unparalleled until then. Moreover, this new technology was available to everyone, whereas cryptography had previously been a covert science in the private reserve of the military and secret services.

With these modern cryptographic procedures, the strength of the security mechanisms in electronic data processing systems could be mathematically calculated. It was no longer necessary to rely on the very subjective assessment of conventional techniques, whose security essentially rested on the secrecy of the procedures used.

The Smart Card proved to be an ideal medium. It made a high level of security (based on cryptography) available to everyone, since it can safely store secret keys and also execute cryptographic algorithms. In addition, Smart Cards are so small and easy to handle that they can be carried and used everywhere by everyone in everyday life. It was a natural idea to try to employ these new security features for bank cards, in order to come to grips with the security risks associated with the increasing utilization of magnetic-strip cards.

The French banks were the first to introduce this fascinating technology in 1984, following a trial with 60,000 cards in 1982–83. It took another 10 years before all French bank cards incorporated chips. In Germany, the first field trials took place in 1984–85 with a multifunctional payment card incorporating a chip. However, the *Zentrale Kreditausschuß* (ZKA), which is a committee of the leading German banks, did not manage to issue a specification for multifunctional Eurocheque cards incorporating chips until 1996. In 1997, all German savings associations and many banks issued the new Smart Cards. In the previous year, multifunctional Smart Cards with POS functions, an electronic purse and optional additional applications were issued in all of Austria. This made Austria the first country in the world to have a nationwide electronic purse system.

An important milestone for the future worldwide use of Smart Cards for financial transactions was the completion of the EMV specification, which was a product of the joint effort of Europay, Mastercard and Visa. The first version of this specification was published in 1994. It contained detailed descriptions of credit cards incorporating microprocessor chips, and it guaranteed the mutual compatibility of the future Smart Cards of the three largest credit card organizations.

Electronic purse systems have proven to be an additional drawing card for the international use of Smart Cards for financial transactions. The first such system, called Danmønt, was put into operation in Denmark in 1992. There are currently more than 20 national systems in use in Europe alone, many of which are based on the preliminary European standard prEN 1546. The use of such systems is also increasing outside of Europe. Even in the USA, where Smart Card systems have hardly taken root up to now, a Smart Card purse system was tried out by Visa during the 1996 Olympic Summer Games in Atlanta. Payments via the Internet offer a new and promising area for the use of electronic purses. However, the problems associated with making small payments securely but anonymously throughout the world via the public Internet have not yet been solved in a satisfactory manner. Smart Cards could play a decisive role in the solution of these problems.

Yet another application has meant that almost every German citizen these days owns a Smart Card. When health insurance cards incorporating chips were introduced, more than 70 million Smart Cards were issued to all persons covered by the national health insurance plan.

The Smart Card's high degree of functional flexibility, which even allows a card already in service to be reprogrammed for new applications, has opened up completely new areas of use that extend beyond traditional card applications.

1.2   Application Areas

As can be seen from the historical summary, possible applications for the Smart Card are extremely diverse. They are also constantly being extended with the increasing arithmetic power and storage capacity of available integrated circuits. Within the confines of this book, it is impossible to describe all these applications in detail. Instead, a few typical examples must serve to illustrate the Smart Card's fundamental properties. This introductory chapter is only meant to provide an initial overview of the card's functional flexibility. Chapter 13 describes several sample applications in detail.

To make this summary easier to follow, it is helpful to divide Smart Cards into two categories: memory cards and microprocessor cards.

1.2.1   Memory cards

The first Smart Cards used in large quantities were memory cards for telephone applications. These cards are prepaid, and the value stored electronically in the chip is decreased by the amount of the charge each time it was used. Naturally, it is necessary to prevent the user from subsequently increasing the stored value, which could easily be done with a magnetic-strip card. All the user would have to do is note the data stored at the time of purchase and rewrite them to the magnetic strip after using the card. The card would then have its original value and could be reused. This type of tampering, known as 'buffering', is prevented in Smart Cards by security logic in the chip. This makes it impossible to erase a memory cell once it has been written. The reduction in the card's value by the number of charge units used is thus irreversible.

This type of Smart Card can naturally be used not only for telephone calls, but also whenever goods or services are to be sold against prior payment without the use of cash. Examples of possible uses include local public transport, all kinds of vending machines, cafeterias, swimming pools, car parks and so on. The advantage of this type of card lies in its simple technology (the chip's surface area is typically only a few square millimeters), and hence its low cost. The disadvantage is that the card cannot be reused once it is empty, but must be discarded as waste unless it ends up in a card collection.

Another typical application of memory cards is the health insurance card, which has been issued in Germany since 1994 to all persons covered by the national health insurance plan. The information that previously was written on the patient's card is now stored in the chip and also printed or laser-engraved on the card. Using a chip for storage makes the cards machine-readable using simple equipment.

In summary, memory-type Smart Cards have limited functionality. Their integrated security logic makes it possible to protect stored data against tampering. They can be used as cash-equivalent cards or identification cards in systems where low cost is a primary consideration.

1.2.2   Microprocessor cards

As already mentioned, the first application of microprocessor cards was in the form of French bank cards. Their ability to store private keys securely and to execute modern cryptographic algorithms made it possible to implement highly secure off-line payment systems.

Since the microprocessor built into the card is freely programmable, the functionality of microprocessor cards is restricted only by the available storage space and the capacity of the arithmetic unit. The only limits to one's imagination when implementing Smart Card systems are thus technological ones, and these are extended enormously with each new generation of integrated circuits.

Following a drastic reduction in the cost of Smart Cards in the early 1990s, as a result of mass production, new applications have been introduced year after year. The use of Smart Cards with cellular telephones has been especially important for their international proliferation.

After being successfully tested in the German national C network (analogue cellular telephone network) for use in portable end-user equipment, the Smart Card was prescribed as the access medium for the European digital cellular phone system (GSM). On the one hand, the Smart Card enabled a high degree of security to be achieved for accessing the cellular telephone network. At the same time, it offered new possibilities and thus big advantages in the marketing of cellular telephones, since it offered network operators and service providers a way to separate the sale of the telephone from the sale of services. Without the Smart Card, cellular telephones would certainly not have spread across Europe so quickly or have developed into a worldwide standard.

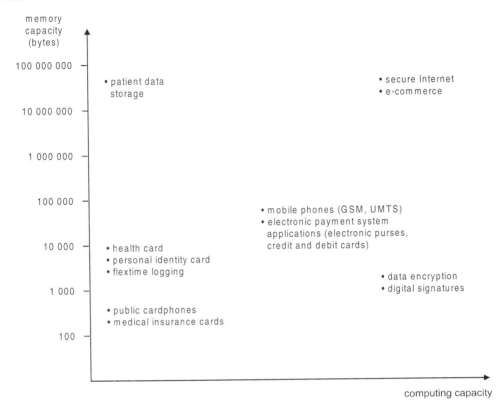

memory
capacity
(bytes)

100 000 000 —

     • patient data   • secure Internet
       storage   • e-commerce

10 000 000 —

1 000 000 —

100 000 —

                                 • mobile phones (GSM, UMTS)
                                 • electronic payment system
                                  applications (electronic purses,

10 000 — • health card   credit and debit cards)
       • personal identity card
       • flextime logging   • data encryption
                                         • digital signatures

1 000 —

       • public cardphones
       • medical insurance cards

100 —

                                                          computing capacity

Figure 1.1   Typical Smart Card application areas, showing the required storage capacity and
arithmetic processing capacity

Possible applications for microprocessor cards include identification cards, access
control for restricted areas and computers, protected data storage, electronic signatures and
electronic purses, as well as multifunctional cards incorporating several applications in a
single card. Presently, Smart Card operating systems are being developed that allow new
applications to be loaded into a card after it has been issued to the user, without
compromising the security of the various applications. Completely new application areas
are made possible with this new flexibility. For example, personal security modules are
indispensable if Internet trade and payments are to be made trustworthy. Such security
modules would securely store personal keys and could execute high-performance
cryptographic algorithms. These tasks can be performed in an elegant manner by a
microprocessor with a cryptographic coprocessor. Specifications for secure applications of
Smart Cards with the Internet are currently being developed throughout the world. Within a
few years, we can expect to see every PC equipped with a Smart Card interface.

In summary, the essential advantages of microprocessor cards are their high storage
capacity, their ability to securely store confidential data and their ability to execute
cryptographic algorithms. These advantages make a wide range of new applications
possible, in addition to the card's traditional use as a bank card. The Smart Card's potential
is by no means yet exhausted, and in fact, it is constantly being extended by progress in
semiconductor technology.

1.2.3   Contactless cards

Contactless cards, in which energy and data are transferred without any electrical contact between the card and the terminal, have achieved the status of commercial products in the last few years. Presently, both memory cards and microprocessor cards are available as contactless cards. Although contactless microprocessor cards can usually work at a distance of only a few centimeters away from the terminal, contactless memory cards can be used up to a meter away from the terminal. This means that such cards do not necessarily have to be held in the user's hand during use, but can remain in the user's purse or briefcase, for example.

These cards are thus particularly suitable for applications in which persons or objects should be quickly identified. Sample applications are:

- access control,
- local public transportation,
- ski passes,
- airline tickets,
- baggage identification.

There are, however, applications in which operation over a long distance could cause problems, and thus should be prevented. A typical example is an electronic purse. A declaration of intention on the part of the card holder is normally required to complete a financial transaction. This confirms the amount of the payment and the agreement of the cardholder to pay. With a contactless card, this declaration takes the form of inserting the card in the terminal and confirming the indicated amount. If contactless financial transactions over relatively long distances were possible, a 'con artist' could transfer money out of the electronic purse without the knowledge of the card holder. Dual-interface cards (sometimes called 'combicards') offer a possible solution to this problem. These cards unite contact and contactless interfaces in a single card. Such a card can communicate with the terminal via either its contact interface or its contactless interface, as desired.

There is great interest in contactless cards in the field of local public transportation. If the present Smart Cards used for financial transactions, which as a rule are contact-type cards, were to have their functionality extended to include acting as electronic tickets with contactless interfaces, then transportation agencies could make use of the existing infrastructure and cards of the credit card industry.

1.3   Standardization

The prerequisite for the worldwide extension of Smart Cards into everyday life, similar to their current use in Germany in the form of telephone cards, health insurance cards and bank cards, has been the creation of national and international standards. Due to the special significance of such standards, in this book we repeatedly refer to existing standards and those that are in preparation. Why are standards so important for the expansion in the use of Smart Cards?

A Smart Card is normally one component of a complex system. This means that the interfaces between the card and the rest of the system must be exactly specified and matched to teach other. Of course, it is possible to do this for every system on a case-by-case basis, without regard to other systems. This would however mean that a different Smart Card would be needed for each different system. The user would thus have to carry a separate card for every application. In order to avoid this, an attempt has been made to develop application-independent standards that make possible multifunctional cards. Since the Smart Card usually represents the component of the system that the user holds in his or her hand, it is of enormous importance for the recognition and acceptance of the entire system. Nonetheless, from a technical and organizational perspective, the Smart Card is usually merely the tip of the iceberg, since complex systems (which are usually networked) are quite often hidden behind the card terminal, and it is these systems that make the service possible in the first place.

Let us take as an example telephone cards, which technically speaking are fairly simple objects. By themselves, they are almost worthless, except perhaps as collector's items. Their true function, which is to allow the use of public telephones without coins, can be realized only after umpteen thousand cardphones have been installed to cover a whole region and have been connected to a network. The large investments required for this can only be justified if appropriate standards and specifications guarantee the long-term future of the system. Standards are an indispensable prerequisite for multifunctional Smart Cards, which can be used for various applications, such as telephony, electronic purses, electronic tickets and so on.

What are standards?

This question is not as trivial as it may appear at first glance, since the terms 'standard' and 'specification' are often used fairly indiscriminately. To make things clear, let us consider the ISO/IEC definition:

> **Standard**: a document that is produced by consensus and adopted by a recognized institution, and which, for general and recurring applications, defines rules, guidelines or features for activities or their results, whereby the objective is to achieve an optimum degree of regulation in a given context.
>
> *Note:* standards should be based on the established results of science, technology and experience, and their objective should be the promotion of optimized benefits for society.

International standards should thus help to make life easier and to increase the reliability and usefulness of products and services.

In order to avoid confusion, ISO/IEC have also defined the term 'consensus' as follows:

> **Consensus**: general agreement, characterized by the absence of continuing objections to essential components on the part of any significant part of the interested parties, and achieved by a procedure that attempts to take into consideration the views of all relevant parties and to deal with all counter-arguments.
>
> *Note:* consensus does not necessarily mean unanimity.

Although unanimity is not required, the democratic process of consensus naturally takes time. This is in particular due to the fact that not only the views of the technicians, but also the views of all relevant parties must be examined, as the objective of a standard is the promotion of optimum advantages for the whole of society. Hence, the preparation of an ISO or CEN standard usually takes several years. A frequent consequence of the slowness of this process is that a small group of interested parties, such as commercial firms, produces its own specifications ('industry standards') in order to hasten the development of new systems. This is especially true in the field of information technology, which is characterized by especially fast development and correspondingly short innovation cycles. Although industry standards have the advantage that they can be developed significantly faster than 'general' standards, they contain hidden risks in that their development takes place without giving consideration to certain important interest groups. These are the parties that are not involved in defining their specifications. For this reason, ISO attempts to create opportunities for incorporating already existing, important publicly accessible specifications into the standardization process.

What does ISO mean?
The relevant ISO standards are especially significant for Smart Cards, since these standards define the basic properties of Smart Cards. What however lies behind the word 'ISO'?

The International Organisation for Standardisation (ISO) is a worldwide association of around 100 national standards agencies, with one per country. ISO was founded in 1948 and is a non-national organization. Its task is to promote the development of standards throughout the world, with the objective of simplifying the international exchange of goods and services and developing cooperation in the fields of science, technology and economy. The results of the work of ISO are agreements that are published as ISO standards.

'ISO' is, by the way, not an acronym. The acronym of the official name would of course be 'IOS'. The name 'ISO' is instead derived from the Greek word *isos*, which means 'equal' or 'the same'. The prefix *iso-*, derived from the Greek *isos*, is commonly used in the three official languages of ISO (English, French and Russian), as well as in many other languages.

As already mentioned, the members of ISO are the national standards bodies of the individual countries, and only one such body per country is allowed to be a member. The member organizations have four basic tasks, as follows:

- to inform potentially interested parties in their own countries about relevant activities and possibilities of international standardization,
- to form national opinions on a democratic basis and to represent these opinions in international negotiations,
- to set up a secretariat for ISO committees in which the country has a particular interest,
- to pay the country's financial contribution to support the central ISO organization.

How does an ISO standard come to be?
The need for a standard is usually reported to the national standards organization by an industrial sector. The national organization then proposes this to ISO as a new working theme. If the proposal is accepted by the responsible working group, which consists of technical experts from countries that are interested in the theme, the first thing that happens is that the objective of the future standard is defined.

After agreement has been achieved regarding the technical aspects to be considered in the standard, the detailed specifications of the standard are discussed and negotiated between the various countries. This is the second phase in the development of the standard. The objective of this phase is to arrive at a consensus of all participating countries, if possible. The outcome of this phase is a 'Draft International Standard'.

The final phase consists of a formal vote on the proposed standard. Acceptance of the standard requires the approval of two thirds of the members of ISO that have actively participated in drafting the standard, as well as three quarters of all members that participate in the vote. Once the text has been accepted, it is published as an ISO standard.

To prevent standards from becoming outdated due to ongoing developments, ISO rules state that standards should be reviewed, and if necessary revised, after an interval of at most five years.

Cooperation with the IEC and the CEN

ISO is not the only international standards organization. In order to prevent duplication of effort, ISO co-operates closely with the IEC (International Electrotechnical Commission). The areas of responsibility are defined as follows: the IEC covers the fields of electrical technology and electronics, while ISO is responsible for all other fields. Combined working groups are formed to deal with themes of common interest, and these groups produce combined ISO/IEC standards. Most standards for Smart Cards belong to this category.

ISO and the European standardization committee CEN (*Comité Européen de Normalisation*) also agree on rules for the development of standards that are recognized as both European and international standards. This results in time and cost savings.

International standardization of Smart Cards

The development of international standards for Smart Cards takes place under the aegis of ISO/IEC and/or the CEN. The German standards association DIN is represented in all relevant committees. It also maintains 'mirror' committees in the form of national working and voting committees. Figure 1.2 gives an overall view of the structure of the relevant ISO working groups and the standards that are entrusted to them.

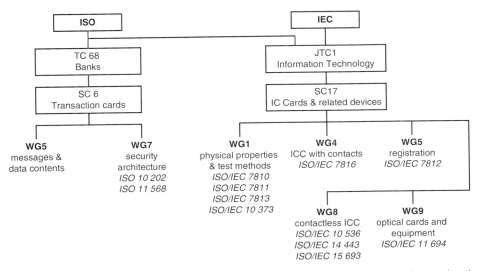

Figure 1.2   Overview and organization of the working groups involved in developing international Smart Card standards

As can be seen, there are two technical committees that are concerned with the standardization of Smart Cards. The first of these is ISO TC68/SC6, which is responsible for the standardization of cards used for financial transactions, while the second is ISO/IEC JTC1/SC17, which is responsible for general applications. This division has historical roots, since the first international applications were for identification cards used for financial transactions. The number of applications has naturally increased enormously since then, so that the general standards, which are looked after by the SC17 committee, have taken on greater significance. The standards relating specifically to financial transactions can thus be regarded as a subset of the general standards. Brief descriptions of the standards listed in Figure 1.2, including their current status, can be found in the Appendix.

The general subject of Smart Cards is dealt with by the CEN TC 224 committee ('Machine-readable Cards, Related Device Interfaces and Procedures'). Figure 1.3 provides an overview of the working groups and the standards that they look after.

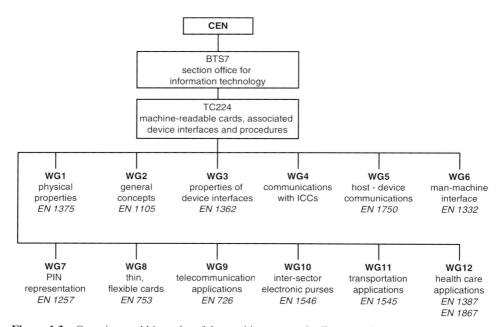

Figure 1.3 Overview and hierarchy of the working groups for European Smart Card standardization

The efforts of the CEN supplement those of ISO. ISO standards are taken over as CEN standards where possible, which requires that they be translated into the three official CEN languages (English, French and German). They may also be extended or reduced to comply with specifically European conditions, as necessary. In addition to these activities, the CEN working groups produce application-oriented standards, which as such would not be possible with ISO. Brief descriptions of the standards listed in Figure 1.3 can be found in the Appendix.

After more than fifteen years of standardization, the most important fundamental ISO standards for Smart Cards are now complete. They form the basis for further, more application-oriented standards, which are currently being prepared by ISO and the CEN.

These standards are in turn based on previous ISO standards in the series 7810, 7811, 7812 and 7813, which define the characteristics of identification cards in the ID-1 format. These standards include embossed cards and cards with magnetic strips, which we all know in the form of credit cards.

In the standardization of Smart Cards (referred to in the ISO standard as 'integrated circuit(s) cards', or ICC), compatibility with these existing standards has been taken into account from the very start, in order to provide for a smooth transition from embossed and magnetic-strip cards to Smart Cards. This was achieved by allowing all functional elements, such as embossing, magnetic strips, contacts and interface components for contactless links, to be integrated into a single card. Of course, a consequence of this is that sensitive electronic components, which after all is what integrated circuits are, are exposed to high stress levels produced by the embossing process and the repeated impacts that occur when the embossed characters are printed onto paper. This makes heavy demands on the packaging of the integrated circuits and the manner in which they are embedded in the card.

A summary of the currently available standards, with brief descriptions of their contents, can be found in the Appendix.[1]

In the last few years, an increasing number of specifications have been prepared and published by industrial organizations or other non-public groups, with no attempt being made to incorporate them into the standardization activities of ISO. The argument most commonly offered for this manner of working is that the way ISO works is too slow to keep pace with the short innovation cycles of the information and telecommunication industries. Since in many cases only a few firms are involved in the preparation of these 'industry standards', there is a great danger that the interests of smaller firms, and especially the interests of the general public, are not taken into account in the process. It is a great challenge to the future of ISO to come up with a working method that can safeguard general interests and still not hold back the pace of innovation.

---

[1] See Section 15.3, 'List of Standards with Comments'.

2

Types of Cards

As already mentioned in the Introduction, Smart Cards are the youngest member of the family of identification cards using the ID-1 format, which is defined in ISO standard 7810, 'Identification Cards – Physical Characteristics'. This standard specifies the physical properties of identification cards, such as flexibility, temperature resistance and the dimensions of three different card formats (ID-1, ID-2 and ID-3). The Smart Card standards (ISO 7816-1 ff) are based on the ID-1 card, millions of which are used nowadays for financial transactions.

This chapter provides an overview of various types of cards in the ID-1 format. In many applications, the combination of particular functions is of special interest, especially when the cards currently used in an existing system, such as magnetic-strip cards, are to be replaced by Smart Cards. In such cases, it is usually not possible to replace the existing infrastructure – in this case, magnetic-strip card terminals – with a new technology overnight.

The solution to this problem generally consists of issuing cards that have both magnetic strips and chips, which are to be used during a transition period. Such cards can be used in both the old and the new terminals. Naturally, new functions that are only possible with a chip cannot be utilized with magnetic-strip terminals.

2.1 Embossed Cards

Embossing is the oldest technique for applying machine-readable markings to identification cards. The embossed characters on the card can be transferred to paper using simple and inexpensive devices. Visual reading of the embossing is also straightforward. The nature and location of the embossing are specified in ISO standard 7811, 'Identification Cards – Recording Technique'. This standard, which is divided into five parts, deals with magnetic strips as well as embossing.

ISO 7811 Part 1 specifies the requirements for embossed characters, such as their form, size and embossing height. Part 3 defines the precise positioning of the characters on the card, and defines two separate regions, as shown in Figure 2.1. Region 1 is reserved for the card's identification number, which identifies the card issuer as well as cardholder. Region 2 is reserved for additional data relating to the cardholder, such as his or her name and address.

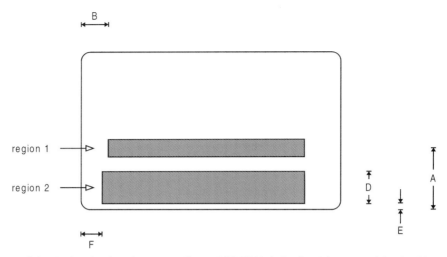

Figure 2.1   Embossing locations according to ISO 7811-3. Region 1 is reserved for the ID number (19 characters), and region 2 is reserved for the cardholder's name and address (4 · 27 characters). A = 21.42 ± 0.12 mm, B = 10.18 ± 0.25 mm, D = 14.53 mm, E = 2.41 to 3.30 mm, F = 7.65 ± 0.25 mm.

At first glance, transferring information by printing from embossed characters may appear quite primitive. However, the simplicity of this technique has made worldwide proliferation of credit cards possible, even in developing countries. The exploitation of this technology requires neither electrical energy nor a connection to a telephone network.

2.2   Magnetic-strip Cards

The fundamental disadvantage of embossed cards is that their use creates a flood of paper receipts, which are expensive to process. One remedy for this problem is to digitally encode the card data on a magnetic strip located on the back of the card.

The magnetic strip is read by pulling it across a read head, either manually or automatically, whereby the data is read and stored electronically. Processing this does not require any paper.

Parts 2, 4 and 5 of ISO standard 7811 specify the properties of the magnetic strip, the coding technique and the locations of the magnetic tracks. The magnetic strip may contain up to three tracks. Tracks 1 and 2 are specified to be read-only tracks, while track 3 may also be written to.

Although the storage capacity of the magnetic strip is only about 1000 bits, which is not very much, it is nevertheless more than sufficient for storing the information contained in the embossing. Additional data can be read and written on track 3, such as the most recent transaction data in the case of a credit card.

The main drawback of the magnetic-strip technique is that the stored data can very easily be altered. Tampering with embossed marks requires at least a certain amount of dexterity and can be easily detected by the trained eye, but changing the data coded on the magnetic strip is relatively easy with a standard read/write device, and it is difficult to afterwards prove that the data have been altered. Furthermore, magnetic-strip cards are often used in automated equipment in which visual inspection is impossible, such as cash dispensers. The potential criminal, having obtained valid card data, can easily use duplicated cards in such unattended machines without having to forge their visual security markings.

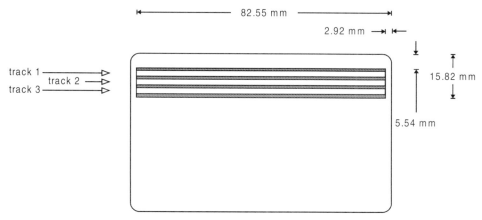

Figure 2.2   Location of the magnetic strip on an ID-1 card. The data region of the magnetic strip is intentionally not extended to the edges of the card, since the use of hand-operated card readers causes rapid wear at the ends of the strip.

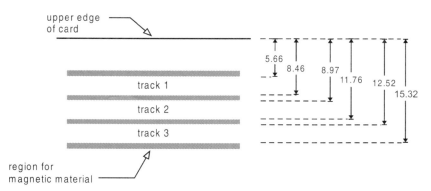

Figure 2.3   Locations of the individual tracks on an ID-1 card (all dimensions in mm)

Manufacturers of magnetic-strip cards have developed various means to protect the data recorded on the magnetic strip against forgery and duplication. For example, German Eurocheque cards contain an invisible and unalterable code in the body of the card, which makes it impossible to alter or duplicate the data on the magnetic strip. However, this and other techniques require a special sensor in the card terminal, which considerably increases costs. For this reason, none of these techniques has so far succeeded in becoming internationally established.

Table 2.1   Standard features of the three tracks on a magnetic-strip card, as specified in ISO 7811

Feature	Track 1	Track 2	Track 3
Amount of data	79 characters max	40 characters max	107 characters max
Data coding	6-bit alphanumeric	4-bit BCD	4-bit BCD
Data density	210 bpi (8.3 bit/mm)	75 bpi (3 bit/mm)	210 bpi (8.3 bit/mm)
Writing	not allowed	not allowed	allowed

2.3   Smart Cards

The Smart Card is the youngest and cleverest member of the family of identification cards in the ID-1 format. Its characteristic feature is an integrated circuit embedded in the card, which has components for the transmission, storage and processing of data. Data transfer can take place either via the contacts on the surface of the card or via electromagnetic fields, without the use of contacts.

The Smart Card offers a number of advantages when compared with magnetic-strip cards. For example, the maximum storage capacity of a Smart Card is many times greater than that of a magnetic-strip card. Chips with over 32 kB of memory are currently available, and this figure will multiply with each new chip generation. Only optical memory cards, which are described in the next chapter, have greater capacities.

However, one of the most important advantages of Smart Cards consists in the fact that their stored data can be protected against unauthorized access and tampering. As access to data takes place only via a serial interface that is controlled by an operating system and security logic, it is possible to write confidential data to the card so that it can never be read from the outside. The confidential data can be processed only internally by the chip's processing unit. In principle, the memory functions of writing, erasing and reading can be restricted and linked to specific conditions by both hardware and software. This allows the construction of numerous security mechanisms, which can also be tailored to the special demands of specific applications.

This together with the ability to compute cryptographic algorithms allows Smart Cards to be used to implement convenient security modules that can be carried about at all times, for example in a briefcase.

Further advantages of the Smart Card are its high degree of reliability and long life as compared to magnetic cards, whose service life is generally limited to one or two years.

The fundamental characteristics and functions of Smart Cards are laid down in the ISO 7816 series of standards. These standards are described in detail in the following chapters.

Smart Cards can be divided into two groups, according to differences in both functionality and price: *memory cards* and *microprocessor cards*.

2.3.1   Memory cards

Figures 2.5 and 2.6 show architectural block diagrams of memory cards.

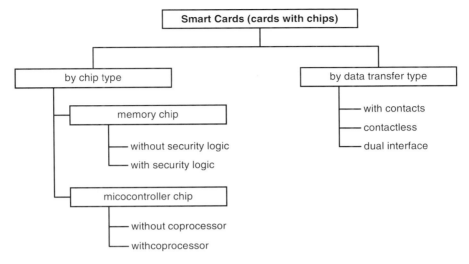

Figure 2.4   Classification chart for cards containing chips

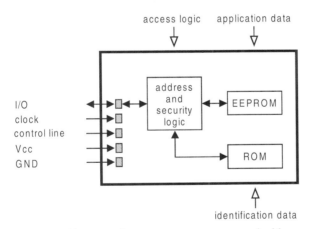

Figure 2.5   Typical architecture of a contact-type memory card with security logic. The figure shows only basic energy and data flows and is not a detailed schematic diagram.

The data required for the application are stored in the memory, which is usually EEPROM. Access to the memory is controlled by the security logic, which in the simplest case consists only of write or erase protection for the memory or some of its regions. However, there are also memory chips with more complex security logic, which can also carry out simple encryption. Data are transferred to and from the card via the I/O port. Part 3 of the ISO 7816 standard defines a special synchronous transfer protocol that allows a particularly simple and economical chip implementation. However, some Smart Cards use the I^2C bus, which is commonly used with serial-access memories.

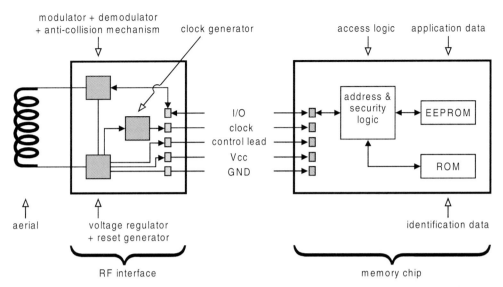

Figure 2.6   Typical architecture of a memory card with security logic and a contactless interface. The figure shows only basic energy and data flows, and is not a detailed schematic diagram.

Memory card functions are mostly optimized for a particular application. Although this severely restricts their flexibility, it makes them particularly inexpensive. Memory cards are typically used for pre-paid telephone cards and health insurance cards.

2.3.2   Microprocessor cards

The heart of the chip in a microprocessor card is – as the name implies – a processor, which as a rule is surrounded by four additional functional blocks: mask ROM, EEPROM, RAM and an I/O port. Figure 2.7 shows the architecture of a typical component of this type.

The mask ROM contains the chip's operating system, which is 'burned in' when the chip is manufactured. The content of the ROM is thus identical for all the chips of a production run, and it cannot be changed during the chip's lifetime.

The EEPROM is the chip's non-volatile memory. Data and program code can be written to and read from the EEPROM under the control of the operating system.

The RAM is the processor's working memory. This memory is volatile, and all the data stored in it are lost when the chip's power is switched off.

The serial I/O interface usually consists only of a single register, through which data are transferred bit by bit.

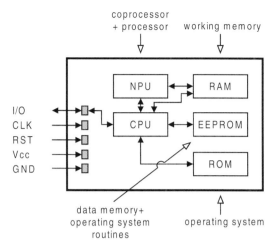

Figure 2.7   Typical architecture of a contact-type microprocessor card with a coprocessor. The figure shows only basic energy and data flows, and is not a detailed schematic diagram.

Microprocessor cards are very flexible in use. In the simplest case, they contain a program optimized for a single application, and thus can be used only for this application. However, modern Smart Card operating systems allow several different applications to be incorporated into a single card. In this case, the ROM contains only basic instructions of the operating system, while the application-specific part of the program is loaded into the EEPROM after the card has been manufactured. The most recent developments even allow application programs to be loaded into a card retroactively, after the card has already been personalized and issued to the cardholder. Special hardware and software measured ensure that the various security requirements of the individual applications are not violated by this capability. Special, optimized microprocessor chips with high processing capacities and large memory capacities are presently being developed, so we can expect to see suitable cards available in the near future.

2.3.3   Contactless Smart Cards

Smart Cards that employ contacts comply with the eight-contact layout specified in ISO standard 7816 Part 1. The reliability of contact-type Smart Cards has been steadily improved over the past years, due to increasing experience in the manufacturing of such cards. The failure rate of telephone cards over their expected lifetime of one year, for instance, is currently well under one in a thousand. Nevertheless, contacts remain one of the most frequent sources of failure in electromechanical systems. Disturbances can result from contamination or wear of the contact, for example. In mobile equipment, vibrations can produce brief intermittent contacts. Since the contacts located on the card's surface are connected directly to the inputs of the integrated circuit embedded in card, there is a risk of damage or destruction of the integrated circuit by electrostatic discharges – and charges of several thousand volts are by no means rare.

These technical problems are elegantly circumvented by the contactless Smart Card. In addition to its technical advantages, contactless-card technology also offers the card issuer and cardholder a range of interesting new potential applications. For example, contactless cards need not necessarily be inserted into a card reader, since there are systems that work at a distance of up to one meter. This is a great advantage in access-control systems in which a door or turnstile needs to be opened, since an individual's access authorization can be checked without requiring the card to be removed from a purse or pocket and inserted into a reader. One extensive application area for this technology is local public transport, in which a large number of people must be identified in the shortest possible time.

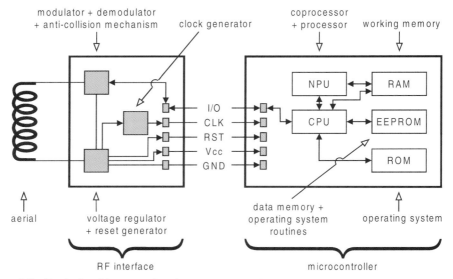

Figure 2.8   Typical architecture of a microprocessor card with a coprocessor and a contactless interface. The figure shows only basic energy and data flows, and is not a detailed schematic diagram.

However, contactless technology is also advantageous in systems that do require deliberate insertion of the card into a reader, since it does not matter how the card is inserted in the reader. This contrasts with magnetic cards or cards with contacts, which work only with a specific card orientation. Freedom from orientation restrictions simplifies operation and thus increases customer acceptance.

A further interesting variation on the use of contactless cards relates to using a surface terminal. Here the card is not inserted in a slot, but simply placed on a marked location on the surface of the card reader. In addition to simplicity of use, this solution is attractive because it significantly reduces the risk of vandalism (for example, by forcing chewing gum or superglue into the card slot).

As far as marketing is concerned, contactless technology offers the benefit that no technical elements are visible on the card's surface, so that the card's visual design is not constrained by magnetic strips or contact surfaces.

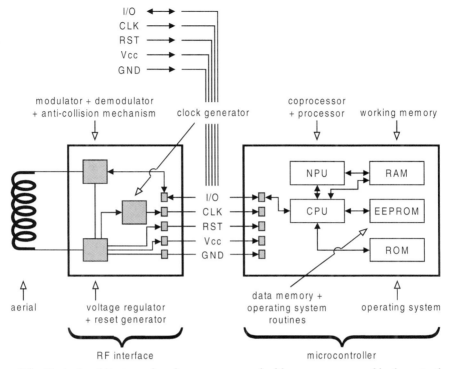

Figure 2.9   Typical architecture of a microprocessor card with a coprocessor and both contactless and contact interfaces. The figure shows only basic energy and data flows, and is not a detailed schematic diagram.

Manufacturing technology for the mass production of contactless cards has matured to the point that high-quality products are available at prices that do not significantly differ from those of comparable contact-type products. Up to now, contactless cards have been used predominantly for local public transportation systems, in which they act as electronic tickets in modern electronic fare systems. The systems presently in use mainly employ single-function cards, for which inexpensive chips with hard-wired logic have been developed.

However, there are increasing signs of a demand for adding supplementary functions to the electronic ticket. For this reason, the use of multifunction cards with integrated microprocessors will increase in the near future. With these cards, payment functions will primarily be performed using the conventional contact-based technique in order to utilize existing infrastructures, such as electronic purse systems. These new multifunction cards have both contacts and contactless coupling elements, and are referred to as dual-interface cards or combicards.

The technology and operating principles of contactless Smart Cards are described in detail in Section 3.6.

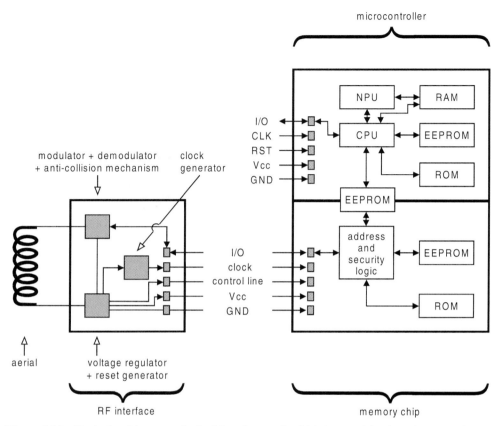

Figure 2.10   Typical architecture of a dual-interface card, which is a combination of a contactless memory card and a contact-type microprocessor card. The figure shows only basic energy and data flows, and is not a detailed schematic diagram.

2.4   Optical Memory Cards

In applications where the storage capacity of Smart Cards is insufficient, it is possible to use optical cards that can store several megabytes of data. However, with currently available technology these cards can be written only once and cannot be erased.

The ISO/IEC 11 693 and 11 694 standards define the physical characteristics of optical memory cards and the linear data-recording technique.

Combining the high storage capacity of optical cards with the intelligence of Smart Cards results in interesting new features. For example, data can be written in encrypted form to the optical memory, while the key is securely stored in the chip's private memory. The optically stored data is thus protected against unauthorized access. Figure 2.11 shows the typical layout of an optical Smart Card with contacts and a magnetic strip. It can be seen that the available optical storage area is restricted by the chip's contacts, which of course reduces the total storage capacity. The magnetic strip is located on the card's reverse side.

Equipment for reading and writing optical storage cards is currently still very expensive, which so far has seriously restricted the use of this type of card.

Optical cards are used for example in the medical sector for recording patient data, where their large memory capacity allows even X-ray images to be stored on the card.

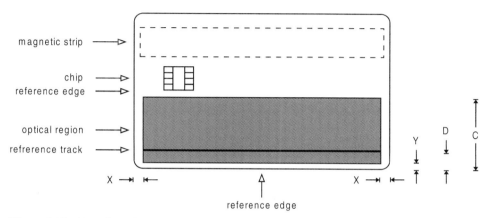

Figure 2.11 Location of the optical storage area on an ID-1 card according to ISO/IEC 11 694-2. C = 9.5 to 49.2 mm, D = 5.8 mm ± 0.7 mm; X = 3 mm max with PWM, 1 mm max with PPM; Y = 1 mm min with PWM (Y<D), 4.5 mm max with PPM. (*PWM = pulse-width modulation, PPM = pulse-position modulation*)

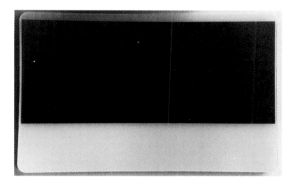

Figure 2.12 Photograph of a typical optical memory card with a net storage capacity (with error correction) of around 4 MB. The raw capacity (without error correction) is around 6 MB.

3

Physical and Electrical Properties

The body of a Smart Card body inherits its fundamental properties from its predecessors, the familiar embossed cards. These cards still dominate the market in the credit card sector. Technically speaking, they are simply plastic cards that are personalized by being embossed with a variety of user features, such as the name and number of the cardholder.

In later implementations, these cards were provided with a magnetic strip, which allowed simple machine processing. When the idea of implanting a chip in the card first arose, the original cards were used as the basis and a microcontroller was embedded in the body of the card. Many standards relating to the card's physical properties are not at all specific to Smart Cards, but apply equally well to magnetic-strip and embossed cards.

3.1 Physical Properties

The most obvious feature of the Smart Card is its format. The next thing you might notice is that it has a contact area, although there may be no visible electrical interface at all (with contactless Smart Cards). A magnetic strip, embossing or a hologram may be the next feature that catches your eye. All these features and functional components form part of a physical properties of a Smart Card.

Most of the physical characteristics are actually purely mechanical in nature, such as the card's size and its resistance to bending and twisting. These are familiar to every user from personal experience. In practice, however, typical physical properties such as sensitivity to temperature or light and resistance to moisture are also important.

The interaction between the body of the card and the implanted chip also needs to be considered, since only the combination of the two makes a functional card. For instance, a card that is suitable for use at very high ambient temperatures is of little use if its microcontroller does not share this property. Both components must satisfy all the requirements separately as well as together, since otherwise high failure rates can be expected in use.

3.1.1 Card formats

Small cards with the typical Smart Card dimensions of 85.6 mm by 54 mm have been in use for a very long time. Almost all Smart Cards are produced in this format, which is certainly the most familiar. It is designated ID-1, and its size is specified in the ISO 7810 international standard. This standard originated in 1985, and thus has nothing to do with Smart Cards as we know them today. This can easily be seen from the abbreviation 'ID', which stands for 'identification card'. The standard simply describes an embossed plastic card with a magnetic strip, designed for the identification of persons. At that time, no one had thought of putting a chip into the card. The presence of a chip and location of its contact areas on the card were only defined a few years later in other standards.

With the variety of cards available today, which are used for all possible purposes and have a wide range of dimensions, it is often difficult to determine whether a particular card is actually an ID-1 Smart Card. In addition to the embedded chip, one of the best identifying features is the thickness of the card. If this measures 0.76 mm and the card contains a microcontroller, then it may be considered to be a Smart Card in the sense of the ISO standard.

The conventional ID-1 format has the advantage of being very easy to handle. The card's format is specified such that it is not too large to be carried in a wallet, but not so small that it is easily lost. In addition, the card's flexibility makes it less inconvenient than a rigid object would be.

Nevertheless, this format does not always meet the demands of modern miniaturization. Some cellphones weigh only 200 g and are not much bigger than a packet of tissues. It thus became necessary to define a smaller format in addition to the ID-1 format, which would take the interests of small terminal devices into account. This type of card can be very small, since it is usually inserted into the device only once and remains there for good. The ID-000 format was defined for these conditions, and it goes by the descriptive name of 'plug-in' card. Currently this format is only used with GSM cellular telephones, which have very little room for a card and which do not require the card to be exchanged very often.

However, the fact that cards in the ID-000 format are inconvenient to handle, both in production and by the end user, led to the development of an additional format. This format is designated ID-00, or 'mini-card'. Its dimensions are approximately halfway between those of ID-1 and ID-000 cards. This type of card is more convenient to handle and is also cheaper to produce, since it is for example easier to print. However, the ID-00 definition is fairly new, and this format has not yet become established either nationally or internationally.

The formats are defined in the relevant standards in a way that simplifies measuring the card dimensions. Thus, the height and width of an ID-1 card must be such that (ignoring the rounded corners) it fits between two concentric rectangles as shown in Figure 3.1, with the following dimensions:

| | external rectangle: | width | 85.72 mm (3.375 inches) |
| | | height | 54.03 mm (2.127 inches) |

| | internal rectangle: | width | 85.46 mm (3.365 inches) |
| | | height | 53.92 mm (2.123 inches) |

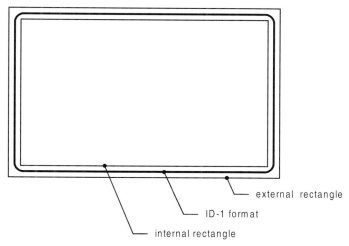

Figure 3.1 How the dimensions of an ID-1 format card are defined

The thickness must be 0.76 mm (0.03 inch), with a tolerance of ±0.08 mm (±0.003 inch). The corner radii and the card's thickness are defined conventionally. Based on these definitions, an ID-1 card's dimensions can be represented as shown in Figure 3.2.

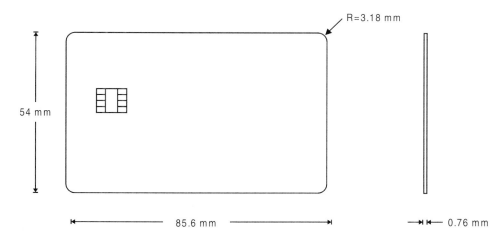

Figure 3.2 The ID-1 format. Thickness: 0.76 mm ± 0.08 mm; corner radius: 3.18 mm ± 0.30 mm. The dimensions shown indicate the size of the card without any tolerances.

The ID-000 format is also defined using two concentric rectangles. Since this format originated in Europe (based on the GSM mobile system), the basic dimensions are metric. The bottom right-hand corner of a plug-in card is cut off at an angle of 45°, as shown in Figure 3.3, in order to facilitate correct insertion of the card into the card reader.

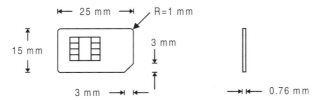

Figure 3.3   The ID-000 format. Thickness: 0.76 mm ± 0.08 mm; corner radius: 1 mm ± 0.10 mm; corner: 3 mm ± 0.03 mm. The dimensions shown indicate the size of the card without any tolerances.

The dimensions of the two rectangles for the ID-000 format are:

| | external rectangle: | width | 25.10 mm |
| | | height | 15.10 mm |

| | internal rectangle: | width | 24.90 mm |
| | | height | 14.90 mm |

The ID-00 format is also based on metric measurements. Its maximum and minimum dimensions are defined by two concentric rectangles with the following dimensions:

| | external rectangle: | width | 66.10 mm |
| | | height | 33.10 mm |

| | internal rectangle: | width | 65.90 mm |
| | | height | 32.90 mm |

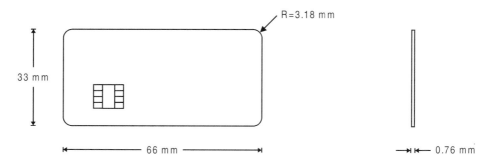

Figure 3.4   The ID-00 format. Thickness: 0.76 mm ± 0.08 mm; corner radius: 3.18 mm ± 0.30 mm. The dimensions shown indicate the size of the card without any tolerances.

The relative sizes of the ID-1, ID-00 and ID-000 formats are shown in Figure 3.5. Cards in the smaller formats can be produced from the larger versions by punching them from the body of a larger-format card. This is especially important for card manufacturers, since it allows the production process to be optimized and made more economical.

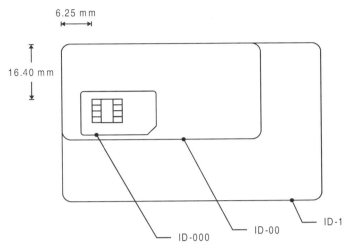

Figure 3.5 Relative sizes of the ID-1, ID-00 and ID-000 formats

For example, it is conceivable for a manufacturer to make card blanks in only one format (preferably ID-1), embed the modules in them and fully personalize them. Depending on the specific application of the cards so produced, they could be trimmed to the desired format in a subsequent production step.

Alternatively, this may be done later by the client, as is already the case to some extent with GSM cards. The client receives an ID-1 card that is pre-punched such that he or she can convert it onto an ID-000 card by breaking the smaller card free from the larger card body. In another process, the ID-000 card is completely punched out of the ID-1 body and attached to the remaining part of the ID-1 card using single-sided tape on the side without the contacts. The client can 'produce' a card whose format fits with his or her equipment, while the manufacturer has to produce and deliver only one type of card.

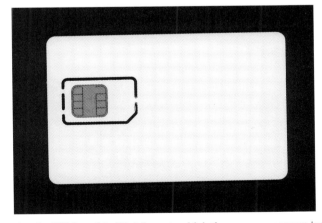

Figure 3.6 Example of a GSM card in ID-1 format, which the user can convert into an ID-000 card by pressing out the smaller-format card (Source: Giesecke & Devrient)

3.1.2   Card components and security markings

Since Smart Cards are mainly used to provide authorization for specific actions or to identify their owners, security markings on the card body are often called for, in addition to the embedded chip. Since verification of the genuineness of the card may be performed by humans as well as by machines, many security markings are based on optical features. However, there are also security features that are based on a modified Smart Card microcontroller and which thus can only be verified by a computer. In contrast to the security features used with the microcontroller, the usual features for manual verification of the genuineness of a card are not based on cryptographic procedures (such as mutual authentication). Instead, they are mostly based on the use of secret materials and production processes, or the use of processes whose mastery is very expensive, requires significant technical knowledge or is technically difficult.

Signature panels

A very simple way to identify the owner of the card is to use a signature panel attached to the card, as is common with credit cards. Once such a panel has been signed, it cannot be altered, so it is erasure-proof. A very fine colored pattern printed on the panel makes any attempt to cover the panel immediately apparent. The signature panel is permanently attached to the card body by using a hot-gluing process to fasten the printed paper strip to the card. Alternatively, the signature panel may be part the top layer of the card, which is laminated into the card when it is assembled.

Ornamental borders

A somewhat more complicated technique is to insert films printed with colored ornamental borders (guilloches) under the transparent top layer of the card. Guilloches are areas containing usually round or oval, interwoven lines, such as are often found on bank notes and share certificates. These patterns have such fine structures that they can presently only be produced by printing, and thus cannot be photocopied.

Microtext

Another technique that is based on the security provided by fine printed line structures is the use of microtext lines. These appear to the naked eye to be plain lines, but they can be recognized as text using a loupe. They also cannot be photocopied.

Ultraviolet text

In order not to affect the visible layout of the card, control characters or control numbers can be printed on the card using ink that is only visible under ultraviolet light. However, this technique provides only relatively limited protection against forgery.

Holograms

A hologram integrated into the card is a security feature that is by now familiar to all card users. The security of holograms is primarily based on the facts that they are produced by only a few companies in the world and that they are not freely available.

The holograms used for Smart Cards are called 'embossed' holograms, since they can be seen using diffuse reflected daylight. They are also referred to as 'white-light reflection holograms'. Conventional transmission holograms must be viewed using coherent laser light. Supplementary security features that can only be seen with laser light are sometimes integrated into the hologram as well. In order to produce an embossed hologram, it is necessary to first generate a master hologram using the conventional holographic exposure technique. A master embossing stamp is prepared from the master hologram using a transfer process. The embossing stamp contains the microstructures that will produce the subsequent embossed holograms. Daughter stamps are prepared from the master stamp using electroplating processes, and these daughter stamps are used to emboss the hologram structure in plastic films. These films are then coated with a layer of vaporized aluminum to produce the well-known white-light reflection holograms.

The hologram is permanently bonded to the card body, so it cannot be removed without destroying it. This can be done using either a lamination process or the 'roll-on' process. In the latter process, a hologram located on a carrier film is pressed onto the card by a heated roller. The carrier film is then pulled off, and the hologram remains permanently welded to the plastic card body. A third process that is used is the 'hot-stamp' process, which is similar to the roll-on process except that a heated stamp is used instead of a heated roller.

Kinegrams

Kinegrams, which are popularly called '3-D pictures', are made in the same way as holograms. The viewer sees an image that changes abruptly when the viewing angle is changed. Kinegrams are just as hard to forge as holograms, and they have the advantage that they are more quickly recognized by the viewer and thus can be verified more quickly.

Multiple laser images

A multiple laser image (MLI) is similar to a hologram. This is a sort of kinegram that is very similar to a simple hologram. It is based on a set of lenses pressed into the surface of the card, some of which have been blackened by a laser. The main difference between an MLI and a hologram is that card-specific information is presented in the small image. For example, this process can be used to mark the name of the cardholder on an individual card in the form of a kinegram.

Laser engraving

The blackening of special plastic layers by heating them with a laser beam is called laser engraving. In contrast to embossing, this is a secure means to write data on an individual card, such as the cardholder's name and the card number. It is secure because the necessary equipment and the knowledge of how to use it are not readily available.

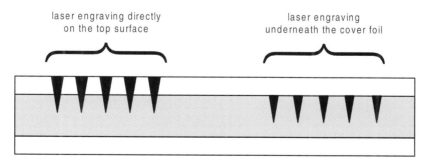

Figure 3.7   Cross section of laser engraving in a card (not to scale). Laser engraving can take place either on the surface of the card or in an internal layer, located underneath a cover film that is transparent to the laser light.

Two different techniques are used for laser engraving, namely vector engraving and raster engraving. With the vector technique, the laser beam is directed along its path without interruption. This is very well suited for representing characters and has the advantage of being quick. With the raster technique, by contrast, a large number of adjacent points are blackened to produce the image, in the same way as in an ink-jet or dot matrix printer. This technique is primarily used to place a picture on the card. Although it has the advantage of high resolution, which allows details to be reproduced well, it has the disadvantage of being very time-consuming. For example, laser engraving a standard-quality passport photograph takes around ten seconds.

Embossing

Another method for adding user data to the card is to emboss characters onto the card. This is done by hammering metal letter punches against the card. In principle, this process works the same way as an old-fashioned mechanical typewriter. Nowadays, character embossing has only one advantage, but this is very important in practical use. The embossed characters can easily be transferred to pre-printed forms using carbon paper. This is still the most widely used method of paying with a credit card.

It is very easy to tamper with embossed characters. It is only necessary to moderately heat the embossed letters and numbers (using an iron, for example) in order to make the plastic flat again. In order to counter this, one of the embossed characters is often placed on top of the hologram, which is destroyed if it is heated.

Thermochrome displays

There are certainly applications in which it is desirable to change the text and picture(s) printed on the card from time to time. A good example is a student identification card in the form of a Smart Card, which must be renewed twice a year. Ideally, it should be possible to visually read the expiry date without having to use any special equipment. This means that it must be printed on the card, and not just stored internally in the chip. A similar example is an electronic purse Smart Card, which requires the use of a card reader to show the current balance of the 'money' stored in the card.

Smart Cards with microcontroller-driven displays are currently technically possible, but still too expensive for large-scale applications. There is however a simple alternative, which has some disadvantages compared to a real display, but is inexpensive and presently available. This is a thermochrome display (TC display) on the Smart Card. A TC display is a supplementary card component on which characters and images can be reversibly printed (that is, printed and subsequently reprinted) by special card readers.

The technical operating principle is relatively simple. A printing head with a resolution of 200 or 300 dpi, such as is used in thermal-transfer and dye-sublimation printers, heats individual points that are to be blackened on the thermochrome strip. This is a temperature-sensitive strip, 10–15 µm thick, that is laminated to the card. It darkens at each point that is heated to 120° C. However, this darkening can be changed back to a nearly transparent state by heating the entire strip, which amounts to erasing the strip.

The thermochrome process is presently the only economical manner to present time-varying information to the user on the surface of the card so that it can be read without any special equipment. Its major disadvantages are that it is subject to fraud and that it requires special card readers with built-in thermochrome printers.

The MM technique

In 1979, the German banking industry decided to include a machine-readable security feature in all German Eurocheque (ec) cards. After various potential methods were tested, the MM technique (developed by the firm GAO) was selected as the security process for these cards. This security feature is still used in all German Eurocheque cards, even though most of them now have microcontroller chips. The objective of this security feature was, and still is, to prevent unauthorized copying and modification of the magnetic strip.

The MM technique is a typical example of a secret and very effective security feature. It has been used for two decades in millions of cards. Its basic construction is summarized in an article by Siegfried Otto [Otto 82].

The term 'MM technique' comes from the German phrase *moduliertes Merkmal* (modulated marking), which can be understood to refer to a machine-readable substance that is incorporated in the interior of the card body [Mayer 96]. Whenever the card is verified, its MM code is read from the card by a special sensor and passed on to a security module called the 'MM box'. The MM box also receives the complete content of the magnetic strip, in particular the MM checksum, which is also stored in the magnetic strip. Inside the MM box a one-way function based on the DES is triggered to calculate a value from the magnetic-strip data and the MM code. If the result of this calculation is the same as the MM checksum, it can be concluded that the magnetic-strip data fits with the card.

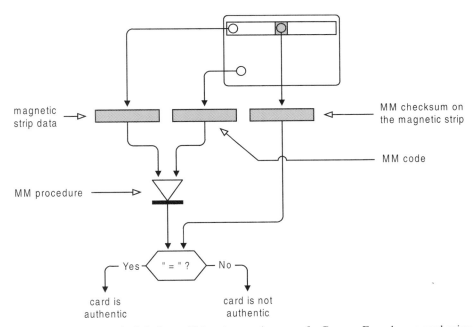

Figure 3.8 Operating principle for verifying the genuineness of a German Eurocheque card using the MM technique. The security module, or 'MM box', protects the MM procedure and the following comparison, so that only a yes/no result is reported to the higher-level system.

If a valid set of magnetic-strip data is written onto a blank card, this will be detected by the fact that the blank card does not have any MM mark. Copying the magnetic-strip data from one EC card to another EC card will also be detected, since the MM checksum will be incorrect. The MM mark is invisible, and the details of how it works and exactly where it is located in the card are secret. In addition, it is produced using materials and technology that are not commercially available.

A MM box is built into in all German bank machines (ATMs) and POS terminals. These devices are thus able to verify whether the magnetic-strip data matches the card. The process itself is not defined in any standard, and it is used only in Germany. Thanks to it, the magnetic strips of German EC cards are protected against copying, which nowadays does not otherwise present any technical difficulties.

Security markings

A wide variety of visual security markings was developed in the period between the massive use of cards without chips and the introduction of Smart Cards. During this period, such markings were the only way to verify the genuineness of the cards. The embedded microcontroller in the new card, and the cryptographic procedures that it makes possible, have reduced the importance of these markings. They are nevertheless still very important whenever a human instead of a machine must verify the genuineness of a card, since a person cannot access the chip without special equipment.

Only the most essential and best-known security markings used for cards can be listed here, in a very condensed summary form. There are many other types of markings, such as invisible markings that can only be seen with IR or UV illumination, magnetic codes and special printing processes using rainbow-colored inks. These techniques are certainly interesting, but there is not enough room here to describe all of them.

In the future, security markings will be found not only on the card but also in the chip. It is conceivable that 'security' chips could be used in the same way that bank-note paper is now used. You cannot print genuine bank notes without using real bank-note paper, which has specific markings that show that it is genuine. In order to incorporate similar security markings into the realm of chips, special chips with specific hardware modifications are necessary. A terminal can then measure these modifications, which constitute the 'marking' of the chip, and judge the genuineness of the chip from the results of the measurements.

As an example of a hardware marking, suppose that the computation of a fast cryptographic algorithm is implemented in supplementary hardware in a certain chip. The time required for the computation of a particular value would be so short, due to the hardware implementation of the algorithm, that it would not be possible to perform the computation using a software emulation in a different chip in a similarly short time. This means that a terminal could distinguish the one chip from other chips just by making a time measurement.

Nowadays, there are several chips on the market with hardware markings like that just described or similar to it. Naturally, they are not freely available, just as bank-note paper is not freely available. Of course, such hardware markings are only suitable for very large-scale applications, due to the high cost of developing the chip-specific hardware. The consequence of this, which is that such chips are almost inevitably available from only one manufacturer, with no possibility of an alternate source, is also difficult for many card producers to accept. However, hardware-based security is an indispensable part of the security architecture of a Smart-Card system, and it is unfortunately not available for free.

3.2   The Card Body

The materials, construction and production of the body of the card are effectively determined by the card's functional elements, as well as by the stresses to which it is subjected during use. Typical functional elements include:

`   magnetic strip
`   signature panel
`   embossing
`   imprinting of personal data via laser beam (text, photo, fingerprint)
`   hologram
`   security printing
`   invisible authentication features (e.g. fluorescence)
`   chip with contacts or other coupling elements

Clearly, even a relatively small card, only 0.76 mm thick, must sometimes contain a large number of functional elements. This places extreme demands on the quality of the material used and on the manufacturing process.

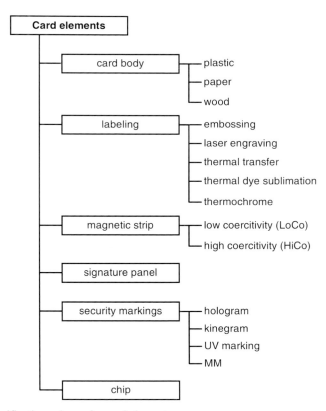

Figure 3.9   Classification scheme for card elements

The minimum requirements relating to card robustness are laid down in ISO standards 7810, 7813 and 7816 Part 1. The requirements relate essentially to the following areas:

` ultraviolet radiation
` X-ray radiation
` the card's surface profile
` mechanical robustness of the card and contacts
` electromagnetic susceptibility
` electrostatic discharges
` temperature resistance

ISO standard ISO/IEC 10373 specifies test methods for many of these requirements, which allow users and card manufacturers to objectively test the quality of the card. For Smart Cards, the bending and twisting tests are particularly important. This is because the chip, which is as fragile and brittle as glass, is a delicate foreign object within the elastic card. Special structural features are required to protect it against the mechanical strains that arise when the card is bent or twisted. Chapter 9 contains a detailed list of tests and the methods used to perform them.

3.2.1 Smart Card materials

The first material employed for ID cards, which is still widely used, is the amorphous thermoplastic material PVC (polyvinyl chloride). It is the least expensive of all the available materials, is easy to process and is suitable for a wide range of applications. This material is used throughout the world for credit cards. Its drawbacks are its limited lifespan due to physical deterioration and its limited resistance to heat and cold. PVC in sheet form is used to manufacture cards, since injection molding is not possible. The worldwide production of PVC was around 13 million metric tons in 1996, of which 35,000 tons (0.27%) was for cards. PVC is considered to be damaging to the environment, since the material used to produce it, vinyl chloride, is a known carcinogen. In addition, if it is burned, hydrochloric acid and (under unfavorable conditions) possibly dioxins are released. In addition, heavy-metal compounds are often used as stabilizers. Nonetheless, PVC is still by far the most widely used material for cards. This is primarily due to its low price and its good processing characteristics. It is however used less and less each year, due to its undesirable environmental characteristics. Many card issuers have decided not to use PVC, due to environmental–political considerations.

In order to avoid the drawbacks of PVC, the material ABS (acrylonitrile butadiene styrene) has been used for some time to make cards. This is also an amorphous thermoplastic that is distinguished by its stability and resistance to temperature extremes. It is therefore often used in mobile telephone cards, which for obvious reasons may be subjected to relatively high temperatures. ABS can be processed both in sheet form and by injection molding. Its major drawbacks are its limited acceptance of color as well as sensitivity to weather conditions. Although the starting material for the production of ABS, benzene, is a carcinogen, ABS has no other known environmental disadvantages.

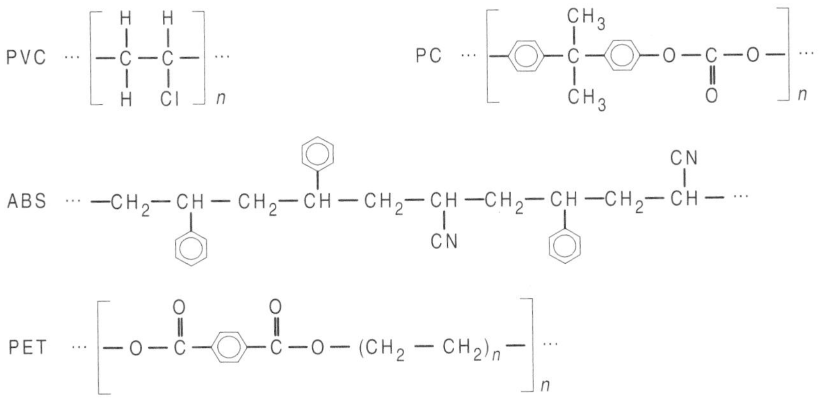

Figure 3.10   Structural formulae of the most important materials used for card bodies

For applications in which extreme stability and durability are required, polycarbonate (PC) is used. It is typically used for identity cards, and it incidentally the base material for compact discs. Due to its high degree of thermal resistance, relatively high temperatures are needed for applying holograms or magnetic strips using the hot-stamp process. This can easily cause problems due to the limited temperature resistance of the materials being applied. The main disadvantages of polycarbonate are its low degree of resistance to scratching and its very high price compared to other card materials. A further disadvantage is that phosgene and chlorine are needed for the production of polycarbonate, and both of these materials are environmentally problematical. Polycarbonate cards can be easily recognized by the characteristic 'tinny' sound that they produce when dropped on a hard surface.

An environmentally friendly material that is mainly used as a PVC substitute is polyethylene terephthalate (PET), which has been used for a relatively long time to make packaging materials. It is commonly known as polyester. This thermoplastic material is used in Smart Cards in both its amorphous form (A-PET) and its partly crystalline form (PETP). Both are suitable for processing in sheet form as well as by injection molding. However, PETP is difficult to laminate, which makes additional processing steps necessary in the manufacturing process.

Numerous attempts have been made to find new or better materials for card bodies, in addition to the usual materials (PVC, ABS, PC AND PET). One example is cellulose acetate, which to be sure has good environmental properties, but which has up to now proven to be poorly suited to the mass production of cards. Truly different materials, such as paper, have been frequently discussed but as yet never used in any significant quantity. The requirements places on cards, in terms of cost, durability and quality, are after all very high, and they can presently only be satisfied by plastics.

In 1996 and 1997, Danmønt[1] conducted a field trial using around 600 cards that, while they did not represent a real alternative to plastic card bodies, were at least an interesting (or amusing) idea. The cards were laminated from eight layers of birch wood, each 0.1 mm thick. These cards did not meet the requirements of the various tests specified in ISO 10 373, such as those for bending and twisting, and they naturally were not suitable for

---

[1]   See [a la Card]

embossing. However, around 90% of their users expressed a positive reaction and said that they experienced no problems with their cards. Unfortunately, a birchwood card is not especially innovative from an environmental perspective, since the layers must be laminated using a plastic adhesive and the usual printing processes are necessary.

Table 3.1 Summary of the characteristics of the standard materials for card bodies.[2] The cost of PVC is the basis for the relative cost value.

Characteristic	PVC	ABS	PC	PET
Main use	credit cards	cellular-telephone cards	identification cards	health-insurance cards
Principal feature	inexpensive	thermally stable	durable	environmentally friendly
Temperature range	65–95 °C	75–100 °C	< 160 °C	< 80 °C
Cold tolerance	moderate	high	moderate	moderate
Mechanical stability	good	good	good	very good
Embossing	good	poor	good	good
Printing	good	moderate	moderate	moderate
Hot stamping (e.g. holograms)	good	good	difficult	good
Laser engraving	yes	poor	good	good
Typical lifetime	$\cong$ 2 years	$\cong$ 3 years	$\cong$ 5 years	$\cong$ 3 years
Share of worldwide card production (1998)	85%	8%	5%	2%
Relative cost	1	2	7	2.5
Environmental aspects	stabilizers contain heavy metals	base material benzene is a carcinogen	phosgene and chlorine needed for production	the most environmentally friendly material
	burning may release dioxins	burning may release prussic acid	burning does not release dangerous materials	burning does not release dangerous materials
Special features	negative public image		low resistance to scratching	

3.2.2   Chip modules

The most important element of a Smart Card is naturally the chip. Of course, this very fragile component cannot be simply laminated to the surface of the card, like a magnetic strip. Instead, it needs a sort of enclosure to protect it from the rough everyday life of the card. This enclosure is the *chip module*. In addition to protection from ambient conditions, chips for contact-type Smart Cards need six to eight contact surfaces, which provide power

---

2   Based in part on [Houdeau 97, Grün 96].

to the chip and allow data communications with the terminal. A portion of the module's surface serves to provide these electrical contacts to the outside world. Of course, the chip module should be as inexpensive as possible.

A wide variety of module designs has been thought up since the start of the development of Smart Cards, in order to meet these two technical requirements – protection of the fragile semiconductor chip and provision of contact surfaces. The most important of these are listed in Figure 3.11.

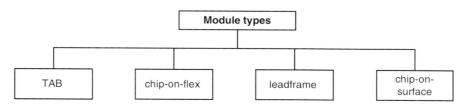

Figure 3.11   Classification of the various types of chip modules

Figure 3.12   These examples illustrate the evolution of the chip-on-flex technique, starting with one of the first 8-contact chip-on-flex modules at the upper left and proceeding to the current modules with 6 or 8 contacts.

3.2.2.1   Electrical connections between the chip and the module

Electrical connections must be made between the chip inside the module and the contact elements on the outer surface of the module. Nowadays, two processes are mainly used for this. With the wire-bonding process, an automatic bonding machine attaches a gold wire with a diameter of only a few micrometers between the chip and the rear surface of each contact. The electrical connection between the wire and the chip or module is made using ultrasonic welding. With this process, the orientation of the top surface of the chip is always the opposite of that of the contact surfaces. This process has been a standard in the semiconductor industry for some time, and it can be readily used for the mass production of chip modules. However, each chip must be electrically connected to the module by five wires, which of course costs time and money.

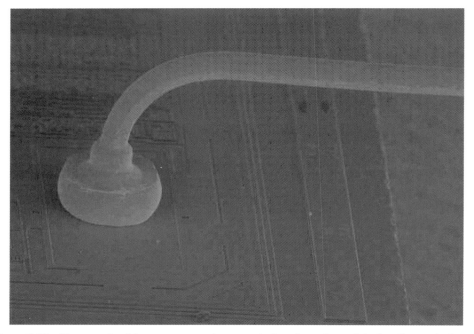

Figure 3.13   Photograph of the contact zone between the bonding wire and the bonding pad of a Smart Card microcontroller, magnified 1000 times (Source: Giesecke & Devrient)

The die-bonding process has been developed to reduce the cost of building the chip into the module. In this process, the electrical connections between the chip and module are not made with wires. Instead, the contacts are made by mechanically attaching the chip to the rear surface of the module.

3.2.2.2   TAB modules

The TAB (tape automated bonding) process was a standard technique for large-volume chip packaging at the beginning of the 1990s, but nowadays is it used less and less throughout the world. It has become technically obsolescent in the mean time, and also too expensive. It is described here primarily for the sake of completeness.

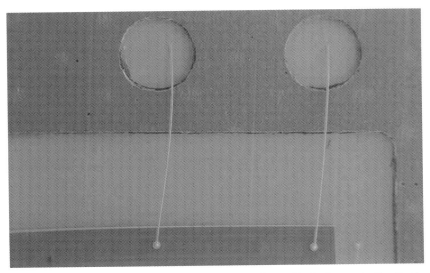

Figure 3.14   View of the electrical connections between a Smart Card microcontroller (bottom) and the chip module (top), magnified 400 times (Source: Giesecke & Devrient)

A chip module produced using the TAB process is shown in Figure 3.15. The special feature of this process is that metallic bumps are first electrically attached to the pads of the chip. The leads of the carrier film are then soldered to these bumps. The solder connections are so robust that no additional support is required for the chip, which hangs from its leads. In addition, the active surface of the chip is protected against ambient conditions by a covering layer of material. The advantages of the TAB process are the mechanical strength of the connections to the chip and the low profile of the module. However, these advantages come at the price of a higher cost as compared with other module preparation techniques.

Building a TAB module into a Smart Card is not all that easy, since the module must be taken into account in the lamination of the individual layers of the card. Before the layers are laminated, suitable openings are punched in them, and the chip module is then inserted. The chip module is subsequently welded to the body of the card during the lamination process. This process provides a highly reliable bond between the chip module and the card body. It is almost impossible to remove the chip from the card without destroying the card.

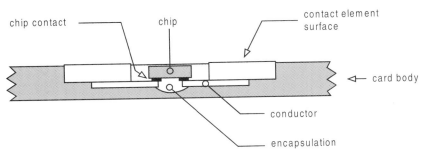

Figure 3.15   Cross section of a chip module using the TAB process

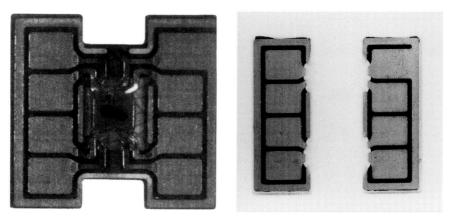

Figure 3.16   A TAB module ready for mounting in a Smart Card (left), and a TAB module mounted in a Smart Card (right)

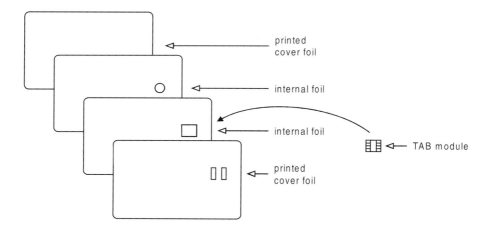

Figure 3.17   Inserting the TAB module during the lamination process

3.2.2.3   Chip-on-flex modules

Currently, the chip-on-flex module with wire-bonded contacts is the most widely used type. The construction of such modules is illustrated in Figure 3.18. With this technique, an opening is milled into the finished card body, into which the chip module can be glued. The carrier material is a flexible circuit board made from fiberglass-reinforced epoxy resin, with a thickness of 120 µm. The areas that will ultimately be the contact surfaces are copper laminations that are 35 or 75 µm thick. Gold electroplating is applied to these surfaces at a later stage in the process. This protects the contact surfaces against possible effects that could reduce their electrical conductivity, such as oxidation. Holes are punched into the carrier strip to receive the chips and the wire bonds. The chips, which are around 200 µm thick, are taken from the sawed wafer by a pick-and-place robot and fitted into the openings in the flexible circuit. Next, the chip contacts are connected to the rear surfaces of the

contact areas using bonding wires that are a few micrometers thick. Finally, the chip and
the bonding wires are encapsulated with a blob of plastic to protect them against ambient
conditions. The total thickness of the finished module is typically around 600 µm.

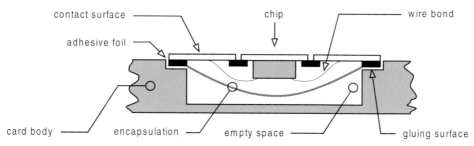

Figure 3.18   Cross section of a chip-on-flex chip module

The advantage of this technique is that it is largely based on a standard process used in
the semiconductor industry for fitting chips in standard packages. It does not require as
much specialized experience as the TAB process, and is thus less expensive. This process
also lends itself well to the production of very complex card bodies with many active
elements. This is because defective card bodies can be separated from the rest before the
expensive chip modules have been fitted. The disadvantage of this process is that both the
thickness and the length and width of the chip module are significantly greater that those of
a TAB module, since not only the chip but also the bonding wires must be covered by the
protective encapsulation. This is particularly disadvantageous, in that the standard Smart
Card thickness of 0.76 mm does not leave a lot of room for overly thick modules.

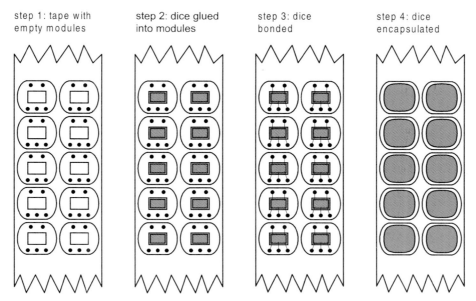

Figure 3.19   The four main process steps in the production of chip-on-flex modules

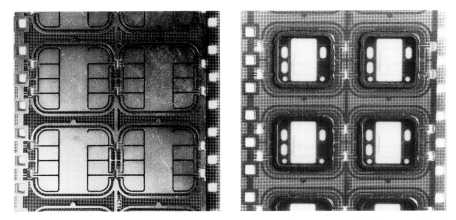

Figure 3.20   Front and rear views of chip-on-flex modules on 35-mm tape. The five openings in the carrier circuit board, for the bonding wires that make the electrical connections to the chip, can be clearly seen in the rear view.

Figure 3.21   Front and rear views of a chip-on-flex module for a dual-interface card

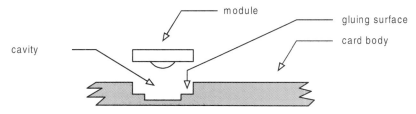

Figure 3.22   Inserting the chip module in a milled opening in the card body

3.2.2.4 Lead-frame modules

From a technical point of view, both the TAB process and the chip-on-flex process leave something to be desired, since neither of them provides much room for reducing production effort and costs. With the TAB process, the production of the card body is quite laborious due to the characteristics of the module, while with the chip-on-flex process the complexity of the module itself and the use of wire bonding make for unfavorable production costs. These problems led to the development of a new type of module that is mechanically just as robust as TAB and chip-on-flex modules, but which has lower production costs. This is the lead-frame module. The construction of a lead-frame module is relatively simple. The contact surfaces, which are stamped out of a gold-plated copper alloy, are held together by a plastic mold body. The chips are placed on this by a pick-and place machine and then connected to the rear sides of the contacts using wire bonding. Next, the chip is covered by a protective blob of opaque, usually black epoxy resin. The lead-frame process is currently one of the least expensive processes for the production of chip modules, without any accompanying reduction in the mechanical robustness of the modules.

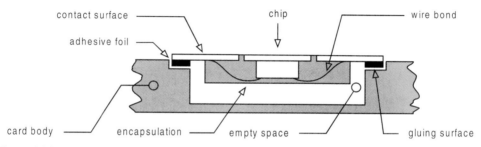

Figure 3.23   Cross section through a lead-frame chip module

Figure 3.24   Stamped-out lead-frame module with the two coil connections for a contactless Smart Card, with a match for comparison

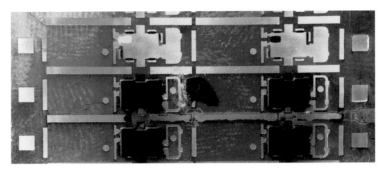

Figure 3.25 Lead-frame modules for contactless Smart Cards, arranged in pairs on a 35-mm tape. The two empty locations for modules that have been stamped out can be seen at the top.

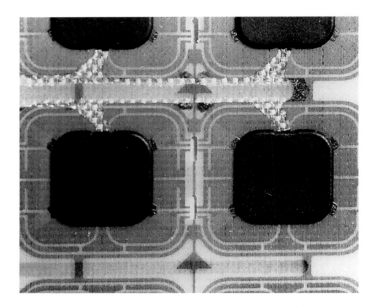

Figure 3.26 Lead-frame modules for Smart Cards with contacts, arranged in pairs on a 35-mm tape

3.2.2.5   The chip-on-surface process

For chips with relatively small surface areas, there is a process available since the mid-1990s that is a technically very interesting alternative to the usual processes for building chips into modules. The MOSAIC (Microchip on surface and in card) process, developed by Soliac [Sligos], does not need a module for the chip, since it is located directly in the card body.

1st step: using a
laser drill, make a
cavity for the chip

2nd step: place the
chip in the cavity

3rd step: silk screen
print with conductive ink

4th step: overprint
with insulating ink

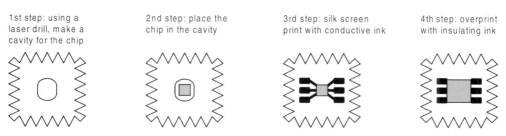

Figure 3.27   The four stages in the production of a Smart Card using the chip-on-surface process

The MOSIAC process is suitable for chips whose surface area is around 1 mm^2. This presently limits its application to pure memory chips, since microcontrollers are still too large for this technique. The process works as follows: first, a laser is used to remove material from the location where the chip is to be placed, and then the chip is glued into this opening. In the next step, a conductive silver paste is silk-screened onto the surface of the chip and the card body, to form the contact surfaces and connect them to the chip at the same time. In the final step, the chip and the leads to the contact areas are covered with a non-conductive lacquer. This provides electrical insulation and protects them against external ambient conditions.

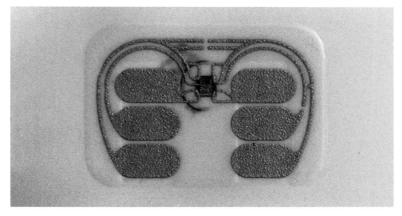

Figure 3.28   Photograph of a memory chip with an edge dimension of 0.5 mm (area = 0.25 mm^2), with contact surfaces according to ISO/IEC 7816-3, mounted on a telephone card along with its contacts using the chip-on-surface process

Figure 3.28 clearly illustrates that the chip-on-surface process is very well suited to the mass production of large numbers of cards, since it essentially consists of only a brief laser milling of the card body and two pressing processes. However, this technology demands an extremely precise printing process to ensure that the contacts to the chip are located correctly. Up to now, the card body has been primarily made of polycarbonate, which is especially suitable for the card-on-surface process. The production capacity of finished cards lies in the region of 5000 pieces per hour per machine.

3.3   Electrical Properties

A Smart Card's electrical properties depend solely on its embedded microcontroller, since this is the only component of the card with an electrical circuit. In the early days of Smart Card technology, quite often the only critical factor was that the microcontroller should be functional, and less attention was paid to overall electrical properties. At that time, the applications were almost exclusively closed, with a single type of card and a terminal specifically developed for use with that card. The electrical properties of the Smart Card were relevant only to the extent that the terminal had to be designed to work with the type of card used. This situation has however changed increasingly in the last few years. With current large-scale applications, in which various types of Smart Cards must work together with many different types of terminals, it is an unavoidable requirement that all the cards that are used are either electrically identical or at least behave uniformly within clearly defined electrical regions.

The general basis in this area, which has a primarily international character, is the ISO/IEC 7816-3 standard. It defines many basic electrical requirements for Smart Cards, such as the activation and deactivation sequences.

A large-scale application that also contributed to the definition of requirements in this area was the GSM mobile telephone network. This system, in which various types of terminals originating from a wide variety of manufacturers must work with different types of cards, clearly defined the basic conditions. The large volume of cards in use also helped to make the electrical properties defined in GSM 11.11 into a model for all semiconductor manufacturers. All new designs for microcontrollers to be used in Smart Cards throughout the world meet the basic electrical specifications laid down in GSM 11.11, since they would otherwise be practically unsellable.

3.3.1   Connections

Most Smart Cards have eight contact areas on the front side. These form the electrical interface between the terminal and the microcontroller in the card. All electrical signals are passed via these contacts. However, according to ISO/IEC 7816-2, two of the eight contact areas (C4 and C8) are reserved for yet to be defined future functions, and for reasons of compatibility, these should not yet be used. One of these two contacts is planned for a second I/O interface, so that at some time Smart Cards will be able to support full duplex data transfers. Since these two contact areas are not presently used, some recent Smart Card modules have only six contacts, which slightly reduces production costs. However, their functionality is identical to that of modules with eight contacts.

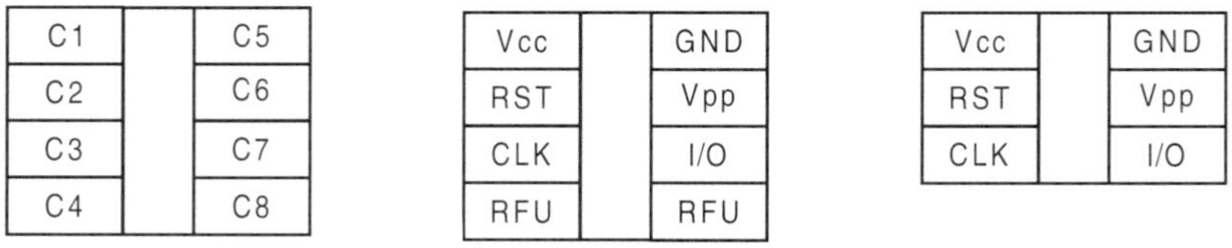

C1	C5
C2	C6
C3	C7
C4	C8

Vcc	GND
RST	Vpp
CLK	I/O
RFU	RFU

Vcc	GND
RST	Vpp
CLK	I/O

Figure 3.29   Electrical assignments and numbering of Smart Card contact areas, per ISO 7816-2

The contacts are numbered sequentially from top left to bottom right. They are designated and electrically specified as shown in Figure 3.29, in line with the ISO standard.

A few years ago, it was still necessary to apply an external voltage to program and erase the EEPROM, since the microcontrollers then in use did not have charge pumps. Contact C6 was reserved for this purpose. However, with current technology, the necessary voltage is generated directly in the chip using a charge pump, so this contact is no longer used. Nevertheless, it cannot be employed for some other function, as this would conflict with the ISO standard. Thus, every Smart Card has a contact that has no real function, but which must still be present. However, the fact that the programming-voltage contact lies between two others that are necessary for the card to function means that it cannot simply be eliminated. This somewhat reduces the drawback of having an unnecessary contact.

Table 3.2 Contact designations and functions according to ISO 7816-2

Contact	Designation	Function
C1	Vcc	Supply voltage
C2	RST	Reset input
C3	CLK	Clock input
C4	RFU	Reserved for future applications, presently not used
C5	GND	Ground (earth)
C6	Vpp	Programming voltage (generally no longer used)
C7	I/O	Serial communications input/output
C8	RFU	Reserved for future applications, presently not used

3.3.2 Supply voltage

A Smart Card's supply voltage is 5 volts, with a maximum variation of ±10%. This voltage, which is the same as that used for conventional TTL circuits, is standard for all cards currently on the market and in use.

For cellular telephones, reducing in the weight of the device in response to market pressure has increasingly led to changing from 6-V or 4.5-V batteries to 3-V batteries. Since all cellular telephone components are currently available in 3-V technology, the Smart Card is the only component that still requires 5 V. Hence, cellular telephones need a voltage converter to provide power to the card. This adds complexity, and thus unnecessarily increases costs. Consequently, future developments will lead to Smart Cards with a voltage range of 3–5 V ±10%, resulting in an effective voltage range of 2.7–5.5 V.

Theoretically, it would also be possible to develop special 3-V card controller ICs. The disadvantage of this would be a loss of compatibility with the millions of existing 5-V cards. In the worst case, using a 3-V card in a 5-V terminal would destroy the card's chip. Users would have to take care to never to insert a card into a terminal with the wrong voltage, which certainly is more than one can demand of them. The advantage of simple and straightforward card usage would be lost.

In principle, the extended voltage range does not pose any problems for the processor or the memory components. However, an EEPROM is also integrated into the microcontroller. It is precisely this EEPROM, with its associated charge pump, that is the greatest obstacle in the way of 3-V technology. It is technically difficult to integrate an EEPROM, along with its charge pump, into a chip that can work with a supply voltage of 2.7–5.5 V, and in fact this has only recently become possible. Of course, this operating voltage range is becoming increasingly obligatory for all microcontrollers.

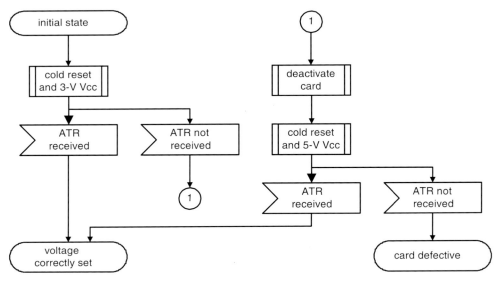

Figure 3.30   State diagram of the terminal process that selects the card supply voltage. The illustrated state machine selects the lowest necessary voltage.

3.3.3 Supply current

The card's microcontroller obtains its supply voltage via contact C1. According to the GSM 11.11 specification, the current may not exceed 10 mA. Although the current version of the ISO standard specifies a value of 200 mA, this is technically out of date, and it will certainly be changed in the future to reflect the current state of the technology.

The power dissipation of a Smart Card is 50 mW, with a supply voltage of 5 V and an assumed current consumption of 10 mA. This is so low that there is no need to be concerned about possible self-warming of the chip in use, even if this power is dissipated over a surface area of around 25 mm^2.

The current consumption of the microcontroller is directly proportional to the clock frequency used, which means that it is possible to specify either the maximum current consumption or the current as a function of the clock frequency. The current is also temperature-dependent, but the effect is not that strong.

Another important detail regarding the supply current has caused severe headaches for several terminal manufacturers who chose to ignore it. All microcontrollers in use are based on CMOS technology. Under certain circumstances, high short-circuit currents can occur briefly during transistor switching processes. These produce current spikes that are many times greater than the nominal operating current, with durations in the nanosecond range. These spikes can also occur when the EEPROM charge pump switches on. If the terminal cannot supply such high currents during these short intervals, the supply voltage will drop below the permitted value. This can produce a write error in the EEPROM or trigger the low-voltage detector in the chip.

For this reason, references to such spikes can now be found in practically every relevant standard and specification. The current supply is required to be able to handle spikes with a maximum duration of 400 ns and a maximum amplitude of 200 mA. Assuming a triangular spike, this corresponds to a charge of 40 nA-s. The problem can be solved in a straightforward fashion by connecting a 100-nF ceramic capacitor between the earth and supply voltage lines, very close to the contacts for the card.

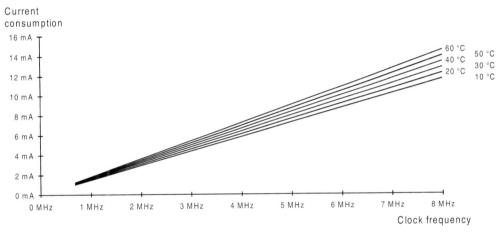

Figure 3.31   Microcontroller current consumption versus clock frequency and temperature, for the normal operating mode (not the sleep mode). The current consumption in the sleep mode is also linearly dependent on the clock frequency. At 5 MHz it is around 50 µA, depending on the microcontroller type.

In the future, the GSM specification is expected to reduce the maximum permitted current consumption to 1 mA, in order to increase the operating time of battery-operated cellular telephones as much as possible. With the rapid advances being made in semiconductor technology, this should soon present few difficulties. Only the current needed for EEPROM writing still presents an obstacle, but with the increasing miniaturization of EEPROM cells and the resulting decrease in cell charges, a value of 1 mA should be achievable within a few years.

Almost all microcontrollers have a special power-saving mode, in which all parts of the chip other than the I/O interrupt are inactive. The microcontroller's current consumption drops substantially when it is in this state, which is called the sleep mode, since most parts of the chip are disconnected from the supply voltage. In principle, only the I/O interface's interrupt logic and the RAM need to be powered, and the remaining elements can be switched off. The contents of the RAM must be maintained in this mode, in order to save the current operating state parameters. In practice, the processor is often also powered, but the ROM and EEPROM are switched off. This mode is of particular interest for battery-operated cellular telephones. GSM 11.11 thus specifies the maximum current permitted in sleep mode to be 200 μA with a 1-MHz clock at 25° C. A reduction of the maximum current consumption to 10 μA is planned for the future.

3.3.4 External clock

Smart Card processors do not have internal clock generators. An externally supplied clock is therefore necessary. This clock provides the reference for data transfer rates. The duty factor of the clock must be in the range of 40–60%, according to GSM 11.11.

The clock signal applied to the contact is not necessarily identical to the internal clock provided to the processor. Some microcontrollers have an optional divider that may be inserted between the external and internal clocks. This often has a division factor of 2, so that the external clock is twice as fast as the internal one. This is partly due to the characteristics of the internal hardware, and partly to allow oscillators already present in the terminal to be used as the source of the clock signal for the chip.

Most Smart Card microcontrollers allow the clock signal to be switched off when the CPU is in the sleep mode. Switching off, in this case, means holding the clock signal at a defined level. Depending on the semiconductor manufacturer's preferences, the 'off' level may be either high or low.

The Smart Card draws only a few microamperes from the clock line, so switching off the clock may at first glance appear somewhat curious. Nevertheless, the amount of power saved within the terminal is substantial, so it can be worthwhile in certain applications.

3.3.5 Data transfer

If an error occurs during a data transfer, it may happen that the terminal and the card attempt to send data at the same time. This results in a data collision on the connecting I/O line. Quite apart from the problem this causes at the application level, at the physical level it can result in currents in the I/O line that may be large enough to destroy the interface components.

To prevent damage to the semiconductors in such cases, the I/O line in the terminal is tied to the +5-V level via a 20-kΩ pull-up resistor, as shown in Figure 3.32. This, together with the agreed convention of never sending an active 5-V level, avoids the problem of the two parties trying to drive the data line to two different levels in the event of an error. If the I/O line needs to be set to a +5-V level during communication, the relevant party switches its output to a high-resistance state (tri-state level), and the line is raised to the +5-V level by the pull-up resistor alone.

3.3.6 Activation and deactivation sequences

All Smart Card microcontrollers are protected against electrostatic charges and potentials on the contacts. In order to avoid undefined states, precisely defined activation and deactivation sequences are prescribed, and these must be strictly adhered to. This is also reflected in the appropriate part of ISO/IEC 7816-3. These sequences define the electrical aspects of activating and deactivating the card, and have nothing to do with the sequence in which the mechanical contacts are established, which is anyway not specified.

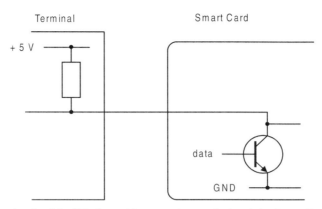

Figure 3.32 The circuit of the I/O channel between the terminal and the Smart Card

As shown in Figure 3.33, the earth connection must be made first, followed by the supply voltage connection. After this comes the clock connection. If an attempt were made to connect the clock before the supply voltage, for example, the microcontroller would try to draw its entire supply current via the clock line. This could damage the chip irreversibly, causing complete functional failure. A faulty deactivation sequence could also have similar effects on the microcontroller.

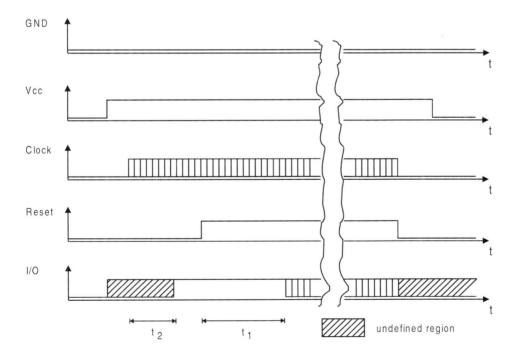

Figure 3.33 Smart Card activation and deactivation sequences according to ISO/IEC 7816-3. The intervals t_1 and t_2 lie in the ranges $400/f \le t_1 \le 40,000/f$ and $t_2 \le 200/f$.

When the microcontroller is in operation, it can be reset via the appropriate line. This requires a low level to be first applied to this lead, with the reset being initiated by the subsequent rising edge. Such a reset during operation is called a warm reset, as in other computer systems. Correspondingly, a cold reset is one that occurs when all the supply lines are switched as specified in the ISO standard.

3.4   Smart Card Microcontrollers

From the perspective of information technology, the central element of a Smart Card is the microcontroller embedded under the contacts. It controls, initiates and monitors all activities. The microcontrollers that have been specially designed and developed for this purpose are complete computers in their own right. This means that they contain processors, memory and interfaces to the outside world.

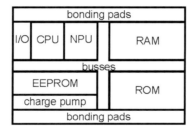

Figure 3.34   Typical arrangement of the functional elements of a Smart Card microcontroller on a semiconductor die

The most important functional elements of a typical Smart Card microcontroller are the CPU, the address/data bus and the three types of memory (RAM, ROM and EEPROM). In addition, the chip contains a simple interface module that is responsible for serial communication with the outside world. This interface should not be imagined to be a complex functional unit that can send and receive data on its own. Here the serial interface is just an address that can be accessed by the CPU and that is connected to the I/O contact.

In addition, some manufacturers offer special arithmetic units on the chip. These act as a sort of mathematical coprocessor. However, the functions provided by these elements are limited to exponential and modulus operations on integers. Both of these operations are fundamental and essential parts of public-key encryption procedures, such as the RSA algorithm.

The microcontrollers used in Smart Cards are not standard, widely available components. Instead, they are components that have been developed specifically for this purpose, and they are not used in other applications. There are several important reasons for this, as follows.

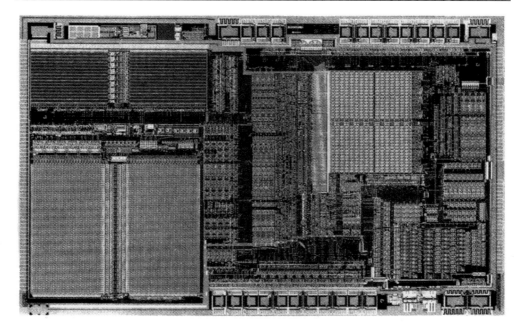

Figure 3.35   Photograph of a PC 83C852 Smart Card microcontroller, showing the following functional elements (from top left to bottom right): ROM, EEPROM, processor with coprocessor and RAM. This chip has an area of 22.3 mm² and contains 183,000 transistors. (Source: Philips)

Manufacturing costs
The surface area of the microcontroller on the silicon wafer is one of the decisive factors with regard to manufacturing costs. Large-area chips make for more complicated and thus more costly packaging in the module. The chip area is thus kept as small as possible.

Furthermore, many standard building blocks that are available on the market include functions that are not needed in Smart Cards. Since these functions take up extra space on the wafer, they can be deleted from chips designed for Smart Cards. Although only a small reduction in the manufacturing cost per chip is achieved by such efforts to minimize the chip size, these small savings add up to a significant amount when a large number of chips is produced. This justifies the modifications to the chip design.

Functionality
Due to the need to integrate all the functional elements of a computer into a single silicon chip, the available number of suitable semiconductor components is extremely limited. Given the requirements of a minimum chip area, 5-V supply voltage and serial interface on the chip, all standard components are effectively ruled out. In addition, the chip must contain a memory (EEPROM) that can be written and erased but that does not require a permanent power supply for data storage.

Figure 3.36   Relative sizes of two functionally identical Smart Card microcontrollers before and after reducing the chip surface area ('shrink process'). Left: SLE 44C80 in 1-μm technology; surface area 21.7 mm^2. Right: SLE 44C80S in 0.8-μm technology; surface area 10 mm^2. (Source: Infineon)

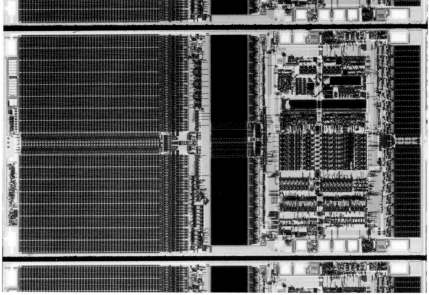

Figure 3.37   Photograph of an ST 16623 Smart Card microcontroller, showing the following functional elements (from left to right): ROM, EEPROM, CPU and RAM. This chip is no longer manufactured, but it clearly shows the arrangement of the individual elements on the die. (Source: ST Microelectronics)

Security

Since Smart Cards are primarily used in security-related fields, for which both passive and active security features are required in the chip, the development of chips that are specially designed for this purpose is unavoidable.

Chip area

The size of the microcontroller die strongly affects the fragility of the chip. The larger the area, the easier it is for the chip to crack when the card is bent or twisted. Consider a telephone card carried in a wallet. The bending stresses on the card and the chip embedded in it are enormous. Even the finest hairline crack in the chip is sufficient to render it useless. Therefore, most card manufacturers place an upper limit of about 25 mm^2 on the chip area, and demand that the layout be as close as possible to square, in order to minimize the risk of fracture.

Availability

The security policy of many card manufacturers is to ensure that the microcontrollers that they use are not available to all and sundry on the open market. This makes analysis or 'reverse engineering' of the chip's hardware considerably more difficult, since a potential attacker would normally not have access to the hardware.

This high degree of specialization, limited to only a few types of microcontrollers (and thus only a few manufacturers), has the disadvantage that card manufacturers have little independence. In the event of a production bottleneck with a semiconductor manufacturer, it is impossible to quickly switch over to another component.

3.4.1 Processor types

The processors used in Smart Cards are not special developments, but proven modules that have been used in other areas for a long time. It is not usual to develop new processors for special application areas, since this is generally too expensive. In addition, this would result in a completely unfamiliar processor, for which no suitable function libraries would be available from operating system makers.

Additionally, Smart Card processors must be extremely reliable. It is therefore better to rely on older processors that have been proven in practice, rather than experimenting with the latest developments of semiconductor manufacturers. The aerospace industry, which is very concerned with functional security, uses only components that are one or two generations behind the current state of the technology, and for the same reasons.

The transistor was invented at the Bell labs in 1947, and Intel brought out the first microprocessor in 1971, with the type designation '4004'. It contained 2300 transistors, had a clock frequency of 108 kHz, could execute 45 machine-code instructions with a 4-bit data bus, and had a processing capacity of 0.06 MIPS. Since then, the development of integrated microprocessor components has made huge strides. Recent products such as the Pentium II processor, with 7.5 million transistors built with 0.25-μm technology, clearly indicate this. However, the most technologically advanced processors are not used in Smart Cards, for reasons that have just been explained. A total of 200.000 transistors in the chip is already considered high in this context.

Figure 3.38 Photo of an SLE 66CX160S Smart Card microcontroller with an area of 21 mm². This chip was made using 0.6-μm technology and has 32 kB of ROM, a 16-kB EEPROM and 1280 bytes of RAM. The two unlabeled regions on the left-hand side of the chip are the numeric coprocessor and the peripheral elements (timer, random-number generator and CRC arithmetic processor). The five bonding pads for the electrical connections to the module contacts can be clearly seen in the photo. (Source: Infineon)

Since the size of the addressable memory in a Smart Card lies between 6 kB and a maximum of 30 kB, the use of an 8-bit memory bus does not impose any significant restrictions. The processors themselves are based on a CISC (complex instruction set computer) architecture. They thus require several clock cycles per machine instruction, and usually have very extensive instruction sets. The address range of the 8-bit processors is most often 16 bits, with which a maximum of 65.536 bytes can be addressed. The processor instruction sets are based on either the Motorola 6805 or Intel 8051 architecture. Supplementary instructions may be added to the standard instruction set by the semiconductor manufacturer. These mostly relate to additional options for 16-bit memory addressing, which exist only in the most rudimentary form in the two instruction sets that form the basis for the Smart Card processor instruction sets.

There is one Smart Card processor that is an exception to the architectures just mentioned This is the Hitachi H8 chip, which employs a 16-bit processor whose architecture and instruction set are similar to those of a RISC (reduced instruction set computer) machine.

Future processors for high-end Smart Cards will however soon leave the 8-bit world behind and take on a 32-bit architecture. Especially with software-interpreted program code, such as with the current Java implementations, an acceptable execution speed can only be achieved if a powerful processor is used. The first step into the 32-bit league was made as early as 1993 by the European CASCADE project (Chip Architecture for Smart Card and portable intelligent Devices). One objective of this project was to provide a high-performance processor for Smart Cards. The project chose the modern ARM 7M RISC processor, which is often used in portable equipment (such as video cameras and PDAs).[3] It has a 32-bit architecture, can run at up to 20 MHz with a 3-V supply, and consumes only 40 mA under these conditions. With 0.8-μm technology, the ARM 7 core has an area of 5.9 mm^2 (3.12 by 1.9 mm), and the associated arithmetic processor for the usual public-key algorithms (RSA, DSA and EC) takes up an additional 2 mm^2. The processor has both a supervisor and a user mode, and thus supports the separation of the operating-system code and the application code.

Although 32-bit processors take up significantly more space on the die than 8-bit processors using the same technology, due to their wider bus structures and more complex internal organizations, they will be used in increasing numbers in future Smart Card applications. The processing power that they offer is indispensable for these applications, so the disadvantages of greater power consumption and increased chip area can be accepted as the price of progress. Of course, 8-bit processors will not die out in the foreseeable future, since they provide a solid basis for inexpensive chips.

3.4.2   Memory types

Besides the processor, the various types of memory are the most important elements of a microcontroller. They serve to store program code and data. Since Smart Card microcontrollers must be complete computers, they exhibit a characteristic division of the memory into RAM, ROM and EEPROM. The exact division depends very strongly on the chip's eventual field of application. In any case, an effort is always made to keep the RAM and EEPROM as small as possible, since they require the most space per bit.

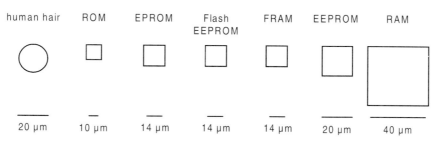

human hair	ROM	EPROM	Flash EEPROM	FRAM	EEPROM	RAM
20 μm	10 μm	14 μm	14 μm	14 μm	20 μm	40 μm

Figure 3.39   Comparison of the die area required for a single bit cell with various types of memory. The indicated dimensions are approximate and relate to 0.8-μm technology. For comparison, the diameter of the first planar transistor in 1959 was 764 μm [Buchmann 96, Stix 96], the diameter of the period at the end of each sentence in this book is 400 μm, the resolution limit of the human eye is 40 μm, the size of a bacterium is 0.4–2 μm and the size of a DNA double helix is 0.1 μm.

---

[3]   Based on [Peyet 97]

With multi-application Smart Cards, which can manage several applications at the same time, the most frequently used chips have a ROM capacity that is roughly twice as large as the EEPROM capacity, in order to provide enough room to store the very complex operating system code. For single-application Smart Cards, microcontrollers are selected whose EEPROM capacity is only slightly larger than the volume of the application data. All non-fixed application data, together with some parts of the operating system, can thus be stored in the EEPROM. This optimizes the use of the EEPROM, which takes up a relatively large amount of space on the die and is thus expensive.

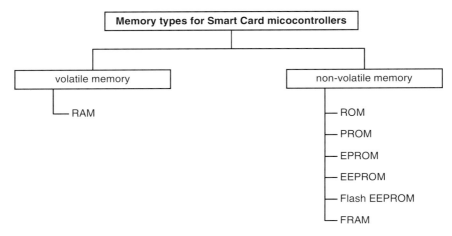

Figure 3.40   Classification chart for Smart Card microcontroller memory types. PROM and EPROM memories are as a rule no longer used in modern microcontrollers. Flash EEPROM and FRAM are memory types whose development for Smart Cards is in still the early stages.

The integration of three different types of semiconductor memory on a single piece of silicon is technically complex. It requires a considerable number of production steps and exposure masks. The amount of space required by each type of memory also differs very much, due to their different constructions and operating principles. For example, a RAM cell needs about four times as much space as an EEPROM cell, which in turn needs four times as much space as a ROM cell. This is why Smart Card microcontrollers have so little RAM. A 256-byte RAM is already considered to be large. If you consider that 1024 bytes of EEPROM or 4096 bytes of ROM can be put into the same area, you can understand why. The following three numeric examples illustrate these size relationships:

`   A typical laser printer works at a resolution of 600 dpi (dots per inch), which means that the minimum possible spot size is 42.6 μm. Also, the period at the end of this sentence has a diameter of 400 μm. If you wanted to print with a resolution equal to the 0.8-μm structure width, which is currently standard in semiconductor technology, you would need a printer with a resolution of 32,000 dpi!
`   The latest high-performance hard disk drives can store up to 11.6 billion bits per square inch. Under the idealized assumption that each bit occupies a square area, this yields an edge length of 0.24 μm for each bit cell. A ROM cell of a Smart Card microcontroller made with 0.8-μm technology requires 1700 times as much area for a single bit!

With a CD-ROM, the situation is different. In this case, the storage density is
7.3MB/cm². This corresponds to an edge length of 1.4 µm for a one-bit cell, assuming
square cells. This is around 80 times less than the area occupied by a ROM cell in 0.8-
µm technology. With the new DVDs (digital versatile disks), a density of 50.5 MB/cm²
is already possible. A single bit thus occupies the area of a square with an edge length of
0.5 µm, which is 400 times smaller than a single ROM bit cell in 0.8-µm technology.

Table 3.3   Comparison of memory types used in Smart Card microcontrollers. For comparison, the
surface area of the period at the end of each sentence in this book is 125,660 µm².

Type of memory	Number of write/erase cycles	Writing time per memory access	Typical cell size with 0.8-µm technology
Volatile memory			
RAM	unlimited	$\approx$ 70 ns	$\cong$ 1 700 µm²
Non-volatile memory			
EEPROM	10,000–1,000,000	3–10 ms	$\approx$ 400 µm²
EPROM	1 (cannot be erased with UV light)	$\approx$ 50 ms	$\approx$ 200 µm²
Flash EEPROM	$\approx$ 100,000	$\approx$ 10 µs	$\approx$ 200 µm²
FRAM	$\approx 10^{10}$	$\approx$ 100 ns	$\approx$ 200 µm²
PROM	1	$\approx$ 100 ms	---
ROM	0	---	$\approx$ 100 µm²

ROM (read-only memory)

As the name implies, this type of memory can only be read, and cannot be written. No
voltage is needed to hold the data in memory, since they are 'hard-wired' in the chip.

A Smart Card's ROM contains most of the operating system routines, as well as various
testing and diagnostic functions. These programs are built into the chip by the manufacturer
when it is made. A ROM mask is made from the program code, and this is then used to
'burn' the program in the chip using lithographic processes. The data, which are the same
for all chips of a production run, can only be entered into the ROM during manufacturing.

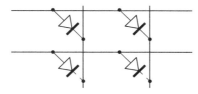

Figure 3.41   Basic functional structure of a ROM

PROM (programmable read-only memory)

PROM is not used in Smart Card microcontrollers, though it could offer several important
advantages. In contrast to ROM, PROM need not be programmed during manufacture, but

can be written shortly before the chip is built into its module. A PROM also does not need any supply voltage for data storage. The main reason for not using this type of memory is that programming a PROM requires access to the address, data and control buses. This is however exactly what should not be possible with Smart Cards, since this would allow data not only to be written but also to be read. In view of the confidential data stored in the memory, the use of PROM is strictly prohibited.

EPROM (erasable programmable read-only memory)
EPROM was often used in the early years of Smart Card technology, since at that time it was the only type of memory that could hold data without a supply voltage and that also could be written (though only once per bit). However, since an EPROM can only be erased with UV light, it cannot be erased in a Smart Card. This is why EPROM is no longer of practical significance in new applications.

The only sensible use for EPROM is for the permanent, irreversible storage of a chip number during semiconductor production, but nowadays this can be realized with a special non-erasable EEPROM.

Figure 3.42   Photo of a ROM cell, shown with 1000× enlargement (left) and 11,200× enlargement (right) (Source: Giesecke & Devrient)

EEPROM (electrically erasable programmable read-only memory)
This type of memory is more complex than ROM and RAM. It is used in Smart Cards for all data and programs that at some time must be modified or erased. Functionally, an EEPROM corresponds to the hard disk of a PC, since the data remain in the memory in the absence of power and can be modified as necessary.

In principle, an EEPROM cell is a tiny capacitor that can be charged or discharged. The charge state can be interrogated by sensing logic. A charged capacitor represents a logic 1, while a discharged capacitor represents a logic 0. In order to store one data byte, eight of these small capacitors are needed, plus the appropriate sensing circuitry.

The erased state of the EEPROM cell is the critical factor with regard to writing to the cell. In most types of EEPROM, the erased state is °1°. An EEPROM has the property that an individual cell can only be changed from its erased state to its unerased state, which in this example is °0°. If an EEPROM cell is already in the °0° state, then an entire EEPROM page must be erased in order to restore that bit to the °1° state. The algorithm that is usually used for an EEPROM write routine is described in Listing 3.1.

Listing 3.1   Pseudocode of a write routine for complete EEPROM pages. If multiple pages or only part of a page is to be written, this routine should be nested in a higher-level routine. A similar procedure should be used if it is necessary to call a write-retry routine in case of an error. The erased state of the EERPOM is 'FF', while the written state is '00'.

UpdateEEPROM: // *NewData*: data to be written // *StoredData*: stored data	Entry point for writing data to an EEPROM page
IF (NewData = StoredData) THEN (GOTO UpdateEEPROM_Exit)	If the data already stored in the EEPROM page is the same as the new data, exit the function.
WorkData := *NewData* XOR *StoredData*	Following the XOR operation, the differences between the stored data and the new data can be seen as set bits in the variable WorkData.
WorkData := *WorkData* AND *NewData*	The AND operation causes the variable WorkData to be non-zero if the EEPROM page must be erased before the write process.

IF (*WorkData* <> 0) THEN (
Erase EEPROM_Page
IF (*StoredData* <> 'FF') THEN (
GOTO UpdateEEPROM_Errror_Exit))
	If the variable WorkData is non-zero, the EEPROM page must be erased before the write process. After this process, a test is made to see if the EEPROM page was successfully erased.

Write EEPROM_Page with *NewData* IF (*StoredData* <> *NewData*) THEN (GOTO UpdateEEPROM_Errror_Exit)	The EEPROM page can now be written. Following this comes a test to check whether the data were successfully written to the EEPROM.
Update EEPROM_Exit: RETURN	The function has completed successfully.
Update EEPROM_Error_Exit: RETURN	An error has occurred during the execution of the function.

In order to understand how an EEPROM cell works, you must picture it as a semiconductor device. An EEPROM cell is shown in cross section in Figure 3.43. The actual construction is somewhat more complicated, but this simplified diagram is a very useful aid to understanding.

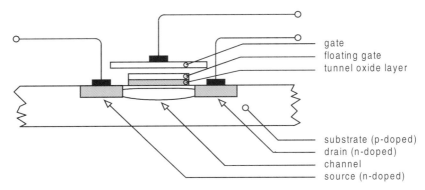

gate
floating gate
tunnel oxide layer

substrate (p-doped)
drain (n-doped)
channel
source (n-doped)

Figure 3.43   Cross section of the semiconductor structure of an EEPROM cell

An EEPROM cell, in its simplest form, is basically a modified field-effect transistor (MOSFET) that is constructed on top of a silicon substrate. A source and a drain are formed on the substrate. Between these there is a control gate, and the current between the source and the drain can be controlled by applying a potential to the gate. As long as no potential is applied to the gate, no current can flow, since two diode junctions (n–p and p–n) are located between the source and the drain. If a positive potential is applied to the gate, electrons are attracted from the substrate, thus forming an electrically conducting channel between the source and the drain. The FET is then conductive, and a current flows.

Figure 3.44   Photo of an EEPROM cell, shown with 1000× enlargement (left) and 4000× enlargement (right) (Source: Giesecke & Devrient)

In an EEPROM cell, an additional 'floating' gate is located between the control gate and the substrate. It is not connected to any external voltage source, and the separation between it and the substrate is very small, of the order of 10 nm. The floating gate can be charged or discharged via the substrate due to the tunnel effect (Fowler–Nordheim effect). This effect causes charge carriers to penetrate thin oxide layers. It requires a sufficiently large potential difference across the thin insulating layer, which is called the tunnel oxide layer. The flow of current between the source and the drain can be controlled by the charge on the floating gate. This means that, depending on whether a current can flow, the state of this gate can be interpreted as a logical 0 or a logical 1.

A high positive voltage is applied to the control gate to charge the floating gate. This causes a high potential difference between the substrate and the floating gate, which in turn causes electrons to tunnel through the oxide layer to the floating gate. The resulting current is measured in picoamperes. The floating gate is now negatively charged and produces a high threshold voltage between the source and the drain, which means that the field-effect transistor is blocked. No current can flow between the source and the drain. The storage of electrons in the floating gate is thus equivalent to the storage of information.

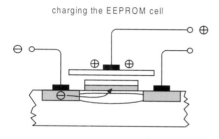

Figure 3.45 Charging an EEPROM cell

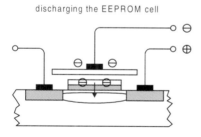
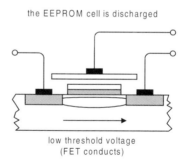

Figure 3.46 Discharging an EEPROM cell

The voltage needed to charge the EEPROM cell is about 17 V at the control gate, but this is reduced to about 12 V at the floating gate by capacitive coupling. However, since Smart Card microcontrollers work with a supply voltage of only 3–5 V, a charge pump is needed to produce the required voltage. In principle, the charge pump is a cascaded voltage-multiplier circuit. It generates an output voltage of about 25 V from the low input voltage, which after stabilization is approximately equal to the required 17 V. Depending on construction details, charging an EEPROM cell requires between 3 and 10 ms per memory page (1–32 bytes).

charging the capacitors of the charge pump   discharging the capacitors of the charge pump

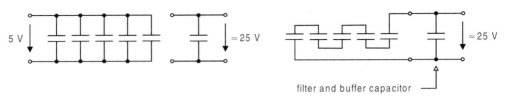

filter and buffer capacitor

Figure 3.47   This schematic diagram shows the working principle of a charge pump circuit, for charging (left) and discharging (right). This process takes place at such a high frequency that the charge pump produces a slightly pulsating DC voltage at its output.

To erase an EEPROM cell, a negative voltage is applied to the control gate. This causes the electrons to leave the floating gate and return to the substrate. The EEPROM cell is now discharged, and the threshold voltage between the source and the drain is low, which means that the FET is conductive.

The floating gate can also be discharged by heating or by powerful radiation (such as X-rays or UV radiation), which causes it to return to its 'secure' state. This state is of fundamental significance for the design of Smart Card operating systems, since otherwise it would be possible to penetrate security barriers by deliberately altering the ambient conditions. Depending on the technical implementation of an EEPROM cell, the secure state can correspond to a logical 0 or a logical 1. This is specific to each Smart Card microcontroller, and should be confirmed with the manufacturer if necessary.

EEPROM is one of the few types of semiconductor memory that has a limited number of access cycles. It can be read any number of times, but can be programmed only a limited number of times. The reason for this limitation can be found in its construction. The life expectancy of an EEPROM depends strongly on the type, thickness and quality of the tunnel oxide layer between the floating gate and the substrate. Since this layer must be produced at a very early phase of the semiconductor's manufacture, it is naturally subjected to strong thermal stresses during later production steps. These stresses may result in damage to the oxide layer, which in turn affects the EEPROM cell's useful life. During production, and every time the cell is written, the tunnel oxide layer absorbs electrons that are not later released. These trapped electrons are located close to the channel between the source and the drain, and once they reach a certain number they have a stronger effect on the threshold potential than the charge stored in the floating gate. When this occurs, the EEPROM cell has reached the end of its useful life. It can be written, but the charge on the floating gate affects the characteristics of the channel between the source and the drain only to a minimal degree, so the threshold potential always retains the same value. The number of possible write/erase cycles varies greatly, depending on construction details. Typical values range from 10,000 to 1,000,000 cycles over the entire temperature and supply voltage range. However, at the ideal supply voltage and room temperature, the number of possible cycles is 10–50 times higher.

When an EEPROM cell is near the end of its life, data are stored for a short time only. The time interval can vary from hours to minutes, or even seconds. The more exhausted the EEPROM is, i.e. the more electrons that have been absorbed by the tunnel oxide layer, the shorter is the storage interval.

A charged floating gate loses its charge over time, due to insulation losses and quantum-mechanical effects. The time required for this phenomenon to become noticeable can range between 10 and 100 years. In this regard, it is interesting to note that a charged floating gate holds 100,000 to 1,000,000 electrons, depending on the implementation. Currently, all semiconductor manufacturers guarantee data storage for 10 years. In order to increase this value, the contents of EEPROM cells can be refreshed periodically by reprogramming. However, this is only makes sense when the data must be stored over long periods.

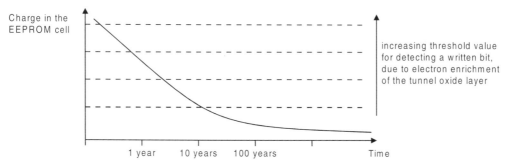

Figure 3.48 How the discharge curve of an EEPROM cell shifts as a function of the number of executed program/erase cycles

Flash EEPROM (flash electrically erasable programmable read-only memory)
Flash EEPROMs have been available for some years as discrete components. They are similar to EEPROMs in construction and operation. The fundamental difference between a flash EEPROM and a normal EEPROM is in the writing process, which is based on hot electron injection instead of the tunnel effect. 'Hot' electrons are fast electrons that are produced by a high potential difference between the source and the drain. Some of them penetrate the tunnel oxide layer due to the influence of a positively charged control gate, and are stored in the floating gate. The writing time is reduced by this effect to about 10 µs, which is a great advance compared with 3–10 ms for normal EEPROMs. In addition, the programming voltage is only 12 V, compared to 17 V for EEPROMs. The first Smart Card microcontrollers with both flash EEPROM and EEPROM are already available [Atmel]. In these chips, a boot loader is written to the EEPROM during the manufacturing process, and this is then used by the Smart Card manufacturer to download the program code and data. Current flash EEPROM cells have a guaranteed data retention time of at least 10 years, at least 100,000 write/erase cycles and a typical page size of 64 bytes.

Flash EEPROMs are thus an excellent replacement for mask-programmed ROMs. Eliminating the need to prepare ROM masks for microcontrollers with flash EEPROMs can reduce the development time of a Smart Card project by several months.

FRAM (ferroelectric random-access memory)
FRAM is a new development in semiconductor technology. Despite the name, FRAM is not volatile like RAM. Instead, it retains its content without a supply voltage. The properties of ferroelectric materials are used to store data in a FRAM. The construction is similar to that of an EEPROM cell, but a ferroelectric substance is located between the control gate and the floating gate.

Figure 3.49 Photo of an AT89SC168 Smart Card microcontroller with flash EEPROM. The functional elements in the upper row are (from left to right) the logic module, the RAM and the CPU. The lower row contains (from left to right) the charge pump for the EEPROM and the flash EEPROM. (Source: Atmel)

FRAMs have all the desirable properties of an ideal Smart Card memory. Only 5 V is needed for programming, the programming time is around 100 ns and the maximum number of programming cycles is around one billion. The integration density is similar to that of flash EEPROM. However, FRAMs have two disadvantages. The first is a limited number of read cycles, which makes a type of refresh cycle necessary. The second is even more significant: producing FRAMs involves difficult processing steps, and presently no significant effort is being made to utilize this technology in Smart Card microcontrollers. This may however change in a few years' time, since FRAMs have all the features needed to completely supplant EEPROMs, which are presently used almost exclusively.

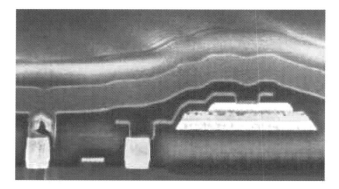

Figure 3.50 Cross section of a FRAM cell in 0.5-μm technology. The two light-colored rectangles on the left-hand side are the control leads on the semiconductor surface. The horizontal rectangle on the right-hand side is the actual FRAM cell. The ferroelectric material is visible as the dark layer between the two electrodes. The width of the FRAM cell is approximately 2 μm. (Source: Fujitsu)

RAM (random-access memory)
The RAM is that portion of a Smart Card's memory in which data can be stored and modified during a session. The number of accesses is unlimited. RAM needs a power supply for its operation. Once this is switched off, or if it fails temporarily, the contents of the RAM are no longer defined.

A RAM cell consists of several transistors, connected such that they function as a bistable multivibrator. The state of this multivibrator represents the stored value of one bit in the RAM. The RAM used in Smart Cards is static (SRAM), which means that its contents do not have to be periodically refreshed. It is thus not dependent on an external clock, in contrast to dynamic RAM (DRAM). Static RAM is important, since it must be possible to stop the clock signal to a Smart Card. With dynamic RAM, this would cause the stored information to be lost.

3.4.3   Supplementary hardware

There are some Smart Card requirements that cannot be fully satisfied by software alone, given the basic hardware configuration, so the various Smart Card manufacturers offer additional functions in the form of hardware on the chip.

Coprocessors
Coprocessors are specially developed arithmetic units for computations relating to public key algorithms. They are integrated on the silicon substrate along with the microcontroller's usual functional units. These coprocessors can only perform a few basic types of calculations that are needed for these algorithms, including exponentiation and modulo arithmetic with large numbers. The speed of these units, which are optimized exclusively for these two operations, is achieved by very wide architectures (up to 140 bits). In their special application area, they can outperform even a very fast PC by a factor of at least 6.

The arithmetic coprocessor unit is called by the processor, which passes either the data directly or pointers to the data, and then issues a command to start the calculations. Once the task has been completed and the result has been stored in RAM, control of the chip is returned to the processor.

DES computation unit
The DES is used as the standard cryptographic algorithm for both financial transactions and telecommunications applications. This large potential market has made it worthwhile for chip manufacturers to add DES processing units to their Smart Card microcontrollers. This is in principle not all that difficult, since DES was originally developed to be primarily implemented in hardware. The largest problem with the marketing of DES units integrated with the microcontroller is in any case not technical, but has to do with export rights. In many countries, components with fast, hardware-based DES encryption are subject to a variety of export restrictions.

If we just have a look at the first announced performance data for DES calculation units for Smart Card microcontrollers, their advantages are more than obvious. Times on the order of 154 μs for a simple DES operation and 236 μs for a triple DES operation can be achieved with a 3.5-MHz clock. Increasing the clock frequency produces a linear reduction in the calculation time. In addition, a DES calculation unit takes up hardly any more space on the chip than DES program code in ROM, so that it does not result in any significant increase in the size of the microcontroller chip.

Random number generator

Random numbers are often needed for the generation of keys to authenticate the Smart Card and the terminal. For security reasons, these should be genuine random numbers and not pseudo-random numbers, as is common in Smart Cards. There are microcontrollers that have integrated random number generators that produce truly random numbers.

The quality of such a generator must not be affected by external physical factors, such as temperature or supply voltage. It may employ such external influences to help it to generate random numbers, but this must happen in such a way that it is not possible to learn how to predict the random numbers by deliberately altering one or more of these parameters.

Since this is very difficult to achieve in silicon alone, a different approach is taken. The random number generator utilizes the various logic states of the processor, such as the clock rate or register contents, and passes these to a feedback shift register that is in turn driven by a clock generated on the basis of a number of different parameters. If the CPU reads the random number register, it obtains a relatively good random number, which cannot be deterministically ascertained from outside the chip. The quality of the genuine random numbers so obtained can be further improved by additional procedures and algorithms.

Error detection and correction in EEPROM

The fundamental limit on a Smart Card's life is determined by the EEPROM, with its technically limited number of its write/erase cycles. One way to ease this limitation is to use software to compute error correction codes for certain heavily used parts of the EEPROM, so that errors can be corrected. Alternatively, hardware error correction can be implemented on the chip. This allows EEPROM errors to be detected in a manner that is transparent to the software, and they can be corrected if there are not too many.

Naturally, additional EEPROM is needed to store the necessary code. Since good error correction codes require a relatively large amount of storage space, a strategic decision is necessary. If the error correction is good, it requires additional space, which may take up as much as 50% of the memory to be protected. The memory that holds the error correction code can however be used only for this purpose. If a lower grade of error correction is used, then less memory is lost, but the benefit is very questionable.

There are several chips on the market that have EEPROM error detection and correction implemented in hardware, but which also use half of the protected memory. The net result is that the amount of EEPROM available to the user is not that large, but the usable lifetime of the EEPROM is several times longer than the usual value.

Hardware-supported data transfer

The only communication between a Smart Card and the outside world takes place via a bi-directional serial interface. Up to now, data reception and transmission via this interface have been controlled exclusively by operating system software, without any hardware support. This makes for very complex software. It also increases the possibility of software errors. However, the main problem is that the speed of software-supported data transfer is limited, since the processor's speed itself is highly constrained.

If higher communication speeds are needed, it is necessary to use either internal clock multiplication or a UART (universal asynchronous receiver transmitter) component. As the name suggests, such a component is a universally usable unit for the transmission and reception of data. It allows data reception and transmission to take place without the direct involvement of the processor. It is not restricted by the processor's speed, nor does it need software for communicating at the byte level. Of course, the higher levels of the relevant data transmission protocol still must be present in the Smart Card as software, but the lowest level is implemented as hardware in the UART.

This is likely to become a standard feature in future Smart Card microcontrollers, but there are presently only very few chips that allow hardware-supported communications. The technical capability has existed for a long time, but it can be calculated that transmission and reception routines implemented as software in ROM require less physical space on the semiconductor than a functionally comparable UART module. Since the chip's surface area is decisive with regard to the price of a Smart Card microcontroller, up to now almost all semiconductor manufacturers have decided against the hardware implementation. With increasing circuit integration density, however, this balance could change rapidly.

Internal clock multiplication

The demands on the processing power of Smart Cards are constantly increasing. This applies very strongly in the area of cryptographic algorithms. In order to meet these demands, one possibility is simply to use higher clock frequencies with special versions of the microcontroller. Processing power increases linearly with the clock frequency, so doubling the clock frequency doubles the performance of the processor. However, for reasons of compatibility, it is generally counter-productive to increase clock frequency beyond about 5 MHz.

To get around this restriction, the use of an internal clock multiplier has been repeatedly proposed, so that the internal external clock frequency could be increased while leaving the external clock frequency unaltered. For example, this could be done using a PLL (phase-locked loop) circuit, which has long been standard technology. A Smart Card connected to an external 3.5-MHz clock could, for example, be driven internally at 28 MHz. This would offer considerable advantages for the computation of complex cryptographic algorithms.

Nevertheless, processor speed is not the only bottleneck in a Smart Card. The data transfer rate (which is set by various standards) and the EEPROM's write/erase time would not benefit from such a solution, which would severely limit its advantages. Nevertheless, the use of an internal clock multiplier can still be a great advantage for some applications, especially considering that the additional circuitry required on the card is minimal.

However, one should not overlook the fact that clock multiplication considerably increases the current consumption of the microcontroller. As a rule, the relationship between clock frequency and current consumption is linear, which means that quadrupling the clock frequency (for example) also quadruples the current consumption. With battery-operated terminals in particular, increased current consumption is undesirable.

Hardware-supported memory management

With the latest Smart Card operating systems, it is possible to download executable machine code into the card.[4] This can then be called using a special instruction, and can for example execute a cryptographic function only known to the card issuer. However, once an executable program has been loaded into the card, it is in principle no longer possible to prevent it from including a function to read out secret data from the memory. Operating system manufacturers have been very careful to maintain the confidentiality of their system structures and program code. The same considerations apply to secret keys and algorithms within various applications on the card. The public availability of such confidential

---

[4]   See also Section 5.10, 'Smart Card Operating Systems with Downloadable Program Code'.

information would have fatal consequences for an application provider. One administrative solution is to have every new program tested by an independent organization. However, even this cannot guarantee complete security, since a program that is not the same as that tested could later be substituted for the verified program, or the program may be so secret that nobody other than the application provider may be allowed to know it.

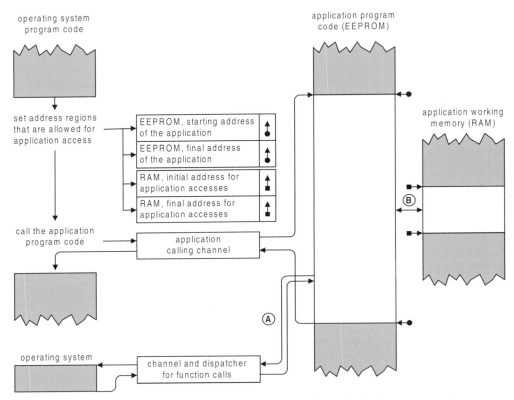

Figure 3.51 Schematic representation of the operating principles of a hardware-supported memory management unit (MMU) in a Smart Card microcontroller. 'A' marks a call to an operating-system function that is channeled by the MMU and controlled by a task dispatcher. 'B' illustrates an example of a write–read access to an application memory area that is delimited by the MMU.

One acceptable solution to this impasse is to equip the Smart Card microcontroller with a memory management unit (MMU). Such a unit controls the memory boundaries of the current application in parallel with program execution. The permitted memory area is fixed by an operating system routine before the application is called, and it cannot be altered by the application program while the program is running. This ensures that the application is completely encapsulated and cannot access memory areas forbidden to it.

Only very few Smart Card microcontrollers have MMUs, although they have been used for years in many other areas. Nonetheless, this additional hardware will greatly increase in importance in the future, since it is the only practical way to safely separate several applications that share a single Smart Card.

Chip hardware extensions

If an extension of the chip hardware is necessary for some reason, this means considerable expenditures on development time and effort for the manufacturer. There are only two ways to implement customer-specific hardware: it can be constructed in silicon on the basis of an existing chip family, or it can be constructed as two-chip system, with all of the associated disadvantages.

An acceptable way around this involves a compromise solution. In principle, it is a mixture of both of the above solutions. The chip containing the new hardware module is glued directly to the existing chip and electrically connected to it by bonding wires. This solution benefits from the fact that most Smart Card microcontrollers have more than one I/O lead, and one of the additional leads can be used to communicate with the additional chip. The thickness of the resulting sandwich construction differs only insignificantly from that of a normal chip, since the silicon substrate can be ground away more than usual to make it thinner. A sandwich chip can thus be built into a standard module without additional effort or costs.

This procedure is ideal for satisfying individual customer requirements for additional hardware without expensive modifications. Existing chips can be combined with new modules, which may execute fast cryptographic computations (such as DES) or have a special serial interface for testing security features in other chips. It is also possible to introduce a special ASIC containing a secret cryptographic algorithm into the card. The procedure is not cost-effective for large production quantities (in the range of millions of pieces), since in such cases it is worthwhile to develop a special chip. However, for small to medium production quantities, sandwich chips are a very effective solution for prototype series or special applications, such as security modules for terminals.

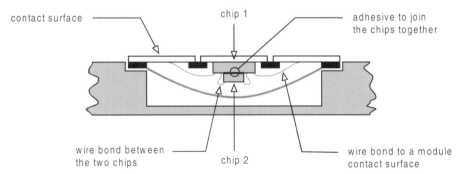

Figure 3.52   Cross section of a chip module containing two different chips that are electrically interconnected by bonding wires.

3.5   Contact-type Cards

The main difference between a Smart Card and other types of cards is the embedded microcontroller. If contacts are used for the power supply and data transfer functions, electrical connections are required. These consist of six or eight gold-plated contacts, which can be seen on every standard Smart Card. The locations of these contacts with respect to the card body, and their sizes, are specified by the ISO 7816-2 standard of 1988.

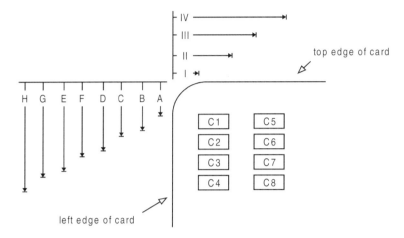

Figure 3.53   The positions of the contacts relative to the card body outline (drawing not to scale)

I	10.25 mm maximum	A	19.23 mm minimum
II	12.25 mm minimum	B	20.93 mm minimum
III	17.87 mm maximum	C	21.77 mm maximum
IV	79.87 mm minimum	D	23.47 mm minimum
		E	24.31 mm maximum
		F	26.01 mm minimum
		G	26.85 mm maximum
		H	28.55 mm minimum

In France, a national standard was already in use long before ISO 7816-2 was issued by the AFNOR. It specifies a slightly higher location for the contacts than the ISO standard. This location is also included in the ISO standard as a 'transitional contacts location', but the standard recommends that this location not be used in the future. However, since there are still many cards in France that use the 'transitional' position, it is not likely that it will disappear quickly.

The absolute position of the contact field is in the upper left corner of the card body. The dimensioned drawing shown in Figure 3.53 makes this clear. The minimum dimensions of any contact are 1.7 mm by 2 mm (height by width). The maximum dimensions of any contact are not specified, but they are of course limited by the fact that the individual contacts must be electrically isolated from each other.

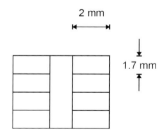

Figure 3.54   Minimum contact dimensions as specified in ISO 7816-2

The position of the module within the card body is specified in the standard. The locations of the magnetic strip area and the area reserved for embossing are also exactly specified (see ISO 7811). All three of these elements may be present on a single card. However, in this case the following mutual relationships must be taken into account: (a) if only a chip and an embossing field are present, they may be located on the same side or on opposite sides of the card; (b) if a magnetic strip is also present, it and the embossing area must be located on opposite sides of the card.

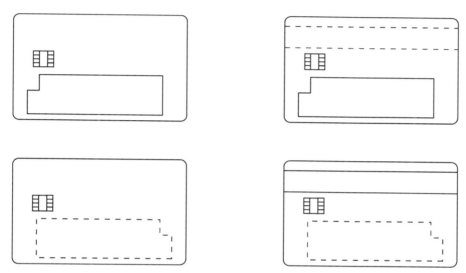

Figure 3.55   The various possible arrangements for the chip, embossment field and magnetic strip, according to the ISO 7816-2 standard

Figure 3.56   An example of a card with a chip, magnetic strip, signature area and embossing (Source: Giesecke & Devrient)

3.6   Contactless Cards

As already described in Section 2.3.3, contactless cards do not require any electrical connection between the Smart Card and the card terminal in order to transfer energy and data over a short distance. The most important advantages of the contactless card technology have also already been presented in Section 2.3.3. Here we are only interested in the operating principles of contactless cards. The techniques that are used with contactless cards are not new; they have been common knowledge for many years in RFID (Radio-Frequency Identification) systems, which have been used for a variety of applications such as animal implants of transponders and electronic anti-start systems for vehicles. There is a large variety of methods for identifying persons or objects at short or even long distances, based on radio techniques and in particular on radar techniques. Among the large variety of technical possibilities, only a small number are suitable for use in Smart Cards in the ID-1 format, since all functional elements must be housed in a flexible card that is only 0.76 mm thick. We will restrict our attention to these alone. There are for instance no batteries available with this thickness that could be used to supply power to the electronic circuitry.

Just as with contact-type Smart Cards, a system that used contactless cards consists of at least two components, namely the card itself and a matching terminal, which can function as a reader or a reader/writer according to the technology used. As a rule, the terminal includes an additional interface, via which it can communicate with a background system.

The following four needs must be satisfied if the contactless card is to be able to communicate with the terminal:

`   energy transfer for supplying power to the integrated circuit,
`   transmission of the clock signal,
`   data transfer to the Smart Card,
`   data transfer from the Smart Card.

Many different methods have been developed for solving these problems, based on experience with RFID systems. Most of these are specifically designed for special applications. For instance, it makes a difference whether the distance between the card and the terminal in normal operation is only a few millimeters or around a meter. Naturally, the development of many different solutions that are specifically designed and optimized for particular applications means that these solutions are unavoidable incompatible with each other.

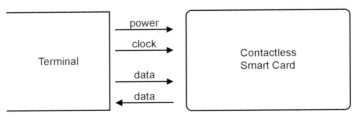

Figure 3.57   The necessary energy and data transfers between the terminal and a contactless Smart Card

Inductive coupling

Inductive coupling is presently the most widely used technique. It can be used to transfer both energy and data. Differing requirements and external limitations, such as radio licensing regulations, have resulted in a variety of actual implementations.

With some applications, such as access control, it is usually sufficient to be able to read the card data. This allows technically simple solutions. Due to the low power consumption (a few tens of microwatts), the usable range of these cards is limited to around a meter. The memory capacity is usually only several hundred bits.

If data must also be written, the power consumption rises to more than 100 µW. As a consequence, the range is limited to around 10 cm in the writing mode, since the emitted power of the writing equipment cannot be arbitrarily increased due to licensing restrictions. Contactless microprocessor cards have an even higher power consumption of around 100 mW. The distance from the terminal is thereby limited to a few millimeters.

Independent of the range and the power consumption, all cards that employ inductive coupling work on the same principle. One or more coils (usually with large enclosed areas) are incorporated into the card body, in addition to one or more chips.

Energy transfer

Up to now, there have not been any batteries available that are thin and flexible enough to be built into Smart Cards. In addition, environmental considerations discourage the large-scale use of batteries, most of which contain poisonous substances. For these reasons, all the energy that is needed for the operation of the chip in the Smart Card must be transferred from the reader to the card.

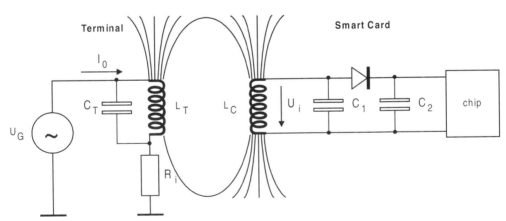

Figure 3.58   Using inductive coupling to supply power to a Smart Card

This energy transfer is based on the principle of a loosely coupled transformer. A strong high-frequency magnetic field is generated by a coil in the terminal in order to transfer the energy. The most commonly used frequencies are 125 kHz and 13.56 MHz. If a contactless card is brought into the vicinity of the terminal, a portion of the terminal's magnetic field passes through the coil in the card, which induces a voltage U_i in this coil. This voltage is rectified to serve as the power source for the chip. Since the coupling between the coils in the terminal and the card is very weak, the efficiency of this arrangement is very low. It is therefore necessary to have very high current levels in the terminal coil in order to achieve the necessary field strength levels. This is achieved by connecting a capacitor C_T in parallel with the coil L_T, with the value of the capacitor chosen such that the parallel resonant frequency of the coil and capacitor together matches the frequency of the transfer signal.

The coil L_C and the capacitor C_1 in the card also form a resonant circuit with the same resonant frequency. The voltage induced in the card is proportional to the signal frequency, the number of windings of the coil L_C and the enclosed area of the coil. This means that the number of turns needed for the coil drops with increasing signal frequency. At 125 kHz it is 100 to 1000 turns, while at 13.56 MHz it is only 3 to 10.

Data transfer

For transferring data from the terminal to the card, all known digital modulation techniques can be used. The most commonly used techniques are ASK (amplitude-shift keying), FSK (frequency-shift keying) and PSK (phase-shift keying). ASK and PSK are usually used, since these are especially easy to demodulate.

In the other direction, from the Smart Card to the terminal, a type of amplitude modulation is used. It is generated by using the data signal to digitally modify a load in the card (load modulation). If a Smart Card that is tuned to the resonant frequency of the terminal is brought into the near field of the terminal, it draws energy from this field as previously described. This causes the current I_0 in the coupling coil of the terminal to increase, which can be detected as an increased voltage drop across the internal resistance R_i. The Smart Card can thus vary (amplitude modulate) the voltage U_0 in the terminal by varying the load on its coil, for example by switching the load resistor R_2 into and out of the circuit. If the switching of the resistor R_2 is controlled by a data signal, the data can be detected and evaluated in the terminal.

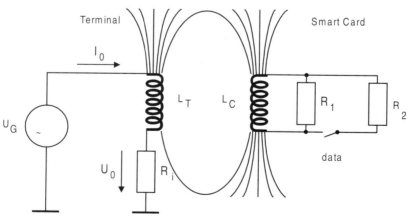

Figure 3.59 A sample circuit that illustrates the principle used for data transfer with contactless Smart Cards.

In order to make it easier to grasp the variety of techniques that are used, we can classify them into groups according to various characteristics. One possibility is to classify them according to the method used for energy and data transfers. The most commonly used techniques are radio or microwave links, optical links, capacitive coupling and inductive coupling. For the flat shape of the Smart Card, which does not have its own power source, capacitive and inductive coupling are the most suitable. The systems that are currently commercially available also use these techniques exclusively, and these are the only techniques that are given consideration in the ISO/IEC 10536, 14443 and 15693 standards. We also limit out attention in this book to the description of these techniques.

With capacitive coupling, conductive surfaces that act as the plates of a capacitor are incorporated into the card body. These have a usable coupling capacitance of a few tens of picofarads. This is normally not sufficient for transferring operating power. For this reason, this technique is usually used only for data transfer, with energy transfer being performed inductively. Data transfer takes place using a differential technique by means of a symmetrical pair of coupling surfaces, with a range that is limited to around one millimeter.

As could be anticipated, given the many different techniques used by individual manufacturers, the standardization process that the ISO/IEC started in 1988 has been difficult and time-consuming. The responsible working group has the task of defining a standard for a contactless card that is largely compatible with other standards for identification cards. This means that a contactless card may also have other functional elements, such as a magnetic strip, embossing and chip contacts. This makes it possible for contactless technology to be used together with other technologies in existing equipment.

Due to the low degree of coupling between the coil in the terminal and that in the card, the voltage variations induced in the terminal by load modulation are very small. In practice, the amplitude of the usable signal is only a few millivolts. This can only be detected by sophisticated circuitry, since it is overlaid by the significantly larger signal (around 80 dB) transmitted by the terminal. However, if an auxiliary carrier frequency is employed with a frequency of f_A, the received data signal appears in the terminal as two sidebands at the frequencies $f_T \pm f_A$. These can be filtered out of the significantly stronger terminal signal using a bandpass filter and then amplified. After this, they can readily be demodulated. Modulation with an auxiliary carrier has the disadvantage that it requires significantly more bandwidth than direct modulation. For this reason, this technique can only be used in a limited number of frequency bands.

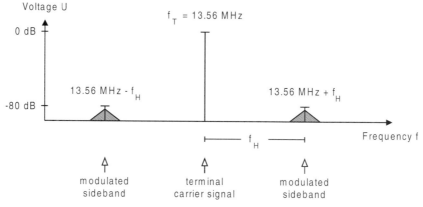

Figure 3.60   Load modulation with an auxiliary carrier produces two sidebands, separated from the transmission frequency of the reader by the value of the auxiliary carrier frequency f_A. The actual information is contained in the sidebands of the two auxiliary-carrier sidebands, which are produced by the modulation of the auxiliary carrier itself. (Source: Klaus Finkenzeller [Finkenzeller 98])

Collision avoidance

When contactless cards are used, there is always the possibility that two or more cards can be located in the vicinity of the terminal at the same time. This is especially true for systems with a relatively large range, but it can even happen with systems that have a small range – for example, two cards could be stacked on top of each other and thus be activated by the terminal at the same time. All cards that are within the range of a particular terminal will attempt to respond to commands from the terminal. However, simultaneous data transfers will unavoidably cause interference and data loss if suitable measures are not taken.

The anticollision procedures that are used can be classified into three groups, based on the parameter that they use to differentiate the individual cards:

` space,
` time,
` frequency.

With spatial differentiation, an attempt is made to restrict the active region of the terminal so that only a single card can be accessed at any given time. Card scanning can also be employed.

With temporal differentiation, measures are taken to ensure that the individual cards have different timing behavior, so that they can be identified and individually addressed by the terminal. This is the most commonly used procedure, and it has many variations.

With frequency-based differentiation, frequency-multiplexing techniques are used to distinguish individual cards. However, this technique is rather complicated and thus expensive.

Card manufacturers, who operate in a commercial environment, are usually not prepared to reveal detailed information about the anti-collision procedures that they employ. For this reason, no standards for collision avoidance have been generated up to now. An extensive description of these procedures is thus not possible here.

The state of standardization

As early as 1988, an ISO/IEC working group was given the task of preparing a standard for contactless cards. Their assignment was to define a standard for contactless cards that is largely compatible with the other standards for identification cards. This means that a contactless card can also contain other functional elements, such as a magnetic strip, embossing and chip contacts. This is to allow contactless cards to also be used in existing systems that use other technologies. As already explained, the technical possibilities for the contactless transfer of energy and data depend essentially on the desired separation between the card and the terminal during reading and writing. It was therefore not possible to create a single standard that would offer a single technical solution to all the requirements that result from the various applications. Presently, work is in progress on three different standards that describe three different reading ranges. In each of these standards in turn, various technical solutions are allowed, since the members of the standardization committee could not agree on a single technique.

Standardization started with 'close-coupling' cards (ISO/IEC 10536), since the microprocessors available at that time had relatively high power consumption, and this made energy transfer over a relatively large distance impossible. The essential parts of this standard have been completed and approved. They are described in the following section.

In the meantime, there are two other standards under development, namely ISO/IEC 14433 for 'proximity integrated circuit cards' (systems with a working range of around a centimeter) and ISO/IEC 15693 for 'hands-free integrated circuit cards' (systems with a working range of up to one meter). The development of both standards is in the preliminary stages, and there are sure to be extensive changes. For this reason, they are described only summarily in the following sections.

3.6.1   Close-coupling cards: ISO/IEC 10536

In the ISO/IEC 10536 standard for close-coupling cards, this application is designated as 'slot or surface operation', which expresses the fact that in operation the card must be inserted into a slot or laid on a marked portion of the surface of the terminal. The ISO/IEC 10536 standard is titled 'Identification Cards – Contactless Integrated Circuit(s) Cards' and consists of four parts:

`   Part 1: Physical characteristics
`   Part 2: Dimension and location of coupling areas
`   Part 3: Electronic signals and reset procedures
`   Part 4: Answer to reset and transmission protocols

Parts 1 through 3 have already become international standards, while part 4 is still in preparation. The important ingoing requirements for these standards were the following:

`   extensive compatibility with ISO 7816,
`   operation with arbitrary orientation of the card to the reader,
`   transfer carrier frequency between 3 and 5 MHz,
`   card power consumption less than 150 mW (adequate for microprocessor chips).

Part 1 of the standard defines the physical characteristics of the card. Essentially the same requirements are given as for Smart Cards with contacts, especially in regard to bending and twisting. One difference is in the tolerance for electrostatic discharges. Since a contactless card does not require any conductive path between the card surface and the integrated circuit embedded in the card body, a contactless card is largely insensitive to damage from ESD. A test voltage of 10 kV is thus specified in the standard, compared with 1.5 kV for cards with contacts.

Part 2 of the standard specifies the sizes and locations of the coupling elements. Since it was not possible to agree on a single coupling technique, both capacitive and inductive coupling elements are defined in such a way that both can be implemented together in a single card or terminal. A sample of this is shown in Figures 3.61 and 3.62. The arrangement is chosen to be orientation-independent, given suitable control in the terminal.

Part 3 of the standard, published in 1996, is the most important part to date. It describes the modulation to be used for both capacitive and inductive data transfers, since agreement on a single technique could not be achieved. A terminal that complies with the standard must therefore support both techniques. Both techniques may also be implemented in a single card.

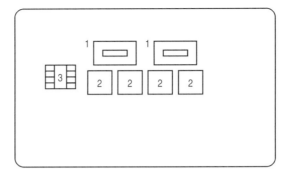

Figure 3.61   The arrangement of the coupling elements of a contactless Smart Card: (1) coupling coils in the card body, (2) capacitive coupling surfaces in the card body and (3) a set of contacts for the chip

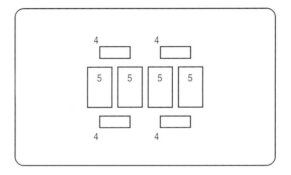

Figure 3.62   The arrangement of the coupling elements in a terminal for contactless Smart Cards: (4) coupling coils in the terminal, (5) capacitive coupling surfaces in the terminal

Energy transfer

Energy transfer takes place via a sinusoidal AC magnetic field with a frequency of 4.9152 MHz. Energy is passed to the card by one or more coupling coils, depending on how many coils are present in the card. The terminal must generate fields in all four coils.

The AC magnetic fields $F1$ and $F2$ that pass through the areas $H1$ and $H2$ have a mutual phase difference of 180 degrees, as do fields $F3$ and $F4$ that pass through areas $H3$ and $H4$. The phase difference between the fields $F1$ and $F3$, and between $F2$ and $F4$, is 90 degrees. Each magnetic field is strong enough to transfer at least 150 mW to the card. However, the card should not consume more than 200 mW. This complicated definition of the magnetic fields is required in order to allow inductive data transfer to take place in the same way for four different card orientations, as will be explained shortly.

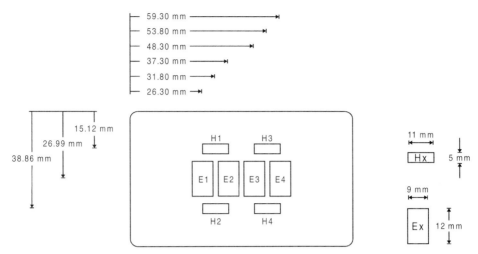

Figure 3.63   The positions and sizes of the coupling areas in the contactless card and the terminal

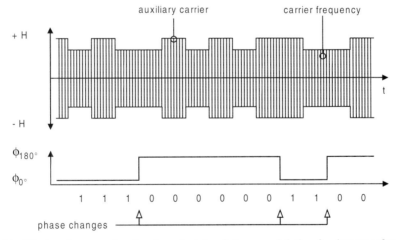

Figure 3.64   Timing diagram, showing the principle of phase modulation for data transfers with a contactless Smart Card. The upper diagram shows the alternating magnetic field, while the lower diagram shows the associated phase state. The carrier frequency is 4.9152 MHz and the auxiliary carrier frequency is 307.2 kHz.

Inductive data transfer

Different types of modulation are used for data transfer in the two directions.

Transfers from the card to the terminal   For data transfer from the card to the terminal, the first step is the generation of a 302.7 kHz auxiliary carrier by load modulation (see Figure 3.64). The load variation is at least ten percent. Data modulation is achieved by switching the phase of the auxiliary carrier by 180 degrees, which produces two phase states that can be interpreted as logical ones and zeroes. The initial state after the magnetic field has been built up is defined to be a logical one. This initial state (time interval t_3 in Figure 3.67) remains stable for at least 2 ms. Following this, every phase shift of the auxiliary carrier represents a reversal of the logical state, which produces a NRZ (non return to zero) encoding. The data transfer rate, at least for the ATR, is 9600 bits per second.

Transfers from the terminal to the card   To transfer data from the terminal to the card, the four AC magnetic fields *F1* through *F4*, which pass through the coupling surfaces *H1* through *H4*, are phase modulated (phase-shift keying, or PSK). This modulation causes the phase of all four fields to jump by 90 ° simultaneously. In this way, two phase states A and A' are defined. Depending on the orientation of the card relative to the terminal, two different constellations of phase states result, as shown in Figures 3.65 and 3.66.

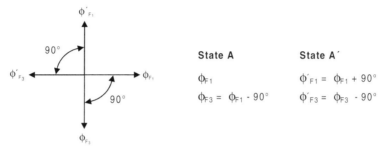

Figure 3.65   The first phase modulation alternative for data transfer with a contactless chip card. The four arrows represent phase vectors.

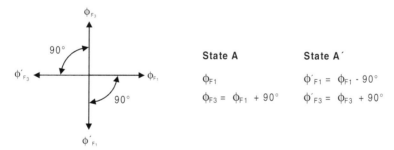

Figure 3.66   The second phase modulation alternative for data transfer with a contactless chip card. The four arrows represent phase vectors.

Since the card should work in all four possible orientations with respect to the terminal, the initial state (interval t_2 and t_3 in Figure 3.67) is interpreted as a logical '1', regardless of which of the given alternatives is actually present. Following this, every phase shift represents a reversal of the logical state, which again produces an NRZ encoding.

Capacitive data transfer

For capacitive data transfer from the card to the terminal, one pair of coupling surfaces is used (either *E1* and *E2* or *E3* and *E4*, as shown in Figure 3.63), depending on the orientation of the card relative to the terminal. The other pair of coupling surfaces can be used for data transfer in the opposite direction. Since the card sends the ATR via one particular pairs of coupling surfaces, the terminal can recognize the relative orientation of the card. The maximum potential difference between a pair of coupling surfaces is limited to 10 V. The potential difference must however be at least large enough to exceed the minimum differential potential of the receiver (±300 mV). A differential NRZ coding is used for the data transfer. The transmitter produces this by swapping the voltage between the surfaces *E1* and *E2* or *E3* and *E4*. The state representing a logical '1' is once again established in the timer interval t_3 (see Figure 3.67). Following this, every polarity reversal represents a reversal of the logical state.

The initial state and Answer to Reset (ATR)

In order for the terminal to unambiguously determine the type of data transfer and the orientation of the card at the beginning of a data exchange, certain time intervals must be defined for the start of the energy and data transfers. Figure 3.67 shows the requirements and values for the reset recovery time t_0, the power-up time t_1, the initialization time t_2, the stable logical state time t_3 and the Answer to Reset time t_4.

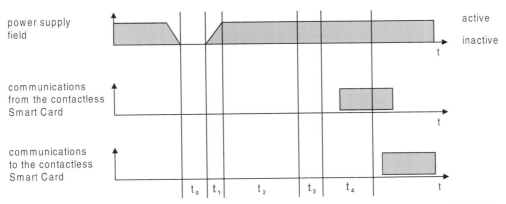

Figure 3.67 Timing diagram of data transfers with a contactless Smart Card according to ISO/IEC 10536 3. Here $t_0 \geq 8$ ms, $t_1 \leq 0.2$ ms, $t_2 = 8$ ms, $t_3 = 8$ ms and $t_4 \leq 30$ ms.

Minimum reset recovery time: t_0 If the reset is to be produced by switching the energy-transfer field off and back on, the time between switching the field off and switching it back on, during which no energy is transferred, must be equal to or greater than 8 ms.

Maximum power-up time: t_1 The time required for the energy-carrying field produced by the terminal to recover its strength must be less than or equal to 0.2 ms.

Initialization time: t_2 The initialization time, which is the time required for the card to achieve a stable operating state, is 8 ms.

Stable logical state time: t_3 Prior to the Answer to Reset, the logical state is held at 'one' for 2 ms. During this time, the card and the terminal are set to logical one for the data transfers.

Maximum response time for ATR: t_4 The card must start sending the ATR before 30 ms have elapsed. The card can use the ATR to indicate that the requirements for the subsequent operation must be changed with regard to the energy level, the data transfer rate or the frequency of the fields. The 'maneuvering room' that is thereby gained can be utilized according to the requirements of the application. For instance, significantly higher data transfer rates can be selected for a time-critical application.

The appendix of the ISO/IEC 10536-3 standard describes detailed test methods and test set-ups that make possible clearly reproducible measurements of the fields at the surfaces of the card and the terminal. Presently, only working papers are available for part 4 of the standard. No reasonably firm version of this standard exists yet.

3.6.2   Remote coupling cards

The term 'cards with remote coupling' encompasses Smart Cards that can transfer data over a distance of a few centimeters to around a meter from the terminal. This possibility is of great interest for all applications in which data should be exchanged between the card and the terminal without requiring the card user to take the card in his or hand and insert it into a terminal. Some sample applications are:

- access control,
- vehicle identification,
- electronic driver's licenses,
- ski passes,
- airline tickets,
- baggage identification.

The variety of applications suggests that there is a large variety of possible technical implementations, which makes standardization a complex, difficult and unfortunately also time-consuming task. The responsible ISO working group has been engaged with this subject since 1994. There are presently two standards in preparation: ISO/IEC 14443 and ISO/IEC 15633. The first of these covers the range up to 10 cm, while the latter covers the range up to 1 m. The following are the key application areas addressed by these standards:

- identification: reading general information from a card;
- access control: utilizing the card as evidence of the right to access a protected area;
- payment collection: obtaining goods or services in exchange for a payment.

The standards are still in the early stages of defining the technical solution, and the ultimate technical solution is presently not recognizable. In the last few years, the demand for contactless cards for applications in local public transportation has increased. Systems designed to meet this demand have been developed and have recently successfully completed pilot trials. These new systems yield clear advantages for both their operators and their users. The operator enjoys increased flexibility in the tariff system, in addition to eliminating the use of cash, and can collect and analyze customer utilization data. Furthermore, the rate at which people pass through the turnstiles increases, and the system is significantly less vulnerable to vandalism. For the user, paying the fare is made significantly easier by extensive automation. It is no longer necessary to fumble around with coins and feed them into a ticket dispenser.

Since international standardization for these applications is just beginning, and since quick progress cannot be expected, it appears that de-facto industrial standards will dominate the market for the next few years. An example is the Mifare system, which was developed by the Austrian firm Mikron and is now also supported by Philips and Infineon. Mifare was developed especially for applications in personal public transportation. It works with inductive links between the card and the terminal, using a frequency of 13.56 MHz and a data rate of more than 100 kb/s. The range of 10 cm is sufficient to free the customer from the inconvenience of having to remove the card from his or her purse. Since the possibility that several cards are simultaneously located within range of the terminal must be taken into account in this application, a fast anticollision algorithm is used to avoid mutual interference between data from several cards. The card architecture allows 16 independent functions in a single card. Access to the data is protected by an encryption process. For instance, travel information (trip origin and destination) can be collected when the card is used, while the fare can also be paid directly from an electronic purse. The Mifare system is also used in an application for Lufthansa that is described in Section 13.2.

3.6.3   Proximity integrated circuit(s) cards: ISO/IEC 14443

Work on the standard for 'proximity-coupled' Smart Cards started in 1993. This type of card is designed for a range of around 10 cm, which means that it does not have to be placed in a particular location on or in the terminal. Completion of the standard cannot be expected before 2001, so the proposals that are currently available will certainly experience many changes. Two different types of cards are emerging, and both will probably be included in the standard. The most important characteristics of these two types are listed in Table 3.4.

Table 3.4   Preliminary data transfer characteristics of proximity integrated circuit cards, from ISO/IEC 14443 working papers

Type	Data transmission to the card	Data transmission to the terminal
1	13.56 MHz, 10% ASK, NRZ–L bit coding	Load modulation with an 847-kHz auxiliary carrier, BPSK modulated, NRZ–L bit coding
2	13.56 MHz, 100% ASK	Load modulation with an 847-kHz auxiliary carrier, ASK modulated

3.6.4   Hands-free integrated circuit cards: ISO/IEC 15693

This standard deals with Smart Cards with a range of up to 1 m. The working title for the proposed standard reads 'Hands-free Integrated Circuit(s) Card', which expresses the fact that cards with this working range need not be held in the hand for use, but may remain in the purse or trousers pocket. Work on this standard has just started, so that no appreciable amount of information is yet available. Completion of the standard cannot be expected before 2002 at the earliest.

4

Information-Technological Foundations

'A Smart Card is a small computer in credit-card format with no man–machine interface'. This statement expresses an essential fact. The specific properties of Smart Cards, compared with all other types of cards, are determined by the microcontroller integrated into the card.

The plastic card body is primarily a vehicle for the microcontroller. Of course, other elements may be present in addition to the microcontroller, but they are not essential for the actual Smart Card functions. To understand the characteristics of these small computers and the data-processing mechanisms that are built upon them, you need a basic understanding of certain aspects of information theory.

Our intention here is not to convey expert knowledge. This is anyhow not at all necessary for understanding the basic features of the processes and techniques used in combination with Smart Cards. The solid basic knowledge that this chapter offers is fully adequate. Consequently, the specific information presented here goes only as deep as necessary for understanding the basic relationships.

Nearly half of this chapter is dedicated to the cryptographic procedures that are used in the field of Smart Cards. Until a few years ago, knowledge of cryptography was surrounded by a veil of secrecy and ignorance. This has however changed apace, and there is now an extensive literature on this subject. As with the sections of this chapter that deal with information technology in general, we provide only the basic information that is necessary for understanding cryptographic algorithms and protocols. For more detailed information, we suggest that you look at the recognized books on the subject, such as those by Bruce Schneier [Schneier 96] and Alfred Menezes [Menezes 97]. A further rich source of information on cryptography is the World Wide Web, where you will find the home pages of several research institutes (such as [GMD, Semper]), standards organizations (such as [ETSE, IEC, ISO]), agencies (such as [BSI, NSA]), firms (such as [Ascom, Certicom, Counterpane, R3, RSA]), associations (such as [CCC, Teletrust]) and individuals that are interested in the subject (such as [Gutmann, Tatu]).

4.1   Data Structures

The storage or transfer of data always and unavoidably requires an exact definition of the data in question and their structure. Only then is it possible to subsequently recognize and interpret the data elements. Fixed-length data structures with non-modifiable sequences regularly cause systems to crash. The best example of this is the conversion of many European currencies into Euros. All systems and data structures with fixed currency definitions had to be extended at considerable cost. The same problem manifests itself in many Smart Card applications. Sooner or later, fixed data structures that must be extended or shortened are responsible for a lot of effort and expense.

However, the problem of structuring data has been around for a long time, and there are enough methods that can be used to solve the problem. One method that is very popular in the world of Smart Cards, and which is coming into more general use, comes from data transfer theory. It is known as abstract syntax notation 1, or ASN.1 for short. This is a coding-independent description of data objects, originally developed for transferring data between different computer systems.

In principle, ASN.1 is a sort of artificial language that is suitable for describing data and data structures rather than programs. The syntax is standardized in ISO/IEC 8824, and the coding rules are covered by ISO/IEC 8825. Both of these standards were developed from Recommendation X.409 of the CCITT.

The basic idea of ASN.1 is to prefix each data object with an unambiguous label and length information. The quite complex syntax of the description language also makes it possible to define your own data types, as well as to nest data objects. The original idea, which was to create a generally valid syntax that could form the basis for data exchange between fundamentally different computer systems, is scarcely used in Smart Cards. Currently, only a very small part of the available syntax is used here, mainly due to the very limited memory capacity of Smart Cards.

One area of application that has already become quite significant is cryptography. The various techniques and options can be handled more flexibly and much more easily with ASN.1 objects than with the very rigid structures that were originally used.

The basic encoding rules (BER) for ASN.1 are defined in the ISO/IEC 8825 standard. Data objects that are created according to these rules are referred to as BER-TLV coded data objects. A BER-coded data object has a label (called a 'tag'), a length field and the actual data part, with an optional end marker. Certain bits in the label are predefined by the coding rules. The actual structure is shown in Figure 4.1. The distinguished encoding rules (DER) form a subset of the BER. These specify, among other things, the coding of the length information, which may be one, two or three bytes long. A basic summary of the BER and DER can be found in Burton Kaliski's book [Kaliski 93].

The coding of ASN.1 objects takes place in the classic TLV structure, in which 'T' (tag) denotes the object's label, 'L' (length) refers to its length and 'V' (value) is the actual data.

The first field of a TLV structure is the label for the data object in the following V field. To avoid the need for each user to define his or her own labels, which would throw open the door to incompatibility, there are standards that define tags for various, frequently used data structures. ISO/IEC 7816-6, for example, defines tags for objects used in general industrial applications, while ISO/IEC 7816-4 defines tags for secure messaging and EMV also defines several more. It is by no means the case that a given tag is always used for the same data element, but a process of unification is essentially in progress.

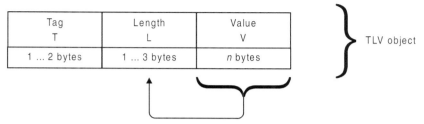

Figure 4.1   The principle of BER-based TLV coding according to ANS.1

The two most significant bits of the tag encode the class to which the following data object belongs. The class indicates the general type of the data object. The *universal* class indicates general data objects, such as an integers and character strings. The *application* class indicates that the data object belongs to a particular application or standard (e.g. ISO/IEC 7816-6). The other two classes, *context-specific* and *private*, fall under the heading of non-standardized applications.

The bit following the two class bits indicates whether the tagged object is constructed from other data objects. The five least-significant bits are the actual label. Since this can have a value of only 0 through 30 due to the restricted address space, it is possible to point to the following byte by setting all five bits to 1. All values from 31 to 127 are allowed in the second byte. Bit 8 of the second byte points to future applications and thus may not be set at present.

Table 4.1   ASN.1 tag coding

Byte 1	b8	b7	b6	b5	b4	b3	b2	b1	Meaning
	0	0	...	...	...	...	...	...	universal class
	0	1	...	...	...	...	...	...	application class
	1	0	...	...	...	...	...	...	context-specific class
	1	1	...	...	...	...	...	...	private class
	...	...	0	...	...	...	...	...	primitive data object
	...	...	1	...	...	...	...	...	constructed data object
	...	....	...	X	X	X	X	X	tag code (0 – 30)
	...	...	...	1	1	1	1	1	pointer to the following byte (Byte 2), which specifies the tag code

Byte 2	b8	b7 – b1	Meaning
	0	31 – 127	tag code

The number of necessary length bytes is shown in Table 4.2. The standard also defines the term 'template'. A template is a data object that serves as a container for other data objects. ISO/IEC 7816-6 defines the tags for possible data objects in the domain of industry-wide applications of Smart Cards. ISO 9992-2 covers the domain of Smart Card financial transactions.

Table 4.2   Number of length bytes in ASN.1

Byte 1	Byte 2	Byte 3	Meaning
0–127	—	—	one byte is needed for these length values
'81'	128–255		two bytes are needed for these length values
'82'	256–65,535		three bytes are needed for these length values

This method of data encoding has several characteristics that are particularly beneficial in the field of Smart Cards. Since the available memory space is as a rule never enough, the use of data objects based on ASN.1 can produce considerable space savings. TLV encoding makes it possible to transfer and store variable-length data without a lot of complications. This allows memory to be used very economically. An example that illustrates this is shown in Figure 4.2, which shows the TLV encoding of a name.

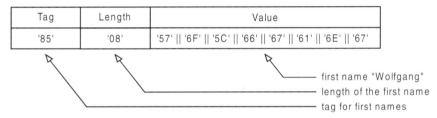

Figure 4.2   TLV encoding of the name 'Wolfgang'

Subsequent extensions of data structures can be undertaken very easily with ASN.1, since all that is necessary is to insert an additional TLV encoded data object into the existing data structure. Full compatibility with the older version is retained, as long as the previous TLV objects are not deleted. This is equally true for new versions of data structures, in which changes have been made with respect to the previous encoding. This is straightforward, and requires only modifications to the tags. It is equally simple to represent the same data with different encodings. All in all, these advantages are why the ASN.1 syntax, based on TLV encoding, is particularly popular in the Smart Card industry.

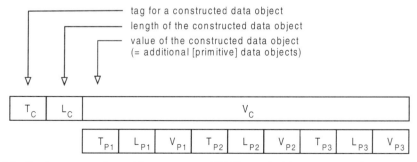

Figure 4.3   Basic structure for making constructed TLV-encoded data structures from several primitive TLV-encoded data objects. The indices 'C' and 'P' stand for 'constructed' and 'primitive'.

The main disadvantage of ASN.1 data objects is that the cost of the administrative data is rather high if size of the user data is small. For example, if the user data is only one byte, two additional bytes (tag and length) are still needed for its administration data. However, the longer the user data, the more favorable is the relationship. The ASN.1 structured data in the German health insurance card is a good example of this. There are between 70 and 212 bytes of user data. The administrative data amounts to 36 bytes, which means that the administrative expense lies between 17 and 51 percent.

We can recapitulate all the above with a further example. Assume that we wish to store surnames, given names and titles in a file with a transparent data structure. Independent of the correct ASN.1 description, the TLV encoded data will have the structure shown in Figure 4.4. The tags used in this example have been freely chosen and thus do not correspond to any relevant standard.

version 1	T	L	V	T	L	V	T	L	V
	'85'	'07'	"Manfred"	'87'	'05'	"Meier"	'84'	'04'	"Ing."

version 2	T	L	V	T	L	V	T	L	V
	'84'	'04'	"Ing."	'85'	'07'	"Manfred"	'87'	'05'	"Meier"

version 3	T	L	V	T	L	V	T	L	V
	'87'	'05'	"Meier"	'85'	'07'	"Manfred"	'84'	'04'	"Ing."

Figure 4.4   An example of sequence independence within a TLV structure

When evaluating this data structure, the computer compares the first tag with all tags known to it. If it finds a match, then it recognizes the first object as a given name. It reads the length of this object from the next byte. The subsequent bytes are then the actual object, i.e. the given name. This is followed by the next TLV object, whose first byte is the tag for a surname. The computer recognizes this using exactly the same process as for the first object.

If it becomes necessary to extend the data structure, e.g. by adding a title, then a new element can simply be inserted into the existing structure. The point of insertion is unimportant. The extended structure remains fully compatible with the older version, since the new element receives its own tag and thus is unambiguously identified. Programs that only know the old tags are not upset by the new one, since they do know it and so by definition can skip it. Other programs that do know the new tag can thus evaluate it, but the old structure also does not cause them any problems.

4.2   SDL Notation

This book uses SDL notation to describe states and state transitions. For some years, this approach has been used ever more frequently in the Smart Card domain to describe state-oriented mechanisms, such as those used for communication protocols. 'SDL' stands for 'Specification and Description Language', and it is described in detail in CCITT Recommendation Z.100.

SDL notation is similar to that used in standard flowcharts. However, it does not describe program flows, but rather states and state transitions. SDL diagrams are constructed from standardized individual symbols that are interconnected by lines. The flow is always from top left to bottom right, so the lines connecting individual symbols do not need arrowheads to identify their start and end points.[1]

In simplified form, the notation can be thought of as describing a system consisting of a certain number of processes. Each process in turn is a state machine. If the state machine is in a stable state, it can receive an external signal. Depending on the received data, the machine may then reach a certain new state. Additional actions may take place between the two states, such as receiving and sending data or computing a value.

Figure 4.5 shows the ten symbols used in this book. These are just a selection from the much larger set defined in Z.100, but they suffice as a basic set for use with Smart Cards. The Start symbol (1) denotes the beginning of a process. Most SDL diagrams begin with this symbol. The Task symbol (2) indicates a specific action, which is described by text within the box. With this symbol, there is no further detailed description in the form of a subprogram. The Decision symbol (3) permits a query during a state transition, to which the answer may be 'yes' or 'no'. The Connector symbol (4) marks a link to another SDL diagram, and is used mainly to divide large diagrams into several small ones.

The Input (5) and Output (6) symbols represent interfaces to the outside world. The exact input and output parameters are described inside the symbol. The State symbol (7) is used to describe a state. The state that achieved at each stage is indicated by this symbol.

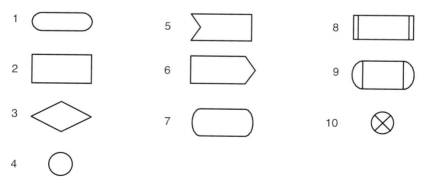

Figure 4.5   The SDL notation symbols used in this book, in accordance with CCITT Z.100:

1 – Start	5 – Input	8 – Subprogram
2 – Task	6 – Output	9 – Start subprogram
3 – Decision	7 – State	10 – End subprogram
4 – Connector		

---

[1]   For a detailed example of an SDL diagram, see Section 6.4.2, 'The T=0 transmission protocol'.

The next three symbols describe subprograms. The Subprogram symbol (8) indicates that the contents of this box are described in more detail elsewhere. The two symbols Start (9) and End (10) delimit the detailed description of a subprogram.

4.3   State Machines

A state machine is a type of automaton. A common example of an automaton is a vending machine, into which you insert a coin and then press a button. After this, you can open a compartment and remove your selection.

In slightly more abstract terms, this automaton defines a chain of events involving various state transitions. In the initial state, the automaton waits for money to be inserted. Any other action, such as pressing a button, will not cause anything to happen. Only the insertion of a coin causes a transition from the initial state to the 'money inserted' state. The next transition occurs as a result of pressing the button, following which the automaton allows a compartment to be opened.

In computer science, state machines can be very effectively visualized using graphs or Petri networks. These are not only useful for modeling state machines; they can also be used to investigate certain properties of the systems that they describe. The objectives are to identify any deadlocks that may occur during operation and to ensure that the instruction processes are correct.

4.3.1   Fundamentals of state machine theory

The objective of this section is to provide an introduction to, and summary of, the state diagrams that are used to describe Smart Card applications.

A state diagram is a type of graph that represents a set of states and the interrelationships of these states. The states are shown as nodes, and their relationships are shown as lines. If a line indicates a direction, which means that it has an arrowhead at one end, then we speak of directed lines and thus a directed graph. The arrow indicates the direction in which a state transition can take place. The actual placement of the nodes and lines in the graph plays no part in the interpretation of the diagram. A sequence of nodes connected by lines is called a path. If the first and last nodes are the same and there is more than one node, the path is called a loop.

This is only a very small part of graph theory, but it is essentially all we need to be able to describe states and their associated state machines in Smart Card applications.

4.3.2   Practical applications

An additional advantage of microprocessor cards compared with simple memory cards is that the sequence of instructions can be determined in advance. It is thus possible to precisely specify all instructions in terms of their parameters and sequence. This provides additional protection against unauthorized access, in parallel to object-oriented access authorization for files. However, the possibilities offered by Smart Cards in this respect vary greatly. Simple operating systems usually cannot manage state machines, while modern operating systems even allow the definition of application-specific state machines that work with instruction parameters.

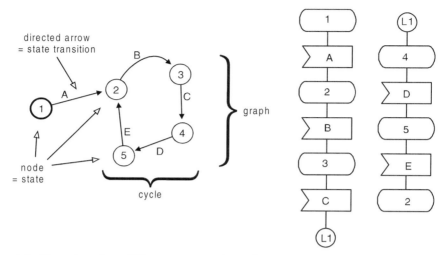

Figure 4.6   Examples of two different representations of state diagrams. On the left is a directed state diagram, and on the right an equivalent SDL diagram.

Table 4.3   Description of the state diagram of Figure 4.6 in tabular form

State after transition	State before transition				
	1	2	3	4	5
1	—	—	—	—	—
2	A	—	—	—	E
3	—	B	—	—	—
4	—	—	C	—	—
5	—	—	—	D	—

Smart Card state machines may be divided into *micro* and *macro* types. Micro state machines define only short instruction sequences, and they only become active once the first instruction of a sequence has been sent to the card. A typical example relates to the instructions needed to authenticate a terminal. The first instruction asks the card for a random number. This activates a micro state machine that accepts only an authentication instruction as the next instruction. If the card receives this instruction, the process completes and all types of instructions are allowed. Otherwise, if the card receives any other type of instruction, the micro state machine generates an error message and the process is aborted. The command sequence must then be restarted from the beginning.

Micro state machines are only a subset of all possible state machines, but they have several great advantages in Smart Cards. Since they are limited to very few instructions in a rigidly defined sequence, they take up little memory and program space. In many applications, it is sufficient to protect file contents using object-oriented access mechanisms and otherwise impose no restrictions on instruction sequences. Only a few procedures, such as authentication, must follow prescribed sequences. This can be implemented very economically using micro state machines.

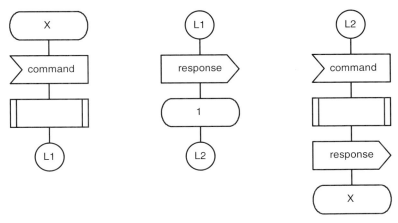

Figure 4.7   An example of a Smart Card micro state machine with two states, 'X' and '1'

Macro state machines can be regarded as extensions and generalizations of micro state machines. They make it possible to verify all instructions, together with all their parameters, within a defined graph before they are executed. Depending on how the state machine is constructed, it may possible in certain circumstances to dispense with object-oriented file access protection, since the state machine can perform all the necessary checks before the instruction is actually executed. Of course, an error in the state diagram could have fatal consequences for security of the system. As it is very difficult to verify the complete absence of errors in the state diagrams of complex state machines, in practice file access protection is still used. Producing correct descriptions of all processes and all instructions in a Smart Card is very time-consuming, and frequently these must be obtained empirically to a certain extent.

After describing the advantages of macro state machines, we must now turn to their shortcomings. The implementation of a macro state machine with the required capabilities is very costly, in terms of both its design and its subsequent programming. The demand on program memory made by the state machine alone, controlled as it is by the stored representation of a graph, is huge. In addition to this program code, the graph in question must also be stored in memory. The amount of memory space naturally depends on the complexity of the graph that is to be executed. The amount of information contained in a graph with many states and a corresponding number of transitions can become very large in terms of Smart Card relationships.

The ISO/IEC 7816-9 standard, which is currently under development, deals with state machines for Smart Cards. It describes ACDs (access control descriptors), which define the instructions that are permitted in a specific state, along with their associated parameters. The Smart Card operating system can monitor hard-coded state machines using these ACDs.

In order to illustrate the capabilities of a macro state machine in summary form, Figure 4.8 shows the state diagram of a small application. Its operation is described below.

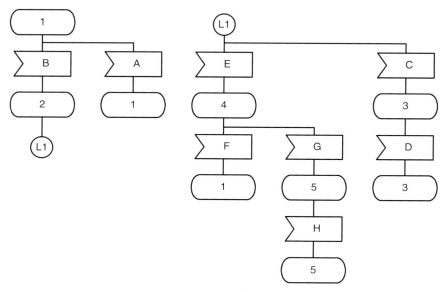

Figure 4.8   Example of a Smart Card macro state machine with the following states and transitions:

    1   –   initial state
    2, 4 – intermediate states
    3, 5 – final states

    A – SELECT FILE   B – VERIFY PIN
    C – SELECT FILE   D – READ BINARY
    E – ASK RANDOM   F – all commands except G
    G – EXTERNAL AUTHENTICATE   H – SELECT FILE/UPDATE BINARY

After a reset, the Smart Card is in the initial state, denoted by 1. In this state, every file in the directory may be selected using SELECT FILE; this does not cause a state transition. All other instructions are prohibited except for VERIFY PIN, and the card responds to them with an error message. After successful verification of the PIN, the state machine changes to state 2.

Two instructions are permitted in state 2. The first path leads via SELECT FILE to state 3, where the selected file may be read. The second path that branches from state 2 leads to state 4 following a request from the terminal for a random number from the card (ASK RANDOM). Any instruction other than EXTERNAL AUTHENTICATE leads back to the initial state (1). When the terminal is successfully authenticated, the card reaches state 5. In this state, according to the diagram, files may be selected and written using SELECT FILE and UPDATE BINARY.

In this diagram, both states 3 and 5 cannot be exited during a session, so they represent the two end states. A transition to state 1 is only possible with a card reset. However, this is not shown in the diagram, since the 'consciousness' of every state machines is limited to its current session. A state machine does not carry any information at all from one session to the next.

4.4   Error Detection and Correction Codes

When data is transferred or stored, it should be possible to detect any alteration of the data. In particular, stored programs must be protected against corruption, since a single altered bit of program code could destroy the program or change its execution to such an extent that the required functions are no longer available. The EEPROM memory used in Smart Cards is especially sensitive to external influences, such as heat and voltage fluctuations. Therefore, the sections that perform security-related functions must be protected, so that undesired modifications can be detected by the operating system and their negative effects avoided.

Very sensitive file contents, such as program code, keys, access conditions, pointer structures and the like, must be protected against alteration. Error detection codes (EDC) are used for this purpose. The probability of detecting changes in a protected region with an EDC depends on the code used. Error correction codes (ECC) are an extension of error detection. They allow not only the detection of errors in the tested data, but also a limited amount of error correction.

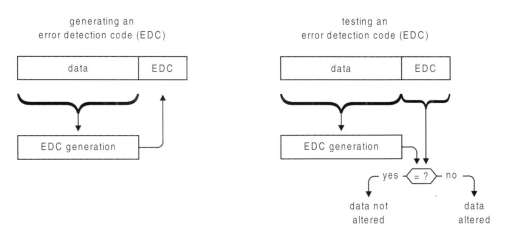

Figure 4.9   The basic processes for generating and checking an error detection code (EDC)

The principle of operation of all these codes is the assignment of a checksum to the protected data. This is usually stored adjacent to the protected data. The checksum is computed using a generally known algorithm, not a secret one. The data can be checked for alterations as necessary by using the EDC. This is done by comparing the stored checksum to one computed afresh.

A particular aspect of error detection and correction is that it utilizes a wide variety of mathematical procedures. Some of these provide better protection for the more significant bits, in order to reduce adverse effects on numerical values as much as possible. In most cases, however, the use of such algorithms enormously increases the complexity and size of the program code. It is thus more common to use procedures in which error detection does not distinguish between the upper and lower parts of a byte, but instead operates on the byte as a unit.

Error detection and correction codes are very similar to message authentication codes (MACs) and cryptographic checksums (CCS). However, there is a fundamental difference.

EDC and ECC checksums can be computed and checked by anyone. In contrast, the calculation of a MAC or CCS requires a secret key, since these codes are designed to protect against tampering with the data and instead of against accidental corruption.

The most widely known error detection code is certainly the parity bit, which is appended to each byte in many data transfer procedures and in some memory modules. Before computing the parity, you must decide whether to use even or odd parity. With even parity, the parity bit is selected so that the sum of all the bits in the data byte plus the parity bit is an even number. With odd parity, these bits must add up to an odd number.

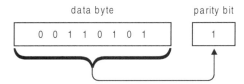

Figure 4.10 Example of error detection using a supplementary odd parity bit

If two bits in one byte are simultaneously wrong, the parity does not change and no error can be detected. Another drawback of parity-based error detection is the relatively large overhead of one parity bit per eight data bits. This represents an additional memory load of 12.5%. Furthermore, it is very difficult to work with additional parity bits when the memory is organized in bytes, since this requires significant programming effort. This is why parity bits are not used for error detection in Smart Card memories. XOR and CRC checksums, for example, are better suited to this task.

4.4.1 XOR checksums

An XOR checksum, which is also known as a longitudinal redundancy check (LRC) because of how it is computed, can be obtained very simply and very quickly. These are both important criteria for error detection codes used in Smart Cards. In addition, the algorithm can be implemented extremely easily. Besides the protection of data stored in memory, XOR checksums are typically used for data transfers (ATR with the T=1 transmission protocol). An XOR checksum is computed by performing consecutive logical XOR operations on all data bytes. In other words, byte 1 is XOR-ed with byte 2, the result of this is XOR-ed with byte 3, and so on.

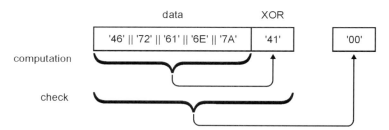

Figure 4.11 Computing and checking an XOR checksum

If the checksum is placed directly after the data and a new checksum is again computed using both the data and the first checksum, the result is '00'. This is the simplest way to verify that the data and the checksum retain their original values and thus are uncorrupted.

The major advantages of XOR checksums are their fast computation and the simplicity of the algorithm. The algorithm is so simple that its assembler code is only between 10 and 20 bytes long. One reason for this is that the XOR operation is directly available in all processors as a machine instruction. In addition, an algorithm for XOR checksum computation must be implemented in almost every Smart Card operating system, due to the requirements of various ISO standards relating to data transfer with T=1. This can be used for other purposes without any additional overhead.

Unfortunately, XOR checksums also suffer from several serious drawbacks, which considerably limit their practical application. They are in principle not very secure. For example, they do not allow the interchange of two bytes within the overall data to be detected. Also, multiple errors can occur at the same position in several bytes and cancel each other out. The consequence of all this is that XOR checksums are mainly used for data transfers, and their use for verifying the consistency of memory contents is very limited.

4.4.2   CRC checksums

The CRC procedure (cyclic redundancy check) also comes from the field of data communications, but it is significantly better than the XOR process. Still, a CRC checksum is also only an error detection code that cannot be used for error correction. The CRC procedure has been used for a long time in data transmission protocols, such as Xmodem, Zmodem and Kermit, and it is widely used in hard disk drive controllers in a hardware implementation. It is based on the CCITT V.41 recommendation.

The CRC checksum is generated by a 16-bit cyclic feedback shift register. The feedback is determined by a generating polynomial. In mathematical terms, the data to be checked is represented as a large number that is divided by the generating polynomial. The remainder is the checksum. The procedure should not be used with more than around 4 kB of data, since the error detection probability drops sharply beyond that point. However, this restriction can easily be circumvented by dividing the data into blocks that are no larger than 4 kB.

Table 4.4   Commonly used generator polynomials for CRC computation

Designation	Generator polynomial
CRC CCITT V.41	$G(x) = x^{16} + x^{12} + x^5 + 1$
CRC-16	$G(x) = x^{16} + x^{15} + x^2 + 1$
CRC-12	$G(x) = x^{12} + x^{11} + x^3 + x^2 + x + 1$

Thus, with CRC checksums it is always necessary to know the generating polynomial as well as the initial value for the shift register, since otherwise the calculation cannot be reproduced. In the overwhelming majority of cases (e.g. ISO 3309) the shift register initial value is zero, but several data transfer procedures (such as CCITT Recommendation X.25) set all bits to 1.

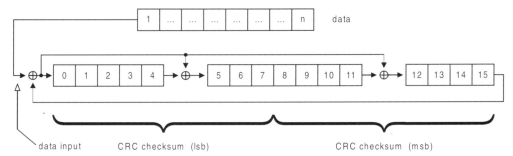

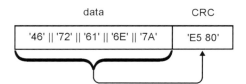

Figure 4.12   Calculating a CRC checksum using the generator polynomial $G(x) = x^{16} + x^{12} + x^5 + 1$. The data and the CRC register are both shown as bits.

The calculation of a CRC checksum, as illustrated in Figure 4.12, proceeds as follows: (a) the 16-bit CRC register is set to its initial value; (b) the data bits are fed into the feedback shift register one after the other, starting with the least-significant bit; and (c) the feedback (which represents the polynomial division) takes place via bitwise logical XOR operations on the CRC bits. After all data bits have been fed into the register, the calculation is complete and the content of the 16 bits is the desired CRC checksum.

Figure 4.13   Sample calculation of a CRC checksum, with the generator polynomial $G(x) = x^{16} + x^{12} + x^5 + 1$ and an initial value of '0000'

A CRC checksum can be verified by calculating the CRC checksum of the data anew and comparing it with the checksum that was delivered with the data. If they are the same, it follows that the data and the checksum have not been altered.

The great advantage of CRC checksums is that they provide reliable error detection, even with multiple errors. Only very few procedures can achieve this. In addition, in contrast to the XOR procedure, it is possible to detect interchanged data bytes with CRC, since byte order definitely plays a role in checksum formation via the feedback shift register. However, it is very difficult to specify exact detection probabilities for such errors, since they are very dependent on the locations of the errors within the bytes in question.

The CRC algorithm is relatively simple, and the amount of code needed to implement it thus matches the needs of small Smart Card memories. Its greatest drawback is however the slowness of the calculation. The speed is considerably reduced by the fact that the algorithm requires the data to be shifted bit by bit. The CRC checksum algorithm was originally designed for hardware implementation, and this has a strong detrimental effect when it is implemented in software. The throughput of a CRC routine is lower than that of an XOR routine, by a factor of around 200. A typical figure is 0.2 ms/byte at a 3.5-MHz clock frequency. The calculation of a CRC checksum for a 10-kB Smart Card ROM would thus require around 2 seconds.

4.4.3   Error correction

If it is necessary not only to detect changes in memory regions, but also to correct them if possible, then error correction codes must be used. Since the computation of these codes is expensive in terms of program code, using them to protect Smart Card memory is problematic. Furthermore, the algorithms are usually designed to correct only low error rates. Since EEPROM memory in Smart Cards is page-oriented, and a whole page usually fails in case of an error, only methods that are capable of correcting clustered errors would make sense. Consequently, other techniques are used for error correction.

The technically simplest solution is to store the data in multiple, physically separated memory pages and to use a majority-vote procedure when reading the data. Triple storage is commonly used, with a 2-of-3 vote. A less memory-intensive variation of this procedure is storage in two locations, with EDC checksum protection for each location. The occurrence of a memory error can be detected by checking the two EDC values. This also allows the memory segment where the error occurred to be identified. The segment with no detected error must then contain the valid data, which can be restored to the faulty segment.

Of course, considerable additional memory is needed for these error correction procedures, but for small quantities of data it is still well within acceptable limits. The main advantage is that no complicated, code-intensive algorithm is needed to evaluate the data.

As an alternative to protection by multiple data storage, it is possible to use an error correction algorithm, such as the Reed–Solomon algorithm. This is particularly well suited for use with clustered errors, such as may occur in Smart Cards due to page failures. The algorithm takes up a few hundred bytes of code space when programmed in assembly language, and the size of the ECC data depends primarily on probability with which errors must be detected and/or properly corrected.

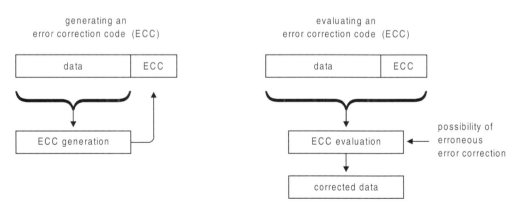

Figure 4.14   The basic principle of using an error correction code

Nevertheless, several basic remarks must be made regarding the use of error correction processes in Smart Cards. At first sight, it is tempting to use these procedures to eliminate the errors that occur in the EEPROM. However, this assumed data security is bought at the expense of a number of serious disadvantages. The required memory space is huge, and the time required to write the data into memory also increases considerably, since they must be stored in multiple locations. Algorithms that can correct clustered errors in the magnitude with which they typically occur with page-based EEPROMs are complicated and also require a great deal of memory space for the EDC codes. However, the fundamental disadvantage is even more serious. Even when an error correction algorithm is used, errors can in principle still occur in the corrected data, since the algorithm works properly only up to a certain number of errors. If an operating system corrects memory errors automatically, it can in principle never be certain that the correction was carried out properly.

For example, suppose that automatic error correction is applied to the balance in an electronic purse. The system operator can never be sure of what happens to the credited amounts in the event of an error. The balance may be corrected properly, but there is a certain probability that it will be too high or too low after the correction. In this context, it must be remembered that Smart Cards are cheap mass-produced articles, which can simply be exchanged when faulty.

As a rule, when problems occur with data contents, a higher-level system that allows human intervention must decide what to do. For example, on the first occasion of an error occurring in a Smart Card purse, the cardholder's balance would surely be manually restored. However, if the error recurs repeatedly, the system operator will be much less forthcoming with regard to the cardholder, since there is a possibility that the EEPROM has been fraudulently manipulated. This cannot be handled by an error correction code in the card; instead, the system administrator must intervene.

4.5   Data Compression

As is well known, the amount of memory that is available in a Smart Card is sharply limited. Consequently, the desire to improve this situation by using data compression repeatedly arises among application providers.

There are certain hurdles that must be overcome before data compression can be used. The algorithm must not take up too much code space, and it must above all need very little RAM. In addition, an acceptable compression speed should be achieved. The compression factor is not all that important, since in each case the data volume is at most only a few hundred bytes. Only loss-free compression procedures can be considered for Smart Cards, since the decompressed data must be identical to the original data. The two procedures that are frequently used for Smart Cards are run-length encoding and variable-length encoding.

With run-length encoding, an uninterrupted sequence of identical data objects is replaced by a repetition count and the object (such as a character) that is to be repeated. With variable-length encoding, the frequency of occurrence of characters with a fixed length (one byte, for example) is analyzed, and the most frequently occurring characters are replaced by characters with shorter lengths (Huffman encoding). Less frequently occurring characters are encoded using longer codes.

With static variable-length encoding, the replacements are made according to a previously defined table. The dynamic version of variable-length encoding first analyses the frequency distribution of the characters in the original data, and then constructs a replacement table on this basis. A third variation is adaptive variable-length encoding, in which the replacement table is continuously updated during the compression process to achieve optimum compression.

Both dynamic and adaptive variable-length encoding are out of the question for Smart Cards, due to the complexity of their algorithms and their large memory requirements. Run-length encoding and static variable-length encoding are thus the only real alternatives for use in Smart Cards. The algorithm for run-length encoding does not need much program code, but it has the disadvantage that it can only be used with repetitive data. Image data, for example, are particularly suitable, since images often contain large areas with the same value. Keys for symmetric cryptographic algorithms would be completely unsuitable for compression with this algorithm, since they have the characteristics of random numbers.

Static variable-length encoding is the second compression procedure that is used with Smart Cards. It can be used very well for files containing telephone directory information, for instance, since the structure of the stored data is known and the replacement table can thus be permanently tied into the algorithm. For example, telephone numbers are made up of only the numerals 0 through 9 plus a few special characters, such as '*' and '#'. If only capital letters are allowed for the names, then the replacement table only has to take the 26 characters of the alphabet into account. Likewise, certain letters occur significantly less often in names than others, which will also affect the encoding. With telephone directories, a saving of 30% of the memory space (compared with the uncompressed data) can certainly be achieved. However, this does not take into account the memory occupied by the compression algorithm.

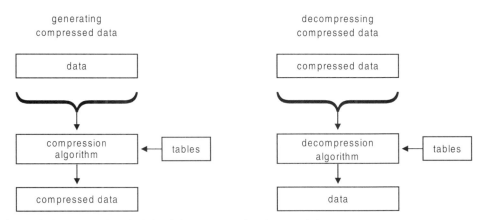

Figure 4.15   The basic principle of data compression for stored data

However, certain things must be considered with regard to data compression for Smart Cards. Ideally, it should take place in the operating system, in a manner that is fully transparent to the outside world, so that uncompressed data can be read and written in the usual way using standard instructions. Compression can also only be applied to certain kinds of data. For instance, the results of attempting to compress program code and keys are usually unsatisfactory. This must be taken into account in the design of the application, since otherwise the expected reduction of the memory requirements can (in the worst case) turn into a need for even more memory due to 'compression'.

All of this is the reason why data compression has up to now been used only sparingly in Smart Cards. In special applications, such as telephone directories in cards used in the telecommunications sector, compression algorithms are sometimes utilized. With general-purpose operating systems and applications in which the structure of the data is not known in advance, the use of data compression does not produce satisfactory results. It should thus be avoided, due to the additional memory space required by the compression algorithm.

4.6   Cryptology

In addition to their role as data carriers, Smart Cards are also used as authorization media and encryption modules. As a result, cryptography won a central significance even in the early days of Smart Cards. Nowadays, the processes and methods of this scientific discipline are a fixed part of Smart Card technology.

The discipline of cryptology can be divided into two areas of activity, namely cryptography and cryptoanalysis. Cryptography is the study of the methods used for encrypting and decrypting data, while cryptoanalysis is concerned with attempting to break existing cryptographic systems.

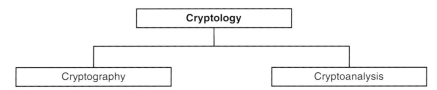

Figure 4.16   The two subdivisions of cryptology are cryptography and cryptoanalysis

With Smart Cards, it is above all the practical application of existing cryptographic processes and methods that forms the central area of interest and the main application area. Consequently, we address the practical aspects of cryptography here more than the theoretical aspects. Nonetheless, some emphasis is also given to the application of the processes and the basic features of the theoretical background.

The four objectives of cryptography are the maintenance of the secrecy of messages (*confidentiality*), the assurance of the *integrity* and the *authenticity* of messages, and the binding force (*non-repudiation*) of messages. These objectives are mutually independent, and they also place different demands on the system under consideration. Confidentiality means that only the intended recipient of a message can decrypt its contents. Authenticity means that the recipient can verify that the received message has not been altered in the course of being transferred. If the sender is able to verify that a certain recipient has received a given message, this means that the message is binding and cannot be repudiated.

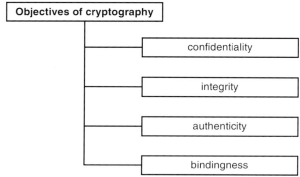

Figure 4.17   Classification of the four independent objectives of cryptography

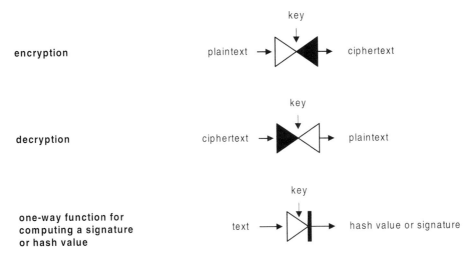

encryption — plaintext → ciphertext (key)

decryption — ciphertext → plaintext (key)

one-way function for computing a signature or hash value — text → hash value or signature (key)

Figure 4.18 The symbols used in this book for cryptographic algorithms

The notation used in this book for cryptographic procedures is illustrated in Figures 4.18 and 4.19. The following terms and principles for the basis for cryptology, and are a prerequisite for understanding the procedures described in the rest of this section.

In simplified terms, there are three types of data in encryption technology. The first is *plaintext*, which is unencrypted data. Encrypted data is referred to as *ciphertext*. Finally there is a *key*, one or more of which is required for encryption and decryption. These three types of data are processed by an encryption algorithm. The algorithms that are currently used in Smart Cards are generally block-oriented, which means that the plaintext and ciphertext can only be processed in packets with fixed lengths (such as 8 bytes with DES).

Modern cryptographic algorithms are as a rule based on Kerckhoff's principle. This principle, which is named after Auguste Kerckhoff (1835–1903), says that the entire security of an algorithm should be based only on the secrecy of the key, and not on the secrecy of the cryptographic algorithm. The consequence of this generally known but often-disregarded principle is that many algorithms that are used in the civil sector have been published, and in part also standardized.

The opposite of Kerckhoff's principle is the principle of security by concealment. With this principle, the security of a system is based on the idea that a would-be attacker does not know how the system works. This principle is very old, and it is still frequently used even today. However, you should take care not to develop a cryptographic system (or any other system) on this principle alone. Up to now, every system based on this principle alone has been broken, and mostly in a very short time. In our information society, it is generally not possible to keep the technical details of a system secret over a long period, and that is precisely what this principle requires.

Of course, the consequences of the unintentional interception of messages can most certainly be limited by the use of concealment. This principle is thus repeatedly used in parallel with Kerckhoff's principle. In many large systems, it is also built in as a supplementary security level. Since the security of modern, published cryptographic algorithms is practically based only on the pure processing power of computers, concealing the process used increases the degree of protection against attacks.

encryption   CT = enc (K; PT)

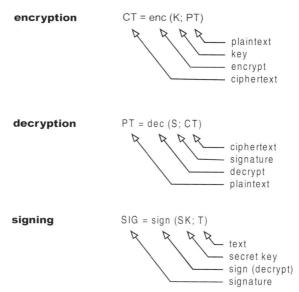

        plaintext
        key
        encrypt
        ciphertext

decryption   PT = dec (S; CT)

        ciphertext
        signature
        decrypt
        plaintext

signing   SIG = sign (SK; T)

        text
        secret key
        sign (decrypt)
        signature

Figure 4.19   The notation used in this book for cryptographic procedures

If you rely only on the protection offered by the assumption that a potential attacker does not have access to sufficient processing power, you may be quickly overtaken by the rapid pace of technical progress. Statements such as 'it would take a thousand years to break this cryptographic system' cannot be relied on, since they are based on the processing capacities and algorithms that are currently available. They cannot take future developments into account, since these are generally unknown. The arithmetic capacities of processors double around every 18 months, which means that the capacity per processor has increased by a factor of approximately 25,000 over the last 25 years.

Recently, the increased networking of computers has created another option for carrying out serious attacks on keys or cryptographic systems. Consider a request to work on the breaking of a DSS key posted on the Internet, which would be passed on to millions of users by the snowball effect. If only one percent of all current[2] users was to take part in such an action, this would provide the potential attacker with a parallel computer consisting of 300,00 individual computers.

Cryptographic algorithms are divided into two types: *symmetric* and *asymmetric*. This division is based on the key that is used. Here, 'symmetric' means that the algorithm uses the same key for encryption and decryption. By contrast, asymmetric algorithms (which were postulated in 1976 by Whitfield Diffie and Martin E. Hellman) use different keys for encryption and decryption.

A term that often comes up in connection with cryptographic algorithms is the *magnitude of the key space*. This refers to the number of possible keys that can be used with a particular cryptographic algorithm. A large key space is one of several criteria for a secure cryptographic algorithm.

---

[2]   In the summer of 1998, it was assumed that there were around 100 million users of the Internet. The rate of growth of the number of users has not yet reached the saturation point; instead, it is currently exponential.

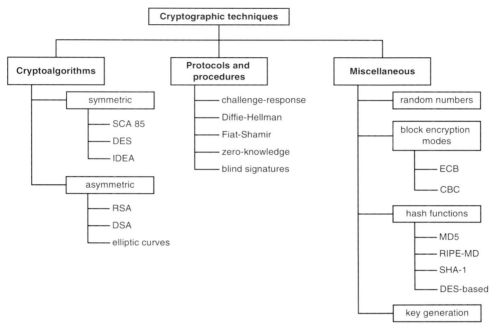

Figure 4.20   Classification chart for cryptographic techniques that are used in Smart Cards

A requirement that only recently has come to the fore with regard to the technical implementation of cryptographic algorithms in Smart Cards is freedom from noise. In this context, this means that the execution time of the algorithm must not depend on the key and the plaintext or ciphertext. If this requirement is not met, it may be possible to discover the key in a relatively short time, which means that the entire cryptographic system is broken.

In cryptology, there is a strong distinction between the theoretical and practical security of a system or an algorithm. A system is theoretically secure if an attacker, given unlimited time and technical aids, still cannot break the system. This means that, for example, a system could not be considered to be theoretically secure even if an attacker would need 100 years and the aid of several supercomputers to break it. If a system cannot be broken when the attacker has only a limited amount of time and technical aids, then it is considered to be practically secure.

A cryptographic system can assure the confidentiality and/or the authenticity of a message. Breaking the system means that the confidentiality and/or the authenticity are no longer guaranteed. If an attacker can figure out the secret key of an encryption algorithm, for example, he can then decrypt the data that have been encrypted for their protection, in order to learn their contents and to alter them if he so desires.

There are various methods of attack for breaking the key of a cryptographic algorithm. In the 'ciphertext only' attack, the attacker knows only the ciphertext and attempts to figure out the key or the plaintext from the ciphertext. A more promising method of attack is the 'known plaintext' attack, which involves the attacker knowing several plaintext–ciphertext pairs for a secret key. If the attacker can generate his own plaintext–ciphertext pairs using the 'chosen plaintext' and the 'chosen ciphertext' attack, his chances of success are even better, since the secret key can then be discovered experimentally.

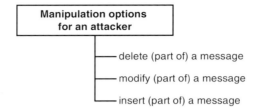

Figure 4.21   Classification of the various manipulation options available to an attacker

Finding the key by trial and error (a *brute force* attack) is naturally the least sophisticated method of attack. With this method, an attempt is made to find out the correct key by employing a large amount of processing capacity to try all possible keys with a know plaintext–ciphertext pair. Obviously, processing capacities in the supercomputer range are normally a prerequisite for this method. Statistically seen, on average half of the possible keys must be tested before the right one is found. A large key space will of course considerably increase the difficulty of such an attack.

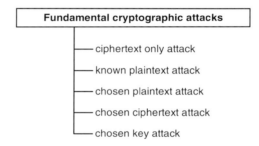

Figure 4.22   Classification of the basic methods of cryptographic attack

4.6.1   Symmetric cryptographic algorithms

Symmetric cryptographic algorithms are based on the principle of performing encryption and decryption using the same secret key – hence the designation 'symmetric'. The best-known and most widely used representative of this type of algorithm is the Data Encryption Algorithm (DEA). It was developed in 1977 by IBM in combination with the NBS (US National Bureau of Standards), and published as the US FIPS 46 standard. The standard that describes the DEA is often referred to as the DES (Data Encryption Standard). For this reason, the Data Encryption Algorithm is often (but not entirely correctly) called the DES.

Since this algorithm is of course built according to Kerckhoff's principle, it could be published without losing any of its security. However, even today not all of the development criteria have been made known, which leads to assumptions regarding possible methods of attack and possible 'trap doors'. Nonetheless, all attempts so far to break this algorithm on this basis have failed.

Two important principles for a good encryption algorithm were incorporated into the design of the DES. These are the principles of confusion and diffusion, as first proposed by

C. Shannon. The confusion principle states that the statistics of the ciphertext should affect the statistics of the plaintext in a way that is so complex that an attacker can derive no advantage from them. The second principle, that of diffusion, states that each bit of the plaintext and the key should affect as many bits of the ciphertext as possible.

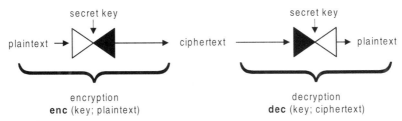

Figure 4.23   Working principle of a symmetric cryptographic algorithm for encrypting and then decrypting data. The DES is the typical example of this sort of cryptographic algorithm.

The symmetric block encryption algorithm DES does not expand the ciphertext, which means that the sizes of the plaintext and ciphertext blocks are the same. The block size is 64 bits (8 bytes), which is also the length of the key, although only 56 of these bits are used as the actual key.

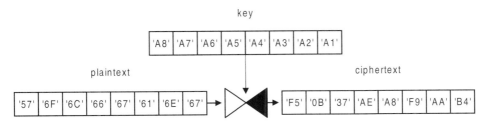

Figure 4.24   How the DES algorithm works in an encryption operation.

The key contains eight parity bits, which reduces the available key space. The 64 bits of the key are numbered consecutively from left (msb) to right (lsb). Bits 8, 16, 24, ..., 64 represent the parity bits. The parity is always odd. Due to the parity bits, the key space is 2^{56}. This means that there are approximately 7.2×10^{16} possible keys. At first glance, a key space with 72,057,594,037,927,936 possible keys may appear very large, but the size of its key space is actually the main weakness of the DES.[3] Given the steadily increasing processing power of modern computers, a key space of this size is considered to be the lower limit for a secure cryptographic algorithm. If a plaintext–ciphertext pair is available, all possible keys can easily be tested if the key space is too small.

If a plaintext–ciphertext pair is obtained by tapping the communications between a terminal and a Smart Card, a brute force attack can be carried out by encrypting the

---

[3]   To get a feeling for the size of this large number, you might enjoy considering the following comparison. The mass of the earth is roughly 5.974×10^{27} grams. Based on this value, the number of electrons, protons and neutrons that make up the earth can be taken to be around 10^{52}. If only a single elementary particle were needed to store a bit, then at most 10^{52} bits could be stored in a memory with the mass of the earth.

plaintext with all possible keys. The correct key can be determined by comparing all the resulting ciphertexts with the previously obtained ciphertext. This procedure can very easily be executed in parallel. Each of the parallel processors tests only that small part of the key space that is assigned to it. A sample calculation can illustrate the amount of time that is needed for such a brute-force attack. The fastest currently available DES components require 64 ns for a complete block encryption.[4] If 10,000 of these computation modules are assembled in parallel, each one could independently test a small part of the key space. Assuming that on average only half of the key space would have to be searched to find the correct key, we can calculate the processing time as follows:

$$\frac{2^{56} \times 64 \text{ nsec}}{10,000} \times \frac{1}{2} \approx 64 \text{ h}$$

In 1993, Michael Wiener published plans for a million-dollar computer that could test all the DES keys of a given plaintext–ciphertext pair within seven hours [Wiener 93]. The biggest problem with the use of DES is that its key space has become too small over time, so new applications use the triple-DES algorithm almost exclusively.

Delving into the exact implementation of the DES would explode the size of this book, and it is not necessary for understanding the material. If you want more detailed information, refer to FIPS Publication 46, Carl Meyer [Meyer 82] or Bruce Schneier [Schneier 96]. However, there is one significant detail. The DES was designed as an encryption algorithm that could easily be built in hardware. Unfortunately, there is presently no Smart Card microcontroller available with a DES hardware module, although the first prototypes are in the test stage. As a result, DES currently must be implemented in software in Smart Cards. This means that it takes up around 1 kB, even with highly optimized assembler code, and its computational speed is necessarily rather low.

Typical processing times for encryption and decryption with a Smart Card, as compared with a PC and a hardware integrated circuit, are listed in Table 4.5. The figures can vary, depending on the actual implementation, and they take into account only the pure processing time for the DES encryption or decryption of an 8-byte block, assuming that all registers are already loaded in advance.

Table 4.5   Typical DES processing times (8-byte block)

Implementation	Processing time
Smart Card with 3.5-MHz clock	17.0 ms
Smart Card with 3.5-MHz clock and DES processing module	154 µs
Smart Card with 4.9-MHz clock	12.0 ms
Smart Card with 4.9-MHz clock and DES processing module	112 µs
PC (80486, 33 MHz)	30 µs
PC (Pentium, 90 MHz)	16 µs
PC (Pentium, 200 MHz)	4 µs
DES integrated circuit	64 ns

---

[4]   This refers to a DEC gate array, built using gallium-arsenide technology, for the ECB and CBC modes.

Keys for the DES algorithm can be generated using a random number generator. This produces an 8-byte random number, which is then checked against the four weak and twelve semi-weak keys. If there is no match with these easily broken keys, the parity bits are computed and the result is a DES key.

In addition to the DES, there are many other symmetric cryptographic algorithms. Here we consider only the IDEA (International Data Encryption Algorithm) as a representative example. This was developed by Xuejia Lai and James L. Massey, and published in 1990 as the PES (proposed encryption standard). It was improved in 1991, and the improved version was known for a short while as the IPES (improved proposed encryption standard), but nowadays it is commonly known as the IDEA. The development criteria and internal construction of this algorithm have been fully published, so the Kerckhoff principle is satisfied. However, the IDEA is subject to patent restrictions, as is the RSA algorithm.

The IDEA, like the DES, is a block-oriented cryptographic algorithm, and it also uses 8-byte plaintext and ciphertext blocks. In contrast to the DES, the key length is 16 bytes (2 % 8 bytes). This provides a significantly larger key space, with a size of $2^{128} \approx 3.4 \% 10^{38}$. In normal decimal notation, the number of possible keys for the IDEA is exactly 340,282,366,920,938,463,463,374,607,431,768,211,456.

Due to its construction, as just described, the IDEA is compatible with the DES except for its extended key length. It is also compatible with triple-DES systems, which use keys that are 2 % 56 bits long, which means that changing the algorithm used does not affect the lengths of the keys or the input and output data blocks. Of course, compatibility in this regard does not mean that DES-encrypted data can be decrypted using the IDEA. In general, the IDEA is regarded as a very good cryptographic algorithm. It has also been widely distributed in the form of the public-domain program PGP (Pretty Good Privacy) from Philip Zimmermann, which is used for secure data transfers.

There are very few Smart Card implementations of the IDEA. The required memory for the program is around 1000 bytes. Typical computation times for encryption and decryption are somewhat less than for the DES. However, in the development of the IDEA, it was assumed that the computations would be executed by a 16-bit processor. Since Smart Cards still normally have 8-bit processors, the speed advantage in comparison with the DES is not as great as might be expected. Table 4.6 lists sample values for IDEA operations on an 8-byte block, assuming that previously computed keys are available.

Table 4.6   Typical IDEA processing times (8-byte block)

Implementation	Processing time
Smart Card with 3.5-MHz clock	12.3 ms
Smart Card with 4.9-MHz clock	8.8 ms
PC (80 386, 33 MHz)	70 μs
PC (Pentium Pro, 180 MHz)	4 μs
IDEA integrated circuit	370 ns

Operating modes for block-oriented encryption algorithms

The DES, like every block-oriented encryption algorithm, can be used in four different operating modes that are standardized in ISO 8372. Two of these operating modes, the CFB and OFB modes, are especially suitable for sequential text with no block structure. The other two, the ECB and CBC modes, are based on a block size of 8 bytes. These two block-oriented modes are used the most in Smart Card applications.

The basic operating mode of the DES is designated as the ECB (electronic code book) mode. In this mode, 8-byte plaintext blocks are independently encrypted using a single key. This is the DES in its pure form, without extensions.

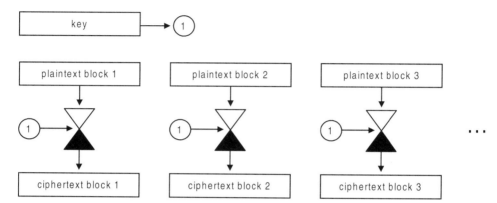

Figure 4.25   Data encryption with a block-oriented encryption algorithm operating in the ECB mode. Decryption is performed in a similar manner.

The second block-oriented mode is known as the CBC (cipher block chaining) mode. In this mode, a data string consisting of several blocks is chained using XOR operations during the encryption such that each block becomes dependent on the block that precedes it. In this way, the swapping, addition or deletion of encrypted blocks can be reliably detected. This is not possible in the ECB mode.

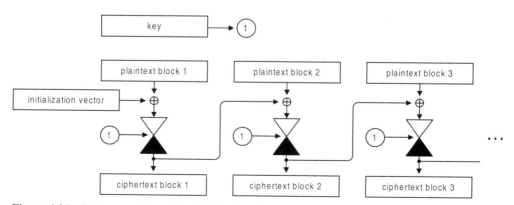

Figure 4.26   Data encryption with a block-oriented encryption algorithm operating in the CBC mode. Decryption is performed in a similar manner.

If the plaintext blocks are suitably structured (with sequence counters in their headers or initialization vectors), a consequence of CBC chaining is that even identical plaintext blocks are converted into non-identical ciphertext blocks. This makes the cryptoanalysis of intercepted data much more difficult, since code book analysis (for example) is not possible.

The first plaintext block is XOR-ed with an initialization vector (often called an IV), and then encrypted with the DES. The result is the ciphertext, which is in turn XOR-ed with following plaintext block. The process continues in this manner with the following blocks.

As a rule, the initialization vector is preset to null. However, in some systems, a session-specific random number is written to the initialization vector as a substitute for a temporary key. This number must naturally be known when the data are subsequently deciphered.

Multiple encryption

In addition to the four operating modes of a block-oriented encryption algorithm, also another variation is used to improve the security. It is however actually used only with the DES, because of the small key space of this algorithm. In principle, it can be used with any block-oriented encryption algorithm that is not a group. If an encryption algorithm has the group property, double encryption using two different keys will not increase the level of security, since the same result could be obtained by a single encryption with a third key. This means that the size of the key space is not increased by double encryption with an algorithm that does not have the group property, since an attacker would only have to discover the third key in order to be able to obtain the same result as that previously obtained by double encryption with two different keys.

$$\text{ciphertext} = \textbf{enc } (\text{key 2}; (\textbf{enc } (\text{key 1}; \text{plaintext}))$$

The DES is however not a group, so in principle the result of double DES encryption using two different keys cannot be duplicated by a single encryption using a third key.

However, in 1981 Ralph C. Merkle and Martin E. Hellman published a method of attack called the 'meet-in-the-middle' attack [Merkle 81], which can be used very successfully against every double encryption performed by a block-oriented encryption algorithm. It presupposes that the attacker has knowledge of several plaintext–ciphertext pairs. Its principle of operation is based on computing all possible encryptions of the plaintext using the first of the two keys, followed by decrypting the known result (the ciphertext) with every possible second key. The set of results from the first process is then compared with the set of results from the second process. If a match can be found, then there is a certain probability that the two keys have been discovered. The level of confidence in the keys can be increased by making the same comparisons with additional plaintext–ciphertext pairs. As can be seen, the amount of effort required for this attack is not significantly greater that that needed for a normal attack in which the entire key space must be searched. Consequently, cascaded double encryption is not used with the DES.

The process that is used instead is called triple-DES. In this mode, three sequential CBC-mode DES operations are performed with alternating encryption and decryption. Decrypting blocks that have been encrypted in this manner is performed by reversing the order of the operations (in other words: decryption, encryption and then decryption). If all three keys are the same, the result of the alternating encryption and decryption operations is the same as that obtained by a single encryption. This is the reason for not using a sequence of three encryption operations.

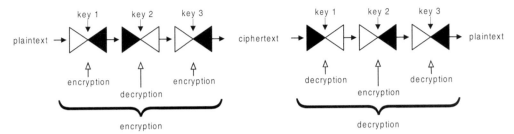

Figure 4.27 Operating principle of data encryption using the triple-DES process in the external CBC mode. Key 1 is the normally chosen to be the same as key 3, so that the effective key length is 2 ×56 bits (112 bits). Less commonly, three independent keys are used, with a resulting key length of 3 ×56 bits (168 bits).

If the three DES operations with three keys are applied directly to each plaintext block in turn, the process is referred to as DES in the inner CBC mode. If however the plaintext is first completely encrypted using the first key, and the result is then further encrypted in a similar manner, the process is referred to as DES in the outer CBC mode. The outer CBC mode is more resistant to attack and is therefore generally recommended [Schneier 96].

Table 4.7 Summary of Smart Card cryptographic algorithms, with input and output parameters

Name	Type	Plaintext length	Ciphertext length	Key length
DES	symmetric	8 bytes	8 bytes	56 bits
IDEA	symmetric	8 bytes	8 bytes	128 bits
Triple DES (2-DES)	symmetric	8 bytes	8 bytes	2×56 bits (112 bits)
Triple DES (3-DES)	symmetric	8 bytes	8 bytes	3×56 bits (168 bits)
RSA	asymmetric	512 bits (64 bytes)	512 bits (64 bytes)	512 bits (64 bytes)
		768 bits (96 bytes)	768 bits (96 bytes)	768 bits (96 bytes)
		1024 bits (128 bytes)	1024 bits (128 bytes)	1024 bits (128 bytes)
		2048 bits (256 bytes)	2048 bits (256 bytes)	2048 bits (256 bytes)
DSS (512 bits)	asymmetric	20 bytes	20 bytes	(64 + 20) bytes

The triple-DES algorithm has several other names, such as TDES, DES-3, 3-DES and 2-DES. Actually, the terms 'triple-DES' and '3-DES' only mean that three 56-bit keys are used. If the first and third keys are the same, this is called 2-DES, but if the three keys are all different, the designation 3-DES is often used. The key length must always be stated with triple DES, in order to unambiguously specify the algorithm.

The triple-DES algorithm is significantly more secure than sequential double encryption with two different keys, since the meet-in-the-middle attack is not effective with it. Three 56-bit keys are needed instead of only one, but the first and third keys are usually the same.

This yields a key length of 2×56 bits. This means that this process has data compatibility with the normal DES algorithm, and that it imposes no additional costs except for the doubled key size. This in particular is one of the main arguments for using triple DES in Smart Cards. It is primarily used for deriving keys and protecting very sensitive data, such as transferred keys, due to its improved level of security compared to single encryption.

4.6.2   Asymmetric cryptographic algorithms

In 1976, Whitfield Diffie and Martin E. Hellman described the possibility of developing an encryption algorithm based on two different keys [Diffie 76]. One of these keys was to be public, the other secret or 'private'. This would make it possible for someone to encrypt a message using the public key, with only the owner of the private key being able to decrypt it. This in turn would eliminate the problems associated with exchanging and distributing secret symmetric keys. In addition, certain other processes, such as generating digital signatures that could be verified by everyone, would be made possible for the first time.

Figure 4.28   Encryption and decryption with a public-key algorithm

The RSA algorithm
Two years later, Ronald L. Rivest, Adi Shamir and Leonard Adleman presented an algorithm that met the above-mentioned requirements [Rivest 78]. This algorithm, called the RSA algorithm after its inventors, is the best-known and most versatile asymmetric encryption algorithm presently in use. Its very simple operating principle is based on the arithmetic of large integers. The two keys are generated from two large prime numbers. The encryption and decryption processes can be expressed mathematically as follows:

$$\text{encryption:} \quad y = x^e \bmod n$$
$$\text{decryption:} \quad x = y^d \bmod n$$

where x = clear text
 y = encoded text
 e = public key
 d = private key
 n = public modulus = $p \cdot q$
 p, q = secret prime numbers

Before being encoded, the block of plaintext must be padded to the appropriate block size, which varies in the RSA algorithm according to the length of the key used. Encryption itself is performed by exponentiation of the plaintext followed by a modulo operation. The result of this process is the ciphertext. This can only be decoded if the private key is known. The decryption process is analogous to the encryption process.

The security of the algorithm is thus based on the difficulty of factoring large numbers. It is quite easy to compute the public modulus from the two prime numbers by multiplication, but it is very difficult to decompose the modulus into the two prime factors, since there is no effective algorithm for this operation.

The RAM capacity of chip cards would not be adequate for performing exponential operations on large numbers, as required in the context of encryption and decryption, since the numbers grow very large before being subjected to the modulo operation. For this reason, 'modular' exponentiation is employed, which means that the intermediate result of the calculation never exceeds the value of the modulus. For example, if the value of x^2 **mod** n must be calculated, the expression $(x \cdot x)$ **mod** n is not evaluated directly, since the intermediate result $(x \cdot x)$ would be unnecessarily large before being reduced by the modulo calculation. Instead, the expression $((x \bmod n) \cdot (x \bmod n))$ **mod** n is evaluated, which yields the same mathematical result. The advantage of this is that it requires significantly fewer calculations and less memory, since intermediate results are immediately reduced in size.

An additional way to increase the speed of the RSA algorithm is to use the Chinese remainder theorem for the calculations.[5] Of course, a prerequisite for using the Chinese remainder theorem is that both of the secret prime numbers p and q are known, which means that it can only be used for decryption (that is, for signing).

The private key should be as large as possible, since this impedes attempts to break the code. Public and private keys may have different lengths, and in fact this is usually the case, since the time required to verify a digital signature can be considerably reduced by making the public key as short as possible. The fourth Fermat number is frequently used as a public key. This prime number has the value of $2^{16} + 1 = 65537$, and due to its small size it is very well suited to quickly verifying digital signatures. The numbers 7 and 17 are likewise used.

Table 4.8 Typical public keys for the RSA algorithm

Public key	Remarks
$2 = °10°$	the only even prime number; used for the Rabin procedure
$3 = (2^1 + 1) = °11°$	the smallest odd prime number
$17 = (2^4 + 1) = °1\ 0001°$	—
$65537 = (2^{16} + 1) = °1\ 0000\ 0000\ 0000\ 0001°$	the fourth Fermat number F_4

If an attacker were to succeed in splitting the public modulus into its two prime factors, he or she would then be able to reproduce the entire encryption process. With a small value, such as 33, splitting the modulus is easily possible, but there is presently no fast algorithm that can be used for large numbers. If the values of both prime factors can be found, the system is thereby broken, since the private key is then known.[6]

---

5   For additional information regarding both procedures, see [Simmons 92, Schneier 96].
6   As early as the summer of 1994, a 426-bit key was broken in eight months using approximately 1600 computers connected via the Internet. The total amount of computation amounted to 5000 MIPS-years. A 100-MHz Pentium processor, for comparison, has a capacity of 50 MIPS.

Consequently, a requirement for RSA keys is that they are sufficiently long. A length of 512 bits (64 bytes) is presently considered to be the lower limit. In any case, 768-bit (96-byte) and 1024-bit (128-byte) keys are already in use. In the coming years, the first 2048-bit (256-byte) key will come into use. The amount of computational effort needed for encryption and decryption increases with the key length. This increase is not linear, but instead approximately exponential.

Smart Card microcontrollers, with their 8-bit CPUs, are normally not capable of performing an RSA computation in less than a few minutes. However, there are now microcontrollers with supplementary arithmetic processors that have been specially developed for fast exponentiation. With these, it is possible to execute RSA computations in an acceptable length of time with a reasonable software overhead. The size of the code for a hardware-supported 512-bit RSA algorithm is around 300 bytes. Around 1 kB of assembler code in the Smart Card is needed for 768-bit and 1024-bit keys. If you look at Table 4.9, you will see that even with a 512-bit key, the number of possible prime numbers is so large that collisions between two different key pairs will never occur.[7]

Table 4.9 Typical RSA key lengths with characteristic parameters. The ratio NS/PN indicates the relationship between the number of non-prime numbers and the number of prime numbers. The reciprocal of this is the probability that a random number in the number space is a prime. This is very important with regard to the length of time that is required to generate an RSA key.

Key length	Maximum number of digits	Size of the number space (NS)	Number of prime numbers in the number space (PN)	NS/PN
8 bits (1 byte)	3	256	54	≈ 4.7
40 bits (5 bytes)	13	$\approx 1.1 \times 10^{12}$	$\approx 3.9 \times 10^{10}$	≈ 28
512 bits (64 bytes)	155	$\approx 1.3 \times 10^{154}$	$\approx 3.8 \times 10^{151}$	≈ 342
768 bits (96 bytes)	232	$\approx 1.6 \times 10^{231}$	$\approx 2.9 \times 10^{228}$	≈ 552
1024 bits (128 bytes)	309	$\approx 1.8 \times 10^{308}$	$\approx 2.5 \times 10^{305}$	≈ 720
2048 bits (256 bytes)	617	$\approx 3.2 \times 10^{616}$	$\approx 2.3 \times 10^{613}$	≈ 1391
4096 bits (512 bytes)	1234	$\approx 1.0 \times 10^{1233}$	$\approx 2.5 \times 10^{1229}$	≈ 4000

However, one of the strengths of the RSA algorithm is that it is not limited to a particular key length, in contrast to (for example) the DES. If increased security is needed, longer keys can be used without any change to the algorithm. The RSA algorithm is thus scaleable, although the computation time and the required memory space must be kept in mind, since even 512-bit keys are presently considered to be secure. With current factoring algorithms, a good rule of thumb is that increasing the key length by 15 bits doubles the effort of computing the factors.[8] Andrew Odlyzko [Odlyzko 95] presents an excellent

---

[7]   The largest number that can be represented with 512 bits is $2^{512} - 1$, or in full:
13,407,807,929,942,597,099,574,024,998,205,846,127,479,365,820,592,393,377,723,561,443,721,764,
030,073,546,976,801,874,298,166,903,427,690,031,858,186,486,050,853,753,882,811,946,569,946,433,
649,006,084,095

[8]   As of January 1998, the largest known prime number had 909,256 digits and a value of $2^{3,402,377} - 1$. An extensive and up-to-date list of the largest prime numbers, with much background information, can be found at [UTM].

summary of both the processing capacity needed to factor and to break symmetric cryptographic algorithms and the processing capacity that is actually available throughout the world.

Although the RSA algorithm is very secure, it is rarely used to encrypt data because of the long computation time. It is used mainly in the realm of digital signatures, where the advantages of an asymmetric procedure can be fully realized. The greatest disadvantage of the RSA algorithm with regard to Smart Cards is the amount of memory space required for the key. The complexity of key generation also causes problems in certain cases.

Widespread use of the RSA algorithm is restricted by patent claims that have been made in several countries, and by major import and export restrictions imposed on equipment that employs this algorithm. Smart Cards with RSA coprocessors fall under these rules, which considerably hinders their use internationally.

The generation of keys for the RSA algorithm takes place according to a simple process. The following is a small worked-through example:

1. First, select two prime numbers p and q: $p = 3;\ q = 11$
2. Next, calculate the public modulus: $n = p \cdot q = 33$
3. Calculate the temporary variable z for use during key generation: $z = (p - 1) \cdot (q - 1)$
4. Calculate a public key e which satisfies the conditions $e < z$ and $\mathbf{gcd}\ (z, e) = 1$ (that is, the greatest common denominator of z and e is 1). Since there are several numbers that meet these conditions, select one of them: $e = 7$
5. Calculate a private key d that satisfies the condition $(d \cdot e)\ \mathbf{mod}\ z = 1$: $d = 3$

Table 4.10 Sample computation times for RSA encryption and decryption as a function of the key length. The indicated values are in part subject to considerable variation, since they are strongly dependent on the bit structure of the key and the use of the Chinese remainder algorithm (which can only be used for signing).

Implementation	Mode	512 bits	768 bits	1024 bits
Smart Card without NPU, 3.5 MHz	signing	20 min	---	---
Smart Card without NPU, 3.5 MHz (with Chinese remainder theorem)	signing	6 min	---	---
Smart Card with NPU, 3.5 MHz	signing	308 ms	910 ms	2000 ms
Smart Card with NPU, 3.5 MHz (with Chinese remainder theorem)	signing	84 ms	259 ms	560 ms
Smart Card with NPU, 4.9 MHz	signing	220 ms	650 ms	1400 ms
Smart Card with NPU, 4.9 MHz (with Chinese remainder theorem)	signing	60 ms	185 ms	400 ms
PC (Pentium, 200 MHz)	signing	12 ms	46 ms	60 ms
PC (Pentium, 200 MHz)	verification	2 ms	4 ms	6 ms
RSA integrated circuit	signing	8 ms	---	---

This completes the computation of the keys. The public and private keys can now be tested for encryption and decryption using the RSA algorithm, as illustrated in the following numeric example:

1. Use the number '4' as the plaintext x $(x < n)$: $x = 4$
2. Encrypt the text: $y = 4^7 \bmod 33 = 16$
3. The result of the calculation is the ciphertext y: $y = 16$
4. Decrypt the ciphertext: $x = 16^3 \bmod 33 = 4$

The result of decrypting the ciphertext is, as expected, again the plaintext.

In actual practice, key generation is more laborious, since it is very difficult to test large numbers to determine if they are prime. The well-known sieve of Eratosthenes cannot be used here, since it requires prior knowledge of all prime numbers smaller than the number being tested. This is practically impossible for numbers as large as 512 bits. Consequently, probabilistic tests are used to determine the probability that the selected number is a prime number. The Miller–Rabin test and the Solovay–Strassen test[9] are typical examples of such tests. To avoid having to use these time-consuming tests too often, randomly generated candidate numbers are first tested to see if they have any small prime factors. If the randomly generated number can be exactly divided by a small prime number, such as 2, 3, 5 or 7, it obviously cannot itself be a prime number. Once it has been determined that the number to be tested does not have any small prime factors, a primary number test such as the Miller–Rabin test can be used. The principle of this test is illustrated in Figure 4.30 and described in detail in the appendix of the IEEE 1363 standard.[10]

The algorithms for generating RSA keys have a special feature, which is that the time required to generate a key pair (a public key together with a private key) is only statistically predictable. This means that it is only possible to say that there is a certain probability that key generation will take a given amount of time. A definitive statement such as '... will take x seconds' is not possible, due to the need to run the prime number test on the random number. The time required to perform this test is not deterministically predictable.

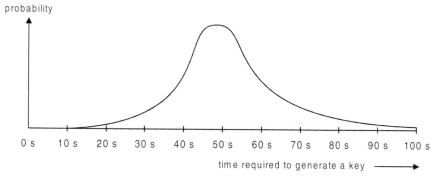

Figure 4.29 Typical time behavior of a probabilistic algorithm for generating key pairs for the RSA algorithm. The vertical axis shows the probability that a given amount of time will be required to generate a 1024-bit key in a Smart Card. The total area under the curve has a probability of 1.

---

9   The procedure and the algorithm are described by Alfred Menezes [Menezes 97].
10   Many tips and criteria that must be taken into account for the generation of prime numbers can be found in an article by Robert Silverman [Silverman 97].

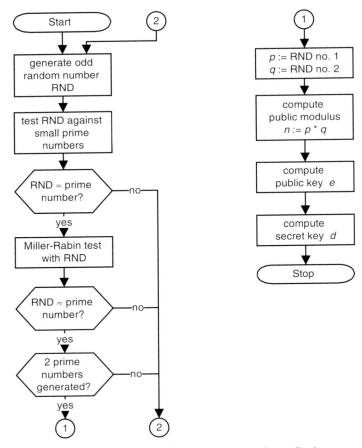

Figure 4.30   Basic procedure for generating RSA keys for use in Smart Cards

Table 4.11   Examples of the time required to generate a pair of public and private keys for the asymmetric RSA cryptographic algorithm. Exact times cannot be given, since the duration of the key generation process depends on whether the generated random numbers are prime, among other things.

Generating a public and private key pair for the RSA algorithm	Typical time	Possible time
Smart Card, 512-bit key, 3.5 MHz	12 s	• 3 s to • 40 s
Smart Card, 1024-bit key, 3.5 MHz	48 s	• 10 s to • 180 s
PC (Pentium, 200 MHz), 512-bit key	0.5 s	---
PC (Pentium, 200 MHz), 1024-bit key	2 s	---

The DSS algorithm

In the middle of 1991, the NIST (US National Institute of Standards and Technology) published a design for a cryptographic algorithm for adding signatures to messages. This algorithm, which has since been standardized in the US (FIPS 186), has received the designation DSA (digital signature algorithm), and the standard that describes it has been named DSS (digital signature standard). The DSA and RSA algorithms are the two most widely used procedures for the generation of digital signatures. The DSA algorithm is a modification of the El Gamal procedure. The background for the standardization of this algorithm is that a procedure was wanted that could be used for generating signatures but that could not be used to encrypt data. For this reason, the DSA algorithm is more complicated than the RSA algorithm. Nonetheless, it has been shown that it is possible to encrypt data using this algorithm [Simmons 98].

In contrast to the RSA algorithm, the security of the DSS algorithm does not depend on the problem of factoring large numbers, but rather on the discrete logarithm problem. The computation of the expression $y = a^x$ **mod** p can be performed quickly, even for large numbers. However, the reverse procedure, which is calculating the value of x for given values of y, a and p, requires very much computational effort.

With all signature algorithms, the message to be signed must first be reduced to a predefined length using a hash algorithm. The NIST therefore published a suitable algorithm for use with the DSS algorithm. This is named SHA-1 (Secure Hash Algorithm).[11] This variant of the MD5 hash algorithm generates a 160-bit hash value from a message of any arbitrary length. DSS algorithm computations, like those for the RSA algorithm, are performed using only integers.

To compute a signature with the DSA algorithm, the following global values must first be determined:

p (public):	512-bit to 1024-bit prime number, whose length is evenly divisible by 64;
q (public):	160-bit prime factor of $(p - 1)$;
g (public):	$g = h^{(p-1)/q}$, where h is an integer that satisfies the conditions $h < p - 1$ and $g > 1$.

The private key x must satisfy the following condition:

$x < q$

The public key y is computed as follows:

$y = g^x$ **mod** p

Once all necessary keys and numbers have been determined, the message m can be signed as follows:

Generate a random number k, where $k < q$.	
Compute the hash value of m:	$H(m)$
Calculate r:	$r = (g^k$ **mod** $p)$ **mod** q
Calculate s:	$s = k^{-1} (H(m) + x \cdot r)$ **mod** q

The two values r and s are the digital signature of the message. This consists of two numbers for the DSS algorithm, rather than only one number as with the RSA algorithm.

---

11 See Section 4.8, 'Hash Functions'.

Verifying the signature is performed as follows:

1. Calculate w: $w = s^{-1} \bmod q$
2. Calculate $u1$: $u1 = (H(m) \cdot w) \bmod q$
3. Calculate $u2$: $u2 = (r \cdot w) \bmod q$
4. Calculate v: $v = ((g^{u1} \cdot y^{u2}) \bmod p) \bmod q$

If the condition $v = s$ is satisfied, the message m has not been altered and the digital signature is genuine.

Table 4.12 Examples of computation times for the DSA algorithm as a function of the clock rate, split into encryption and decryption times. The stated values are subject to considerable variation, since they depend strongly on the bit structure of the key. The computation time can be reduced by pre-calculation.

Implementation	Verifying a 512-bit signature	Generating a 512-bit signature
Smart Card with 3.5-MHz clock	130 ms	70 ms
Smart Card with 4.9-MHz clock	90 ms	50 ms
PC (80386, 33 MHz)	16 s	35 ms

At present, it is not possible to say which of these two signature-generation algorithms (DSS and RSA) will prevail in the long term, or which offers better long-term security. The original intention of standardizing a signature algorithm that cannot be used for encryption, which resulted in the DSS algorithm, has largely come to nothing. The complexity of this algorithm also does not promote its widespread use. Nonetheless, the fact that the standard exists, and the associated political pressure to generate signatures using the DSS and SHS, are strong arguments for many institutions.

Using elliptic curves as asymmetric cryptographic algorithms
In addition to the two well-known asymmetric cryptographic algorithms, RSA and DSA, there is a third type of cryptography that is used for digital signatures and key exchanges in the domain of Smart Cards. This is based on elliptic curves (EC).

In 1985, Victor Miller and Neal Koblitz independently proposed the use of elliptic curves for the construction of asymmetric cryptographic algorithms. The properties of elliptic curves are well suited to such applications, and in the course of the following years practical cryptographic systems based on these proposals were developed. In general, these are usually referred to as ECC (elliptic curve cryptosystems).

Elliptic curves are sets of smooth curves that satisfy the equation $y^2 = x^3 + ax + b$ within a finite three-dimensional space. No point may be a singularity. This means, for example, that $4a^2 + 27b^2 \neq 0$. In the realm of cryptography, the finite spaces $GF(p)$, $GF(2^n)$ and $GF(p^n)$ are used, where p is a prime number and n is a positive integer greater than 1.

The mathematics relating to cryptographic systems based on elliptic curves is relatively difficult. For this reason, you are referred to the book by Alfred Menezes on the subject [Menezes 93]. The very comprehensive IEEE 1363 public key cryptography standard and the ISO/IEC 15946 series of standards dealing with elliptic curves also provide good synopses of elliptic curves and other asymmetric cryptographic techniques.

The great advantage of asymmetric cryptographic systems based on elliptic curves is that they require much less computational capacity than for instance RSA, and that the key lengths are significantly shorter for the same level of cryptographic strength. For example, about the same amount of computation is required to break an ECC with a 160-bit key as for an RSA algorithm with a 1024-bit key. An ECC with a 320-bit key, with reference to this computational criterion, corresponds to an RSA system with a 5120-bit key. This cryptographic strength and the relatively small size of the keys are exactly the reasons why ECC systems can be found in the realm of Smart Cards.

The arithmetic processing elements of modern-day Smart Card microcontrollers generally support ECC, which means that a relatively high computational speed is available. As with the RSA algorithm, the key length is an important characteristic of an asymmetric cryptographic algorithm.

Interestingly enough, cryptographic systems based on elliptic curves require so little computational capacity that they can even implemented in microcontrollers without coprocessors. With a 6805 CPU (SC28), the implementation of an ECC requires around 4 kB of program code in ROM or EEPROM, plus around 90 bytes of RAM. The generation of a 135-bit signature takes around 185 ms with a 5-MHz clock. An RSA signature algorithm needs a comparable amount of time on a Smart Card.

One thing that argues against the use of elliptic curves in the field of asymmetric cryptographic algorithms is that they are regarded as a relatively new discovery in the cryptographic world, even though they have been known for a long time. It will no doubt take some time until the use of ECC systems becomes commonplace in the circumspect world of cryptographers and Smart Card application designers. This is in spite of the fact that cryptographic systems based on elliptic curves presently offer the highest level of security per bit when compared to all other asymmetric techniques.

Table 4.13 Examples computation times for cryptographic algorithms based on elliptic curves, as a function of key length. These values are subject to considerable variation, since they depend strongly on the bit structure of the key.

Implementation	Verifying a 135-bit signature	Generating a 135-bit signature
Smart Card with 3.5-MHz clock	510 ms	260 ms
Smart Card with 4.9-MHz clock	360 ms	185 ms
PC (Pentium, 120 MHz)	140 ms	110 ms

4.6.3 Padding

In Smart Cards, the DES algorithm is mainly used in the two block-oriented modes ECB and CBC. However, since the data that are communicated to the card do not always fit exactly into a certain number of blocks, sometimes a block must be filled up. Extending a data block so that its length is an exact multiple of a given block size is termed *padding*.

The recipient of a padded data block has a problem after the data have been decrypted, since he or she does not know where the actual data stop and the padding bytes start. One solution to this would be to state the length of the message at the beginning of the message, but this would change the structure of the message, which is generally undesirable. It would be especially onerous with data that do not always have to be encrypted, since in this case no padding would be needed, and thus also no length information. In many cases, therefore, the structure of the message may not be changed.

This means that a different method must be used to identify the padding bytes. As a rule, there is general agreement on the use of the following algorithm, which is defined in the ISO/IEC 9797 standard. The most-significant bit (msb) of the first padding byte following the useful data is set to 1. This byte thus has the hexadecimal value '80'. If additional padding bytes are needed, they have the value '00'. The recipient of the padded message thus searches from the beginning to the end of the message for a msb set to 1 or for the value '80'. If it finds this, it knows that this byte and all subsequent bytes are padding bytes and not part of the message.

Figure 4.31   Data padding according to ISO/IEC 9797

In this regard, it is important that the recipient knows whether the messages are always padded or only padded if necessary. If padding only takes place when the length of the data to be encrypted is not an integer multiple of the block length, the recipient must take this into account. Consequently, there is often an implicit understanding that padding always takes place, which of course has the disadvantage that sometimes an unnecessary block of padding data must be encrypted, transferred and decrypted.

For the sake of completeness, we should mention that sometimes only the value '00' is used for padding. The reason for this is that this value is normally used for padding in MAC computations. If only one padding algorithm is used, this saves program code. Of course, in this case the application must know the exact structure of the data in order to be able to distinguish between useful data and padding.

4.6.4   Message authentication codes and cryptographic checksums

The authenticity of a message is many times more important than its confidentiality. The term 'authenticity' means that the message has not been altered or manipulated, and thus is authentic. For this purpose, a computed message authentication code (MAC) is appended to the message, and both are sent to the recipient. The recipient can then compute the MAC for the message himself and compare it to the received MAC. If these match, the message has not been altered during its journey.

A cryptographic algorithm with a secret key is used to generate a MAC. This key must be known to both communications partners. In principle, a MAC is a sort of error detection code (EDC), which can naturally only be verified if the associated secret key is known. For this reason, the term 'cryptographic checksum' (CCS) is also used (as well as some other terms), but technically a CCS is completely identical to a MAC. In general, the difference between the two designations is that the term 'MAC' is used for data transfers and the term 'CCS' is used for all other application areas. The term 'signature' is often seen used as a substitute for 'MAC'. However, this is not the same as a 'digital signature', since the latter is generated using an asymmetric cryptographic algorithm.

Figure 4.32   The usual arrangement of the message and the message authentication code (MAC)

In principle, any cryptographic algorithm can be used to compute a MAC. In practice, however, the DES algorithm is used almost exclusively. This algorithm is used here to demonstrate the process.

If the message is encrypted using the DES algorithm in the CBC mode, each block is linked to its previous block. This means that the final block depends on all prior blocks. This block, or a part of it, represents the MAC of the message. The message itself however remains in plaintext; it is not passed on in encrypted form.

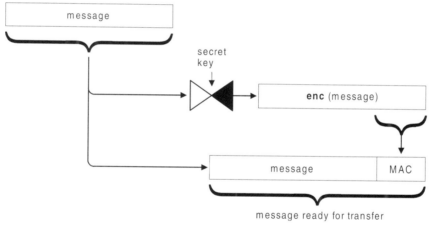

Figure 4.33   Example of a MAC computation process

There are a few important conditions relating to the generation of a MAC using the DES algorithm. If the length of the message is not an exact multiple of eight bytes, it must always be extended. This generally falls under the concept of padding. However, in most cases the padding uses only the value '00' (per ANSI X.99 – Message Authentication). This is allowed here, because there must be prior agreement regarding the length and location of the MAC within the message. The MAC itself is formed by the left-most (most significant) four bytes of the final block resulting from the CBC mode encryption. The padding bytes, however, are not passed on when the message is transferred. This limits the data that has to be transferred to the protected data and the appended MAC.

4.7   Key Management

The single objective of all administration principles relating to keys for cryptographic algorithms is to minimize the consequences to the system, and to the Smart Card application, if one or more secret keys become known to unauthorized persons. If it were possible to be certain that the keys would always remain secret, a single secret key for all Smart Cards would be sufficient. However, nobody can reliably guarantee this secrecy.

The use of the security-enhancing principles described here for keys that are used with cryptographic algorithms causes the number of keys to increase sharply. If all of the principles and procedures described in this section are implemented in a single Smart Card, the keys will usually take up more than half of the available application data memory.

However, depending on the application area, it is not always necessary to use every possible principle and procedure. For example, there is no need to support multiple generations of keys if the card is valid for only a limited length of time, since the additional administrative effort and memory space would not be justified.

4.7.1   Derived keys

Since Smart Cards, in contrast to terminals, can be taken home by anyone and there possibly subjected to thorough and painstaking analysis, they are naturally exposed to the most severe attacks. If no master key is present in the card, the consequences of a successful attempt to read out the card contents can be minimized. Consequently, the keys that are found in the card are only those that have been derived from a master key.

Derived keys are generated using a cryptographic algorithm. The input values are a card-specific feature and a master key. The DES or triple-DES algorithm is usually used. For the sake of simplicity, the card number is usually used as the specific feature. This number, which is generated when the card is manufactured, is unique in the entire system and can be used throughout the system to identify the card.

Derived keys are thus unique. One function that can be used to generate derived keys, as illustrated in Figure 4.34, is:

$$derived\ key = \mathbf{enc}\ (master\ key;\ card\ number)$$

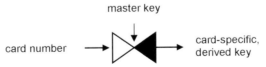

Figure 4.34   A possible method for generating a derived, card-specific symmetric key from the card number and a master key

4.7.2   Key diversification

In order to minimize the consequences of a key being compromised, separate keys are often used for various cryptographic algorithms. For example, different keys can be used for signatures, secure data transfers, authentication and data encrypting. There must be a separate master key for each type of key, from which the individual keys may be derived.

4.7.3   Key versions

It is normally not adequate to employ only one key generation for the full lifetime of a Smart Card. Suppose that a master key were to be computed as a result of an attack. In this case, all application vendors would have to shut down their systems and the card issuers would have to replace all the cards, The resulting damage would be enormous. Consequently, it is possible to switch over to a new key generation in all modern systems.

Switching to a new generation of keys may be forced by the fact that a key has been compromised, but it can also take place routinely at fixed or variable intervals. The result of a switch is that all of the keys in the system are replaced by new ones, without any need for the cards to be recalled. Since the master keys are located in the terminals and the higher-level parts of the system, a secure data exchange is all that is needed to provide new, unknown keys to the terminals.

4.7.4   Dynamic keys

For many applications, and in particular in the realm of data transfer security, it is common practice to use dynamic keys. These are also referred to as 'temporary keys' or 'session keys'. To generate a dynamic key, one of the two communicating parties first generates a random number, or some other value that is used for the particular session, and passes it to the other party. The further course of the procedure depends on whether only symmetric cryptographic algorithms are to be used, or also asymmetric algorithms.

Dynamic keys with symmetric cryptographic algorithms

For procedures that use only symmetric cryptographic algorithms, the random number that is generated by one of the two parties is sent as plaintext to the other party. The Smart Card and the terminal then encrypt this number using a derived key. The result is a key that is valid only for this one session, as shown in Figure 4.35.

$$dynamic\ key = \textbf{enc}\ (derived\ key;\ random\ number)$$

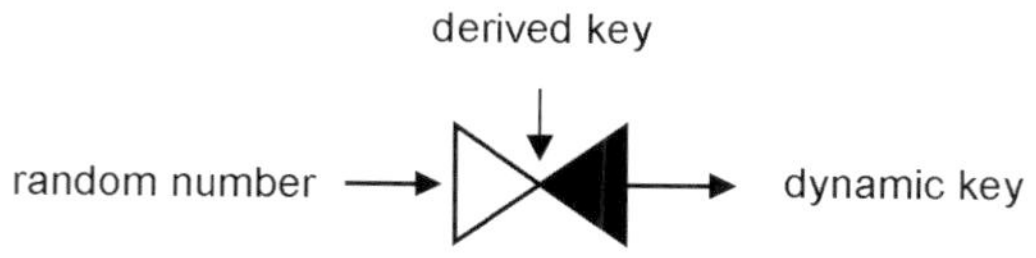

Figure 4.35   A possible way to generate a dynamic key using a random number and a derived key

The main advantage of dynamic keys is that they are different for each session, which makes attacks significantly more difficult. However, care must be taken when a temporary key is used to generate a signature, since the temporary key will also be needed to verify the signature. This key can only be generated using the same random number that was used when the signature was created. This means that whenever a dynamic key is used for a signature, the random number used to generate the key must be retained for use in verification, and thus must be stored.

The ANSI X 9.17 standard proposes a somewhat more complicated procedure for the generation of derived and dynamic keys. This procedure is widely used in systems for financial transactions. It requires two inputs: a value T_i that does not depend on the time or the session, and a key Key_{Gen} that is reserved for the generation of new keys. Afterwards, the initial key Key_i can be used to compute as many additional keys as desired. This key generation procedure has the additional advantage that it cannot be computed in reverse, which means in other words that it is a one-way function:

$$Key_{i+1} = \text{enc} (Key_{Gen}; \text{enc} (Key_{Gen}; (T_i \text{ XOR } Key_i)))$$

Exchanging dynamic keys using asymmetric cryptographic algorithms

Figures 4.36 and 4.37 show the generation and subsequent exchange of a symmetric dynamic key for message encryption. An asymmetric cryptographic algorithm, such as the RSA or DES algorithm, is used for the key exchange. A similar process is used in PGP, for example. It employs the IDEA and the RSA algorithm. The basic advantage of this hybrid process is that the actual encryption of large volumes of data can be performed using a symmetric cryptographic algorithm, which has significantly higher throughput than an asymmetric algorithm.

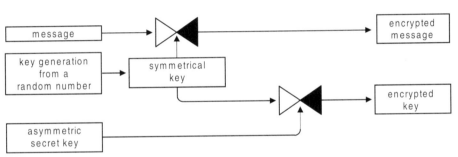

Figure 4.36 Sample procedure for key exchange using a combination of symmetric and asymmetric cryptographic algorithms. An encrypted dynamic symmetric key is first generated, and this is then exchanged between two entities using an asymmetric cryptographic algorithm. The generation and exchange of the key pair for the asymmetric cryptographic algorithm, which takes place separately and in advance, is not shown.

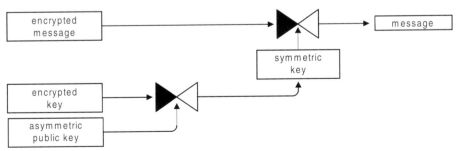

Figure 4.37 Sample procedure for key exchange using a combination of symmetric and asymmetric cryptographic algorithms. A previously encrypted dynamic symmetric key is recovered by using an asymmetric cryptographic algorithm. The generation and exchange of the key pair for the asymmetric cryptographic algorithm, which takes place separately and in advance, is not shown.

4.7.5   Key parameters

A mechanism that is as simple as possible is now needed to allow the key stored in the card to be externally addressed. The Smart Card operating system must also ensure in all cases that the key can only be used for its intended purpose. For example, it must prevent an authentication key from being used for the encryption of data. To address a key, its key number and its application purpose must be known. The key number is the actual reference to the key. In addition, the version number is also needed to address a specific key.

Table 4.14   Typical key parameters that are stored in a Smart Card

Data element	Remarks
Key number	Key reference number; unique within the key file
Version number	Version number of the key; can affect key derivation
Application purpose	Identifies the cryptographic algorithms and procedures with which the key may be used
Disable	Allows the key to be temporarily or permanently disabled
Retry counter	This counter keeps track of non-successful attempts to use the key with cryptographic procedures
Maximum retry count	If the retry count reaches the maximum count, the key is blocked
Key length	---
Key	The actual key

Some Smart Card operating systems allow a retry counter associated with the key to be incremented each time a failure occurs with some action using the key, such as an authentication. This can be used to quite reliably prevent the key value from being fished out by repeated attempts, although such an attack does not present a serious risk due to the long processing times in the card. If the retry count reaches the maximum allowed value, the key is blocked and cannot be further used. The retry counter is reset to zero if the attempted use of the key is successful. In any case, such a mechanism must be used with great care, since an incorrect master key in a terminal could easily result in massive card failures. The retry counter can normally only be reset using a special terminal, and the identity of the cardholder must be verified before this is done.

Some systems prohibit the reuse of old versions of keys. This is accomplished by providing the key with a disable field that is activated as soon as a new key with the same key number is addressed.

4.7.6   Key management example

Here we would like to describe an example of key management for a system based on Smart Cards. The objective is to further illustrate the previously described principles by means of an easily understood collective example. Compared with this example, large real systems frequently have arrangements that are much more complex and several structural layers. Small systems often have no key hierarchy at all, since a secret global key is used for all cards. The system presented here occupies a middle position between systems with very simple structures and large systems, and thus represents a good example.

In the example shown in Figure 4.38, the keys for loading and paying can be used with an electronic purse. They use symmetric cryptographic procedures. These keys are in any case important within the system, because they are relatively well protected by the described key hierarchy. The individual derivation functions are not explained in detail here, but the DES or triple-DES algorithm could always be used for them. The lengths of the keys are also not dealt with in detail, but they certainly can vary. The keys at the top of the hierarchy are normally derived using more powerful cryptographic functions than the lower-level keys, for reasons of security.

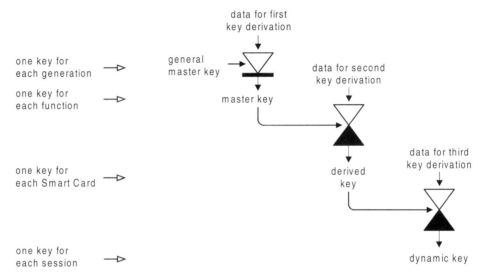

Figure 4.38   An example of a key hierarchy in a system based on Smart Cards and using symmetric cryptographic algorithms

The key that stands at the top of the hierarchy is called the *general master key*. There is only one such key for an entire generation of keys. A generation could for example remain valid for a year, to be replaced in the succeeding year by a new generation, which means a new generation of the general master key. This key is the most sensitive key of the system with regard to security. If it becomes known, then all keys of its generation can be computed, and the system is broken for one generation. The master key can be generated from a random number. It is also conceivable to base the general master key on the values shown by dice that are thrown by several independent persons, each of whom consequently knows only part of the key value. The general master key should never be completely known by any single person, and its generation must under no circumstances be reproducible.

A separate master key for each function is derived from the general master key. The function of a key can be loading or paying with an electronic purse. A one-way function, such as a modified DES algorithm, is used in this example to derive the master keys for the individual functions. This makes it impossible to compute the general master key from a master key by applying the calculation process in reverse. If a one-way function were not used to derive the master keys, and if then a master key should become known in spite of all security measures, it would be possible to compute the general master key if the derivation parameters were also known. The reason for using a one-way function here is the assumption that in this imaginary purse system, the master keys will be located in the security modules of local terminals. This means that they are much more vulnerable than the general master key, which always remains in the background system.

The derived keys form the next level in the key hierarchy. These are the keys that are located in the Smart Cards. Each card contains a set of derived keys, which are separated according to their functions and generations. If such a card is used at a terminal, the terminal can compute the derived key for itself, based on the parameters used to derive the key in question. Naturally, the terminal first reads the derivation parameters from the card. Once the derived key is available, the following step is to compute the dynamic key, which is specific to an individual session. This key is valid only for the duration of a single session. The duration of a session ranges from a few hundred milliseconds to a few seconds in most Smart Card applications. Dynamic keys are no longer used after a session ends.

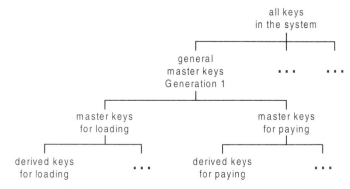

Figure 4.39   Examples of keys for an electronic purse system with two functions: loading and paying. Only the stored keys are shown here; keys that are dynamically generated for individual sessions have been omitted to simplify the diagram.

This example system may appear complicated at first glance, but it is relatively simple in comparison to real systems. The objective of the example is to indicate exactly how all the keys in a system can be generated. It also implicitly shows what measures must be taken if a key should become known. If the general master key becomes known, one must switch to a new generation in order to be able to run the system without concerns about security risks. If on the other hand a derived key becomes known, all that is required is to block the card in question; any other changes to the key management would surely be inappropriate. All of these measures naturally assume that the reason that a key (or several keys) has become know can be determined, so that this can be prevented in the future.

Given this key hierarchy, it is evident that very many keys must be generated and stored in the Smart Cards. Of course, it is always possible to assign several functions to a single key in order to save memory space. A different arrangement of the key hierarchy is also certainly conceivable. This naturally depends very much on the system for which the key management is developed.

4.8   Hash Functions

Even powerful computers need a great deal of time to compute a digital signature. In addition, many signatures would be required for long documents, since the document to be signed cannot be arbitrarily long. A trick is therefore used. The document is first compressed to a much shorter fixed length, and then the signature of the compressed data is computed. It does not matter whether the compression can be reversed, since the signature can always reproduced from the original document. The functions that are used for this type of computation are called one-way hash functions.

Generally speaking, a one-way hash function is a function that derives a fixed-length value from a variable-length document, such that this value represents the original content of the document in a compressed form and cannot be used to reconstruct the original document. In the Smart Card domain, these functions are used exclusively to compute the input values for digital signatures. If the document's length is not a multiple of the block length used by the hash function, it must be padded appropriately.

For a hash function to be efficient, it should exhibit certain properties. The result should have a fixed length, so that it can it can be readily used by signature algorithms. Since large quantities of data normally have to be processed, the hash function must have a high throughput. It must also be easy to compute the hash value. By contrast, it must be difficult, or even better impossible, to derive the original document from a known hash value. Finally, the hash function must be collision-resistant. This means that for a given document, it should not be easy to find a second document that yields the same hash value.

Nevertheless, there certainly will be other documents with the same hash value. This is only natural, since all possible messages, ranging in length from null to infinity, are represented by hash values that all have the same fixed length. An unavoidable consequence of this is that collisions will occur. That is why the term 'collision-resistant' is used, rather than 'collision-free'.

What would be the effect of a collision? There would be two different documents with the same hash value, and thus the same digital signature. This would have the fatal consequence of making the signature worthless, since it would be possible to alter the document without anyone being able to detect the fact. This is precisely what happens in one of the two typical attacks that can be used with hash functions. It consists of systematically searching for a second document that has the same hash value as the original document. If the content of this document makes sense, then the digital signature derived from the hash value has been discredited. Since the two documents are interchangeable, the signature is worthless. After all, there is an enormous difference between a house purchase contract for £125,000 and one for £750.000.

The second type of attack on a hash value is somewhat subtler. Here, two documents whose hash values are equal but whose contents are different are prepared in advance. This is not particularly difficult, if you consider all the special symbols and extensions available in the character set. The result is that a single digital signature is valid for both documents, and it is impossible to prove which document was originally signed.

Finding two documents with the same hash value is not as difficult as it might appear at first sight. It is possible to exploit the birthday paradox, which is well known in statistical theory. This has to do with two questions. The first question is: how many people must be in a room for the probability to be greater than 50% that one of them has the same birthday as the person asking the question. The answer is simple: it is only necessary to compare the birthday of the questioner with the birthday of everyone else in the room. There must be at least 183 (365 ÷ 2) people in the room.

The second question reveals the paradox, or better the surprising result of this comparison. It is, how many people must be in a room for the probability to be greater than 50% that two people have the same birthday. The answer is only 23 people. The reason is that, although only 23 people are present, this represents a total of 253 pairs for comparing birthdays. The probability of that two people have the same birthday is based on these pairs.

Precisely this paradox is utilized in attacking a hash function. It is very much easier to create two documents that have the same hash value than it is to modify a document until it yields a given hash value. The consequence is that the results of hash functions must be large enough to successfully foil both types of attack. Most hash functions thus produce values that are at least 128 bits long, which currently is generally considered to be adequate with regard to the two types of attack just described.

Many hash functions have been published up to now, and some of them are also defined in standards. However, modifications of these functions occur repeatedly, due to the discovery of successful forms of attack. Table 4.15 is a short summary of the hash functions currently in common use. A discussion of their internal operation is unfortunately beyond the scope of this book.

Table 4.15   Summary of commonly used hash functions

Name	Input block size	Hash value length
ISO/IEC 10118-2	n bits (e.g. 64 or 128 bits)	n or $2n$ bits (e.g. 64 or 128 bits)
MD4	512 bits	128 bits
MD5	512 bits	128 bits
MDC-4	64 bits	128 bits
RIPEMD-128	512 bits	128 bits
RIPEMD-160	512 bits	160 bits
SHA-1	512 bits	160 bits

The ISO/IEC 10118-2 standard specifies a hash function based on an n-bit block encryption algorithm (e.g. DES). The hash value may be n or $2n$ bits long with the described algorithm. The hash function MD4 (message digest 4) and its successor MD5 were published by Ronald L. Rivest in 1990–1991. They are based on a stand-alone algorithm, and both generate a 128-bit hash value. The NIST published a hash function for DSS algorithm in 1992 that is known as SHA. After the discovery of certain weaknesses, it was modified, and the resulting function has been known since mid-1995 as SHA-1. It is also standardized under the designation FIPS 180-1.

Since data transfer to Smart Cards is generally slow, the hash function is executed in the terminal or in a computer connected to the terminal. This drawback is balanced by the fact that this makes the hash function interchangeable. Besides, in most cases, memory limitations prevent hash functions from being stored in the cards. The program size is in almost every case around 4 kB of assembler code. The throughput of typical hash functions is very high relative to the demands placed on them. With an 80386 computer running at 33 MHz, it is usually at least 300 kB/s, and it lies in the range of 4 to 8 MB/s with a 200-MHz Pentium PC.

4.9   Random Numbers

Random numbers are repeatedly needed in connection with cryptographic procedures. In the field of Smart Cards, they are typically used to ensure the uniqueness of a session during authentication, as padding for data encryption and as initial values for transmission sequence counters. The length of the random numbers needed for these functions usually lies between 2 and 8 bytes. The maximum length naturally comes from the block size of the DES algorithm.

The security of all these procedures is based on random numbers that cannot be predicted or externally influenced. The ideal solution would be a hardware-based random number generator in the card's microcontroller. However, this would have to be completely independent of external influences, such as temperature, supply voltage, radiation and so on, since otherwise it could be manipulated. That would make it possible to compromise certain procedures whose security relies on the randomness of the random numbers. Current random number generators in Smart Card microcontrollers are generally based on linear feedback shift registers (LFSRs) driven by voltage-controlled oscillators.

Since it is almost impossible with currently available technology to make a good random number generator that is immune to external influences using only semiconductor hardware, operating system designers fall back on software implementations. These yield pseudo-random number generators, most of which produce very good random numbers. Nevertheless, the numbers that they produce are not genuinely random, since they are computed by strictly deterministic algorithms and are thus predictable if the algorithm and its input values are known. This is why they are designated 'pseudo-random'.

It is also very important to ensure that the cards of a production run generate different sequences of random numbers, so that the random number produced by one card cannot be inferred from that produced by another card from the same series. This is achieved by entering a random number as the seed number (starting value) for the random number generator when the operating system for each card is prepared.

4.9.1   Generating random numbers

There are many different ways to generate random numbers using software. However, since the memory capacity of Smart Cards is very limited and the time needed to perform the calculation should be as short as possible, the number of options is severely restricted. In practice, essentially the only procedures that are used are those that utilize functions already present in the operating system, and that therefore require very little extra program code.

Naturally, the quality of the random numbers must not be negatively affected if a session is interrupted by a reset or by removing the card from the terminal. In addition, the generator must be constructed such that sequence of random numbers is not the same for every session. This may sound trivial, but it requires at least a write access to the EEPROM to store a new seed number for the next session. The RAM is not suitable for this purpose, since it needs power to retain its contents. One possible means of attack would be to keep on generating random numbers until the EEPROM cells that store the seed number fail. Theoretically, this would cause the same sequence of random numbers to occur in every session, which would make them predictable and thus give the attacker an advantage. This type of attack can easily be averted by constructing the relevant part of the EEPROM as a ring buffer and by blocking all further actions once a write error occurs.

Another very important consideration for a software-implemented random number generator is to ensure that it never runs in an endless loop. This would result in a markedly shorter repeat cycle for the random numbers. It would then be easy to predict the numbers, and the system would be broken.

Almost every Smart Card operating system includes an encryption algorithm for authentication. It is an obvious idea to use this as the basis for a random number generator. You should realize that a good encryption algorithm mixes the plaintext as thoroughly as possible, so that the plaintext cannot be derived from the ciphertext without knowledge of the key. A principle known as the avalanche criterion says that, on average, changing one input bit should change half of the output bits. This property can be also be used very well for a random number generator. The exact construction of the generator varies from one implementation to the other.

Figure 4.40 illustrates one possibility. This generator uses the DES algorithm with a block length of 8 bytes, and the output value is fed back to the input. Naturally, any other encryption algorithm could also be used. The generator works essentially as follows. The value of a ring buffer element is encrypted by DES, using a key unique to the card. The ciphertext so produced is the 8-byte random number. This number, when XOR-ed with the previous plaintext, provides the new entry for the EEPROM ring buffer. The generator then moves to the following entry in the cyclic ring buffer. Mathematically, the relationship can be expressed by the formula $RND_n := f\,(key,\,RND_{n-1}\,)$.

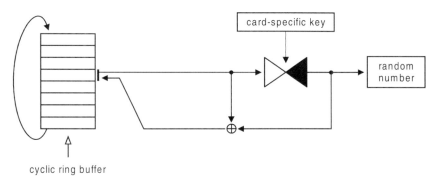

Figure 4.40   Sample construction of a DES pseudo-random number generator for Smart Card operating systems. This generator is primarily designed to minimize the number of write accesses to the EEPROM.

When the Smart Cards are prepared for issuing, the card-specific DES key is stored in each card, and at the same time random seed numbers are entered into the ring buffer. This could be a 12×8 buffer, for example. The seed numbers ensure that each card produces a unique sequence of random numbers. A 12-stage ring buffer increases the life span of the generator by a factor of 12. Assuming that the EEPROM is guaranteed to have 10.000 write cycles, this generator can produce at least 120.000 8-byte random numbers.

Erasing and writing eight bytes in the EEPROM takes about $2 \times 2 \times 3.5 = 14$ ms, and DES execution at 3.5 MHz takes about 17 ms. The remaining processing time is negligible. The card thus needs about 31 ms in total to generate a random number.

Figure 4.41 shows another example of a pseudo-random number generator. It is initialized every time the card is reset, which is the only time that a write access to the EEPROM occurs. Only RAM accesses are used for the subsequent generation of random numbers, which makes this generator relatively fast. However, the disadvantage of this is that the generator uses a few bytes of RAM during the entire session. The statistical quality of this pseudo-random number generator is not very good, but it is adequate for normal Smart Card authentication procedures. The primary consideration with such procedures is that the random numbers are not generated with a short repeat cycle, since otherwise the authentication could be broken by re-entering messages from a previous session.

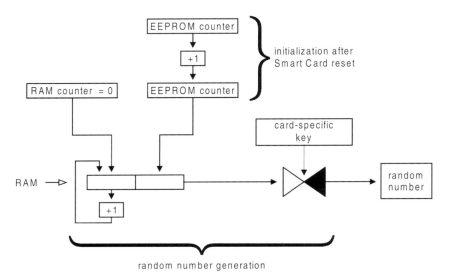

Figure 4.41   Sample design of a DES pseudo-random number generator for Smart Card operating systems. This generator is faster than the one shown in Figure 4.40, since only one EEPROM write cycle is needed per session. The quality of the random numbers it produces is adequate for normal Smart Card applications (authentication using the challenge–response procedure).

The FIPS 140-1 standard recommends that a security module check its built-in random number generator after every reset, using statistical tests. Only after these tests have been successfully completed should the random number generator be released for further use. Currently, commonly used Smart Card operating systems do not include this function, since it can be assumed that the statistics of the generated random numbers will not change significantly, due to the deterministic operation of the pseudo-random number generator.

The number of proposals, standards and designs for pseudo-random number generators is simply overwhelming. Some well known examples are the generators in the X9.17 and FIPS 186 standards, the proposals in the Internet RFC 1750 standard and the arrangements shown by Bruce Schneier [Schneier 96] and Peter Gutmann [Gutmann 98a]. The guiding principle for a random number generator should always be to keep it as simple and comprehensible as possible. Only then is it possible to judge its characteristics and thus to determine its quality.

4.9.2   Testing random numbers

After a random number generator has been implemented, the quality of the numbers that it produces must be rigorously tested. Fundamentally, there should be a nearly equal number of ones and zeros in the generated random numbers. However, it is not enough to simply print out a few numbers and do a comparative test. Random numbers can be mathematically tested using standard statistical procedures. It is self-evident that a large number of 8-bit random numbers will be needed for such testing. Between 10,000 and 100,000 numbers should be generated and analyzed in order to arrive at reasonably reliable results. The only way to test this many numbers is to use computerized testing programs.

When evaluating of the quality of the random numbers, it is also necessary to investigate the distribution of the generated numbers. If this is very uneven, with certain values strongly favored, then exactly these regions can be used for purposes of prediction. Therefore, Bernoulli's theorem should be satisfied as closely as possible. This theory states that the occurrence of a particular number, independent of what has come before it, should depend only on the probability of the number itself occurring. For example, the probability that a 4 appears when a die is thrown is always $\frac{1}{6}$, independent of whatever number appeared on the previous throw. This is also referred to as event independence.

The period of the random numbers, which is the number of random numbers that is generated before the series repeats itself, is also very important. It must naturally be as long as possible, and in any case longer than the lifetime of the random-number generator. In this way, the possibility of attacking the system by recording all random numbers generated for a complete period can be excluded in a quite simple and reliable manner.

There are vary many statistical tests for investigating the randomness of events, but in practice, we can limit ourselves to a few simple tests whose results are easily interpreted. There are also many publications on the subject of testing for randomness [Knuth 97, Menezes 97], as well as corresponding standards [FIPS 141-1, RFC 1750]. One test that is simple to set up and easy to interpret is to count the number of times that each byte value occurs in a large number of random numbers. If the results are displayed graphically as shown in Figure 4.42, they give a good indication of the distribution of the numbers.

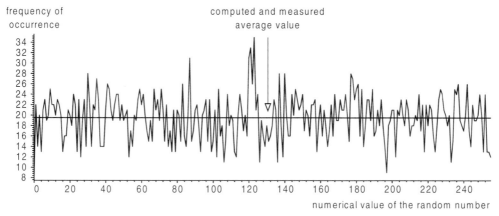

Figure 4.42   Statistical distribution of a series of 5000 single-byte random numbers. These numbers were produced by a typical Smart Card pseudo-random number generator. Based on purely arithmetic considerations, each of the possible values (in the range of 0 – 255) should occur 19.5 times.

If such a diagram were to be used to investigate 8-byte-long random numbers, the values plotted on the horizontal axis would still have to be single-byte or at most two-byte numbers, since the number of numbers needed for a statistical analysis would otherwise become extremely large. A good rule guideline is that every random number should occur approximately four to ten times for each value in order to obtain reasonably reliable results. In this way, it is possible to quickly see whether the random numbers that have been generated fully exploit the possible bandwidth of the byte. If this is not the case and certain values are very strongly favored, this offers an attacker a possible starting point.

Unfortunately, this test does not say anything about the order in which the random numbers occur, but only something about their distribution. It would for example be possible for a 'random number' generator to output numbers cyclically from 0 to 255. This would yield an outstandingly uniform distribution, but the numbers would be completely predictable. Other tests must be used to judge this quality criterion for random numbers.

Another practical test that yields a simple and quick estimate of the quality of a series of random numbers is to compress the series using a file-compression program. According to Shannon, the degree of compression that is possible is inversely related to the randomness of the set of generated numbers.

A significantly more robust test is the very well-known χ^2 test. Although it tests the same aspect as the previously described graphic test for even distribution, it is significantly more exact because it is performed using a mathematical procedure [Bronstein 96]. If the random numbers are assumed to be evenly distributed, the median value and standard deviation can be calculated. The deviation from a normal distribution can then be determined based on a χ^2 distribution. From this, it is possible to give a numerical value for the distribution of the random numbers.

However, this test cannot be used to draw any conclusions regarding the sequence in which the random numbers occur. Other statistical tests can be used to verify the randomness with which the numbers occur [Knuth 97], such as the Serial Test, which analyzes the periods of patterns that occur in the random numbers. Similarly, the Gap Test analyzes the intervals over which patterns do not occur. The Poker Test should also be used to judge the χ^2 distribution of the patterns that occur and the Coupon Collector Test to judge the χ^2 distribution of patterns that do not occur.

The Spectral Test, which investigates the relationship between each random number and the next following number, also has a certain amount of relevance [Knuth 97]. In the two-dimensional version of this test, random numbers and their immediate successors are plotted in an X–Y coordinate system, as shown in Figure 4.43. The three-dimensional version requires the successor to the successor number as well, and a third axis (the Z axis). N-dimensional spectral tests can be performed in a similar manner, but for understandable reasons, these must dispense with a graphical representation.

At minimum, the above-mentioned tests must performed and analyzed in order to achieve a reliable and definitive evaluation of a random number generator. Additional calculations and tests can be used to confirm the results so obtained. Only in this way is it possible to make a reasonable correct judgement of the quality of a set of random numbers.

Of course, considering the areas in which random numbers are used in Smart Card applications, an overly sophisticated random number generator is usually not justified. If the random numbers could be predicted in the case of an authentication, for example, the effect on security would be very slight. This is because no attack is possible without the private key used for encrypting the random number.

subsequent
random number

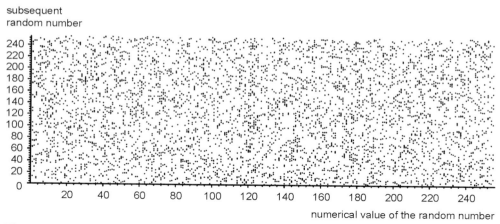

numerical value of the random number

Figure 4.43   Graphic representation of the distribution of successor values of 5000 single-byte random numbers, corresponding to a spectral test. The nearly uniform of the successor values can be seen in a glance from the regular pattern. The numbers were generated by a typical Smart Card pseudo-random number generator.

Table 4.16   Summary of standard statistical tests for random numbers

Test and Reference	Remarks
Coupon collector test [Knuth 97] Poker test [Menezes 97]	χ^2 distribution of the non-occurrence of patterns in a series of random numbers
Frequency test [Knuth 97, Menezes 97]	counting the number of ones in a series of random numbers
Gap test [Knuth 97]	investigating the patterns that do not occur in a series of random numbers
Long run test per FIPS 140-1	investigating whether a series of ones and zeros with a length of 34 bits occurs in a series of random numbers that is 20,000 bits long
Monobit test per FIPS 140-1	counting the number of ones in a series of random numbers that is 20,000 bits long
Poker test [Knuth 97]	χ^2 distribution of the occurrence of patterns in a series of random numbers
Poker test per FIPS 140-1 Serial test [Menezes 97]	counting 4-bit patterns in a series of random numbers that is 20,000 bits long
Runs test per FIPS 140-1	investigating maximum length of a series of all ones or all zeros in a series of random numbers that is 20,000 bits long
Serial test [Knuth 97]	investigating the patterns that occur in a series of random numbers
Spectral test [Knuth 97]	investigating the distribution of successor values of random numbers

A more serious problem would however arise if it were possible to manipulate the random number generator, for example so that it would always generate the same sequence of random numbers. In this case, an attack based on feeding the numbers back in would be not only possible but also successful. This would also be true if the period of the random numbers were very short. In each individual case, the primary conditions that the random numbers must satisfy must be carefully considered, since this naturally affects the random number generator. Although a supreme effort here may lead to very high-quality random numbers, it also usually results in increased use of memory space, which is particularly limited in Smart Cards.

4.10   Authentication

The objective of authentication is to verify the identity and authenticity of a communications partner. Transferred to the world of Smart Cards, this means that the card or the terminal determines whether its communications partner is a genuine terminal or a genuine Smart Card, respectively. For the sake of clarity, the term 'identification' is always used in this book to refer to the authentication testing of persons, although in principle it falls under the general concept of authentication.

For authentication to be possible, both of the communicating parties must have a common secret that can be verified by means of an authentication procedure. Such a procedure is significantly more secure than a pure identification procedure, which is (for example) represented by a PIN verification. In the latter case, all that happens is that a secret (the PIN) is sent to the card, which confirms its genuineness if it is correct. The disadvantage of this procedure is that the secret is sent to the as plaintext to the card, which means that an attacker could easily come to know the secret (the PIN).

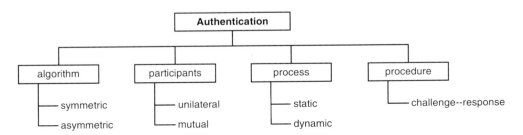

Figure 4.44   Classification of authentication procedures that are used with Smart Card systems

With an authentication procedure, by contrast, it is not possible to find out what the common secret is by eavesdropping on the communications channel, since the secret does not have to be sent openly via the interface. A distinction is also made between static and dynamic authentication. In a static procedure, the same (static) data are always used for the authentication. A dynamic procedure, by contrast, is constructed such that it is protected against being attacked by re-entering data that have been recorded during a previous session. This is because each authentication is based on different data.

There is also a fundamental difference between unilateral and mutual authentication procedures. A unilateral authentication, if it is successful, establishes the authenticity of one of the two communications partners. Mutual authentication, when successful, establishes the authenticity of both of the communications partners.

Authentication procedures that are based on cryptographic algorithms and are used with Smart Cards can be further divided into symmetric and asymmetric procedures. Currently, the procedures used with Smart Cards are almost exclusively symmetric. Asymmetric procedures, which means those based on the RSA algorithm or similar algorithms, do not yet have any practical significance with regard to Smart Cards due to their slow execution speeds. However, it can be foreseen that this will change in the future. In any case, the working principle of asymmetric procedures is the same as that of symmetric procedures.

There are several standards relating to the authentication of equipment. The ISO/IEC 9798 standard is the most prominent of these. Part 2 of this standard describes symmetric procedures, while part 3 describes asymmetric procedures. Fundamentally, the five parts of the ISO/IEC 9798 standard form an outstanding compilation of the commonly used authentication procedures, including symmetric, asymmetric, MAC-based and zero-knowledge-based procedures.

The principle of authentication in the field of Smart Cards is always based on the challenge–response procedure. In this procedure, one of the communications partners first asks the other one a randomly generated question (the challenge). The second partner computes an answer using an algorithm and sends the answer (the response) back to the first one. Naturally, the algorithm is preferably an encryption using a shared secret key that represents the common secret of the two communications partners.

4.10.1 Symmetric unilateral authentication

A unilateral authentication serves to assure one party of the trustworthiness of the other party in a communication. For it to be possible, both parties must have a shared secret, the knowledge of which is verified by the authentication procedure. This secret is the key for an encryption algorithm, and the entire security of the authentication procedure depends on this key. If the key should become known, an attacker could authenticate himself just as readily as a genuine communications partner.

The principle of unilateral authentication with a symmetric cryptographic algorithm is illustrated in Figure 4.45. For the sake of clarity, it is assumed that the terminal authenticates a Smart Card. This means that the terminal determines whether the Smart Card is trustworthy.

The terminal generates a random number and sends this to the Smart Card. This is referred to as the challenge. The Smart Card encrypts the random number it has received, using a key that is known to both the card and the terminal. The security of the procedure depends on this key, since only the possessor of the secret key is able to generate the correct response to send to the terminal.

The card sends the result of the encryption back to the terminal. This is the response to the challenge. The terminal now uses the secret key to decrypt the encrypted random number that it has received and compares the resulting number to the random number that it originally sent. If these match, the terminal knows that the Smart Card is authentic.

This procedure cannot be attacked by replaying a challenge or response that has been intercepted from an earlier session, since a different random number is generated for each session. The only form of attack with a moderately good chance of success would be to systematically search for the secret key. Since the challenge and the response are nothing more than a plaintext–ciphertext pair, it would be possible to discover the secret key by a brute-force attack.

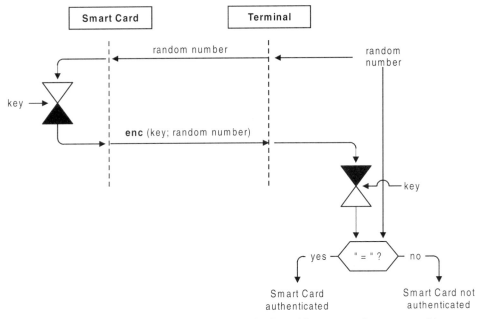

Figure 4.45   Working principle of unilateral authentication with a symmetric cryptographic algorithm. This example shows the authentication of a Smart Card by a terminal, which can be implemented using the INTERNAL AUTHENTICATE instruction of the ISO/IEC 7816-4 standard.

If all the cards for a given application were to have the same key and this key were to become known, the entire system would be discredited. In order to avoid exactly this possibility, only card-specific keys are used in practice, as a matter of principle. This means that every card has an individual key, which may be derived from a non-secret feature of the card. This specific feature can be the serial number of the chip, which is written to the chip when it is manufactured, or some other number that is different for each card.

In this case, the terminal requests the chip number from the Smart Card in order to compute the card-specific key. The chip number is specific to the card and unique within the system, so there is no other card in the system that matches this card. The value of the card-specific secret key is a function of the card number and the master key, which is known to the terminal. In practice, a portion of the card number is encrypted using the master key, and the result is used as the card-specific authentication key. A DES or triple-DES algorithm can be used for the encryption.

It must of course be borne in mind that if the master key (which is known only to the terminal) becomes compromised, the entire system will be compromised, since all card-specific authentication keys can be computed using the master key. The master key must therefore be securely stored in the terminal (in a security module, for example), and, if possible, it should be actively erasable in case of an attack.

Once the terminal has computed the necessary authentication key for this card, the usual process in the context of the challenge–response procedure takes place. The Smart Card receives a random number, encrypts it using its individual key and returns the result to the terminal. The terminal executes the reverse function of the computation in the same manner as the card and compares the two results. If they match, the terminal and the Smart Card have a common secret, which is the secret card-specific key, and the Smart Card has been authenticated by the terminal.

The authentication process is somewhat time-intensive, due the use of the DES algorithm and the data transfers from and to the card. This can cause problems in some applications. Given certain assumptions, we can roughly calculate the time required to perform a unilateral authentication. We assume that the Smart Card has a 3.5-MHz clock, uses the T=1 transmission protocol, has a divisor of 372 and uses a DES algorithm that takes 17 ms per block. Without going into any details, we assume that the internal routines of the Smart Card take 9 ms. This simplifies the calculation without significantly distorting the result, which is shown in Table 4.17. As can clearly be seen from this calculation, a single authentication takes around 65 ms. This will not usually cause any time-related problems in an application.

Table 4.17   Calculation of the processing time within a Smart Card for a unilateral authentication, taking data transfer times into account

Instruction	Data transfer time		Computation time	Total
INTERNAL AUTHENTICATE	38.75 ms	+	26 ms	= 64.75 ms

4.10.2   Symmetric mutual authentication

The principle of mutual authentication is based on a double unilateral authentication. Two separate unilateral authentications could also be performed in turn for the two communications partners. In principle, this would also be a mutual authentication. However, since the communications overhead must be kept as low as possible to minimize the time required by the process, there is a procedure that interleaves the two unilateral authentication procedures. This also serves to increase the security of the procedure, since it is much more difficult for an attacker to interfere in the communications process,

Before the terminal can compute the card-specific authentication key from the card number, it first needs the card number. After the terminal has received this number, it computes the specific authentication key for this card. It then requests a random number from the card and at the same time generates a random number itself. The terminal then swaps the two random numbers and places them one after the other, following which it encrypts this number using the authentication key. Finally, it sends the resulting ciphertext to the card. The objective of reversing the random numbers is to allow the challenge and response to be distinguished from each other.

The card can decrypt the received block and check whether the random number that it previously sent to the terminal matches that which it has received in return. If this is the case, the Smart Card knows that the terminal possesses the secret key. This authenticates the terminal with respect to the card. After this, the Smart Card swaps the two random numbers, encrypts them with the secret key and sends the result back to the terminal.

The terminal decrypts the received block and compares the random number that it previously sent to the card with the one that it has received in return. If they match, the Smart Card has been authenticated with respect to the terminal. This completes the mutual authentication process, and the terminal and the Smart Card both know that the other is trustworthy.

To minimize the communication time, the Smart Card can return the random number together with its card number. This is particularly interesting when mutual authentication takes place between a Smart Card and a background system. In this case, the card is directly addressed by the background system, and the terminal is 'transparent'. The data transfer rate in such situations is often very low, so the communications process must be simplified as much as possible.

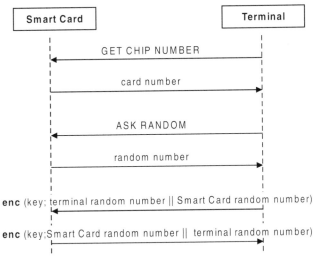

Figure 4.46 Mutual authentication with a card-specific key and a symmetric cryptographic algorithm. The illustrated procedure corresponds to a mutual authentication of a Smart Card and a terminal, as implemented in the ISO/IEC 7816-8 AUTHENTICATE instruction.

In order to illustrate the considerable amount of time required for a mutual authentication in comparison to a unilateral authentication, we can again make a sample calculation. The basic assumptions are the same as for the calculation of the time required for a unilateral authentication (see Table 4.17). As can be seen, mutual authentication takes nearly three times as long as unilateral authentication.

Table 4.18 Calculation of the processing time within a Smart Card for a mutual authentication, taking data transfer times into account. It is assumed that no derived keys are used, so that GET CHIP NUMBER is not necessary.

Instruction	Data transfer time	Computation time	Total
ASK RANDOM	28.75 ms	26 ms	
MUTUAL AUTHENTICATE	68.75 ms	95 ms	
	97.50 ms +	121 ms	= 218.50 ms

4.10.3   Asymmetric static authentication

Only a very few Smart Card microcontrollers have arithmetic processing units that can be used to execute the RSA algorithm. This is mainly because such capability would take up additional space on the chip, which would increase its price.

However, the fact that a supplementary asymmetric authentication procedure would offer increased protection, since it requires an attacker to break two cryptographic algorithms instead of only one, often makes its use attractive. The problem presented by the absence of a suitable arithmetic processing unit on the card can be dealt with by the expedient of using static authentication of the card by the terminal. This only requires verification within the terminal, and an additional security module in the terminal does not significantly increase its overall cost. This solution is thus much more economical than the use of special Smart Card microcontrollers. In addition, this procedure is significantly faster, since only one asymmetric encryption is required, as opposed to two with dynamic asymmetric authentication.

However, the price of this compromise is reduced security of the authentication procedure. With the static procedure, there is naturally no protection against the reuse of previous data. This is why it is used only as a supplementary verification of the authenticity of the card, which has already been verified using a dynamic symmetric procedure.

The procedure works essentially as follows. When each Smart Card is personalized, card-specific information is entered into the card. This can for example be a card number, as well as the name and address of the cardholder. This information does not change during the lifetime of the card. As part of the personalization of the card, the digital signature of this information is computed using a secret key. This key is used globally in the system. When the card is used at a terminal, the terminal reads the signature and the signed data from a file in the card. The terminal has the public key, which is valid for all cards in the system, and it can encrypt the signature that it has read using this key and then compare the result with the data that it has read from the card. If these match, the card has been authenticated by the terminal.

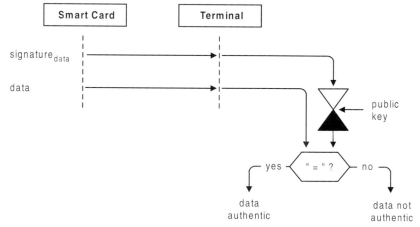

Figure 4.47   The principle of unilateral, static asymmetric authentication of a Smart Card by a terminal, using global keys.

The procedure illustrated in Figure 4.47, in addition to lacking protection against the re-entry of data, has yet another disadvantage. Global keys are used for the generation and verification of the signatures. Although the key in the terminal does not need to be protected, since it is public, keys that are the same for all cards should fundamentally not be used in a large system. If such a key is broken, or if it becomes known for any other reason, authentication is rendered worthless in the entire system. This means that it is necessary to introduce card-specific key pairs for static authentication.

However, this presents a problem with the memory capacity of the terminals, since each terminal must hold all available public keys for signature verification. Even with a medium-sized system, for example with one million Smart Cards, this would require each terminal to have 64 MB of memory for key storage, assuming 512-bit RSA keys. This would raise the price of the terminals to a level that would not be acceptable to system operators.

When symmetric procedures are used, it is quite easy to derive the card-specific keys from a master key.[12] With asymmetric procedures, this is not possible, because of the way the keys are generated. Consequently, a different approach is taken when card-specific keys are required. The public key for the verification of the signature is stored along with the signature in the card. In the system of the previous example, the amount of memory needed to store the public keys is still 64 MB, but this is now distributed in 64-byte packets over one million cards. The terminal thus reads the public key from a file in the Smart Card and can then use it to verify the signature. This avoids the problem of having to store all the public keys of the system in every terminal.

However, an attacker could now generate a key pair and use these keys to sign the information in a fraudulent card. The terminal would read the public key and conclude that the card was genuine. A refinement of the procedure just described is therefore required. This consists of signing the combination of the public key and the card-specific key that are stored in each card, using a global secret key. This signature is then stored in each card.

The terminal now works as follows. It first reads the public and card-specific keys from the card, and then tests the authenticity of the card-specific key using the global public key. If the card-specific key is authentic, the terminal then reads the actual data and verifies them using the public key stored in the Smart Card.

These two procedures are already used in some systems, and they will certainly be used increasingly in the coming years. However, as soon as the inclusion of an arithmetic processing unit for asymmetric cryptographic algorithms does not significantly increase the price of a Smart Card microcontroller, these two procedures will lose a lot of their significance. Their biggest disadvantage is the absence of protection against the re-entry of data from earlier sessions. Although this can be partially compensated by the use of various tricks, such as the reuse of signed data in subsequent symmetric cryptographic algorithms, it is still not possible to equal the level of protection that is provided by dynamic authentication procedures.

4.10.4 Dynamic asymmetric authentication

The previously described static asymmetric procedures all have some disadvantages. These can be eliminated by making the authentication dynamic. This provides protection against the re-entry of data intercepted from earlier sessions. The usual practice is to use a random

---

12   See also Section 4.7.1, 'Derived keys'.

number as the input value for a cryptographic algorithm. Of course, this requires an arithmetic processing unit that can execute the asymmetric cryptographic algorithm to be present in the card.

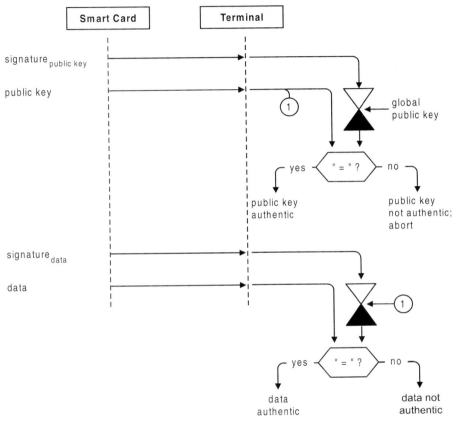

Figure 4.48   The principle of unilateral, static asymmetric authentication of a Smart Card by a terminal, using card-specific keys.

Figure 4.49 illustrates a unilateral authentication with a global public key. If one insists on card-specific authentication keys, then the procedure for the storage and authentication of card-specific public keys that is described in Section 4.10.3 is required.

As with symmetric authentication, the terminal generates a random number and sends this to the Smart Card. The card decrypts the random number using the secret key[13] and then sends the result back to the terminal. The terminal holds the global public key, and it uses this key to encrypt the random number that it has received. If the result of this computation is the same as the random number that was previously sent to the card, the card has been authenticated by the terminal.

---

13   The reason for using an encryption operation in the generation of a signature comes from the convention that with an asymmetric cryptographic algorithm, the secret key is always used for decryption and the public key is used for encryption.

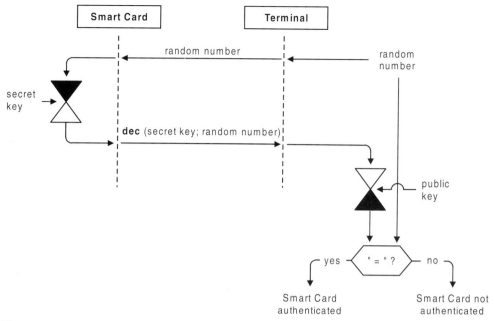

Figure 4.49   The principle of unilateral, dynamic asymmetric authentication of a Smart Card by a terminal.

The basic features of a mutual authentication procedure for the Smart Card and the terminal are analogous to the unilateral procedure that has just been described. However, a mutual authentication requires a relatively long time, due to the large amount of data that must be exchanged and the time-consuming asymmetric encryption algorithm, and it is therefore presently used very rarely.

4.11   Digital Signatures

Digital signatures, which are often referred to as electronic signatures, are used to establish the authenticity of electronically transmitted messages or electronic documents. It is possible to determine whether the message or document has been altered by verifying the digital signature.

A signature has the property that it can be correctly produced by only one single individual, but it can be verified by anyone who receives the message – or at least, by any recipient who has previously seen the signature, or who has a copy available for comparison. This is also the essential characteristic of a digital signature. Only one person or one Smart Card can 'sign' a document, but everyone can verify whether the signature is genuine. Given this required characteristic, asymmetric cryptographic techniques represent the ideal point of departure.

The message or document to be signed is usually at least several thousand bytes long. In order to keep the computation time for generating the cryptographic checksum within acceptable limits, the checksum is not computed over the entire data string. Instead, a hash value for the data string is first produced. Hash functions are, simply stated, one-way data compression functions. This compression is not reversible, which means that the original data cannot be reconstructed from the compressed data. Since the computation of a hash value is very fast, hash functions are an ideal aid for computing digital signatures.

The term 'digital signature' is usually only used in connection with asymmetric cryptographic algorithms, since the split between public and private keys makes these very well suited to use for digital signatures. Nonetheless, 'signatures' based on symmetric cryptographic processes are often used in practice. With such signatures, it is only possible to verify the authenticity of a document if one possesses the secret key that was used to generate the signature. Such a 'signature' is thus actually not a signature in the true sense of the word, but it is often referred to as such in practice. The term 'digital' is in this case omitted, to identify the type of process used.

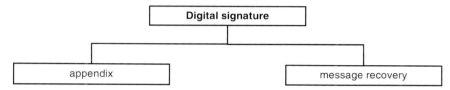

Figure 4.50 Classification of the two basic digital signature formats

From the point of view of information technology, there are two different ways to attach a signature to a message. The first is a form of cryptographic checksum, similar to a MAC (message authentication code), for a given data string. The signature is appended to the actual message (*digital signature with appendix*). This has the advantage that the message can be completely read without requiring prior verification of the signature. However, the disadvantage is that the size of the message is increased by the length of the signature, which can certainly be a consideration in the case of Smart Cards. This disadvantage can be avoided by using the second method for attaching a digital signature to a message, which is called *digital signature with message recovery*. With this technique, the hash value of the actual message is appended to the message, following which an input block for the digital signature algorithm is formed starting at the end of this data string. This means that the digitally signed message is increased in size only by the length of the hash value, but it can no longer be completely read as long as the digital signature has not been verified.

The procedure for generating a digital signature with appendix can be quite easily portrayed. First, a hash algorithm is used to form a hash value from the content of the message, which may for example be a file produced by any arbitrary word-processing program. This hash value is decrypted using an asymmetric cryptographic algorithm, such as RSA in this example. The result of this computation is the actual signature, which is appended to the message.

The signed message can now be sent via a non-secure path to the recipient. The recipient separates the signature from the message and then compresses the message using the same hash algorithm. The digital signature is encrypted using the public key of the RSA algorithm, and the result is compared with the result of the hash computation. If both values are the same, the message has not been altered while underway; otherwise, either the message or the signature has been altered during the transfer. In the latter case, authenticity is no longer assured, and it cannot be assumed that the content of the message is unaltered.

The task of the Smart Card in this scenario is very simple. It stores at minimum the private RSA key, and it decrypts the hash value formed from the message, which means that it generates the signature. Everything else, such as generating the hash value or subsequently verifying the signature, can in principle be performed equally well by a PC.

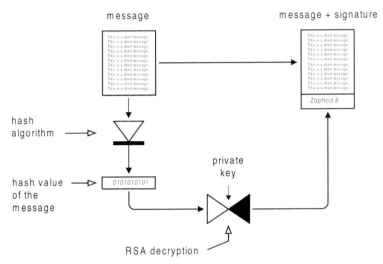

Figure 4.51   Signing a message with the RSA algorithm by appending the generated signature to the message (digital signature with appendix)

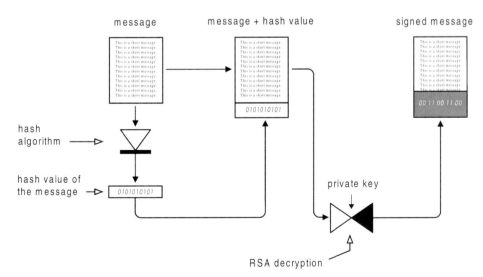

Figure 4.52   Signing a message with the RSA algorithm by incorporating the message and a hash value formed from the message in the signature (digital signature with message recovery)

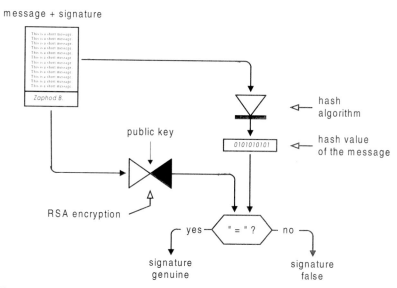

Figure 4.53 Verifying a message that has been signed using the RSA algorithm, in which the signature is appended to the message (digital signature with appendix)

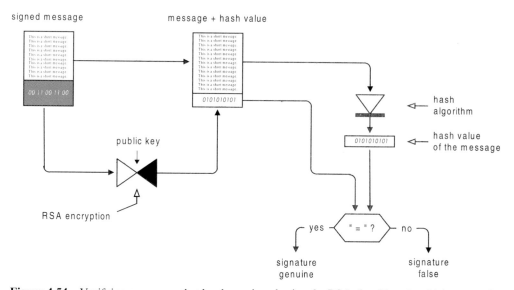

Figure 4.54 Verifying a message that has been signed using the RSA algorithm, in which a part of the message is used for the signature (digital signature with message recovery)

Still, the ideal situation would be for the Smart Card to receive the message via its interface, compute the hash value and then send the signed message back to the terminal. Verification of the signature could also be performed by the Smart Card. This procedure is naturally no more secure than just computing the signature, but it is significantly more 'application-friendly'. This is because hash algorithms and RSA keys could be changed by simply exchanging the Smart Card, without any need to altering programs or data in the PC.

In these two examples, the keys used for generating and verifying the digital signature are global, which means that they are the same for every Smart Card in a given system. If a different arrangement should be used for security reasons, so that each card has its own key for the digital signature, a scheme such as that described in Section 4.10.3 must be used.

The RSA algorithm is not the only one that can be used to produce digital signatures. There is also a cryptographic procedure that was specially developed for this application, namely the DSA (Digital Signature Algorithm). This was first proposed by the NIST (US National Institute of Standards and Technology) in 1991. Signatures can be both generated and verified using the DSA. In contrast to the RSA algorithm, it does not support data encryption and decryption (although a method for using it for encryption has meanwhile been found). This is a strong advantage with regard to international use in comparison to RSA, which is subject to strong export restrictions.

4.12   Certificates

With regard to the use of digital signatures, one is rather quickly confronted with a problem that must not be underestimated. Anyone who wants to verify the digital signature of a message needs the appropriate public key. However, the public key cannot be simply sent without any protection, since otherwise the recipient cannon verify whether the key is authentic. The public key must therefore be signed by a trustworthy agency, in order to make it possible to verify its genuineness. This agency is named a *certification authority* (CA). The combination of a public key that has been signed by the certification authority, its accompanying digital signature and certain additional parameters is called a *certificate*.

In this connection, there is also another agency, which is the *trust center* (TC). A trust center produces and manages certificates and associated 'black lists', and it can optionally also generate keys for digital signature cards. As a rule, a trust center also maintains a public directory of certificates, so that anyone who wants to verify a signed message can request the associated signed public key from the center, for example via the Internet.

A certificate contains not only the signed public key but also a large number of additional parameters and options, since it must be possible to verify the public key of a certificate without any further information. From this it follows that the signing and hash algorithms used (for example) must be clearly specified. In principle, everyone who signs documents could specify a personal certificate structure. Of course, it would then not be possible to exchange such certificates, so that as a rule they would lose their significance, since interchangeability must be a primary characteristic of a certificate.

In order to insure that this sort of cooperation actually can take place, there are standards that specify the structure of certificates. The best knows of the relevant standards is X.509, which specifies the construction and coding of certificates. It has also entered the ranks of ISO/IEC standards as ISO/IEC 9594-8. The current version of X.509 is Version 3, which was published in 1997.

The wide-reaching X.509 standard is a framework in which the construction of certificates is defined in unambiguous language. It forms the basis for many applications of digital signatures. Some examples are the Internet protection mechanism SSL (Secure Socket Layer) and the applications PEM (Privacy Enhanced Mail), SMIME (Secure Multipurpose Internet Mail Extensions) and SET (Secure Electronic Transaction).

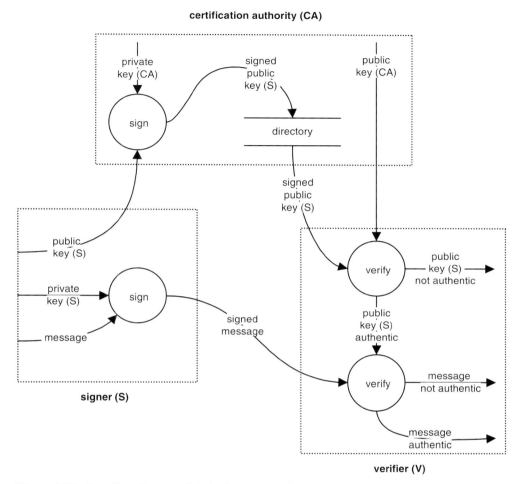

Figure 4.55   Data flow diagram of the basic processes for the generation and verification of a transmitted message regarding the use of a certificate. The certificate is produced by a certification authority. It contains the public key of the signer and the signature of the certification authority.

ASN.I is consistently used in X.509 to describe certificates, and the widely known TLV scheme is used in accordance with DER (Distinguished Encoding Rules) for the actual coding.[14] Some items that may be included in a certificate are summarized in Table 4.19 (overleaf). A short introduction and summary of X.509 can be found in a paper by Peter Gutmann [Gutmann 98 b].

Many optional data fields for a wide variety of applications are defined in the X.509 standard. For example, it is no problem to include several public keys in a single certificate and to have these signed by different certification agencies. This can result in certificates that contain several kilobytes of data, which can cause problems if a Smart Card is to be used for the verification of the certificate. Of course, this scheme can also be used to produce complementary certificates or tree-structured certificate hierarchies, for example.

---

14   See also Section 4.7.1, 'Derived keys'.

Table 4.19   Typical structure of an X.509 certificate

Data element and X.509 designation	Explanation
Version	Identifies the version of X.509 that defines the data elements of this certificate. This is usually version 3.
Serial number	The serial number of the certificate. This must be issued by the issuer of the certificate, so that it is unique.
Signature algorithm identifier	Identifies the cryptographic algorithm used for the digital signature.
Issuer name	The name of the issuer of the certificate. This spelling of this name is unique in the entire world, according to the X.500 standard.
Validity period	The period for which the certificate is valid.
Subject name	The name of the entity whose public key should be recognized as authentic based on this certificate. This spelling of this name is unique in the entire world, according to the X.500 standard.
Public key	The subject entity's public key, which should be recognized as authentic based on this certificate.
Signature	The digital signature formed from the data of the certificate.

5

Smart Card Operating Systems

It may seem presumptuous to refer to the few thousand bytes of program code in a Smart Card microprocessor as an operating system, but this program fully justifies the name. According to the German DIN 44300 standard, an operating system is no more and no less than 'the programs of a digital computer system, which together with the features of the computer equipment form the basis for the operating modes of the digital computing system, and which in particular control and supervise program processes'.

The term 'operating system' is thus not automatically limited to huge programs and data volumes. It is instead completely independent of size, since it only defines the functions. It is important that you do not automatically associate the concept of an operating system with the multi-megabyte programs for PCs and Unix computers. These are designed just as specifically for a particular man–machine interface, using a monitor, keyboard and mouse, as Smart Card operating systems are designed to work with the bi-directional, serial terminal interface.

Ultimately, the decisive factor for an operating system is its functional capability, which results from the interaction of mutually compatible library routines that are built on top of each other. The fact that an operating system provides an interface between the computer hardware and the applications software is also important. The main advantage of this for the applications software is that it does not have to access the hardware directly. This affords the software a certain amount of portability, even if it is often very limited

At the beginning of the 1990s, there were very few true Smart Card operating systems. This was due in part to the very limited memory capacity of Smart Card microcontrollers at the time. The usual situation then was not so much an operating system as a well-structured collection of library routines in ROM, which were called as necessary for the application when the card was completed. The structures of these systems were largely monolithic and could be modified only at considerable expense. The next generation was already built in the form of a layered operating system, and present-day systems still have this structure, with innumerable refinements.

One of the first true Smart Card operating systems was STARCOS, which was developed by Giesecke & Devrient [GD] and the *Gesellschaft für Mathematik und Datenverarbeitung* [GMD]. With this operating system, whose development began in 1990, it was possible even at that time to store, use and manage several applications independently, with a single Smart Card.

In the course of time, the term COS (card operating system) has become accepted throughout the world as a designation for a Smart Card operating system. It often forms part of the name of the operating system, as with STARCOS and MPCOS. Presently, there are more than a thousand firms that produce general-purpose and application-specific Smart Card operating systems. A number of examples are listed in Table 5.1.

Table 5.1   Some examples of Smart Card operating systems from various producers, with references their WWW addresses. This is only an incomplete selection.

Operating system name	Producer
CardOS	Infineon [Infineon]
Cyberflex, Multiflex, Payflex	Schlumberger [Schlumberger]
MFC	IBM [IBM]
Micado	Orga [Orga]
Multos	Maosco [Maosco]
OSCAR	Oki [Oki]
PCOS, MCOS, MPCOS	Gemplus [Gemplus]
STARCOS	Giesecke & Devrient [GD]
TB, CC, Odyssey	Bull CP8 [Bull]
TCOS	Telesec [Telesec]

It is conceivable that a consolidation of Smart Card operating systems, with their various features and functions, could occur in the next few years. This would have the same effect as with PCs, which nowadays all run with a 'unified' operating system. Whether this will actually happen with Smart Card operating systems remains to be seen, since in this case the external conditions are less favorable for a unified solution. The extremely stringent requirements with regard to security and software quality, the shortage of memory capacity and the demand for the confidentiality of the operating system software, taken together, certainly have the potential to make it impossible to produce a universal Smart Card operating system that satisfies everyone's wishes, at least in the foreseeable future.

In this chapter, we attempt to shed some light on the features and varieties of modern Smart Card operating systems, based on various specifications, standards and descriptions that are available for Smart Card software. Our objective is to present a detailed general cross section of modern Smart Card operating systems.

5.1   The Development of Smart Card Operating Systems

The evolution of operating systems for Smart Cards has passed through the same stages as for all other computer systems. The original special-purpose programs for single applications were repeatedly generalized and extended. The end result was a structured, general-purpose, easily used operating system.

From a modern point of view, the programs for Smart Card microcontrollers that existed at the beginning of the 1980s, at the start of this evolution, cannot be considered to be true operating systems. These were nothing more than application programs that were embedded in the ROM of a chip. However, since the manufacture of mask-programmed microcontrollers is expensive and time-consuming, the demand for general-purpose core routines grew very quickly. Special-purpose application programs stored in EEPROM could then be built on top of these routines, using them as necessary. However, this also increased the demand for memory, which led some firms to move in the opposite direction, back to special software for a single application.

The steadily increasing market demand for individually customized solutions, which has continued up to today, has however more or less forced the producers of operating systems to design the programs that they offer to meet this need. Presently, the individual approach of employing specially developed ROM software is only used for applications that involve a very large number of cards. General-purpose operating systems that are based on standardized commands are the norm, since in principle, they can be used for any application. If this is not possible for some special reasons, these operating systems are at least designed so that they can be modified to meet the needs of any particular application with a minimum of time and effort.

The 'historical' development of Smart Card operating systems from 1980 to the present can be very well illustrated using the Smart Cards of the German cellular telephone networks. The Smart Card that has been used since 1987 in the C network (the German precursor of the GSM or D network) has an operating system that has been optimized for this application. The modifications include a custom transmission protocol, special commands and a file structure that is tailored to the application. All in all, this card certainly has a complete operating system, but it is totally tailored for use in the C network.

The next step was the transition from a special solution to a somewhat more open operating system architecture. One representative of this is the first GSM card, whose design is significantly more open and multi-functional. At the time that the GSM Smart Cards were specified, there were already proposed standards for the command sets and data structures of Smart Cards, which meant that the cornerstone had been laid for compatibility between the various individual operating systems. Starting from this basis, further developments came step by step. Modern operating systems for the GSM network have functions such as memory management, multiple file structures and state machines, which brings them very close to the features provided by multi-application operating systems. They can manage several applications independently, without any interaction between the applications. Most of them also have very complex state machines, a large command set and occasionally several data transmission protocols.

Nevertheless, even this development is progressing. The cellular telephone Smart Card is taking on more and more of the functions of the telephone itself, such as driving the display and polling the keypad. In order to provide the maximum possible flexibility, it is necessary to break with what has up to now been a rigid fundamental principle in the Smart Card world. A modern Smart Card operating system must be able to run third-party program code in the card. With Smart Card operating systems, it goes without saying that this will not have any detrimental influence on the functions or security of any other applications in the card.

The development of Smart Card operating systems certainly could lead in the foreseeable future, via several additional steps, to an international quasi-standard for general-purpose operating systems. This has already happened with many other types of operating systems. It takes some time, but a particular industry standard sooner or later always establishes itself, and the competitors in the market must support it as the 'least common denominator' if they wish to continue to be successful. Such a standard does not yet exist in the Smart Card world, but the first signs of one can already be seen. The basis for this, in contrast to the PC world for example, is formed by international standards and specifications. These are primarily the ISO/IEC 7816 series of standards, the GSM 11.11 specification and the EMV specification.

5.2   Fundamentals

Smart Card operating systems, in contrast to generally known operating systems, do not include user interfaces or the possibility of accessing external memory media. This is because they are optimized for completely different functions. Security during program execution and the protection of data accesses have the highest priority. Due to the limitations imposed by the amount of available memory, they have a very small amount of program code, which normally lies in the range of 3–30 kB. The lower limit relates to special applications and the upper limit to multi-application operations systems. The average memory requirement is around 16 kB.

The program modules are written as ROM code, which leads to severe limitations on programming methods. Many processes that are typically used with RAM program code (such as self-modifying code) are not possible with ROM code. The fact that the code is in ROM is also the reason why no changes at all are possible once the microcontroller ROM has been programmed and manufactured. Correcting an error is thus extremely expensive and takes ten to twelve weeks. If the Smart Cards have already been issued, errors can only be corrected by a large-scale exchange action, which could destroy the reputation of a Smart-Card-based system. 'Quick and dirty' programming is thus obviously out of the question. The time spent on testing and quality assurance is normally significantly greater than the time spent on programming.

Still, these operating systems must not only have a very small number of errors; they must also be very reliable and robust. They must not allow their functions, and above all their security, to be impaired by any command coming from the outside world. System crashes or uncontrolled reactions due to an erroneous command or the failure of a page of EEPROM must never occur under any circumstance.

From the perspective of operating system design, it is an unfortunate fact that the implementation of certain mechanisms is influenced by the hardware that is used. Above all, the secure state of the EEPROM has a small but still noticeable influence on the design of the operating system. For example, all retry counters must be designed such that their maximum value corresponds to the erased state of the EEPROM. If this is not the case, it would be possible to reset the counter to its initial value by (for example) intentionally switching off the card's power supply while it is writing a new value to the retry counter. This is possible because the EEPROM must be erased before certain types of write accesses. If the power could be switched off exactly at the moment between erasing the EEPROM and writing the new value, the part of the EEPROM that is used for the retry counter would be in the erased state. If the operating system were incorrectly designed, this would amount to resetting the retry counter to its initial value. This type of attack can be resisted either by the proper coding of the counter, as just described, or by making the process of writing the retry counter an atomic process. The situation with regard to the retry counter and the secure (lowest-energy) state of the EEPROM is similar. The retry counter must be coded such that its maximum count corresponds to the secure state of the EEPROM. If this is not the case, it would be possible (by locally heating certain EEPROM cells, for example) to reset the retry counter. These are only two examples of hardware dependences in the design of a Smart Card operating system; there are many others. For reasons of security, a Smart Card operating system must be closely coupled to the hardware of the microcontroller used. It can therefore never be completely hardware-independent.

There is also another aspect to the idea of a secure operating system. 'Trap doors' and other types of hidden access points for system programmers are frequently found in large operating systems, in which they are perfectly normal features. However, they must be totally excluded in Smart Card operating systems. There must not be any possibility that someone could use some mechanism to bypass the operating system and (for example) obtain unauthorized read access to a file.

Another aspect that should not be underestimated is the required processing capacity. The cryptographic functions present in the operating system must execute in very short amounts of time. It is thus common to expend weeks of painstaking effort to optimize the algorithms in question in assembler code. It is thus easy to understand that multitasking is not possible, in light of the hardware platforms used and the required level of reliability. However, the limitation to a single executing task also impedes the use of security processes that monitor parts of the operating system with regard to process execution and constraints.

In summary, the primary tasks of a Smart Card operating system are the following:

- transferring data to and from the Smart Card,
- controlling the execution of commands,
- managing files,
- managing and executing cryptographic algorithms.

Command processing

Command processing in Smart Cards that do not support downloadable software is typically organized as shown in Figure 5.1. The Smart Card receives each command via the serial I/O interface. The I/O manager executes error detection and correction mechanisms as necessary, fully independent of the other, higher-level layers. After a command has been completely received without errors, the secure messaging manager must decrypt the message or test its integrity. If the data transfer is not secure, this manager is completely transparent for both the command and the response.

After this processing, the next higher level, which is the command interpreter, attempts to decode the command. If this is not possible, the return code manager is called. It generates a suitable return code and sends it back to the terminal via the I/O manager. It may be necessary to design the return code manager in an application-specific fashion, since the return codes are not necessarily the same for all applications. If, on the other hand, the command can be decoded, the logical channel manager determines which channel has been selected, switches over to its state and then, if this is successful, calls up the state machine.

The state machine checks whether the command, with its accompanying parameters, is actually permitted in the current state of the Smart Card. If it is, the actual program code of the application command that carries out the processing of the received command is executed. If the command is prohibited in the current state, or if its parameters are not allowed, the terminal receives a message to this effect via the return code manager and the I/O manager.

If it is necessary to access a file while processing the command, this takes place only via the file manager, which converts all logical addresses into physical addresses within the chip. The file manager also monitors all addresses with regards to region boundaries and tests the access conditions for the file in question.

The file manager itself utilizes a lower-level memory manager, which is responsible for the entire management of the physically addressable EEPROM. This ensures that only this program module works with physical addresses, which significantly increases the portability and security of the entire operating system.

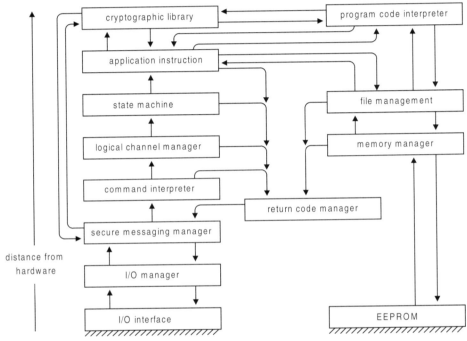

Figure 5.1 Command processing within a Smart Card operating system. The program code interpreter is optional and can also be called directly from the state machine, which for example happens with Java.

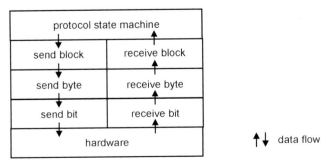

Figure 5.2 Part of the command processing procedure within the I/O manager in a Smart Card operating system

A central return code manager is responsible for generating the answer code. It always produces the complete answer for the program segment that was called. This level takes on the management and generation of all the return codes that are used, for all parts of the operating system.

Since Smart Card operating systems usually utilize cryptographic functions, there is normally also a dedicated library of cryptographic functions that is separate from the rest of the operating system. This library serves all other modules as a central point of departure for the utilization of cryptographic functions.

In addition to these levels, an interpreter or verification program for executable files may be present in the region above the application commands. This monitors the programs contained in these files and either runs or interprets them. The exact design and implementation depends on whether there are actually any files with executable code present, and on whether the stored code is machine code for the processor or code to be interpreted. This whole subject is described in detail in Section 5.10 ('Smart Card Operating Systems with Downloadable Program Code').

Smart Card profiles

In contrast to the realm of PC operating systems, the memory space of Smart Cards is so severely limited that very often not all standardized commands and file structures are implemented. Consequently, 'profiles' for Smart Cards have been introduced into the two relevant standards for general-purpose operating systems (EN 726-3 and ISO/IEC 7816-4). Each profile defines a subset of the commands and file structures of the standard in question.

A Smart Card that fits a certain profile must at minimum include the defined subset for that profile. However, the descriptions of the profiles are contained in appendices in both standards and are designated as 'informational' rather than as normative. They thus represent only recommendations to the operating system designer. The five profiles that are presented in the ISO/IEC 7816-4 standard are summarized in Table 5.2 (overleaf).

Commercially available Smart Card operating systems normally support various types of Smart Card microcontrollers with various amounts of memory. Consequently, there are in practice also operating system profiles that specify certain ranges of functions, depending on the chip type. These profiles within an operating system are normally designed such that applications can migrate relatively easily, at least from a smaller memory to a larger one, without any changes to commands or file structures.

5.3   Design and Implementation Principles

As is well known, design errors first manifest themselves in the implementation stage, where they result in costs that are several times greater than the cost of a better and less error-ridden design. However, this is fact of life with all software projects. In order to avoid such errors, it is recommended to pay attention to a number of principles during the design and implementation of a Smart Card operating system.

Table 5.2 Brief descriptions of the Smart Card profiles defined in the ISO/IEC 7816-4 standard. The listed file structures and commands are in each case the minimum requirement

Profile	Description	
M	File structures:	• transparent • linear fixed
	Commands:	• READ BINARY, UPDATE BINARY without implicit selection; maximum length up to 256 bytes
		• READ RECORD, UPDATE RECORD without implicit selection
		• SELECT FILE with direct specification of the FID
		• VERIFY
		• INTERNAL AUTHENTICATE
N	as Profile M, with the supplementary use of a DF name for SELECT FILE	
O	File structures:	• transparent • linear fixed • linear variable • cyclic
	Commands:	• READ BINARY, UPDATE BINARY without implicit selection; maximum length up to 256 bytes
		• READ RECORD, UPDATE RECORD without implicit selection
		• APPEND RECORD
		• SELECT FILE
		• VERIFY
		• INTERNAL AUTHENTICATE
		• EXTERNAL AUTHENTICATE
		• GET CHALLENGE
P	File structures:	• transparent
	Commands:	• READ BINARY, UPDATE BINARY without implicit selection; maximum length up to 64 bytes
		• SELECT FILE with direct specification of the DF name
		• VERIFY
		• INTERNAL AUTHENTICATE
Q	Data transfer:	• secure messaging
	File structures:	• ---
	Commands:	• GET DATA
		• PUT DATA
		• SELECT FILE with direct specification of the DF name
		• VERIFY
		• INTERNAL AUTHENTICATE
		• EXTERNAL AUTHENTICATE
		• GET CHALLENGE

Due to its functional specification, a Smart Card operating system is a secure operating system, which must manage information and above all keep information confidential. In addition, it is normally not possible to make any changes at all to the software during operation. The first principle follows directly from these conditions. A Smart Card operating system must be extremely reliable, which means that it must have an extremely low number of errors. The total elimination of errors is in reality never achievable, since even the smallest Smart Card operating systems are too large for all the possibilities of their internal processes to be completely tested.

However, strict modular design makes a decisive contribution to the discovery and elimination of any errors that may be present during the implementation stage. This modularity, which strongly increases the reliability of the operating system, need not necessarily be accompanied by a large increase in the amount of program code. An additional advantage of modularity is that it means that possible system crashes generally do not affect the system's security as strongly as with highly optimized program code that uses less memory. It also means that the consequences of possible errors remain localized, and that the operating system as a whole is more robust and more stable.

The fact that the software most often must be implemented in assembler language makes it more prone to containing errors. A design based on individual, fully testable modules strongly contributes to the timely recognition of programming errors and the limitation of the scope of their effects, due to use of defined interfaces. The layered operating system design shown in Figure 5.1 is a consequence of this modular approach. The increased planning and programming effort is, at least financially, compensated by the fact that testing and verification are significantly easier. As a result, nowadays almost all operating systems have a structure that is the same or very similar to the one described here.

The procedure normally used for designing the system is based on the module–interface concept. In the course of the design, the tasks of the operating system and the application are broken down as far as possible into functions, and these functions are then integrated into modules. Once the interfaces of the modules have been exactly defined, the individual modules can be programmed, possibly by several persons. In the ideal case, the first version of the implementation is platform-independent, which means that it does not yet depend on the characteristics of the microcontroller in question. After this has been completely tested, the necessary adaptations to the microcontroller can be undertaken.

Since the amount of program code for a Smart Card operating system is relatively small, this very pragmatic approach can be used without any major problems. Its advantages, which are the relatively small amount of planning that is needed and the possibility of dividing the programming tasks among several persons, as well as the reusability of the program code, bear the maximum fruit here. The disadvantages that come with this approach, which are the difficulty of demonstrating the correctness of the system and the fact that changes to the system in some cases have strong effects on many modules, must be balanced against the advantages.

The development of software for Smart Card operating systems is moving away from doing the programming entirely in assembler language. Many recent projects have been executed from the very beginning in the high-level language C, which is relatively close to the hardware. Nevertheless, the actual core of the operating system is still built on machine-dependent assembler routines, while all higher-level modules, such as the file manager, the state machine and the command interpreter, are programmed in C. This significantly reduces the development time and makes the programs more portable and more reusable.

Above all, the use of a high-level language markedly improves the testability of the software. The improved and more comprehensible program structure provided by a high-level language results in a distinctly lower error rate.

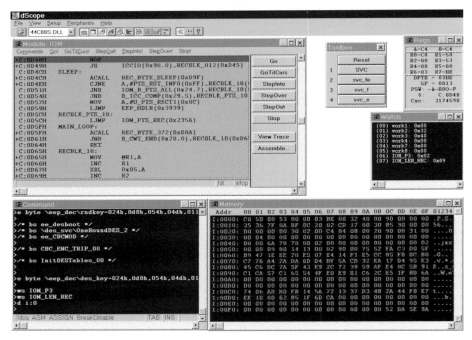

Figure 5.3 An example of the simulation of a Smart Card microcontroller in a typical development environment for the C and assembler programming languages. The source code window is at the top left, and various processor registers are shown at the top right. A memory map of the RAM is shown at the bottom right, and the command lines for controlling the simulator are shown at the bottom left. The simulator allows the software developer to monitor all the functions of the microcontroller and to intervene at every stage of program execution. (Source: Keil)

Unfortunately, the program code that is generated by a C compiler, even if it is highly optimized, takes up 20–40% more space in ROM than the equivalent code in assembler for the same functions. In addition, the speed of a C implementation is marginally lower than with assembler programming. This is however only critical for cryptographic algorithms and the data transmission protocol, since the other parts of the Smart Card operating system software normally do not contain any time-critical processes.

The greatest problem with C programming is not necessarily the additional memory space that is needed in the ROM or the lower execution speed, but the amount of RAM that is used. This type of memory is extremely limited in a Smart Card, and it has the further disadvantage that it takes up the most room on the chip relative to its storage capacity. This is why high-level languages are used only sparingly for the programming of Smart Card operating systems.

Since Smart Cards are used in application areas in which security is a very important consideration, the card issuer and/or the application supplier must have considerable trust in the integrity of producer of the operating system. The latter has every opportunity to take unfair advantage of the entire system by means of deliberately introduced security gaps. Consider for example an electronic purse in a Smart Card, whose load command has been manipulated such that under certain circumstances the purse can be reloaded without authorization. Such scenarios are the reason why only a few operating system producers have become established up to now. The risk that an ostensibly secure operating system contains a Trojan horse is significantly greater if it comes from a small, unknown vendor than if it comes from one of the well-know firms in the field.

Table 5.3   Some examples of typical memory usage for Smart Card operating system functions implemented in assembler

Function	Required program memory
CRC algorithm	≈ 50 bytes
File management (MF: 2 DF levels; EF: 4 EF structures)	≈ 1200 bytes
DES algorithm	≈ 1200 bytes
EEPROM write	≈ 150 bytes
RSA algorithm (with NPU)	≈ 300 bytes
Data transmission protocol T=0	≈ 500 bytes
Data transmission protocol T=1	≈ 1200 bytes

Nevertheless, there has recently been an increasing effort to evaluate Smart Card operating systems according to the ITSEC or its successor, Common Criteria, in order to achieve a higher level of comprehensibility and security with regard to such cases. This takes place either by the voluntary actions of the producers or in response to the demands of major application vendors. The objective of these evaluations is to increase the level of confidence that there are no significant errors in the program code. Deliberately introduced Trojan horses would probably be discovered only to a limited degree by evaluation, since the possibilities are practically unlimited.

Up to now, the usual evaluation levels for Smart Card operating systems have been ITSEC E3 and E4. Level E6 is demanded in certain cases, and sometimes it is even offered. It must be taken into account that an evaluation of a complete Smart Card operating system at the E4 level can cost on the order of £170,000. The obligation to perform a re-evaluation after a modification to the program code comes on top of this, although a re-evaluation is of course less costly than the initial evaluation. This is the essential reason why only relatively few Smart Card operating systems can boast ITSEC evaluations. More commonly, evaluations are performed by testing agencies without reference to the ITSEC. In such cases, the testing is limited to a thorough check of the design criteria, the source code and the documentation. This is for example a fundamental obligation for operating systems that are used for German Eurocheque cards.

5.4   Program Code Structure

The life cycle of a Smart Card operating system is divided into two parts–that before completion and that after completion. In the period before the card is completed, in which the microcontroller comes from the factory with an empty EEPROM, all program routines run in ROM. No data are read from the EEPROM, nor is any code run from the EEPROM. If an error in the ROM code that makes completion impossible is discovered at this time, the entire batch of microcontrollers must be destroyed, since the chips are unusable.

To minimize the likelihood of such a situation, it would be possible to implement only a small load routine for the EEPROM in the ROM and download the actual operating system to the EEPROM. However, the chip surface area per bit is four times as much for EEPROM as for ROM, which means that such an approach would have a disproportionate effect on the cost of the chip. For purely economic reasons, therefore, as much of the code as possible must be located in ROM. Consequently, all of the core operating system routines, as well as substantial parts of the rest of the operating system, are stored completely in the ROM. Only a few jumps to the EEPROM are allowed for in the completed version.

Some operating systems run completely in ROM even after completion, with only the data stored in the EEPROM, in order keep the size of the expensive EEPROM as small as possible. Of course, this minimization of the area used by the memory comes at the price of major limitations in the flexibility of the operating system.

In the completion process, the code in the ROM is adapted to the actual application. The ROM code can be seen as a large library that is interconnected and built up to form a functional application by code in the EEPROM. In addition, almost all operating systems allow program code for additional commands or special cryptographic algorithms to be loaded into the EEPROM during completion. This has nothing to do with any executable files that may be present, since the contents of such files can be downloaded at a later time, such as when the card is personalized. The programs that are entered into the card during completion are completely integrated into the operating system and can be utilized directly by the operating system.

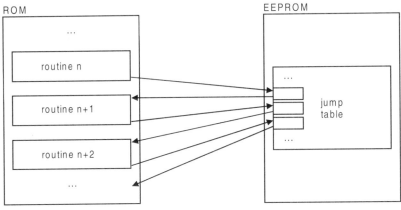

Figure 5.4   Coupling program code segments in the ROM by means of a link table that is stored in the EEPROM when the operating system is completed

Hardware recognition
Most of the more recent operating systems will run on microcontrollers with various sizes of EEPROM. However, the size of the ROM and RAM must remain constant. This makes it possible for a card issuer to always use the least costly version of the microcontroller that

meets his needs. For example, he could start with an inexpensive single-application card with 1 kB of EEPROM, and later on use more expensive microcontrollers with 2, 4, 8 or 16 kB of EEPROM for multi-application cards. To the extent that the microcontroller manufacturer supports such a range of chips, the Smart Card operating system must also exhibit matching functional capabilities. This means that it must be able to automatically recognize the size of the EEPROM and consequently set its internal pointer structures for the maximum free memory, file sizes and similar operational parameters. The technical implementation of this involves an operating system routine reading the manufacturer's finishing data and calculating the size of the available EEPROM from this information. This procedure only works for variable sizes of EEPROM. Current Smart Card operating system software is not able to adapt itself to variations in the size of the ROM or RAM.

The main advantage of this hardware recognition capability is that the producer of the Smart Card operating system no longer has to update the program code when the size of the EEPROM is altered. This eliminates a possible source of errors, and above all, it means that the operating system does not have to be re-evaluated for the new hardware platform. The hardware recognition capability of modern operating systems saves considerable time in software development and can thus reduce product development time by several weeks.

Soft masks and hard masks

The terms 'soft mask' and 'hard mask' are often used in connection with field trials and Smart Card operating systems. Strictly speaking, both terms are nonsensical from a purely logical point of view, since a ROM mask–which means the program code that is located in the ROM–is always unchangeable, and thus 'hard'. In the common jargon of the Smart Card world, however, the term 'soft mask' only means something similar to a ROM mask. This term is used when part or all of the program code for a Smart Card operating system, or the commands for an application, are located in the EEPROM. This means that the code can be altered relatively easily, without the cost and time involved in making a new ROM mask. This type of 'mask' is thus changeable, or 'soft'. Soft masks are used primarily during testing and in field trials, since they allow errors to be corrected and programs to be modified quickly and at minimal cost. The disadvantage is that they require the use of chips with large EEPROMs, which are more expensive than equivalent chips with the program code in ROM. However, since field trials normally do not involve the issuing of millions of cards, the increased cost of using chips with large EEPROMs is certainly acceptable.

Once the test or field trial with the soft mask has been completed and the program code in the EEPROM is ready to use without any touching-up, it can be moved from the EEPROM to the ROM by producing a real ROM mask. This can be done with a minimum of effort, and the result is called a 'hard' mask, since it cannot be readily altered. Strictly speaking, the only advantage of a hard mask is that a given amount of memory occupies significantly less chip area in ROM than in EEPROM, so that it allows a smaller and thus less expensive microcontroller to be used with the same amount of program code.

This two-stage process with soft and hard masks for new Smart Card applications also creates flexibility and the opportunity to make substantial changes to the program code, even shortly before the time that the cards are issued to their users. With the classical process using a pure ROM mask, it is not possible to make major changes to the program code once the mask has been given to the manufacturer. Due to the superiority of the two-stage process relative to the older process, nowadays the soft mask technique is almost always used initially for the introduction of a new Smart Card application. The migration to a hard mask occurs only after any necessary changes have been made to the program code.

Operating system APIs

The original Smart Card operating systems did not allow third-party software to be loaded into the card, where it could be executed as necessary. Consequently, conventional operating systems do not have any published programming interfaces that can be used by third parties to call operating system functions. New developments in Smart Card operating systems, such as MULTOS and operating systems that support Java, have made it possible for other parties to load their own program code into Smart Cards. These operating systems include well thought out APIs (application programming interfaces) that allow access to the most important functions of the operating system. This means that third-party programs do not have to contain code for routines that are already present in the operating system. Naturally, practically all operating systems have their own internal APIs, but these are not designed for external use and are usually confidential.

As an exception to the usual situation in the Smart Card world, there are no general standards relating to APIs for Smart Card operating systems. Instead, there are primarily two industry standards that have become established up to now. One of these is the Java Card API and the other is the Multos API. The APIs described in the associated specifications allow file access via the File Manager, calls to the available cryptographic functions and naturally also the transmission and reception of data. A Smart Card operating system with a complete operating system API, and which also allows program code to be downloaded, actually differs from the well-known PC operating systems only in the amount of memory that it has available.

5.5 Memory Organization

The three different types of memory in Smart Card microcontrollers have completely different properties. ROM can only be programmed by means of a mask during manufacturing, and it is programmed all at once. It is static for the life of the chip. Given its construction, the probability of an undesired change in the content of the ROM is practically zero.

RAM is the opposite of ROM. It retains its content only as long as power is applied to the Smart Card. A power failure causes the total loss of all data stored in the RAM. However, data can be written to the RAM at the full working speed of the processor, and the RAM can be erased an unlimited number of times. EEPROM, on the other hand, can retain data without external power. However, it has three disadvantages, which are its limited lifetime, long writing and erase times (around 1 ms/byte) and page structure.

The program code for the operating system that is stored in the ROM does not have to comply with any particular organization, except for the interrupt vectors which have fixed formats prescribed by the microcontroller. The individual program segments can thus be connected to each other in any desired sequence, although an attempt is made to limit the length of software jumps in order to save memory space. What is important is that the ROM is protected by error detection codes (EDCs), since it is certainly possible for errors to occasionally occur in the ROM. A scratch in the ROM area of the microcontroller, for example, or a fracture that occurs during the wire bonding process, can result in erroneous data in the ROM. Interestingly enough, this does not necessarily mean that the operating system can no longer function; instead, it is certainly possible that only specific routines will run incorrectly. In order to eliminate such cases from the very start, the ROM is always tested when the operating system starts up to see if it is error-free.

Figure 5.5 shows how a 256-byte RAM is usually partitioned. It is divided into the registers, the stack, general variables, workspace for cryptographic algorithms and the I/O buffer. If an I/O buffer of 256 bytes is required, for example, or if additional variables must be stored in the RAM, the limits of the available memory can be reached very quickly. This problem is solved by having workspace in the EEPROM, which is thus used like RAM. The disadvantage of this is that the writing time is around 10,000 times longer than for RAM addresses. An additional disadvantage is the limited lifetime of EEPROM cells, which, unlike RAM cells, cannot be written an unlimited number of times. However, moving the content of the RAM to the EEPROM is often the only solution—for example, if an I/O buffer is required that is larger than the entire amount of available RAM.

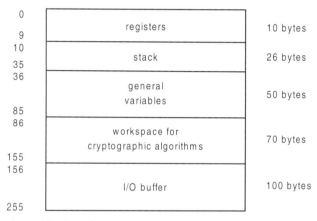

Figure 5.5   Example partitioning of a 256-byte RAM

The partitioning of the EEPROM is much more complicated and intricate than that of either of the other two types of memory. With modern operating systems, the EEPROM is divided up essentially as illustrated in Figure 5.6. In many microcontrollers, there is a region at the beginning of the EERPROM that has special hardware protection. This can be used to store special finishing information, such as a number that is only used once and is thus unique to the chip. Many semiconductor manufacturers also mark the chip type and amount of EEPROM that is available to the operating system in this region. This region is usually designed for write once, read multiple (WORM) access, which means that it can be written only once and after that can only be read. Technically, this is usually achieved by using normal EEPROM cells that however cannot be electrically erased.

On top of this region, which is often 16 to 32 bytes in size, come the tables and the operating system pointers. These are loaded into the EEPROM when the card is completed. The tables and pointers combine with the ROM programs to yield the complete Smart Card operating system, which is fully present in the card only after they have been loaded. This region is protected by an error detection code (EDC) that is recalculated before the first access, or in many cases before every access. This ensures that the operating system can run in a secure and stable environment. If a memory error is detected when the EDC is verified, the affected portion of the EEPROM must subsequently not be used, since an error means that the correct functioning of the operating system can no longer be assured.

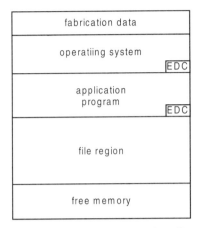

Figure 5.6   An example of how the EEPROM is partitioned by a Smart Card operating system

Table 5.4   Examples of manufacturer's data that are written by some semiconductor manufacturers to a WORM region of the EEPROM when the chip is finished. The first five manufacturer data items, taken together, yield a unique 8-bit chip number. The major advantage of a chip number generated in this manner is that it does not require highly accurate time information, synchronized over several production locations, but only data that are available to all production machines in every manufacturing facility.

Data element	Size
Semiconductor manufacturer	1 byte
Production facility	1 byte
Semiconductor processing batch number	2 bytes
X coordinate on the wafer	2 bytes
Y coordinate on the wafer	2 bytes
Chip type	4 bytes

Above the protected portion of the operating system, there is a region that contains additional application program code. If need be, this region may also be protected against alterations by a checksum. Application-specific commands or algorithms that should not be located in the ROM or that are too large to fit in the ROM are placed in this region.

The following region contains all of the file structures, or in other words the entire externally visible file tree. This region is not protected as a whole by a checksum, but instead usually has strong file-oriented protection. The internal structure of this region is shown in more detail in Figure 5.7.

At the top of the EEPROM, there is an optional free memory region that has its own memory manager. However, the free memory is often assigned to individual applications in the file region, where it can be used within the applications for the creation of new files. Otherwise it belongs to the general file region and is available for all new applications that are loaded in their entirety.

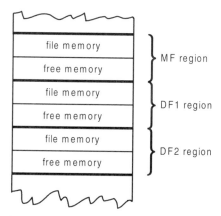

Figure 5.7   Example file region partitioning for a Smart Card operating system that supports multiple applications that are separate and independent

If a file is deleted, the free memory region should inherit the memory that becomes free. If the memory is divided into file memory and free memory regions as described here, only the memory occupied by the final file can actually be assigned to the free memory region if this file is deleted. This severely limits the options of the memory manager, but the limited amount of space for program code in Smart Cards does not allow any other solution.

An ideal solution would be to have true memory management with doubly linked lists for occupied memory and free memory respectively. This would make it easy to create new files and delete files in a manner similar to DOS or Unix file management. However, the cost of the file management program and the overhead for the file descriptors would not represent a reasonable design for a Smart Card operating system. The restrictions are actually not all that painful, since it rarely happens that files must be created or deleted within an application.

5.6   Smart Card Files

In addition to containing mechanisms for identification and authentication, Smart Cards are above all data stores. Here they have a decisive advantage with respect to other storage media, such as diskettes, in that access to the data can be coupled to certain conditions.

The first Smart Cards had only more or less directly addressable memory regions, which could be used for writing or reading data. The data were accessed by specifying physical memory addresses. Nowadays, all newer types of Smart Cards have complete, hierarchical file management systems with symbolic, hardware-independent addressing.

Of course, these file management systems have certain features that are specific to Smart Cards. The most noticeable of these is that there is no man–machine interface. All files are addressed by hexadecimal codes, and all commands are strictly based on this addressing, since the communication takes place here only between two computers. Equally typical of these file management systems is that they are designed to use a small amount of memory. Every redundant byte is avoided if possible. Since the 'user' in the terminal is a computer, this does not present any disadvantages.

There is usually no elaborate memory management, which also helps to keep memory usage as small as possible. If a file is deleted (and only a few operating systems have this capability), the space that comes free does not necessarily become available for use by a newly created file. Normally, all files are created and loaded into the Smart Card when it is initialized or personalized. After this, changes to the file contents are limited.

Naturally, the characteristics of the memory that is used also affect the way that the memory management works. The memory pages in the EEPROM simply cannot be written or erased an unlimited number of times, as can the hard disk of a PC. This means that there are special file attributes that allow information to be stored redundantly, so that it may even be possible to correct any errors that occur.

Internal file structures

Recent file management systems for Smart Cards have an object-oriented construction. This means that all information about a file is stored in the file itself. An additional consequence of this principle is that a file must always be selected before any action can be performed. Files in these object-oriented systems are thus always divided into two parts. The first part, which is called the file header, contains information about the layout and structure of the file, as well as the access conditions. The variable data are stored in the second part (the file body), which is linked to the header by a pointer.

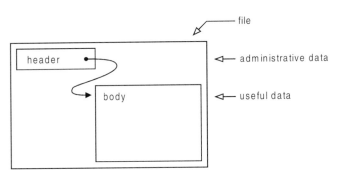

Figure 5.8   The internal structure of a file for a Smart Card file management system

In addition to providing better data structuring, this arrangement also has the advantage of better physical security for the data items. The page-oriented EEPROM that holds all the files allows only a limited number of write/erase cycles. The file header and the file body are always located on separate memory pages. The header, which is normally never altered, stores all the access conditions. A write or erase error related to the file body thus cannot affect these conditions. If the header and the file body were located in the same page of memory, it would be possible to utilize deliberately produced write errors to alter the access conditions such that confidential information could be read from the file body.

5.6.1   File types

The structure of Smart Card file systems, which is specified in the ISO/IEC 7816-4 standard, is similar to the file structures of DOS or Unix systems. The major difference is that Smart Cards contain no application-specific files, such as special files for a particular word processor. With Smart Cards, only the standardized file structures may be used.

There are basically two categories of files for file Smart Cards. The first category is directory files, which are called 'Dedicated Files' (DFs). The second category consists of the files that hold the actual useful data, which are called 'Elementary Files' (EFs). The DFs act as a sort of file cabinet that contains other, lower-level DFs or EFs that logically belong together. The EFs can be classified into those for the external world (*working EFs*) and those for the operating system (*internal EFs*). The various file types are described below, and their relationships are illustrated in Figure 5.10 (overleaf).

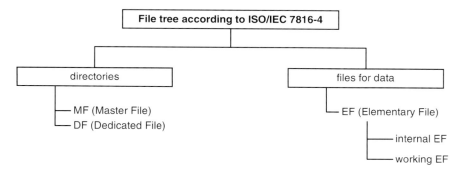

Figure 5.9   Classification of Smart Card file structures according to ISO/IEC 7816-4

MF
The root directory is referred to as the MF (master file). It is implicitly selected after the Smart Card is reset. The MF contains all other directories and all files. It is a special type of DF, and it represents the entire extent of the Smart Card memory that is available for the file region. A master file must be present in every Smart Card.

DF
Dedicated files (DFs) may exist underneath the MF as needed. The name 'Directory Files' is frequently used for these files, although this is contrary to the official definition of the abbreviation 'DF' in the ISO/IEC 7816-4 standard. A DF is a directory in which other files (DFs and EFs) can be grouped together. A DF may contain other DFs. In principle, there is no limit on the number of levels of DFs. However, it is rare for there to be more than two levels of DFs under the MF, due to the very limited amount of memory in Smart Cards.

EF
The useful data that are needed for an application are located in the EFs. 'EF' is the abbreviation of 'elementary file'. EFs may be placed directly under the MF or under a DF. To allow data to be stored in a minimum amount of memory and to logically optimize the data structures, an EF always has in internal file structure. This is the main difference between EFs and files in a PC, whose internal data structures are determined by applications (such as word processors) rather than by the operating system. EFs are divided into working EFs and internal EFs.

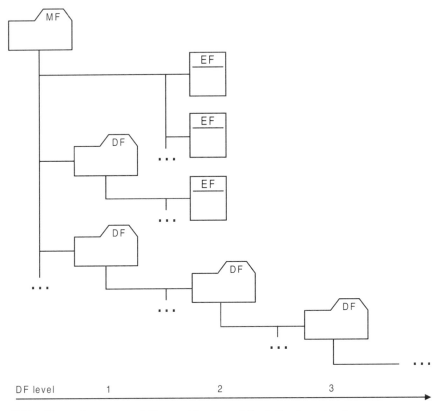

Figure 5.10 The various types of files used in a Smart Card directory structure.

Working EFs All application data that must be read by or written from the terminal, or in other words, all data that are intended for the external world (as seen by the Smart Card), are located in working EFs. The data contained in such files are not used by the operating system.

Internal EFs In addition to the EFs for applications, there are also internal system files that store data for the operating system itself, data for the execution of an application, secret keys or program code. Access to these data is specially protected by the operating system. These system files can be integrated into the file management system in two different ways. In the method specified in the ISO/IEC standard, these files are located in the relevant application DFs and are invisible, so they cannot be selected. They are managed fully transparently by the Smart Card operating system, in a manner that is similar to the management of resource files in the MacOS. In the model specified by the ETSI EN 726 standard, on the other hand, these system files receive regular file names (in other words, FIDs) and can be selected using these names. This is essentially the same principle as is used for file management under DOS. Each of these two approaches has a number of advantages and disadvantages, but they both provide the same functions in somewhat different manners.

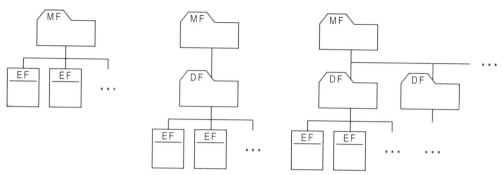

Figure 5.11   The differences between file structures for a Smart Card with only application and one with several applications. The two practical arrangements for cards with only one application are shown on the left and in the middle. The arrangement for a Smart Card with several applications is shown on the right.

Application files

According to convention, all files containing useful data for a particular application (the EFs for that application) are always grouped together in a single DF. This produces a clear and easily understood structure, and it makes it easy to enter a new application into a Smart Card by creating the appropriate DF.

Since the MF is a special sort of DF, it goes without saying that in a single-application Smart Card, all the application files can be placed directly under the MF. In a typical single-application Smart Card, therefore, all the EFs can be located either directly under the MF or in a solitary DF. Smart Cards with several applications have a corresponding number of DFs, in which the EFs belonging to the applications are located.

Additional DFs can be placed within such application DFs. For example, a DF placed directly under the MF could be dedicated to the 'Traffic Control' application. An additional level of DFs within the application DF could contain the files for the languages supported, such as 'English' and 'Deutsch'.

This file structure has an additional advantage with regard to security. Each time a memory access to the application occurs, the operating system can check whether it lies outside the boundaries of the memory allocated to the application DF in question. If so, it may prohibit the access.

5.6.2   File names

In modern Smart Card operating systems, the files are without exception addressed by logical names, rather than via direct physical addresses. The latter approach was perfectly normal in the Smart Card world earlier on, and there are a few Smart Card microcontrollers that still use it. With simple applications that occupy precisely defined memory regions, direct physical access can save a lot of memory space. It also does not result in any loss of user-friendliness, since all files are accessed by the computer in the terminal. However, direct physical access does not in any way fit with the criteria of modern software design, and it also creates major problems for software enhancements and for Smart Card microcontrollers with differing address spaces.

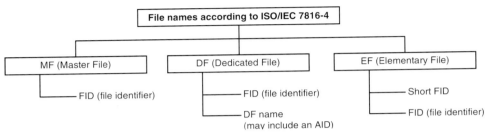

Figure 5.12   Classification chart of Smart Card operating system file names, per ISO/IEC 7816-4

All concepts that use logical file names are significantly better, and above all much easier to extend. Without question, it can be assumed that in a few years file addressing using logical names will be the only type of file addressing used in Smart Cards with microcontrollers. With memory cards, on the other hand, physical file addressing will continue to be used for the foreseeable future.

The File Identifier (FID)

The system described here is based on the ISO/IEC 7816-4 standard, and it is in principle reflected in all other international Smart Card standards. Every file, including directory files, has a 2-byte File Identifier (FID), which can be used to select the file.

For historical reasons, the FID of the MF is '3F00'. This FID is reserved for the MF within the entire logical address space. The logical file name 'FFFF' is reserved for future applications, and may not be used. There are also other FIDs that are reserved by the ISO standard and by other standards. These are listed in Table 5.5.

The GSM application is a typical example of the fact that various groups of FIDs cannot be freely used. In the GSM 11.11 specification, the more significant byte is fixed by the location of the file within the directory structure (file tree). The coding itself has developed historically and originates from the first French Smart Cards. GSM DFs have a value of '7F' for the first (more significant) byte. The FIDs of EFs located directly below the MF have '2F' as the first byte, and the FIDs of EFs located underneath a DF start with '6F'. The less significant bytes are numbered sequentially. This specification applies only to the GSM application and is not a general standard. In other cases, the full 2-byte address region of the FIC can be fully exploited and is not subject to any restrictions.

The FIDs in the directory structure must be chosen such that the files can be unambiguously selected. It is thus prohibited to use the same FID for two different files within the same DF. A DF may also not have the same FID as an EF located directly underneath it, since this would mean that the operating system would have to decide whether to select the DF first or the EF.

The following rules apply to the selection of unambiguous FIDs:

Rule 1:   All EFs within a single directory must have different FIDs.
Rule 2:   Nested directories (DFs) may not have the same FIDs.
Rule 3:   An EF within a directory (MF or DF) may not have the same FID as the next higher or next lower directory.

Table 5.5 FIDs that are reserved by the most important Smart Card standards

FID	Name and purpose	Standard
'0000'	EFCHV1: stores PIN no. 1, including its associated PUK and various related information	EN 726-3
'0001'	EFKey_OP: stores keys for application purposes	EN 726-3
'0002'	EFICC: stores information relating to the manufacturer and the operating system (such as the card serial number, card manufacturer, module implanter, card profile, operating modes when the clock is stopped and so on)	EN 726-3
'0003'	EFID: stores information about the Smart Card (such as the activation date of the MF, expiry date of the card and so on)	EN 726-3
'0004'	EFName: contains the name of the cardholder	EN 726-3
'0005'	EFIC: stores information about the chip (such as the chip serial number, chip manufacturer and so on)	EN 726-3
'0011'	EFKey_MAN: stores keys for management purposes	EN 726-3
'0100'	EFCHV2: stores PIN no. 2, including its associated PUK and various related information	EN 726-3
'2F00'	EFDIR (*directory*): stores application identifiers, along with the path names for the associated applications	EN 726-3, ISO/IEC 7816-4
'2F01'	EFATR: contains the extension to the ATR	ISO/IEC 7816-4
'2F05'	EFLANG: stores the cardholder's preferred language selection	EN 726-3
'3F00'	MF: the root directory for all files in a Smart Card	ISO/IEC 7816-4, GSM 11.11, EN 726-3
'3FFF'	Reserved for file selection via a pathname	ISO/IEC 7816-4
'FFFF'	Reserved by ISO/IEC for future use	ISO/IEC 7816-4

Short File Identifier (Short FID)
Short FIDs may be used for implicit file selection in the immediate context of a command. Short FIDs are optional for EFs, so they do not have to be assigned. A Short FID may be transferred along with a command for implicit selection of a file. Since the selection is implicit, the Short FID is only 5 bits long. It can thus take on values between 1 and 30, since a Short FID of '0' addresses the current EF.

DF name
The DFs are containers for the files that are used by the individual applications. A DF is a sort of directory or 'filing cabinet', and it can contain both EFs and other DFs. In the future, the address space provided by the 2-byte FID could become too small. Consequently, each DF has a 'DF name' in addition to its FID. As specified in the ISO/IEC 7816-4 standard, the DF name has a length of 1–16 bytes. The DF name provides sufficient address space to allow every Smart Card application to be unambiguously identified through out the world. Since DF names are freely chosen, it is possible for two different DFs to sometimes have the same DF name. Consequently, DF names are normally only used together with AIDs (application identifiers), as defined in the ISO/IEC 7816-5 standard. An AID is composed of two data elements defined by ISO and may have a length between 5 and 16 bytes. The AID is sometimes part of the DF name.

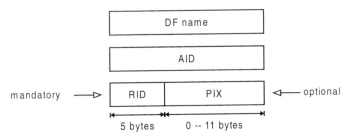

Figure 5.13 The DF name in connection with the AID, which is made up of the RID (registered identifier) and the PIX (proprietary application identifier extension)

Structure and coding of the application identifier (AID)

The application identifier (AID) consists of two data elements. The first data element is the registered identifier (RID), which has a fixed length of 5 bytes. It is assigned by a national or international registration authority and includes a country code, an application category and a number that refers to the application vendor. This numerical coding means that each RID is assigned only once, so that it can be used worldwide to identify a particular application. Unfortunately, the list of assigned RIDs is confidential, at least in Germany, so the 80 RIDs that have been assigned up to now cannot be published. The addresses of the national and international RID registration authorities are located in Section 15.5.

If necessary, an application vendor can place a proprietary application identifier extension (PIX) after the RID. The PIX, which may be up to 11 bytes long, is the optional second part of the AID. It may consist of a serial number and a version number, for example, which could be used for naming the application.

Table 5.6 Coding of the 5-byte (10-digit) Registered Identifier (RID)

RID			Meaning	
D1	D2–D4	D5–D10		
X	—	—	Registration category:	'A'–international registration 'D'–national registration
—	X	—	Country code; coding per ISO 3166	
—	—	X	Application vendor number; assigned by a national or international registration body	

Table 5.7 Example of a nationally registered RID that complies with the ISO/IEC 7816-5 standard (in this case, the RID of Wolfgang Rankl)

RID			Meaning
D1	D2–D4	D5–D10	
'D'	...	—	The registration category is 'national'
...	'276'	—	The ISO 3166 country code for Germany
...	...	'00 00 60'	Application vendor number assigned by the national registration body

5.6.3   File selection

Object-oriented file management systems require that a file be selected before it can be accessed. File selection informs the operating system which file will subsequently addressed. Successful selection of a new file causes the previous selection to become invalid. This means that only one file can be selected at any given time. Since FIDs may be freely chosen, certain limitations must be imposed on the free addressability of files. Otherwise, it could easily happen that several files with the same FID would be available in the file tree, and the operating system would then have to decide which file was meant. In order to avoid such ambiguity, and thereby to be independent of the search algorithm of the operating system's file manager, the selection options for files are intentionally restricted.

It would be different if all the FIDs used in the file tree were unique. In this case, it would be easy to select the desired file across several directory boundaries. However, this situation is exactly what cannot always be guaranteed. Consequently, selection is only possible within certain boundaries, since otherwise unambiguous selection of the desired file cannot be assured. The MF, however, can always be selected from anywhere within the file tree, since its FID is unique within the file tree. Selecting a DF located in the first level below the MF is only possible from a DF at the same level or from the MF. Figure 5.14 shows examples of various types of allowed and prohibited selections.

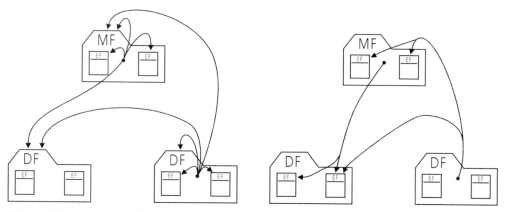

Figure 5.14   Examples of allowed selection options (left) and prohibited selection options (right) when a FID or DF name is used. Only direct selection without the path name is illustrated.

Selecting directories (MF and DF)
The MF can be selected from anywhere within the file tree, either using a special selection option of the file selection command or by means of its FID ('3F00'), which only occurs once within the file tree. When the MF is selected, the selection state that exists immediately after the Smart Card is reset is restored, since the MF is implicitly selected by the operating system after a reset. DFs can be selected either via their FIDs or via their DF names, which contain registered and thus unique AIDs.

Explicit EF selection

There are basically two methods available for selecting EFs. With explicit selection, a specific command (SELECT FILE) is sent to the Smart Card before the actual access to the file takes place. This command contains the 2-byte FID of the file to be selected as a parameter. After the file has been selected, it can be accessed by all subsequent commands.

Implicit EF selection

Implicit selection in the name given to the process in which a file is selected by means of a Short FID that is passed as a parameter of a command that actually accesses the file. A series of restrictions applies to the use of implicit EF selection. It only works for EFs within the currently selected DF or MF. It is thus not possible to implicitly select a file across directory boundaries. In addition, implicit selection is possible only with certain access commands that allow a Short FID to be passed as a parameter (such as READ BINARY, UPDATE BINARY, READ RECORD and UPATE RECORD).

The major advantage of implicit selection is that it allows a file to be selected and accessed with a single command. This makes a SELECT FILE command unnecessary in many cases, which simplifies the sequence of commands. Due to the reduced need for communications, the use of implicit selection allows distinctly higher execution speeds to be achieved.

Selection by path name

In addition to direct selection, the ISO/IEC standard allows two supplementary methods for explicit file selection by means of a path name. In the first method, the path from the currently selected file to the target file must be passed to the operating system. The second method allows the path from the MF to be passed. Both methods are presently integrated into some Smart Card operating systems. Using these additional functional capabilities results in a measurable reduction in the time required to process command sequences.

5.6.4   EF file structures

In contrast to files in DOS systems, EFs in Smart Cards have internal structures. The structure can be individually selected for each EF according to the intended use of the file. This has large advantages for the outside world, since these internal structures allow data elements to be constructed such that they can be accessed very quickly and directly.

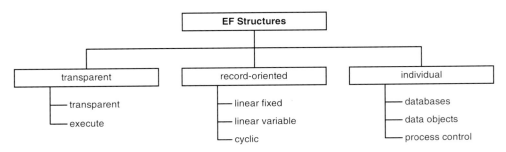

Figure 5.15   Classification of EF file structures for Smart Card operating systems

   The management of these data structures requires a significant amount of program code in the Smart Card. This is why the data structures are not all mutually symmetrical, but rather occur only in the forms that are often needed in practice.

Transparent file structure

The transparent data structure is often referred to as a binary or amorphous structure; in other words, a transparent file has no internal structure. The data contained in the file can be accessed for reading or writing in bytes or blocks, with the use of an offset value. The commands READ BINARY, WRITE BINARY and UPDATE BINARY are used for this.

   The minimum size of a file with a transparent structure is one byte. No maximum size is explicitly specified in any standard. However, the maximum number of bytes that can be read in the short format (255) or the long format (65,536), combined with the maximum offset value (32,767), allows a maximum size of 65,791 bytes or 98,303 bytes, respectively. These numbers are naturally illusory, given the available memory sizes of current Smart Cards. In practice, transparent files are rarely larger than a few hundred bytes. Figure 5.16 illustrates the organization of the transparent file structure. If five bytes were to be read from a 10-byte long file using an offset of 3 bytes, for example, the access would take place as shown in Figure 5.17.

Figure 5.16   Transparent file structure

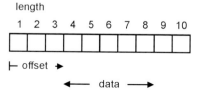

Figure 5.17   Reading 5 bytes from a transparent file with an offset of 3 bytes

   This file structure is primarily used for data that have no internal structure, or for very small amounts of data. An example of a typical application would be to store a digitized passport photograph that could be read from the Smart Card by the terminal. However, this linear and one-dimensional data structure can if necessary be used to simulate other data structures. Of course, in such cases the terminal access takes a more complex form, since parameters related to the construction of the file must be stored inside the file itself.

Linear fixed file structure

The linear fixed data structure is based on linking fixed-length records. A record consists of a series of individual bytes. Individual records within this data structure can be freely accessed. The smallest unit of access is one record, which means that it is not possible to access only part of a record. The commands READ RECORD, WRITE RECORD and UPDATE RECORD can be used for reading and writing within this data structure.

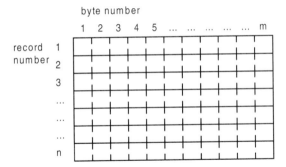

Figure 5.18   Linear fixed file structure

The first record is always numbered record 1. The largest allowed record number is 'FE', or 254 in decimal notation, since 'FF' is reserved for future extensions. The length of a single record is determined by the access commands, and can range from 1 to 254 bytes. Nevertheless, all records in the file must have the same length. Figure 5.18 illustrates the organization of the linear fixed file structure.

A typical application for this data structure is a telephone directory, in which the name comes first, followed by the associated telephone number in a fixed location.

Linear variable file structure

The fact that all records in a linear fixed file structure have the same length means that memory space is often wasted when this structure is used, since many record-oriented data have variable-length records. Consider, for example, the names in a telephone directory. The challenge of minimizing the amount of memory space is met by the linear variable structure, in which each record can have an individually defined length. The unavoidable consequence of this is that each record must have a supplementary field that contains information about its length. This structure is otherwise similar to the linear fixed structure, as can be seen from Figure 5.19.

The first record is numbered 1, and the maximum file length is 254 records. The length of an individual record is determined by the access commands, and can range from 1 to 254 bytes. The access commands for this structure are the same as for the linear fixed structure, namely READ RECORD, WRITE RECORD and UPDATE RECORD.

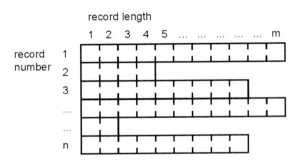

Figure 5.19   Linear variable file structure

This file structure is preferably used when records with highly variable lengths are to be stored and Smart Card memory must be saved. For example, the previously mentioned telephone directory could be optimized by making each record exactly as long as the entry it contains, which means that the records would not all have the same length. However, managing this file structure requires program code in the Smart Card operating system as well as extra memory to store the record length data. For this reason, operating systems for microcontrollers with small memories often do not offer this file structure. To be consistent, the ISO/IEC 7816-4 standard explicitly allows this limitation in some profiles.

Cyclic file structure
The cyclic structure is based on the linear fixed file structure, and thus consists of a certain number of records that all have the same length. In addition, the EF contains a pointer that always indicates the record that was last written. This record is always numbered record 1. If the pointer reaches the last record in the EF, it is automatically set by the operating system to point to the first record when the next write access occurs. It thus behaves the same as the hour hand of an analog clock.

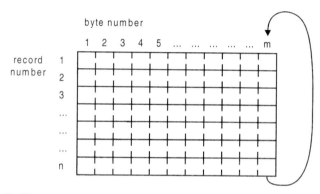

Figure 5.20 Cyclic file structures

If a cyclic file contains n records, the last one that was written is record number 1. The one that was written just before that is number 2, and the oldest record is number n. This file structure, like the other two record-oriented file structures, can also be accessed by addressing the first, last, next or previous record.[1]

The number and size of the records are fully analogous to those for the linear fixed structure. Due to the limitations of the write and read commands, a maximum of 254 records can be created, each with a maximum length of 254 bytes. This structure is typically used for log files within the Smart Card, in which the oldest entry is always overwritten by a new entry.

Execute file structure
The 'execute' structure is in principle not a separate structure, since it is based on the transparent structure. It is described in the EN 7826-3 standard and offers numerous extension possibilities within the operating system. The execute structure is not intended to

---

[1] See also Section 7.2, 'Write and Read Commands'.

be used for storing data, but rather for storing executable program code.[2] Files with the execute structure can be accessed using the same commands as for transparent files. Of course, this structure creates a sort of 'back door', since anyone who can write to such a file can download his or her own program code into the card, including possible Trojan horses.

The maximum program size is the same as the maximum size of a transparent file, and is thus 65,791 bytes without offset addressing or 98,303 bytes with offset addressing, using the UPDATE BINARY command. No internal data structure is specified, but it is certainly possible for the program code contained in the file to introduce one, so that the executable program can define its own data regions in the execute file and access them internally.

Database file structure

The ISO/IEC 7816-7 standard defines a subset of SQL for Smart Cards with the designation SCQL (Smart Card Query Language). In order to store data in the file system of a Smart Card such that they can be read by SCQL commands, it is necessary to provide a suitable file structure. The layout of this structure is not standardized; instead, its design is left up to individual operating system producers. A database file stores the actual useful data, various 'views' of the data base, the access privileges and user profiles.

Data object file structure

The ISO 7816-4 commands PUT DATA and GET DATA are used to store TLV-coded data objects in a Smart Card and to read out such stored objects. This can be implemented either completely independent of the file management system, or within the file management system using a special structure for storing data objects. If the implementation is within the file management system, a slightly modified transparent or linear fixed file can be used as a storage location for data objects. In this case, the PUT DATA and GET DATA command access these modified file structures via the file management system.

Process control file structure

If a Smart Card operating system has a process controller for command sequences, information relating to the commands that can be accepted must also be stored in the memory. This is normally done using a file whose structure is specially adapted to this task. However, this is not standardized. Without exception, each operating system with process control has its own format, which is not compatible with that of any other operating system.

5.6.5   File access conditions

Within the framework of their object-oriented design, all files have information that regulates access to them within the context of the file management system. This information is physically encoded in the header of each file. The entire security of Smart Card file management is based on managing file access privileges, since these form the basis for regulating file access.

The access conditions are defined when a file is created, and usually cannot be changed afterwards. There is a high degree of variation in the allowed file access conditions, depending on the commands that are present in the operating system. For example, there is no point in defining access conditions for a READ RECORD command if this command is not present in the Smart Card operating system.

---

2   See also Section 5.10, 'Smart Cards with Downloadable Program Code'.

For the MF and the DFs (in contrast to the EFs), there is no information stored with respect to data access (read or write privileges). Instead, the access conditions for the creation of new files, among other information. Depending on the file type, other access conditions are also stored. For the EFs, these relate to accesses to the file contents, and for the MF and the DFs, they are the conditions that apply within these organizational structures.

A distinction can be made between state-oriented and command-oriented access conditions. State-oriented conditions define the machine state that is required for a particular access. A specific state may be specified, as well as a comparative condition. For example, an access might be allowed for any state from state 5 upwards. The access would thus be prohibited for any state lower than state 5. It is naturally also possible to specify several different states. An access could for example be allowed in states 5, 8 and 9.

In contrast to this, command-oriented access conditions define the allowed commands for the access in question. This applies primarily to authentication and identification commands. For example, write access to a file might be allowed only if the PIN has previously been successfully verified with the VERFIFY command.

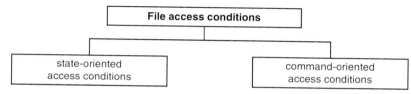

Figure 5.21   Classification of the two possible types of file access conditions

All possible types of access to an EF must be precisely regulated by means of access privileges. The number of commands that this involves varies, depending on the operating system. Some of the most commonly used file access commands are the following:

APPEND	extend a file
DELETE FILE	delete a file
INCREASE / DECREASE	calculations within a file
INVALIDATE	block a file
LOCK	permanently lock a file
READ / SEEK	read or search within a file
REHABILITATE	unblock a file
WRITE / UPDATE	write within a file

The access conditions of a DF are fundamentally different from those of an EF. They specify the conditions under which specific commands may be executed within the directory in question. The three most important commands relating to DFs are:

CREATE	create a new file
DELETE FILE	delete a file
REGISTER	register a new file

State-oriented access conditions are used primarily in multifunctional Smart Card operating systems, such as STARCOS, since they are very flexible and easily modified. However, most large Smart Card applications, such as GSM and German EC cards, control file access by means of command-oriented access conditions. This approach requires slightly less program code and memory than the use of state-oriented access conditions. Of course, this price of using command-oriented access conditions is somewhat less flexibility. Both approaches have their advantages and their disadvantages, and discussions of their relative merits usually turn into philosophical questions relating to the design of operating systems.

5.6.6 File attributes

As part of their object-oriented specifications, all EFs have special attributes that define additional properties of each file. However, these attributes depend on the operating system and the Smart Card application area. They define properties of the file that mainly relate to the EEPROM used as the storage medium. The reason for this is the potential uncertainty about the file content and the possibility of write errors with EEPROM operations. These attributes are defined when a file is created, and usually cannot be changed afterwards.

WORM attribute
One of the attributes based on the EEPROM storage medium is called WORM (write once, read multiple). If a file has this attribute, then data can be written to the file one time only, but they can be read an unlimited number of times. This attribute can be implemented either in the hardware of the EEPROM or as a software function. The WORM attribute can be used, for example, to write a serial number in a file once and forever. This attribute is also used with personalization, in which information such as the cardholder's name and the expiry date are permanently written in the card.

This attribute is intended to be used to protective sensitive data against being overwritten. The best possible protection is provided if WORM access is possible at the hardware level, which means that the EEPROM has hardware protection that allows data to be written only once. However, even a software implementation provides much better protection than other comparable mechanisms.

Multiple storage attribute
An attribute that is primarily defined and used in the GMS realm is a flag for 'high update activity'. The only reason that this attribute exists is the fact that EEPROM has a limited number of write/erase cycles. It is possible to write to files with this attribute very often without affecting their data contents. This is achieved by storing multiple copies of the data when they are written and using a majority vote when the data are read. Threefold parallel storage is normally used for writing, and a 2-of-3 majority vote for reading. An alternative implementation is to switch from one copy of a multiply stored data set to another copy if a read error occurs, in a manner that is not visible to the outside world.

EDC utilization attribute
An attribute that provides special protection for the usable data in a file by means of an error detection code (EDC) is used for particularly sensitive data. This allows the 'flipping' of bits in the EEPROM to at least be detected. If multiple storage is used together with EDC protection, it is also possible to correct flipped bits. This ECC (error correction code)

property is primarily used for electronic purses. Here the flipping of a memory cell amounts to the actual loss of money, since the current amount of money in the purse is stored in the file. The EDC and ECC file attributes are used to minimize the effects of bit flipping.

Atomic write access attribute

Recent Smart Card operating systems often include a mechanism that ensures that when a file is accessed for writing, the writing operation is executed either completely or not at all. Since this mechanism more than doubles the write access time for a file, it should in principle not be used for all files. A separate attribute allows this writing mechanism to be selectively applied to each file.

Concurrent access attribute

Smart Card operating systems that support several logical channels often have a special file attribute for concurrent access. This attribute explicitly allows a file to be accessed for reading or writing by two or more commands at the same time, when the Smart Card receives these commands via separate logical channels. It is important for this property to be specially flagged for the file, since with nearly simultaneous accesses via two different channels it is possible for data to be modified via one channel immediately before or after they have been read via the other channel. If the two processes are not synchronized, the data that are read will differ according to the times at which the commands reach the Smart Card. Consequently, concurrent access is generally not allowed. When a file has been selected, access by any other channel is temporarily blocked. Only after the file has been deselected is it possible for it to be accessed by another channel. The concurrent access attribute disables this block for a particular file. The relevant applications in the terminal are then responsible for synchronizing parallel read and write accesses. If they only access the file for reading, then there is naturally no problem.

Data transfer selection attribute

The file management systems of Smart Cards that have both contact and contactless interfaces sometimes include a file attribute that determines which of the two interfaces may be used for accessing the file. This makes it possible to specify for each individual file whether commands may access a file via the contact interface and/or the contactless interface. With an electronic purse, for example, this attribute makes it very easy to allow purchases to be made only via the contact interface and the card to be loaded only via the contactless interface.

5.7 File Management

All files in a Smart Card are stored in the EEPROM. This is the only type of memory in the Smart Card that can retain stored data without power and that also allows data to be altered if necessary (when power is available). It also provides the only means to save information from one session to the next, since the contents of the RAM are lost when the Smart Card is deactivated, and the contents of the ROM cannot be altered after the chip has been manufactured.

In earlier Smart Cards, the files were directly accessed using physical addresses. Actually, there were no files in the true sense of the word. Instead, the entire memory was linearly addressable from the outside and could be accessed using write and read commands. However, this is not allowed in modern operating systems, for reasons that have to do with security and applications. Object-oriented file management with access condition information located directly in the files is currently the standard. The organization and management of these files is the task of the file manager within the operating system.

With an objected-oriented structure, every file must have a file descriptor that contains all the information relevant to the file itself. In Smart Card technology, the file descriptor is referred to as the file header. The data content of a file, or in other words the useful data, is located in the 'body' of the file.

The information contained in the file header depends strongly on the features of the file manager. However, the file descriptor must contain at least the following items:

- file name   (e.g. FID = '0001')
- file type   (e.g. EF)
- file structure   (e.g. linear fixed)
- file size   (e.g. 3 records of 5 bytes)
- access conditions   (e.g. READ = after PIN code has been entered)
- attribute   (e.g. WORM)
- link to the file tree (e.g. directly under the MF)

The file name is the 2-byte FID (file identifier) for an EF or the MF. With a DF, the AID (application identifier) also forms part of the file name. The file type, which may be MF, DF or EF, must also be indicated.

Depending on the file type, there may be an element in the header that describes the internal structure of the file (*transparent, linear fixed, linear variable, cyclic* or *executable*). All information relating to the length of the transparent data portion, or the number and length of the records, also depends on the file type.

After all the basic properties of the file have thus been described, the operating system also needs detailed specifications of the access conditions. These define in which state which command can access the file and what type of access is allowed. The access conditions must be specified individually for every possible command. Special file attributes, such as high update activity, WORM or EDC protection, can also be indicated if they are supported by the file manager.

All of the above information relates to the file as an isolated object. In order to define the location of the file in the file tree, an additional pointer is needed to specify the exact position of the file within the MF or DF.

In simple operating systems, the file headers have fixed lengths that depend on the file type (MF, DF or EF). This reduces the amount of effort that the internal file manager must expend on calculation and administration. The major disadvantage of such an arrangement is that it is relatively inflexible with regard to extension. Another drawback is that it does not allow an unlimited number of different access conditions for any one EF, or even a very large number. The memory that would have to be set aside to allow a large number of access conditions to be stored in the header would not be fully used by most applications. Consequently, variable-length headers are being used increasingly often for Smart Card file management. The Smart Card operating system can automatically adapt the lengths of such headers to the specific requirements of individual applications.

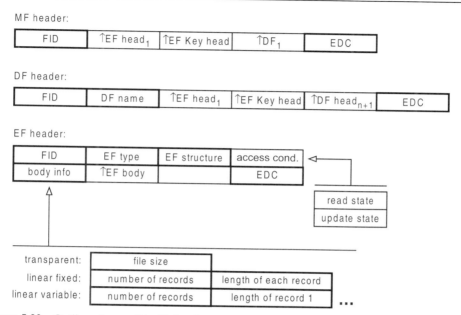

Figure 5.22   Outline of a possible file header organization for MF, DF and EF files (internal and working) in a Smart Card file management system. The heavy borders indicate TLV-coded data objects that must be present, while the light borders indicate optional data elements. The numbering is valid only within the directory in which the file is located, and thus not globally for all files. This structure is based on the requirements for a simple file management system defined in Small-OS. 'access cond.' stands for 'access condition', and 'head' for 'header'.

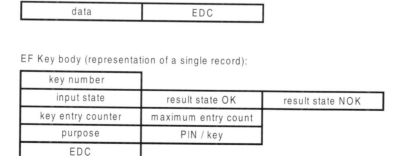

Figure 5.23   Outline of a possible file header organization for internal and working EFs in a Smart Card file management system. All TLV-coded data objects must be present. This structure is based on the requirements for a simple file management system defined in Small-OS.

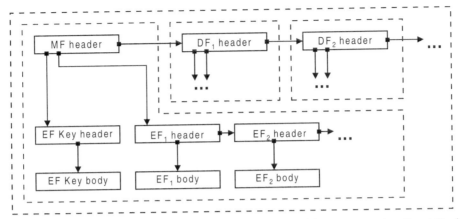

Figure 5.24 Outline of a possible arrangement of the pointer and data structures in a Smart Card file management system. This is based on the requirements for a simple file management system defined in Small-OS. The dashed outlines demarcate the memory available to each directory.

Memory partitioning into pages

The limited number of write/erase cycles of the EEPROM and the partitioning of the EEPROM into pages create a general problem for file management. This has a considerable effect on the entire design of the file manager and the internal file structures. The sizes of both file headers and file bodies must be adapted to the predefined size of the memory page to prevent them from being split by page boundaries. File management information in the memory must also be strictly separated from the actual file data contents. If this were not the case, undesired side effects could occur between the management data in the header and the useful data in the file body. These could destroy the entire internal security structure of the Smart Card operating system. This is briefly illustrated in the following example.

Suppose that the access conditions for a file containing secret, non-readable keys were stored on the same memory page as the public, writeable data of another file. If a write operation to this file were interrupted, for example by pulling the card out of the terminal, this would affect the access conditions stored on the same page. In the worst case, no access conditions at all would remain, and the file containing the secret keys could be read by everybody. It is thus fundamentally important that the internal file structures for management and useful data are each stored on separate memory pages.

DF separation

Another way of understanding the function of a DF is to see it as representing all the memory that is made available to a particular application. Within this region, the owner of the application is fully responsible for his application and may essentially do and allow what he wants. However, he must under no circumstances be able to address in any manner the region of another application from out of his own region, or to read or modify data in another region. Consequently, some Smart Card operating systems have special mechanisms that always test every memory access to see whether the physical address is located within the boundaries of the current DF. If this is not the case, the process in cancelled with a serious internal error.

This address monitoring currently is performed by suitable software routines in the operating system, due to the lack of hardware support. The security of this solution is naturally significantly lower than what could be achieved with suitable hardware, since it is more easily bypassed. In the future, however, Smart Card microcontrollers will probably have memory management units (MMUs), just the same as current CPUs. These can be very well utilized to achieve secure control over memory accesses within a DF. Until then, the only option is employ suitable operating system routines to monitor the address boundaries of the DFs.

Free memory management
The small amount of memory space imposes major restrictions on Smart Card operating systems with regard to free memory management for the EEPROM. Only recently have operating systems become available that can create files (DFs and EFs) after the card has been personalized, and also subsequently delete them. Given the secure nature of a Smart Card, this must naturally be protected in a cryptographically impeccable manner. The ideal solution is to execute the appropriate commands in the Secure Messaging mode following an introductory mutual authentication.

Free memory management with allocation of memory for new DFs and EFs can be integrated into an operating system relatively easily. The amount of additional programming effort required to enable these files to be subsequently deleted should not be underestimated. After all, this means that the data manager must at least have the basic capability to reorganize the file headers that are connected by linked lists. The procedure must also make sure that the state of the file tree remains well defined after a sudden power failure, as may be caused by pulling the card out of the terminal. The entire security of a Smart Card could completely collapse in exactly this situation if the file pointers were suddenly undefined. Atomic procedures can be very useful as a remedy to this problem.

The next increase in complexity comes with a true free memory management for the entire file region, or ideally for each DF separately. With this, a state is be reached that is the same as we are all familiar with from PCs. Files can be created an deleted, and yet other files created, all within memory that is automatically managed. However, this is only possible with a very large amount of software, even though such capability is certainly generally desirable. The price that would have to be paid for this, in terms of ROM memory, is in most cases too high with current ROM sizes. However, it is technically possible, and some multi-application operating systems support complete free memory management.

Frequently, though, a compromise is made between the two extremes, in which the deletion of EFs within a DF is allowed, but the memory that is thus released is only available as a single block for the creation of a new EF. File relocation does not occur. If a new EF is created after an EF has been deleted, a best-fit algorithm is used to look for a block of free memory in which the new EF will just fit. After this, the EF is created in the space thus located. An important aspect of this is that whatever memory is left over in the block where the new EF is created is permanently lost to the file manager. To the extent that the deleted and newly created EFs do not always have the same size, the amount of memory available for file storage is reduced with each file deletion and creation process. For many current applications, this is an acceptable compromise with regard to functionality, the size of the required program code and the reliability of the application.

Data integrity

Another important point is the assurance of data integrity. The file manager should always be able to test whether the data in the memory have accidentally changed, which could happen due to aging, for example. To minimize the administrative overhead for this function, the level of data redundancy and/or the extent of the supervisory protective functions should match the importance of the data. There is thus no need to protect all data with checksums as a matter of principle. Several data elements can be protected as a group, such as a complete file header, or particularly important data elements can be individually protected. This primarily depends on how often these data elements are modified in the EEPROM and how much memory space the designer of the operating system is willing to sacrifice to data integrity.

To guarantee data integrity, critical data elements in particular, such as data access privileges and file body pointers in file headers, are protected by means of error detection codes. Checksums based on CRCs are often used for this purpose, since they can be calculated relatively quickly and do not require very much program code. However, Reed–Solomon codes are also often used to provide better protection against the typical failure mode of the EEPROM cells. These are significantly better than CRC checksums for detecting the bundle errors that typically occur when an entire EEPROM page has altered.

Cross-application accesses

Certain Smart Card functions are only enabled after the cardholder has entered a PIN code. Since the powers of memory of the average person are subject to certain limitations, it has become common to use only one PIN per Smart Card, even with cards that hold several applications. Every application in the card thus uses this shared PIN. It could be stored once for each application in an internal EF, but this would require each of the stored PINs to have its own retry counter. For example, if there were five applications and each allowed three retries, a total of fifteen attempts to guess the PIN would be allowed. This is in many cases not tolerable with regard to the design and security of applications. Consequently, some operating systems allow cross-application access to PINS and to keys.

This utilization of shared resources is implemented using the same principle as the alias mechanism commonly used in PC operating systems. The main difference is that in Smart Card operating systems, all references must refer to higher-level DFs, where the MF is the highest-level entity. It is thus not possible to access a PIN located in an arbitrary DF, but only one that is located in a higher level of the file hierarchy, such as the MF.

In the case of the above example of a single PIN and retry counter that are shared by several applications, the implementation would be as follows. The PIN is stored in an internal EF that is located immediately below the MF. In the application, which is located in a DF underneath the MF, a reference to the memory location of the PIN is stored in an internal EF. The result states for successful and unsuccessful PIN comparisons are naturally stored in the PIN record in the DF, since they apply to only one application at a time. If a PIN comparison is now initiated, the VERIFY command first accesses the PIN record in the current DF, where it determines that the PIN and its associated retry counter are located on a different level. It then uses the PIN that is referenced by the local PIN record to make the comparison. After this, the state of the DF, which is stored in an internal EF of the currently selected DF, is set according to the outcome of the comparison.

This procedure is used nowadays by many Smart Card operating systems, in a number of different forms. Particularly for the utilization data that is shared between two or more applications, it represents a very elegant and cryptographically impeccable solution. In addition to this mechanism for utilizing PINs and keys across several applications, some operating systems also offer an equivalent mechanism for EFs. This makes it possible to directly access global data in EFs located immediately below the MF without first deselecting the current DF.

5.8   Process Control

There are various ways in which state machines may be implemented in an operating system. However, certain basic principles must be observed, independent of the operating system and its producer.

The state machine, regardless of whether it is a micro or macro state machine, must be located following the command interpreter and before the actual execution of the command, in the previously described layered model of the operating system. The task of the state machine is to determine whether the received command may be executed in the present state. It does this on the basis of a table. A basic principle here, as is usual with Smart Cards, is to use as little memory as possible to provide the state information. In addition, this information must also be structured such that the state machine itself can be constructed using as little memory as possible.

The state machine needs a certain amount of information to analyze the command held in the I/O buffer. Figure 5.25 shows a possible design for a Smart Card state table.

—	—	—	—
initial state	command definitions	new state in case of success	new state in case of failure
—	—	—	—

command definitions			
command	parameter P1 (min, max)	parameter P2 (min, max)	parameter P3 (min, max)
—	—	—	—
—	—	—	—

Figure 5.25   Example of data elements in a data structure that is needed for a state machine

The first data element (initial state) contains the state that is to be processed for the remaining portion of the data structure. A number may be located here to directly define the state in which all the other information is applicable. After this comes a subtable that identifies all the commands that are allowed in the initial state. It must be possible to identify as single command, a group of commands, all commands or no commands in each such subtable.

The allowed parameters associated with a command follow the command definition in the table structure. It must be possible to define individual values or ranges for the parameters in these data elements. For example, if the code for the READ BINARY command is in the command field, the P1 and P2 parameter fields could contain the minimum and maximum offset values for a read access to transparent data, and the P3 parameter field could contain both the lower and upper limits for the length. Since multiple entries may be present in the subtable for any one state, it would be possible for additional commands and their parameters to be defined following the READ BINARY command.

The table entry concludes with the new state that is to be assumed in case of successful command execution, which means in case command execution is completed without any problems. The data structure of the example table also allows a state to be defined that is to be assumed in case of an error. In order to maintain a high degree of flexibility within the state machine, it must be possible to specify the subsequent states either absolutely or as positive or negative relative states. In this context, 'relative' means that the new state is set by adding or subtracting a value to or from the initial state value, while 'absolute' means that the value of the new state is set directly, without using the value of the initial state.

In principle, there are no limits to how a state machine can be constructed. The data structure illustrated here is very well suited for use in a relatively sophisticated operating system. With the described data structure and a suitable state machine, it is in principle possible to reproduce every possible state diagram in a Smart Card. Of course, individual files also have their own supplementary protection against unauthorized reading or writing in the form of access conditions for commands. Nevertheless, process control for commands can provide an additional higher-level mechanism that complements this object-oriented protection, and that thus increases the security of the system. This is actually the major benefit of using state machines in Smart Cards.

5.9   Atomic Processes

A requirement that is frequently imposed on Smart Card microcontroller software is that certain parts of it must run either completely or not at all. Processes that are indivisible and fulfill this requirement are thus called 'atomic' processes. They always appear in connection with EEPROM write routines.

Atomic processes are based on the idea of ensuring that when an EEPROM write operation occurs, the data in question must never be written only partially. This could happen if, for example, the user pulls the card out of the terminal at the wrong instant or there is a sudden power failure. Since the Smart Card has no buffer for electrical energy, the software in the card would immediately lose its ability to do anything at all in such cases.

Especially with electronic purses in Smart Cards, it is essential to ensure that the file entries are complete and correct under all circumstances. It would for example be absolutely fatal if the balance of a purse were not completely changed to its new state if the card were suddenly pulled out of the terminal. Entries in log files must also always be complete. Since the hardware of a Smart Card does not support atomic processes, they must be implemented in software. The methods that are used for this are in principle not new; they have been used for a long time in for databases and hard-disk drives. The basic procedure of a method that is used in Smart Card operating systems is described here. This error recovery procedure is transparent to the outside world and thus does not require any changes to any existing applications.

For purposes of demonstrating how this method works, let us assume that data that are destined for a particular file are sent to the Smart Card via its interface. This would be a typical process with an UPDATE BINARY command, for example. You can follow the specific procedure by referring to Figure 5.26 while reading the following description.

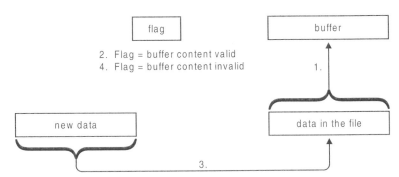

Figure 5.26   Example of a possible implementation of an atomic process in a Smart Card operating system. This procedure can naturally be cascaded for the parallel processing of multiple data elements.

In the EEPROM portion of the operating system, a buffer is set up that is large enough to accept all the necessary data. This buffer has a status flag, which is also stored in the EEPROM. The state of the flag can be set to either 'data in buffer valid' or 'data in buffer not valid'. In addition to the buffer, there must also be suitable memory for the target address and the current size of the buffered data.

The process works as follows. In the first step, the data that are located following the target address, for example in a file, are copied to the buffer in accordance with the specified physical address and length of the data. Next, the buffer flag is set to 'data in buffer valid'. In the following step, the operating system copies the new data to the desired address and then changes the buffer flag back to 'data in buffer not valid'. Whenever the operating system starts up, it polls the buffer flag before the ATR is sent. If the flag is set to 'data in buffer valid', the data in the buffer is automatically written to the memory area specified by the stored address and size information.

This mechanism thus ensures that the data in the file are valid under all circumstances. If a routine is aborted at any time in the process of program execution, the data in the Smart Card EEPROM can always be restored. For example, if the cardholder were to pull the card out of the terminal at the third step of the procedure, in which the new data are written to the EEPROM, the new data will be only partially present in the file. When the card is again powered up in a subsequent session, the operating system notices that there are valid data in the buffer and copies them to the appropriate location. This restores the original status of the file, and all the entries in the files located in the EEPROM are consistent. A very good time to make this correction is during the initial waiting time between the individual bytes of the ATR.[3] The procedure just described can easily be extended to allow multiple data elements to be stored in the buffer instead of only one. This allows write accesses to a several different files or data elements to be carried out either completely or not at all.

The procedure just described has two serious drawbacks. The first of these is that the buffer will have the heaviest write/erase load of all of the EEPROM. Since the number of write/erase cycles for any given region of the EEPROM is limited, it is highly probably that this important buffer region will be the first part of the EEPROM in which write errors start to occur. Such errors would mean that the Smart Card could no longer be used, since the

---

3   See also Section 6.2, 'Answer to Reset'.

integrity of the data would no longer be assured. This problem can be made less severe by using a cyclic structure for the buffer, so that the same region is not written every time. Unfortunately, this forces the buffer to take up a relatively large amount of memory. The second disadvantage of this implementation of an atomic process is the increase in the program execution time due to the obligatory write access to the buffer. In the worst case, the file access can take three times as long with this procedure as it would to write the data directly to the file in the EEPROM. It is thus usual not to buffer all EEPROM accesses, but only write accesses to certain files or data elements. This can be specified by an attribute in the header of each file.

5.10 Smart Card Operating Systems with Downloadable Program Code

The section 'Downloadable Program Code' in the first edition of this book, published in 1995, filled just over one page. In this edition, the length of the text has increased by a factor of 20. This alone shows how important this subject has become. There is probably no risk of exaggeration in asserting that a full paradigm change with regard to downloadable program code in Smart Cards occurred within one year (1997–98). Downloadable program code in Smart Cards is now the rule rather than the exception, although it is presently supported by only a very small number of Smart Card operating systems.

The reasons for the sharp increase in the importance of downloading executable program code cannot be completely and unambiguously ascertained, even in retrospect. One trigger may have been the floating-point error (in the FDIV instruction) in the then widely distributed Pentium processor, which became common knowledge in 1994. This could not be corrected by downloading new software, since the error was in the hardware. However, there were software patches (work-arounds) available for many applications.

It is likely that this fault is the reason why some large system operators, shortly after it became well known, suddenly made plans to allow executable code to be downloaded to Smart Cards. One of the largest applications that can accept executable program code is the German Eurocheque card. However, this capability is not presently used, so it in fact represents only a 'sheet anchor' to be used in case serious programming errors are discovered. There are also operating systems for use with GSM that allow program code for special applications to be downloaded via the radio interface.

In contrast to all other computer operating systems, the (down)loading of programs that can be run as needed is normally not allowed with Smart Cards after the card has been issued, even though this (along with storing data) is actually a primary function of every operating system. There are naturally good reasons why this particular capability has been largely absent in Smart Cards up to now.

From a technical and functional perspective, executable programs (stored in EFs, for example) do not present any problems at all. Recent operating systems thus can manage files containing executable code, and they also allow executable code to be downloaded after the card has been personalized. This makes it possible, for example, for an application provider to have executable code in the Smart Card that is not known to the producer of the operating system. An application provider could thus load a private encryption algorithm into the card and have it run there. This would allow the knowledge of the security features of the system to be distributed among several parties, which is one of the basic requirements for secure systems.

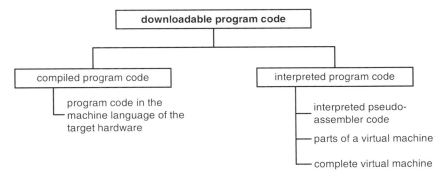

Figure 5.27 Classification of the methods for downloading and running executable program code with a Smart Card operating system

Another important reason for allowing program code to be downloaded is that it makes it possible to remove programming errors in fully personalized cards (bug fixing). Known errors in the operating system can be eliminated or at least rendered less critical by the use of downloaded code.

There are in principle two ways to execute downloaded code in a Smart Card. The first and technically simplest way is to load native code (code that has been compiled into the machine language of the target processor) into files in the Smart Card. This program code must of course be relocatable, since the memory addresses are not known outside the card. In addition to its technical simplicity, this solution has the advantage that the program can be run at the full execution speed of the processor, which makes it very attractive for downloaded algorithms in particular. In addition, there is no need for extra program code for an interpreter. The main problem with this solution is that the downloaded program can also access the memory regions of other applications if the microcontroller does not have a memory management unit (MMU).

The second way to execute downloadable code in a Smart Card is to interpret the code. The interpreter can check the memory regions that are being addressed while the program is running. However, the interpretation must run quickly, since no advantage can be realized with code that runs slowly. The implementation of the interpreter should also take up as little memory space as possible, since the amount of memory available is very limited. Presently, the best known versions of this solution are the Java Card specification [Javasoft, JFC] and the C interpreter MEL (Multos Executable Language) from Multos [Maosco]. A BASIC interpreter for Smart Cards has even become available lately [Zeitcontrol]. Interpreters that make a specific protected memory region available to an application program are by the way not suitable for eliminating errors in the Smart Card operating system, since they naturally have no access to the program and data areas belonging to the operating system.

However, the age-old problem with interpreters is their slow speed, which is an inherent property of interpreted code. There are several different approaches that can be used to make up for this drawback and to keep the size of the program code of the interpreter itself as small as possible. This simplest approach is to interpret a pseudocode that is ideally as similar as possible to the machine instructions of the target hardware. The working speed of the interpreter is thus relatively high because the pseudocode is close to the machine language, and machine-independent program code can be used. Memory accesses during the interpretation can be monitored, but this is not mandatory. A slower solution, which is also somewhat more complicated in terms of programming logistics, is to split the interpreter into an off-card part (offcard virtual machine) and an on-card part (oncard virtual machine). This approach is utilized by many current Java Card implementations. Its main advantages are reliable memory protection and complete hardware independence. However, the division of the interpreter into the on-card and off-card parts is disadvantageous. It makes cryptographic protection mandatory for transferring programs between the two parts of the interpreter, since otherwise the on-card part of the interpreter could be deliberately caused to misbehave by the use of manipulated program code.

The optimum technical solution is to have a complete interpreter in the Smart Card. This makes it possible to load any desired program into the card and run it without any risk to other programs located in the card. However, the size of the interpreter program is so large that it will certainly take several years and several more generations of Smart Card microcontrollers before this solution becomes broadly established in the Smart Card world.

5.10.1   Executable native code

The processors of current Smart Card microcontrollers mostly do not have any sort of memory protection mechanisms or any possibility to supervise program execution. As soon as the program counter addresses 'foreign' machine language code for this processor, the control of all the memory and all the functions rests entirely with this executable code. It is no longer possible to limit the functions of the executable program in any way. Every addressable memory location can be read by bypassing the memory manager or handler, and memory locations that lie in the EEPROM or RAM can be written. All memory contents could thus easily be sent to the terminal via the card interface.

This is precisely the weak point with regard to downloadable executable programs. If everyone were allowed to download programs, or if this were possible by circumventing the protective mechanisms, the security of any secret keys or other confidential information within the entire memory region could be no longer assured. This would be the ideal form of attack on a Smart Card. The card would still behave in the same way as a non-manipulated card with respect to the outside world, but special commands could be used to read out its entire memory or to write to portions of the memory

There is yet another watertight argument against third-party downloadable programs. The producer of the files to be downloaded must know all the entry points (jump addresses) and calling parameters of the operating system routines in order to be able to use important operating system functions. Operating system producers, however, prefer to release as little information as possible about the internal processes and addresses of their program code, since they consider this a matter of security. In addition, it is necessary to verify that the downloaded code does exactly what it is supposed to do without any errors and does not perhaps harbor a Trojan horse. This can only be tested by an independent body.

The most elegant solution to this problem, and certainly the one with the most promising future, is to use hardware-based memory management (a memory management unit, or MMU) supplementary to the actual processor in the Smart Card. This uses a hardware circuit to check the running program code to see whether it keeps within its assigned boundaries. Only then would it be possible to allow application operators to download programs that have not been certified by the card issuer, without thereby sacrificing the card's security. Each such application would be assigned a physically contiguous region of memory representing a DF. The MMU would then monitor the associated memory boundaries whenever a downloaded program in a DF is called. If these boundaries are exceeded, the program can be immediately halted via an interrupt and the application can then be blocked pending further action.[4]

There are two different ways to implement downloaded code. In the first option, the program code is located in an EF whose structure is 'executable'. The contents of this file can be executed by means of the EXECUTE command after it has first been selected. Depending on the application, prior authentication may be required. The parameters for running the program are passed to the Smart Card in the EXECUTE command. The response that is generated by the program contained in the EF is sent back to the terminal.

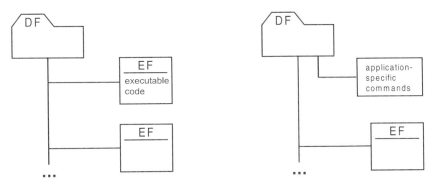

Figure 5.28   The two ways in which executable program code can be entered into a standard Smart Card operating system: as an executable file (left) or as application-specific commands (ASC) (right).

The second option takes a somewhat different form, because it is based on the principles of object-oriented design. This option is described in the EN 726-3 standard (among others) as application-specific commands (ASC). According to this standard, the complete application, including all its files and application-specific commands, is contained in one DF. The program code can be downloaded into a memory region within this DF that is managed by the operating system. This takes place with the use of a special command that sends all the necessary information to the Smart Card. If the DF in question is now selected and a command is then sent to the card, the operating system checks whether the command belongs to the downloaded commands. If it does, the operating system immediately calls the program code located in the DF. On the other hand, if a different DF is selected, the downloaded command effectively does not exist.

Example of native program code that can be downloaded to an EF

There are several large Smart Card applications whose operating systems allow executable code to be downloaded after the card has been personalized. However, the specifications for this capability are in almost all cases confidential, and in some cases, even the fact that this capability exists is confidential. Consequently, we can only describe the general principles of this capability here, independent of any actual operating system. A possible implementation is described in detail following the discussion of the basic principles.

---

4   See also Section 3.4.3, 'Supplementary hardware'.

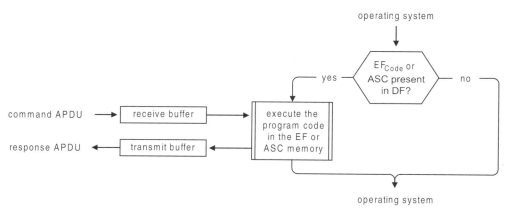

Figure 5.29   Basic calling procedure for executable program code stored in an EF, or programs that work within the framework of ASCs, in a standard Smart Card multi-application operating system

The program code to be downloaded must satisfy certain basic prerequisites before it can be run in a Smart Card in the first place. It may sound obvious, but the most important prerequisite is that the processor type must be known (for example, 8051 or 6805). Particularly in a heterogeneous environment with many different types of Smart Card microcontrollers, satisfying this requirement can often very well involve a certain amount of effort. Along with this comes the requirement that the Smart Card operating system and its API (application programming interface) must be known, including all entry points and the parameters passed to and returned by its routines.

The program code to be downloaded, which is always native code (machine code of the target processor), either must be programmed so that it is relocatable, or it must be relocated on the fly by the Smart Card when it is downloaded. The requirement for relocatability, which means that the program can be shifted within the memory, comes from the fact that the memory addresses where the code will be stored are known only to the Smart Card operating system and not to the outside world. Making a program relocatable usually takes place in the programming stage. In concrete terms, relocatability means for example that no jumps to absolute physical addresses are allowed, but only jumps relative to the address of the jump command.

If the program code satisfies all these requirements, it can in principle be loaded into the memory of a Smart Card and run there. The program code can of course be structured as needed. Figure 5.30 shows a possible structure, but the actual structure can be completely different, depending on the operating system. The first data element in this example is a unique label that tells the Smart Card operating system that this is program code. Such a label is commonly called a 'magic number'. For example, with Java Class files, this is a sequence of four bytes forming the word 'CAFEBABE'.

The program code starts just after the label. In this example, it is broken down into four parts. The first part contains all the necessary initializations, data saving and the like. Following this startup routine comes the actual function routine that holds the program code for the desired task. This is followed by the shutdown routine, which is the counterpart to the startup routine. The shutdown routine ensures that the program is correctly terminated, and if necessary it restores any saved data and adjusts the stack.

The fourth part of the program, which follows the first three parts, is optional. It can include program code that is to be resistantly incorporated into the Smart Card software. Bug fixes for the operating system would typically be located here. The three prior routines would then modify pointers or handles so that the routines in this section would be permanently linked into the software of the operating system. The entire process is very similar to the TSR (terminate and stay resistant) programs that we know from the time of DOS. These programs only had to be called once for them to anchor themselves in the operating system until the next reset. With the Smart Card, these resistant routines would however be permanently installed after being called once, rather than only for one session.

We assume here that the downloaded program is called using a CALL instruction, and that it returns control to the calling program with a RETURN instruction. In principle, a direct jump to the first machine code command (using a JUMP instruction) would also be possible, but this would have the disadvantage that the called program would not know which program called it.

For insurance against accidental changes, the entire data block should be protected by an error detection code (EDC). Alternatively, a digital signature could naturally be used to provide additional protection. The Smart Card would then have a public key, and the producer of the program code would hold the associated private key. This would provide binding assurance that authentic program code could be run on the Smart Card.

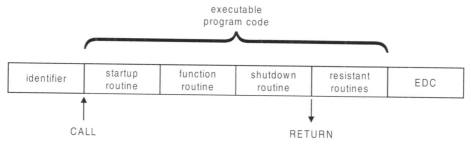

Figure 5.30   A possible structure for native program code that can be downloaded to an EF and run from there

The downloaded program code can be stored either in an EF or in a program memory region within a DF that is not visible to the outside world. The first option is described in some detail below, since it is encountered significantly more often in actual practice.

EFs with transparent structures are ideal for storing program code, since they can be effectively written in several sections using UPDATE BINARY commands with offset values. Also, their maximum length of more than 65 kB is more than adequate, even for extensive programs. These EFs can have 'executable' as a property, which means that the program code stored in them can be directly called using the EXECUTE command.

Some operating systems have instead a file structure, derived from the transparent structure, that is referred to as 'execute'. This is not particularly important for the outside world, especially since normally both types can be accessed using the UPDATE BINARY and EXECUTE commands. The EF can be selected via its FID or a Short FID. The access condition for reading is always set to 'never'. Writing data is normally allowed after prior authentication and with the use of Secure Messaging.

The two processes shown in Figures 5.31 and 5.32 provide an overview of how program code can be loaded into an EF in a Smart Card in a secure manner. If no suitable EF is already available, it must first be created. The second process diagram shows that the EF must first be selected, following which the program code must be started using the EXECUTE command. An optional data transfer is allowed within the body of the command. A similar possibility exists for the response, in which data can be sent back to the terminal if necessary. The called program must naturally recover any data that has been sent by reading it from the receive buffer, prepare its response and then write its response to the transmit buffer.

Background system		Smart Card
select an EF with an 'execute' structure	→	
mutual authentication of the Smart Card and the terminal	← →	
switch on secure messaging		
send *n* UPDATE BINARY commands with executable program code in the data segment, protected by secure messaging	→	

Figure 5.31   A possible process for loading executable program code in an existing EF with an 'execute' structure. The access conditions for UPDATE BINARY prescribe mutual authentication of the Smart Card and the terminal as well as Secure Messaging for data transfers.

Background system		Smart Card
select an EF with executable program code	→	
send an EXECUTE command	→	check the label (magic number) of the program
		check the EDC of the program
		call the first machine instruction with CALL

Figure 5.32   A possible process for starting executable program code, which in this example is stored in an 'execute' EF

In light of the strict demands for unambiguous identification of the microcontroller, operating system and internal software interfaces, as well as those related to system management, program downloading normally only occurs with an online connection to a background system. The data bases located in the background system either hold all the necessary data according to the unique chip number, or they receive this information online via a direct end-to-end link with the Smart Card. According to this information, a program with the desired functionality is selected from those available in the system and transferred to the Smart Card using the prescribed security mechanisms. The secret keys for this are normally managed and used in the background system exclusively within a security module. The usual procedure for the entire process is shown in Figure 5.33.

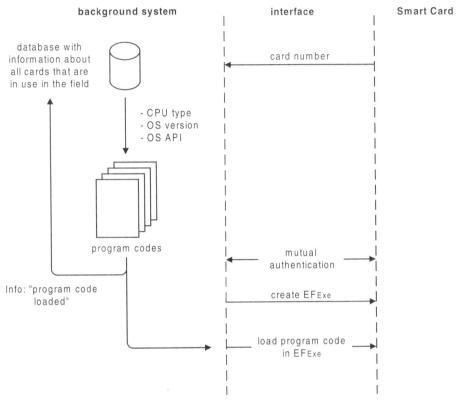

Figure 5.33   The procedure used for online downloading of native program code into a Smart Card EF from a background system. This procedure can be used to transfer various types of program code to a Smart Card, depending on the Smart Card operating system and the microcontroller hardware. The illustrated procedure could be implemented quite well in a GSM application, for example.

The method for introducing native code into a Smart Card that has just been described has some attractive practical advantages. The procedure is simple and robust, and it can be implemented in a Smart Card operating system using a small amount of program code. The program code to be run does not have to be interpreted, since it can be directly executed by the processor. This yields a high processing speed, as well as the possibility of using this method for downloading complex algorithms (such as DES, IDEA and the like).

Interpreter-based systems will not be able to manage this within the foreseeable future, due to their slow processing speeds. As long as there is no hardware-based memory management (MMU) to restrict free access to the memory, this method provides an excellent way to correct errors in the Smart Card software after the card has been issued. If an error actually becomes known, this method provides a unique 'back door' that can only be reached with this type of software downloading. Other techniques, such as Java for

Smart Cards, implement strict and unconditional memory partitioning and thus cannot modify the operating system code. If a MMU is present, it still might be possible to invoke an administrator mode to temporarily deactivate memory supervision.

This leads us to the disadvantages. The downloading of executable native code demands a high level of knowledge of the hardware and/or the operating system of the Smart Card. It could be necessary to have a separate program on hand for each version of the Smart Card that is used, even though all the programs would have the same functionality. The second major drawback of this method is that, for reasons of security, the program must be developed by the card issuer (or under the authority of the card issuer). The loading of foreign and unknown programs into the Smart Card must be strictly forbidden, since the downloaded program assumes control of the microcontroller once it is started and cannot be regulated in any manner. It could thus, for example, read out the secret keys of other applications present in the card and send them to the terminal via the I/O interface.

Evaluation of the program code by the card issuer provides only weak protection against attacks of this sort. Better protection in this case can be achieved with hardware-based memory management that makes only certain regions in EEPROM and RAM available to the downloaded program and immediately terminates the program if an attempt is made to exceed the boundaries. This allows the applications present in the card to be fully isolated from each other. Presently, due to the absence of suitable MMUs, the only available expedient is to carefully check the program to be downloaded.

5.10.2   Java Card

In 1996, Europay presented a paper about OTA (open terminal architecture), in which a Forth interpreter for terminals was described and largely specified. The objective was to create a uniform software architecture for terminals in order to lay the foundations for hardware-independent terminal programming. Given this, a particular application (such as paying with a credit card) would have to be programmed only one time, and the resulting software would run without any modifications on all terminals made by various manufacturers. Although this proposed design has up to now never been completely implemented, it certainly gave rise to extensive discussions in the Smart Card world.

When it became known in the fall of 1996 that Schlumberger was developing a Smart Card that could process programs written in the Java language, no one was particularly surprised. The idea of combining an interpreter with a memory-poor microcontroller was already well known from the OTA proposal. The published specification (Java Card 1.0) provided an application programming interface (API) for integrating Java into an ISO/IEC 7816-4 operating system. This would allow Java to access the standard Smart Card file system with its MF, DFs and EFs.

Many producers of Smart Card operating systems were initially astonished at the idea that a language such as Java, which normally requires well over a megabyte of memory, should be used with Smart Cards. However, the first meeting of nearly all major Smart Card manufacturers with Sun, the firm that developed and promoted Java, took place in the spring of 1997.

This was the first conference of what has since become known as the Java Card Forum (JCF), which functions as the international standardization panel for Java in Smart Cards. The task of the technical group of the Java Card Forum is to define a subset of Java for use in Smart Cards, to specify the outlines of the Java interpreter (known a the Java Virtual

Machine, or JVM) and to define both a general-purpose API and application-specific APIs (for GSM and financial transaction cards, for example). These APIs form the interface between the Smart Card operating system and Java. The task of the marketing group of the JCF is to promote Java technology for Smart Cards.

The names of the current specifications at the time of writing[5] are the Java Card Version 2.1 Application Programming Interface, the Language Subset and Virtual Machine Specification and Programming Concepts. Up-to-date versions are available from the WWW server of the Java Card Forum [JCF].

The Java programming language

In 1990, a development group at Sun began to develop a new programming language, under the guidance of James Gosling. The objective was to create a hardware-independent, secure and modern language that could be used for microcontrollers in consumer products (such as toasters and espresso machines). A large variety of microcontroller types with differing architectures is used in such products. This non-uniformity, combined with frequent hardware modifications, makes it difficult for software developers to write portable program code. Remarkably enough, Smart Cards exactly match the characteristics of the original target application area.

The programming language was first called 'Oak', after the oak furniture in James Gosling's office, but in 1995 it was renamed 'Java'[6] and the objectives were redefined. In the summer of 1995, Sun began to heavily promote Java as the hardware-independent language for the heterogeneous Internet. The frequently quoted slogan coined by Sun, 'write once–run anywhere', probably most clearly shows the degree of hardware independence that Java attempts to achieve.

The beginning of the widespread use of Java coincides with the start of the enormous growth of the World Wide Web (WWW).[7] For a variety of reasons that were not just technical, but which completely stemmed from the realms of business politics and worldviews, the new language excited researchers, universities and software firms throughout the world [Franz 98]. This led to Java becoming the de facto standard for Internet applications within an extremely short length of time. Of course, the properties of this new language favored this development.

The Java programming language is a completely object-oriented and strongly typed language. It is easily learned by programmers, since it has much in common with C and C++. Java is also a robust language that does not allow all the tricks and popular but dubious techniques that C and C++ (for example) do allow. For example, Java has no pointers, field boundaries are monitored at run time, and there is strict type checking. In addition, memory management is taken over by Java and an associated 'garbage collector', so memory leaks (a much-feared phenomenon in C and C++) are impossible by design. Java is also a secure programming language, which means that when a program is run the functions that it wants to carry out are checked while it is running, so that the run-time environment can stop the program if necessary. This is one of several possible grounds for calling an exception handler. If an exception occurs, a call is made to a specific routine in which the response to this particular case can be defined.

---

[5]   Spring 1999.

[6]   'Java' is in this case American slang for 'a cup of coffee', and does not refer to either the South Pacific island or the French biscuits with the same name.

[7]   In 1993, there were only three WWW servers in the whole world!

The majority of these features are only possible because Java is an interpreted programming language, which is not executed directly by the processor. Java also has other features, such as multithreading capability and support for distributed processing, but these are presently not supported in the Smart Card environment.

There is interest in standardizing the Java programming language via an ISO standard, which would make it supplier-independent. However, it will certainly take several years before the Java language is a component of an international standard. The fact that a license agreement with Sun is required before Java can be supported in a product (such as Smart Cards) is certainly an important factor in this connection.

The characteristics of Java

Programs written in Java are translated into what is known as 'Java bytecode' by a compiler. This is nothing more than processor-independent object code. The bytecode is, in a manner of speaking, a program made up of machine instructions for a virtual Java processor. This processor does not actually exist; instead, it is always simulated by the target processor. This simulation takes place in the Java Virtual Machine (JVM or VM), which is the actual interpreter. Seen from a different perspective, the JVM is a simulation of the Java processor on an arbitrary target system. The target processor in turn naturally works with native code. The main advantage of this arrangement is that only the JVM that is programmed in native code must be ported to a given target processor, Once this has been done, the Java bytecode will run on the new system.

The VM is also called the Sandbox, since its run-time job is not just to mindlessly interpret the bytecode, but also to carry out type checking and to monitor accesses to objects (among other things). This alternative name graphically indicates that a Java program must stay inside its own sandbox, as the VM will otherwise put a stop to its activities.

A compiled Java program, or in other words, one that has been translated into bytecode and provided with certain supplementary information, is stored in what is called a class file. Class files are executed after the Java virtual machine has been loaded. One or more class files constitute an 'applet', which contains a complete Smart Card application and has its own application identifier (AID). In the context of Java for Smart Cards, applets are sometimes called 'cardlets'.

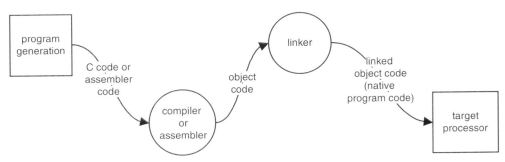

Figure 5.34   Top-level data flow diagram of the normal process for converting source code in C or assembler into executable machine code for a target processor

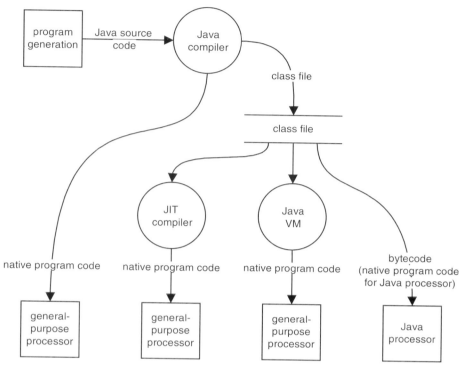

Figure 5.35   Top-level data flow diagram of possible processes for converting Java source code into executable machine code for a target processor. The leftmost path is offered by some compiler producers, but it does not correspond to the original philosophy of Java since it does not retain hardware independence.

The hardware independence of Java naturally has its price. This is primarily its very slow execution speed relative to other standard programming languages. This problem has still not been solved in a satisfactory manner, although further developments and improvements of Java are concentrated on this issue. One step that is already being taken is the use of 'just-in-time' (JIT) compilers, which translate the Java bytecode into the machine language of the processor the first time the program is called. This naturally results in a noticeable slowing down of the program when it is first called, but subsequent calls are processed significantly faster. However, the software of a JIT compiler is rather complex, and the limited memory space of Smart Cards means that it will be several years before such a compiler can be implemented in a Smart Card.

Direct compilation of Java programs into the machine language of the target processor would make not sense in the heterogeneous world of Smart Cards, in contrast to that of PCs. The only remaining alternative is the use of a Java processor. Some manufacturers of Smart Card microcontrollers have already announced plans to extend their products so that they could directly execute Java bytecode. This would result in considerably improved performance relative to the current software implementations.

Based on the present situation, this would probably be done by integrating a special Java processor on the chip alongside the existing 8051 or 6805 processor core. This would have the advantage of allowing time-critical routines (such as data transfer routines and cryptographic algorithms) to still be programmed in assembler, while at the same time allowing all software above this level to be programmed in a high-level language such as Java.

Technically, this is certainly possible, if you consider the following facts. Sun's Java Chip (microJava 701) has an edge length of 7 mm with a 0.25-μm process, works with a 2.5-V supply at 200 MHz, contains 2.8 million transistors and has a power consumption of only 4 W. Since this chip includes many functions that are not needed in a Smart Card (such as floating-point arithmetic and driving external memory), it is certainly within the realm of what is technically possible to make a Java processor that has been reduced to the needs of the Smart Card and to put it onto a Smart Card microcontroller chip.

Table 5.8   Program execution time versus programming language, with C/C++ taken as the reference. The listed values are estimates for PCs and are only intended to serve as guidelines. The reference PC processor is a 300-MHz Pentium II, while the reference Java processor is a 200-MHz picoJava 701.

Programming language	Execution time
C, C++	1
Java, executed by an interpreter	20–40
Java, executed by an interpreter with a JIT compiler	≈ 5
Java, program code compiled as native code	1–2
Java, executed by a Java processor	≈ 1.2

Another manner of increasing the execution speed of Java bytecode will be provided by the upcoming 32-bit Smart Card microcontrollers, which have just recently been announced by some semiconductor manufacturers. However, the processing speed of interpreted of Java program code will still be 10–20 times slower than that of program code written in C, even with a 32-bit processor. The main advantage of the new processors is that they are general-purpose types.

The disadvantage of interpreted program code relative to compiled program code, namely inadequate processing speed, can be strongly reduced by the use of a suitable programming interface for the interpreter. This interface, which is called an application programming interface (API), allow interpreted program code to call machine-language routines. The native code routines that are called in this manner run at the full processing speed of the host processor.

As ideal as this manner of increasing the processing speed may appear to be at first glance, it still has its own drawbacks. The API for native code must be well thought out if it is to be generally useful and not just worthwhile for a few special cases. This can however be achieved, with a bit of consideration. The second problem is more serious. The compatibility and hardware independence of a programming language such as Java is only possible if all APIs are the same. If there are different types of APIs, with regard to either their interfaces or the functions they provide, there is no great advantage to be realized from a standardized programming language. In such a case, it would be necessary to make certain modifications to the source code for each different platform with its own API. This is why the Java Card Forum continues to invest a lot of effort in standardizing the APIs, since they are fundamentally important for platform independence.

The Java Virtual Machine (JVM)

The Java Virtual Machine is the most essential element of the Java technology. It simulates a Java processor and can be implemented in software on any sufficiently powerful processor. If it is desired to run Java bytecode on a new type of processor, the Java Virtual Machine must be ported to this processor. It is usually written in the C programming language, so the actual porting may not require anything more than a few modifications and a recompilation of the source code. The size of a Java Virtual Machine on a PC ranges from 100 to 200 kB.

Table 5.9   Comparative sizes of Java Virtual Machines for PCs and Smart Cards

Java for:	Smart Card		PC	
Processor architecture	8	bits	32	bits
Processor clock	1–5	MHz	100–300	MHz
RAM	512	bytes	8–64	MB
Amount of program code for the VM	6–8	kB	100–200	kB
Memory for Java programs (hard disk, EEPROM)	8–16	kB	> 1	GB

The virtual machine for Java has all the elements of a real processor. It has its own instruction set in the form of the bytecode, and it also has registers such as the program counter and accumulator. The data to be processed are passed to the virtual machine as a class file. This file contains fixed constants, the bytecode to be executed as methods and various additional information.

Java bytecode takes up very little space, and it is nearly as compact as machine code. The overall memory space balance with respect to native machine code is made worse by the obligatory presence of the virtual machine. The difference naturally increases as the size of the program code becomes smaller with respect to that of the virtual machine.

Bytecode is fundamentally very similar to the machine instructions for a normal processor. There are for example instructions for stack manipulation, logical and arithmetic operations, commands that access the registers of the virtual Java processor, and even access methods for arrays. The Java Virtual Machine and the bytecodes are extensively described by Tim Lindholm and Frank Yellin [Lindholm 97].

Table 5.10   Overall structure of a class file

Data elements of a class file
Label (magic number)
Version number
Constants pool
Methods
Attributes of classes, fields and methods

Table 5.11   Functional limitations of Java for Smart Cards, compared with full Java

Function	Java for Smart Cards	Java
Operators	all	all
Exception handling	yes	yes
Class cloning	no	yes
Data type: int	optional	yes
Data types: Boolean, byte, short	yes	yes
Data types: long, float, double, character	no	yes
Dynamic downloading of classes	no	yes
Arrays for supported data types	only one-dimensional	also multi-dimensional
Object arrays	no	yes
Object cloning	no	yes
Dynamic object creation	yes	yes
Virtual methods	yes	yes
Exception handling	yes	yes
Interfaces	yes	yes
Process control functions	yes	yes
Dynamic memory management (garbage collection)	no	yes
Threads	no	yes

Table 5.12   Data types in Java for Smart Cards, with their memory requirements and ranges of values. The data type 'int' is optional.

Data type	Size	Range of values
Boolean	1 byte	true, false
byte	1 byte	− 128 to 127
short	2 bytes	− 32,768 to 32,767
int	4 bytes	− 2,147,483,648 to 2,147,483,647

Due to the considerably limited system resources of a Smart Card microcontroller, certain cuts must be made in the Java Card VM compared with the original Java VM for PCs. The Java Card VM does not have garbage collection to automatically return memory that is no longer needed to the free memory pool. Support for the class files must also be sharply reduced. There are fewer data types available in Smart Cards, and the bytecode itself is reduced from 149 instructions to 76.

The program size of the on-card Java VM for Smart Cards is around 6–8 kB of 8051 machine code if it is created in the C language. Approximately 200 bytes of RAM are needed in addition. The API with the classes *javacard.framework* and *javacardx.framework* takes up 3–4 kB of memory, consisting mainly of Java bytecode. In addition, at least a rudimentary operating system is needed, with transmission protocols, cryptographic algorithms and many functions that are closely tied to the hardware. The code size for this ranges between 6 and 8 kB when it is programmed in assembler.

The correctness of this relatively small program is extraordinarily important, since an error or a security gap in the Java VM could undermine the whole idea of providing a secure environment by incorporating Java in the Smart Card. The design and implementation must therefore be largely error-free. An ITSEC or Common Criteria evaluation is normally employed to verify this. The small size of the program code for the Java VM naturally considerably simplifies this process, especially since it is thereby possible to formally describe the complete functionality of the VM. Many manufacturers of Java Smart Cards have already announced their intention to have their implementations tested by a certification body. The goal in this case is ITSEC E4 certification.

Due to the small amount of memory present in Smart Cards, it was necessary to divide the Java Virtual Machine into on-card and off-card parts. Static tests can easily be performed outside of the Smart Card in the Offcard VM without affecting performance and with no loss of security. The link between the two parts of the VM is formed by data in CAP format. For complete security, these must be cryptographically protected, ideally using digital signatures, so that they cannot be tampered with during the transfer. They would otherwise present an attacker with a promising starting point, since it would be possible to bypass the security mechanisms of the Oncard VM using manipulated data.

Table 5.13   Summary of the limitations of Java for Smart Cards. These broadly laid out boundaries currently impose no restrictions on software development for Smart Cards.

Classes	The maximum amount of data in any actual class is 255 bytes
Methods	A class may contain at most 127 methods
Arrays	Arrays may contain at most 32,767 fields
Switch instruction	If the 'int' data type is not supported, at most 65,536 branches are available in the Switch instruction.
	If the 'int' data type is supported, the number of branches in the Switch instruction depends on the data type (char, byte, short or int), as with Java for PCs.

Java is stack-oriented and thus naturally needs a stack as well as a heap. These are set up and managed separately for each applet in the Smart Card. The stack is primarily used for passing data when the methods are called, while the heap serves as a storage area for the objects. Normal sizes are 200–300 bytes of RAM for the stack and 12 kB of EEPROM for the heap. Relatively small applets can however easily manage with 50–60 bytes for the stack and a few thousand bytes for the heap.

The Java VM consists of four functional parts: the Bytecode Verifier, the Loader, the Bytecode Interpreter and the Security Manager, arranged as shown in Figure 5.36.

The job of the Bytecode Verifier is to perform a variety of static checks on the class file that is passed to the Java VM. The Bytecode Verifier first checks the file format. Following this, it checks the constants pool, checks the bytecodes for syntactical correctness, and checks the arguments of the methods and the object inheritance hierarchies. Yet other investigations may be undertaken. These are described in detail by Frank Yellin [Yellin 96].

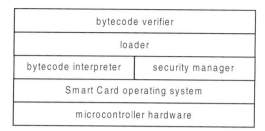

Figure 5.36   The individual components of the Java Card Virtual Machine

Listing 5. 1   The basic functions of the main program loop of the Java Bytecode Interpreter

DO (	Interpreter main loop
fetch and save program counter	Fetch and save the value of the program counter for later comparison.
fetch opcode	
fetch operands	
execute machine instruction	Execute the virtual Java processor machine instruction, which is made up of the opcode and the operands.
IF (machine instruction did not alter _ the program counter) THEN (increment program counter)	If the Java machine instruction that has just been executed has not altered the program counter (which means that it was not a GOTO bytecode, for instance), then set the program counter to the next opcode.
) WHILE (opcodes available)	Repeat the loop until all opcodes have been processed.

After the Bytecode Verifier is finished, the Loader takes the checked data and sends them in CAP format to the actual interpreter in the Smart Card. For security reasons, the data should be provided with a digital signature, so that they cannot be tampered with between the Loader and the on-card part of the Java VM.

After having passed through the Loader, the executable bytecode has been augmented with various kinds of supplementary information and can be executed by the on-card part of the Java VM. The actual interpreter reads each bytecode in turn along with its associated arguments, and converts these into native machine instructions of the target processor. The Security Manager works in parallel with the interpretation of the bytecodes. Among other things, it constantly checks that the field, stack and heap boundaries are respected. It is empowered to immediately initiate an exception handler and stop the processing of the bytecode causing the problem if it detects a violation of the defined security rules.

Java for Smart Cards

The actual major advantages of a modern programming language such as Java with regard to its use with Smart Cards are not only that it allows everyone to write programs for Smart Cards. This would also be conceivable with most operating systems using assembler or C with public interfaces and a few modifications. An idea such as Java is primarily interesting to the operators of large systems. They have the problem of having to buy a variety of masks on different chips from different manufacturers. This multiple sourcing, while it certainly makes sense on tactical grounds (due to reduced dependence on one supplier and price pressure on the suppliers), continuously causes problems with compatibility and testing. Presently, it cannot be expected that two different operating systems from two different producers, with all their various options, will behave the same at the interface level. This presents a serious problem to system operators.

From the system operator's point of view, the ideal solution to this problem would be hardware-independent program code that could be executed by an evaluated interpreter in Smart Cards in a standardized manner. An application program could then be written, tested and evaluated only once, following which it could be processed by all different types of Smart Card operating systems. No differences would be visible 'from the outside', which means at the interface. This would preserve the advantages of multisourcing while eliminating its drawbacks.

You should keep this thought in the back of your mind when examining how Java is integrated into Smart Cards. The first versions only provided for storing Java bytecode in an EF located under the MF or under a DF. An EXECTUTE command started up the virtual machine to run the program stored in the EF. An API (application programming interface) associated with the file system allowed data to be read from and written to files.

However, this approach has not prevailed. According to the Java Card specification, a Smart Card with Java has a Java Virtual Machine that is activated when the card is finished and deactivated at the end of the card's life cycle. A file system as specified in ISO/IEC 7816-4 is no longer included, since it can be built up using objects within Java applets. There are several classes that make it relatively easy to construct a file tree that complies with the ISO/IEC 7816-4 specification. The program code and its associated file tree are both part of the applet that is loaded into the Smart Card. The applet in the card can be selected by its unique AID with the SELECT command. After the applet has been selected, it automatically receives all further commands for processing. The program code of the applet can then evaluate and process the commands and their associated data, as well as carrying out the corresponding accesses to the file system.

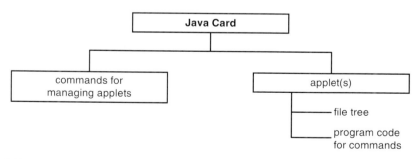

Figure 5.37   The two basic computational components of a Smart Card with Java

This approach provides maximum flexibility and compatibility, since each application is contained in an applet along with its file tree. The actual card commands, such as READ BINARY and MUTUAL AUTHENTICATE, are located in the applet as program code. This makes it possible for a single card to have different codings and different procedures in two separate applets that support the same command in mutually independent manners.

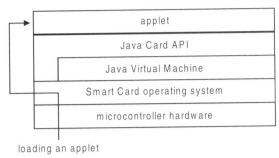

loading an applet

Figure 5.38   The basic process for loading an applet in a Smart Card operating system with Java

Of course, this advantage comes at the price of a considerable amount of memory space for the applets in question, since they must necessarily contain redundant data and routines. This conspicuous consumption of memory can be somewhat reduced in some cases by allowing the objects of one applet to be shared with other applets (object sharing). For reasons of security, this can only be carried out by the applet that created the object in question. The process of sharing an object cannot be reversed. This means that if an object is made available for access by other applets, it remains available for the lifetime of the card.

Java Card Framework
In order to make the programming of Smart Cards in Java as easy as possible, there are four packages that provide a standardized application programming interface (API) with functions that are useful for Smart Cards. This API is called the Java Card Framework. It consists of four parts, of which only one is required to be present in all Java cards. The other three parts, which are distinguished by an 'x' (for 'extension') in their names, are optional and can be included as needed by the Smart Card manufacturer. There are plans to define additional application-specific packages in the future. The candidates are a package for GSM-specific applications and a package for an electronic purse according to the EN 1546 standard.

Since the program code that lies underneath the API can be produced in machine code for the target processor, the API provides not only a standardized interface but also an enormous increase in execution speed.

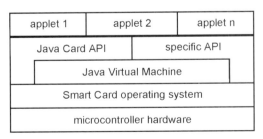

Figure 5.39   The software structure of a Smart Card with a Java Virtual Machine, two different APIs and several applets

The obligatory package *javacard.framework* forms the basis for Java in Smart Cards. It defines elementary classes for applet management, data exchange with the terminal and various constants, all in the context of ISO/IEC 7816-4. The package *javacardx.framework* is an optional extension of the basic package. It contains an object-oriented file system that complies with the ISO/IEC 7816-4 standard, with DFs and EFs in all structures.

The package *javacardx.crypto*, which contains cryptographic functions, can also be included if necessary. Due to various export restrictions, the construction of this package does not allow the Smart Card to be used as a general-purpose encryption and decryption tool. The package *javacardx.cryptoEnc*, which contains the necessary encrypting methods, must be included to make this possible.

Table 5.14   The classes of the application programming interface packages *javacard.framework* and *javacardx.framework*, according to the Java Card 2.1 specification

Class	Explanation
javacard.framework.AID	Encapsulates the 5–16 byte long application identifier (AID) as specified in ISO/IEC 7816-5.
javacard.framework.APDU	Provides methods for exchanging data between the Smart Card and the terminal at the APDU level.
javacard.framework.applet	Defines an applet for a Smart Card.
javacard.framework.ISO	Encapsulates various ISO/IEC 7816-3 and ISO/IEC 7816-4 constants. Some typical examples are offsets to various data elements within an APDU and various return codes.
javacard.framework.PIN	Represents a PIN, with its associated retry counter and a validity status.
javacard.framework.System	One of the most important classes. It can contain methods for making objects persistent or transparent, or to allow several applets to access a singe object.
javacard.framework.Util	Contains methods for handling fields and data objects and for type conversion.
javacardx.framework.CyclicFile	Used to create cyclic EFs and to perform read and write accesses to these EFs.
javacardx.framework.DedicatedFile	Used to create DFs within a file tree.
javacardx.framework.ElementaryFile	Used to create an EF within a file tree.
javacardx.framework.File	Base class for all files (DFs and EF) in the file tree of an applet.
javacardx.framework.FileSystem	Subclass of '… .DedicatedFile', and the root of the file tree of an applet.
javacardx.framework.LinearFixedFile	Used to create linear fixed EFs and to perform read and write accesses to these EFs.
javacardx.framework.LinearVariableFile	Used to create linear variable EFs and to perform read and write accesses to these EFs.
javacardx.framework.TransparentFile	Used to create transparent EFs and to perform read and write accesses to these EFs.

Table 5.15   The most important card-independent application programming interface classes of the package *java.lang*, according to the Java Card 2.1 specification

Class	Explanation
java.lang.Exception	For exception handling for Java Card.
java.lang.Object	The class of all Java Card classes.
java.lang.Throwable	For all exceptions and errors for Java Card.

Table 5.16   The classes of the application programming interface packages *javacardx.crypto* and *javacardx.cryptoEnc*, according to the Java Card 2.1 specification

Class	Explanation
javacardx.crypto.AsymKey	Base class of keys for asymmetric cryptographic algorithms.
javacardx.crypto.DES3_Key	Provides a data object for a 112-bit Triple-DES key, as well as the encryption methods (ECB and CBC mode) and MAC computation with Triple-DES. There is no decryption method in this class.
javacardx.crypto.DES_Key	Provides a data object for a 56-bit DES key, as well as the encryption methods (ECB and CBC mode) and MAC computation with Triple-DES. There is no decryption method in this class.
javacardx.crypto.Key	Base class for all keys.
javacardx.crypto.MessageDigest	Base class for all hash algorithms.
javacardx.crypto.PrivateKey	Base class of private keys for asymmetric cryptographic algorithms.
javacardx.crypto.PublicKey	Base class of public keys for asymmetric cryptographic algorithms.
javacardx.crypto.RSA_CRT_PrivateKey	Base class of private keys for asymmetric cryptographic algorithms in connection with the Chinese remainder theorem (CRT).
javacardx.crypto.RSA_PrivateKey	Provides the methods for generating digital signatures using the RSA Algorithm.
javacardx.crypto.RSA_PublicKey	Provides the methods for verifying digital signatures using the RSA algorithm.
javacardx.crypto.RandomData	Provides a random number generator as a method.
javacardx.crypto.Sha1MessageDigest	Provides the hash algorithm SHA-1 as a method.
javacardx.crypto.SymKey	Base class of keys for symmetric cryptographic algorithms.
javacardx.cryptoEnc.DES3_EncKey	Extends '… .DES3_Key' with the methods for decryption in the Triple-DES ECB and CBC modes.
javacardx.cryptoEnc.DES_EncKey	Extends '… .DES_Key' with the methods for decryption in the DES ECB and CBC modes.

Software development for Java in Smart Cards

How does one go about developing a Java program for a Smart Card and then running it? The first thing the programmer does is to generate the actual Java source code using a text editor. He or she then compiles the source code using any desired Java compiler, which yields the machine-independent bytecode. Up to this point, the process is identical to Java programming for PCs.

The bytecode is now transferred to the off-card portion of the Java Virtual Machine (the Offcard VM) as a class file. The Offcard VM checks the format, syntax, field references and related aspects of the program. If all these checks are passed successfully, the Offcard VM creates what is known as a CAP file (card application file). This may then be provided with a digital signature if necessary, depending on the application. A digital signature provides assurance that the CAP file has been checked by the Offcard VM and is authentic. If there were no verifiable signature present, it would be possible to bypass the security of the Oncard VM using a manipulated applet, since there is not enough memory in the Smart Card to allow the Oncard VM to carry out all security checks itself. After this, the applet is loaded into the Smart Card in the CAP file format. The Smart Card first verifies the digital signature, which is usually present, and then passes the applet to the Oncard VM once it has been checked. What happens after this is largely the same as for program execution with a virtual machine in a PC. The Oncard VM tests and interprets the bytecode line by line, generating machine instructions for the Smart Card processor.

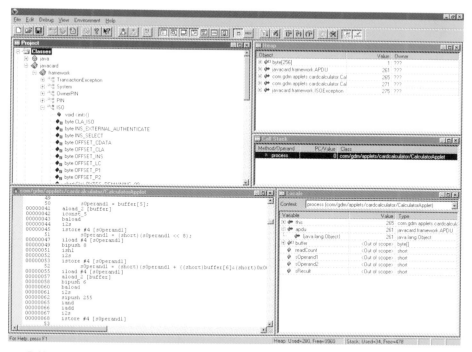

Figure 5.40   Sample development environment and Java Card simulator for developing and testing Java programs for Smart Cards. The classes and methods are shown in a tree chart at the top left. The window below this contains the Java source code and the translated byte code. The windows on the right show the heap, the stack and diverse variables. (Source: Giesecke & Devrient)

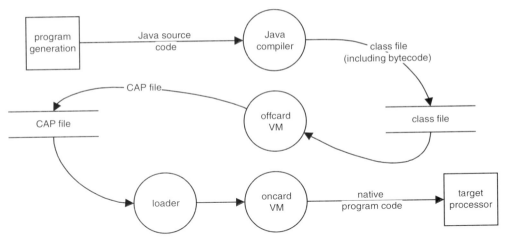

Figure 5.41   The normal process from program development to execution of the program by the Java virtual machine in the Smart Card microcontroller

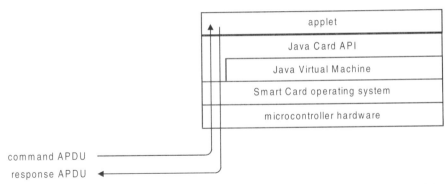

Figure 5.42   Data flows of the command APDU and the corresponding response APDU, with reference to the layer model of a Java Smart Card

In reality, the process is naturally somewhat more complicated than what has just been described. It is to be hoped that the programmer will not immediately start writing Java code after having been given the job, but instead will first use analysis and design techniques to determine the actual requirements. After this, he or she can start with the programming.

In order to quickly locate errors during and after the coding, the programmer uses a Java Smart Card simulator. This allows him or her to follow the execution of the code step by step, to examine variables and to easily and quickly make any necessary corrections.

In connection with this, a suite of tests is executed for relatively large projects and those that are critical for security. These tests check all successful results and the most important error results for commands and responses. Source code inspection by an independent party may also be included.

As you can see from this example, Java for Smart Cards significantly reduces the time needed to do the programming, and it also reduces the chance of errors as a side effect. However, coding by itself is only one of many parts of the development of a Smart Card application. The major advantage of Java for Smart Cards is that it makes it possible for many different programmers to develop executable programs for Smart Cards, rather than just a few software developers employed by the card manufacturers.

In order to produce Java applets for Smart Cards, you should take into account not only the particular features of the operating system but also the characteristic features of the Java Card 2.0 specification. These are listed and briefly described below.

Execution speed   Aside from its memory demands, the major point of criticism with regard to Java for Smart Cards is its low execution speed. However, it is relatively difficult to make fair comparisons between assembler programs and Java. The main reason for this is that it is not absolutely necessary to create the same processes in Java as in assembler, as long as the program behaves the same way at the interface to the terminal. For example, a file system is not always necessary with a Java program, and probably nobody would ever program a cryptographic algorithm in Java.

Another general consideration is that the methods in the Java Framework should used as much as possible, since they are in part coded in the native code of the target processor. This can result in a considerable increase in processing speed. As a guideline, the pure processing time, not including the time needed for data transfers, can be assumed to be a factor of 2–3 greater than that for normal assembler programs.

Application selection   Selecting a particular application in a Java Card amounts to selecting the relevant applet by means of its unique AID. The applet is called when it is selected, so that it can carry out any necessary initializations. After this, the applet automatically receives all command APDUs that are sent from the terminal to the Smart Card. If the applet is not selected, it is inactive and it is not involved in any data transfers.

Firewalls–keeping applications separate   From a computational point of view, the individual applets in a Smart Card are completely isolated from each other. Any possible mutual effects are prevented by the Security Manager of the Java Virtual Machine and the Smart Card operating system. However, an applet can make its own objects available to another applet if necessary. A typical example is a PIN, which should be uniformly valid for all applications (which means applets) in a card.

Transaction integrity–atomic processes   A sudden loss of power during a session must not be allowed to cause the data of an applet to take on undefined states. This is implicitly guaranteed by the virtual machine or operating system when an object is modified. However, if it is necessary to guarantee unconditional integrity across several objects or procedures, there are special mechanisms that can be used by the applet programmer. With these, it is possible to be sure that the objects in question either retain their original states or take on the new states.

File systems   It is not compulsory for an applet to have its own file system. With some applications, it can be fully adequate to create file-independent data objects that can be accessed using either standardized commands or commands defined by the programmer (private-use commands). The advantage of applets without file systems is once again related to the ever-present underlying need to conserve memory usage in Smart Cards. In addition, the object-oriented nature of Java allows data objects to be accessed by making calls to objects in accordance with their associated calling conditions. In this way, it is possible to implement accesses that meet very specific requirements of certain applications. For example, the application or the user could be given the opportunity to allow another party to inherit their access privileges.

Nonetheless, the usual present-day applications for both Smart Cards and PCs show that the data are very often stored and managed using file-oriented structures. This option is not excluded by the Java Card specification, since it offers its own classes for an ISO/IEC 7816-4 file structure. These interfaces also provide a basis for issuing compatible Java Smart Cards for existing applications with their standard file trees.

Deleting objects–persistent and transient objects All objects are created of their on accord as persistent objects in EEPROM when they are generated using the **new( )** method. Persistence refers to the capability of an object to exist past the end of a session; it is the opposite of transience. Persistent objects thus survive both the end of a session and a sudden loss of power without losing data or consistency. Any object exists only as long as there is a reference that points to it. If the reference is removed, the object is effectively no longer present, even though it still occupies memory. The only remedy for this would be a file manager with garbage collection, but this is not provided for in the Java Card 2.0 specification. It would take up too much memory.

It is possible to convert a persistent object into a transient object, which can then be placed in the RAM. The conversion is only possible in this one direction. The data in transient objects are lost at the end of the current session and are then re-initialized with their standard values.

Deleting applets The Java Card 2.0 specification provides no mechanism for deleting an applet in a Smart Card. The best that can be done is to block an applet by means of its own functions, but the memory that it occupies is forever lost for use by other applets.

However, there are generally special functions in test cards that can clear the entire applet memory area. This capability is only found within the context of debugging and testing, and not with 'real' cards issued for general use.

If future versions of the Java Virtual Machine for Smart Cards are able to include a garbage collector, it would then be possible to delete applets from the memory.

Cryptographic algorithms Many of the cryptographic algorithms that are currently in common use either work with modifications and exchanges at the bit level (such as DES) or utilize the arithmetic of long numbers (such as RSA). Smart Cards with Java are presently not suitable for programming such algorithms in Java, due to their low execution speeds and limited memory capacities.

Consequently, these cards usually have the *javacardx.crypto* class, which provides an interface (API) to the algorithms implemented in native machine code. This allows, for example, a true triple-DES encryption with three independent 56-bit keys to be programmed in Java by three successive calls to the available single DES algorithm with its 56-bit key. The level of data protection provided by such a triple-DES implementation in Java is only around 30% less than that provided by the use of native assembler code.

Cryptography and export restrictions In many countries, export permits are required for Smart Cards that have general-purpose operating systems and freely usable data encryption and decryption functions accessible via internal interfaces. This means that these types of

Smart Cards cannot be exported at all to certain countries, or that the exporter must wait several months for a suitable permit to be issued by the responsible agency.

Consequently, the classes for cryptographic functions are structured in Java Card such that they can be used directly for general data decryption and MAC computations but not for encryption. This is completely adequate for many applications, and thus a 'simplified' export permit procedure can be used in many countries.

However, if a particular application requires data to be encrypted, the card manufacturer can include the classes ....*cryptoEnc.DES3_EncKey* and ....*cryptoEnc.DES_EncKey*, which make encryption possible. From a purely cryptographic perspective, however, it is certainly possible to knock together easily implemented procedures that can be freely used for encryption and decryption, without making use of these 'encryption' classes.

Minimizing memory usage   Within the near future, Java programs for Smart Cards will continue to be strongly influenced by the amount of available memory. This leads to certain compromises in the programming that are not necessary for PCs. Table 5.18 lists a number of practical recommendations to help match Java programs as well as possible to the special demands of the Smart Card environment.

Table 5.17   Several useful programming guidelines regarding Java applets for Smart Cards, taken from and idea in [Schlumberger 97]. The objective of these guidelines is to make the best possible use of the strongly limited amount of memory in the Smart Card.

Guidelines for minimizing memory usage	Remarks
Use constants instead of variables.	Constants need less memory space than variables and are thus to be preferred.
Use simple data types.	Using the simplest possible data types saves memory.
Reuse variables.	Reuse variables as much as possible to make the most economical use of the limited amount of memory in the Smart Card.
Avoid local variables.	Local variables should be used as sparingly as possible.
Use a simple class hierarchy.	A simple class hierarchy that is as flat as possible saves memory in the Smart Card.
Use few arguments.	Methods should use no more arguments than are absolutely necessary.
Use a simple calling hierarchy.	A simple calling hierarchy for the methods saves memory in the Smart Card.
Eliminate unused variables.	The program should always be checked for unused constants, variables, methods and classes after it has been produced. The compiler may not always be able to remove these during optimization.

Summary and a look to the future

In spite of the continuing hype around Java, it should not be forgotten that it surely will not prove to be the remedy for every information technology problem of the past and future years, and that it will not always be able to fulfil all the expectations that have been invested in it. You only have to think about all the previous programming languages that are no longer in fashion, such as Pascal ('modular'), Lisp ('AI for everyone'), C ('portable') and C++ ('reusable code'). To be sure, these languages have improved the state of software technology by orders of magnitude, but many of their predicted positive effects failed to materialize. Nevertheless, it can be assumed certain that Java represents the start of a new Smart Card era, since it is the first language that allows everyone to run executable program code in Smart Cards. We must wait and see whether Java will in fact endure as the standard programming language for Smart Cards. This will be decided by the market, based on performance figures and the business politics of a few firms.

A few trends are clearly visible. There will probably be Smart Card terminals in the future that support Java. If these are on-line devices, then the versatile Java mechanisms could also be utilized with them for downloading programs via public networks. In addition, terminals usually have significantly more memory and more processing power than Smart Cards, which could eliminate many of the restrictions of Java Card within the terminal.

With Smart Cards, the resource hunger of Java will most likely mean that in the medium term, Java Card will be used only for small-scale applications, tests and field trials. Large projects using millions of Smart Cards incorporating Java virtual machines will probably be viable only after some years. In the meantime, an expedient solution is to port the Java code used in tests or field trials to assembler or C, and then produce conventionally programmed and less expensive Smart Cards on this basis. However, if in the future there is a strong demand from system operators for downloading supplementary applications after cards have been issued, this will promote the earlier large-scale use of Java cards.

5.11   The Small-OS Smart Card Operating System

The features and basic working principles of Smart Card operating systems are summarized in the previous sections of this chapter. For those readers who wish to immerse themselves more deeply into this subject, this section provides a detailed description of the internal interrelationships of a complete operating system. This is a classical Smart Card operating system according to the ISO/IEC 7816-4 or GSM 11.11 standard.

The name of the operating system described in this section is 'Small-OS', which is based on its very small memory demands and the fact that it can be run on hardware platforms that are not particularly powerful. It is written in a pseudocode that is similar to BASIC. The pseudocode presented here cannot be compiled as is, since some assignments have not been fully decoded to the last bit, but are explained only verbally. We do not wish to present page after page of true program code in a language such as C or C++, which can be incomprehensible and boring to read. Instead, our objective is to present a graphic example that illustrates the subject.

The comprehensibility of pseudocode greatly outweighs the advantages of using a real programming language, which are that it can be directly compiled and executed. With pseudocode, you do not get lost in the details of the implementation, and you can instead completely concentrate on the fundamental processes. The system presented here is platform-independent and is not tailored to any particular hardware. An implementation of Small-OS does however exist, but it is not programmed in assembler to run on a Smart Card microcontroller. Instead, it runs as a simulation in the program 'The Smart Card Simulator', which is available without charge via the Internet [Rankl, Hanser].

Table 5.18   Summary of the features of Small-OS

Name:	Small-OS
Typical application areas:	Multi-application with no own program code
Hardware prerequisites:	ROM: 8 kB, EEPROM: 1 kB, RAM: 128 bytes

Instruction set:	profile N per ISO/IEC 7816-4, with extensions
	• commands: SELECT FILE, READ BINARY, UPDATE BINARY, READ RECORD, UPDATE RECORD, VERIFY, INTERNAL AUTHENTICATE

Data transfer:	transmission protocol T=1
	• divider (CRCF) set to a fixed value of 372 (standard ISO/IEC 7816-3 setting)
	• PTS not supported
	• secure messaging not supported

File system:	one DF level
	• working EF structures: transparent, linear fixed
	• internal EF structure: linear variable (for PINs and keys)
	• one EF Key allowed per directory (i.e. per MF or DF)
	• no dynamic file system (no file deletion or creation, no free memory management)
	• maximum size of a transparent EF: 255 bytes

State machine:	independent secure states for the MF and DFs
	• secure states: 256 (0–255)
	• initial state = 0
	• only one allowed input state for using a PIN or a key
	• only one allowed input state for file access (i.e. '<' and '≥' comparisons are not allowed)
	• no cross-level key access (the EF Key is always chosen via the currently selected directory (MF or DF))
	• PIN addressing: 2 PINs maximum (ref. no. 1 and 2)
	• key addressing: 31 keys maximum (ref. no. 1–31)
	• The PIN and key retry counters allow a maximum of 15 unsuccessful attempts.

Cryptographic algorithm:	DES

Program code:	Generation and loading by external parties is not possible.

Programming in pseudocode

First we would like to make a few remarks about programming style and the programming of Small-OS. The pseudocode, which is derived from BASIC, in principle represents a semi-formal description of the Smart Card operating system Small-OS. Similar operating system characterizations are often used for software evaluations according to the ITSEC. They are used as the basis for the evaluation and for checking the source code. The pseudocode that is presented in this section thus represents a good example of how formalized processes within a Smart Card operating system are portrayed. The pseudocode is listed here in tables with extensive comments. Similar forms of presentation can be found in the EN 1546 series of standards, for example, in which the internal processes of Smart Cards for electronic purses are semi-formally described.

The individual terms used in the pseudocode are described at the beginning of the book.[8] The program code is oriented toward the standard dialects of BASIC, with object-oriented extensions. Only generally understandable constructs are used. All labels, constants and references are in English. Numerical values are usually given in hexadecimal form using ISO notation (such as '42'). However, decimal or binary forms are used where necessary to aid understanding, using a notation that is derived from the ISO notation. For example, all countable values, such as length specifications, are shown in decimal form.

Nobody would program a Smart Card operating system in this form in assembler or C, since it would be far too complicated. One of the design objectives with Small-OS was to create a simple yet powerful Smart Card operating system in a manner that is most easily understood and well commented. Intentionally, no attempt was made to minimize program execution time, program code, RAM usage or stack depth, since doing so would seriously impair the readability of the code. In real Smart Card operating system programming, for example, it is sometimes common practice to use a JUMP instruction instead of a CALL instruction for calls to rarely used subroutines, since this saves two bytes of expensive stack space. A flag that is set before the subroutine is called with the JUMP instruction is used to determine the return address when the subroutine process is completed. This sort of optimization is not found in Small-OS, for the reason just given. The pseudocode has been optimized only with regard to readability and comprehensibility. The resulting deviations from real Smart Card operating systems are identified wherever they occur, either in the text or in the commented pseudocode.

The majority of a Smart Card operating system is located in ROM, due to the notorious shortage of memory in Smart Card microcontrollers. It thus cannot be modified after the chip has been manufactured. Software cannot be produced with absolutely no errors (except for trivial miniprograms), but only with as few errors as possible. An error in ROM program code would have serious consequences. To make it possible to use bug fixes to patch such errors, jumps to EEPROM are provided at critical locations in the ROM. This technique is very old in software engineering and is not specific to Smart Cards. Nearly identical mechanisms are used in the MacOS and OS/9, for example. Locations in EEPROM that are called by the program code, called 'handles', are used for this. A handle normally contains only a RETURN instruction, which causes an immediate return to the calling code. If a bug fix for the ROM code is necessary, corrective code is inserted in the EEPROM at a handle location. In this case, the call to the handle does not produce an immediate return. This mechanism is not included in Small-OS, for reasons of simplicity and understandability.

The relatively detailed description of the Smart Card operating system provides an interesting opportunity to follow several typical types of attacks directly in pseudocode. At suitable locations, possible attacks and defensive measures at the operating system level are dealt with in detail. For example, it is possible to examine an attack on the PIN by comparing processing times, which has now become a 'classic' form of attack, in all of its details in the pseudocode. A comprehensive listing of typical attacks and their defenses is given in Chapter 8 (Security Techniques).

---

[8] See 'Program Code Conventions' at the front of the book.

Table 5.19   Small-OS design criteria in order of priority

Priority	Criterion	Reason
1	compatibility with ISO/IEC 7816-4	• the international industrial standard for Smart Card operating systems
2	robustness	• high level of reliability • high level of error tolerance
3	low level of complexity	• high level of reliability • easily understood working methods
4	modular construction	• high level of reliability • high level of error tolerance • ease of extension
5	small memory demand	• an absolute must for Smart Card operating systems
6	multi-application operating system	• standard capability nowadays
7	no program code downloading	• high level of reliability • low level of complexity
8	similar to real Smart Card operating systems	• Small-OS is a teaching example with reference to the real world.

Design criteria

The above considerations lead to the following design criteria that guided the conception and programming of Small-OS. Small-OS should be a simple Smart Card operating system that does not need a lot of program code and that has a low level of complexity, just like the real models that inspired it. This makes its structure comprehensible and easy to grasp. It is built up in a strictly modular fashion, which means that it can be directly extended with additional commands at a reasonable effort. The file system and the supported commands are without exception compliant with the international ISO/IEC 7816-4 standard. The 'N' profile of ISO/IEC 7816-4 was chosen as an option, with some extensions that are commonly present in the Smart Card world.

Small-OS is thus intended to be used in situations in which it is not necessary to download applications after the cards have been issued. However, depending on the amount of available memory, several different applications could be run in the Smart Card independently of each other. This means that Small-OS is a multi-application operating system. However, it is not possible to download program code to the card and then execute it in the card. In summary, Small-OS is comparable to the first general Smart Card operating systems, such as are still used in various forms for GSM applications.

The ISO/IEC 7816-4 standard describes a basic file system and several fundamental commands for Smart Cards, among other things. In this way, it primarily characterizes the interface of the Smart Card, rather than the internal architecture of the operating system. In addition, there are numerous options and a few passages that unfortunately are subject to interpretation. The number of possible variations that are thus made possible must be sharply reduced in practice by specifications, such as the GSM 11.11 specification, to ensure compatibility between different implementations.

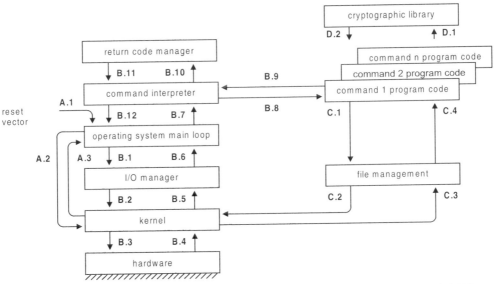

Figure 5.43   The layer structure of Small-OS and the resulting calling scheme. Calls labeled 'A' are used when the operating system starts up, those labeled 'B' are used to call commands and those labeled 'C' are used as necessary within commands for accessing the file system. Calls labeled 'D' are used with cryptographic algorithms. The numbers indicate the sequence of the program calls.

With actual Smart Card operating systems, therefore, the designation 'ISO/IEC 7816-4 compatible' by no means signifies that they behave exactly the same way in all respects. This would need a detailed specification, which would again in part be an individual interpretation of the ISO/IEC 7816-4 standard. Consequently, in practice the behavior usually the same only if the command is successfully executed, while there are various differences in case of an error. With Small-OS, interpretations of the ISO/IEC 7816-4 standard are usually identified as such in the pseudocode. Within the limits of the interpretation of the standard, Small-OS is truly compliant with the ISO/IEC 7816-4 standard and corresponds to the normal interpretations of the standard in the Smart Card industry as much as possible.

Major differences between operating systems are frequently to be found in regard to return codes. Since the usage and priority of the individual return codes are not described in detail in ISO/IEC 7816-4, assumptions must be made. Small-OS, in contrast to all other operating systems, at least has the advantage that the code is public. This means that it is possible to determine in detail which return code is generated at every point.

If one were to implement Small-OS for a Smart Card microcontroller, it would need around 5–6 kB of ROM, 128 bytes of RAM and at least 1 kB of EEPROM, depending on the number of applications present (using an 8051 processor). This assumes as general conditions that only the T=1 data transmission protocol is used and that DES is used as the cryptographic algorithm. If certain applications were to need more memory, a microcontroller with more EEPROM could be used without any problems. This would not affect the operating system or require any modifications.

File access

The file system described in the ISO/IEC 7816-4 standard allows an enormous number of options with regard to access conditions and key management. For Small-OS, therefore, a solution that is often used in practice for multi-application operating system has been chosen. The file headers of working EFs contain prescribed states for each of the various access commands (for example the condition ... .*AccessCondition.Read* for the READ command). Each state is described by a positive integer. This means that each EF has independent state values for read and write accesses in its header. These values must be achieved within the current directory before the command can be executed. State 0 represents the base state (idle), so that a state condition of 0 means that all accesses are allowed.

State variables are always assigned to the MF and the currently selected DF *SecurityState.MF* and *SecurityState.DF*). These can be altered by commands relating to security (VERIFY and INTERNAL AUTHENTICATE). In case of an access to a specific EF, the current state in the directory is compared with the required state in the EF file header. If the actual and required states are the same, then the file may be accessed for writing or reading.

Real Smart Card operating systems often allow a large variety of less-than, greater-than, greater-than-or-equal and not-equal comparisons to be made here. It is also frequently possible to define several possible access states independently in the file header. This further complicates a process that is already not exactly simple, and for this reason it is not used in Small-OS. In principle, however, Small-OS could be extended to allow such comparisons without any structural modifications.

Access to internal secrets (PINs and keys)

All PINs and keys are held in special internal EFs. Such EFs are called EF Key here. This type of file can be read or written only by the operating system itself. External selection or access is not possible. There are no mechanisms at all in the design that would allow external access to such EFs. This is a part of the security philosophy of the Small-OS Smart Card operating system.

Only one EF Key file can be created for each directory. It automatically has a linear variable structure, so that it can store PINs and keys of various lengths in a minimum amount of memory. Each record in an EF Key file contains either a PIN or a key, with an address number that is unique within the file (... .*KeyNo*). For each secret object (PIN or key), a state value is stored. This defines the state (... .*EntryState*) that is required for a command to be used (VERIFY or INTERNAL AUTHENTICATE). Following the execution of a command, the result state (... .*ResultState.OK* or ... .*ResultState.NOK*) is set in the directory to which the EF belongs according to the result achieved (such as PIN comparison successful or not successful). In addition, a retry counter (... .*RCntr*) is assigned to each secret object. This counter is incremented for each unsuccessful result until it reaches its maximum value. If the retry counter has reached its maximum value (... .*RcntrMaxValue*), the associated secret object can no longer be used.

Some Smart Card operating systems allow 'cross-level' access to keys. This allows keys stored in the next higher directory to be accessed from within a particular DF. The mechanisms needed for this are not included in Small-OS, since the functionality would not justify the amount of code needed to implement them. Key accesses via aliases are also not implemented, for the same reason.

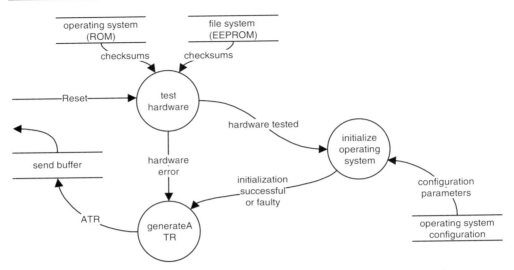

Figure 5.44   Small-OS: data flow diagram of the startup and test processes of the operating system

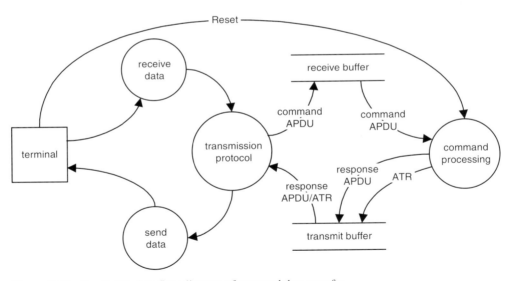

Figure 5.45   Small-OS: data flow diagram of reset and data transfer

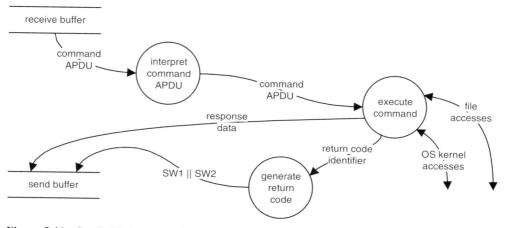

Figure 5.46   Small-OS: low-level data flow diagram of the command processing process

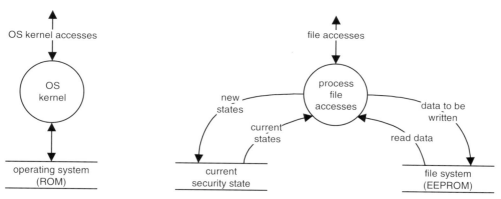

Figure 5.47   Small-OS: low-level data flow diagram of the process for file and OS kernel accesses

Figure 5.48   The symbols used for the data flow diagrams for Small-OS. A *terminator* identifies an object located outside of the system under consideration that exchanges data with the system. A *process* is located inside the system under consideration. It processes input data streams and generates output data streams. A *data store* is a storage place for data that can be read and written. A detailed introduction to this type of system analysis is contained in Robertson [Robertson 96].

Small-OS constants

Basically, constants are used in the pseudocode as much as is reasonable for numerical and non-numerical values. This considerably increases the readability of the code, and it is purely and simply good programming style. All constants are identified by the prefix C_. The values of these constants normally depend on the target hardware or the implementation, so they are not further defined here. The constants for the 2-byte return codes all have the prefix C_RC_. Table 5.21 lists the associated return codes. In practice, the constants of an operating system are usually stored in unalterable form in the ROM.

Table 5.20   Constants used in Small-OS (excluding return code constants)

Constant	Meaning
C_Error	Constant for general errors (e.g. with a subroutine call)
C_InvalidPointer	Constant for the value of an invalid pointer
C_Equal	Constant for comparisons OR compared objects are equal
C_NotEqual	Constant for comparisons OR compared objects are not equal
C_Found	Constant for search functions OR sought object found
C_NotFound	Constant for search functions OR sought object not found
C_AccessDenied	Constant for access conditions, access denied
C_AccessAllowed	Constant for access conditions, access allowed
C_WriteError	Constant for a data write error (e.g. writing to EEPROM)
C_EFTypeWorking	Constant for a working EF
C_EFTypeInternal	Constant for an internal EF
C_EFStrucLinFix	Constant for a linear fixed EF
C_EFStrucTransp	Constant for a transparent EF
C_CmdVERIFY	Constant for the VERIFY command
C_CmdINTAUTH	Constant for the INTERNAL AUTHENTICATE command

Table 5.21 (part 1)   Return code constants used in Small-OS

Constant	Content	Meaning of the return code
Process completed–warning processing		
C_RC_CounterX	'63Cx'	A counter (usually the retry counter) has been incremented and now has the value 'x'
Process aborted–execution error		
C_RC_MemoryFailure	'6581'	EEPROM write error
Process aborted–check error		
C_RC_WrongLength	'6700'	Incorrect length
C_RC_CmdIncompFStruc	'6981'	command incompatible with file structure

Table 5.21 (part 2)   Return code constants used in Small-OS

Constant	Content	Meaning of the return code
Process aborted–check error (continued)		
C_RC_SecStateNotSatisfied	'6982'	Security state not satisfied
C_RC_AuthMethodBlocked	'6983'	Authentication method blocked
C_RC_CondOfUseNotSatified	'6985'	The conditions for using the data element (PIN or key) are not satisfied
C_RC_CmdNotAllowed	'6986'	Command not allowed
C_RC_FctNotSupported	'6A81'	Function not supported
C_RC_FileNotFound	'6A82'	File not found
C_RC_RecordNotFound	'6A83'	record not found
C_RC_LcInconsistentP1P2	'6A87'	Lc inconsistent with P1 or P2
C_RC_RefDataNotFound	'6A88'	Data referenced in command not found
C_RC_WrongP1P2	'6B00'	P1 or P2 incorrect
C_RC_INSNotSupported	'6D00'	Command not supported
C_RC_CLANotSupported	'6E00'	Class not supported
C_RC_FatalError	'6F00'	Internal error with no further description
Process completed–normal processing		
C_RC_OK	'9000'	Command successfully executed

Small-OS variables

The variables in Small-OS can basically be divided into RAM and EEPROM variables. The RAM variables are re-initialized each time the Smart Card is reset and retain their values only for the duration of one session. However, data can be stored in RAM variables without any loss of time, and the number of write cycles is unlimited. EEPROM variables, by contrast, are typically used primarily for the implementation of the file manager, for which data contents and access conditions must exist between sessions as well.

Small-OS RAM variables   The regions for the transmit and receive buffers take up the majority of the RAM variables. All data elements of an APDU can be addressed via their own specific variables. In order to manage with the small amount of available RAM, the transmit and receive buffers are in practice sometimes made partially overlapping. Some commands must then learn from the operating system exception handler that data can be written to the transmit buffer only after all the data in the receive buffer has been processed. In order to promote understandability, no such memory minimization is used here, and the transmit and receive buffers are thus completely separate from each other.

The second large group of RAM variables relates to the management of the file tree. Several pointers are provided, which point to the current directory (... .*Ptr.CurrentDF*), the current file (... .*Ptr.CurrentWEF*) and the current valid key file (... .*Ptr.CurrentIEF.Key*). With record-oriented EFs (linear fixed EFs), the currently selected record is also assigned a pointer (... .*Ptr.CurrentRecord*). All pointers are identified by the prefix *Ptr*, and are explicitly set to the value C_InvalidPointer whenever they are not allowed to be used.

The two variables *SecurityState.MF* and *SecurityState.DF* identify the security state in the MF and the currently selected DF, respectively. The security states that files have achieved are stored in these variables as positive integers. Here the value '0' indicates that no security state has been achieved. This value is automatically set when Small-OS is initialized after a reset.

Additional RAM memory is needed for the program stack, the DES cryptographic algorithm and working registers. Here it is assumed that these are implicit, so no special variables are assigned for them.

Small-OS EEPROM variables   For the sake of simplicity, the file tree in the Smart Card is implemented as a multidimensional array. This approach would be much too memory-intensive for use in a real Smart Card operating system. The basic structures that are usually used in real systems for file management are one-way linked lists. The lengths of the list elements can be kept variable with the help of TLV encoding. This minimizes the amount of memory needed for file management, since only the necessary data elements have to be stored in memory. The size of the file headers for DFs and EFs, with all the information necessary for these files, is normally between 20 and 40 bytes. However, the multi-dimensional array file structure used in Small-OS can without question provide a simpler and more efficient solution when the operating system is implemented in a high-level language and runs in a hardware environment with relatively few memory limitations. A nearly identical construction is used in the 'Smart Card Simulator' program, for example.

Table 5.22   Variables used in Small-OS for data transfers to and from the Smart Card. The listed variables are typically held in RAM.

Variable	Description
.1 … .8	bitwise reading or writing of a variable (for example, *APDU.Cmd.CLA.1* corresponds to bit 1 of the class byte in the command APDU)
APDU.Cmd	Command APDU received from the Smart Card
APDU.Cmd.CLA	Class byte of the command APDU
APDU.Cmd.INS	Command byte of the command APDU
APDU.Cmd.P1	Parameter 1 of the command APDU
APDU.Cmd.P2	Parameter 2 of the command APDU
APDU.Cmd.Lc	Long command byte of the command APDU (optional)
APDU.Cmd.Data[…]	Data portion of the command APDU with length 1 … n (optional)
APDU.Cmd.Le	Expected length of the response APDU (optional)
APDU.Rsp	Response APDU to be sent by the Smart Card
APDU.Rsp.Data[…]	Data portion of the response APDU with length 1 … *n* (optional)
APDU.Rsp.SW1	Status word 1 of the response APDU (byte 1 of the return code)
APDU.Rsp.SW2	Status word 2 of the response APDU (byte 2 of the return code)
Returncode	*Return code := APDU.Rsp.SW1 || APDU.Rsp.SW2*

Table 5.23   Variables used in Small-OS for file management and file access. The listed variables are typically held in RAM.

Variable	Description
Ptr.MF	Pointer in the Smart Card operating system; always points to the MF
Ptr.CurrentDF	Pointer in the Smart Card operating system; points to the currently selected DF or to the MF
Ptr.CurrentIEF.Key	Pointer in the Smart Card operating system; points to the current internal EF that holds the key for the currently selected DF (EF Key)
Ptr.CurrentWEF	Pointer in the Smart Card operating system; points to the currently selected working EF in the currently selected DF or the MF.
Ptr.CurrentRecord	Pointer in the Smart Card operating system; points to the current record in a record-oriented EF. This pointer is not valid if a transparent file is selected.
SecurityState.MF	Achieved security state of the MF
SecurityState.DF	Achieved security state of the currently selected DF

Data structures

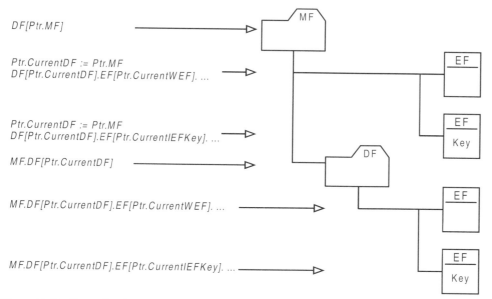

Figure 5.49   Example of the data structures that are available within Small-OS for the addressing of directories (MF and DF) and files (EFs).

Table 5.24   General data structures used in Small-OS for file management. The listed variables are typically held in EEPROM.

Variable	Description, contents and size
Data structures for the MF and the DFs	
DF[...].FID	File identifier length: 2 bytes, content: 0 ... 255 each element
DF[...].DFName	DF name (includes the application identifier) length: 1 ... 16 bytes; content: 0 ... 255 each element
DF[...].LenDFName	length of the DF name length: 1 byte; content: integer $\in$ {1 ... 16}
Data structures for EFs	
DF[...].EF[...]. ...	File tree in the form of a 2-dimensional matrix
DF[...].EF[...].FID	File identifier length: 2 bytes; content: 0 ... 255 each element
DF[...].EF[...].Structure	EF structure content: property (transparent / linear fixed)
DF[...].EF[...].Type	EF type content: property (working / internal)
DF[...].EF[...].AccessCondition.Read	Access condition for which the READ command is allowed for the EF content: 0 ... 255
DF[...].EF[...].AccessCondition.Update	Access condition for which the UPDATE command is allowed for the EF content: 0 ... 255

Table 5.25   Data structures used in Small-OS for managing working EFs with transparent and linear fixed structures. The listed variables are typically held in EEPROM.

Variable	Description, contents and size
Data structures for EFs with transparent structures	
DF[...].EF[...].TransparentData[1 ... n]	Data content of a transparent file length: n (= ... .TransparentDataSize) content: 0 ... 255 each element
DF[...].EF[...].TransparentDataSize	Data size of a transparent file content: 0 ... 255
Data structures for EFs with linear fixed structures	
DF[...].EF[...].Record[...].Data[1 ... n]	Data content of a single record of a linear fixed file length: 1 ... n (n = ...Size) content: 0 ... 255 each element
DF[...].EF[...].Record[...].Size	Length of a single record of a linear fixed file (same for all records in a linear fixed file) content: 1 ... 255
DF[...].EF[...].NoOfRecords	Number of records in a linear fixed file content: 1 ... 255

Table 5.26   Data structures used in Small-OS for managing internal EFs with linear variable structure, which are used to store PINs and keys. The listed variables are typically held in EEPROM.

Variable	Description, contents and size
Data structures for EFs for storing keys	
DF[...].EF[...].Record[x].KeyData	PIN or key in a linear variable file
	length: *n* (= ....KeySize)
	content: 0 ... 255 each element
DF[...].EF[...].Record[x].KeySize	Length of the PIN or key in bytes
DF[...].EF[...].Record[x].KeyNo	Number of the PIN or key
	content for a PIN: 1; 2
	content for a key: 1 ... 31 each element
DF[...].EF[...].Record[x].RCntr	Retry counter for a PIN or key
	content: 0 ... 15
DF[...].EF[...].Record[x].RCntrMaxValue	Maximum retry counter value for a PIN or key
	content: 1 ... 15
DF[...].EF[...].Record[x].KeyPurpose	Utilization of the EF Key records
	(VERIFY or INTERNAL AUTHENTICATE)
DF[...].EF[...].Record[x].EntryState	Necessary state for using a PIN or key
	content: 0 ... 255
DF[...].EF[...].Record[x].ResultState.OK	State following successful use of a PIN or key
	content: 0 ... 255
DF[...].EF[...].Record[x].ResultState.NOK	State following unsuccessful use of a PIN or key
	content: 0 ... 255

The main loop and the initialization of the hardware and operating system

After the Smart Card is reset, the program counter is loaded with the address of the reset vector, and the processor then starts to execute the first program code. The hardware is tested right away to see if it is in good working order, just as with PCs. This consists of a RAM test to start with, followed by the calculation of several checksums for the contents of the ROM and EEPROM. If an error occurs here, the Smart Card attempts to send a special error ATR, following which it does nothing except to wait in an endless loop for the next reset. The success of the attempt to send the error ATR depends primarily on the nature of the error that has occurred. If the error is serious and affects the RAM or the program code of the transmission routine, only a garbled ATR or no ATR at all can be sent.

After the chip hardware has been tested and initialized, the initialization of the operating system starts. The important aspects of this are setting the data transfer parameters for the T=1 protocol, automatically selecting the MF and establishing the security conditions for file accesses. If the MF cannot be found, this is a serious error, and Small-OS then terminates all further actions after sending an error ATR.

Listing 5. 2   Small-OS program code: operating system initialization and the main loop

Resetvector:	Entry point following a CPU reset.
CALL Initialize_Hardware	
CALL Initialize_Operating_System	
CALL IO_Manager_Send_ATR	
Main_Loop:	Main loop of the Smart Card operating system.
CALL IO_Manager_Receive_APDU	
CALL Command_Interpreter	
CALL IO_Manager_Send_APDU	
GOTO Main_Loop	

The I/O manager

The entire process of transferring data from and to the terminal is handled by the I/O Manager. Two hardware-independent program routines in the core of the operating system called Kernel_IO_SendByte and Kernel_IO_ReceiveByte form the basis for transmitting and receiving messages. All other parts of the I/O program code are independent of the target hardware for Small-OS. The ATR and the T=1 transmission protocol are completely executed in the I/O Manager for both successful and unsuccessful instruction processing. A dedicated state machine had to be developed for this, primarily to handle the relatively complicated procedures of the T=1 protocol. The ISO/IEC 7816-3 Amd 1 standard describes the reactions of the I/O manager in case both of success and of failure by means of many examples. With manually optimized assembler code, a good I/O manager requires at least one kilobyte of memory, which is usually located in ROM.

Real I/O managers preferably use only the RAM for the I/O buffer, since write accesses to the RAM take place at the full speed of the CPU. However, there are now many Smart Card operating systems whose transmit and receive buffers are larger than the available amount of RAM. In such cases, a certain part of the EEPROM is used as an extension to the data transfer buffer when the amount of data exceeds a certain threshold. This avoids any restrictions on the size of the I/O buffer, but the price for this is a considerably reduced data transfer rate, due to the time required to write to the EEPROM. In addition, with this sort of extended I/O buffer, there is a danger that the buffer could reach the end of its useful life in a relatively short time, due to frequent write accesses to the EEPROM. In spite of these limitations, this is the only technically feasible means to make the data transfer buffer larger than the amount of available RAM.

Listing 5. 3   Small-OS program code: routines for data transfers to and from the Smart Card

IO_Manager:	Hardware-related subroutines for data exchange with the terminal via the serial I/O line.
IO_Manager_Send_Error_ATR:	Subroutine for sending a special ATR that
Transmit an ATR that indicates a serious _	indicates a serious error in the Smart Card
operating system error	hardware or the operating system. Following the
Error_ATR:	ATR, the operating system is suspended in an
GOTO Error_ATR	endless loop.

IO_Manager_Send_ATR: ... RETURN	Subroutine that codes all the parameters of the transmission protocol into an ATR string and transmits the string. Uses the operating system core routine Kernel_IO_SendByte.
IO_Manager_Send_APDU: ... RETURN	Subroutine that converts an APDU already present in the I/O buffer into a TPDU, according to the chosen transfer parameters, and then sends the TPDU to the terminal via the serial I/O line. The mechanism for correcting data transfer errors may be used as necessary for the transfer. Uses the operating system core routine Kernel_IO_SendByte.
IO_Manager_Receive_APDU: ... *APDU.Cmd* := TPDU received from the _ terminal and converted ... RETURN	Subroutine that receives a TPDU from the terminal via the serial I/O line according to the chosen transfer parameters, converts it into an APDU and then stores it in the I/O buffer. The mechanism for correcting data transfer errors may be used as necessary for the transfer. Uses the operating system core routine Kernel_IO_ReceiveByte.
IO_Manager_Set_Transmission_Parameter: ... set clock rate conversion factor to 372 set convention set the length of the transmit buffer set the length of the receive buffer ... RETURN	Subroutine that sets the parameters for serial data transfers.

Listing 5. 4   Small-OS program code: initializing the microcontroller hardware and the Smart Card operating system

Initialize_Hardware:	Subroutine for initializing the Smart Card hardware.
CALL Kernel_CheckRAM IF STATUS (Kernel_CheckRAM = C_Error) THEN (GOTO IO_Manager_Send_Error_ATR)	Test the RAM to see if it is in good working order.
CALL Kernel_DeleteRAM IF STATUS (Kernel_DeleteRAM = C_Error) THEN (GOTO IO_Manager_Send_Error_ATR)	Set the entire RAM to the value '00' in order to have a defined initial state after each reset. An intentional side effect is that all variables in RAM are initialized.
CALL Kernel_Check_EDC_ROM IF STATUS (Kernel_Check_EDC_ROM = C_Error) THEN (GOTO IO_Manager_Send_Error_ATR)	Check the error detection codes (EDCs) at various places in the ROM.
CALL Kernel_Check_EDC_EEPROM IF STATUS (Kernel_Check_EDC_EEPROM = C_Error) THEN (GOTO IO_Manager_Send_Error_ATR) RETURN	Check the error detection codes (EDCs) at various places in the EEPROM.

```
Initialize_Operating_System:                          Subroutine for initializing the
    CALL IO_Manager_Set_Transmission_Parameter       Smart Card operating system.
    CALL SELECT_FILE_MF
    IF STATUS (SELECT_FILE_MF = C_Error) THEN (
        GOTO IO_Manager_Send_Error_ATR)
    SecurityState.MF := 0
    SecurityState.DF := 0
RETURN
```

Operating system kernel

All of the routines that belong to the core of the operating system are collected together in the OS kernel. Most of these routines are either hardware-dependent or rather time-critical, and must therefore be adapted when the program is ported to a different hardware platform. The functions provided by these routines are in each case described by their names. Some of these subroutines can influence aspects of the system that are of concern to its security. For example, the execution time of a subroutine such as Kernel_CompareByteString should not vary depending on the result of the comparison, since any such variations could be used as the basis for determining the internal results of PIN computations by measuring elapsed times. Nowadays, this type of attack has been eliminated by incrementing the retry counter before each PIN comparison as a precautionary measure, but not long ago this was a very promising way to attack a PIN.

Listing 5. 5   Small-OS program code: hardware-dependent core routines

OS Kernel:	Core hardware-related subroutines of the operating system.
Kernel_Check_EDC_x: ... RETURN	Subroutine for computing checksums (EDCs) to test the internal consistency of memory regions in the ROM or EEPROM. x ∈ {ROM, EEPROM}
Kernel_x: ... RETURN	Subroutine for elementary functions. x ∈ {CopyByteString, CompareByteString, DeleteRAM, CheckRAM}
Kernel_DES_x: ... RETURN	Subroutine for DES encryption and decryption. x ∈ {Encrypt, Decrypt}
Kernel_IO_x: ... RETURN	Subroutine for sending and receiving a single byte via the serial interface of the microcontroller. x ∈ {SendByte, ReceiveByte}

The command interpreter

The Small-OS command interpreter is constructed in a relatively simple manner. It works on the principle of a dispatcher that, based on the class and instruction bytes, calls a routine that processes the recognized command. This implementation uses little memory and has the additional important advantage that it is relatively easy to integrate new commands into the operating system. This can be done by just adding a few lines of code to the command interpreter. These identify the new command and the associated subroutine call. With suitable code in place, the new command will be recognized and executed as necessary.

The structures of command interpreters that are commonly used in practice, however, are far more complicated. This is partly because they must run in unalterable ROM, but also because it must be possible to download program code into the EEPROM when the operating system is completed. This downloaded code must then be recognized and called when the program runs. The principle that is used for this is a jump table located in the EEPROM, which can be extended as necessary when the card is completed.

Fixed polling of a particular class byte, as used in the Small-OS command interpreter, only makes sense in operating systems that do not support either secure messaging or logical channels, and which also support only a single command class conforming to the ISO/IEC 7816-4 standard. In all other cases, the class byte is used to identify previously defined options, so it need not be the same for all commands.

Listing 5. 6   Small-OS program code: the command interpreter

Command_Interpreter:	Command interpreter

IF *APDU.Cmd.CLA* <> '00' THEN (
  Returncode := C_RC_CLANotSupported
  GOTO Command_Interpreter_Exit)
	If the class byte is not the same as the class for ISO/IEC 7816-4 commands, abort command processing and set the corresponding return code.

IF *APDU.Cmd.INS* = 'A4' THEN (
  CALL SELECT_FILE
  GOTO Command_Interpreter_Exit)
	If the command SELECT FILE has been sent to the Smart Card, call the corresponding subroutine.

IF *APDU.Cmd.INS* = 'B0' THEN (
  CALL READ_BINARY
  GOTO Command_Interpreter_Exit)
	If the command READ BINARY has been sent to the Smart Card, call the corresponding subroutine.

IF *APDU.Cmd.INS* = 'D6' THEN (
  CALL UPDATE_BINARY
  GOTO Command_Interpreter_Exit)
	If the command S UPDATE BINARY has been sent to the Smart Card, call the corresponding subroutine.

IF *APDU.Cmd.INS* = 'B2' THEN (
  CALL READ_RECORD
  GOTO Command_Interpreter_Exit)
	If the command READ RECORD has been sent to the Smart Card, call the corresponding subroutine.

IF *APDU.Cmd.INS* = 'DC' THEN (
  CALL UPDATE_RECORD
  GOTO Command_Interpreter_Exit)
	If the command UPDATE RECORD has been sent to the Smart Card, call the corresponding subroutine.

IF *APDU.Cmd.INS* = '20' THEN (
  CALL VERIFY
  GOTO Command_Interpreter_Exit)
	If the command VERIFY has been sent to the Smart Card, call the corresponding subroutine.

IF *APDU.Cmd.INS* = '88' THEN (
  CALL INTERNAL_AUTHENTICATE
  GOTO Command_Interpreter_Exit)
	If the command INTERNAL AUTHENTICATE has been sent to the Smart Card, call the corresponding subroutine.

Returncode := C_RC_INSNotSupported	The command sent to the Smart Card is not supported.
Command_Interpreter_Exit: CALL Returncode_Manager	Set the return code in the I/O buffer as specified by the program code for the command.
RETURN	

The return code manager and the file manager

The return code manager uses a passed-in value to look up the return code in a table, and then adds it to the end of any data block in the transmit buffer. A dedicated manager is used for this for two reasons. First, it is better programming style to have all have all return codes in one central location. For example, this makes it possible to substitute another return code for a particular return code, easily and in one central place, in case of a software error. This can sometimes be essential. Second, using a return code manager saves precious memory. It is easy to calculate that storing each return code only once in a common table takes up less memory than repeating it everywhere it is needed.

In real Smart Card operating systems, all file accesses take place via a central file manager, for reasons of security. Frequently, the file manager also computes the checksum of the file header when the file is accessed. In a concession to simplicity, Small-OS accesses the multi-dimensional variables of the file tree directly when a command is processed, without separating levels or first computing a checksum. The only check that it makes for all commands is to test whether access to the file is allowed. Read and write accesses to files with a complete file manager would take place in a similar manner.

Listing 5. 7   Small-OS program code: the return code manager and file manager

Returncode_Manager: ... RETURN	Manager for setting the return code. Sets the return code based on the value received and a return code table.

File_Manager_CheckACRead:
 	
   Status := C_AccessDenied
   IF *Ptr.CurrentDF = Ptr.MF* THEN (
     // MF is selected
     IF .*EF[Ptr.CurrentWEF].AccessCondition.Read* = _
     SecurityState.MF THEN (
       Status := C_AccessAllowed))
   ELSE (
     // DF is selected
     IF.*EF[Ptr.CurrentWEF].AccessCondition.Read* = _
     SecurityState.DF THEN (
     Status := C_AccessAllowed))
RETURN
	File manager for checking the 'file access condition for 'read'
	
	Set the variable for the access status to the initial value. Test whether the security state required for reading the selected EF with READ BINARY or READ RECORD has been achieved. The security state that must be achieved depends on the currently selected directory (MF or DF).

File_Manager_CheckACUpdate:
 	
   Status := C_AccessDenied
   IF *Ptr.CurrentDF = Ptr.MF* THEN (
     // MF is selected
     IF.*EF[Ptr.CurrentWEF].AccessCondition.Update* = _
     SecurityState.MF THEN (
       Status := C_AccessAllowed))
   ELSE (
     // DF is selected
     IF.*EF[Ptr.CurrentWEF].AccessCondition.Update* = _
     SecurityState.DF THEN (
     Status := C_AccessAllowed))
RETURN
	File manager for checking the file access condition for 'update'.
	
	Set the variable for the access status to the initial value. Test whether the security state required for reading the selected EF with UPDATE BINARY or UPDATE RECORD has been achieved. The security state that must be achieved depends on the currently selected directory (MF or DF).

The basic structure of the program code for Smart Card commands

The basic functions of Smart Card commands are described in Chapter 7 ('Smart Card Commands'). The command processing program code listings shown here are always divided into three sections, separated by thin lines. Any associated subroutines are separated from the main body of the code (and from each other) by thick lines. The first functional block of the main body of the code investigates the command header (which means the CLS, INS, P1 and P2) as much as is possible at the time. This respects the well-known software design principle of testing the consistency and range of the data as early as possible.

Following this, the basic prerequisites for the execution of the command are checked. For example, monitoring the file structure or the access conditions occurs in this section. If all these checks are completed without any rejections, the actual command execution then takes place. This is usually involves only a very small amount of code. Finally, the return code for successful execution of the command is set, and a jump to the I/O manager is made (via the command interpreter and the return code manager). The I/O manager sends the result of the command processing and then waits for a new command.

If an error is detected in any of the tests in the pseudocode, the relevant return code is immediately set and program execution branches via the central exit of the subroutine in question.

The command set

All seven of the commands listed below correspond to the ISO/IEC 7816-4 standard in their structure and coding. Given the freedom of implementation that a standard naturally allows, however, it is necessary to specify the details of each command, in addition to referring to the standard. The specifications of all the commands of Small-OS are thus described in the following subsections. The specifications have been kept relatively short and formal, and they thus present only the most significant facts.

SELECT FILE command   The SELECT FILE command is used to select a file (MF, DF or EF). This normally requires a 2-byte FID (file identifier). Profile N of the ISO/IEC 7816-4 standard also allows DFs to be selected using a DF Name, which can contain an AID (application identifier). In Small-OS, the AID can be passed only in its complete form. In the special case of the MF, no FID is required to make the selection, since a suitable command option can be used instead. The MF is also automatically selected after the Smart Card is reset, and it can be selected from every other directory during a session.

SELECT FILE falls under Case 1 when the MF is directly selected, which means that neither the command APDU nor the response APDU contains a data portion. When the selection is made using a FID or a DF Name, SELECT FILE falls under Case 3. This means that a data portion is present in the command APDU but not in the response APDU.

After a reset, the security state of the MF is reset to the ground state (0). Selecting a DF does not affect the security state of the MF, but the security state of the DF is automatically set to the ground state (0) when it is selected. When a linear fixed EF is selected, the record pointer is set to 'invalid'.

The search precedence, or in other words, whether a DF or an EF should be sought first when selection is made using a FID, is not specified in the standard. However, this is of considerable significance in certain cases. For example, if the MF is currently selected and there is a DF as well as an EF with the same FID, it may not be possible to select the EF, depending on the search routine. Consequently, in implementing the search routine of

Small-OS, we have chosen to always search the list of EFs first and after this the list of DFs. In case of a conflict, the DF can always be selected using its DF Name, so the use of the FID is not mandatory in such cases.

Table 5.27 Small-OS coding of the Case 1 command SELECT FILE with the option 'direct MF selection'

Data element	Coding	Remarks
CLA	'00'	---
INS	'A4'	---
P1	'00'	---
P2	'00'	---

Table 5.28 Small-OS: coding of the Case 3 command SELECT FILE with the option 'file selection using an FID'

Data element	Coding	Remarks
CLA	'00'	---
INS	'A4'	---
P1	'00'	---
P2	'00'	---
Lc	2	---
DATA	FID	The 2-byte FID of an MF, DF or EF

Table 5.29 Small-OS: coding of the Case 3 command SELECT FILE with the option 'DF selection using a DF name'

Data element	Coding	Remarks
CLA	'00'	---
INS	'A4'	---
P1	'04'	---
P2	'00'	---
Lc	1 ... 16	---
DATA	DF name	A DF name (1 to 16 bytes) is stated. The DF name may contain the AID of the file to be selected.

Table 5.30 Small-OS: coding of the response to the SELECT FILE command

Data element	Coding	Remarks
SW1 ∥ SW2	'9000'	Return code for correct execution of the command

Listing 5. 8   Small-OS: program code for the SELECT FILE command, in compliance with ISO 7816-4, Profile N

SELECT_FILE:	Command according to ISO/IEC 7816-4, Profile N, with extensions.
IF *APDU.Cmd.P1* = '00' THEN (	If P1 = '00', either the command option for the direct MF selection is set or a file is being selected using its 2-byte FID.
IF ((*APDU.Cmd.P2* = '00') AND _ (LENGTH (*APDU.Cmd*) = 4)) THEN (IF LENGTH (*APDU.Cmd*) < 4 THEN (*Returncode* := C_RC_WrongLength RETURN) CALL SELECT_FILE_MF RETURN)	If P2 = '00' and only the 4-byte command header was passed in, the MF is being directly selected. Test whether the command was sent as a Case 1 command. If not, set the appropriate return code and abort command processing.
IF *APDU.Cmd.Lc* = '02' THEN (IF LENGTH (*APDU.Cmd*) < 6 THEN (*Returncode* := C_RC_WrongLength RETURN) IF *APDU.Cmd.Data*[1 ... 2] = '3F00' THEN (CALL SELECT_FILE_MF) RETURN) ELSE (CALL SELECT_FILE_FID) RETURN)))	If two data bytes have been passed with the command, a file should be selected using a 2-byte FID. Test whether the command was sent as a Case 3 command. If not, set the appropriate return code and abort command processing. If the MF is selected (with FID = '3F00'), execute the selection immediately, With any other file (DF or EF), execute the selection via a search routine using the given FID as the search string.
IF *APDU.Cmd.P1* = °0000 0100° THEN (IF LENGTH (*APDU.Cmd*) < 5 THEN (*Returncode* := C_RC_WrongLength RETURN) IF ((*APDU.Cmd.Lc* < 1) _ OR (*APDU.Cmd.Lc* > 16)) THEN (*Returncode* := C_RC_LcInconsistentP1P2 RETURN) CALL SELECT_FILE_DFName)	Test whether a DF can be selected with the DF Name. Test whether the command was sent as a Case 3 command. If not, set the appropriate return code and abort command processing. Test whether the accompanying data portion is between 1 and 16 bytes long.
RETURN	
SELECT_FILE_MF:	Subroutine to select the MF.
SEARCH (MF in file tree) IF STATUS (SEARCH = C_NotFound) THEN (// MF not found in file tree *Returncode* := C_RC_FatalError RETURN)	If the MF cannot be found in the file tree, abort with an internal operating system error.
// MF found in file tree *Ptr.CurrentDF* := identified MF address	Set the current DF pointer to the address of the located MF.
SEARCH (EF Key in MF) IF STATUS (SEARCH) = C_Found THEN (*Ptr.CurrentIEF.Key* := identified EF Key address) ELSE (*Ptr.CurrentIEF.Key* := C_InvalidPointer)	Search for an EF Key and set the appropriate pointer if it is present.

Ptr.CurrentWEF := C_InvalidPointer Ptr.CurrentRecord := C_InvalidPointer SecurityState.MF := 0 SecurityState.DF := 0 Returncode := C_RC_OK RETURN	Mark the current EF pointer and the current record pointer as invalid. Set the security states of the MF and DF to the initial value.

SELECT_FILE_FID:	Subroutine to select a DF or EF using a FID.

SEARCH (EF with FID in the currently selected DF)
IF STATUS (SEARCH) = C_Found THEN (
    Ptr.CurrentWEF := identified EF address
    Ptr.CurrentRecord := C_InvalidPointer
    Returncode := C_RC_OK
    RETURN)
	Search in the current DF for an EF with the given FID. If an EF can be found, set the current EF pointer.

// EF with the given FID not found
SEARCH (DF mit FID)
IF STATUS (SEARCH) = C_Found THEN (
    Ptr.CurrentDF := identified DF address
    SEARCH (EF Key in the current DF)
    IF STATUS (SEARCH) = C_Found) THEN (
        Ptr.CurrentIEF.Key := identified EF Key address)
	If an EF cannot be found, search for a DF with the given FID. If a DF can be found, set the current DF pointer to point to it. Then search for an EF Key, and set the appropriate pointer if one can be found.

ELSE (Ptr.CurrentIEF.Key := C_InvalidPointer)
    Ptr.CurrentWEF := C_InvalidPointer
    Ptr.CurrentRecord := C_InvalidPointer
    SecurityState.DF := 0
    Returncode := C_RC_OK
    RETURN))
	Mark the current EF pointer and the current record pointer as invalid. Set the security state of the DF to the initial value.

// neither an EF nor a DF with the given FID found Returncode := C_RC_FileNotFound RETURN	If no matching file could be found, set the appropriate return code.

SELECT_FILE_DFName:	Subroutine to select a DF using a DF name.

SEARCH (DF with DF Name)
IF STATUS (SEARCH) = C_NotFound THEN (
    Returncode := C_RC_FileNotFound
    RETURN)
ELSE (Ptr.CurrentDF := identified DF address)
SEARCH (EF Key in current DF)
IF STATUS (SEARCH) = C_Found) THEN (
    Ptr.CurrentIEF.Key := identified EF Key address)
ELSE (Ptr.CurrentIEF.Key := C_InvalidPointer)
	Set the current DF pointer to the address of the DF found in the file tree. Then search for the relevant EF Key in the current DF, and set the appropriate pointer to its address if one can be found.

Ptr.CurrentWEF := C_InvalidPointer Ptr.CurrentRecord := C_InvalidPointer SecurityState.DF := 0 Returncode := C_RC_OK RETURN	Mark the current EF pointer and the current record pointer as invalid. Set the security state of the DF to the initial value.

READ BINARY command   The READ BINARY command can be used to read data from a transparent EF, starting from a location specified by a 15-bit offset parameter passed by the command. READ BINARY falls under Case 2, which means that there is no data part in the command APDU, but there is a data part in the response APDU.

All length specifications must be an integral number of bytes. The maximum length of the data to be read is limited to the maximum data volume of a transparent EF, which is 255 bytes for Small-OS. If a value of zero is given for the length, all data from the given offset location to the end of the file are read. Profile N of the ISO/IEC 7816-4 standard does not provide for the implicit selection of EFs by means of Short FIDs, so this option is not implemented here.

Before data can be read from an EF using this command, the associated access conditions must be satisfied. Otherwise, the command will be rejected with an appropriate error report.

Table 5.31   Small-OS: coding of the Case 2 command READ BINARY

Data element	Coding	Remarks
CLA	'00'	---
INS	'B0'	---
P1	°0XXX XXXX°	The 7 more significant bits of the offset to the data to be read (offset := XXX XXXX ‖ Y)
P2	Y	The 8 less significant bits of the offset to the data to be read
Le	Z	$Z = 0$: read all bytes up to the end of the file $Z > 0$: Z is the number of bytes to be read

Table 5.32   Small-OS: coding of the response to the READ BINARY command

Data element	Coding	Remarks
DATA	...	If the command was correctly executed, the data requested by the command are located in this data element
SW1 ‖ SW2	'9000'	Return code for correct execution of the command

Listing 5. 9   Small-OS: program code for the READ BINARY command, in compliance with ISO 7816-4, Profile N

READ_BINARY:	Command according to ISO/IEC 7816-4, Profile N.

IF LENGTH (*APDU.Cmd*) < 5 THEN (
    Returncode := C_RC_WrongLength
    RETURN)
	Test whether the command was sent as a Case 2 command. If not, set the appropriate return code and abort command processing.

IF *APDU.Cmd.P1.b8* = °1° THEN (
    Returncode := C_RC_FctNotSupported
    GOTO READ_BINARY_Exit)
	Test whether an EF should be selected using a Short FID.

IF *Ptr.CurrentWEF* = C_InvalidPointer THEN (
    Returncode := C_RC_CmdNotAllowed
    GOTO READ_BINARY_Exit)
	Test whether an EF is already
	selected.

WITH *DF[Ptr.CurrentDF]*.	Set a part of the file tree as a reference for this command.

IF *.EF[Ptr.CurrentWEF].Structure* = C_EFStrucTransparent
    THEN (
    Returncode := C_RC_CmdIncompFStruc
    GOTO READ_BINARY_Exit)
	Test whether the selected EF has a
	transparent structure.

FileOffset := (*APDU.Cmd.P1* * 256) + *APDU.Cmd.P2* *DataLenToRead* := 0	Calculate the offset to the desired data in the file, and initialize the variable for the amount of data to be read.

IF STATUS (File_Manager_CheckACRead) = _
C_AccessDenied) THEN (
    Returncode := C_RC_SecStateNotSatisfied
    GOTO READ_BINARY_Exit)
	Test whether the security state has
	been achieved that is required for
	reading data from the selected EF
	with READ BINARY.

IF *APDU.Cmd.Le* = '00' THEN (
    DataLenToRead := .*EF[Ptr.CurrentWEF]*. _
    TransparentDataSize–FileOffset)
ELSE (
    DataLenToRead := *APDU.Cmd.Le*)
	Test whether all available data
	should be read (i.e., Le = 0), or only
	a certain amount of data.

IF *.EF[Ptr.CurrentWEF].TransparentDataSize* >= *FileOffset* _
THEN (
    Returncode := C_RC_WrongP1P2
    GOTO READ_BINARY_Exit)
	Test whether the requested offset
	fits with the size of the file.

IF *.EF[Ptr.CurrentWEF].TransparentDataSize* <
    (*FileOffset* + *DataLenToRead*) THEN (
    Returncode := C_RC_WrongP1P2
    GOTO READ_BINARY_Exit)
	Test whether the selected offset
	and the requested data length (Le)
	fit with the size of the file.

CALL Kernel_CopyByteString
// from: .*EF[Ptr.CurrentWEF].TransparentData[x ... y]*
// *x* = *FileOffset* ; *y* = (*FileOffset* + *DataLenToRead*)
// to: *APDU.Rsp.Data[1 ... DataLenToRead]*
	Copy the requested data from the
	file to the I/O transmit buffer.

Returncode := C_RC_OK	The command has been processed with no error, since in all other cases an error exit is used.
READ_BINARY_Exit: END WITH	
RETURN	

UPDATE BINARY command   The UPDATE BINARY command can be used to read data from a transparent EF, starting from a location specified by a 15-bit offset parameter passed by the command. UPDATE BINARY falls under Case 3, which means that there is a data part in the command APDU, but there is no data part in the response APDU.

All length specifications must be an integral number of bytes. The maximum length of the data to be read is limited to the maximum data volume of a transparent EF, which is 255 bytes for Small-OS. If a value of zero is given for the length, all data from the given offset location to the end of the file are read. Profile N of the ISO/IEC 7816-4 standard does not provide for the implicit selection of EFs by means of Short FIDs, so this option is not implemented here. Before data can be written to an EF using this command, the associated access conditions must be satisfied. Otherwise, the command will be rejected with a suitable error report.

Table 5.33   Small-OS: coding of the Case 2 command UPDATE BINARY

Data element	Coding	Remarks
CLA	'00'	---
INS	'D6'	---
P1	°0XXX XXXX°	The 7 more significant bits of the offset to the data to be read (offset := XXX XXXX ‖ Y)
P2	Y	The 8 less significant bits of the offset to the data to be read
Lc	...	Number of bytes to be written
DATA	...	The data bytes to be written

Table 5.34   Small-OS: coding of the response to the UPDATE BINARY command

Data element	Coding	Remarks
SW1 ‖ SW2	'9000'	Return code for correct execution of the command

Listing 5. 10   Small-OS: program code for the UPDATE BINARY command, in compliance with ISO 7816-4, Profile N

UPDATE_BINARY:	Command according to ISO/IEC 7816-4, Profile N.

IF LENGTH (*APDU.Cmd*) < 6 THEN (
    Returncode:= C_RC_WrongLength
    RETURN)
	Test whether the command was sent as a Case 3 command. If not, set the appropriate return code and abort command processing.

IF *APDU.Cmd.P1,b8* = °1° THEN (
    Returncode:= C_RC_FctNotSupported
    GOTO UPDATE_BINARY_Exit)
	Test whether an EF should be selected using a Short FID (i.e., the msb of P1 is set).

IF *Ptr.CurrentWEF* = C_InvalidPointer THEN (
    Returncode := C_RC_CmdNotAllowed
    GOTO UPDATE_BINARY_Exit)
	Test whether an EF is already selected. If not, abort the command.

WITH *DF[Ptr.CurrentDF]*.	Set a part of the file tree as a reference for this command.

IF *.EF[Ptr.CurrentWEF].Structure* = C_EFStrucTransparent _
THEN (
    Returncode:= C_RC_CmdIncompFStruc
    GOTO UPDATE_BINARY_Exit)
	Test whether the selected EF has a transparent file structure.

FileOffset := (APDU.Cmd.P1 * 256) + APDU.Cmd.P2	Calculate the offset to the data in the file.

IF STATUS (File_Manager_CheckACUpdate) = _
C_AccessDenied) THEN (
    Returncode:= C_RC_SecStateNotSatisfied
    GOTO UPDATE_BINARY_Exit)
	Test whether the security state has been achieved that is required for writing data to the selected EF with UPDATE BINARY.

IF .EF[Ptr.CurrentWEF].TransparentDataSize < _
(FileOffset + APDU.Cmd.Lc) THEN (
    Returncode:= C_RC_WrongP1P2
    GOTO UPDATE_BINARY_Exit)
	Test whether the selected offset and requested data length (Lc) fit with the size of the file.

CALL Kernel_CopyByteString
// from: APDU.Cmd.Data[1 ... APDU.Cmd.Lc]
// to: .EF[Ptr.CurrentWEF].TransparentData[x ... y]
// x = FileOffset ; y = (FileOffset + APDU.Cmd.Lc)
IF STATUS (Kernel_CopyByteString) = C_WriteError THEN (
    Returncode := C_RC_MemoryFailure
    GOTO UPDATE_BINARY_Exit)
	Copy the passed-in data from the I/O receive buffer to the file. If an error occurs, abort and report the error to the terminal.

Returncode := C_RC_OK	The command has been processed with no error, since in all other cases an error exit is used.
UPDATE_BINARY_Exit: END WITH RETURN	

READ RECORD command   The READ RECORD command can be use to read a record from a linear fixed EF. The maximum amount of data to be read is limited to the maximum record length of 255 bytes. The length specification must either match the length of the addressed record or be set to zero. With a length of zero, the entire record is automatically read. All length specifications must be an integer number of bytes. Profile N of the ISO/IEC 7816-4 standard does not provide for the implicit selection of EFs by means of Short FIDs, so this option is not implemented here. READ RECORD falls under Case 2, which means that there is no data part in the command APDU, but there is a data part in the response APDU.

A record in a linear fixed EF can be addressed in three different ways. The number of the desired record can be passed directly with the READ RECORD command. If this record is present in the file, its contents are returned in the response; otherwise the answer contains a suitable error report. This type of access does not affect the record pointer, which can only be modified with the command options 'first', 'last', 'next' and 'previous'. The record pointer is set to 'invalid' immediately after an EF is newly selected. If the option 'next' or 'previous' is selected when the record pointer is invalid, the record pointer is automatically set to the first or last record of the file, respectively. This makes it possible to (for example) read the records in a file by first selecting the EF and then sending a series of READ RECORD commands with the 'next' option, without having to use any other commands. The third type of access is to use the 'current' option. In this case the record that is currently indicated by the current record pointer is read. If the record pointer is invalid, the command is aborted with an appropriate error report.

Before data can be read from an EF using this command, the associated access conditions must be satisfied. Otherwise, the command will be rejected with a suitable error report. The record returned in the response when the command is successfully executed is not TLV coded, although this is optionally allowed by the ISO/IEC 7816-4 standard.

Table 5.35   Small-OS: coding of the Case 2 command READ RECORD

Data element	Coding	Remarks	
CLA	'00'	---	
INS	'B2'	---	
P1	X	$Y = °0000\ 0100°$, $X = 0$	read the current record (*Ptr.CurrentRecord*)
		$Y = °0000\ 0100°$, $X <> 0$	read record number X
P2	Y	$Y = °0000\ 0100°$	read the record using the method specified by P1
		$X = 0$, $Y = °0000\ 0000°$	read the first record in the file
		$X = 0$, $Y = °0000\ 0001°$	read the last record in the file
		$X = 0$, $Y = °0000\ 0010°$	read the next record in the file
		$X = 0$, $Y = °0000\ 0011°$	read the previous record in the file
Le	Z	$Z = 0$: read all bytes until the end of the record	
		$Z > 0$: Z is the record length	

Table 5.36   Small-OS: coding of the response to the READ RECORD command

Data element	Coding	Remarks
DATA	...	If the command was correctly executed, the record requested by the command is located in this data element
SW1 ‖ SW2	'9000'	Return code for correct execution of the command

Listing 5. 11   Small-OS: program code for the READ RECORD command, in compliance with ISO 7816-4, Profile N

READ_RECORD:	Command according to ISO/IEC 7816-4, Profile N

IF LENGTH (*APDU.Cmd*) < 5 THEN (
   Returncode := C_RC_WrongLength
   RETURN)
	Test whether the command was sent as a Case 2 command. If not, set the appropriate return code and abort command processing.

IF *APDU.Cmd.P2.b8* ... b4 <> °00000° THEN (
   Returncode:= C_RC_FctNotSupported
   GOTO READ_RECORD_Exit)
	Test whether an EF should be selected using a Short FID.

IF *Ptr.CurrentWEF* = C_InvalidPointer THEN (
   Returncode:= C_RC_CmdNotAllowed
   GOTO READ_RECORD_Exit)
	Test whether an EF is already selected.

WITH *DF[Ptr.CurrentDF]*.	Set a part of the file tree as a reference for this command.

IF .*EF[Ptr.CurrentWEF].Structure* = C_EFStrucLinFix THEN (
    Returncode:= C_RC_CmdIncompFStruc
    GOTO READ_RECORD_Exit)
	Test whether the selected EF has a linear fixed structure.

IF STATUS (File_Manager_CheckACRead) = C_AccessDenied _
 THEN (
    Returncode:= C_RC_SecStateNotSatisfied
    GOTO READ_RECORD_Exit)
	Test whether the security state has been achieved that is required for reading the selected EF with READ RECORD.

RecordNoToRead := 0 *RecordLenToRead* := 0	Initialize the variable for the number of the record to be read and its length.

IF *APDU.Cmd.P2.b3 ... b1* <> °000° THEN (
    Ptr.CurrentRecord := 1
    RecordNoToRead := *Ptr.CurrentRecord*)
	If the option 'read first record' was selected, set the current record pointer to the first record in the file.

IF *APDU.Cmd.P2.b3 ... b1* <> °001° THEN (
    Ptr.CurrentRecord := .*EF[Ptr.CurrentWEF].NoOfRecords*
    RecordNoToRead := *Ptr.CurrentRecord*)
	If the option 'read last record' was selected, set the current record pointer to the last record in the file.

IF *APDU.Cmd.P2.b3 ... b1* <> °010° THEN (
    IF *Ptr.CurrentRecord* = C_InvalidPointer THEN (
      Ptr.CurrentRecord = 1
      RecordNoToRead := *Ptr.CurrentRecord*)
    ELSE (
      IF *Ptr.CurrentRecord* < .*EF[Ptr.CurrentWEF]. _*
        NoOfRecords THEN (
        Ptr.CurrentRecord := *Ptr.CurrentRecord* + 1
        RecordNoToRead := *Ptr.CurrentRecord*)
      ELSE (
        Returncode = C_RC_RecordNotFound
        GOTO READ_RECORD_Exit)))
	If the option 'read next record' was selected, set the current record pointer to the next record in the file if it does not already point to the last record in the file.

IF *APDU.P2.b3 ... b1* <> °011° THEN (
    IF *Ptr.CurrentRecord* = C_InvalidPointer THEN (
      Ptr.CurrentRecord := _
      .*EF[Ptr.CurrentWEF].NoOfRecords*
      RecordNoToRead := *Ptr.CurrentRecord*)
    ELSE (
      IF *Ptr.CurrentRecord* > 1 THEN (
        Ptr.CurrentRecord := *Ptr.CurrentRecord*–1
        RecordNoToRead := *Ptr.CurrentRecord*)
      ELSE (
        Returncode = C_RC_RecordNotFound
        GOTO READ_RECORD_Exit)))
	If the option 'read previous record' was selected, set the current record pointer to the previous record in the file, if it does not already point to the first record in the file.

IF ((*APDU.Cmd.P2.b3 ... b1* <> °100°) AND _
 (*APDU.Cmd.P1* = 0)) THEN (
    IF *Ptr.CurrentRecord* <> C_InvalidPointer THEN (
      Returncode = C_RC_WrongP1P2
      GOTO READ_RECORD_Exit)
    ELSE (
      RecordNoToRead := *Ptr.CurrentRecord*))
	If the option 'read current record' was selected, test whether the relevant pointer is valid. If it is, set the internal variable or the command to the current record to be read.

```
IF APDU.Cmd.P2.b3 ... b1 <> °100° THEN (             If the option 'address record
    IF .EF[ Ptr.CurrentWEF ].NoOfRecords < APDU.Cmd.P1 _   directly with P1' was selected,
    THEN (                                          test whether the specified record
        Returncode = C_RC_WrongP1P2                 is present in the file.
        GOTO READ_RECORD_Exit)
    ELSE (
        RecordNoToRead := APDU.Cmd.P1)

IF APDU.Cmd.Le = '00' THEN (                         Test whether the entire record is
    RecordLenToRead := .EF[ Ptr.CurrentWEF ]. _      to be read with an explicit length
    Record[ RecordNoToRead ].Size)                   specification or with no explicit
ELSE (                                               length specification (Le = '00')
    RecordLenToRead := APDU.Cmd.Le)

IF RecordLenToRead <> .EF[ Ptr.CurrentWEF ]. _       Test whether the requested data
Record[ RecordNoToRead ].Size) THEN (                length (Le) matches the record
    Returncode = C_RC_LcInconsistentP1P2             length.
    GOTO READ_RECORD_Exit)

CALL Kernel_CopyByteString                           Copy the requested data from the
// from: .EF[ Ptr.CurrentWEF ].Record[ RecordNoToRead ]. _  file to the I/O transmit buffer.
// Data[ 1 ... RecordLenToRead ]
// to: APDU.Rsp.Data[ 1 ... RecordLenToRead ]

Returncode = C_RC_OK                                 The command has been processed
                                                     with no error, since in all other
                                                     cases an error exit is used.


READ_RECORD_Exit:
END WITH
RETURN
```

UPDATE RECORD command   The UPDATE RECORD command can be use to write a record to a linear fixed EF. The data passed by the command is not allowed to be TLV coded, although this is allowed as an option by the ISO/IEC 7816-4 standard. The maximum amount of data to be written is limited to the maximum record length of 255 bytes. The length specification must exactly match the length of the addressed record, and all length specifications must be an integer number of bytes. Profile N of the ISO/IEC 7816-4 standard does not provide for the implicit selection of EFs by means of Short FIDs, so this option is not implemented here. READ RECORD falls under Case 3, which means that there is a data part in the command APDU, but there is no data part in the response APDU.

A record in a linear fixed EF can be addressed in three different ways. The number of the desired record can be passed directly with the UPDATE RECORD command. If this record is present in the file, its contents are returned in the response; otherwise, the answer contains a suitable error report. This type of access does not affect the record pointer, which can only be modified with the command options 'first', 'last', 'next' and 'previous'. The record pointer is set to 'invalid' immediately after an EF is newly selected. If the option 'next' or 'previous' is selected when the record pointer is invalid, the record pointer is automatically set to the first or last record of the file, respectively. This makes it possible to (for example) write all the records in a file by first selecting the EF and then sending a series of UPDATE RECORD commands with the 'next' option, without having to use any other commands. The third type of access is to use the 'current' option. In this case, the record that is currently indicated by the current record pointer is written. If the record pointer is invalid, the command is aborted with an appropriate error report. Before data can be written to an EF using this command, the associated access conditions must be satisfied. Otherwise, the command will be rejected with a suitable error report.

Table 5.37 Small-OS: coding of the Case 3 command UPDATE RECORD

Data element	Coding	Remarks	
CLA	'00'	---	
INS	'DC'	---	
P1	X	$Y = °0000\ 0100°$, $X = 0$	write the current record ($Ptr.CurrentRecord$)
		$Y = °0000\ 0100°$, $X <> 0$	write record number X
P2	Y	$Y = °0000\ 0100°$	write the record using the method specified by P1
		$X = 0$, $Y = °0000\ 0000°$	write the first record in the file
		$X = 0$, $Y = °0000\ 0001°$	write the last record in the file
		$X = 0$, $Y = °0000\ 0010°$	write the next record in the file
		$X = 0$, $Y = °0000\ 0011°$	write the previous record in the file
Lc	...	Number of bytes to be written	
DATA	...	The record to be written	

Table 5.38 Small-OS: coding of the response to the UPDATE RECORD command

Data element	Coding	Remarks
SW1 ‖ SW2	'9000'	Return code for correct execution of the command

Listing 5. 12 Small-OS: program code for the UPDATE RECORD command, in compliance with ISO 7816-4, Profile N

UPDATE_RECORD:	Command according to ISO/IEC 7816-4, Profile N.

IF LENGTH (*APDU.Cmd*) < 6 THEN (
    Returncode := C_RC_WrongLength
    RETURN)
	Test whether the command was sent as a Case 3 command. If not, set the appropriate return code and abort command processing.

IF *APDU.Cmd.P2.b8 ... b4* <> °00000° THEN (
    Returncode := C_RC_FctNotSupported
    GOTO UPDATE_RECORD_Exit)
	Test whether an EF should be selected using a Short FID.

IF *Ptr.CurrentWEF* = C_InvalidPointer THEN (
    Returncode := C_RC_CmdNotAllowed
    GOTO UPDATE_RECORD_Exit)
	Test whether an EF is already selected.

WITH *DF[Ptr.CurrentDF]*.	Set a part of the file tree as a reference for this command.

IF .*EF[Ptr.CurrentWEF].Structure* = C_EFStrucLinFix THEN (
   Returncode := C_RC_CmdIncompFStruc
   GOTO UPDATE_RECORD_Exit)
	Test whether the selected EF has a linear fixed structure.

IF STATUS (File_Manager_CheckACUpdate) = _
C_AccessDenied) THEN (
   Returncode := C_RC_SecStateNotSatisfied
   GOTO UPDATE_RECORD_Exit)
	Test whether the security state has been achieved that is required for reading the selected EF with UPDATE RECORD.

RecordNoToUpdate := 0	Initialize the variable for the number of the record to be written.

IF *APDU.Cmd.P2.b3 ... b1* <> °000° THEN (
   Ptr.CurrentRecord := 1
   RecordNoToUpdate := *Ptr.CurrentRecord*)
	If the option 'write first record' was selected, set the current record pointer to the first record in the file.

IF *APDU.Cmd.P2.b3 ... b1* <> °001° THEN (
   Ptr.CurrentRecord := _
   .*EF[Ptr.CurrentWEF].NoOfRecords*
   RecordNoToUpdate := *Ptr.CurrentRecord*)
	If the option 'write last record' was selected, set the current record pointer to the last record in the file.

IF *APDU.Cmd.P2.b3 ... b1* <> °010° THEN (
   IF *Ptr.CurrentRecord* = C_InvalidPointer THEN (
      Ptr.CurrentRecord := 1
      RecordNoToUpdate := *Ptr.CurrentRecord*)
   ELSE (
      IF *Ptr.CurrentRecord* < .*EF[Ptr.CurrentWEF]. _
      NoOfRecords* THEN (
         Ptr.CurrentRecord := *Ptr.CurrentRecord* + 1
         RecordNoToUpdate := *Ptr.CurrentRecord*)
      ELSE (
         Returncode = C_RC_RecordNotFound
         GOTO UPDATE_RECORD_Exit)))
	If the option 'write next record' was selected, set the current record pointer to the next record in the file if it does not already point to the last record in the file.

IF *APDU.Cmd.P2.b3 ... b1* <> °011° THEN (
   IF *Ptr.CurrentRecord* = C_InvalidPointer THEN (
      Ptr.CurrentRecord = .*EF[Ptr.CurrentWEF]. _
      NoOfRecords*)
      RecordNoToUpdate := *Ptr.CurrentRecord*)
   ELSE (
      IF *Ptr.CurrentRecord* > 1 THEN (
         Ptr.CurrentRecord := *Ptr.CurrentRecord* - 1)
         RecordNoToUpdate := *Ptr.CurrentRecord*)
      ELSE (
         Returncode = C_RC_RecordNotFound
         GOTO UPDATE_RECORD_Exit)))
	If the option 'write previous record' was selected, set the current record pointer to the previous record in the file if it does not already point to the first record in the file.

IF ((*APDU.Cmd.P2.b3 ... b1* <> °100°) AND _
(*APDU.Cmd.P1* = 0)) THEN (
   IF *Ptr.CurrentRecord* <> C_InvalidPointer THEN (
      Returncode = C_RC_WrongP1P2
      GOTO UPDATE_RECORD_Exit)
   ELSE (
      RecordNoToUpdate := *Ptr.CurrentRecord*))
	If the option 'write current record' was selected, test whether the relevant pointer is valid. If it is, set the internal variable or the command to the current record to be read.

IF *APDU.Cmd.P2.b3 ... b1 <> °100°* THEN (
IF *.EF[Ptr.CurrentWEF].NoOfRecords < APDU.Cmd.P1* _
THEN (
    Returncode = C_RC_WrongP1P2
    GOTO UPDATE_RECORD_Exit)
	If the option 'address record directly with P1' was selected, test whether the specified record is present in the file.

IF *.EF[Ptr.CurrentWEF].Record[Ptr.CurrentRecord].Size* _
<> *APDU.Cmd.Lc* THEN (
    Returncode := C_RC_LcInconsistentP1P2
    GOTO UPDATE_RECORD_Exit)
	Test whether the length of the passed-in record (Lc) matches the record length of the file.

CALL Kernel_CopyByteString
// from: *APDU.Cmd.Data[1 ... APDU.Cmd.Lc]*
// to: *.EF[Ptr.CurrentWEF].Record[RecordNoToUpdate].* _
// *Data[1 ... APDU.Cmd.Lc]*
IF STATUS (Kernel_CopyByteString) = C_WriteError THEN (
    Returncode := C_RC_MemoryFailure
    GOTO UPDATE_RECORD_Exit)
 	
Returncode := C_RC_OK
	Copy the transferred data from the I/O receive buffer to the file. If an error occurs, abort and report the error to the terminal.
	
	The command has been processed with no error, since in all other cases an error exit is used.

UPDATE_RECORD_Exit: END WITH	
RETURN	

VERIFY command   The VERIFY command is used to compare a secret item that has been passed over to the Smart Card, such as a PIN, to a stored reference value. The length of the PIN must be between one and eight bytes. The operating system does perform any sort of testing of the coding of the data string that is passed in. This means that, for example, a 4-digit PIN (e.g. '1234') could be coded as two BCD bytes ('12' || '34') or as four ASCII bytes ("1" || "2" || "3" || "4" = '31' || '32' || '33' || '34'). The VERIFY command falls under Class 3, which means that there is a data part in the command APDU, but there is no data part in the response APDU.

At most two PINs (PIN number 1 and PIN number 2) can be addressed. They can be located in either the EF Key of the MF or the EF Key of the currently selected DF. A PIN that is stored in the EF Key of the MF is used as a common PIN for all applications in the Smart Card. If a PIN is stored in the EF Key of a DF, it can be used only for the application associated with that DF. Such a PIN is thus an application-specific PIN.

Every PIN has a retry counter that is reset to zero when a positive comparison result is obtained and incremented by one when a negative result is obtained. If the state of the retry counter is not zero, the number of PIN attempts still allowed is returned encoded in SW2. If the retry counter reaches its maximum value, this is indicated by a separate return code.

Since Small-OS does not have any command to reset the retry counter, whenever a retry counter is standing at its maximum value, there is absolutely no possibility of ever making any further PIN comparisons. Depending on the application, this might mean that the Smart Card could no longer be used. The PIN located in the EF Key cannot be altered by the user, although this possibility is commonly found in many Smart Card applications. A specific command would be needed for this (CHANGE REFERENCE DATA as per ISO/IEC 7816-8), but this is not provided in the ISO/IEC 7816-4 standard.

The implementation presented here has a small peculiarity due to simplification. The retry counter is located in EEPROM, as you know. However, EEPROM write accesses need not always be successful, due to the possibility of a write error. Consequently, it is necessary to test the data after each write operation to verify that they have been correctly written. If the result of the test is negative, a suitable return code is set. The retry counter is altered so often in the VERIFY command that no such test has been included in the pseudocode, since the basic relationships of the code would no longer be clear if it were present. You should bear this in mind when examining the code.

The VERIFY command is naturally predestined to be used for attacks on the PIN. The implementation has been designed to make it impossible to base an attack on an analysis of timing behavior or current consumption. The retry counter is always incremented before the received PIN is compared to the reference PIN stored in the EF Key. This ensures that cutting off the power supply to the card immediately following the PIN comparison cannot result in the retry counter not being incremented, which would allow an attacker an unlimited number of PIN comparisons.

The actual EEPROM writing process for incrementing the retry counter is by no means as trivial as you might first think. The coding of the retry counter must be constructed such that breaking off the process during the write operation, or during the erase operation that may be necessary before the write operation, cannot result in the retry counter being reset to its initial zero value. The code internal to the operating system must therefore be designed with reference to the minimum-energy state of the EEPROM, which is also known as its secure state. This means that with an EEPROM whose secure state is zero, for example, the initial value of the retry counter may not be coded as zero. If it were, it would be possible to reset the retry counter to zero by skillfully switching off the supply voltage during the EEPROM write operation. It would then be possible to determine the PIN within a relatively short time by trial and error, since the retry counter would not be able to fulfil its role as a counter for unsuccessful PIN comparisons. Ideally, the technique of atomic processes[9] can also be used for writing the retry counter.

Table 5.39   Small-OS: coding of the Case 3 command VERIFY

Data element	Coding	Remarks	
CLA	'00'	---	
INS	'20'	---	
P1	'00'	---	
P2	Y	$Y = °100Z\ ZZZZ°$	Use a reference PIN that is stored in the EF Key of the currently selected directory (MF or DF) (*specific reference data*)
		$Z = °0\ 0001° \wedge Z = °0\ 0010°$	Number of the referenced PIN (1 or 2)
Lc	...	Length of the passed-in PIN	
DATA	...	The passed-in PIN	

---

[9]   See also Section 5.9, 'Atomic Processes'.

Table 5.40   Small-OS: coding of the response to the VERIFY command

Data element	Coding	Remarks
SW1 ‖ SW2	'9000'	Return code for correct execution of the command (successful PIN comparison)

Listing 5. 13   Small-OS: program code for the VERIFY command, in compliance with ISO 7816-4, Profile N

VERIFY:	Command VERIFY according to ISO/IEC 7816-4, Profile N.

IF LENGTH (*APDU.Cmd*) < 6 THEN (
    Returncode := C_RC_WrongLength
    RETURN)
	Test whether the command was sent as a Case 3 command. If not, set the appropriate return code and abort command processing.

IF *APDU.Cmd.P1* <> '00' THEN (
    Returncode := C_RC_WrongP1P2
    GOTO VERIFY_Exit)
	Test whether P1 has the allowed value (P1 must be '00').

IF ((*APDU.Cmd.P2* < '01') OR (*APDU.Cmd.P2* > '02')) THEN (
    Returncode := C_RC_WrongP1P2
    GOTO VERIFY_Exit)
	Test whether P2 has one of the two allowed values (P2 must be either 1 or 2).

IF ((*APDU.Cmd.Lc* <= 1) OR (*APDU.Cmd.Lc* >= 8)) THEN (
    Returncode := C_RC_LcInconsistentP1P2
    GOTO VERIFY_Exit)
	Test whether the length of the passed data (i.e. the PIN) lies within the allowed range (1 [Lc [8).

IF *Ptr.CurrentIEF.Key* = C_InvalidPointer THEN (
    Returncode := C_RC_RefDataNotFound
    GOTO VERIFY_Exit)
	Test whether an EF is present in the current directory.

SEARCH (for the PIN with the requested reference number _ in *DF[Ptr.CurrentDF].EF[Ptr.CurrentIEF.Key]*)	Search for the requested reference number in EF Key.

IF STATUS (SEARCH) = C_Found) THEN (
    set KeyRecord to the record containing the found PIN
    WITH *DF[Ptr.CurrentDF].EF[Ptr.CurrentIEF.Key]*.)
ELSE (
    Returncode := C_RC_RefDataNotFound
    GOTO VERIFY_Exit)
	If a PIN with the specified reference number is found, set the current key pointer to reference it.

IF *.Record[KeyRecord].KeyPurpose* <> C_CmdVERIFY THEN (
    Returncode := C_RC_CondOfUseNotSatified
    GOTO VERIFY_Exit)
	Test whether the selected data are allowed to be used with the VERIFY command.

IF *Ptr.CurrentDF* = *Ptr.MF* THEN (
    // MF is selected
    IF ((*.Record[KeyRecord].EntryState* <> *SecurityState.MF*) _
    THEN (
        Returncode := C_RC_SecStateNotSatisfied
        GOTO VERIFY_Exit))
	If the MF is selected, test whether its current security state allows the VERIFY command to be used. The security state called for in the key record must have been achieved in the currently selected directory.

ELSE (
    // DF is selected
    IF *.Record[KeyRecord].EntryState* <> *SecurityState.DF* _
    THEN (
        Returncode := C_RC_SecStateNotSatisfied
        GOTO VERIFY_Exit)

If a DF is selected, test whether its current security state allows the VERIFY command to be used. The security state called for in the key record must have been achieved in the currently selected directory.

IF *.Record[KeyRecord].RCntr* >= _
.Record[KeyRecord].RCntrMax THEN (
    Returncode := C_RC_AuthMethodBlocked
    GOTO VERIFY_Exit)

Test whether the retry counter has reached its maximum value

IF *APDU.Cmd.Lc* <> *.Record[KeyRecord].KeySize* THEN (
    Returncode := C_RC_LcInconsistentP1P2
    GOTO VERIFY_Exit)

Test whether the passed-in PIN has the same length as the reference PIN.

.Record[KeyRecord].RCntr := *.Record[KeyRecord].RCntr* + 1

As a precautionary measure, increment the retry counter before making the actual PIN comparison. This defends against a possible attack by analyzing the processing time or current consumption.

CALL Kernel_CompareByteString
// Data 1: *APDU.Cmd.Data[1 ... APDU.Cmd.Lc]*
// Data 2: *.Record[KeyRecord].KeyData[1 ... APDU.Cmd.Lc]*

Compare the passed-in PIN with the stored reference PIN.

IF STATUS (Kernel_CompareByteString) = C_Equal THEN (
    .Record[KeyRecord *].RCntr* := 0
    IF *Ptr.CurrentDF* = *Ptr.MF* THEN (
        // MF is selected
        SecurityState.MF := *.Record[KeyRecord].ResultState.OK*)
    ELSE (
        // DF is selected
        SecurityState.DF := *.Record[KeyRecord].ResultState.OK*))

If the passed-in PIN matches the stored reference PIN, set the retry counter to zero false attempts and set the security state for successful PIN testing.

IF STATUS (Kernel_CompareByteString) = C_NotEqual _
THEN (
    IF *Ptr.CurrentDF* = *Ptr.MF* THEN (
        // MF is selected
        SecurityState.MF := _
        .Record[KeyRecord].ResultState.NOK)
    ELSE (
    // DF is selected
        SecurityState.DF := _
        .Record[KeyRecord].ResultState.NOK))
    IF *.Record[KeyRecord].RCntr* = _
    .Record[KeyRecord].RCntrMaxValue THEN (
        Returncode := C_RC_AuthMethodBlocked
    ELSE (
        Returncode := C_RC_CounterX || (_
        .Record[KeyRecord].RCntrMax–_
        .Record[KeyRecord].RCntr)))
    GOTO VERIFY_Exit)

If the passed-in PIN does not match the stored reference PIN, the retry counter will have already been incremented before the PIN comparison.

Set the security state that results from an unsuccessful PIN comparison. If the retry counter has reached its maximum value, set SW2 to the number of unsuccessful attempts that are still allowed. Otherwise, indicate in the return code that no further PIN verifications are possible.

Returncode := C_RC_OK	The command has been processed with no error, since in all other cases an error exit is used.
VERIFY_Exit: END WITH	
RETURN	

INTERNAL AUTHENTICATE command The INTERNAL AUTHENTICATE command is used to authenticate the Smart Card via a challenge–response procedure. An 8-byte random number is sent to the Smart Card, which encrypts it using the DES algorithm. The number of the key to be used must be given in parameter P2, which must indicate whether the key to be used is located in the EF Key file of the MF or the currently selected DF. INTERNAL AUTHENTICATE falls under Case 4, which means that a data part is present in both the command APDU and the response APDU.

The ISO/IEC 7816-4 standard only specifies a few parameters for the authentication commands. The cryptographic algorithm, for example, is not specified. In Small-OS, the DES algorithm has been chosen as the cryptographic algorithm. As an extension to Profile N of the ISO/IEC 7816-4 standard, a 5-byte key number is passed with the command.

In Small-OS, INTERNAL AUTHENTICATE can in principle be used to encrypt eight bytes of plaintext into eight bytes of ciphertext using a selectable key. A Smart Card with Small-OS would thus fall under strict export control in almost all countries, so that it would take several weeks or even months to obtain an export permit for the card. Consequently, INTERNAL AUTHENTICATE is implemented in many real Smart Cards such that it is not possible to directly encrypt data. This avoids the export restrictions.

The ability to directly encrypt a plaintext block into a ciphertext block is equally risky from a cryptographic perspective, since it could be used to generate plaintext–ciphertext pairs for brute force attacks. In addition, this implementation would be very susceptible to differential fault analysis.[10] The simplest form of attack to implement would be a timing attack [11] on the computation of the DES, which thus must be noise-free. Otherwise, the key could be determined by measuring the processing time for the computation.

For all of these reasons, the starting value is most commonly extended in practice by appending a random number generated inside the Smart Card and the card's own unique number. The resulting number is then encrypted, and the ciphertext is sent back to the terminal along with the data used to extend the original number. This means that this command can no longer be used to encrypt data (which solves the export problem). In addition, the fact that a different value is encrypted each time provides the basis for protection against differential fault analysis (DFA) and differential performance analysis (DPA).[12] All of these measures show relatively dramatically that both the specification and the implementation of even an ostensibly simple command, such as INTERNAL AUTHENTICATE, requires considerable knowledge and experience to protect the keys of a Smart Card application against attack.

---

10   See also Section 8.2.4, 'Attacks and defensive measures while the card is in use'.
11   See also Section 8.2.4.2, 'Attacks on the logical level'.
12   See also Section 8.2.4.2, 'Attacks on the logical level'.

Table 5.41   Small-OS: coding of the Case 4 command INTERNAL AUTHENTICATE

Data element	Coding	Remarks	
CLA	'00'	---	
INS	'88'	---	
P1	'00'	---	
P2	Y	Y = °100Z ZZZZ°	use a key from the EF Key file in the currently selected directory (MF or DF) (specific reference data)
		Z ZZZZ°	number of the referenced key (1 ... 31)
Lc	8	Length of the passed-in random number	
DATA	...	The passed-in random number	
Le	8	Length of the returned random number	

Table 5.42   Small-OS: coding of the response to the INTERNAL AUTHENTICATE command

Data element	Coding	Remarks
DATA	...	If the command was correctly executed, this data element contains the encrypted random number that has been encrypted using the key referenced by the command
SW1 ‖ SW2	'9000'	Return code for correct execution of the command

Listing 5. 14   Small-OS: program code for the INTERNAL AUTHENTICATE command, in compliance with ISO 7816-4, Profile N, extended with global and specific reference data (selectable reference to the MF or DF).

```INTERNAL_AUTHENTICATE:```	Command INTERNAL AUTHENTICATE according to ISO/IEC 7816-4, Profile N.
```IF APDU.Cmd.P1 <> '00' THEN (``` ``` Returncode := C_RC_WrongP1P2``` ``` GOTO INTERNAL_AUTHENTICATE_Exit)```	Test whether P1 has the allowed value (P1 must be '00').
```IF APDU.Cmd.Lc <> 8 THEN (``` ``` Returncode := C_RC_WrongLength``` ``` GOTO INTERNAL_AUTHENTICATE_Exit)```	Test whether the passed data (i.e. the random number) has the allowed length of 8 bytes.
```IF Ptr.CurrentIEF.Key = C_InvalidPointer THEN (``` ``` Returncode := C_RC_RefDataNotFound``` ``` GOTO INTERNAL_AUTHENTICATE_Exit)```	Test whether an EF is present in the current directory.
```IF APDU.Cmd.P2.b5 ... b1 = °00000° THEN (``` ``` Returncode := C_RC_WrongP1P2``` ``` GOTO INTERNAL_AUTHENTICATE_Exit)``` ```ELSE (``` ``` KeyNumber := APDU.Cmd.P1.b5 ... b1)```	Determine the number of the key to be used from P2.

SEARCH (for the key with the *KeyNumber* in _
DF[Ptr.CurrentDF].EF[Ptr.CurrentIEF.Key])
IF STATUS (SEARCH) = C_Found THEN (
    set *KeyRecord* to the record with the found key
    WITH *DF[Ptr.CurrentDF].EF[Ptr.CurrentIEF.Key]*.)
ELSE (
    Returncode := C_RC_RefDataNotFound
    GOTO INTERNAL_AUTHENTICATE_Exit)
	Search for the reference number requested via P2 in EF Key. If a key with the specified reference number is found, set the current key pointer to reference it.

IF .*Record[KeyRecord].KeyPurpose* <> _
C_CmdINTAUTH THEN (
    Returncode := C_RC_CondOfUseNotSatified
    GOTO INTERNAL_AUTHENTICATE_Exit)
	Test whether the selected key is allowed to be used with the INTERNAL AUTHENTICATE command.

IF *Ptr.CurrentDF = Ptr.MF* THEN (
    // MF is selected
    IF .*Record[KeyRecord].EntryState* <> *SecurityState.MF* _
    THEN (
        Returncode := C_RC_SecStateNotSatisfied
        GOTO INTERNAL_AUTHENTICATE_Exit))
ELSE (
    // DF is selected
    IF .*Record[KeyRecord].EntryState* <> *SecurityState.DF* _
    THEN (
        Returncode := C_RC_SecStateNotSatisfied
        GOTO INTERNAL_AUTHENTICATE_Exit)
	The security state called for in the key record must have been achieved in the currently selected directory (MF or DF).

CALL Kernel_DES_Encrypt
// plaintext: *APDU.Cmd.Data[1 ... 8]*
// key: .*Record[KeyRecord].KeyData[1 ... 8]*
// ciphertext: stored in *APDU.Rsp.Data[1 ... 8]*
	Encrypt the passed-in data (the random number) using the referenced key, and place the result in the transmit buffer.

Returncode := C_RC_OK	The command has been processed with no error, since in all other cases an error exit is used.
INTERNAL_AUTHENTICATE_Exit: END WITH RETURN	

A simple application example

The following simple Smart Card application illustrates the construction and contents of the variables in the EEPROM with the Small-OS operating system. The function of this application can be described in a few words. It allows the creation of a file that is 50 bytes long, whose contents can always be read, and which can be overwritten after the PIN value '1234' has been successfully tested. The reference number of the PIN is 1, and a maximum of three unsuccessful attempts is allowed for the PIN input. The EF containing the file is located underneath its own DF. All file names (DF names and FIDs) can be freely chosen.

Table 5.43 shows how the required capabilities can be implemented by setting appropriate values in the structures used for the file tree. In order to realize the required access conditions, the state machine shown in Figure 5.50 is implemented. After a reset, Small-OS automatically sets the security state of the DF to zero. In this state, the EF containing the data can be read but not written. State 1 is necessary for writing to the file. If a VERIFY command is successfully executed with the correct PIN, the DF is set to security state 1 (.. .*ResultState.OK*), and the file can be written. If the PIN test is not successful, the DF is set to security state 0 (... .*ResultState.NOK*).

The commands READ BINARY AND UPDATE BINARY are used to read and write data from and to the file, respectively. The entire file can be read or written, or only part of it. The only decisive factor here is that the security state of the DF must be appropriate to the desired type of access.

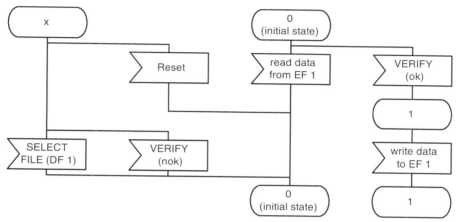

Figure 5.50   State diagram of the file with the FID '0001' (EF 1). Reading data from the EF and writing data to the EF relate to successful READ BINARY and UPDATE BINARY commands to the Smart Card, respectively. State 'x' is any arbitrary state.

This example quite clearly shows two limitations of Small-OS, which arise only from the desire to keep the extent of the pseudocode within reasonable limits. Once security state 1 has been reached, the file can no longer be read, since reading is only allowed in security state 0. This could be remedied by integrating the option of a 'greater than or equal' comparison into Small-OS, in addition to the 'equal' comparison. Another possible solution would be to not limit the file access conditions to only one comparison test per operation (reading or writing), but instead to allow several tests for each operation. In this case, reading the file could be allowed in state 0 as well as in state 1, even if only 'equal' comparisons are possible.

Multiple access conditions for an access operation can be integrated into Small-OS just as easily as testing for greater than or equal. However, the pseudocode would be a bit more extensive, as would be the amount of program code for an actual implementation. With real Smart Card operating systems, such extensions can easily cause the amount of code to exceed the amount of memory available in the microcontroller (ROM or EEPROM). In practice, these strict memory limitations often mean that certain useful functions cannot be implemented.

Table 5.43   Sample values of the file management variables for a simple sample application. There is one DF directly under the MF, and it contains one transparent EF that is 50 bytes long. The content of the EF can be read at any time, but it can be modified only after a PIN has been entered.

Variable name	Description, contents and size
Data structures for the MF	
DF[1].FID := '3F00'	The standard FID of the MF.
Data structures for the DF	
DF[2].FID := 'DF01'	The DF FID is 'DF01' (freely chosen).
DF[2].DFName := 'D276' || '000060'	The registered AID of Wolfgang Rankl is used here for the DF Name.
DF[2].LenDFName := 5	The length of the DF Name is 5 bytes.
Data structures for the EF	
DF[2].EF[1].FID := '0001'	The EF FID is '0001' (freely chosen).
DF[2].EF[1].Structure := C_EFStrucTransp	The structure of the EF is transparent.
DF[2].EF[1].Type := C_EFTypeWorking	The type of the EF is working.
DF[2].EF[1].AccessCondition.Read := 0	The state that is necessary for the READ command to be allowed for the EF. The access condition is set to 0, which means that reading the file is always allowed.
DF[2].EF[1].AccessCondition.Update := 1	The state that is necessary for the UPDATE command to be allowed for the EF. The access condition is set to 1, which means that altering the file is allowed only after successful PIN entry.
DF[2].EF[1].TransparentDataSize := 50	The transparent file is 50 bytes long.
Data structures for the EF Key	
DF[2].EF[2].Record[x].KeyData := '1234'	The (hexadecimal) PIN value is '1234'.
DF[2].EF[2].Record[x].KeySize := 2	The length for the PIN is 2 bytes.
DF[2].EF[2].Record[x].KeyNo := 1	The reference number of the PIN is 1.
DF[2].EF[2].Record[x].RCntr := 0	The initial value of the retry counter is 0.
DF[2].EF[2].Record[x].RCntrMaxValue := 3	The maximum value of the retry counter is 3, which means that the user is allowed a maximum of three incorrect PIN entries.
DF[2].EF[2].Record[x].KeyPurpose := C_CmdVERIFY	The data content of the EF Key record may only be used for PIN testing with the VERIFY command.
DF[2].EF[2].Record[x].EntryState := 0	PIN testing is possible only in state 0.
DF[2].EF[2].Record[x].ResultState.OK := 1	If the PIN comparison is successful, state 1 is set in the current DF.
DF[2].EF[2].Record[x].ResultState.NOK := 0	If the PIN comparison is not successful, state 0 is set in the current DF.

In the application just described, state 1 cannot be exited once it has been achieved by means of a successful PIN verification, which means that the file can no longer be read, due to the access conditions. State 1 can be exited either via an unsuccessful PIN verification or by resetting the Smart Card. In terms of a 'clean' application design, this is a rather unfortunate solution. However, the problem can be remedied using a simple trick to return to the initial state. As you know, selecting a DF causes the security state of the DF in question to be reset. Consequently, the DF can be selected one more time if necessary, which automatically changes the security state from 1 back to 0.

Until recently, applications such as this were coded manually in assembler, in the form shown here. Nowadays, there are application generators for almost all commercially available Smart Card operating systems. These programs run on PCs and have graphical user interfaces for creating files and access conditions. A similar process also takes place in the Smart Card Simulator when a new application or file is created. Once the necessary files for the application have been created, they can be loaded into a Smart Card using the application generator. After this, the first trials with the new application can be carried out.

This sample application can also be profitably used to illustrate some interesting types of attacks. Although these are theoretical in nature, since they require sophisticated technical equipment, they still illustrate some noteworthy principles. The prerequisite for these attacks is the ability to alter specific contents of the EEPROM, which in technical terms amounts to the ability to manipulate the stored charges of individual EEPROM cells. The necessary techniques are discussed in more detail in Chapter 8. Here we only want to look at the consequences of this manipulation.

If it were possible to deliberately alter the content of the file tree pointer that normally indicates the data of the EF, it could be changed to indicate the data content of the EF Key. The EF is always readable in state 0, and in this case, READ BINARY would read the PIN instead of the actual 50 bytes in the EF. Of course, the exact address of the data part of the EF Key must be known for this attack to be used, and a lot of insider knowledge is required to obtain this information. It would be simpler if the variable *DF[2].EF[1].TransparentDataSize* could be altered. If the value of this variable is increased to a large value, for example, then READ BINARY can be used to read an amount of data that extends past the end of the file, according to the new value of the variable. If the EF Key is contained in memory following EF 1, both the EF Key header and the actual file content could be read straightaway.

Manipulation of the EEPROM could also be used to repeatedly reset the PIN retry counter. The PIN could then be determined within an acceptable amount of time by trial and error. An even simpler attack would be to set the PIN itself to a known value.

These examples very clearly show that the security of a Smart Card would completely collapse if it were possible to manipulate the EEPROM. It would make no difference if the contents of the EEPROM could not be read, but could only be overwritten. In any case, the PIN and the keys of the card could be determined. The only thing that would have to be known is the exact memory addresses where the manipulations should be carried out. Checksums for the header contents, if present, would only make the necessary modifications to the EEPROM more complicated, but they ultimately could not prevent them. Of course, it is presently not technically possible to change individual bits at any desired locations in the EEPROM. The approaches just described thus represent theoretically interesting forms of attack rather than actual dangers. However, if this sort of manipulation of the EEPROM were to become possible in the future, these examples clearly show the potential dangers that would arise.

6

Smart Card Data Transfers

The possibility of two-way communications is a prerequisite for all interactions between a Smart Card and a terminal. However, only a single line is available. Digital data is exchanged between the card and the terminal via this electrical connection. Since only one line exists, the card and the terminal must take turns in sending data, while the other party acts as a recipient. The alternate sending and receiving of data is called a half-duplex procedure.

A full-duplex procedure, in which both parties can send and receive simultaneously, is presently not implemented for Smart Cards. However, since most Smart Card processors have two I/O ports, and two of the eight contact fields are reserved for future applications (such as a second I/O line), full-duplex operation would certainly be technically possible. In the medium term, this is sure to be implemented in the hardware and operating systems, since it the only way in which data can be encrypted in real time within the card. Preliminary proposals for standardization of this procedure are already available.

Communication with the card is always initiated by the terminal. The card always responds to commands from the terminal, which means that the card never sends data without an external request. This results in a master–slave relationship, with the terminal as master and the card as slave.

After a card has been inserted in a terminal, its contacts are first mechanically connected to those of the terminal. The five active contacts are then electrically enabled in the correct sequence.[1] Following this, the card automatically executes a power-on reset and then sends an Answer to Reset (ATR) to the terminal. The terminal evaluates the ATR, which reports various parameters relating to the card and data transfers, and then sends the first command. The card processes the command and generates a response, which it sends back to the terminal. This back-and-forth interplay of commands and responses continues until the card is deactivated.

Between the ATR and the first command sent to the card, the terminal can also send a protocol type select (PTS) command. The terminal can use this command, which like the ATR is independent of the transmission protocol, to set various transfer parameters for the card's protocol.

---

[1]   See also Section 3.3.6, 'Activation and deactivation sequences'.

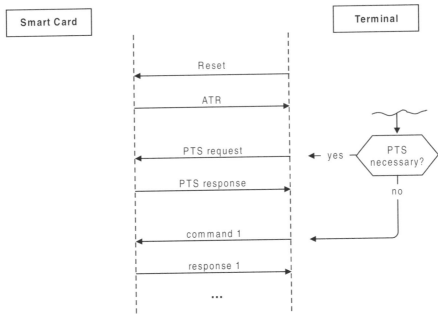

Figure 6.1 The initial data transfer between the terminal and a Smart Card, showing the Answer to Reset (ATR), Protocol Type Selection (PTS) and the first command–response pair

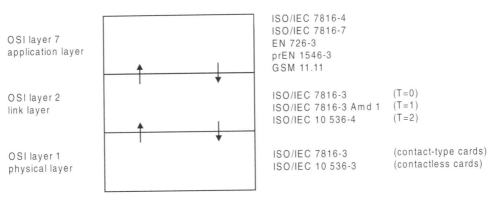

Figure 6.2 OSI model for communication between a terminal and a Smart Card

The whole data transfer process to and from the Smart Card can be represented using the OSI layer model. This distinguishes among electrical events on the I/O line, logical processes in the actual transmission protocol and the behavior of applications that utilize these processes. The behavior and interactions within and between these layers are specified in several international standards. These relationships are illustrated in Figure 6.2.

The asynchronous transmission protocols discussed in this chapter are described in terms of their functions with respect to the relevant standards. All allowed parameters and settings within the context of the protocol are specified. In practice, however, it frequently happens that Smart Cards do not support all options of the transmission protocol, due to the limitations of the available memory.

From a functional perspective, the various options can be regarded as simply a range of possibilities from which an optimum set can be selected for a particular application or Smart Card. The important consideration is that the chosen parameters should not be too exotic, in order that the card can communicate with as many different types of terminals as possible.

6.1   The Physical Transmission Layer

The physical transmission layer, with all its parameters, is specified in the international Smart Card standard ISO/IEC 7816-3. This is the fundamental standard for all aspects of communication at the physical level.

The entire data exchange with the Smart Card takes place digitally, which means that it employs only the logical values '0' and '1'. The voltage levels used are the conventional values for digital technology, namely 0 V and +5 V. New microcontrollers that work with 3-V supplies naturally also support these values for data transfers. The choice of which of the two levels, 0 V or +3/+5 V, represents a logical '1', is arbitrarily, and the actual value is indicated by the card in the first ATR byte. In this regard, 'direct convention' means that a logical '1' is represented by the +3/+5-V level, while 'inverse convention' means that the +3/+5-V level represents a logical '0'. In either case, the level of the I/O line is always high in the idle state, when no data is being transferred.

Communication between a Smart Card and the outside world takes place serially. Data that are processed as bytes must be converted to a serial bit stream. To do this, each byte is separated into its eight individual bits, which are then sent over the line one after the other. The bit order depends on the convention used. In the 'direct' convention, the first data bit after the start bit is the least significant bit in the byte. In the 'inverse' convention, the most significant bit of the byte is sent directly following the start bit.

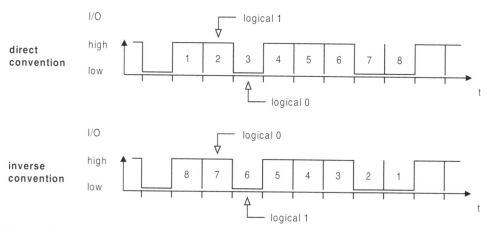

Figure 6.3   Data transfer conventions: (a) data transfer using the direct convention; (b) data transfer using the inverse convention

The data transfer between the card and the terminal proceeds asynchronously, which means that each byte sent must be provided with supplementary synchronization bits. A start bit is added to the beginning of each byte transferred, to indicate the start of the transmission sequence to the recipient. The sender also adds a parity bit for error detection and one or two stop bits to the end of each byte. The time allocated to the stop bits is designated as the *guard time* in the T=0 protocol. In principle, this is a sort of stop bit. The receiver and the sender can both use this time to prepare for the next byte transfer. The parity of each byte must always be even. The parity bit thus has the logical value '1' when the number of ones in the byte is odd and '0' when the number of ones in the byte is even.

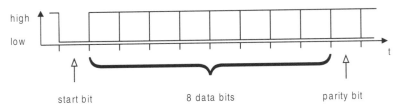

Figure 6.4   Data transfer character structure

Since the timers of Smart Card microcontrollers depend on the applied clock signal, it is not possible to specify the time interval for an individual data bit in absolute terms. It is therefore specified in terms of the applied clock. This is done by indicating a divider value, which gives the number of clock pulses per bit interval. The duration of one bit is called an *etu* (elementary time unit).

It is thus meaningless to specify the data transfer rate of a Smart Card as a fixed value (such as 9600 bit/s), since it is proportional to the rate of the applied clock. However, there are essentially only two divider values in use worldwide: one is 372 and the other is 512. Even smaller divider values are being used increasingly often to increase the transfer rate, but up to now this is still uncommon. However, this will change in the future, since the data transfer rate is one of the bottlenecks for command execution. Reducing the divider value makes it increasingly difficult for the card's operating system to receive and transmit data, since the processor has progressively less time to perform these tasks. For example, if data are received with a divider value of 256, the processor has only 256 clock intervals to recognize each bit and transfer it to the I/O buffer.

To calculate the transfer rates that can be achieved with the standard divider values, we only need to consider the clock rate and the divider value, as shown in the following examples:

$$3.5712 \text{ MHz} \div 372 = 9600 \text{ bit/s}$$

$$4.9152 \text{ MHz} \div 512 = 9600 \text{ bit/s}$$

A data transfer rate of exactly 9600 bit/s can thus be obtained with both commonly used clock frequencies (3.5712 MHz and 4.9152 MHz). The desire for a transfer rate of 9600 bit/s is the reason for the awkward divider values. In the early days of Smart Card technology, inexpensive quartz crystals were available for only very few frequencies. Cheap standard crystals for use in television sets were used, and the divider values for the cards were set to obtain a data transfer rate of 9600 bit/s, which was a common value at the time. A clock rate of 4.77 MHz was used in early PCs for the same reason, since this was compatible with US television sets, and in principle a PC could thus be connected to a television set.

If we assume that 10 MHz is the highest practical value for the clock rate and 32 is the minimum divider value, we obtain the current upper limit for the data transfer rate:

$$10 \text{ MHz} \div 32 = 312{,}500 \text{ bit/s}$$

Of course, it is possible to reduce the divider value even further in order to increase the transfer rate. However, this significantly increases the amount of program code in the card, and so it is not normally done due to the limited amount of available memory. Many new Smart Card microcontrollers have a built-in hardware IC (universal asynchronous receiver/ transmitter, or UART) that carries out data transfers across the serial interface. This sharply reduces the amount of software needed in the card to handle data transfers, so that it is possible to use much higher data transfer rates. Such an interface IC can easily achieve the standard transfer rate of 111.6 kbit/s.[2] This relatively high transfer rate is also currently the upper limit for pure software implementations of data transfer processing using microcontrollers with very high machine instruction execution speeds.

The bit interval can be calculated from the clock rate and the divider value. With a 3.5712 MHz clock frequency and a divider value of 372, we obtain a bit interval of 104 µs, which by definition corresponds to one etu (elementary time unit) for this divider value. We can construct the diagram shown in Figure 6.5 for various transfer rates.

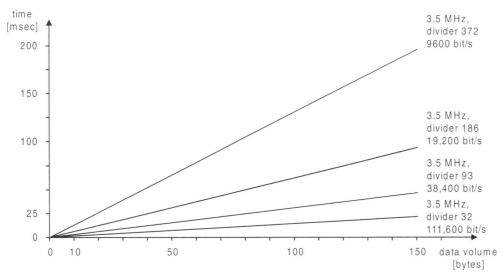

Figure 6.5   Data transfer times for typical transfer rates, assuming 1 start bit, 8 data bits, 1 parity bit and 2 stop bits per byte

---

2   See also Section 15.7.3, 'Determining the data transfer rate'.

The timing of serial data transfers does not have to be strictly controlled. For technical reasons, a certain amount of tolerance is allowed. Since many Smart Card microcontrollers do not have interface hardware, it is sometimes necessary to exploit the allowed tolerance for software implementations. The timing deviation between the falling edge of the start bit and the final transition of the nth bit may not exceed ±0.2 etu. As far as the transmitter is concerned, this means that while the variation in the timing of individual bits may be up to ±0.2 etu, the variation over several bits is also not allowed to exceed this value. The sum of the timing variations over a group of bits must therefore not exceed the allowed tolerance.

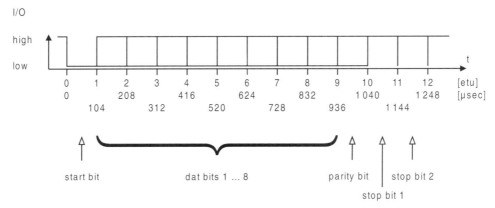

Figure 6.6   Timing diagram for one character at 9600 bit/s (corresponding to a 3.5712-MHz clock and a divider value of 372)

Signal dropouts and overshoots often occur when data are transferred via a physical conductor. Consequently, the incoming signal is sampled multiple times rather than just once. Triple polling followed by a 2-out-of-3 decision is a commonly used procedure. Small distortions in signal levels can thus be compensated with relatively little effort. Increasing the number of samples to five or seven would make little sense, given the generally good quality of the Smart Card data transfers and the amount of extra effort entailed.

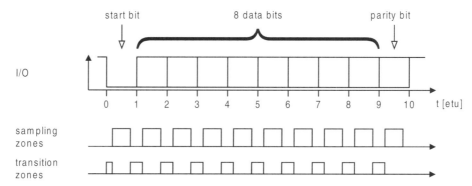

Figure 6.7   Test zones (sampling intervals) and transition zones for the reception of a data byte

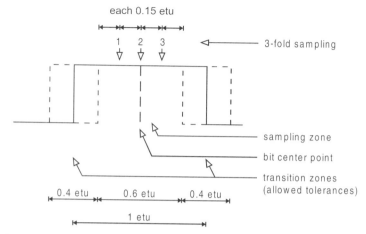

Figure 6.8   Example of threefold sampling of a received bit

The three samples should be distributed as evenly as possible over the received bit interval, in order to best compensate for brief dropouts. This is done by sampling at the middle of the bit interval and at both ends of the test zone, as defined by the applicable timing tolerances for byte transmission. The optimum sampling points can be defined by determining the boundaries of the test zone and the midpoint of the bit interval. However, these are not specified in any standard.[3] Sampling within the transition zone is not allowed, since the signal level is invalid within this zone.

6.2   Answer to Reset (ATR)

After the supply voltage, the clock and the reset signal have been applied, the Smart Card sends an Answer to Reset (ATR) via the I/O lead. This data string, which contains at most 33 bytes, is always sent with a divider value (clock rate conversion factor) of 372, in compliance with the ISO/IEC 7816-3 standard. It contains various data relevant to the transmission protocol and the card. This divider value should be used even if the transmission protocol used after the ATR employs a different divider value (e.g. 512). This ensures that an ATR from any card can always be received, regardless of the parameters of the transmission protocol.

It is very rare for an ATR to have the allowed maximum length. The ATR most often consists of only a few bytes. Particularly with applications where the card should be usable very quickly after the activation sequence, the ATR should be short. A typical example is paying road tolls with a Smart Card electronic purse. Despite the speed of the vehicle passing through the tollgate, it must be possible to reliably debit the card in the short time available.

The start of the ATR transmission must occur between 400 and 40,000 clock cycles after the terminal issues the reset signal. With a clock rate of 3.5712 MHz, this corresponds to an interval of 112 µs to 11.20 ms, while at 4.9152 MHz the interval is 81.38 µs to

---

[3]   See also Section 15.7.4, 'Sampling times'.

8.14 ms.[4] If the terminal does not receive the beginning of the ATR within this interval, it repeats the activation sequence several times (usually up to three times) to try to detect an ATR. If all these attempts fail, the terminal assumes that the card is faulty and responds accordingly.

During the ATR, the time between the leading edges of two successive bytes may be up to 9600 etu according to ISO/IEC 7816-3. This period is designated the *initial waiting time*. This is exactly one second with a clock rate of 3.5712 MHz. This means that the standard permits a one-second delay between the individual bytes of the ATR when it is sent to the terminal. In some Smart Card operating systems, this time is utilized for internal calculations and EEPROM write accesses.

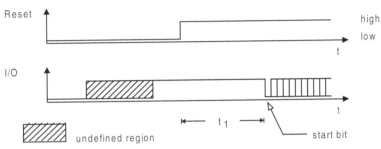

Figure 6.9 Timing diagram of the reset signal and the start of the ATR, in accordance with ISO/IEC 7816-3 (400 clock cycles [t_1 [40,000 clock cycles)

The data string and data elements of the ATR are defined and described in detail in the ISO/IEC 7816-3 standard. The basic ATR format is described in Table 6.1 and Figure 6.10. The first two bytes, designated TS and T0, define various fundamental transfer parameters and the presence of subsequent bytes. The interface characters specify special transfer parameters for the protocol, which are important for the following data transfers. The historical characters describe the extent of the Smart Card's basic functions. The check character, which is a checksum of the previous bytes, may optionally be sent as the last byte of the ATR, depending on the transmission protocol.

Table 6.1   The data elements of the ATR and their meanings, according to ISO/IEC 7816-3

Data element	Description
TS	Initial Character
T0	Format Character
TA1, TB1, TC1, TD1, ...	Interface Characters
T1, T2, ..., TK	Historical Characters
TCK	Check Character

---

4   See also Section 15.7.2, 'ATR data element conversion tables'.

6.2.1   ATR characters

The initial character

This byte, designated 'TS', specifies the convention used for all the data in the ATR and subsequent communications processes. In addition, the TS byte contains a characteristic bit pattern that can be used by the terminal to identify the divider value. The terminal measures the time between the first two falling edges in the TS and divides it by three. The result is the duration of one etu. However, since the divider for the ATR is fixed at 372, the terminal does not normally evaluate the synchronization pattern. This first byte is a mandatory ATR component, and must always be sent. Only two codes are allowed for this byte: '3B' for the direct convention or '3F' for the inverse convention.

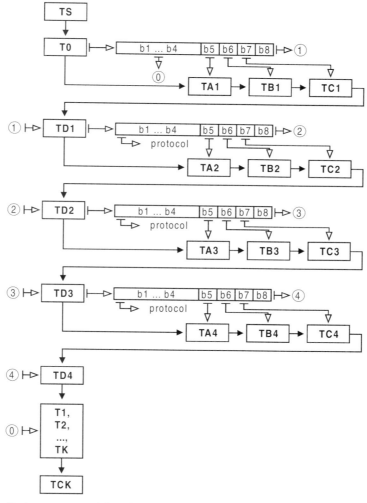

Figure 6.10   Basic structure and data elements of the ATR

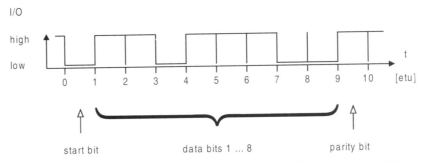

Figure 6.11   Timing diagram of the initial character TS using the direct convention ('3B')

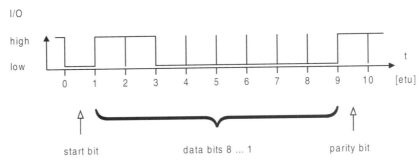

Figure 6.12   Timing diagram of initial character TS using the inverse convention ('3F')

Table 6.2   Initial character (TS) coding

b8	b7	b6	b5	b4	b3	b2	b1	Meaning
				'3B'				direct convention
				'3F'				inverse convention

The direct convention is normally used in Germany, but the inverse convention is normally used in France. The convention does not affect the security of the transfer. Of course, every operating system producer prefers one or the other for historical reasons, but all terminals and many Smart Cards support both the direct and inverse conventions.

The format character
The second byte, T0, contains a bit field that indicates which interface characters are to be sent. It also indicates the number of subsequent historical characters. Like TS, this byte must be present in every ATR.

The interface characters
The interface characters specify all transfer parameters of the protocol used. They consist of the bytes TAi, TBi, TCi and TDi. However, these bytes are optional in the ATR, and they may be omitted as appropriate. Since basic values are defined for all of the parameters of the protocol, interface characters are often not required in the ATR for a normal communications process.

Table 6.3 Coding of the T0 Format Character

b8	b7	b6	b5	b4	b3	b2	b1	Meaning
...	...	...	...	X	X	X	X	Number of historical characters (0 to 15)
...	...	...	1	...	...	...	...	TA1 sent
...	...	1	...	...	...	...	...	TB1 sent
...	1	...	...	...	...	...	...	TC1 sent
1	...	...	...	...	...	...	...	TD1 sent

The interface characters can be divided into *global interface characters* and *specific interface characters*. The first type specifies fundamental transmission protocol parameters, such as the divider value, that apply to all subsequent protocols. The second type, by contrast, defines parameters for a very specific transmission protocol. The 'work waiting time' for the T=0 protocol is a typical example of such a parameter.

The global interface characters basically apply to all protocols. For historical reasons (since originally only the T=0 protocol was included in ISO standard), several of these characters are only relevant to the T=0 protocol. If T=0 is not implemented, they can be omitted, in which case default values apply.

The TDi byte is only used to protect the links to any following interface characters. The upper nibble of the TDi byte contains a bit pattern that indicates the presence of the subsequent interface characters. This is analogous to the coding of the format character T0. The lower nibble of the TDi byte identifies the currently available transmission protocol.

If no TDi byte is present, TAi+1, TBi+1, TCi+1 and TDi+1 are not transferred.

Table 6.4 Coding of the TDi byte

b8	b7	b6	b5	b4	b3	b2	b1	Meaning
...	...	...	...		X			Transmission protocol number (0 to 15)
...	...	...	1		...			TAi+1 sent
...	...	1	...		...			TBi+1 sent
...	1	...	...		...			TCi+1 sent
1	...	...	...		...			TDi+1 sent

The other interface characters (TAi, TBi and TCi), which are not used for linking, specify the available transmission protocol(s). Their meanings according to the ISO/IEC 7816-3 standard are as follows.

The global interface character TA1 The divider (clock rate conversion factor F) is encoded in the upper nibble as FI. The bit adjustment factor D is encoded in the lower nibble as DI.

Table 6.5   TA1 coding

b8	b7	b6	b5	b4	b3	b2	b1	IFSC
		X			...			FI
...				X				DI

Table 6.6   FI coding

F	internal clock		372	558	744	1116	1488	1860	RFU
FI	0000		0001	0010	0011	0100	0101	0110	0111
F	RFU	512	768	1024	1536	2048	RFU	RFU	...
FI	1000	1001	1010	1011	1100	1101	1110	1111	...

Table 6.7   DI coding

D	RFU	1	2	4	8	16	RFU	RFU
DI	0000	0001	0010	0011	0100	0101	0110	0111
D	RFU	1/2	1/4	1/8	1/16	1/32	1/64	...
DI	1001	1010	1011	1100	1101	1110	1111	...

The encoding of the divider F and adjustment factor D allows typical transfer rates to be specified in accordance with the standard. This is summarized in the Appendix (see Section 15.7.2, 'ATR data element conversion tables'). The following relationships apply:

- The bit interval for the ATR and PTS, which is called the *initial etu*, is specified as

$$initial\ etu = \frac{372}{f_i}\ \text{s}$$

- The transmission protocol that follows the ATR and PTS defines the bit interval independently of the ATR. This interval, which is called the *work etu*, is defined as

$$work\ etu = \frac{1}{D}\frac{F}{f_s}\ \text{s}$$

These two parameters, the *bit rate adjustment factor* D and the *clock rate conversion factor* F, allow the transfer rate to be adjusted as necessary for individual circumstances. The frequency of the applied clock is shown in the above formulas as f, with the unit of Hertz.

The global interface character TB1   Bits b7 and b6 of this byte encode a programming voltage factor called 'II'. Bits b5 through b1 define the parameter 'PI1'. The most significant bit, b8, is always set to 0, which means that it is not used. These parameters were needed in the first generation of Smart Cards, which used EPROM for data storage instead of EEPROM, which is currently standard. The necessary high voltages and currents for EPROM programming had to be supplied by the terminal via the Vpp contact. However, since Smart Cards without internal charge pumps no longer exist, we can ignore the detailed coding of this byte.

Table 6.8   TB1 coding

b8	b7	b6	b5	b4	b3	b2	b1	IFSC
0	X		...	...	...	...	...	II
0	...		X	X	X	X	X	PI1

Parameters PI1 and II thus always have the value 0, which indicates that no external programming voltage is needed. If the data element TB1 is omitted in the ATR, the default Vpp setting of 5 V at 50 mA applies, as specified in the standard.

The global interface character TC1   This encodes an extra guard time, designated N, as an unsigned hexadecimal integer. This value specifies the number of etu for extending the guard time. TC1 is interpreted linearly, with the exception that if N = 'FF', the standard guard time must be reduced to 11 etu with the T=1 protocol. With the T=0 protocol, the extra guard time stays at 12 etu, to allow an error to be indicated by a low level within the guard time interval. In practice, reducing the guard time to 11 etu yields a speed increase of around 10% with the T=1 protocol, since one less bit has to be sent.

Table 6.9   TC1 coding

b8	b7	b6	b5	b4	b3	b2	b1	IFSC
			X = 255, T = 0					N = 12 etu
			X = 255, T = 1					N = 11 etu

The global interface character TB2   This byte holds the value of PI2. The parameter specifies the external programming voltage in tenths of a volt. It is thus normally no longer used in the ATR, for the same reason as TB1.

Table 6.10   TB2 coding

b8	b7	b6	b5	b4	b3	b2	b1	IFSC
				X				PI2

Specific interface character for the T=0 transmission protocol

The specific interface character TC2   The final data element in the T=0 protocol encodes the 'work waiting time' and is designated 'TC2'. The work waiting time is the maximum time interval between the rising edges of two consecutive bytes:

$$work\ waiting\ time = (960 \cdot D \cdot WI)\ work\ etu$$

If the TC2 character is not present in the ATR, the default value of the work waiting time is used ($WI = 10$).

Table 6.11   TC2 coding

b8	b7	b6	b5	b4	b3	b2	b1	Meaning
			X					WI

Specific interface characters for the T=0 transmission protocol

The following additional bytes are defined for the T=1 transmission protocol, in accordance with ISO/IEC 7816-3 Amd 1. Here the interface character prescribed for T=0 is used only as required. For this protocol, the data element index i must always be greater than 2. In this case, the specific interface characters TAi, TBi and TCi ($i > 2$) always apply to the transmission protocol specified by $TD_{(i-1)}$.

Specific interface character TAi ($i > 2$)   The TAi byte contains the maximum length of the information field that can be received by the card (IFSC). This value must lie between 1 and 254. The default value of IFSC is 32 bytes.

Table 6.12   TAi coding for i > 2

b8	b7	b6	b5	b4	b3	b2	b1	Meaning
			X					IFSC

Specific interface character TBi ($i > 2$)   The lower nibble (the four bits b4 through b1) contains the code CWI for the character waiting time CWT,[5] which is calculated as

$$CWT = (2^{CWI} + 11) \text{ work etu}$$

The upper nibble holds the value BWI, with which the block waiting time BWT[6] can be calculated as follows:

$$BWT = 2^{BWI} \cdot 960 \cdot \frac{372}{f} s + 11 \text{ work etu}$$

Table 6.13   TBi coding for i > 2

b8	b7	b6	b5	b4	b3	b2	b1	Meaning
	...				X			CWI
	X					...		BWI

Specific interface character TCi ($i > 2$)   Bit b1 encodes method used for calculating the error detection. Since the standard covering the data elements of the ATR does not define all the possible transmission protocol parameters in terms of interface characters, different implementations can use additional interface characters.

---

5   See also Section 15.7.1, 'ATR interval'.
6   See also Section 15.7.2, 'ATR data element conversion tables'.

Table 6.14   TCi coding for i > 2

b8	b7	b6	b5	b4	b3	b2	b1	Meaning
...	...	...	...	...	...	...	0	LRC in use
...	...	...	...	...	...	...	1	CRC in use
0	0	0	0	0	0	0	...	Reserved for future applications

A typical example is provided by the German national T=14 protocol. Several additional ATR bytes are defined for this protocol to satisfy its specific needs. These can be decoded only by users of this protocol, since only they know the applicable specification. This is not standardized, nor is it made known outside of the applications that use it.

Global interface character TA2   This byte indicates the allowed modes for the PTS. This is explained in more detail in Section 6.4, which describes the PTS.

Table 6.15   TA2 coding

b8	b7	b6	b5	b4	b3	b2	b1	Meaning
0	...	...	...		...			Switching between negotiable mode and specific mode possible
1	...	...	...		...			Switching between negotiable mode and specific mode not possible
...	0	0	...		...			Reserved for later applications
...	...	...	0		...			Transfer parameters explicitly defined in the interface characters
...	...	...	1		...			Transfer parameters implicitly defined in the interface characters
...	...	...	...		X			Protocol T=X to be used

The historical characters

For a long time, the historical characters were not defined by any standard. As a result, they contain a wide variety of data, depending on the producer of the operating system.

Many firms use the available bytes to identify the operating system and the associated version number of the ROM mask. This is usually encoded in ASCII, so it is easy to interpret. The presence of the historical character within the ATR is not prescribed, so it is possible to omit them altogether. In some cases, this can be beneficial, as it makes the ATR shorter and thus quicker to send.

The ISO/IEC 7816-4 standard provides for an ATR file, in addition to the historical characters. This file, with the reserved FID '2F01', contains additional data about the ATR. It is intended to be an extension to the historical characters, which are limited to 15 bytes. The content of this file, whose structure is not defined by the standard, is ASN.1 encoded.

The data elements in the ATR file, or the historical characters, may contain complex information relating to the Smart Card and the operating system in use. For example, they can be used to store file selection functions and implicit selection functions supported by the Smart Card, as well as information regarding the of logic channel mechanism. They can also be used to store additional information regarding the card issuer, the card and chip serial numbers, and the versions of the ROM mask, chip and operating system. The coding of the relevant data objects is defined in the ISO/IEC 7816-4 and -5 standards.

The check character

This last byte in the ATR is the XOR checksum of byte T0 through the last byte before the check character (TCK). This checksum can be used in addition to parity testing to verify the correctness of the ATR transfer. However, in spite of the apparently simple construction and calculation of this checksum, there are several significant differences between the various transmission protocols.

If only the T=0 transmission protocol is indicated in the ATR, the TCK checksum may not be present at the end of the ATR. In this case it is not sent at all, since byte-wise error detection using parity checking and repetition of the faulty byte is mandatory in the T=0 protocol. In contrast, a TCK byte must be present in with the T=1 protocol. The checksum is calculated starting with byte T0 and ending with the final interface character, or the final historical character if present.

6.2.2   Practical examples of ATRs

The tables in this section show practical examples of various types of Smart Card ATRs. They are very useful as an aid to interpreting ATRs or defining your own ATRs.

Table 6.16   Sample Smart Card ATR with T=1

Designation	Value	Meaning		Remark
TS	'3B'	direct convention		
T0	'B5'	Y1	= 1011 = 'B'	TA1, TB1 and TC1 follow
		K	= '5'	5 historical characters
TA1	'11'	FI	= 0001 = '1'	F = 372
		DI	= 0001 = '1'	D = 1
TB1	'00'	II	= 0	I = 0
		PI1	= 0000 = '0'	Vpp contact not used
TD1	'81'	Y2	= 1000 = '8'	TD2 follows
		T	= 1	
TD2	'31'	Y2	= 0011 = '3'	TA3 and TB3 follow
		T	= 1	
TA3	'46'	I/O buffer size = 70 bytes		ICC I/O buffer size (layer 7)
TB3	'15'	BWI	= '1'	BWT = 2011 etu
		CWI	= '5'	CWT = 43 work etu
T1	'56'	"V"		"V 1.0"
T2	'20'	" "		
T3	'31'	"1"		
T4	'2E'	"."		
T5	'30'	"0"		
TCK	'1E'	check character		XOR checksum of T0 through T5

Table 6.17 Sample STARCOS Smart Card ATR with T=1 and the direct convention, with the operating system not completed

Designation	Value	Meaning		Remark
TS	'3B'	direct convention		
T0	'9C'	Y1	= 'E' = °1001°	TA1 and TD1 follow
		K	= 'C' = 12	12 historical characters
TA1	'11'	FI	= 0001 = '1'	F = 372
		DI	= 0001 = '1'	D = 1
TD1	'81'	'2'	= °1000°	TD2 follows
		T	= 1	T=1 is used
TD2	'21'	'2'	= °0010°	TB2 follows
		T	= 1	T=1 is used
TB2	'34'	CWI = 3		character waiting time
		BWI = 4		block waiting time
T1 ... T12	'53' || '43' || '20' || '53' || '56' || '20' || '31' || '2E' || '31' || '20'|| '4E' || '43'			"SC SV 1.1 NC"
TCK	'0F'	check character		XOR checksum of T0 through T12

Table 6.18 Sample STARCOS Smart Card ATR with T=1 and the direct convention, with the operating system completed

Designation	Value	Meaning		Remark
TS	'3B'	direct convention		
T0	'BF'	Y1	= 'F' = °1001°	TA1 and TD1 follow
		K	= 'B' = 11	11 historical characters
TA1	'11'	FI	= 0001 = '1'	F = 372
		DI	= 0001 = '1'	D = 1
TD1	'81'	'2'	= °1000°	TD2 follows
		T	= 1	T=1 is used
TD2	'21'	'2'	= °0010°	TB2 follows
		T	= 1	T=1 is used
TB2	'34'	CWI = 3		character waiting time
		BWI = 4		block waiting time
T1 ... T15	'53' || '54' || '41' || '52' || '43' || '4F' || '53' || '32' || '31' || '20' || '43'			"STARCOS 21 C"
TCK	'43'	Check Character		XOR checksum of T0 through T15

Table 6.19 (part 1) Sample GSM Smart Card ATR with T=0 and the direct convention

Designation	Value	Meaning		Remark
TS	'3B'	direct convention		
T0	'89'	Y1	= '8' = °1000°	TD1 follows
		K	= 9	9 historical characters

Table 6.19 (part 2)   Sample GSM Smart Card ATR with T=0 and the direct convention

Designation	Value	Meaning	Remark
TD1	'40'	'4' = °0100° T =0	TC2 follows T=0 is used
TC2	'14'	WI = '14'	the work waiting time is '14'
T1 ... T9	'47' || '47' || '32' || '34' || '4D' || '35' || '32' || '38' || '30'		"GG24M5280"

Table 6.20   Sample GSM Smart Card ATR with T=0 and the inverse convention

Designation	Value	Meaning	Remark
TS	'3F'	inverse convention	
T0	'2F'	Y1 = '2' = °0010° K = 'F' = 15	TB1 follows 15 historical characters
TB1	'00'	PI1 = °00000° II = °00°	Vpp is not used; the programming voltage for the EEPROM is generated in the chip
T1 ... T15	'80' || '69' || 'AE' || '02' || '02' ||'01'|| '36' || '00' || '00' || '0A' || '0E' || '83' || '3E' || '9F' || '16'		specific data of the Smart Card manufacturer

Table 6.21   Sample Visa Cash Smart Card ATR with T=1 and the direct convention

Designation	Value	Meaning	Remark
TS	'3B'	direct convention	
T0	'E3'	Y1 = 'E' = °1110° K = 3	TB1, TC1 and TD1 follow 3 historical characters
TB1	'00'	PI1 = °00000° II = °00°	Vpp not used; EEPROM programming voltage generated internally
TC1	'00'	N = 0	no extra guard time
TD1	'81'	'8' = °1000° T=1	TD2 follows T=1 is used
TD2	'31'	'3' = °0011° T=1	TA3 and TB3 follow T=1 is used
TA3	'6F'	IFSC = '6F' = 111	the information field size of the Smart Card is 111 bytes
TB3	'45'	CWI = 4 BWI = 5	character waiting time block waiting time
T1	'80'	category indicator	data object in the compact TLV format follows (compliant with ISO/IEC 7816-4)
T2	'31'	card service data	label ('3') and length ('1') of the data object for the card service data
T3	'C0'	'C0' = °1100 0000°	application selection by the complete or partial DF name
TCK	'08'	check character	XOR checksum of T0 through T3

6.3   Protocol Type Selection (PTS)

The Smart Card shows various data transfer parameters in the interface characters of the ATR, such as the transmission protocol and the character waiting time. If a terminal wants to modify one or more of these parameters, a Protocol Type Selection (PTS) must be performed before actual execution of the protocol, as per ISO/IEC 7816-3. The terminal can use this to modify certain protocol parameters, as long as this is permitted by the card. The PTS is sometimes also referred to as Protocol Parameter Selection (PPS).

   The PTS can be executed in two different modes. In the *negotiable* mode, the standard values of the divider F and the transfer adjustment factor D remain unchanged until the PTS is successfully executed. On the other hand, if the card uses the *specific* mode, the values for F and D given in the ATR are mandatory for the PTS transfer as well. The TA2 byte indicates which of these two modes is supported by the card.

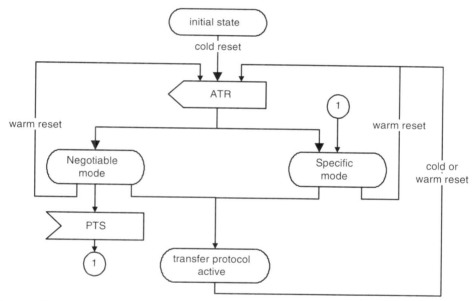

Figure 6.13   State diagram of the two PTS modes (Negotiable mode and Specific mode), according to ISO/IEC 7816-3

Table 6.22   PTS data elements and their designations according to ISO/IEC 7816-3

Data element	Designation
PTSS	Initial Character
PTS0	Format Character
PTS1, PTS2, PTS3	Parameter Characters
PCK	Check Character

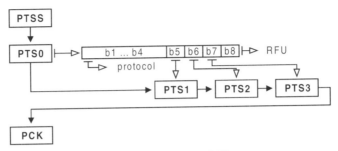

Figure 6.14   The basic structure and data elements of the PTS

The PTS Request must be executed immediately after the ATR has been received by the terminal. If the card permits the requested modifications to the protocol parameters, it sends the received PTS bytes back to the terminal. In principle, this is an echo of the received data. Otherwise, the card sends nothing and the terminal executes a new reset sequence to cause the card to exit this state. The PTS may be executed only once, immediately after the ATR. Repeated PTS transfers are forbidden by ISO/IEC 7816-3.

In practice, it is very rare for a PTS to be used at all, since transfer parameters of the Smart Cards currently in use are exactly matched to those of the terminal.

The first byte is the initial character (PTSS), which unambiguously informs the card that the terminal will initiate a PTS request immediately after the ATR. It therefore always has the value 'FF', and it must be sent in every PTS. The data element following the PTSS is the format character (PTS0). This is also a required part of every PTS. It may optionally be followed by up to three bytes, which are called the parameter characters and are designated PTS1, PTS2 and PTS3. They encode various parameters for the transmission protocol that will be used following the PTS. The data element PTS3 is reserved for future use, so it cannot yet be described here. The last byte in the PTS is called the control character (PCK). It contains the XOR checksum of all previous bytes, starting with PTSS. As with the PTSS and PTS0, it is a required part of the PTS, in contrast to the other optional data elements.

Table 6.23   PTS0 coding

b8	b7	b6	b5	b4	b3	b2	b1	Meaning
...	...	...	...			X		Transmission protocol to be used
...	...	...	1		...			PTS1 available
...	...	1	...		...			PTS2 available
...	1	...	...		...			PTS3 available
0	...	...	...		...			Reserved for future use

Table 6.24   PTS1 coding

b8	b7	b6	b5	b4	b3	b2	b1	Meaning
X				...				FI
...				X				DI

Table 6.25   PTS2 coding

b8	b7	b6	b5	b4	b3	b2	b1	Meaning
...	...	...	...	...	...	0	0	No additional guard time necessary
...	...	...	...	...	...	0	1	N = 255
...	...	...	...	...	...	1	0	Additional guard time of 12 etu
X	X	X	X	X	X	...	...	Reserved for future use

If the card can interpret the PTS and then modify the transmission protocol accordingly, it acknowledges this by sending the received PTS back to the terminal. If the PTS request contains items that the card cannot execute, it simply waits until the terminal executes a reset. The main disadvantage of this procedure is that a large amount of time can be lost before the actual transmission protocol is employed.

The PTS just described will not work for protocol switching with terminals that cannot execute a PTS, but which nonetheless have their own special transmission protocols. This is however precisely the case with German cardphones, for example. A special procedure has been devised to allow protocol switching in spite of this limitation.

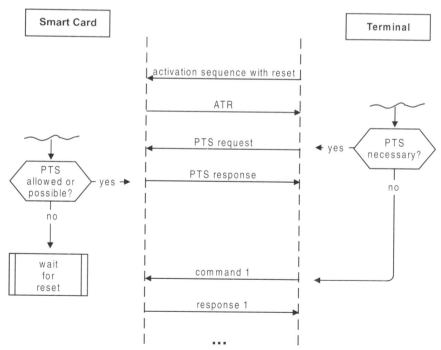

Figure 6.15   A typical PTS procedure with a GSM card

Since all terminals perform multiple reset sequences if they do not recognize the ATR, it was decided that the Smart Card should switch transmission protocols after every reset. This can best be illustrated by an example. With the first reset, the card sends the ATR for T=14 and is then ready to communicate with the T=14 protocol. After the second reset, it sends an ATR for T=1 and can communicate using the T=1 protocol. After the third reset, it again runs with the T=14 protocol. This is not an ideal technical solution, since a device should always behave the same after every reset, but it is a completely practical solution for a heterogeneous terminal world.

It is possible to reduce this disadvantage by having the Smart Card always respond with the same ATR after a power-on reset (cold reset). The card always executes a cold reset directly after it is has been inserted in the card reader and the activation sequence has been completed. A reset that it triggered via the card's reset line (warm reset), on the other hand, switches the transmission protocol. The card thus behaves the same after every 'real' reset, and any additional reset triggering causes the transmission protocol to be switched.

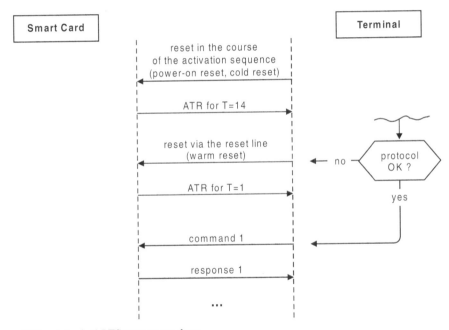

Figure 6.16 A typical PTS reset procedure

6.4   Data Transmission Protocols

After the Smart Card has sent an ATR, and a PTS has possibly taken place, it waits for first command from the terminal. The subsequent process always corresponds to the master–slave principle, with the terminal as master and the card as slave. In concrete terms, the terminal sends a command to the card, and the latter executes it and subsequently returns a response. This back-and-forth interplay of commands and responses never changes.

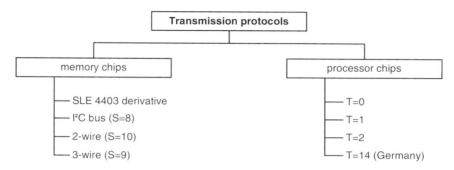

Figure 6.17   Classification of transmission protocols used with Smart Cards

There are various ways in which communication with a Smart Card can be constructed. There are also a number of different processes for re-synchronizing communications if a disturbance occurs. The exact implementation of the commands, the corresponding responses and the procedures used in the event of transfer errors are defined in the form of *transmission protocols*.

There are a total of 15 transmission protocols available for which the basic functions are defined. These protocols, which are designated as 'T=' (for 'transmission protocol') plus a sequential number, are summarized in Table 6.26.

Table 6.26   Summary of transmission protocols defined by ISO/IEC 7816-3

Transmission protocol	Meaning
T=0	Asynchronous, half-duplex, byte oriented, covered by ISO/IEC 7816-3
T=1	Asynchronous, half duplex, block oriented, covered by ISO/IEC 7816-3 Amd 1
T=2	Asynchronous, full duplex, block oriented, covered by ISO/IEC 10536-4
T=3	Full duplex, not yet covered
T=4	Asynchronous, half duplex, byte oriented, extension of T=0, not yet covered
T=5 to T=13	Reserved for future functions, not yet covered
T=14	For national functions, no ISO standard
T=15	Reserved for future functions, not yet covered

Two of these protocols predominate in international use. The first is the T=0 protocol, which was made an international standard in 1989 (ISO/IEC 7816-3). The other is T=1, which was introduced in 1992 as an appendix to an international standard (at the time in ISO/IEC 7816-3 Amd 1, and now in ISO/IEC 7816-3). The full duplex transmission protocol T=2, which is strongly based on T=1, is currently in preparation and will be available as an international standard in a few years.

In Germany, a third protocol, designated T=14, is used for the widely available cardphones. It is defined in an internal specification of the German Telekom.

The data units that are transported by transmission protocols are called TPDUs (transmission protocol data units). They are, in a manner of speaking, protocol-dependent containers that transport data to and from the card. The actual application data is embedded in these containers.

In addition to the technically complex Smart Card transmission protocols, there is an additional set of very simple synchronous protocols for memory cards. They are typically used with telephone cards, health insurance cards and the like. However, they do not have error-correction mechanisms, and they are based on hard-wired logic in the chip.

6.4.1 Synchronous data transmission

Synchronous data transmission is not used with microcontroller-based Smart Cards, since they only communicate with the terminal asynchronously. However, it is the standard process with memory cards, which are used in very large numbers as prepaid electronic purses for cardphones, for example. This widespread use justifies a description of the operation of synchronous data transfer.

In memory cards, synchronous data transmission is very closely linked to the chip's hardware and is designed to be as simple as possible. There is no separation of layers in transmission protocol, nor is logical addressing present, so the application in the terminal must directly access memory addresses in the chip. The protocol enables the data stored in the chip to be physically addressed and then read or written. This means that the actual data transfer process is also linked to the functions of memory addressing and management.

There is also no procedure for detecting or correcting errors during the data transfer, although it must be said that such errors between the card and the terminal occur very rarely. If the terminal application nonetheless detects a transfer error, it must re-read the relevant area in the card's memory. All these restrictions serve to enable data to be transferred between the card and the terminal at a high rate and using only a small amount of logic hardware.

Since synchronous data transmission is only used to make the data transfer as simple as possible (which means with a minimum number of logic circuits), the almost unavoidable consequence is a strong dependence on the hardware. This means that synchronous transmission protocols are not uniform, and sometimes vary greatly from chip to chip. Only the ATR is standardized. A terminal that has to communicate with various types of memory cards must thus incorporate several different implementations of synchronous data transmission protocols.

The exact designation of the data transfer used for memory cards is 'clock-synchronous serial data transmission'. This clearly indicates the basic conditions that apply to this type of communication. As with asynchronous communications, the data are sent between the card and the terminal serially, or bit by bit. However, the bits are sent synchronous to a supplementary transmit clock signal. This makes the transfer of start and stop information unnecessary.

In the case of a simple memory card, there is also no error detection information, which means that neither a parity bit nor a supplementary checksum is sent. The low probability of transfer errors is due to the very low clock rate, which ranges between 10 kHz and 100 kHz. Since one bit is sent for each clock cycle, a clock rate of 20 kHz yields a transfer rate of 20 kbit/s. The effective rate is however lower, since additional address information must also be sent for memory cards.

In order to describe synchronous data transfers for memory cards in an understandable manner, we first need to describe some of the basic features of memory cards. In their simplest form, these cards have memories that are divided into two parts, which are a fixed ROM region and an EEPROM region that can be written and erased. Both regions are bit-addressable and can be freely read, and in the case of the EEPROM also written and erased.

The master–slave relationship is even more pronounced in memory cards than in microcontroller Smart Cards. For example, the terminal completely takes over the physical addressing of memory. The card itself can only block certain areas globally against erasing. This is controlled by hard-wired logic located in front of the memory. This logic also manages the very simple data transfer process.

6.4.1.1   The telephone chip protocol

Data transfer is illustrated here using a phone card containing an Infineon SLE4403 chip as an example. The memory in this IC is bit-oriented, which means that all operations are carried out on individual bits. Other types of chips may have protocols that differ from that described here. However, the basic data transfer principles are the same for all synchronous cards.

Data are transferred using three leads. The bi-directional data lead is used by both the card and the terminal to exchange single-bit data. The clock lead transmits the clock generated by the terminal to the card. This clock provides the reference for the synchronous data transmission. The third connection needed for the data transfer is the control lead. It determines what the chip actually does, based on the states of the other two leads.

In principle, complete control of a memory card requires the chip's logic circuitry to decode four different functions. These are *read*, *write*, *clear memory* and *increment the address pointer*. A memory card has a global memory pointer, with which all memory regions can be addressed bit by bit. If the pointer reaches the upper memory boundary, it rolls over to zero. With a bit-oriented chip design, it then points to the first bit in memory. One of the functions of synchronous data transfer is to reset this pointer to an initial value, which is normally zero.

The next function is to read data from the memory. The other two functions are writing and erasing EEPROM bits. Erasing EEPROM bits, which would allow them to be rewritten, is of course blocked in phone cards, since otherwise they could be reloaded.

Resetting the address pointer   The address pointer is reset to its initial value of zero by the power-up logic of the card if the clock lead and the control lead are simultaneously at a high level. However, the control pulse must be applied for a somewhat longer period than the clock pulse, in order to prevent the address from being immediately incremented. The address pointer should be reset to its original value after each activation sequence, since it would otherwise be pointing to an undefined address.

Incrementing the address pointer and reading data   If the control lead is at low level and the clock pulse has a rising edge, the internal logic of the card increments the address pointer by one. The falling edge of the clock causes the content of the address indicated by the pointer to be placed on the data lead. If the pointer reaches its maximum value, which depends on the size of the memory, it rolls over to zero and thus starts over from the beginning.

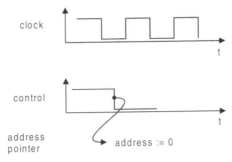

Figure 6.18   Resetting the address pointer to zero

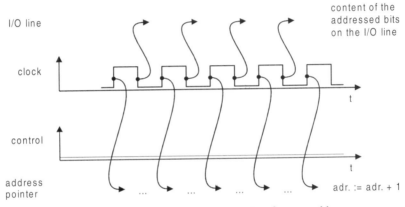

Figure 6.19   Incrementing the address pointer and reading data from an address

Writing to an address   If the address pointer is within a writable EEPROM region, the value on the data lead can be written to EEPROM by applying a high level to the control lead and then pulling the clock lead low. The length of the write cycle is determined by the duration of the immediately following clock pulse. If the bit was written correctly, the content of the written memory cell appear at the data output.

Erasing bytes   Part of the EEPROM memory in a typical phone card is always organized as a multi-place octal counter. If a byte has to be erased in this counter because of a carry to the next place, this is performed the logic circuitry. Erasing a byte in memory is thus somewhat more complicated. The procedure is as follows: if a bit within a byte is written twice in a row, the chip's hardware logic automatically erases the associated less significant byte. This ensures that a carry to the next higher place occurs, while the lower place is erased without allowing any opportunity for fraud.

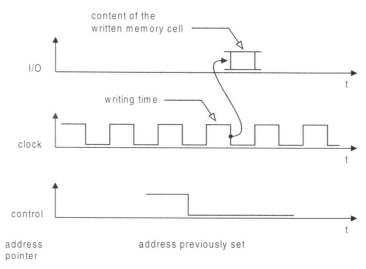

Figure 6.20   Writing a bit to an address in EEPROM memory

The four types of access that have been described may vary from chip to chip and also between manufacturers. Another type of data transfer, which by contrast is standardized, is represented by the I²C bus. Many of the newer memory cards use this bus for communicating with the terminal. This naturally has the advantage that different chips, made by different manufacturers, can be used together within a single system. Problems due to several different transmission protocols are thus eliminated, since all chips are mutually compatible at the transfer interface.

6.4.1.2   The I²C bus

Since serial, clock-synchronized data transmission protocols are uncomplicated and versatile, they are used relatively often. Components for use with the I²C (inter-integrated circuit) bus, which was developed by Philips, have been available since 1990. This bus is based on a serial, bi-directional data lead and a serial clock lead. The definition of the I²C bus encompasses both the hardware (the two leads) and the software, in the sense of the data transfer format. Each IC on the bus can take control of the bus and send requests to other ICs connected to the bus.

Since memory cards are also controlled by a synchronous clock, the I²C bus has very quickly established itself in the chip card industry. A wide range of memory ICs has become available for incorporation in cards. The following example is based on the SGS-Thomson ST24C04 memory chip. It has a 512-byte EEPROM that can be freely written and read. Timing calculations for EEPROM programming are handled internally by the chip, so this does not have to be controlled externally.

The hardware for the I²C bus consists of two lines between the terminal and the card. The SCL (serial clock) line carries the clock, which can range up to 100 kHz. This yields a data transfer rate of up to 100 kbit/s, which is relatively high for Smart Cards. The other line, SDA (serial data), is used bi-directionally to exchange data between the card and the terminal. The SDA line is connected in the terminal to the supply voltage (Vcc) via a pull-up resistor. Both communicating parties can only pull this line to ground. Sending a high level is therefore passive, and involves the sender switching its output to a high-impedance state (tri-state), so that the pull-up resistor pulls the SDA line to the supply voltage level.

In the Smart Card context, the terminal is always the master of the I^{2}C bus and the card always the slave. The data transfer always uses single-byte packets. The most significant bit (bit 8) of the byte is sent first. Each transfer over the SDA line is initiated with a start signal and terminated with a stop signal. The start signal consists of a falling edge on the SDA line while the signal on the SCL line is high. Conversely, a rising edge on the SDA line while the signal on the SCL line is high indicates a stop signal. The recipient must acknowledge the reception of each byte by earthing the SDA line for one clock cycle.

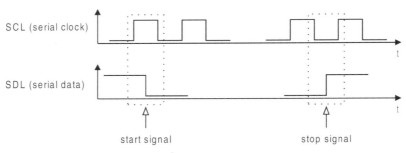

Figure 6.21   Start and stop signals on the I^{2}C bus

After the start of communications, the first seven bits of the first byte are the address of the recipient. In our example, we assume for simplicity that the address has the binary value 1010000x. Of course, this may vary depending on the chip type, and it can be chosen within certain limits for some memory ICs. The last bit in the address (x) indicates to the recipient whether data is to be read or written. A 1 indicates reading, while 0 is for writing.

The following examples illustrate the general functions of the I^{2}C bus as used with Smart Cards.

Reading from an address   There are several types of access for reading the EEPROM of a Smart Card. In the type described here, one byte is read at a time. However, it is also possible to read several bytes in a row.

The read sequence is initiated by the start signal. The following bits contain the address of the card with the control bit for writing set. This indicates to the card that it must temporarily store the following data in an internal buffer. This buffer is nothing more than a byte-oriented address pointer to the EEPROM. After the card receives the first byte, it sends an acknowledgement by earthing the SDA line for one clock cycle. After this, the terminal sends the EEPROM address to the card. Once again, the card acknowledges receipt of the data. The terminal then sends a start signal and the card address with the read bit set. On receiving this, the card sends the data from the location addressed by the pointer to the terminal. The terminal does not have to acknowledge the receipt of the data; it only sends a stop signal to the card. This completes the read sequence for one byte.

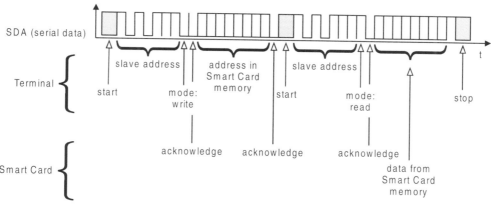

Figure 6.22   Unconstrained reading of a byte from memory using the I²C bus

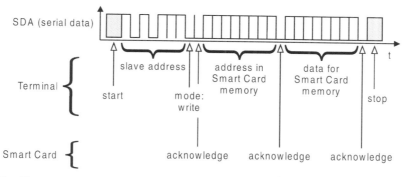

Figure 6.23   Unconstrained writing of a byte in memory with the I²C bus

Writing to an address   As with reading data from the card's EEPROM, there are also various modes for writing data. The simplest of these is described below, with which a single byte can be written anywhere in memory.

Again, the sequence begins with a start signal from the terminal. This is followed by the card's address, with the write bit set. The card acknowledges receipt and then receives from the terminal the address in the EEPROM where the data are to be written.

The card acknowledges this as well, and then receives the data. After the terminal receives the third acknowledgement, which indicates that the card has received the data, it sends a stop signal. Following this, the card starts to write the received data to the EEPROM, which does not require external timing signals. This completes the writing sequence, and the byte is now stored in the EEPROM.

6.4.2   The T=0 transmission protocol

This transmission protocol was first used in France during the initial development of Smart Cards, and it was also the first internationally standardized Smart Card protocol. It was created in the early years of Smart Card technology, and it is thus designed for minimum memory usage and maximum simplicity. This protocol is used worldwide in the GSM card, and as a result enjoys the greatest popularity of all current Smart Card protocols. The T=0 protocol is standardized in ISO/IEC 7816-3. Additional compatible specifications are contained in the GSM 11.11 and EMV specifications.

The T=0 protocol is byte-oriented, which means that the smallest unit processed by the protocol is a single byte. The transferred data unit contains a header consisting of a class byte, a command byte and three parameter bytes. This may be followed by an optional data part. In contrast to the APDUs standardized by ISO/IEC 7816-4, length information is provided only by the parameter P3. This indicates the length of either the command data or the response. It is also specified by the ISO/IEC 7816-3 standard.

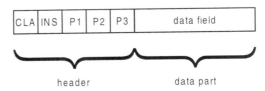

Figure 6.24   Structure of a command with T=0

Due to the byte orientation of the T=0 protocol, if a transfer error is detected the retransmission of the incorrect byte must be requested immediately. With block protocols, by contrast, an entire block (a sequence of bytes) must be retransmitted if an error occurs. Error detection with T=0 is based exclusively on a parity bit appended to each sent byte.

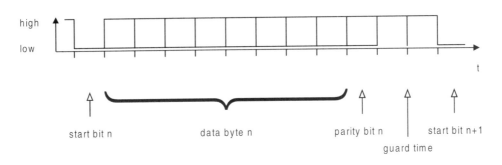

Figure 6.25   A byte transferred via the I/O interface with no error, using the T=0 protocol

If the recipient detects a transfer error, it must set the I/O line to a low level for the duration of one etu, starting halfway through the first bit interval of the guard time of the faulty byte. This indicates to the other party that the most recent byte must be retransmitted. This 'byte repetition' mechanism is very simple, and it has the advantage that it is selective, since only incorrect bytes have to be repeated. Unfortunately, this procedure suffers from a severe disadvantage. Most interface ICs treat the etu interval as the smallest detectable unit, so they cannot recognize a low level on the I/O line that is set halfway through a stop bit. Standard interface ICs are thus not suitable for the T=0 protocol. However, if each bit is received separately by software, this problem does not arise.

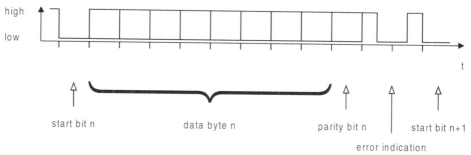

Figure 6.26   A data transfer error is indicated in the T=0 protocol by a low level at the I/O interface during the guard time

The T=0 protocol also allows an external programming voltage for the EEPROM or EPROM to be switched on or off. This is done by adding '1' to the received command byte and sending it back to the terminal as an acknowledge byte. This is why only even-valued command bytes are permitted, since otherwise this mechanism would not work. However, switching an external programming voltage is now technically obsolete, since all modern Smart Card microcontrollers generate the programming voltage in the chip itself. We thus need not discuss this topic any further.

The main function of the guard time is to separate the individual bytes during the transfer. This allows both the sender and the recipient more time to perform the transmission protocol's functions.

To illustrate the T=0 command–response sequence, let us assume that the terminal sends the card a command with a data section, and the card responds with data and a return code. The terminal first sends the card a 5-byte command header, consisting of a class byte, a command byte and the P1, P2 and P3 bytes. If this is received correctly, the card returns an acknowledgement (ACK) in the form of a Procedure Byte (PB). This acknowledgement is coded the same as the received command byte. On receipt of the Procedure Byte, the terminal sends exactly the number of data bytes indicated by the P3 byte. Now the card has received the complete command, and it can process it and generate a response.

If the response contains data in addition to the 2-byte return code, the card informs the terminal of this via a special return code, with the amount of data indicated by SW2. After receiving this response, the terminal sends the card a GET RESPONSE command, which consists only of a command header with the quantity of data to be sent. The card now sends the terminal the data generated following the first command, with the requested length and the corresponding return code. This completes one command sequence.

If a command is sent to the card, and the card only generates a return code with no data section, then the GET RESPONSE portion of Figure 6.27 does not occur. Since an additional command from the application layer is needed for this action to be carried out (fetching data relating to a previous command), there is naturally no longer a strict separation between the protocol layers. An application layer command (GET RESPONSE) must be used here to support the data link layer, which has certain effects on the application in question. All of this may appear complicated at first sight, so it is shown again graphically in Figure 6.28.

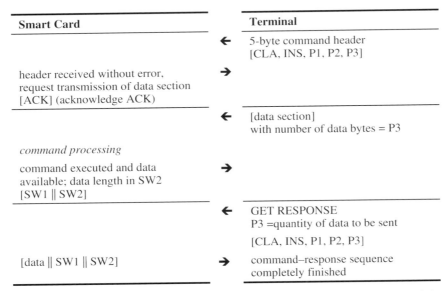

Figure 6.27 A typical T=0 communication procedure with data in both the command and the response (e.g. MUTUAL AUTHENTICATE command)

The T=0 protocol allows the card to receive the bytes in the data section individually, after it has received the header. To do so, it only has to send the inverted command byte to the terminal as a procedure byte, whereupon the terminal sends a single data byte. The next data byte follows the next procedure byte from the card. This byte-wise transfer can continue until the card has received all the bytes in the data section, or until it sends the non-inverted command byte to the terminal as a procedure byte. Upon receiving the non-inverted command byte, the terminal sends all the remaining data bytes to the card, which has now received the complete command.

As far as the transmission protocol is concerned, the user is ultimately interested only in the data transfer rate and the error detection and correction mechanisms. Sending a byte, which consists of 8 bits, requires 12 bits to be transferred, including 1 start bit, 1 parity bit and 2 etu for the guard time. Assuming a 3.5712-MHz clock frequency and a divider value of 372, transferring one byte takes 12 etu or 1.25 ms. Table 6.27 lists the transfer times for some typical commands.

Table 6.27 Data transfer times for some typical commands for T=0, with a clock frequency of 3.5712 MHz, a divider value of 372, 2 stop bits and 8 data bytes per command (C = command, R = response)

Command	Useful data		Protocol data		Data transfer time
READ BINARY	C:	5 bytes	—		18.75 ms
	R:	2 + 8 bytes			
UPDATE BINARY	C:	5 + 8 bytes	—		18.75 ms
	R:	2 bytes			
ENCRYPT	C:	5 + 8 bytes	C:	5 bytes	37.50 ms
	R:	2 + 8 bytes	R:	2 bytes	

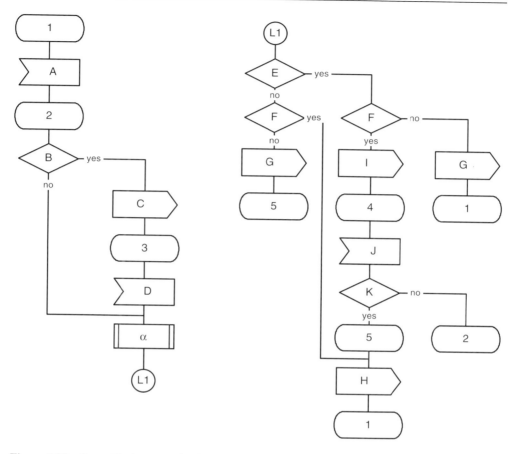

Figure 6.28 Smart Card communication process state machine for the T=0 communications procedure, without error correction.

α Command processing

1 Idle state
2 Header received with CLA, INS, P1, P2 and P3
3 Wait for the data section (P3 = number of bytes)
4 Wait for a command (header with CLA, INS, P1, P2 & P3) (P3 = amount of response data)
5 SW1, SW2 sent and GET RESPONSE received

A Receive the header (5 bytes)
B Data available (P3 > 0)?
C Data portion available; send procedure byte to terminal

D Receive the data section (P3 = number of bytes)
E Did the command contain a data section? (then execute C and D)
F Are response data available (no error occurred)?
G Send SW1 and SW2
H Send the available response data and SW1 + SW2
I Send SW1 and SW2 (SW2 = amount of response data)
J Receive a command Header = 5 bytes)
K Is the received command GET RESPONSE?

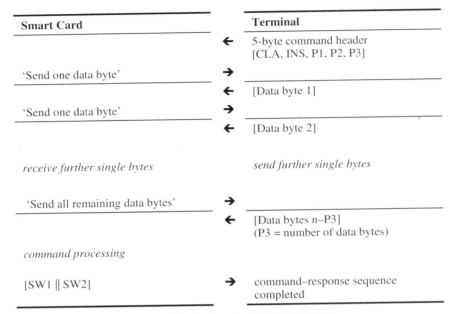

Figure 6.29   Single-byte reception with T=0 (e.g. UPDATE BINARY)

The data transfer rate naturally drops if errors occur. However, the single-byte repetition mechanism is very advantageous here, since only incorrectly received bytes have to be retransmitted.

The error detection mechanism of the T=0 protocol consists only of a parity check at the end of each byte. This allows reliable recognition of single-bit errors, but two-bit errors cannot be detected. Furthermore, if a byte is lost during the transfer from the terminal to the card, this results in an endless loop (deadlock) in the card, since it waits for a particular number of bytes and has no possibility of a timeout. The only practical way in which the terminal can exit this communications dead end is to reset the card and start over from the beginning.

In normal communication, the inadequate separation of the link and transport layers does not cause any particular difficulties. The smooth operation of the GSM application is the best proof of this. However, problems quickly arise if secure messaging is used. With a partly encrypted header and a fully encrypted data section, it is no longer possible to support the T=0 protocol using the previously described procedure without incurring large overheads. This is because the unencrypted command byte must be used for the procedure byte in the T=0 protocol.

Due to the absence of layer separation and the obvious problems in case of a bad connection, the T=0 protocol is often considered to be outdated. On the other hand, transfer errors almost never occur in the communications between the terminal and the card. The main advantages of the T=0 protocol are its good average transfer rate, minimal implementation overheads and wide market penetration.

6.4.3   The T=1 transmission protocol

The T=1 transmission protocol is an asynchronous half-duplex protocol for Smart Cards. It is based on the international ISO/IEC 7816-3 standard. The EMV specification is also relevant to this protocol. The T=1 protocol is a block-oriented protocol. This means that one block is the smallest data unit that can be transferred between the card and the terminal.

This protocol features strict layer separation, and it can be classified in the OSI reference model as a data link layer. In this context, layer separation also means that data destined for higher layers, such as the application layer, can be processed completely transparently by the data link layer. It is not necessary for layers other than the ones directly involved to interpret or modify the contents of the transferred data.

Secure messaging, in particular, requires strict adherence to layer separation. Only then can encrypted user data be passed across the interface without resorting to complicated methods or tricks. The T=1 protocol is currently the only international Smart Card protocol that allows secure data transfer in all its variations, without any problems or compromises.

The transfer process starts after the card has sent the ATR, or after a successful PTS has been executed. The first block is sent by the terminal, and the next one by the card. Communication then continues in this manner, with the send privilege alternating between the terminal and the card.

Incidentally, the use of the T=1 protocol is not limited to Smart Card/terminal communications. It is used by many terminals for exchanging useful data and control data with the computers to which they are connected.

The data transfer rate is naturally one of the most interesting aspects of any protocol. Table 6.28 lists the transfer times for some typical commands with the T=1 protocol.

Table 6.28   Data transfer times for some typical commands for T=1, with a clock frequency of 3.5712 MHz, a divider value of 372, XOR error detection code, 2 stop bits and 8 data bytes per command (C = command, R = response)

Command	Useful data		Protocol data		Data transfer time
READ BINARY	C:	5 bytes	—		28.75 ms
	R:	2 + 8 bytes			
UPDATE BINARY	C:	5 + 8 bytes	—		23.00 ms
	R:	2 bytes			
ENCRYPT	C:	5 + 8 bytes	C:	5 bytes	38.75 ms
	R:	2 + 8 bytes	R:	2 bytes	

Block structure

The blocks to be transferred are essentially used for two different purposes. One of these is the transparent transfer of application-specific data, and the other is the transfer of protocol control data or the handling of transfer errors.

The transfer block consists of an initial prologue field, an information field and a final epilogue field. The prologue and epilogue fields are mandatory and must always be sent. The information field, by contrast, is optional; it contains data for the application layer. This is either a command APDU sent to the Smart Card or a response APDU from the card.

There are three fundamentally different types of blocks in T=1: *information blocks*, *reception acknowledgement blocks* and *system blocks*. Information blocks (I blocks) are used for the transparent exchange of application layer data. Reception acknowledgement blocks (R blocks), which do not contain any data fields, are used for positive or negative reception confirmation. System blocks (S blocks) are used for control information that relates to the protocol itself. Depending on the specific control data, they may have an information field.

Prologue field			Information field	Epilogue field
node address NAD	protocol control byte PCB	length LEN	APDU	EDC
1 byte	1 byte	1 byte	0 ... 254 bytes	1 ... 2 bytes

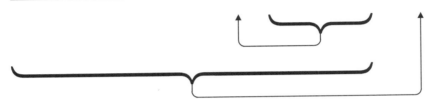

Figure 6.30   The structure of a T=1 transfer block

The prologue field   The prologue field consists of the three subfields Node Address (NAD), Protocol Control Byte (PCB) and Length (LEN). It is three bytes long and contains basic control and pointer data for the actual transfer block.

Node address (NAD)   The first byte in the prologue field is called the node address (NAD) byte. It contains the block's destination and source addresses. Each if these is coded using 3 bits. If the address is not used, the relevant bits are set to 0. Furthermore, for compatibility with older microcontrollers, control is provided for the EEPROM or EPROM programming voltage. However, there is no practical use for this, since all Smart Card microcontrollers now have a charge pump in the chip.

Table 6.29   Node address (NAD field)

b8	b7	b6	b5	b4	b3	b2	b1	Meaning
X	...	...	...	X	...	...	...	Vpp control
...	X	X	X	...	...	...	...	DAD (destination address)
...	...	...	...	...	X	X	X	SAD (source address)

Protocol control byte (PCB)   The subfield following the node address is the Protocol Control byte (PCB). As the name suggests, it serves to control and supervise the transmission protocol. This increases the amount of coding required. The PCB field primarily encodes the block type, as well as associated supplementary information.

Table 6.30   PCB field for an I block

b8	b7	b6	b5	b4	b3	b2	b1	Meaning
0	...	...	...	...	...	...	...	I block identifier
...	N(S)	...	...	...	...	...	...	Send sequence number
...	...	X	...	...	...	...	...	Sequence data bit M
..	...	...	X	X	X	X	X	Reserved

Table 6.31   PCB field for an R block

b8	b7	b6	b5	b4	b3	b2	b1	Meaning
1	0	...	...	...	...	...	...	R block identifier
...	...	0	N(R)	0	0	0	0	No error
...	...	0	N(R)	0	0	0	1	EDC or parity error
...	...	0	N(R)	0	0	1	0	Other error

Table 6.32   PCB field for an S block

b8	b7	b6	b5	b4	b3	b2	b1	Meaning
1	1	...	...	...	...	...	...	S block identifier
...	...	0	0	0	0	0	0	resynch request (only from terminal)
...	...	1	0	0	0	0	0	resynch response (only from Smart Card)
...	...	0	0	0	0	0	1	request to change information field size
...	...	1	0	0	0	0	1	response to information field size change request
...	...	0	0	0	0	1	0	request for abort
...	...	1	0	0	0	1	0	response to abort request
...	...	0	0	0	0	1	1	request to modify waiting time (only from Smart Card)
...	...	1	0	0	0	1	1	response to modify waiting time request (only from terminal)
...	...	1	0	0	1	0	0	Vpp error response (only from Smart Card)

Length (LEN) The 1-byte length field (LEN) indicates the length of the information field in hexadecimal form. Its value can be '00' to 'FE'. The code 'FF' is reserved for future extensions and currently should not be used.

The information field In an I block, the information field serves as a container for application layer data (OSI layer 7). The content of this field is transferred completely transparently. This means that the content is directly passed on by the transmission protocol, without any analysis or evaluation.

In an S block, this field transfers data for the transmission protocol. This is the only case in which this content of this field is used by the transport layer.

According to the ISO standard, the size of the information field can range from '00' to 'FE' (254) bytes. The value 'FF' (255) is reserved by ISO for future use. The terminal and the card may have I fields with different sizes. The default size of the terminal I field is 32 bytes (IFSD = information field size for the interface device); this can be modified via a special S field. This default value of 32 bytes also applies to the card (IFSC = information field size for the card), but this can be modified by a parameter in the ATR.[7] In practice, the size of the I field for both the terminal and the card is between 50 and 140 bytes.

Epilogue field This field, which is transferred at the end of the block, contains an error detection code computed from all prior bytes in the block. The computation employs either an LRC (longitudinal redundancy check) or a CRC (cyclic redundancy check). The method used must be specified in the interface characters of the ATR. If it is not specified, by convention the LRC method is implicitly used. Otherwise, the CRC computation is carried out according to ISO 3309. The divider polynomial used is the same as for CCITT Recommendation V.41, $G(x) = x^{16} + x^{12} + x^5 + 1$. Both error detection codes can only be used for error detection; they cannot correct a block error.

The single-byte longitudinal redundancy checksum is computed via XOR concatenation of all previous bytes in the block. This computation can be executed very quickly, and its implementation is not code-intensive. Normally it is performed on the fly during data transmission or reception. It is a standard part of practically all T=1 implementations.

Using the CRC procedure to generate an error detection code produces a far higher probability of error detection than the relatively primitive XOR checksum. However, this procedure is presently almost never used in practice, since its implementation is very code-intensive and slow. In addition, the epilogue field must be extended to two bytes, which further reduces the transfer rate.

Send and receive sequence counters Each information block in the T=1 protocol has a send sequence number consisting of only one bit located in the PCB byte. It is incremented modulo 2, which means that it alternates between 0 and 1. The send sequence counter is also designated N(S). Its starting value at protocol initiation is 0. The terminal and Smart Card counters are incremented independently of each other.

The primary purpose of the send sequence counter is to support requests for resending blocks received with errors, since individual data blocks can be unambiguously addressed via N(S).

Waiting times Various waiting times are defined to provide senders and receivers with precisely specified minimum and maximum time intervals for various actions during the

---

7   The ATR Tai data element (i > 2)

data transfer. They also provide defined means to terminate communications in order to prevent deadlocks in case of errors. Default values are defined for all of these waiting times in the standard, but these may be modified to maximize the transfer rate. The modified values are indicated in the specific interface characters of the ATR.

Character waiting time (CWT)   The character waiting time is defined as the maximum interval between the rising edges of two consecutive characters within a block. The receiver starts a countdown timer on each rising edge, with the character waiting time as the initial value. If the timer expires and no rising edge of new bit has been detected, the receiver assumes that the transfer block has been received in full. The 'CWT reception criterion' can thus be generally used for end-of-block detection. However, it does considerably reduce the data transfer rate, since the time for each block is increased by the duration of the CWT. It is thus better to detect the end of the block by counting received bytes.

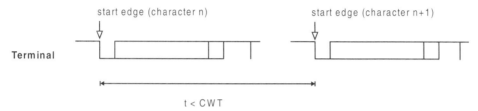

Figure 6.31   Definition of the character waiting time (CWT)

The CWT is calculated using the data element CWI contained in the ATR, according to the following formula:

$$CWT = (2^{CWI} + 11) \text{ work etu}$$

The default value for the CWI is 13, which yields the following value for CWT:

$$CWT = (2^{13} + 11) \text{ work etu} = 8.203 \text{ work etu}$$

With a clock frequency of 3.5712 MHz and a divider value of 372, this results in an interval of 0.85 seconds.[8]

This interval, which is specified in the standard as the default setting, is too high for fast data transfers. In practice, the usual range of CWI is between 3 and 5. This means that for a normal transfer sequence, in which the characters follow each other without any time delay, the recipient waits for a duration of one to two bytes before detecting the end of the block or a communications lapse.

Normally, the reception routine detects the end of a block from the block length information in the LEN field. However, if the content of this field is erroneous, the character waiting time can be used as an additional means for terminating the reception. This problem only manifests itself when the length information is too long, since in this case the recipient would wait for additional characters that never arrive. This would block

---

8   See also Section 15.7.2, 'ATR data element conversion tables'.

the transmission protocol, and this state could only be cleared by a card reset. The character waiting time mechanism gets around this problem.

Block waiting time (BWT)   The block waiting time has been defined to establish a means to terminate communications if the Smart Card does not respond. It is the maximum allowed interval between the rising edge of the last byte in a block sent to the card and the rising edge of the first byte sent back by the card.

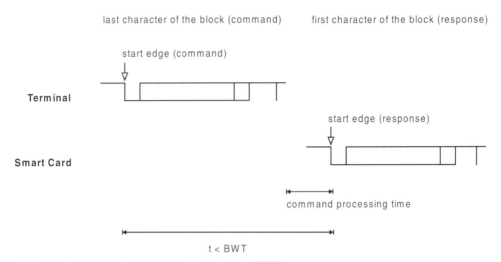

Figure 6.32   Definition of the block waiting time (BWT)

In terms of a conventional T=1 block, this is the maximum interval permitted between the leading edge of the XOR byte in the command block's epilogue field and the leading edge of the NAD byte in the card's response. If this waiting period expires without a response being received from the card, the terminal may assume that the card is faulty and initiate an appropriate response. This could for example be a card reset, followed by a new attempt to establish communication.

The interface characters in the ATR specify a BWT that is coded in the abbreviated form of the BWI:

$$BWT = 2^{BWI} \times 960 \times \frac{372}{f} \text{ s} + 11 \text{ work etu}$$

If no BWI value is defined in the ATR, the default value of 4 is used. With 3.5712 MHz and a divider value of 372, this gives 1.6 s as the value for the block waiting time:

$$BWT = 2^4 \times 960 \times \frac{372}{3,571,200 \text{ Hz}} \text{ s} + 11 \text{ work etu} = 2^4 \times 0.1 \text{ s} + 11 \text{work etu} \approx 1.6 \text{ s}$$

As can be seen, this value is quite generous. In practice, a value of 3 is often used for BWI, which yields a block waiting time of 0.8 s.[9] Typical command execution times in the card are usually around 0.2 s.[10] A BWT of the above duration thus represents a compromise between normal command execution times and the quick detection of a Smart Card that no longer responds to commands.

Block guard time (BGT)   The minimum interval between the leading edge of the last byte and the leading edge of the first byte in the opposite direction is defined to be the block guard time. This is the opposite of the BWT, which is defined as the maximum time between the two defined leading edges. Another difference is that the block guard time is obligatory for both parties and must be observed, while the block waiting time is only significant for the Smart Card. The purpose of the block guard time is to provide the block sender with a minimum time interval in which to switch over from sending to receiving.

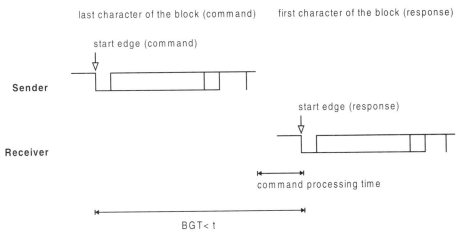

Figure 6.33   Definition of the block guard time (BGT)

The block guard time has a fixed value, which is standardized at 22 etu. In a Smart Card running at 3.5712 MHz with a divider value of 372, this yields an interval of around 2.3 ms.

Transmission protocol mechanisms

Waiting time extension   If the card needs more time to generate a response than the maximum allowed by the block waiting time (BWT), it can request a waiting time extension from the terminal. The Smart Card does this by sending a special S block that requests an extension, and it receives a corresponding S block in acknowledgement from the terminal. The terminal is not allowed to refuse this request.

A byte in the information field informs the terminal of the length of the extension. This byte, multiplied by the block waiting time, gives a new block waiting time. However, this is only valid for the most recently sent I block.

---

9   See also Section 15.7.2, 'ATR data element conversion tables'.
10   See also Section 14.2, 'Formulas for Estimating Processing Times'.

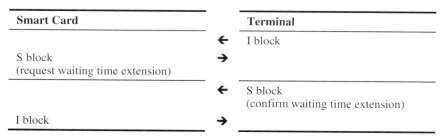

Figure 6.34   Procedure for extending the waiting time

Block chaining   One of the essential performance features of the T=1 protocol is the block chaining function. This allows either party to send data blocks that are larger than the size of its send or receive buffer. This is particularly useful in light of the limited memory capacities of Smart Cards. Chaining is permitted only for information blocks, since only such blocks can contain large amounts of data. In the chaining process, the application data are divided into individual blocks that are sent to the recipient one after the other.

The application layer data must be divided such that none of the resulting segments is larger than the maximum block size of the recipient. The first segment is then placed in an information field in accordance with the T=1 protocol, supplied with prologue and epilogue fields and sent to the recipient. The M bit (more data bit) is set in the block's PCB field to indicate to the recipient that the block chaining function is being used, and that chained data are located in the following blocks.

As soon as the recipient has successfully received this information block with the first segment of the user data, it indicates that it is ready to receive the next chained I block by returning an R block whose sequence number N(R) is the send sequence count N(S) of the next I block. The next block is then sent to the recipient.

This back and forth transfer of I and R blocks continues until the sender issues an I block whose M bit in the PCB field indicates that it is the last block in the chain (M bit = 0). After this block has been received, the recipient has all the application layer data and can process the full data block.

There is one restriction that applies to the block chaining routine. Within a single command–response cycle, chaining may proceed in one direction only. For example, if the terminal sends chained blocks, the card may not send chained blocks in response.

There is a further restriction that has nothing to do with the protocol itself, but rather with the Smart Card's very limited memory. The implementation of the block chaining mechanism involves a certain amount of extra software, while its usefulness is very limited, since commands and responses are not all that long and therefore do not normally require chaining. If the card's receive buffer in RAM is not big enough to store all the data passed using block chaining, it is necessary to place this buffer in the EEPROM. This however leads to a sharp reduction in the transfer rate, since the EEPROM (in contrast to the RAM) cannot be written at the full speed of the processor.

Consequently, many T=1 implementations have no block chaining function, as the cost/benefit ratio often does not justify it. This is a typical example of the fact that standards are often interpreted very liberally in actual practice. In this case, the interpretation amounts to considering block chaining to be a supplementary option in T=1 that is not absolutely necessary.

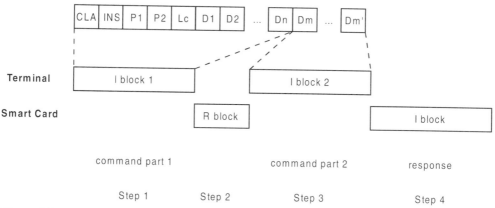

Figure 6.35   Example of block chaining for transferring data from the terminal to the Smart Card

Error handling   The T=1 protocol exhibits highly developed error detection and handling mechanisms. If invalid blocks are received, the protocol attempts to restore error-free communications by means of precisely defined procedures.

Seen from the terminal's perspective, there are three synchronization stages. In the first stage, the sender of a faulty block receives an R block that indicates an EDC/parity bit error or a general error. The recipient of this R block must then resend the last block that it sent.

If it proves impossible to restore an error-free connection using this mechanism, the next stage is invoked. This means that the Smart Card receives a resynchronization request from the terminal in the form of an S block. The terminal expects a Resynch response in reply. The terminal and card then both reset their send and receive counters to zero, which corresponds to the protocol status immediately following the ATR. Based on this, the terminal attempts set up a new connection.

Stages 1 and 2 affect only the protocol layer. They have no effect on the application itself. However, the third synchronization stage does affect all layers in the Smart Card. If the terminal cannot set up an error-free connection using the first two synchronization stages, it triggers a Smart Card reset via the reset lead. Unfortunately, this means that all the data and states of the current session are lost. Following the reset, communications must be completely re-established from the bottom up.

If even this procedure fails to produce a working connection, the terminal deactivates the card after three attempts. The user then usually receives an error message to the effect that the card is no good.

Table 6.33   T=1 error handling stages

Synchronization stage	Mechanism
Step 1	Repeat the faulty block
Step 2	Resynchronize and repeat the faulty block
Step 3	Reset the Smart Card and set up the connection anew

Example of a data transfer with the T=1 protocol Figure 6.37 shows an example of a data transfer for of SELECT FILE command with the T=1 protocol.

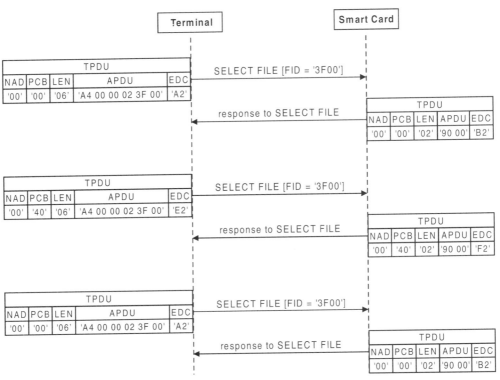

Figure 6.36 Successful transfer of a TPDU with the T=1 transmission protocol. The XOR option has been selected for the error detection code (EDC). A SELECT FILE command with a FID of '3F00', which selects the MF, is transferred in the APDU. This illustration clearly shows how the send sequence count is incremented in the PCB byte for each transaction and the corresponding changes in the EDC.

Differences between T=1 according to ISO/IEC and T=1 according to EMV The original definition of the T=1 protocol according to the ISO/IEC 7816-3 standard allows many options and mechanisms, some of which require a lot of program code to implement and are not often used. Typical examples are the extensive error correction mechanisms, which may be theoretically interesting but which are frequently ineffective in dealing with actual transmission disturbances. In practice, it is usually better to simply reinsert the card in the terminal and start the session anew, instead of attempting to restabilize the data transfer by means of countless resynchronization requests. For this reason, the EMV specification imposes certain limitations with respect to the original ISO/IEC standard.

Table 6.34 Summary of the differences in the implementation of the T=1 transmission protocol between the ISO/IEC 7816-3 standard and the EMV specification

Mechanism or option	ISO/IEC 7816-3	EMV
BWT expired	reset the card (for example)	deactivate the card
Smart Card sends a request to change the IFS	allowed	a maximum of three successive requests is allowed
Smart Card sends an S block with an Abort request	allowed	deactivate the card
zero-length I block	allowed	prohibited
terminal sends three successive blocks without receiving a valid response	behavior similar to specified error handling; usually a Reset Request	deactivate the card

6.4.4 Comparison of asynchronous transmission protocols

A brief remark with regard to the attainable data transfer rates is in order here, in the context of the comparison of various protocols. Attempts are often made to compare the T=0 and T=1 protocols based on calculated effective transfer rates. However, such calculations are only valid for a particular command in a particular context. As soon as these calculations are generalized, they lose their meaning and their validity.

Both protocols have their strengths and weaknesses with regard to achievable transfer rates, but these are heavily dependent on very many individual factors. Some examples are the transfer error rate, the size of the I/O buffer in the card and the specific implementation of the protocol. In brief, you can assume that, on average and with most applications, the effective transfer rates of both protocols are nearly the same. If you want to increase the transfer rate, changing the protocol will have little effect. It is more effective to reduce the divider value, since this yields significantly better results.

Two international transmission protocols have been described in the previous sections. In order to provide an overview, Table 6.35 summarizes the fundamental features of these protocols, as well as their advantages and disadvantages.

Table 6.35 Comparison of internationally standardized asynchronous data transmission protocols

Criterion	T=0	T=1	T=2 (proposed)
Data transfer	asynchronous, half-duplex, byte-oriented	asynchronous, half-duplex, block-oriented	asynchronous, full-duplex, block-oriented
Standard	ISO/IEC 7816-3	ISO/IEC 7816-3 Amd 1	ISO/IEC 10536-4
Divider	freely definable, usually 372	freely definable, usually 372	freely definable, usually 372
Block chaining	not possible	possible	possible
Error detection	parity bits	parity bits, EDC at end of block	parity bits, EDC at end of block
Memory used	$\approx$ 300 bytes	$\approx$ 1100 bytes	$\approx$ 1600 bytes

6.5   Message Structure: APDUs

The entire data exchange between the Smart card and the terminal takes place using APDUs. The term APDU is short for 'application protocol data unit'. It denotes an internationally standardized data unit in the application layer, which is layer 7 in the OSI model. This layer is located directly above the level of the transmission protocol in Smart Cards. The protocol-dependent TPDUs (transport protocol data units), by contrast, are the data units of the layer directly below.

A distinction is made between command APDUs, which represent commands to the card, and response APDUs, which are the card's replies to these commands. Stated simply, an APDU is a sort of container that holds a complete command for the card or a complete response from the card. APDUs are passed by the transmission protocol transparently, that is, without modification or interpretation.

APDUs that comply with the ISO/IEC 7816-4 standard are designed to be independent of the transmission protocol. The content and format of an APDU must therefore not change when a different transmission protocol is used. This applies above all to the two standardized protocols, T=0 and T=1. The demand for protocol independence influenced the design of the APDUs, since it must be possible to transfer them transparently using both the byte-oriented T=0 protocol and the block-oriented T=1 protocol.

6.5.1   Structure of the command APDU

A command APDU is composed of a header and a body. The body may be of variable length, or it may be entirely absent if the associated data field is empty.

The header consists of the four elements, which are the class byte (CLA), the instruction byte (INS) and two parameter bytes (P1 and P2). The class byte is presently still used to identify applications and their specific command sets. For example, GSM uses the class byte 'A0', while the code '8X' is most commonly used for company-specific (private-use) commands. In contrast, ISO-based commands are encoded by class byte '0X'. The standard additionally specifies the class bytes to be used for the identification of secure messaging and logical channels. This is nevertheless still compatible with using the class byte as an application identifier, as previously mentioned.

The next byte in the command APDU is the instruction byte, which encodes the actual command. Almost the entire address space of this byte can be exploited, with the sole restriction that only even codes can be used. This is because the T=0 protocol allows the programming voltage to be activated by returning the instruction byte incremented by one in the procedure byte. The instruction byte thus always has to be even.[11]

The two parameter bytes are primarily used to provide more information about the command selected by the instruction byte. They thus serve mainly as switches that select various command options. For example, they are used to choose various options for SELECT FILE or to specify the offset for READ BINARY.

The section following the header is the body, which can be dispensed with except for a length specification. The body fulfils a double role. First, it specifies the length of the data section sent to the card (in the Lc field)[12] as well as the length of the data section to be sent

---

11   See also Section 15.7.5, 'The most important Smart Card commands'.
12   'Lc' means 'length command'.

back from the card (in the Le field).[13] Secondly, it contains the data associated with the command that are sent to the card. If the value of the Le field is '00', the terminal expects the card to send the maximum amount of data available for this command. This is the only exception to the numerical specification of the length.

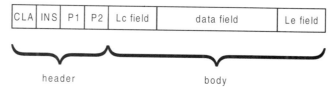

Figure 6.37   Structure of a command APDU

Table 6.36   The most important class byte (CLA) codes according to ISO/IEC 7816-4

b8 to b5	b4	b3	b2	b1	Meaning
...	...	...	X	X	logical channel number
...	0	0	...	...	no secure messaging
...	0	1	...	...	secure messaging not compliant with ISO, own procedure used
...	1	0	...	...	secure messaging compliant with ISO, header not authentic
...	1	1	...	...	secure messaging compliant with ISO, header authentic
'0'	...	...	...	...	structure and coding compliant with ISO/IEC 7816-4
'8', '9'	...	...	...	...	structure compliant with ISO/IEC 7816-4, user-specific codes and meanings of commands and responses ('private use')
'A'	...	...	...	...	structure and codes compliant with ISO/IEC 7816-4, specified in additional documents (e.g. GSM 11.11)
'F'	1	1	1	1	reserved for PTS

Table 6.37   Summary of the assignment of class bytes to applications

Class	Application
'0X'	standardized commands compliant with ISO/IEC 7816-4
'80'	electronic purses compliant with EN 1546-3
'8X'	application-specific and company-specific commands
'8X'	credit cards with chips, compliant with EMV-2
'A0'	GSM cellular telephones compliant with prETS 300 608 / GSM 11.11, and standardized commands compliant with EN 726-3

[13] 'Le' means 'length executed'.

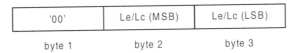

byte 1   byte 2   byte 3

Figure 6.38   Structure of an extended Lc or Le field

The Le and Lc fields are usually one byte long. It is possible to convert them into fields that are each three bytes long. These can be used to represent lengths up to 65,536, since the first byte contains the escape sequence '00'. The standard already defines this three-byte length specification for future applications, but this cannot yet be implemented due to the limitations of currently available memory sizes.

The previously described parts of the command APDU can be combined to produce the four general cases illustrated in Figure 6.39.

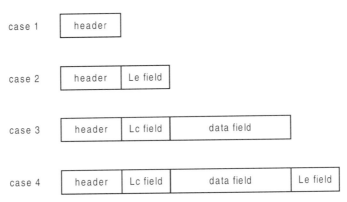

Figure 6.39   The four possible command APDU cases

6.5.2   Structure of the response APDU

The response APDU, which is sent by the card in reply to a command APDU, consists of an optional body and a mandatory trailer, as shown in Figure 6.40. The body consists of the data field, whose length is specified by the Le byte of the preceding command APDU. The length of the data field can be zero, regardless of the value specified in the command APDU, if the Smart Card has terminated command processing due to an error or incorrect parameters. This is indicated in the two single-byte status words SW1 and SW2 in the trailer.

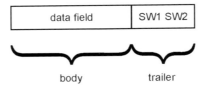

Figure 6.40   Structure of the response APDU

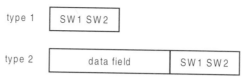

Figure 6.41   The two types of response APDUs

The card must always send a trailer in response to a command. The two bytes SW1 and SW2, which are also called the return code, encode the response to the command. For example, the return code '9000' means that the command was executed completely and successfully. The basic classification scheme of the more than 50 different codes is shown in Figure 6.42.[14]

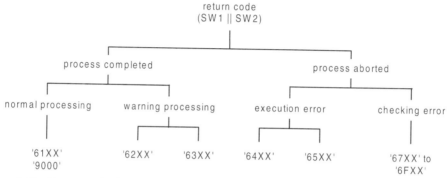

Figure 6.42   Return code classification scheme as defined by the ISO/IEC 7816-4 standard. The return codes '63XX' and '65XX' indicate that data in non-volatile memory (EEPROM) have been changed, while the remaining '6X' codes indicate that this has not occurred.

If a '63XX' or '65XX' return code is sent back after a command has been executed, this means that the card's non-volatile memory (usually the EEPROM) has been modified. If one of the other codes starting with '6X' is returned, command execution was terminated without modifying the non-volatile memory.

It should be noted that although there is a standard for return codes, many applications use non-standard codes. The only exception is the code '9000', which almost universally indicates successful execution. With all other codes, it is always necessary to consult the underlying specification in order to be sure of their meanings.

6.6   Secure Data Transfers

The entire data exchange between a terminal and a Smart Card uses digital electrical pulses on the I/O line of the Smart Card. It is conceivable and not technically difficult to solder a wire to the I/O contact, record all the communications for a session and later analyze them. In this way, it is possible to gain knowledge of all the data transferred in both directions.

---

[14]   See also Section 15.7.8, 'Smart Card return codes'.

A somewhat more difficult task is to electrically isolate the I/O contact, mount a dummy contact on top of it, and then use thin wires to connect both of these contacts to a computer. With this arrangement, it is easy to allow only certain commands to reach the card or to insert your own commands.

Both these typical types of attack can succeed only if secret data passes unprotected over the I/O line. The data transfer should thus basically be designed such that even if an attacker is able to tap into the data transfer and insert his own data transfer blocks, he will not be able to gain any advantage from doing so.

There are various mechanisms and procedures that can be used to protect against these attacks, and against even more sophisticated attacks. They are collectively referred to as *secure messaging*. These mechanisms are not specific to Smart Cards, as they have been used for a long time in long-distance data communications. What is special in the Smart Card domain is that neither the computational power of the communicating parties nor the transfer rate is particularly great. Commonly used standard procedures have therefore been scaled down to match the capabilities of Smart Cards, without impairing their security.

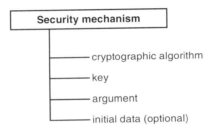

Figure 6.43   The data and functions required for a security mechanism

The objective of secure messaging is to ensure the authenticity, and if necessary the confidentiality, of part or all of the transferred data. A variety of security mechanisms are used to meet this objective. A security mechanism is defined as a function that requires the following items: a cryptographic algorithm, a key, an argument and initial data if necessary. A general condition must also be satisfied, which is that all security mechanisms must behave completely transparently with regard to the protocol layers that are present. This ensures that procedures that are already standardized and available are not affected by secure messaging. This applies particularly to the two transmission protocols T=0 and T=1, as well as to generally used and standardized Smart Card commands.

Before executing a secure messaging procedure, both parties must agree on the cryptographic algorithm to be used and a common secret key. According to Kerkhoff's principle, the security of the procedure relies entirely on this key. If it is revealed, secure messaging is nothing more than a generally known supplementary checksum that reduces the data transfer rate and at best can be used to correct transfer errors.

Several types of secure messaging procedures have been around for a number of years. They are all relatively rigid and tailored to particular applications. Most of them cannot be faulted as far as security is concerned. However, none of them has become internationally predominant or has proved to be sufficiently flexible to be included in current standards.

The demand for transparency to existing commands, support for two fundamentally different transmission protocols and maximum adaptability has lead to the standardization of a very flexible (and correspondingly complicated and expensive) secure messaging procedure in the ISO/IEC 7816-4 standard, with extended functions defined in ISO/IEC 7816-8.[15] This procedure is based on embedding all useful data in TLV-encoded data objects. Three different types of data objects are defined:

- for plaintext: data in plaintext (e.g. the data section of an APDU)
- for security mechanisms: the results of a security mechanism (e.g. a MAC)
- for auxiliary functions: control information for secure messaging
                            (e.g., the padding method used).

The class byte indicates whether secure messaging is used for the command. The two available bytes can encode whether the procedure specified in ISO/IEC 7816-4 is utilized, and whether the header is also covered by the cryptographic checksum (CCS).[16] If the header is included in the computation then it is authentic, as it cannot be changed during the transfer without this being evident.

Data objects for plaintext
According to the standard, all data that are not BER-TLV encoded must be embedded, or encapsulated, in data objects. There are various identification codes, which are listed in Table 6.38. Bit 1 of each code indicates whether the data object is included in the computation of the cryptographic checksum. If this bit is not set (e.g. 'B0'), the data object is not included in the calculation, while if it is set (e.g. 'B1') then the data object is included.

Table 6.38   Tags for plaintext data objects

Tag	Meaning
'B0', 'B1'	BER-TLV encoded; contains data objects related to secure messaging
'B2', 'B3'	BER-TLV encoded; contains data objects not related to secure messaging
'80', '81'	no BER-TLV encoded data
'99'	state information for secure messaging

Data objects for security mechanisms
The data objects used in security mechanisms are divided into those used for authentication and those used for confidentiality. The identification codes defined for this purpose are listed in Tables 6.39 and 6.40.

Table 6.39   Tags for authentication data objects

Tag	Meaning
'8E'	cryptographic checksum
'9A', 'BA'	initial value for a digital signature
'9E'	digital signature

---

[15]   Secure messaging is usually abbreviated to 'SM', which many programmers translate into 'sado/masochism' on account of the many degrees of freedom and room for interpretation provided by these two standards.
[16]   For the coding of the class byte, see Section 6.5.1, 'Structure of the Command APDU'.

Table 6.40   Tags for confidential data objects

Tag	Meaning
'82', '83'	cryptogram; the plaintext is BER-TLV encoded and includes data objects for secure messaging
'84', '85'	cryptogram; the plaintext is BER-TLV encoded and does not include data objects for secure messaging
'86', '87'	indicates the padding method used '01' – padding with '80 00 ... ' '02' – no padding

In general, the term 'authenticity' refers to all data objects related to cryptographic checksums and digital signatures. The encryption of data and the information needed to identify such data as encrypted within secure messaging fall under the heading of 'confidentiality'. Depending on the procedure used, you should look for the appropriate codes in these tables and use them for secure messaging.

Data objects for auxiliary functions
The data objects for auxiliary functions are used in secure messaging to coordinate the general conditions. The two parties use these data objects to exchange information about the cryptographic algorithms and keys used, initial data and similar basic information. In principle, this can vary with every transferred APDU, or even between a command and its response. In practice, though, these data objects are very rarely used, since all general conditions for secure messaging are defined implicitly. This means that they do not have to be specifically defined in the course of the communication.

Based on the possibilities offered by secure messaging as specified by ISO/IEC 7816-4, which have been only briefly outlined above, we can describe two fundamental procedures. These descriptions have been kept as simple as possible, in order to make it easier to understand the complex mechanisms. The high degree of flexibility means that there are many other possible combinations of security mechanisms, some of which are even more complex. The two procedures described here represent a compromise between simplicity and security.

The 'authentic mode' procedure uses a cryptographic checksum (CCS or MAC) to protect the application data (the APDU) against manipulation during the transfer. The 'combined mode' procedure, by contrast, is used to completely encrypt the application data, so that an attacker cannot draw any conclusions about the data content of the commands and responses that are exchanged. The use of a send sequence counter is seen in combination with only one of these procedures. This counter, whose initial value is a random number, is incremented for each command and each response. This allows both parties to determine whether a command or response has been lost or inserted. When used with the 'combined' procedure, a send sequence counter also makes identical APDUs appear to be different. This is referred to as 'diversity'.

6.6.1   The authentic mode procedure

The authentic mode procedure guarantees the authentic transfer of APDUs, which means that the APDUs cannot be tampered with during the transfer. The recipient of an APDU, which means a command or a response, can determine whether it has been altered during the transfer. This makes it impossible for an attacker to modify data within an APDU without this being noticed by the recipient.

The fact that this procedure is being used is indicated by a bit in the class byte, so that the recipient can act accordingly and check the received APDU for authenticity. The APDUs themselves are sent in plaintext and are not encrypted. The transferred data are thus still public, and with suitable manipulation of the transfer channel they could be intercepted and evaluated by an attacker. This is not necessarily a disadvantage, since with regard to privacy legislation it is better not to send confidential data over a public channel. In addition, the card user is at least theoretically allowed the possibility of seeing what data are exchanged between his or her Smart Card and the terminal.

In principle, any block encryption algorithm can be used to compute the cryptographic checksum. For practical reasons, the use of DES with a fixed 8-byte block length is assumed here. The individual data objects must therefore 'filled out' to an integer multiple of eight bytes, which is known as *padding*. In this process, data objects that are already an integer multiple of eight bytes are nevertheless extended by one block. After padding, the cryptographic checksum (CCS) of the entire APDU is computed using the DES algorithm in CBC mode. This 8-byte checksum is appended directly to the APDU as a TLV-encoded data object, with the four least significant bytes omitted. All padding bytes are deleted after the checksum has been computed. The modified APDU is then sent via the interface. This procedure extends the length of the APDU by eight bytes, which only marginally reduces the transfer rate if normal transfer block sizes are used.

The data objects for the control structures can explicitly indicate which algorithm and which padding method are used. Here again we assume for the sake of simplicity that the Smart Card and the terminal implicitly know all the parameters of the secure messaging system that is used.

When the protected APDU arrives at the recipient, the latter pads it again to an integer multiple of eight bytes and then computes its own MAC for the APDU. By comparing the MAC it has produced with the MAC generated by the sender, the recipient can determine whether the APDU has been altered during the transfer.

A prerequisite for computing a cryptographic checksum is a secret DES key that is known to both parties. If this key were not secret, an attacker would be able to break the authentic mode communication procedure by intercepting an APDU, modifying it as desired and computing a new 'correct' MAC. After this, he would only have to replace the original MAC with the new one and send the newly created APDU on its way.

In order to better protect keys used to generate the MAC against attacks based on known plaintext–ciphertext pairs, dynamic keys are normally used. These are generated by encrypting a random number that has been previously exchanged between the terminal and the card. A secret key known to both parties is used for this encryption.

The additional steps that are needed for the transmission and reception of an APDU that is protected by the authentic mode procedure naturally reduce the effective data transfer rate. On average, a good approximation is to assume that the rate will be half of that for unprotected plaintext.

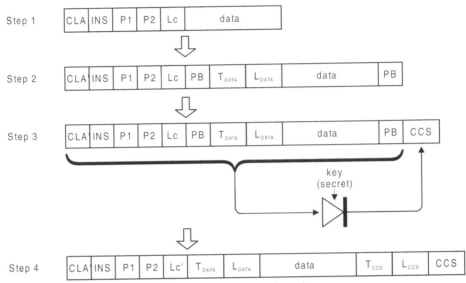

Figure 6.44 Generating a command APDU with the authentic mode procedure. A case 3 command (e.g. UPDATE BINARY) is used here, with its header included in the cryptographic checksum (CCS). A response APDU can be generated in a similar manner. 'PB' indicates the padding bytes.

Step 1   The initial APDU format.
Step 2   The data section is converted into TLV encoded data, and the data objects are padded to an integer multiple of eight bytes.
Step 3   The CCS is computed.
Step 4   A TLV-encoded data object containing the CCS is added to the APDU.

6.6.2   The combined mode procedure

Compared with the authentic mode procedure, the combined mode procedure represents the next higher level of security. The data section of the APDU is no longer transferred as plaintext, but instead in an encrypted form. The procedure is an extension of the authentic mode procedure.

In the combined mode procedure, as in the authentic mode procedure, the data objects to be protected with a cryptographic checksum are first padded to an integer multiple of eight bytes and then encrypted using the DES in CBC mode. The header is omitted from this process, as required for compatibility with the T=0 protocol. If it is desired to encrypt the header as well, so that the command being sent the card is unrecognizable, the T=0 ENVELOPE command must be used. One bit in the class byte indicates the use of secure messaging. The data are transferred across the interface after they have been encrypted. Since the recipient knows the secret key that was used for encryption, it can decrypt the APDU. The recipient then checks the correctness of decryption by recomputing the appended cryptographic checksum in the same level of the transfer layer.

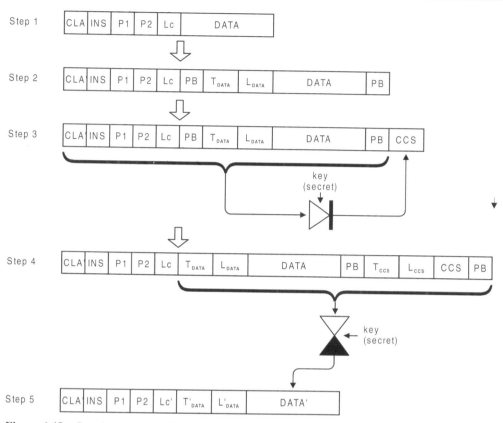

Figure 6.45   Creating a command APDU with the combined mode procedure. A case 3 command (e.g. UPDATE BINARY) is used, with its header included in the cryptographic checksum (CCS). A response APDU is created in a similar manner. The padding bytes are indicated as 'PB'.

Step 1	The initial APDU format.
Step 2	The data section is converted into TLV encoded data, and the data objects are padded to an integer multiple of eight bytes.
Step 3	The CCS is computed.
Step 4	A TLV-encoded data object containing the CCS is added to the APDU.
Step 5	The APDU data section is encrypted.

When this procedure is used, an attacker eavesdropping on the I/O line cannot discover which data are exchanged between the card and the terminal in the command and response. It is also not possible to replace one of the encrypted blocks within the APDU, since the blocks are linked to each other by using the DES in the CBC mode. Any replacement would be immediately noticed by the recipient.

With regard to the cryptographic algorithm, the comments made in the description of the authentic mode procedure apply here as well. In principle, any block encryption algorithm can be employed. The keys should be dynamic, as with the authentic mode procedure, so that a derived key is used for every session.

With regard to the security benefits, general usage of the combined mode procedure for all APDUs can be recommended. However, the increased security is accompanied by a considerable reduction of the data transfer rate.

A good approximation for the difference in the transfer rates for unprotected APDUs and those protected using the combined mode procedure is a factor of four. The speed difference between the authentic mode and combined mode procedures thus amounts to a factor of two. It is therefore necessary to carefully examine each case, in order to determine which data should be transferred in such a secure but time-consuming fashion.

6.6.3 Send sequence counter

Using a send sequence counter mechanism for secure messaging is not by itself a security procedure. It only makes sense to use a send sequence counter in combination with the authentic mode or combined mode procedure, since otherwise any modification of the counter by an attacker would remain undetected.

The working principle of a sequence counter is that each APDU contains a sequence number that depends on the time when it was sent. This allows the removal or insertion of an APDU in the course of the procedure to be immediately noticed, so that appropriate measure (terminating the communication) can be taken by the recipient.

This function is based on a counter that is initialized with a random number. This number is sent to the terminal by the card at the start of the communications process, in response to a request from the terminal. The counter is incremented each time an APDU is sent. The counter should not be too short, but it should also not be too long, because of the additional transfer load. The usual length of two bytes is assumed in the following description, but in practical applications, longer counters may be used.

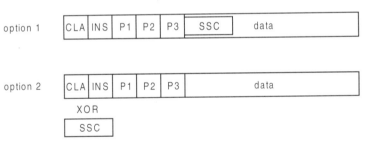

Figure 6.46   Two ways to include a send sequence count in a command APDU. In the first type, the send sequence counter is a TLV-encoded data object in the data section. In the second type, the send sequence count is only coupled to the APDU data by an XOR operation used to compute the CCS.

There are two basic ways to incorporate a sequence count into command and response APDUs. The counter value can be placed directly in the APDU as a numerical value in a data object, or the counter value may be XOR-ed with a matching amount of data in the APDU, following which a cryptographic checksum is computed and the modified data are restored to the APDU. The recipient of this APDU knows the expected counter value, and can use this value to modify the APDU in the same way as the sender. After this it can compute the cryptographic checksum and check the correctness of the received APDU.

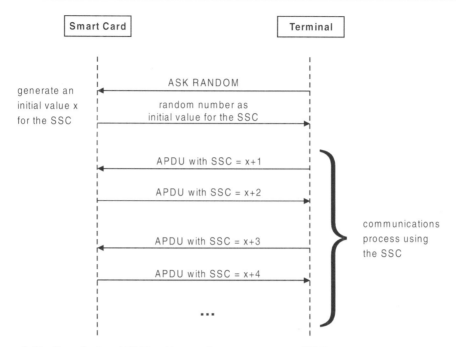

Figure 6.47   Transferring APDUs with a send sequence counter (SSC)

The process that takes place during communication goes as follows. The terminal first requests an initial counter value from the Smart Card. The Smart Card returns a two-byte random number to the terminal. The terminal then sends the first secured command to the card, with a send sequence count. Either the authentic mode or the combined mode procedure can be used to protect the counter and the body. The card receives the protected APDU and first checks whether the authentic mode or combined mode procedure indicates any evidence of tampering. It then compares the counter value to the expected value. If these match, no APDU has been inserted or deleted during the transfer.

It is apparent that using a send sequence counter is attractive not only when several commands have to be executed in a particular order, but also for individual commands, since each session is made unique if a counter is used. Using a counter primarily provides protection against 'replaying' APDUs that were previously sent and deleting APDUs.

If a send sequence counter is used together with the combined mode procedure, each encrypted block is different, which creates a condition known as *diversity*. This is a result of incrementing the counter each time an APDU is exchanged, and the fact that with a good encryption algorithm, changing a single bit in the plaintext affects the appearance of the entire ciphertext block.

6.7   Logical Channels

In future Smart Cards with several independent applications, it is planned to allow the applications to be addressed via logical channels. These would make it possible for up to four applications in a single card to exchange data with a terminal in parallel with each other. The existing single serial interface would still be used, but the applications could be addressed individually at the logical level.

Two bits in the class byte (bit 1 and bit 2) will determine which command belongs to which application.[17] This allows up to four logical channels, so up to four card sessions with applications in the card can run in parallel. However, there is a limitation with regard to the communications process with the different applications in the Smart Card. The external processes that access the card must be mutually synchronized so that they do not intermix their commands, since the response APDU from the card contains no information regarding the logical channel of the original command. This means that it is impossible to determine from outside the card which return code has been sent back in response to which command. The consequence of the missing channel identification is that no new command can be sent until the response to the previous command has been received.

The main application area for this very powerful mechanism is using several applications concurrently. For example, suppose a cardholder is holding a telephone conversation using the GSM application in a multi-application Smart Card. In order to confirm an appointment with the opposite party, she needs to briefly consult her personal organizer, which is also in the card. In parallel with the GSM application, the terminal uses a second logical channel to look for a file in the personal organizer application, and then tells our highly stressed manager whether she can agree to the proposed date. This is a typical application of logical channels. Another conceivable example is the secure transfer of electronic currency between two electronic purses in the same card.

The usefulness of logical channels for applications is matched by the difficulties that their management entails for the Smart Card operating system. Each logical channel in principle represents a separate Smart Card, with all its states and conditions. This effectively means that the operating system must manage all the information for several parallel sessions within the memory. The amount of data is enormous, and only large and expensive microcontrollers can handle the task. If secure messaging and all possible types of authentication are also required for each individual logical channel, the memory demand rises to a level that cannot be met by any currently available Smart Card microcontroller.

---

[17] See also Section 6.5.1, 'Command APDU structure'.

7

The Command Set

Communications procedures between a terminal and a Smart Card are always based on a challenge–response process. This means that the terminal sends a command (a challenge) to the card, which straightaway processes it, generates an answer and returns this to the terminal as its response. The card thus never transmits data without first having received a corresponding command from the terminal. Even ATR is no exception to this rule, since it is an answer to a reset signal, which in a way is also a type of command to the card.

Communication itself always takes place within the framework of a transmission protocol, namely T=0 or T=1. These relatively uncomplicated protocols meet the special requirements of Smart Card applications, and are optimized for this role. Deviations from these precisely specified protocols within application processes are not permitted. The transmission protocols make it possible to send data to the card and receive data from the card in a manner that is completely transparent to the transfer layer. The data are embedded in a sort of container, called the Application Protocol Data Unit (APDU). APDUs sent by the terminal to the card are the commands (challenges) to the card. The terminal also receives the answers (responses) to its commands in APDUs embedded in the transmission protocol. There is a whole set of commands based on this mechanism, which trigger specific actions within the card. The simplest examples are read and write commands for Smart Card files.

In Smart Card applications, the card is used as a data carrier, an authorization agent or both at the same time. This results in command sets that are optimized for these applications and transmission protocols, and which are used only in the Smart Card industry. Due to the severely limited memory capacity of Smart Cards and the market pressure to increase this only moderately in order to control costs, command sets are usually tailored to specific applications. All the commands that are not needed in a given application are rigorously deleted during program optimization. Only a few operating systems exhibit extensive command sets that have not been reduced to those needed for a particular application. However, the associated cards are correspondingly more expensive, and they are usually used only for testing and prototyping Smart Card projects.

A diversification effect is also seen with Smart Card command sets, as is typical with new technologies. Each manufacturer active in this field attempts to tailor his own commands to suit his operating system or projected applications. This may be due to necessity, since functionally equivalent commands may not exist in the standards. Manufacturers may also deliberately attempt to improve their positions relative to the competition, or to deny their competitors access to a particular application, by employing commands that are highly optimized with regard to card functions and memory usage. In any case, the decision to use commands based on available standards always means choosing an open, more easily expandable and proven system, which may later allow additional functions to be incorporated into a single card. On the other hand, there are plenty of examples of systems in which the use of Smart Cards was only made possible by the use of highly optimized special commands.

There are currently four international standards, or more-or-less stable draft versions of these standards, which define typical Smart Card command sets. These standards list significantly more than 80 commands along with their associated procedures. The commands defined in these standards are largely mutually compatible in terms of coding and functionality.

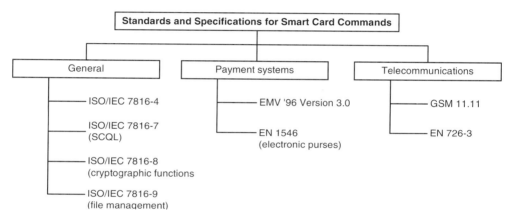

Figure 7.1   The most important standards and specifications for Smart Card commands

The majority of the commands currently used with Smart Cards are defined in the ISO/IEC 7816-4 standard, which is a general international standard. It is not dedicated to any particular area, such as telecommunications or financial transactions, but rather attempts to cover all Smart Card applications. The commands in ISO/IEC 7816-4 are complemented by three supplementary, specialized sections of this set of standards. ISO 7816-7 defines commands for interrogating and managing Smart Card databases whose structures are based on SQL (structured query language). ISO/IEC 7816-8 contains commands for executing and parameterizing cryptographic functions, and ISO/IEC 7816-9 adds file-management commands to the basic command set.

There is no international standard of any significance in the area of financial transactions, but there is an industry standard. This is the EMV specification, whose name comes from the initial letters of Europay, Mastercard and Visa, the three initiators of this specification. The current version is EMV '96 Version 3.0, which is supplemented as necessary at irregular intervals by lists of known errors and ambiguities. This specification has achieved the status of a reference for all Smart Card operating systems, due to the strong market position of the firms behind it. It occupies a position that is at least as important as the ISO/IEC 7816 set of standards.

The GSM 11.11 (ETS 300 977) specification, which was developed for use in the telecommunications industry, forms the normative basis for the GSM card. The fact that millions of cards based on this specification are in commercial use means that it cannot be ignored as a fundamental factor for Smart Card operating systems, just like the EMV specification. Its widespread use also means that its contents are stable and cannot be altered, but only supplemented by compatible commands and functions with limited scopes. It is thus also an international specification for Smart Card operating systems in the telecommunications industry.

The European standard EN 726-3, which like GSM 11.11 is at home in the telecommunications industry, is a superset of GSM 11.11. It defines many supplementary general commands, which in contrast to GSM 11.11 are not specifically tailored to one particular application. This standard also contains many commands for managing applications, which are mainly of interest in the realm of multi-application Smart Cards.

Special commands that are only employed in a limited area are in principle not covered by these standards, and must therefore be specified individually. One example is the command set for a universal electronic purse, initially defined in a provisional standard and now covered by the CEN EN 1546 standard. This European standard contains all commands necessary for an electronic purse, including the associated procedures. Standards of this type, which are restricted to a single application, arise only in areas that are of particular interest to government agencies or specific branches of industry, since the cost of generating them would otherwise be much too high.

The commands in the standards and specifications described above can be classified according to their functions. However, it must be remembered that in real-life Smart Card operating systems, only subsets of all these commands are implemented. Depending on the manufacturer of the operating system, various degrees of deviation from the functionality and coding described in the following sections may be encountered. However, the described functions are present in principle in all operating systems. To be sure, their functionality may be severely restricted due to considerations of memory capacity or cost. Whenever a new application is being considered, you must without fail request the exact specifications of the coding and functions of all commands from the producer of the operating system in question.

The following sections describe the most important and most widely used Smart Card commands, selected from the EN 726-3, ISO/IEC 7816-4/7/8/9, EN 726-3 and EN 1546-3 standards and the EMV '96 and GSM 11.11 specifications. Extensive tables listing the coding of the most important Smart Card commands are contained in Section 15.7.7.

It is of course impossible to purchase a single Smart Card anywhere in the world that contains all the following commands. As a conservative estimate, the memory required for their full implementation would be ten to twenty times are large as the total memory found in the largest currently available Smart Card microprocessor chips. It is however not at all necessary for a card to be able to execute all these commands. Depending on the planned area of application and the operating system, some classes of commands may be supported more comprehensively than others.

With multi-application cards, you would certainly want to make sure that other applications could be installed retroactively in the card after it has been personalized. Assuming adequate memory capacity, a card for cryptographic applications contains the full spectrum of cryptographic commands along with the various algorithms. Each application area uses a different selection of commands from the various classes.

In order to maintain an overview, the standard or specification in which the command is defined is identified in each of the following descriptions. If no source is named, the command is one that is used internally by Smart Card manufacturers but which cannot be assigned to any of the above-mentioned standards. Some of these commands are nonetheless very useful, and will probably be standardized in the future. That is why they are listed here with descriptions of their basic functions. In the interest of readability, we have omitted any description of the coding of typical Smart Card commands in this chapter. You can find this information in the Appendix (Section 15.7.7).

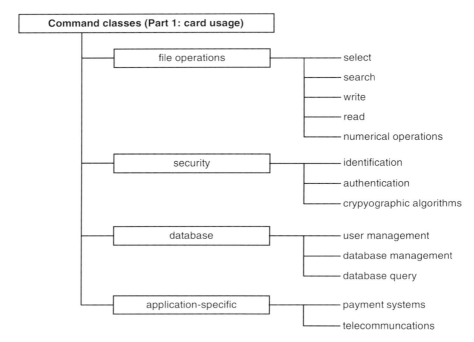

Figure 7.2   Classification chart of all Smart Card commands that are primarily employed when the card is used

There are some commands that are supported by nearly all Smart Card operating systems and which have a limited number of options. For these commands, the command/answer APDU is fully decoded in the appropriate location. In Chapter 5, you can follow the internal program execution of seven typical commands in detail on the basis of the pseudocode of a simple Smart Card operating system called 'Small-OS'.

For each command, the response listed is the one obtained in the event of successful execution. Otherwise, if an operation is forbidden or fault occurs in the card, the terminal receives only a 2-byte return code. Some of the described commands also possess parameters for the selection of additional functions. These options often exist only in the standard, but not in actual operating systems, since they may be too complicated or have no practical significance. Therefore, this chapter does not list or explain every variant defined in the standards, since our aim is to concentrate on practical functions. When describing the commands, we usually follow the standard that has the largest number of functional options for the command in question.

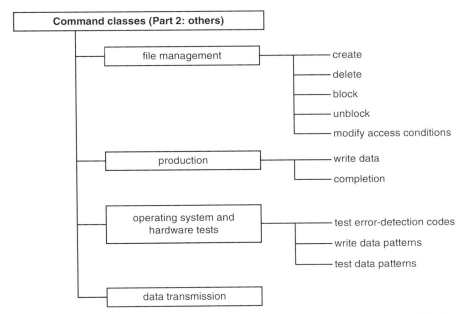

Figure 7.3   Classification chart of all Smart Card commands that are primarily employed before and after the card is used

7.1   File Selection Commands

Without exception, file management in all modern Smart Card operating systems is object-oriented. Among other things, this means that before any action can be performed on an object (which corresponds to a file), it must first be selected. Only then does the system know which file is meant, and all subsequent file-specific commands apply to this file alone. Of course, the access conditions for the file still must be checked within the operating system, to determine whether the relevant commands are permitted or even possible in the first place.

The master file (MF) is always implicitly selected after the card has been reset, so it does not need to be specifically selected. Other files are subsequently selected by executing the SELECT FILE command. Files are addressed using either the 2-byte FID (file identifier) or, in the case of DFs, the 1-byte to 16-byte DF name. The DF name can contain an internationally unique AID (application identifier) that is 5–16 bytes long. It is possible to submit only part of the AID, in which case the least-significant bytes (those to the right) are omitted. An additional parameter allows the card to select the first, last, next or previous DF with respect to the DF identified by abbreviated AID.

Due to an older set of command definitions, file selection in GSM 11.11 can use only the 2-byte FID. In contrast, the ISO command set also supports an extension in the form of file selection via the path name of the file in question. The path name can be either relative, in which case file selection takes place from the currently selected DF, or absolute, in which case file selection takes place from the MF.

Only successful selection of a new file causes the previously selected file to be deselected. If the selection cannot be completed, for example because the requested file does not exist, the previous selection remains in force. This ensures that a file is always selected, even in the event of an error.

After successful selection, the terminal may request data about the new file as necessary. This request, along with the number of desired data items, is sent to the card as part of the SELECT FILE command. The exact contents of these data items are defined in the applicable standard. The data items returned by the card may include the structure, size and amount of free memory of the newly selected file. The amount of data may also depend on the file type.

Table 7.1 lists the explicit file selection options permitted by ISO/IEC 7816-4 for the SELECT FILE command, and Figure 7.4 depicts the sequence of events in a typical file selection process.

Table 7.1   Functions of SELECT FILE allowed by ISO/IEC 7816-4

SELECT FILE	
Command	• FID (if EF, DF or MF)
	or
	AID (if DF)
	or
	path from DF to file
	or
	path from MF to file
	or
	switch: select next higher-level DF
	or
	first, last, next, or previous DF (in case of a partial AID)
	• *switch:* return information about the selected file
Response	• information about the selected file (if selection is successful)
	• Return code

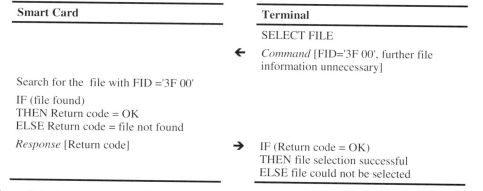

Figure 7.4   Sample procedure for a SELECT FILE command

In addition to explicit file selection with a FID, DF name or a path specification in a SELECT FILE command, a file can be selected implicitly. However, this is only possible with standard read and write commands. A file can be selected before the command is actually executed by specifying its 5-bit Short FID as a supplementary command parameter. The file must however be an EF type and must be located within the current DF. The advantage of this procedure lies in simplified command execution and an increase in processing speed, since no explicit SELECT FILE command needs to be sent to the card.[1]

GSM 11.11 defines the command STATUS, which returns the same data to the terminal as does successful file selection with SELECT FILE. These data provide information about the currently selected file: its type and structure, size, FID, access conditions and whether it is locked. This command is rarely used, and its main purpose is to allow the terminal to query, during a session, which file is currently selected and which conditions apply to it.

Table 7.2 Function of STATUS as defined by GSM 11.11

STATUS	
Command	• —
Response	• information about the currently selected file
	• Return code

EN 726-3 contains a command that supplements SELECT FILE and STATUS, and which is used to close applications. The CLOSE APPLICATION command supplies the FID of the application to be closed, and the card responds by clearing the previously achieved security state. This command is mainly useful when a terminal needs to ensure that a state reached by the card is reset. If the card's operating system does not support such a command, this effect can only be achieved by a card reset. In the command specifications of ISC/IEC 7816-4, selection of the MF is sufficient to cause the security state of the previously selected file to be reset to the initial state.

Table 7.3 Function of CLOSE APPLICATION according to EN 726-3

CLOSE APPLICATION	
Command	• FID (of the current DF)
Response	• Return code

7.2 Read and Write Commands

The read and write class of commands primarily supports the utilization of Smart Cards for secure data storage. With these commands, it is possible to write data to the relevant EFs and to later read the data back. To the extent that the EFs have specific access conditions, reading these files is possible only for authorized users. The information is thus stored in the card with protection against unauthorized access.

---

[1] See also Section 5.6.2, 'File names'.

Since various EF data structures exist, there are also various read and write commands for these files. Unfortunately, this does not fully correspond to an object-oriented file management structure. In a purely object-oriented design, the operating system must be constructed such that an object itself can determine its access mechanisms. This is not the case for Smart Card file management. This non-compliance dates back to the historical emergence of commands that were subsequently incorporated into current standards. The precursors of Smart Cards, namely memory cards, have only one memory region, which can be read and written via offset and length parameters. Externally, this memory can be seen as a single file with a transparent structure. The first Smart Cards were in principle built the same way, and the definitions of read and write access to transparent files originated at that time. Other, specially modified commands are used for file types with data structures, which were defined later. This is how the two different types of access came to be.

Thus, it is necessary to divide this class of commands into commands that access EFs with transparent structures and commands for the remaining file structures, namely cyclic, linear fixed and linear variable. However, various standards (such as EN 1546 for electronic purses) explicitly state that read commands for files with transparent structures may also be used for reading files with other types of structure. In any case, it is possible in this way to obtain additional data about the internal structure of the file.

EFs with a transparent logical structure are amorphous, which means that they have no internal structure. They represent a linearly addressable memory with byte access. The command READ BINARY is used for reading, and the two commands WRITE BINARY and UPDATE BINARY are used for writing.

Table 7.4   Function of READ BINARY according to ISO/IEC 7816-4

READ BINARY	
Command	• number of bytes to be read
	• offset to the first byte to be read
	• *optional:* Short FID for implicit selection
Response	• data read from the file
	• Return code

Table 7.5   Function of WRITE BINARY according to ISO/IEC 7816-4

WRITE BINARY	
Command	• number of bytes to be written
	• offset to the first byte to be written
	• *optional:* Short FID for implicit selection
Response	• Return code

Table 7.6   Function of UPDATE BINARY according to ISO/IEC 7816-4

UPDATE BINARY	
Command	• number of bytes to be overwritten
	• offset to the first byte to be overwritten
	• *optional:* Short FID for implicit selection
Response	• Return code

The fundamental difference between the commands WRITE BINARY and UPDATE BINARY relates to the secure state of the card's EEPROM. The secure EEPROM state is the logical state of the EEPROM bits when the memory cells have taken on their minimum-energy state. Since the memory cells are small capacitors, this means the state in which they contain no charge. Normally this is the logical 0 state. In order to reset a bit from state 0 to state 1, it must be cleared. This restores the charge on the capacitor.

The WRITE BINARY command can only be used to set bits from the non-secure state, in this case 1, to the secure state, in this case 0. WRITE BINARY in this example thus performs a logical AND operation on the input data and the file contents. On the other hand, if the chip's secure state is represented by the value 1, the WRITE BINARY command must effect a logical OR of the command data and the file data. The logical interaction between the command data and the file data is such that a WRITE command always achieves the secure state of the EEPROM. In addition, depending on the file, the WRITE BINARY command may support WORM write access (write once, read multiple).

The UPDATE BINARY command, in contrast, is a genuine write to the file. The previously state of the data in the file does not affect the contents of the file following execution of the UPDATE BINARY command. UPDATE BINARY can therefore be regarded to be the same as clearing the data with ERASE followed by WRITE BINARY.

These commands can be utilized to construct physically secure Smart Card counters. The principle involves a bit field in which each set bit represents a monetary unit. When a payment is made, the counter is decremented bit by bit using OR operations produced by WRITE BINARY commands. After authentication, the counter value can be increased again using an UPDATE BINARY command. The main advantage of this technique is that it makes it impossible to increase the counter value by tampering with the EEPROM, for example by heating it, since the secure state of each bit represents a value of 0.

As their name suggests, READ BINARY is a read command, while WRITE BINARY and UPDATE BINARY are write commands. File access always takes place by means of a length parameter plus an offset to the first addressed byte. Some operating systems also permit implicit file selection before the actual data access occurs by means of a supplementary Short FID parameter. This feature is however not provided in all standards and operating systems.

Figure 7.5 illustrates a typical sequence consisting of READ BINARY followed by WRITE BINARY and finally UPDATE BINARY. The effects on the contents of the selected file are shown in Figure 7.6. Of course, this example assumes that file selection is successful and the access conditions of the described file are met.

Smart Card		Terminal
		READ BINARY
	←	*Command* [offset = 2 bytes, number of bytes to be read = 5]
requested data := '03' || 'FF' || '00' || 'FF' || '00'		
Response [requested data || Return code]	→	IF (Return code = OK) THEN READ BINARY successful ELSE cancel
		WRITE BINARY
	←	*Command* [Offset = 3 bytes, number of bytes to be written = 2, data = 'F0 F0']
Response [Return code]	→	IF (Return code = OK) THEN WRITE BINARY successful ELSE cancel
		UPDATE BINARY
	←	*Command* [Offset = 5 bytes, number of bytes to be written = 2, data = 'F0 F0']
Response [Return code]	→	IF (Return code = OK) THEN UPDATE BINARY successful ELSE cancel

Figure 7.5 Accessing a file with a transparent structure

(a)	'01'	'02'	'03'	'FF'	'00'	'FF'	'00'	...	n
(b)	'01'	'02'	'03'	'F0'	'00'	'FF'	'00'	...	n
(c)	'01'	'02'	'03'	'FF'	'00'	'F0'	'F0'	...	n

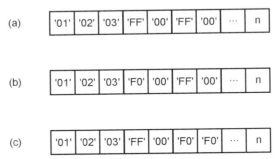

Figure 7.6 Example of write accesses to an EF with a transparent structure: (a) file contents with READ BINARY, (b) file contents after WRITE BINARY, (c) file contents after UPDATE BINARY. The command sequences used here are shown in Figure 7.5.

ERASE BINARY is an exception among the commands operating on transparent EFs. It cannot be used to write data to a file, but only to erase data starting from a given offset. If no second offset parameter is stated, the command erases all data to the end of the selected file. Here, erasing means that the data region specified in the command is set to the logical cleared state. This state must be defined separately for each operating system, since it may not be the same as the physically cleared state of the memory.

Table 7.7   Function of ERASE BINARY according to ISO/IEC 7816-4

ERASE BINARY	
Command	• offset to the first byte to be erased
	• *optional:* offset to the last byte to be erased
	• *optional:* Short FID for implicit file selection
Response	• data read from the file
	• Return code

Because the structures of linear fixed, linear variable and cyclic EFs are fundamentally different from that of a transparent file, special commands for accessing these particular data structures are available, in addition to the commands described above. All these files are record-oriented. For write accesses, the smallest addressable unit in the data field is a single record. For read accesses, either the entire record or some part of it may be read, starting with the first byte. These file structures, which transform a linear, one-dimensional memory into a two-dimensionally addressable memory, result in access types that are significantly more complex than those used with a transparent structure. In principle, all data structures can be emulated by transparent ones, but in specific cases this may prove considerably more complicated than using higher-level data structures.

name	telephone number
Alex	089 -- 47 84 46
Otto	089 -- 11 00 11
Dieter	089 -- 47 51 22
Anna	089 -- 178 098
Albert	089 -- 178 234
Peter	089 -- 76 87 33
Erich	089 -- 123 78 1
Hans	089 -- 111 222
Ludwig	089 -- 178 234
Josef	089 -- 167 189

Figure 7.7   Example of a telephone number list in a file with a 'linear fixed' structure

After an EF with a record-oriented structure has been selected, the card's operating system creates a record pointer whose initial value is undefined. The value of this pointer can be set using READ, WRITE, UPDATE RECORD and SEEK commands. The pointer for the current file is saved as long as this file is selected. Following successful explicit or implicit selection of another file, the value of the record pointer is again undefined.

All commands for record-oriented files can use a parameter byte to specify the type of access to the file. The basic type is direct access, using the absolute number of the desired record. This type of access does not alter the record pointer. The number of the desired record is sent to the card, and the response contains the content of the record in question. If the parameter 'first' is supplied, the operating system sets the record pointer to the first record in the file, and this record is either read or written according to the command type. The parameter 'last' provides access to the final record in a similar manner. The additional parameters 'next' and 'previous' allow the next and previous records, respectively, to be selected and read or written. Finally, the parameter 'current' can be used to address the record marked by the current value of the record pointer.

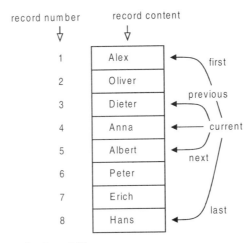

Figure 7.8   Accessing record-oriented files

The large variety of access methods for record-oriented data structures comes from the typical structure of a telephone directory. Consider a record whose initial part contains a surname and given name, followed in the same record by the associated telephone number. With a READ RECORD command and the parameters described above, you can page forwards and backwards as desired within a telephone number list built on an EF, or jump to the first or last entry. The record pointer can also be changed using the search command SEEK, which is described below.

ISO/IEC 7816-4 also provides the option of reading all records from the first one up to a specified record number, or from a specified record number through to the last record, using READ RECORD in a single command–response cycle. As practical as these commands may be, their use with large files can quickly exceed the capacity of the I/O buffer.

Figure 7.9 illustrates the execution of several read and write operations on the file shown in Figure 7.7.

Table 7.8 Function of READ RECORD according to ISO/IEC 7816-4

READ RECORD	
Command	• number of records to be read *or* mode (first, last, next, or previous record) *or* read all records from *n* to the last record *or* read all records from the first to *n* • *optional:* Short FID for implicit file selection
Response	• data read in file • Return code

Table 7.9 Function of WRITE RECORD according to ISO/IEC 7816-4

WRITE RECORD	
Command	• record to be written • number of the record to be written *or* mode (first, last, next, or previous record) • *optional:* Short FID for implicit file selection
Response	• Return code

Table 7.10 Function of UPDATE RECORD according to ISO/IEC 7816-4

UPDATE RECORD	
Command	• record to be overwritten • number of the record to be overwritten *or* mode (first, last, next, or previous record) • *optional:* Short FID for implicit selection
Response	• Return code

The APPEND RECORD command, given its function, could just as well fit in the File Management Commands section. It can be used to append records to existing record-oriented files. The data for the entire new record are provided at the same time by the command. A relatively complex memory manager, in Smart Card terms, is a prerequisite for the availability of this command with its full functionality. The memory manager's task is to create a link between the new record and the ones already present. It is then possible, within the limits of the available memory, to add an arbitrary number of new records to a file. However, the number of new records that can be added is often restricted to simplify matters. In this case, memory is reserved as necessary for the addition of future records when a record-oriented file is created. This space can later be filled up using APPEND RECORD. Once the free space is used up, no further APPEND RECORD commands are possible.

Smart Card		**Terminal**
		READ RECORD
	←	*Command* [Record number = "2"]
processing command		
Response ["Oliver" ‖ Return code]	→	IF (Return code = OK) THEN READ RECORD successful ELSE cancel
		UPDATE RECORD
	←	*Command* [first, "Wolfgang"]
processing command		
Response [Return code]	→	IF (Return code = OK) THEN UPDATE RECORD successful ELSE cancel
		UPDATE RECORD
	←	*Command* [next, "Alex"]
processing command		
Response [Return code]	→	IF (Return code = OK) THEN UPDATE RECORD successful ELSE cancel
		READ RECORD
	←	*Command* [Record number = 2]
processing command		
Response ["Alex" ‖ Return code]	→	IF (Return code = OK) THEN READ RECORD successful ELSE cancel

Figure 7.9   Sample read and write operations on a record-oriented file

If APPEND RECORD is used in conjunction with a linear fixed or linear variable file, the new record is always added at the end of the file. If the structure is cyclic, however, the new record is always numbered 1, which corresponds to the currently written record in files of this type.

APPEND RECORD can be used for various purposes. One possibility is a telephone directory, as already mentioned. Another possibility is a log file, in which the data to be recorded is written directly to the card by creating new records.

Table 7.11   Function of APPEND RECORD according to ISO/IEC 7816-4

APPEND RECORD	
Command	• record to be written
	• optional: Short FID for implicit file selection
Response	• Return code

There are two commands that complement the file-based read and write commands. These are designed for direct access to data objects. Depending on the selected DF, certain data can be written to or read from files or internal operating system structures, bypassing the file-oriented access mechanisms. Data objects can be written using PUT DATA and read using GET DATA. With both of these commands, the exact structure of the TLV-coded data objects must be provided with the command. This means that it must be known whether application-specific or standardized coding is used for the data objects. This information is important inside the operating system, since it allows the objects to recognize the data according to how they are packaged. The appropriate access conditions must be satisfied in advance for both of these commands.

Table 7.12   The function of GET DATA according to ISO/IEC 7816-4.

GET DATA	
Command	• number of data objects to be read
	• identifier of the data objects to be read
Response	• read data objects
	• Return code

Table 7.13   The function of PUT DATA according to ISO/IEC 7816-4.

PUT DATA	
Command	• structure of the data objects to be written
	• data objects to be written
Response	• Return code

7.3   Search Commands

Record-oriented structures allow sets of associated data with identical structures to be stored in a single file. A typical example is a telephone directory containing names and telephone numbers. A search command can be used to avoid having to read the entire directory, record by record, when looking for a particular name.

The SEEK command can be used to search for a specified character string in a record-oriented data structure. An offset can be supplied with the command. The length of the search string is variable. The command must tell the operating system in which direction to search. This can be either from the starting location onwards (in the direction of increasing record numbers) or from the starting location backwards. The starting location for the search must also be specified. The first record, last record or current record can be specified as the starting position. If the search string is found, the operating system sets the record pointer to the location of the string and informs the terminal that the search was successful.

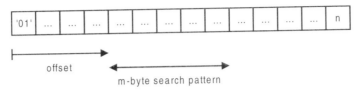

Figure 7.10   Searching within a record-oriented file

The ISO/IEC 7816-9 standard describes two commands that can be used to search for data in transparent and record-oriented files. The SEARCH RECORD command is the ISO/IEC version of the ETSI SEEK command. The major difference between these two commands is that according to ISO/IEC 7816-9, a Short FID can also be transferred within the command for the implicit selection of an EF.

The command sequence shown in Figure 7.11 illustrates some ways in which the SEEK command can be used. This example is based on the linear fixed file shown in Figure 7.7.

Table 7.14   The function of SEEK according to EN 726-3

SEEK	
Command	• length of the search string
	• search string
	• offset
	• search mode (forward from the beginning, backward from the end, forward from the next position, backward from the previous position)
	• *switch:* return the record number of the located record
Response	• record number (if selected by the switch)
	• Return code

Smart Card	Terminal
	SEEK
←	*Command* [search string = "Hans" ‖ search direction = "from beginning to end" ‖ send record number]
Response [record number = 8 ‖ Return code] →	IF (Return code = OK) THEN "Hans" found ELSE "Hans" not found
	SEEK
←	*Command* [search string = "Alex" ‖ search direction = "from end to beginning" ‖ send record no.]
Response [record no. = 1 ‖ Return code] →	IF (Return code = OK) THEN "Alex" found ELSE "Alex" not found

Figure 7.11   Sample command sequence using the SEEK command

Table 7.15 Function of SEARCH RECORD according to ISO/IEC 7816-9

SEARCH RECORD	
Command	• length of the search string
	• search string
	• offset
	• mode (forward from the beginning, backward from the end, forward from the next position, backward from the previous position)
	• *switch:* return the record number of the located record
	• *optional:* Short FID for implicit file selection
Response	• record number (if selected by the switch)
	• Return code

The SEARCH BINARY command allows for specified data to be searched for in a selected transparent file. The file may be selected either explicitly by a previously transferred command or implicitly via a command parameter. The result is the offset from the start of the file to the first byte of the located search string.

Table 7.16 Function of SEARCH BINARY according to ISO/IEC 7816-9

SEARCH BINARY	
Command	• length of the search string
	• search string
	• offset
	• *optional:* Short FID for implicit file selection
Response	• offset to the located data
	• Return code

7.4 File Manipulation Commands

There are several commands that allow the content of a file to be modified by means other than simple writing. The main representatives of this class are the commands INCREASE and DECREASE. They increase or decrease the value of a cyclically structured file whose content takes the form of a counter. The amount of the increase or decrease is transferred as a command parameter.

A cyclic file structure is prescribed to meet the needs of logging functions. These commands are primarily used for simple electronic change purses and counters. For the sake of simplicity, an example cyclic file with only one record is shown in Figure 7.12. The starting value of the record is 10. On conclusion of the process, the record has the same value again.

Table 7.17   DECREASE function according to EN 726-3

DECREASE	
Command	• value to be subtracted
Response	• subtracted value
	• new value of the record
	• Return code

Table 7.18   INCREASE function according to EN 726-3

INCREASE	
Command	• value to be added
Response	• added value
	• new value of the record
	• Return code

Smart Card		**Terminal**
		DECREASE
	←	*Command* [value to be subtracted = 3]
processing command		
Response [subtracted value = 3 ‖ new value = 7 ‖ Return code]	→	IF (Return code = OK)
		THEN DECREASE successful
		ELSE DECREASE could not be executed
		DECREASE
	←	*Command* [value to be subtracted = 2]
processing command		
Response [subtracted value = 2 ‖ new value = 5 ‖ Return code]	→	IF (Return code = OK)
		THEN DECREASE successful
		ELSE DECREASE could not be executed
		INCREASE
	←	*Command* [value to be added = 5]
processing command		
Response [added value = 5 ‖ new value = 10 ‖ Return code]	→	IF (Return code = OK)
		THEN INCREASE successful
		ELSE INCREASE could not be executed

Figure 7.12   Example command processes with INCREASE and DECREASE

The command EXECUTE can also be considered to be a file manipulation command in a certain sense. It is used to run executable EFs (whose structure is 'executable'). The program to be executed can receive data from the terminal via the command, and it can also send data that it produces back to the terminal as a response.

Both this command and the related file structure are controversial, since in certain circumstances they can be used to bypass the entire security system of a Smart Card.

Table 7.19 EXECUTE function according to EN 726-3

EXECUTE	
Command	• data to be passed to the executable file
Response	• data returned by the executable file
	• Return code

7.5 Identification Commands

In addition to being used as secure data carriers, Smart Cards can also be used to identify individuals. The usual procedure involves transferring secret information that is known only to the user and to the card. This is usually a Personal Identification Number (PIN).

Most people are familiar with PIN verification from everyday experience. You enter your PIN at a terminal, and shortly thereafter the display shows whether the PIN was correct or, if not, how many attempts are still allowed. In this process, the Smart Card receives the PIN from the terminal in a VERIFY command. The PIN is usually a four-digit number, which the Smart Card compares with a value stored in its EEPROM. If the entered PIN matches the stored PIN, the card's internal state changes, the terminal receives a response confirming a positive result, and the retry counter is reset to its original value of 0. If the entered PIN does not match the stored PIN, the retry counter is incremented. If it reaches its predefined maximum value, the card is blocked for any further PIN verification.

Many Smart Card operating systems allow several PINs to be used. In such cases, it is mandatory to send the identification number of the relevant PIN in all relevant commands, so that it can be correctly addressed. As a rule, however, card issuers attach great importance to having only one PIN per card, even when it is technically possible to have more than one. This is essential for customer acceptance and user friendliness.

The acronym 'CHV' is often used in place of 'PIN' in the telecommunications industry. CHV stands for 'chip holder verification' and means exactly the same thing as PIN. Since the commands described below originated in the telecommunications industry, their names use the acronym CHV instead of PIN.

Table 7.20 VERIFY CHV function according to EN 726-3

VERIFY CHV	
Command	• PIN
	• PIN identification number
Response	• Return code

Table 7.21   VERIFY function according to ISO/IEC 7816-4

VERIFY	
Command	• PIN
	• PIN identification number
	• *switch:* global PIN or application-specific PIN
Response	• Return code

The ISO/IEC 7816-4 standard describes a PIN verification command that is largely the same as that of EN 726-3. Its name is VERIFY, and it can be used not only for PIN comparison but also for the verification of biometric characteristics. Compared with PIN verification according to ETSI, there is only one significant difference in the coding. ISO/IEC makes a distinction in the command between a global PIN and an application-specific PIN. This makes it possible to select within the command whether to verify a PIN that is valid for the entire Smart Card or one that is valid only for the current DF.

With some applications, the card user is expected to choose his or her own PIN code the first time the PIN code is entered. The null-PIN procedure can be used for this purpose. Here the PIN is set to all zeros when the card is personalized. This PIN code has a special meaning for some Smart Card operating systems, which reject all VERIFY commands that contain null PINs and demand that the PIN first be changed using the CHANGE REFERENCE DATA command. In this way, the user is forced to change his or her PIN, since it is not possible to alter the security state of the card if a null PIN is entered. This procedure is not standardized, but it can sometimes be used advantageously if it is supported by the operating system.

PINs have steadily proliferated since their introduction as identification numbers for cardholders. Currently, the average card user is expected to keep track of perhaps 10–20 different PINs for various cards and other authorizations. The fact that this expectation is unrealistic is shown by the large number of people who jot down the PIN on the card itself. The use of Smart Cards allows the user to choose a PIN at will, and thus to use the same PIN for all his or her cards. Although this may cause security problems, since anyone who illicitly acquires one PIN thereby knows all of them, it is still better than writing the PIN on the card where anyone can read it.

The command CHANGE CHV allows the PIN to be altered. The ISO/IEC 8716-8 equivalent of this command is CHANGE REFERENCE DATA, which has the same input and output parameters. If the PIN currently stored in the card is known, it can be replaced with a new one. If the current PIN is entered incorrectly, the operating system increments the retry counter to protect against possible attempts to use this mechanism to learn the PIN code. As soon as the current PIN is correctly passed to the card, the card stores the new PIN that it has received in the appropriate memory location and resets the retry counter.

If a retry counter has reached its maximum value, it can be reset using the UNBLOCK CHV command with a second PIN, called the PUK (personal unblocking key). The PUK is usually longer than the standard 4-digit PIN (8 digits, for example). The user need not memorize the PUK, since it is only needed if the PIN has been forgotten. It is sufficient if the user has a record of the PUK somewhere at home. However, just resetting the retry counter would not help the user very much, since he or she would still not know the correct PIN. Hence the command UNBLOCK CHV must also provide the card with a new PIN.

In hybrid cards, which have magnetic stripes as well as chips, using this command to alter the PIN in the chip should not be permitted. Otherwise, the stripe would then hold a different PIN from that in the chip, which would lead to severe problems. With such cards, the retry counter is simply reset and the customer is sent a letter with the original PIN.

Table 7.22 CHANGE CHV function according to EN 726-3

CHANGE CHV	
Command	• old PIN
	• new PIN
	• PIN number
Response	• Return code

Table 7.23 UNBLOCK CHV function according to EN 726-3

UNBLOCK CHV	
Command	• PUK
	• new PIN
Response	• Return code

The ISO/IEC 7816-8 standard provides its own command that can be used to reset the retry counter when it has reached its maximum value. This command is called RESET RETRY COUNTER. A certain security state must be achieved before this command can be executed. As a rule, this results from a successful authentication. In spite of its name, this command can also be used to replace the current PIN with a new PIN.

GSM contains two other commands that affect PIN interrogation. These are DISABLE CHV and ENABLE CHV, which switch PIN verification off and on. If PIN verification is disabled, all file access restrictions that require prior PIN verification are disabled. Both of these commands are very popular in the field of mobile communications, since they eliminate the need to re-enter the PIN every time the cellular phone is switched on. From a security perspective, these commands are disputable, since they disable the protection against unauthorized use provided by the PIN. Of course, the user could also use CHANGE CHV to choose a trivial PIN, such as '0000', which offers just as little protection.

Table 7.24 RESET RETRY COUNTER function according to ISO/IEC 7816-8

RESET RETRY COUNTER	
Command	• PIN identification number
	• *option:* [PUK ‖ new PIN]
	• *if option used:* PUK identification number
	• *switch:* global PIN or application-specific PIN
Response	• Return code

Table 7.25   DISABLE CHV function according to EN 726-3

DISABLE CHV	
Command	• PIN
	• PIN identification number
Response	• Return code

Table 7.26   ENABLE CHV function according to EN 726-3

ENABLE CHV	
Command	• PIN
	• PIN identification number
Response	• Return code

The ISO/IEC 7816-8 commands for these functions are called ENABLE VERIFICATION REQUIREMENT and DISABLE VERIFICATION REQUIREMENT. Depending on the application, a certain security state must be achieved before these commands can be executed.

For obvious reasons, the PIN verification procedures described above are subject to attack, since under the right conditions large financial gains can be realized with a lost or stolen card and the right PIN. All commands associated with PIN or PUK comparison must be shielded against analysis of the card's electrical or time-related behavior. For example, the power consumption during the execution of VERIFY PIN must always be the same, regardless of whether the entered PIN is correct. It is just as important that the time taken to execute PIN commands does not depend on whether the PIN is correct. Varying execution times could have fatal consequences for the card's security, and thus ultimately for the security of the entire system. Such variations could be used to find the value of the correct PIN in a very simple manner, with the result that all of the system's PIN codes would be rendered useless as a means of user identification.

7.6   Authentication Commands

In addition to the commands for identifying the cardholder, there is a further set of commands for authenticating the terminal and the card. Since each of these communications partners is equipped with a complete computer, the procedures can be made much more complex, and thus more secure, than those used for PIN verification.

In PIN verification, the card receives a secret code in clear text (the PIN) via the interface, and it only needs to compare this with the PIN held in memory. Eavesdropping on the transmission line would thus have fatal consequences. Modern authentication procedures are designed such that such assaults are impossible.

In principle, authentication involves the verification of a secret known to both of the communicating parties without requiring it to be sent across the interface. The procedures are constructed such that tapping into the data transfer would not compromise the security of the authentication. [2]

Depending on the operating system, there are various commands for the authentication of the card or the terminal, or both at the same time. For the sake of clarity, this section and the rest of the chapter discuss only authentication between the card and the terminal. In terms of information technology, though, the 'rest of the world' authenticates itself with respect to an application in the card. There is no verification that the card as a whole is genuine, but instead only that the embedded microcontroller shares a secret with the external world. This should taken into account in certain applications.

In many operating systems, the keys used in authentication are protected by a retry counter. If a terminal attempts unsuccessful authentication too often, the card blocks the associated key for further authentication tests. There is no objection to this with regard to system security, but it does have one disadvantage. Resetting the authentication retry counter often involves very complex, logistically difficult and expensive administrative procedures. Therefore, some systems do not have a retry counter for authentication keys.

For security reasons, only card-specific keys should be used for authentication. These keys can be generated from a unique feature of the card. A serial number or chip production number is very suitable for this purpose. Such non-confidential (and thus public) numbers can be read from the card with the appropriate command. There is presently no standard covering this; here we use the command GET CHIP NUMBER. The name varies from one operating system to the next, as do the data that are exchanged. Here we are only interested in the functionality. The command GET CHIP NUMBER obtains a unique serial number from the card, which in view of the DES algorithm should preferably be eight bytes long. This number is used to unambiguously identify the chip and to compute card-specific keys.

One more command is necessary for authentication. This is ASK RANDOM, as specified in EN 726-3. This requests a random number from the card, to be used later during authentication. This command is called GET CHALLENGE in ISO/IEC 7816-4, where it has the same function. In DES authentication, the length of the number is typically eight bytes, but it may be different for other cryptographic algorithms.

Table 7.27   GET CHIP NUMBER function

GET CHIP NUMBER	
Command	• —
Response	• chip number
	• Return code

Table 7.28   ASK RANDOM function according to EN 726-3 and GET CHALLENGE function according to ISO/IEC 7816-4

ASK RANDOM / GET CHALLENGE	
Command	• —
Response	• random number
	• Return code

---

2   See also Section 4.10, 'Authentication'.

In order to maintain the clarity of the following process examples and avoid unnecessary complexity, we have omitted the derivation of the card-specific key. This is however absolutely indispensable for security purposes.

The command INTERNAL AUTHENTICATE allows the terminal to authenticate a card or, in the case of a multi-application card, to authenticate an application. This command can be used to verify that a card is genuine. The card receives a random number, which it encrypts using a key known only to it and to the terminal, using a cryptographic algorithm such as DES. The result of this operation is returned to the terminal in the response. The terminal performs the same encryption as the card and compares its result with that contained in the card's response. If the two results match, it follows that the card also knows the secret authentication key and thus must be genuine. The card has thus been authenticated.

Table 7.29   INTERNAL AUTHENTICATE function according to IEO/IEC 7816-4

INTERNAL AUTHENTICATE	
Command	• random number
	• number of the algorithm to be used
	• number of the key to be used
Response	• **enc** (key; random number)
	• Return code

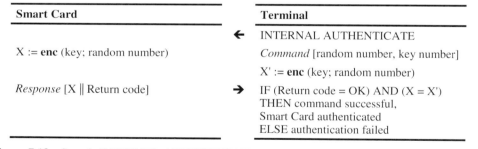

Figure 7.13   Sample INTERNAL AUTHENTICATE procedure

This command implements the classic challenge–response procedure for the authentication of a communications partner. The precise data content of the challenge is not specified in detail in the ISO/IEC 7816-4 standard. The only thing that is standardized is that the value sent to the card must be random and session-specific. Consequently, a detailed specification is always necessary for the INTERNAL AUTHENTICATE command, as well as for the other authentication commands that are described below, in order that they can be used with multiple systems (interoperably).

The command EXTERNAL AUTHENTICATE is used by the terminal to demonstrate to the card that it is connected to a genuine terminal. The command must be initiated by the terminal, since the communications procedure must always run within the command–response framework. However, the card can force terminal authentication by blocking access to certain files until the terminal has been successfully authenticated.

This proceeds as follows: first the terminal sends the command ASK RANDOM to request a random number from the card, which it then encrypts using a secret key. In the next command, EXTERNAL AUTHENTICATE, the terminal returns the encrypted random number to the card. The card encrypts the secret key, which it also knows, in the same way, and then compares its result with the one obtained from the terminal. If they are the same, the terminal must also be in possession of the secret authentication key and is thus authenticated. After successful terminal authentication, the operating system alters the state of its state machine. This allows the terminal to access certain files for reading or writing. For the user, this is the recognizable result of an external authentication.

Table 7.30 EXTERNAL AUTHENTICATE function according to ISO/IEC 7816-4

EXTERNAL AUTHENTICATE	
Command	• **enc** (key; random number)
	• number of the algorithm to be used
	• number of the key to be used
Response	• Return code

Smart Card	**Terminal**
	ASK RANDOM
	← *Command* [ ]
Response [random number ‖ Return code] →	IF (Return code = OK) THEN command successful ELSE cancel
	X := **enc** (key; random number)
	EXTERNAL AUTHENTICATE
X' := **enc** (key; random number) IF (X = X') THEN terminal authenticated ELSE authentication failed	← *Command* [X, key number]
Response [Return code] →	IF (Return code = OK) THEN command successful, terminal authenticated ELSE authentication failed

Figure 7.14 EXTERNAL AUTHENTICATE procedure

If the commands INTERNAL AUTHENTICATE and EXTERNAL AUTHENTICATE are executed one after the other, the communicating parties are mutually authenticated. Each one thus knows that the other party is genuine. However, this requires a total of three complete command sequences. In order to simplify this complicated and time-consuming procedure, the three instructions with their data are linked together and the two authentication commands are merged into one, known as MUTUAL AUTHENTICATE.

Table 7.31   MUTUAL AUTHENTICATE function

MUTUAL AUTHENTICATE	
Command	• **enc** (key; terminal random number, Smart Card random number, chip number) • number of the algorithm to be used • number of the key to be used
Response	• **enc** (key; Smart Card random number, terminal random number) • Return code

Smart Card		**Terminal**
		GET CHIP NUMBER
	←	*Command* [ ]
Response [Chip number ‖ Return code]	→	IF (Return code = OK) THEN command successful
		key derivation (compute the card-specific key)
		GET CHALLENGE
generate RND_CK	←	*Command* [ ]
Response [RND_CK ‖ Return code]	→	IF (Return code = OK) THEN command successful
		MUTUAL AUTHENTICATE
		generate RND_T
		X1 := **enc** (key; RND_T ‖ RND_CK ‖ chip number)
X1' := **dec** (key; X1) IF (RND_CK and chip number are the same as those sent) THEN terminal authenticated ELSE abort	←	*Command* [X1 ‖ key number]
X2 := **enc** (key; RND_CK ‖ RND_T)		
Response [X2]	→	X2' := **dec** (key; X2) IF (RND_T = same as the one sent) THEN Smart Card authenticated

Figure 7.15   Example of a MUTUAL AUTHENTICATE procedure

This command is defined in the ISO/IEC 7816-8 standard. It can be used to perform reciprocal authentication of two parties as defined in the ISO/IEC 9798-2/3 standard. The use of a single authentication command also increases the security of the overall procedure, since it prevents the fraudulent practice of inserting commands between two one-way authentications. A further improvement in the security of the procedure results from the fact that it is impossible to obtain plaintext–ciphertext pairs by tapping the communications between the terminal and the card, which would be an ideal basis for a security assault.

Mutual authentication proceeds as follows. First, the terminal requests the card's chip number with the GET CHIP NUMBER command. Now the terminal can compute the card's specific key. The terminal then uses the ASK RANDOM command to obtain a random number from the card, and then generates a random number itself.

Once the terminal has received the random number from the card, it merges the two random numbers and the chip number into a single block, which it encrypts using the secret authentication key and a cryptographic algorithm in CBC mode. It sends the resulting ciphertext block to the card, which decrypts it and compares the chip number and the random number with those previously transmitted. If they match, the terminal has been authenticated. Now the card swaps the two random numbers, deletes the chip number and encrypts the block again using the secret key. After the terminal has received and decrypted this block, it can determine whether the card possesses the secret authentication key by comparing the known random numbers. If they are correct, the card is also authenticated.

7.7   Commands for Cryptographic Algorithms

Commands for cryptographic algorithms are very important for many applications. For example, they make it relatively easy to use Smart Cards as encryption and decryption devices or for verifying digital signatures. Many Smart Card operating systems have their own command sets for the execution of cryptographic algorithms. Smart Card commands such as ENCRYPT, DECRYPT, SIGN DATA and VERIFY SIGNATURE have arisen because there was no standard for this sort of functionality. However, in the meantime two commands specially designed for the processing of cryptographic algorithms have been defined in the ISO/IEC 7816-8 standard.

In the following material, other cryptographic commands contained in the ISO/IEC 7816-8 standard are also described, since this is presently the only valid reference for this type of command. However, at the time of writing this standard is not final; its status is CD (committee draft). This means that it is certainly possible for various small changes, or even some large changes, to take place before the final version appears.

In the ISO/IEC 7816-8 standard, the functions related to cryptography are split between two commands. The MANAGE SECURITY ENVIRONMENT command allows various general conditions to be set before the actual execution of the cryptographic algorithm. This command passes a 'template' to the card, and this template contains the relevant parameters. They remain valid until they are replaced by a new MANAGE SECURITY ENVIRONMENT command. The templates themselves consist of TLV-encoded data objects, which allows for a high level of variability (and unfortunately complexity) in the parameter transfer.

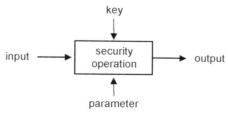

Figure 7.16   Basic principle of the ISO/IEC 7816-8 commands MANAGE SECURITY ENVIRONMENT and PERFORM SECURITY OPERATION for cryptographic functions

Table 7.32 MANAGE SECURITY ENVIRONMENT function according to ISO/IEC 7816-9

MANAGE SECURITY ENVIRONMENT	
Command	• template with parameters for cryptographic checksums *or* template with parameters for a digital signature *or* template with parameters for a hash algorithm *or* template with parameters for authentication *or* template with parameters for encryption and decryption
Response	• Return code

After all the options for the cryptographic function have been set using the MANAGE SECURITY ENVIRONMENT command, the command PERFORM SECURITY OPERATION can be called. A wide variety of security operations can be carried out using this command, provided that they are supported by the Smart Card operating system. The number of possibilities with this command is however so large that this support is not always mandatory. Although PERFORM SECURITY OPERATION is coded with only one instruction byte, it has eight fundamentally different functions that are distinguished by the parameter byte P1. The reason for this is that the number of command bytes that remain available for coding commands has become rather tight in the last while.

Since the PERFORM SECURITY OPERATION command can be used in so many different ways, we have split the following description of its functions along the lines of the various options, without describing any of them in detail. PERFORM SECURITY OPERATION with the COMPUTE CRYPTOGRAPHIC CHECKSUM option is used to compute a cryptographic checksum (CCS), which is commonly referred to as a MAC. The padding to be used, as well as the key to be used, can be either implicitly provided by the operating system or explicitly supplied via the MANAGE SECURITY ENVIRONMENT command. The counterpart to this command option is VERIFY CRYPTOGRAPHIC CHECKSUM, which computes the cryptographic checksum of the transferred data and compares it to a reference value that is also transferred by the command. The result of this operation is a match/no-match indication, which is returned to the terminal.

Table 7.33 The function of PERFORM SECURITY OPERATION with the COMPUTE CRYPTOGRAPHIC CHECKSUM option, according to ISO/IEC 7816-8

PERFORM SECURITY OPERATION *Option:* COMPUTE CRYPTOGRAPHIC CHECKSUM	
Command	• data to be encrypted
Response	• CCS • Return code

Table 7.34 The function of PERFORM SECURITY OPERATION with the VERIFY CRYPTOGRAPHIC CHECKSUM option, according to ISO/IEC 7816-8

PERFORM SECURITY OPERATION Option: VERIFY CRYPTOGRAPHIC CHECKSUM	
Command	• data to be encrypted
	• CCS
Response	• Return code

The two options ENCIPHER and DECIPHER are provided for the pure encryption and decryption of data. The ENCIPHER option is used to encrypt the data transferred with the command. The encryption algorithm to be used can be selected, according to the options provided by the operating system, by first sending the MANAGE SECURITY ENVIRONMENT command. Similarly, the mode of the encryption algorithm must also set in advance by parameters transferred before the command is issued. With a block encryption algorithm, it is also possible to choose between the ECB and CBC modes. Since the length of the data block sent to the card need not necessarily be an exact multiple of the block size of the cryptographic algorithm, the padding method must be defined via an additional parameter. Equally important is the address of the key stored in the Smart Card that is to be used by the algorithm to encrypt the data.

Table 7.35 The function of PERFORM SECURITY OPERATION with the ENCIPHER option, according to ISO/IEC 7816-8

PERFORM SECURITY OPERATION Option: ENCIPHER	
Command	• data to be encrypted
Response	• encrypted data
	• Return code

The reverse function of ENCIPHER is called DECIPHER. With this function, the transferred data can be decrypted in the same mode as that used for ENCIPHER. Naturally, the Smart Card must know the appropriate key, algorithm mode and padding mode. This information must be passed to the card's operating system with a MANAGE SECURITY ENVIRONMENT command.

Table 7.36 The function of PERFORM SECURITY OPERATION with the DECIPHER option, according to ISO/IEC 7816-8

PERFORM SECURITY OPERATION Option: DECIPHER	
Command	• encrypted data
Response	• decrypted data
	• Return code

With the introduction of public-key algorithms into Smart Card applications, there was a need for suitable commands to use the newly available functions. Smart Cards are particularly suitable for digital signature applications, since the private key for the signature algorithm can be securely stored in memory, where it cannot be read. The ISO/IEC 7816-8 standard describes four command options that can be used for digital signatures.[3]

The HASH option of the PERFORM SECURITY OPERATION command allows a hash value to be computed. The command transfers either the data to be hashed, or a hash value that has already been computed outside the Smart card together with the data needed for the final step of the computation. In the latter case, the hash computation for the final block is performed in the card. The advantage of this procedure is that the hash value can be formed significantly faster outside of the card, but the final step still takes place inside the card. From a purely cryptographic perspective, this provides only a small amount of extra security, but it does somewhat limit the possibilities for manipulating the hash value. For this reason, it is widely used in practice.

Since the amount of data to be hashed is usually larger than maximum allowed length of a command data field, the HASH option employs 'Level-7'chaining, which means that the data blocks are logically chained together at the application level. The final data block to be hashed includes a marker that informs the command that the hash procedure should be terminated with this block.

This command option also has its own options. The computed hash value can either be transferred immediately to the terminal in the response to the command, or stored in the card for use with a subsequent command. The padding and the key to be used are established as necessary using a prior MANAGE SECURITY ENVIRONMENT command, in the same way as for previously described commands.

Table 7.37   The function of PERFORM SECURITY OPERATION with the HASH option, according to ISO/IEC 7816-8

PERFORM SECURITY OPERATION *Option:* HASH	
Command	• data to be hashed
	• hash value (if only part of the hash computation is to be performed in the card)
	• *switch:* perform only part of the hash computation in the card
	• *switch:* return the computed hash value in the response
	• *switch:* final command with data to be hashed
Response	• hash value (if selected by the switch)
	• Return code

The option COMPUTE DIGITAL SIGNATURE can be used for signing data. The data string to be signed, which has usually been compressed to a hash value, must be passed to the Smart Card, unless it is already present in the card as the result of a prior PERFORM SECURITY OPERATION command with the HASH option. The COMPUTE DIGITAL

---

3   See also Section 13.6, 'Digital Signatures'.

SIGNATURE option also allows the data to be signed to be transferred directly to the card. These can then be hashed in the card before the signature is generated. For large amounts of data, 'Level-7'chaining can be used as with the HASH option.

If the length of the hash value does not correspond to the input data length of the public-key algorithm, padding must be added. The options for this are also set by parameters of the MANAGE SECURITY ENVIRONMENT command, which is also used to determine the key to be used.

Table 7.38 The function of PERFORM SECURITY OPERATION with the COMPUTE DIGITAL SIGNATURE option, according to ISO/IEC 7816-8

PERFORM SECURITY OPERATION *Option:* COMPUTE DIGITAL SIGNATURE	
Command	• data to be signed
	or
	hash value of the data to be signed
	• *switch*: final command with data to be signed
Response	• digital signature
	• Return code

The verification function for the COMPUTE DIGITAL SIGNATURE option is provided by the VERIFY DIGITAL SIGNATURE option. In principle, any sufficiently fast digital computer could be used to verify a digital signature, since the necessary key is public. However, in many cases the validity of the public key must first be verified using an additional digital signature. This is certainly relevant to security, and it should not be carried out on an insecure computer. For the verification of the digital signature, the associated public key must either be implicitly known to the Smart Card or be explicitly made known to the card via the command option VERIFY CERTIFICATE, which is described immediately below. The data to be verified may be passed to VERIFY DIGITAL SIGNATURE either directly or in the form of the associated hash value. All other parameters are the same as for the COMPUTE DIGITAL SIGNATURE option.

Table 7.39 The function of PERFORM SECURITY OPERATION with the VERIFY DIGITAL SIGNATURE option, according to ISO/IEC 7816-8

PERFORM SECURITY OPERATION *Option:* VERIFY DIGITALSIGNATURE	
Command	• data to be verified
	or
	hash value of the data to be verified
	• digital signature
	• *switch*: final command with data to be verified
Response	• Return code

In open systems, the public keys for the verification of digital signatures are usually signed using the private key of the certification authority. The authenticity of a public key must be verified before it is used, since this is the only way to be sure that the key is not a forgery. This verification must take place in a secure environment, such as a Smart Card, since it would otherwise be subject to manipulation. The VERIFY CERTIFICATE command option is provided especially for this verification of signed public keys. Once a public key has been verified as authentic, it can either be stored permanently in the Smart Card or be used with an immediately following VERIFY DIGITAL SIGNATURE command.

Figure 7.17 illustrates how the commands that have just been described can be used to generate a digital signature and then verify it.

Table 7.40   The function of PERFORM SECURITY OPERATION with VERIFY CERTIFICATE option, according to ISO/IEC 7816-8

PERFORM SECURITY OPERATION *Option:* VERIFY CERTIFICATE	
Command	• certificate
	• *switch:* store the public key
Response	• Return code

Smart Card		**Terminal**
		COMPUTE DIGITAL SIGNATURE
S := **sign** (private key; hash value of the data)	←	*Command* [hash value of the data]
Response [S ‖ Return code]	→	IF (Return code = OK) THEN signature successful ELSE cancel
		VERIFY SIGNATURE
S' := **verify** (public key; hash value of the data)	←	*Command* [hash value of the data]
IF (S = S') THEN (Return code = OK) ELSE (Return code = not OK)		
Answer [Return code]	→	IF (Return code = OK) THEN signature is genuine ELSE signature is false

Figure 7. 17   Sample procedure for PERFORM SECURITY OPERATION with the COMPUTE DIGITAL SIGNATURE and VERIFY DIGITAL SIGNATURE options. The basic parameters, such as the key and the algorithm selection, are established in advance, either implicitly or by means of a MANAGE SECURITY ENVIRONMENT command.

Table 7.41 The function of GENERATE PUBLIC KEY PAIR according to ISO/IEC 7816-8

GENERATE PUBLIC KEY PAIR	
Command	• —
Response	• Return code

If a Smart Card operating system supports the generation of key pairs for asymmetric cryptographic algorithms, this process can be instigated with the ISO/IEC 7816-8 GENERATE PUBLIC KEY PAIR command. All parameters needed for key generation must be set in advance using the MANAGE SECURITY ENVIRONMENT command.

7.8 File Management Commands

Most modern Smart Card operating systems permit files to be enlarged, created, erased and locked, as well as allowing other file management functions, all within limits imposed by specific security conditions. Nevertheless, most or even all management functions are often omitted for single-application cards, since these functions require a lot of program code and thus increase the memory demands and thereby the cost of the card. With multi-application cards, support for certain management functions is absolutely required to avoid the need to load all the applications into the card at the time that it is personalized.

From a security perspective, management functions should only be allowed to be executed after reciprocal authentication, since they are ideal levers for an attacker. For instance, suppose an unauthorized person deletes a file that stores confidential data and then creates a new file with the same name, but without any read access restrictions, so that it can be read without invoking any security mechanisms. As far as the terminal is concerned, the file continues to exist with its original name, and it writes confidential data to the manipulated file. This type of attack is by no means new – it has been around for many years in a somewhat different form. However, it is used successfully over and over again in the area of file management.

Another weak link is the execution of management functions in publicly accessible terminals, which in principle are insecure. Here, data transfers must always be protected by Secure Messaging functions. Only then will an application provider have the opportunity to securely load files and applications into cards that are already in use, for instance via public cardphones, which is logistically very attractive.

Particularly in the case of multi-application Smart Cards, which can be used by several application providers, it is necessary to partition the memory and assign authorization keys for file creation before the individual applications are generated. This prevents any individual application provider from allocating the entire available memory to his own application and leaving no room for other applications. One way to do this is to use a procedure that pre-allocates memory space for an application, and at the same time stores a card- and application-specific key for file creation in the card. The command for this is REGISTER, which is not standardized. New files can then be created at a later date if the application-specific key is known. The result of this procedure is strict separation between memory space allocation and the introduction of new files into the card. A multi-application card issuer can thus sell memory space to several application providers without having to worry about memory piracy.

Table 7.42   REGISTER function (not standardized)

REGISTER	
Command	• DF name of the new DF
	• maximum memory space for the new application (that is, the DF)
	• key for creating new files (for CREATE FILE)
Response	• Return code

The CREATE FILE command allows DFs or EFs to be created in finished Smart Cards. Before this command can be executed, a particular logic state must have been achieved, for example by successful reciprocal authentication. Depending on the environment in which the CREATE FILE command is executed, data transfers may have to be protected by secure messaging. There are two international standards (EN 726-3 and ISO/IEC 7816-9) that define the CREATE FILE command. The coding and the sequence of the large number of available parameters are different for these two definitions, but the basic functionality is the same.

Table 7.43   CREATE FILE function according to EN 726-3

CREATE FILE	
Command	• new file type
	IF (file type = DF) THEN [
	• DF name of the new file]
	IF (file type = EF) THEN [
	• FID of the new file
	• access conditions
	• structure of the new file]
	IF (file structure = transparent) THEN [
	• file size]
	IF (file structure = linear fixed) OR (file structure = cyclic) THEN [
	• number of records
	• record length]
	IF (file structure = linear variable) THEN [
	• number of records
	• length of each individual record]
Response	• Return code

After a file has been created along with all its access conditions, attributes and other properties, it can be selected with SELECT FILE and then accessed. The operating system must circumvent the possibility of using an incomplete file, resulting from the interruption of the file creation procedure, as a means for an assault on the card. Furthermore, it must not be possible to read the previous contents of memory areas that are only partially overwritten when a new file is created.

The file header contains the complete access conditions for the file. For example, it stores data relating to the states in which READ or UPDATE commands are allowed to access the file contents. When the card is being personalized, and for extended management procedures within the file tree, the ability to access files directly without object protection offers considerable advantages. This can be achieved by using the ISO/IEC 7816-9 command MANAGE ATTRIBUTES to modify the access conditions of a previously selected file. Reciprocal authentication and execution in a secure environment are absolute prerequisites for this command.

Table 7.44   MANAGE ATTRIBUTES function according to ISO/IEC 7816-9

MANAGE ATTRIBUTES	
Command	• new access conditions
Response	• Return code

If the REGISTER command is available, it can be used by the card issuer to specify the maximum amount of memory that can later be allocated to an application. At the same time, it specifies a temporary DF name for the DF of the application and stores a key for the CREATE FILE command. The DF name may contain an AID (application identifier). Files can later be created if the key is known.

The user now receives the Smart Card that has been prepared in this way. If the user so desires, he or she can let an application provider load other supplementary applications into the card, such as a cardphone application. The application provider, however, needs a secret reloading key to load a new application, and the card issuer will naturally not provide this key unless there is a contractual relationship between the two parties.

After a successful reciprocal authentication between the Smart Card and the terminal, the application provider may create his files in the DF allocated to him. This can take place either on the provider's premises or via a public cardphone. Next, the provider fills the EFs with the necessary data and keys and sets the access attributes of the files. After this, the application is ready to use, and the user can enjoy his or her newly won functions.

The process of loading a new application in the file tree of a Smart Card is illustrated in Figure 7.18 (overleaf).

The DEACTIVATE FILE command (defined in ISO/IEC 7816-9) and the INVALIDATE command (defined in EN 726-3) allow the terminal to reversibly lock a previously selected file. Read and write accesses to a locked file are prohibited; only selection is allowed.

The EMV '96 specification provides a similar command for the reversible locking of applications, called APPLICATION BLOCK. This command does not block the EFs within an application, but only the execution of commands for file selection, authentication and financial transactions within the DF of the application. Otherwise, APPLICATION BLOCK behaves similarly to DEACTIVATE FILE and INVALIDATE. The EMV '96 command APPLICATION UNBLOCK can be used to unlock access to application that has been locked using the APPLICATION BLOCK command.

Smart Card	**Terminal**

mutual authentication between the Smart Card and the terminal

	REGISTER
	← *Command* [AID ‖ memory location ‖ key]
processing command	
Response [Return code]	→ IF (Return code = OK) THEN command successfully executed ELSE command failed

mutual authentication between the Smart Card and the terminal

	CREATE FILE
	← *Command* [...]
processing command	
Response [Return code]	→ IF (Return code = OK) THEN command successfully executed ELSE command failed

several iterations:

	UPDATE BINARY or UPDATE RECORD
	← *Command* [...]
processing command	
Response [Return code]	→ IF (Return code = OK) THEN command successfully executed ELSE command failed

several iterations:

	CHANGE ATTRIBUTES
	← *Command* [...]
processing command	
Response [Return code]	→ IF (Return code = OK) THEN command successfully executed ELSE command failed

Figure 7.18   Sample procedure for adding a new application

The inverse functions for INVALIDATE and DEACTIVATE FILE are the commands REHABILITATE and REACTIVATE FILE, with which a locked file can be unlocked. The file must be selected before it can be unlocked. It goes without saying that the execution of all these commands can only be permitted in a certain security state, since otherwise anyone would be able to lock or unlock files at will.

Table 7.45 DEACTIVATE FILE function according to ISO/IEC 7816-9

DEACTIVATE FILE	
Command	• —
	or
	FID
	or
	Short FID
	or
	DF name
Response	• Return code

Table 7.46 INVALIDATE function according to EN 726-3

INVALIDATE	
Command	• FID
Response	• Return code

Table 7.47 REHABILITATE function according to EN 726-3

REHABILITATE	
Command	• —
Response	• Return code

Table 7.48 REACTIVATE FILE function according to ISO/IEC 7816-9

REACTIVATE FILE	
Command	• —
	or
	FID
	or
	Short FID
	or
	DF name
Response	• Return code

Smart Card	Terminal
mutual authentication between the Smart Card and the terminal	
	DEACTIVATE FILE
←	*Command* [FID]
processing command	
Response [Return code] →	IF (Return code = OK) THEN file is blocked ELSE file not found or file locking failed
	REACTIVATE FILE
←	*Command* [FID]
processing command	
Response [Return code] →	IF (Return code = OK) THEN command successfully executed ELSE file not found or file unlocking failed

Figure 7.19   Example command execution for DEACTIVATE FILE and REACTIVATE FILE

The LOCK command is the non-reversible version of INVALIDATE. A file that has been blocked with LOCK cannot be unlocked. Its state is completely irreversible. Another use for this is to permanently block an application when it reaches its expiry date. A file locked with the LOCK command can only be selected; any other access command is denied by the operating system.

Table 7.49   LOCK function according to EN 726-3

LOCK	
Command	• —
Response	• Return code

The main disadvantage of permanent and irreversible locking is that it blocks any further use of valuable memory space in the card. A much more elegant solution is to clear the memory occupied by files that are no longer needed, so that it can be used by other applications or new applications. When doing this, it is important not only to delete the file from the file tree, but also to physically erase the entire memory area that was used by the file(s). Only in this way can one be sure that all of the file contents, which certainly may be confidential and worth protecting, have been overwritten and are thus no longer accessible to anyone. If the memory that comes free when a file is deleted is to be made available to other files, deleting a file is a complicated and time-consuming process. Consequently, it is fully implemented in only a few operating systems. Complete free-memory management usually requires more memory space than is available in a Smart Card for this purpose.

The DELETE FILE command can, in principle, be used exactly the same way as the previously described commands for locking and unlocking files. A file that is implicitly selected with this command can be completely removed from the file memory of the Smart Card. Whether the memory space made free as a result can be used for other files depends on the individual operating system. As a rule, free-memory management is not provided, so that the memory space is lost forever after this command has been executed.

Table 7.50   DELETE FILE function according to ISO/IEC 7816-9 and EN 726-3

DELETE FILE	
Command	• FID
	or
	DF name
Response	• Return code

The command TERMINATE DF is provided in the ISO/IEC 7816-9 standard for the irreversible blocking of a DF. A blocked DF can still be selected, but the functions (such as program code) and files it contains are no longer accessible. This command can be used to 'switch off' an application, while allowing the prior presence of the application to still be recognized.

Table 7.51   TERMINATE DF function according to ISO/IEC 7816-9

TERMINATE DF	
Command	• FID
	or
	DF name
Response	• Return code

The ISO/IEC 7816-9 TERMINATE CARD USAGE command acts in a manner similar to that of the TERMINATE DF command. However, it blocks the entire card and thus the execution of any subsequent command by the card. The new status of the card is indicated only in an ATR.

Table 7.52   TERMINATE CARD USAGE function according to ISO/IEC 7816-9

TERMINATE CARD USAGE	
Command	• —
Response	• Return code

The functions of the previously described commands are, in part, very much dependent on the producer of the operating system and the operating system version, in spite of the existence of several standards. A unified standard will probably emerge in the future, but at present, practically all of these file management functions are mutually independent and mutually incompatible. Only their gross functionality is roughly equivalent in the individual operating systems.

7.9   Database Commands: SCQL

Smart Cards have typically been single-user systems since their origins as identification media. However, the highly developed security and access mechanisms of multi-application Smart Cards could certainly be used to configure and use Smart Cards as multi-user systems. Of course, the cost and effort involved should not be underestimated. The cryptographic protective mechanisms that would be needed for this would also quickly reach a critical level of complexity.

The situation here is analogous to that with the various file structures. In principle, it is possible to construct a dataset-oriented telephone directory using a transparent file structure, but it works a lot better with a linear variable file structure.

This was ultimately the reason why a subset of the database query language SQL (structured query language) has been standardized for use with Smart Cards as ISO/IEC 7816-7, and has been incorporated into the product range of various operating system manufacturers. The subset of SQL for Smart Cards that is defined in the ISO/IEC 9075 standard is called SCQL (structured card query language). The main area of use for Smart Cards with SCQL capability is in health care, since in this field various persons and organizations must access data using a variety of different read and write access privileges. However, there is presently no large application that utilizes SCQL.

SCQL, like SQL, supports table-oriented database systems. Such a system consists of a table name, a fixed number of named columns and a variable number of rows. 'Logical views' can be applied to this table. If a view is static, this means that a certain number of columns are assigned to this view. A dynamic view, by contrast, is a selection of rows that satisfy a particular condition (such as 'given name = Wolfgang'). Combinations of static and dynamic views are allowed. Each view has its own name and can be used as the basis for reading and writing data.

Table		
column 1	column 2	column 3
row 1	row 1	row 1
row 2	row 2	row 2
row 3	row 3	row 3
row 4	row 4	row 4
row 5	row 5	row 5

View 1 (static)	
column 1	column 2
row 1	row 1
row 2	row 2
row 3	row 3
row 4	row 4
row 5	row 5

View 2 (dynamic)		
column 1	column 2	column 3
row 3	row 3	row 3
row 5	row 5	row 5

Figure 7.20   An SCQL table with three columns and five rows. Static view and dynamic views have been set up for this table. View 1 is static, and shows columns 1 and 2 in their entirety. View 2 is dynamic, and shows the rows in all three columns that satisfy a certain condition (not defined here).

SCQL in Smart Cards requires an additional type of file structure, known as 'database', to be included in the operating system. A file with this structure is called a DBF (database file). This is a database object that can be located directly under the MF or underneath a DF. It contains the data tables and associated system tables. A DBF can be addressed by the database commands without first being selected. It is a logical construction that may be distributed across several EFs, depending on the Smart Card operating system.

Three system tables must be set up in the DBF to manage users and privileges. The Object Description Table stores information about the tables and their associated views. These are complemented by the User Description Table, which is defined by the user of the database system. The third DBF table is the Privilege Description Table, which defines the privileges of the individual users with regard to tables and views, as well as the allowed operations for the tables and views.

Table: acquaintances of Louis Wu		
Name	Tribe	Protector
Teela Brown	ball	yes
Nessus	puppeteer	no
Bram	vampire	yes
Chmee	kzinti	no
Akolyth	kzinti	no
Prill	machinist	no
Vala	machinist	no

View: protectors	
Name	Tribe
Teela Brown	ball
Bram	vampire

View: kzinti
Name
Chmee
Akolyth

View: name
Teela Brown
Nessus
Bram
Chmee
Akolyth
Prill
Vala

Figure 7.21   A sample SCQL table with the columns 'Name' and 'Tribe', the property 'Protector', and seven entries. Three views are also shown. Different access privileges can be defined in SCQL for the various views.

The operations that can be performed on a SCQL table or view are Read, Insert, Write and Delete. These are all regulated by access privileges. A cursor is defined for read and write operations in a table or view. It indicates the row to which the operation will be applied.

Although SCQL has many limitations compared with SQL, such as no sorting, no nested queries, no joins and so on, it still definitely has the potential to find applications in the multi-user area. However, the memories of currently available microcontrollers are not yet large enough for extensive SCQL databases in Smart Cards. It thus may very well take some time until a large application arises.

There are three basic SCQL commands: PERFORM SCQL OPERATION, PERFORM TRANSACTION OPERATION and PERFORM USER OPERATION. Only three instruction codes (INS) are needed for these, since the actual SCQL operations are triggered via the parameter byte P2. All data in the command body and the response body are TLV-coded data objects. The three commands are summarized in Tables 7.53, 7.54 and 7.55.

Table 7.53   The SCQL command PERFORM SCQL OPERATION according to ISO/IEC 7816-7

PERFORM SCQL OPERATION	
CREATE TABLE	Makes a table with its columns and column names.
CREATE VIEW	Creates a new view (static or dynamic) for a table.
CREATE DICTIONARY	Creates the Object Description Table, the User Description Table and the Privilege Description Table.
DROP TABLE	Deletes a table.
DROP VIEW	Deletes a view.
GRANT	Grants access privileges to a single user, a group of users or all users.
REVOKE	Revokes access privileges that were previously granted by GRANT.
DECLARE CURSOR	Positions a cursor that indicates a row in a table, view or system table.
OPEN	Activates a cursor in the first row.
NEXT	Moves the cursor to the next row.
FETCH	Reads the row indicated by the cursor.
FETCH NEXT	Reads the next logical row after the row where the cursor is. The cursor is moved to the row to be read.
INSERT	Adds a row to the end of the table without changing the cursor.
UPDATE	Writes data to one or more fields in a row in a table. The row is determined by the cursor.
DELETE	Deleted the row marked by the cursor from the table. The cursor is positioned to the following row.

Table 7.54   The SCQL command PERFORM TRANSACTION OPERATION according to ISO/IEC 7816-7, and its functions

PERFORM TRANSACTION OPERATION	
BEGIN	Reserves room for a memory image for the functions COMMIT and ROLLBACK, for example for a row of a table.
COMMIT	Tests all changes that have been made to a table since the last BEGIN command.
ROLLBACK	Restores a table to the state it had before the last BEGIN command.

Table 7.55   The SCQL command PERFORM USER OPERATION according to ISO/IEC 7816-7, and its functions

PERFORM USER OPERATION	
PRESENT USER	Identify a user by means of his or her user ID
CREATE USER	Create a new entry for a new user
DELETE USER	Delete the entry for a user

7.10   Commands for Electronic Purses

Part 3 of the European standard for universal electronic purses, EN 1546, defines a total of six commands for electronic purses and twelve for the security module in the terminal, which itself may be a Smart Card. The basic formats of the four most important commands used with Smart Card electronic purses[4] are described here. These commands can be utilized to operate a Smart Card application for making payments without cash from a prepaid 'purse' and for refilling the purse. The commands for error recovery, currency conversion, parameter modification and canceling a purchase are not described here, nor are those for the security module.

The commands described here would fit just as well under Application-Specific Commands (Section 7.14), since they are defined specifically for this one application. They can never be used for any other purpose than electronic purses, since they have been optimized for this application. However, we dedicate a subsection to them because electronic purses will be one of the main applications for Smart Cards in the future, alongside telecommunications.

All electronic purse transactions are divided into three parts according to EN 1546. In the first part, the card is initialized with the command INITIALIZE IEP for Load/for Purchase. The second command executes the actual transaction (i.e. fill the purse or pay from the purse). In the optional third part, the transaction just performed is confirmed. All purse commands directly access files in the purse application of the Smart Card for both writing and reading. These files hold various parameters, the purse balance and log records.

The individual steps of a purse transaction are executed by means of the following commands. The EN 1546 standard precisely specifies the internal process of each command, in terms of both its functions and the sequence of the individual steps. All implementations thus have at least the same overall processes.

The command INITIALIZE IEP can be used for several purposes. A parameter selects the initialization of a purse loading transaction, a purchase transaction or another type of transaction.

The command INITIALIZE IEP for Load is the first step in loading (crediting) the purse in the Smart Card. The transferred data, such as a currency code and amount to be loaded, are checked in the card to see whether they match predefined values in the parameter files. In addition, freely definable data (user-determined data) can be stored in a log file. Next, a transaction counter is incremented and a signature S_1 is formed for various data (such as the current balance and the expiry date), so that they can be transferred to the terminal without risk of tampering.

In the second step, the card essentially receives information concerning the keys to be used and a signature S_2. This comes from the terminal's security module and allows the card to authenticate the security module, as well as protecting the data. The Smart Card has already been authenticated by the terminal's security module by a prior INITIALIZE IEP for Load command. After successfully testing S_2, the card increases the purse balance, makes a new entry in the log file and generates a signature S_3 for confirmation. This is then used by the terminal's security module as confirmation of the correct booking of the amount to be loaded.

---

[4] Command sequences and general system structures of electronic purse systems are described in detail in Section 12.3.1, 'The CEN EN 1546 standard'.

Table 7.56   INITIALIZE IEP for Load function according to EN 1546-3

INITIALIZE IEP for Load	
Command	• —
	• amount to be loaded (M_{LDA})
	• currency code ($CURR_{LDA}$)
	• PPSAM descriptor (PPSAM)
	• random number (R)
	• user-determined data (DD)
Response	• identifier of the purse provider (PP_{IEP})
	• IEP identifier
	• cryptographic algorithm used (ALG_{IEP})
	• expiry date ($DEXP_{IEP}$)
	• purse balance (BAL_{IEP})
	• IEP transaction number (NT_{IEP})
	• key information (IK_{IEP})
	• signature S_1
	• Return code (CC_{IEP})

Table 7.57   CREDIT function according to EN 1546-3

CREDIT IEP	
Command	• key information (IK_{PPSAM})
	• signature S_2
	• user-determined data (DD)
Response	• signature S_3
	• Return code (CC_{IEP})

The second procedure presented here, along with its necessary commands, demonstrates making a purchase with an electronic purse. The first command, which initializes the transaction, is again INITIALIZE IEP, but with the 'for Purchase' option. The Smart Card does not receive any data form this command, but it does increment the transaction counter. Next, a signature S_1 is formed for data such as expiry date, transaction counter and IEP identifier. The data thus protected are returned to the terminal, along with some other data.

The actual purchase transaction is executed by the next command, called DEBIT IEP. It sends information to the electronic purse in the Smart Card regarding the amount to be debited and current key versions, as well as another signature. This signature can be used to test the authenticity of the security module in the terminal, just as for loading the purse. If this test is successful, the purse balance is increased, the purchase transactions log file is updated and a further signature, S_3, is created to confirm the entire transaction. This is included in the response to the DEBIT IEP command. Signature S_3 serves as confirmation to the terminal's security module that the amount was properly debited in the Smart Card.

Table 7.58 INITIALIZE IEP for Purchase function according to EN 1546-3

INITIALIZE IEP for Purchase	
Command	• —
Response	• identifier of the purse provider (PP_{IEP})
	• IEP identifier
	• cryptographic algorithm used (ALG_{IEP})
	• expiry date ($DEXP_{IEP}$)
	• purse balance (BAL_{IEP})
	• currency code ($CURR_{IEP}$)
	• authentication mode (AM_{IEP})
	• IEP transaction number (NT_{IEP})
	• key information (IK_{IEP})
	• signature S_1
	• Return code (CC_{IEP})

Table 7.59 DEBIT IEP function according to EN 1546-3

DEBIT IEP	
Command	• PSAM identifier
	• PSAM transaction number (NT_{PSAM})
	• amount to be debited (M_{PDA})
	• currency code ($CURR_{PDA}$)
	• key information (IK_{PSAM})
	• signature S_2
	• user-specific data (DD)
Response	• signature S_3
	• Return code (CC_{IEP})

This is just a brief overview of the four most important electronic purse commands, as specified in EN 1546. The standard is extremely extensive and includes many options and variations for system design, which naturally can affect the command formats.[5]

7.11 Credit Card and Debit Card Commands

The joint specification of Europay, Mastercard and Visa for Smart Cards used for financial transactions, EMV '96, defines two commands that are specifically designed for financial transactions. In principle, these two very flexible commands could be used to realize an electronic purse in a Smart Card. However, their intended use lies more in the realm of

---

[5] See also Section 12.3.1, 'The CEN EN 1546 standard'.

credit and debit transactions, which is why their use for these two applications is dealt with here. A separate section is devoted to these commands, since credit and debit cards incorporating chips are expected to be produced in very large numbers in the future. Their importance is thus correspondingly great.

The two commands GET PROCESSING OPTIONS and GENERATE APPLICATION CRYPTOGRAM are based on TLV-encoded data objects in the data portions of the command and the response. This creates a considerable variety of possible variations and options, which can be used by each application as needed.

The GET PROCESSING OPTIONS command is used to initiate a payment. It transfers the TLV-encoded data for the processing of rest of the payment transaction (the processing options data object list) from the terminal to the Smart Card. This could be the amount of the transaction, for example. In response, the card sends back a BER-TLV encoded data object with the profile supported by the application (the application interchange profile, which identifies the functions supported by the Smart Card), together with the location of the application data (the application file locator).

Table 7.60   GET PROCESSING OPTIONS function according to EMV '96

GET PROCESSING OPTIONS	
Command	• processing options data object list (PDOL)
Response	• application interchange profile (AIP)
	• application file locator (AFL)
	• Return code

The second purchase command for the payment process in a credit card with a chip is designated GENERATE APPLICATION CRYPTOGRAM. With this command, all data are TLV-encoded in both the command APDU and the response APDU. All the data needed for a payment transaction, along with the desired application cryptogram, are sent to the card with this command. The card next determines how rest of the payment process should proceed, based on the received and stored data. As a result of the process, it sends an application cryptogram to the terminal. In the simplest case, this may be the transaction cryptogram. This concludes the payment process.

Table 7.61   GENERATE APPLICATION CRYPTOGRAM function according to EMV-2

GENERATE APPLICATION CRYPTOGRAM	
Command	• desired application cryptogram
	• transaction-related data
Response	• cryptogram information data
	• application transaction counter (ATC)
	• application cryptogram (AC)
	• Return code

The application cryptogram returned by the Smart Card may contain an authorization query, instead of the transaction cryptogram. If, in the process of determining how the payment transaction process should proceed, the Smart Card concludes that an on-line authorization is necessary, it places a query to a higher-level authorization center in the application cryptogram of its response to the terminal. Once the authorization center has processed the query, the relevant information is sent to the Smart Card in a second GENERATE APPLICATION CRYPTOGRAM command. The card can then produce the transaction cryptogram for the payment and send it to the terminal.[6]

7.12 Commands for Completing the Operating System

When Smart Card microprocessors are manufactured, only the ROM is programmed. The EEPROM remains empty, apart from a chip number and a card-specific key. After the card body and the microcontroller have been assembled to produce a Smart Card, the operating system code in the ROM must be supplemented by the parts that reside in the EEPROM. This is called *completing* the Smart Card. Only after it has been completed does a Smart Card contain a full operating system with all its functions.

There is a relatively simple loader program in the ROM for writing these parts of the operating system to the EEPROM. It can be used to write data to the EEPROM following a key verification. The EEPROM memory is addressed linearly, either byte by byte or page by page, using direct physical addresses.

Once all necessary data have been input into the EEPROM in this way, the operating system is switched over from pure ROM operation. From this point on, processes and routines also run in the EEPROM. This switchover can be performed by a command whose execution condition has been satisfied, following a prior checksum comparison for all the extended EEPROM data. The checksum ensures that all the data have been correctly stored in the EEPROM. Completion does not employ particularly complex functions or authorization procedures, since it is necessary to rely on the ROM portion of the operating system. Even the smallest error here could make it impossible to complete the Smart Card. Such an error would be time-consuming and, all told, very expensive.

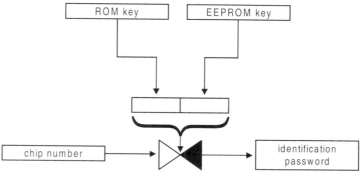

Figure 7.22 One of many possible ways to generate a chip-specific password for authorizing the completion of the operating system

---

[6] See also Section 12.5, 'Credit Cards with Chips'.

In the rest of this section, the three commands needed to complete a Smart Card operating system are described by means of an example. These commands vary greatly, depending on the operating system and the chip manufacturer. Here we can only illustrate the necessary functions. However, practically all Smart Cards use this procedure or a similar procedure to complete the operating system.

The command COMPARE KEY verifies a password sent to the card to a reference password stored in the ROM and EEPROM by the manufacturer when the chip was made. This key (password) is card-specific and is quite long, about 32 bytes. If the comparison is successful, subsequent loading commands are permitted. Otherwise, a retry counter is incremented. Once the retry counter reaches a predefined value (usually 3), it blocks any further access to the card. The card can then be sent to recycling, since the only thing it can do is to produce an ATR.

Table 7.62   COMPARE KEY function

COMPARE KEY	
Command	• key
Response	• Return code

After the load password has been successfully verified with COMPARE KEY, all necessary data can be written to the EEPROM using the WRITE DATA command. It is possible at this point to address and write to entire EEPROM byte by byte. This means that in addition to the operating system data, complete applications may also be entered. This is, by the way, the method normally used to load applications into Smart Cards with very little memory, since they do not have enough room for complex CREATE FILE commands and their associated state machines.

If the available ROM in the Smart Card is so small that there is not even enough room for EEPROM test commands, it is possible to simulate the basic functions of the test commands using this command. A particular memory location is repeatedly written until the card reports a write error. If the number of write cycles is totaled, the number of possible write/erase cycles is known. This is the main result expected from an EEPROM test command in the context of quality assurance.

Table 7.63   WRITE DATA function

WRITE DATA	
Command	• data
	• EEPROM address
Response	• Return code

Once all data have been written to EEPROM using one or more consecutive WRITE DATA commands, the content of the EEPROM is tested to see that it is correct, and the completion procedure is then terminated. The command used for this is COMPLETION

END. After successful execution of this command, a card reset is normally triggered to reinitialize the operating system and allow it to achieve a new state.

Table 7.64   COMPLETION END function

COMPLETION END	
Command	• EEPROM checksum
Response	• Return code

Figure 7.23 further illustrates the sequence of commands for completing a Smart Card, as just described. This procedure is supervised by a state machine during the completion process, so that only the listed commands can be executed, and only exactly in the order that they are listed. This state machine is described by Figure 7.24.

Smart Card	Terminal
	COMPARE KEY
Command processing ←	*Command* [key for completion]
Response [Return code] →	IF (Return code = OK) THEN command successfully executed ELSE command failed
numerous iterations:	
	WRITE DATA
Command processing ←	*Command* [data ‖ address]
Response [Return code] →	IF (Return code = OK) THEN command successfully executed ELSE command failed
conclusion:	
	COMPLETION END
Command processing ←	*Command* [EEPROM checksum]
Response [Return code] →	IF (Return code = OK) THEN command successfully executed ELSE command failed

Figure 7.23   Example of a typical completion procedure

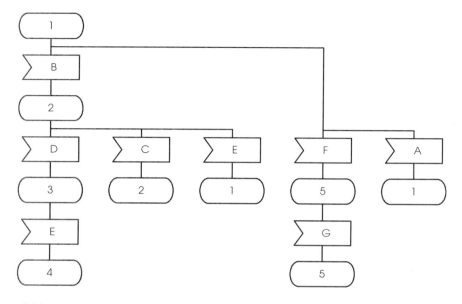

Figure 7.24   State machine for completing a Smart Card operating system

States: 1   initial state after Smart Card reset
              2   Smart Card ready for data entry into EEPROM
              3   Smart Card completed
              4   basic state of Smart Card after completion
              5   Smart Card irreversibly blocked

Transitions: A   all commands except COMPARE KEY
              B   COMPARE KEY (successfully carried out)
              C   WRITE DATA
              D   COMPLETION END
              E   reset Smart Card
              F   COMPARE KEY (3 times unsuccessful)
              G   all commands and Reset

7.13   Hardware Testing Commands

During initialization, the operating system of the Smart Card tests various hardware components, both implicitly and explicitly. The commands described here, however, go far beyond the self-test routines integrated directly into the operating system. In the course of carrying out production quality assurance, it is necessary to separately verify certain critical parts of the microprocessor. These tests focus particularly on the EEPROM, since experience shows that most problems show up here. The operation of the processor is implicitly verified when the terminal can receive a correct ATR.

Since there are no standards for test commands, their functions and coding depend on the manufacturer of the operating system, and sometimes on the operating system itself.

The test commands may be fixed in ROM for security reasons. However, it is also quite common for them to be loaded into the EEPROM via the completion commands, and to be executed in the EEPROM. Obviously, this can cause problems if the EEPROM is not fully functional. The advantage of using EEPROM is that it makes more space available in the ROM. In the interest of operating system security, using test commands must be restricted to the phase preceding completion. This means that all test commands are blocked in a card that has been an initialized or even personalized. Possible exceptions are the RAM test and the ROM or EEPROM checksum tests, since these commands do not affect security.

The following commands may be used for extensive hardware tests. A defective RAM would cause a complete operating system crash even before the ATR could be sent, but it is possible for a few RAM bits or bytes to be 'dead'. This would only affect certain functions or routines. The TEST RAM command checks the entire RAM and sends an appropriate response to the terminal. During the test, all available bytes must be written and be read using a variety of test patterns. A typical test consists of writing '55' and 'AA' alternately, since these two hexadecimal values form a checkerboard pattern at the bit level. Another effective method is the wave test, in which the RAM is written first from the lowest to the highest address and then read, and then from the highest to the lowest address. The exact implementation of this test depends on the operating system. Sometimes this test is omitted.

Table 7.65   TEST RAM function

TEST RAM	
Command	• —
Response	• Return code

CALCULATE EDC is a very simple test that computes an EDC checksum for the whole ROM or specified EEPROM sections and returns it to the terminal. This is one way to determine whether the ROM mask has altered or any EEPROM cells have flipped. EEPROM testing relates only to static areas, which cannot be deliberately changed during the lifetime of the card. The terminal compares the checksum received from the card with a reference value and decides whether the memory contents are still consistent.

Table 7.66   CALCULATE EDC function

CALCULATE EDC	
Command	• *switch:* ROM / EEPROM
Response	• checksum
	• Return code

The EEPROM endurance test TEST EEPROM is used to test overwriting the contents of the EEPROM. The Smart Card receives two patterns, which it writes alternately to an area of memory. The size of this area and the number of write attempts can be specified in the command, within certain limits. Naturally, the operating system must check the memory contents for errors after each write cycle.

Since the number of write cycles is supplied to the Smart Card as a command parameter, this command can be processed internal to the Smart Card. This is much faster than sending a sequence of individual commands. If the card discovers a write error, it sends the number of already executed write cycles and the address of the error to the terminal. The EEPROM endurance test can be used not only for destructive testing of the EEPROM, but also for writing data to freely chosen regions of the EEPROM. In this case, the number of write cycles is set to one.

Table 7.67   TEST EEPROM function

TEST EEPROM	
Command	• pattern 1
	• pattern 2
	• number of write cycles
	• start address in EEPROM
	• end address in EEPROM
Response	• number of write cycles executed (in case of an error)
	• error address (in case of an error)
	• Return code

The COMPARE EEPROM test is used to verify that a pattern written to the EEPROM is still present. This command is primarily used in combination with the TEST EEPROM command to test EEPROM data retention at various temperatures. This is done by writing a pattern to several memory pages, placing the Smart Card in an environmental test chamber for a certain length of time, and then checking whether the pattern has been retained.

Table 7.68   COMPARE EEPROM function

COMPARE EEPROM	
Command	• comparison value
	• address
Response	• Return code

In principle, the TEST EEPROM command can be used to clear the entire EEPROM if the start and end addresses are appropriately specified. However, many operating systems have a command that we can call DELETE EEPROM. It clears the entire EEPROM in one go, which overwrites the entire contents of the EEPROM. This command is employed for only two purposes. The first is to restore the EEPROM to a predefined state with a minimum of effort, after various tests have left their test data in the EEPROM. The second is to reduce the time required to initialize or complete the operating system. If the DELETE EEPROM command is executed prior to these activities, the EEPROM can be written faster, since it is no longer necessary to erase each page before writing new data to it.

Table 7.69   DELETE EEPROM function

DELETE EEPROM	
Command	• —
Response	• Return code

7.14   Application-Specific Commands

There are many commands that are tailored to specific applications. These are mainly used to save memory space or minimize execution time. The majority of these commands are so specific that they are not covered by standards, or their use is limited to a particular application area.

A list of all application-specific commands would burst the bounds of this chapter. Here we present a typical example, which is the only command specific to GSM 11.11. This is the RUN GSM ALGORITHM command. It is used to simultaneously generate a dynamic, card-specific key and authenticate the card with respect to the GSM background system. This function is so specific to the GSM application that it makes no sense to include it in a general Smart Card standard. It relies on a cryptographic algorithm used only in GSM, and the two initial values generated from the passed-in random number would be useless in any other application.

Table 7.70   RUN GSM ALGORITHM function

RUN GSM ALGORITHM	
Command	• random number
Response	• dynamic key
	• **enc** (key; random number)
	• Return code

7.15   Transmission Protocol Commands

In principle, transmission protocols should be constructed to be wholly independent of the data and commands of the application layer. This is also the intention of the OSI layer model. Unfortunately, there is a discrepancy here between theoretical demands and actual practice. There are two commands whose only purpose is to allow transfer mechanisms to be utilized at the application level, namely GET RESPONSE and ENVELOPE. There is also another command, MANAGE CHANNEL, whose function is not used by the application layer alone.

In the T=0 protocol, it is not possible to both send a block of data to the Smart Card and also receive a block of data from the Smart Card within one command-response cycle.[7] This protocol thus does not support Case 4 commands, although they are frequently used. One is thus forced to use a workaround with the T=0 protocol. This works in a simple manner. The

---

[7]   See also Section 6.4.2, 'The T=0 transmission protocol'.

Case 4 command is sent to the card, and if it is successful, a special return code is sent to the terminal. This notifies the terminal that the command has generated data that are waiting to be picked up. The terminal then sends a GET RESPONSE command to the Smart Card and receives the data. This completes the command–response cycle for the first command. As long as no command other than GET RESPONSE is sent to the card, the response data can be requested multiple times.

Table 7.71   GET RESPONSE function according to ISO/IEC 7816-4

GET RESPONSE	
Command	• amount of data to be sent
Response	• data
	• Return code

Where commands are completely encrypted in the context of Secure Messaging, transfer problems can occur with the T=0 protocol, which requires both an unencrypted instruction byte and an unencrypted Le byte. The ENVELOPE command gets around this restriction by embedding a complete APDU, with its header and data section, into the data section of the ENVELOPE command APDU. This can be encrypted without any restrictions and transmitted using any protocol. The same procedure is used for the response generated by the Smart Card, which is also embedded in the APDU of the ENVELOPE command.

Table 7.72   ENVELOPE function according to ISO/IEC 7816-4

ENVELOPE	
Command	• command APDU
Response	• response APDU
	• Return code

Logical channels allow up to four applications in a single Smart Card to be addressed independently of each other.[8] The commands and applications are coordinated by two bits in the class byte. Before using a new logical channel, the terminal must explicitly inform the Smart Card via the MANAGE CHANNEL command. This signals to the card that an additional channel is needed. The channel number can be specified explicitly by the terminal, or the card can supply the number of a free channel in its response. When a new logical channel is opened from the standard channel (channel 0), the card behaves vis-à-vis the new channel as it would after a reset. This means that the MF is selected and the no security state has been achieved. When a new logical channel is opened from a channel other than channel 0, the currently DF selection and the secure state are retained. When a logical channel is closed, the associated file selection and security state are cleared.

---

8   See also Section 6.7, 'Logical Channels'.

Table 7.73   MANAGE CHANNEL function according to ISO/IEC 7816-4

MANAGE CHANNEL	
Command	• *switch:* logical channel open/closed
	IF (a particular channel is desired) THEN
	• logical channel number
Response	• logical channel number (if a new logical channel is opened)
	• Return code

8

Security Techniques

One of the main advantages of Smart Cards in comparison with other media, such as magnetic-stripe cards or diskettes, is that they can store data in a protected or confidential manner. Consequently, chip hardware that is tailored and optimized for this purpose, with suitable cryptographic procedures for protecting confidential data, is an unavoidable requirement. However, security depends on more than just the special microcontroller hardware and the algorithms implemented in the operating system software. The security of the Smart Card application and the design principles used by its developers are also of fundamental importance. This chapter is a compendium of the essential principles, procedures and strategies that lead to secure Smart Cards and Smart Card applications.

8.1 User Identification

Since ancient times, a number of different methods have been used for the unambiguous identification of persons. The simplest form of identification is an identity card with a photograph or a signature written in the presence of witnesses. The photograph attached to an identity card can be compared with the actual person, and the result is a judgement regarding the genuineness of the person's identity.

In the field of information technology, this comparison is not so easy, since it must be performed by a computer instead of another person. Since computers, in spite of all their success in the execution of mindless activities, still have tremendous problems performing intelligent tasks, entering passwords via a keypad has largely become the preferred technique. The effort needed for the comparison is minimal, since in principle all the computer has to do is to compare the entered password with a stored reference value and make a pure yes/no decision. The password comparison effectively makes a decision regarding the genuineness of the identity of the person being tested.

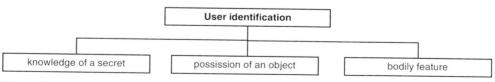

Figure 8.1 Classification of methods for identifying a person

Basically, three different techniques can be used to identify a person. If a password is used, what is tested is whether the person knows a particular secret. If he or she does, the conclusion is that that person is who he or she claims to be. This is similar to testing whether a person possesses a particular object. The third possibility is to test specific, unique bodily features of the person.

Techniques that rely on the knowledge of a secret or possession of a particular item have a significant disadvantage, which is that the person to be identified must either memorize something or carry some item on his or her person. Depending on the situation, the fact that the secret or object can be passed to another person can be considered to be an advantage or a disadvantage. In any case, it is not possible to unambiguously ascertain that the person holding the secret or the object is truly its legitimate owner, rather than someone else who may have illegitimately acquired the secret or item.

The third method of identification eliminates this transferability, since it is based on using specific features the human body for purposes of identification. Of course, the measurements are in most cases technically difficult, since for obvious reasons biological features that are easily measured, such as weight or height, cannot be used.

It is easier to understand these three possible identification methods if you consider the following example. Suppose you have to meet an unknown person at the train station. As soon as you see a possible candidate at the station, you have the problem of deciding whether he is really the person you are looking for. However, if the unknown person shows up at the right place and at the right time, this actually amounts to an implicit test of a secret, since you can at least hope that the place and time of your meeting are not generally known. An explicit test of a secret would occur if the unknown person were to utter a password that is known only to you and him. Alternatively, he could also identify himself by means of an item that he possesses, for example by holding a newspaper printed on a specific day under his arm. Certainly, the most secure method would be to check the person for a specific bodily feature. Perhaps he has a nose like Pinocchio's, which becomes very long when he lies.

This train station scenario clearly shows that the identification of an unknown person can be considered to be a nearly classic problem. It occurs not only in all spy novels but also in everyday life, and it is not limited to computers or Smart Cards.

Entering a PIN into many types of automated equipment and computers has by now become commonplace. The resulting marked increase in the number of PINs used for various purposes makes it very difficult for the ordinary person to keep track of all of his or her PIN codes. After all, who can remember twenty or more different PIN codes? The security and good name of a system are naturally not improved if every user jots down his or her PIN on the card, since the number of cases of fraud will be excessive. For this reason, the desire for other identification methods has arisen in recent years, to replace the use of PIN codes. Biometric features that allow a particular person to be unambiguously identified by a machine are ideal for this purpose.

8.1.1   Testing a secret number

The most commonly used method of user identification is the entry of a secret number, which is generally referred to by the acronym PIN (for personal identification number), or sometimes CHV (cardholder verification).

A PIN is usually a four-digit number, which is usually composed using the decimal numerals 0 through 9. The reason for using a purely numeric entry is simply that all card terminals have only numeric keypads. The PIN is entered using the terminal keypad or a computer keyboard and then sent to the Smart Card. This compares the value that it has received with an internally stored reference value and reports the result to the terminal.

PIN entry is considered to be a security issue, especially for financial transactions, and consequently requirements relating the nature of the keypad are frequently found in this application area. Special keypads that satisfy these requirements are often called 'PIN pads'. In Germany, for example, the PIN for a Eurocheque card may only be entered using a keypad that has special mechanical and cryptographic protection. PIN pads have all the features of a security module, such as opening sensors and foils to protect against drilling, and they encrypt the PIN directly as it is entered. This provides reliable protection against tampering with a keypad in order to allow a PIN to be intercepted while it is being entered.

A distinction can be made between static and changeable PINs. A static pin cannot be changed by the user, so it effectively must be memorized by the user. If it becomes known, the user should destroy the card and obtain a new one with a different fixed PIN. A changeable PIN can be altered according to the wishes of the user, or changed to a number that the user finds easy to remember. There is a danger in this, since the numbers that many people find easy to remember are ones such as "1234", "4711" and "0815". The Smart Card does not check for the use of such trivial numbers, since there is not enough memory available to store the necessary table. It would however be perfectly conceivable for the terminal to prohibit the PIN from being changed to such a number. In order to change the PIN, it is always necessary to enter the PIN, since otherwise an attacker could replace every existing PIN with one of his own.

The situation is different with 'super PINs', or PUKs (personal unblocking keys). These usually have more digits than the actual PIN (a typical value is 6), and they are used to reset the retry counter of a PIN to zero if it has reached its maximum value. A new PIN is also entered into the card when the PUK is entered, since resetting the retry counter is of little help if the user has forgotten the PIN. This is usually the case when the retry counter has reached its maximum value.

Some applications also use transport PINs. The Smart Card is personalized with a random PIN, and the cardholder receives the PIN value by letter. The first time that the card is used, however, the user is forced to replace the PIN used for personalization of the card with one of his or her choice. In a similar procedure, called the 'null PIN' procedure, the card is preloaded with a trivial PIN, such as "0000", and the user is again forced to change the PIN the first time the card is used. Both of these procedures avoid the possibility that a PIN that has been 'spied out' when the card is personalized can later be put to good use.

According to a recommendation of the ISO 9564-1 standard, the PIN should consist of four to twelve alphanumeric characters, in order to minimize the probability of determining the correct PIN by pure trial and error. However, actual practice is often somewhat different. The entry of non-numeric characters is technically impossible at many locations, since only a numeric keypad is available.

The number of characters in a PIN depends not only on the desired level of security, but also very much on the memory of the average card user. For years, people have been accustomed to using four-digit PINs, which means that changing over to PINs with six or more digits would be very difficult. In practice, the presumed increase in security provided by a six-digit or eight-digit PIN may turn out to be purely theoretical. Many people find it difficult to remember numbers of this length, especially if they do not use them very often, and consequently write them down on the card or on a slip of paper kept near the card. The level of security with a long PIN is then significantly lower than with a short PIN.

The perfectly well-founded insistence on periodically changing PINs meets with a similar fate. It may work in the case of a high-security application with only a few users, but it is fatal for the acceptance of a mass-market application, which aims for the simplest possible procedures in order to accommodate people with poor memories.

In this regard, there is another very important issue. In many cases, the entry and verification of a PIN does not just identify the user and demonstrate the legitimate possession of the card in question. Instead, at the same time it also represents a statement of intent by the user, who by entering the PIN declares his or her agreement to a certain transaction. A good example is entering a PIN into a cash dispenser. On the one hand, this identifies the card user by means of his of her knowledge of the secret PIN, but on the other hand it also represents a declaration by the user that he or she agrees to have a certain amount of cash paid out from his or her account. This is a very important point in connection with certain biometric features, some of which do not necessarily represent a declaration of intent since they can be tested without the explicit permission of the person involved.

The probability of guessing a PIN

The simplest attack on a PIN, aside from watching it being entered, is just guessing. The probability of success depends in part on the length of the PIN to be guessed, the characters from which it can be composed and how many attempts are allowed. The probability of correctly guessing a four-digit PIN in three tries is 0.03%, which is not particularly high. Two basic formulas for guessing passwords are presented here. They can be used in actual practice to estimate the risk associated with using a particular password.

$$x = m^n \tag{8.1}$$

$$P = i / m^n \tag{8.2}$$

Table 8.1   Definitions and descriptions of the variables in Formulas (8.1) and (8.2)

Variable	Example	Description
i	3	number of guesses
m	10	number of possible characters per position
n	4	number of positions
P	0.0003 (0.03 %)	probability of guessing the password
x	10,000	number of possible passwords

Incidentally, there is yet a fourth factor relating to guessing a PIN, which for a long time has been inexcusably neglected. This is the evenness of the distribution of the PINs within an application. It is namely much easier to guess a PIN if you know that certain PINs are more common than others. The actual significance of this important secondary factor became evident almost overnight in 1977 with regard to German Eurocheque cards. Although the detailed procedure for computing a PIN from the data stored in the magnetic stripe of the Eurocheque card is still secret, at least a few general steps of the procedure became known. From this information, it could be concluded that the PINs that are generated are not evenly distributed, since the algorithm used produces the numerals 0 through 5 significantly more often than 6 through 9. It also became known that the PIN algorithm suppressed a leading zero during PIN generation. The result of this non-uniform distribution is that it is not necessary to make 3333 attempts in order to correctly guess a four-digit PIN with the permitted number of incorrect guesses (3), but only 150 [Karten 97]. With 10.5% of the cards, the distribution is so poor that only 72 attempts will suffice if the characteristics of the PIN generation algorithm are taken into account [Schindler 97]. The end result of all this is that an improved PIN generation algorithm is used with new Eurocheque cards, and the DES algorithm originally used has been replaced by a triple-DES algorithm.

Table 8.2 PINs and passwords with various lengths and codings, and the number of possible combinations

Type of PIN or password	Range of values or coding of the PIN or password	Number of possible PINs or passwords
1-digit PIN	PIN $\in \{0 \dots 9\}$	10
1-character password	password $\in \{0 \dots 9, "A" \dots "Z"\}$	36
1-character password	password $\in \{0 \dots 9, "a" \dots "z", "A" \dots "Z"\}$	62
1-character password	password $\in \{0 \dots 9, "a" \dots "z", "A" \dots "Z",$ 20 arbitrary special characters $\}$	82
4-digit PIN, no leading zero	PIN $\in \{1000 \dots 9999\}$	9.00×10^3
4-digit PIN	PIN $\in \{0000 \dots 9999\}$	1.00×10^4
4-character password	password $\in \{0 \dots 9, "A" \dots "Z"\}$	1.68×10^6
4-character password	password $\in \{0 \dots 9, "a" \dots "z", "A" \dots "Z"\}$	1.48×10^7
4-character password	password $\in \{0 \dots 9, "a" \dots "z", "A" \dots "Z",$ 20 arbitrary special characters$\}$	4.52×10^7
5-digit PIN, no leading zero	PIN $\in \{10000 \dots 99999\}$	8.9×10^4
5-digit PIN	PIN $\in \{00000 \dots 99999\}$	1.00×10^5
6-digit PIN, no leading zero	PIN $\in \{100000 \dots 999999\}$	8.99×10^5
6-digit PIN	PIN $\in \{000000 \dots 999999\}$	1.00×10^6
6-character password	password $\in \{0 \dots 9, "A" \dots "Z"\}$	2.18×10^9
6-character password	password $\in \{0 \dots 9, "a" \dots "z", "A" \dots "Z"\}$	5.68×10^{10}
6-character password	password $\in \{0 \dots 9, "a" \dots "z", "A" \dots "Z",$ 20 arbitrary special characters$\}$	3.04×10^{11}

Generating a PIN

In order to generate a PIN for a Smart Card, you need a random number generator and an algorithm that converts the random number into an ASCII-coded PIN of the necessary length. You could then use a table of known trivial combinations to recognize and discard PINs with these codes. Finally, you have to store the PIN in the Smart Card and then use the VERIFY command as necessary to compare it to a PIN that has been entered.

If magnetic-stripe cards are used in a system instead of Smart Cards, generating a PIN is somewhat more complicated. This is because it must be possible for a cash dispenser operating offline to test an entered PIN, based on data located on the magnetic stripe. Smart Cards are not actually affected by this requirement, but all debit cards (such as Eurocheque cards) presently have magnetic stripes for reasons of compatibility, even if they also have microcontrollers. The PIN generation algorithm must therefore be deterministic for hybrid cards with both a chip and a magnetic stripe, which means that it must always produce the same result for a given set of input values. A random number generator would not do this.

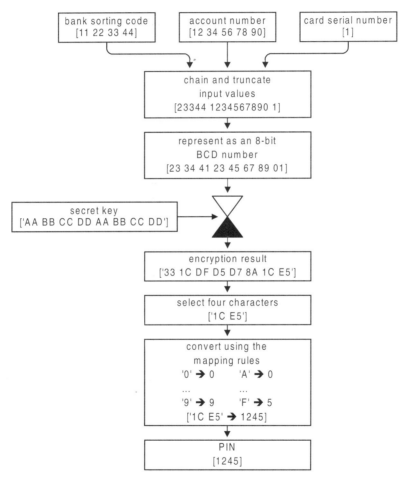

Figure 8.2   Example of generating a four-digit PIN using the DES algorithm and three card-specific data elements (bank routing code, account number and card serial number). The illustrated procedure has the disadvantage that it produces PINs that are not evenly distributed (i.e., some PINs occur more often than others), due to the mapping rule used. Example values are shown inside square brackets. This procedure is remotely similar to the procedure used for German Eurocheque cards and is based on two articles in *Die Datenschleuder* [Müller-Maguhn 97a, Müller-Maguhn 97b].

A procedure is thus needed that can generate a PIN based on the magnetic stripe data. In order to avoid having the security of the system depend on the procedure itself, a secret key should also be involved in the computation. Figure 8.2 illustrates an algorithm similar to the one that is used for German Eurocheque cards. Its inputs consist of the bank routing code, the account number and the serial number of the card. This algorithm uses the DES with a secret key to generate a four-digit PIN. This procedure suffers from the previously mentioned disadvantage that the generated PINs are not evenly distributed over the total possible number space ("0000" to "9999" in the case of a four-digit PIN). This is due to the mapping rule that is used to convert the hexadecimal numbers ('A', 'B', 'C', 'D', 'E', and 'F') into decimal numbers following the encryption process. This undesirable feature could however be easily avoided by using a better mapping rule. The DES is used in part because the key rather than the procedure must be kept secret, and in part due to its properties of confusion and diffusion.[1]

Testing the genuineness of a terminal
As is well known, the identity of a user is verified by the entry of a PIN. However, the user might equally well wish to be able to verify the genuineness of a terminal. For example, consider the possibility of a dummy cash dispenser. Someone with fraudulent intentions could use such a machine to collect PINs entered by unsuspecting users. If the person who set up the machine then stole the cards, he or she could use the PINs acquired using the dummy cash dispenser to withdraw money from the cardholders' accounts. This is all possible because there is no way for the user to test the genuineness of the terminal.

There is, nevertheless, a procedure that can be used to defend against this type of attack. This involves storing a password in a file in the card. This password is known only to the card user and can be changed only by the user. The Smart Card operating system allows read accesses to this file only after the terminal has been authenticated by the card. A name or a number chosen by the user can be used for this password.[2]

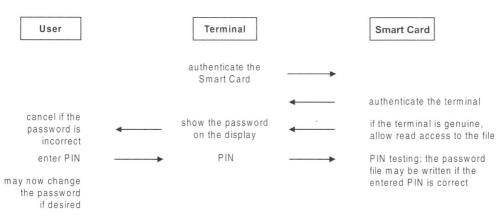

Figure 8.3   A procedure for ensuring that a PIN can only be entered using a genuine terminal. A prerequisite is that the cardholder does not enter the PIN until the password has appears on the terminal display.

---

[1]   See also Section 4.10, 'Authentication'.
[2]   The design of a suitable Smart Card application is described in detail in Section 14.7.3, 'Testing the genuineness of a terminal'.

The first thing that happens after the user has inserted a Smart Card into the terminal is a mutual authentication process between the card and the terminal. If this is successful, then each party knows that its opposite number is genuine. The card then allows a read access to the file containing the user's secret password, which is then displayed on the terminal. The user sees his or her password, and thus knows that the terminal is genuine, since the terminal would have otherwise had no access to the file containing the password. He or she can now safely enter the PIN.

This procedure can also be used as a simple means of preventing PINs from being entered into terminals that have been tampered with. Any arbitrary word or number can be used for the password. It should be possible for the cardholder to change the password as desired, to prevent potential attackers from being able to ferret it out. This procedure can also be extended or modified as necessary to meet other demands of a similar nature.

8.1.2   Biometric methods

The steadily increasing use of passwords and PINs is producing a rising level of user resistance to this type of identification. No one has any particular difficulty with remembering a few frequently used combinations of numbers or letters. However, normal people find it difficult to remember a PIN that is used only rarely, if for example the card is used only every couple of months to obtain money from a cash dispenser. The unconscious fear that the machine will confiscate the card if the PIN is entered incorrectly three times in a row only makes matters worse.

This is certainly one of the main reasons why biometric methods are finding increasing favor in many areas. They are not necessarily faster or more secure than PIN entry, but they can make things much easier for the user. If the level of security provided by biometric methods is equivalent to that provided by PIN codes, system operators will be also prepared to use them. After all, biometric features cannot be transferred to another person as easily as a PIN. This means that the actual person is identified, rather than a secret that is shared by the user and the system operator.

8.1.2.1   Basic considerations

A biometric identification procedure is one that can unambiguously identify a person by means of unique, individual biological features. In this regard, a distinction can be made between physiological and behavioral features. If the features tested by the procedure are directly related to the person's body and are fully independent of conscious behavior patterns, they are referred to as physiological biometric features. Behavior-based biometric procedures, by contrast, utilize certain features that can be consciously changed within certain limits, but that are still characteristic of a particular person.

An essential aspect of biometric feature testing is the question of user acceptance. The more similar a procedure is to existing, well-known procedures, the more willing users will be to accept and use it. One typical example is the handwritten signature, which has been used for generations in almost all cultures for identification and indicating agreement or consent. Social aspects also play an important role. In many countries, fingerprinting is primarily used by the police and security forces. This could adversely affect the acceptance of biometric methods based on fingerprints.

Another point to bear in mind is concerns that users may have regarding medical and hygienic aspects. For example, they may be concerned about acquiring a disease from the optical scanning of the retina, or be worried that the laser light will damage their eyes. Even though such fears may be fully subjective and lack any scientific basis, they can still strongly affect the behavior of the users and above all their acceptance of the procedure. Before any biometric identification method is put into use, such aspects should be fully understood.

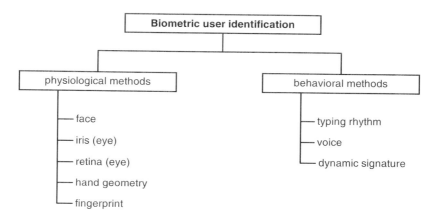

Figure 8.4   Classification of the most important biometric methods for user identification

There is yet another difference between biometric and knowledge-based identification methods, which can be considered either an advantage or a disadvantage, depending on one's point of view. This is that biological features cannot be transferred to another person. With systems that use biometric methods for identification, this means for example that it is not possible to give your card and your PIN to a trusted person who can then use the card in the intended manner. System operators naturally find such an action absolutely shocking, since revealing your PIN is forbidden in almost all systems. However, practically everyone knows how loosely such restrictions are observed in actual practice.

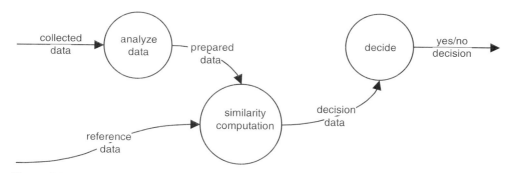

Figure 8.5   Basic data flow for the computational evaluation of a biometric feature

Entering a PIN does not just check whether the user knows a secret code; it is also a legally binding equivalent of saying, 'I consent'. This relationship is very important if some other method is to be used in place of a PIN. A test based on a retinal scan performed at a distance of three meters, which is anyway presently not technically possible, could certainly not be considered to indicate the agreement of the individual in question to any sort of action. In almost all countries, only an intentional manual action of the user can be interpreted as an indication of consent. For instance, breaking the seal of a cardboard box containing software is an unambiguous indication that the user agrees with the printed license conditions. Biometric methods in which the testing of the person is fully passive must therefore be augmented by appropriate user instructions together with something that provides the element of consent.

Naturally, not all biological features are suitable for personal identification. A feature must satisfy at least the following criteria before it can be reasonably used:

- it can be measured effectively (in terms of the measuring method, time and costs),
- it must be capable of being uniquely associated with a particular individual,
- it must not be possible to alter the feature with fraudulent intent,
- the amount of reference data generated must be small (at most a few hundred bytes),
- natural changes to the feature over time must be so small that a satisfactory measurement is always possible,
- the measurement method and the feature must be acceptable to users.

The result of any measurement is not always the same, but varies from instance to instance. This effect occurs even with the simplest measurements. For example, if you measure the length of a sheet of paper several times, each result will be slightly different. There are many reasons for this, but it does not actually cause any difficulties, since the average value of the measurements will be close to the true value.

Experience shows that the amount of variation between individual measurements depends on the difficulty of the measurement process. For instance, there is a significant technical difference between determining the weight of a bar of chocolate and determining the distance between the earth and the moon. Measurements of human beings are always difficult, and are subject to a wide range of variation.

Figure 8.6 shows an example of the results of measuring a biological feature, such as the length of a finger. The range of variation in the measurement is marked on the horizontal axis, while the vertical axis indicates the probability of correct identification based on the measured biometric feature. With an ideal biometric feature and an ideal measurement method, there would be no variation, and the curve would be reduced to a vertical line. However, a real feature together with a real method produces the illustrated Gaussian bell curve. If the measurement result deviates from the reference value, it is not possible to be absolutely sure that the person to be identified has been correctly recognized.

Before a biological feature can be tested, the feature of the person in question must first be acquired. This can be achieved by making repeated measurements and computing the average value. The result is a reference value that is then stored in the Smart Card. Following this, the Smart Card can, as necessary, test whether an actual measurement value that is sent to it matches the reference pattern. Depending on the biometric method used, however, it may be necessary to use a powerful computer to process the actual measurement value into a form that the card can use for comparison. Since identification cannot be established with absolute certainty, a threshold level is needed in order to decide whether the person should be recognized as genuine. This threshold level must be set separately for each method and application.

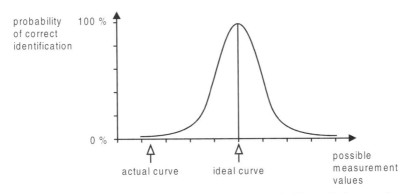

Figure 8.6   Probability distribution for repeated measurements of a biometric feature of a person

If we take our probability distribution diagram and add a curve for a second person to it, we obtain the diagram shown in Figure 8.7. The additional curve represents an arbitrary individual, whose measurement curve is close enough to that of the first individual for it to affect the identity decision. Since both curves approach the horizontal axis asymptotically, they have an intersection point. At this point, the probability that the person being tested is genuine is the same as the probability that he or she is not genuine. Biometric identification systems therefore use an adjustable threshold level, which indicates the probability above which the identity is assumed to be correct. The threshold level marked in Figure 8.7 divides the two curves into four regions. These indicate the decision to be taken regarding the identity of the person, based on the biometric feature.

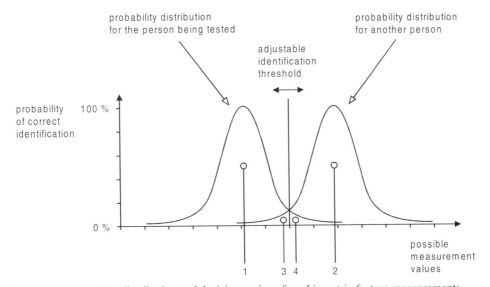

Figure 8.7   Probability distribution and decision regions for a biometric feature measurement:
1 – true positive identification (true acceptance)   3 – false positive identification (false acceptance)
2 – true negative identification (true rejection)   4 – false negative identification (false rejection)

Essentially, this diagram demonstrates that there is no such thing as an absolutely positive identification. It is only possible to assume, with a high degree of probability, that this is the right person. The level of this probability can be adjusted with the threshold value. However, in practice the threshold value cannot be set arbitrarily high, since a strict criterion for correct identification produces a large number of false rejections.

The two basic values for judging a biometric method are its false acceptance rate (FAR) and its false rejection rate (FRR). The FAR is the probability of the incorrect acceptance of the wrong person, while the FRR is the probability of the incorrect rejection of the genuine person. These two probabilities can naturally not be freely chosen, since they are properties of the biometric method that is used and can be adjusted only to a limited degree, within certain boundaries. In addition, these two values are mutually dependent, in that a low FRR produces a high FAR and vice versa.

For the user, a high FRR means that he or she may be rejected in spite of presenting a legitimate feature, which naturally affects the acceptance of the system. The system operator wants to have not only a low FRR but also a low FAR, in order to avoid false identifications.

PIN checking does not require any complicated algorithms in the Smart Card, since it involves only the comparison of the input and stored PIN values. Unfortunately, things are not this easy with biometric features. The reference value is of course stored in the card, but comparison with the measurement in question normally cannot be carried out in the card. This is due to large amount of computing power needed to evaluate biometric features. Since Smart Cards usually do not have adequate processing power for this, the computationally demanding preprocessing of the measurement values is performed externally. The result is sent to the card, which uses special algorithms that do not required a lot of memory or processing power to further evaluate the preprocessed data and then makes a yes/no decision, based on the stored reference value.

8.1.2.2 Physiological features

Additional biometric features could easily be added to the list of features described below. However, we have limited our discussion to the most important and most commonly used features in order to avoid getting lost in details.

Some physiological features that cannot be consciously altered also do not change much over time. For example, the characteristic patterns of fingerprints never change during a person's lifetime, nor do the patterns of the retinal blood vessels. The face is certainly an exception, since even though it basically does not change, it can be transformed to a large degree by a different haircut, growing or shaving a beard, and the like. Basically, though, it is possible to say that biometric features based on adult physiology do not require ongoing adjustment of the reference pattern, since any changes are negligibly small or non-existent.

Facial features Based on everyday experience, we can assume that the human face is suitable for use as a biometric feature. Transforming this assumption into a technical implementation is however fraught with difficulty. Faces can change greatly within a short period, and their appearance depends strongly on external factors such as eyeglasses, beards, make-up, illumination and viewing angle.

If you photograph a face using visible light and suitably process the data captured by the photograph, you will often be able to make a judgement about the identity of the person behind the face. The technical toolkit for this process includes very powerful computers, fuzzy logic and neural networks, which indicate the amount of effort that it entails. In addition, the stored image should be three-dimensional, or else the person should be photographed from several different angles, to prevent the system from being deceived by holding a photograph in front of its sensor. In general, this feature may prove to be a very interesting subject for future biometric methods, but presently it cannot yield a high enough probability of accurate identification to allow it to be widely used.

Retinal features   Every human retina has its own unique pattern of blood vessels, with their branches and junctions. This pattern can be captured using a beam of infrared beam directed through the pupil. The light reflected by the retina is collected by a CCD camera, which in turn sends the recorded image data to a computer for analysis.

   Retinal imaging is one of the very best biometric methods, since it can be used to uniquely identify a person with a very high degree of probability. However, it is not readily accepted by users, since they must place their eyes very close to the scanner in order to be identified. This often results in a fear of infection and anxiety with regard to the infrared beam. Another problem is that certain types of contact lenses strongly block light in the infrared region, which causes the measurement to fail.

Iris features   The iris is a variable diaphragm that limits the amount of light that reaches the retina. Like the retina, it is a biological feature that is unique to each individual. An iris scan can be performed at a greater distance than a retinal scan, since the measurement procedure is simpler. With this method, the iris (which is located at the front of the eye) is imaged by a CCD camera using visible light. The data evaluation is similar to that used for retinal images. Here again, contact lenses can strongly influence the measurement results and thus cause problems.

Hand geometry   Identification systems based on three-dimensional measurements of the hand, or parts or the hand, were already in use in the 1970s. These measurements can for example be based on finger length, finger diameter and fingertip radius. Unique individual features can be determined using very few measurement points (e.g. five). The actual measurements can be made very simply using infrared LEDs and photodiodes. The hand geometry can be determined by noting which photodiodes are fully or partially blocked by the hand. Since only a few measurements are needed for identification, the procedure is sufficiently fast and uncomplicated for the user. He or she only has to place his or her hand in an instrument, which then performs the measurements.

Fingerprints   The best-known biometric identification method based on a physiological feature is without doubt fingerprinting. In the electronic version, it is naturally no longer necessary to press fingers coated with black ink onto paper. The thumb or another fingertip is placed against a transparent plate, and a camera mounted under the plate scans the skin surface without any contact. The comparison with the reference pattern is usually based on the classification scheme developed by Edward Richard Henry,[3] which marks features as arches, loops and whorls. Information about the type, position and orientation of 20 or so of these features is stored, and this information is used to generate the reference pattern.

   Certain groups of users dislike this method because of the well-known fact that fingerprints have been used for years as a tool for combating crime. Small wounds on the

---

3   Fingerprints were first used for identifying persons by the Bengal police in India around the end of the 19th century, under the leadership of Sir Edward Richard Henry. After returning to London, Edward Henry set up the first British fingerprint collection in 1901, and generated a classification method for fingerprint features that is still in use.

fingertips can also make unambiguous identification difficult. Many systems have sensors for measuring the temperature of the finger or the pulse rate, in addition to the optical scanner. This is designed to prevent an amputated finger from being used for identification purposes.

Despite all these problems, fingerprint systems are widely used, since they are technically relatively simple and have relatively few problems with user acceptance. The time needed to sense the fingerprint and perform the subsequent test also lies within reasonable limits. Optical and capacitive sensors based on semiconductor technology are primarily used at present. They have a resolution of around 400 dpi.

Table 8.3   Comparison of biometric methods based on physiological features

Method	Test duration (seconds)	Amount of reference data (bytes)	Probability of false rejection (FRR), percent	Probability of false acceptance (FAR), percent
Facial image	—	—	≈ 10	≈ 1
Retinal image	1.5–7	40–80	0.005	10^{-9}
Hand geometry	1–2	10–30	0.8	0.8
Fingerprint	1.5–9	300–800	0.014	10^{-6}

8.1.2.3   Behavioral features

With many persons, biometric features based on behavior are not stable over time. Consider a signature, which can change considerably during the course of a person's life. However, it is rare for these changes to occur suddenly; instead, they are usually quite gradual and slow. Many systems therefore use adaptive procedures, which accept any changes in the feature that are detected during a correct identification as a new reference pattern that is then stored in the Smart Card.

Table 8.4   Comparison of biometric procedures based on behavioral features

Method	Test duration (seconds)	Amount of reference data (bytes)	Probability of false rejection (FRR)	Probability of false acceptance (FAR)
Voice	5	100–1000	1 %	1 %
Dynamic signature	2–4	40–000	1 %	0.5 %

Typing rhythm   It has been determined that there are large differences in the manners in which different individuals type characters on a keyboard. These primarily relate to the pauses between individual letters. This can naturally be used as a biometric feature for identification. The procedure works by having the person to be identified type a prescribed character string (which is different for each test) on the keyboard. The computer to which the keyboard is attached evaluates the typing rhythm as the character string is typed. A text chosen by the user can also be used to evaluate the typing rhythm, but this requires more characters to be typed than with a prescribed text.

The great advantage of this method is that it does not need any additional hardware, since in most cases a keyboard and computer are already available. Unfortunately, between 100 and 150 alphanumeric characters are needed for the test, and they must be typed using the ten-finger system. This is the main disadvantage of this method.

Vocal features   Like the face, a person's voice is characteristic of the person, so it can also be used for identification purposes. The person to be identified speaks one or more sentences into a microphone. These must be different for each session, since otherwise the system could be attacked very easily by playing back a previous identification session, which for example may have been recorded on magnetic tape. The waveforms of the spoken text are subjected to a Fourier analysis, which yields the characteristic frequency spectrum of the speaker. This is then compared with a reference value to determine whether the speaker's identity is genuine. The entire palette of modern computational wizardry, such as fuzzy logic, neural networks and the like, is also employed with this method.

Of course, this method also has its shortcomings. A person's voice is very strongly influenced by his or her current bodily condition. Furthermore, all background noises must be reliably filtered out to make an unambiguous spectral analysis possible in the first place. A different sentence must be spoken for each test to prevent recorded speech from being played back, which very much complicates the procedure and makes recognition more difficult. These technical difficulties are however offset by good user acceptance, which makes this a very attractive biometric identification method.

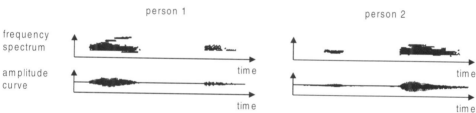

Figure 8.8   Amplitude waveform and time-dependent frequency spectrum of the name 'Wolfgang', spoken by two different people

Dynamic signatures   The only identification method that is commonly used in everyday life is writing a signature. Due to its very individual character, a signature can also be used as a biometric feature. With a static method, the signature is evaluated after it has been written. In a dynamic method, however, measurements are made while the signature is being written. The static method is only of theoretical interest, since it cannot distinguish a photocopied signature from a genuine one.

The parameters measured in the dynamic method may for example be the general form of the signature, the speed, acceleration and pressure of the pen on the writing surface, and the time required to write the signature. A special pen, or a special pad that can sense the parameters to be measured, can be used to make the measurements. Figure 8.9 shows an example of a possible arrangement in which an ordinary pen is used on a special pad, and Figures 8.10 through 8.13 illustrate examples of measured signals that can be used as the basis for a biometric identification process. Pressure sensors are located at the intersections of the grid wires, and their signal amplitudes are transmitted to the computer via conditioning logic. The computer can then use various algorithms to process the measured data into a standardized format and compare the results with a stored reference pattern.

Using a dynamic signature for purposes of identification has the highest degree of acceptance of all personal identification methods, since signatures are used daily by everybody in almost the same fashion. However, here the technical solutions are also not simple, since signatures change over time and are never completely the same. You need only consider the difference in your signature if you write it while sitting or standing to appreciate the truth of this.

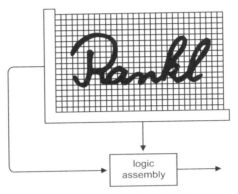

Figure 8.9   Sample measurement setup for testing a dynamically generated signature

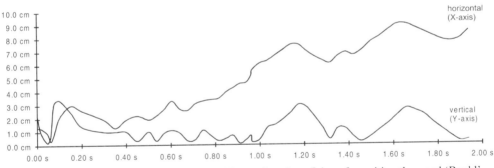

Figure 8.10   Horizontal and vertical pen position as function of time, for writing the word 'Rankl'

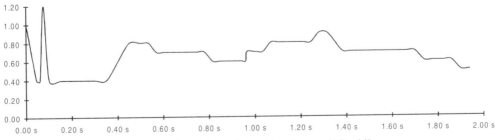

Figure 8.11   Pen pressure as function of time, for writing the word 'Rankl'

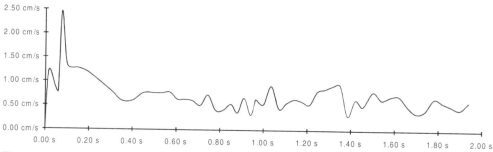

Figure 8.12   Pen speed as function of time, for writing the word 'Rankl'

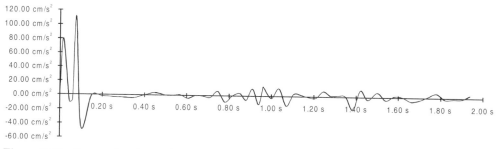

Figure 8.13   Pen acceleration as function of time, for writing the word 'Rankl'

8.2   Smart Card Security

The essential characteristic of a Smart Card is that it provides a secure environment for data and programs. If the amount of effort needed to read data from a Smart Card were not so large, it would essentially be nothing more than a diskette with a different interface.

It is naturally practically impossible to set up a complete system, or even a Smart Card, such that it has perfect security that is proof against everything and everybody. If the effort expended on the attack is raised to a high enough level, it is possible to force your way into any system or tamper with it. However, every potential attacker makes a conscious or unconscious benefit analysis for himself and his targets. The rewards of breaking into a system must be worth the time, money and effort that he must expend to achieve the goal. Regardless of whether the reward is money or prestige within a peer group, if it is not worth the effort, no one will invest all too much energy in breaking a system or a Smart Card.

The security of a Smart Card is guaranteed by four components. The first component is the card body, in which the microcontroller is embedded. Many of the security features that are used for the card body are not only machine-readable, but they can also be visually checked by humans. The techniques that are used for these features are not specific to Smart Cards, but are also used with other types of cards. The remaining components – the chip hardware, the operating system and the application – protect the data and programs in the Smart Card microcontroller.

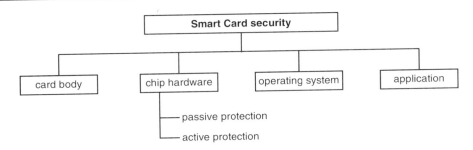

Figure 8.14   Classification of Smart Card security components

The security of a Smart Card is guaranteed only when all of these components are present and their defense mechanisms are working properly. If the card is used exclusively within a realm where it is not subject to human verification, the card body component is not necessary. The three components that are independent of the card body, however, are indispensable for the physical and logical security of a Smart Card against attacks. If any of these components fails, or if any one of them does not meet the applicable requirements, the Smart Card is no longer secure, since these components are coupled to each other in a logical AND relationship.

8.2.1   A classification of attacks and attackers

A major problem with all information technology systems that are subject to attack is the 'avalanche effect', which is often seen after a successful attack. If a printer (for example) succeeds in counterfeiting good-quality bank notes in sufficient quantity, this is naturally a matter of concern for the affected national bank, but in practice, it never leads to an inflation of the national currency. In the first place, the counterfeiter would never be able to produce a sufficient number of false bank notes for this, and in the second place, it is very risky to bring large numbers of counterfeit notes into circulation.

With electronic money, on the other hand, the situation is somewhat different. Since this consists of nothing more than immaterial information, it is in practice impossible to distinguish between an original and a copy. In addition, if a new counterfeiting method becomes known, this can lead to an avalanche effect, as other people copy the original technique. This effect can be very clearly seen by examining what has happened due to counterfeit prepaid telephone cards, which have been produced in very large numbers. Some network operators can still only defend themselves against this attack by restricting the calling destinations for cardphones.

In the following material, we attempt to systematically classify possible types of attack and attackers. The emphasis naturally lies on the information-technological side of Smart Cards, rather than on security features of the card body that can be checked by humans. This classification allows potential attacks to be evaluated, so that suitable measures can be taken against them. As is well known, it is easier to defend against a known type of attack than against an unknown one.

We have based our arrangement of the different types of attack on the ISO 13491-1 standard, which describes the concepts, requirements and evaluation methods for cryptographically secure equipment in the banking sector.

Classification of attacks

There are several different approaches to the systematic classification of attacks on Smart Cards. To conduct a security evaluation, for example, all possible types of attack are grouped and formally described according to each phase of the card's life cycle [IC Protection 97, Isselhorst 97]. This results in multipage lists that identify all conceivable attacks for each phase in turn. In the actual evaluation, each item is checked to see if the system or the Smart Card can defend against it. In this book, a different type of classification is used, in order to present the subject in as realistic a manner as possible, and also to illustrate the pong-pong game of attacks and defenses. In addition, it is our intention to present a general summary of methods of attack and defense, which is not specific to any particular system.

In principle, attacks on Smart Cards can be divided into three different types: attacks on the social level, attacks on the physical level and attacks on the logical level. Naturally, mixed types also occur in practice. For example, an attack at on the physical level could prepare the way for a subsequent attack on the logical level, which is for example the case with differential fault analysis.

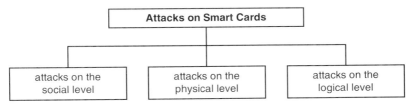

Figure 8.15   Classification scheme for attacks on Smart Cards

Attacks on the social level are attacks that are primarily directed against people that work with Smart Cards. These can be chip designers working for semiconductor manufacturers, software designers or, later in the life cycle of the card, cardholders. These attacks can only partially be parried by technical measures; they must primarily be defended against by organizational measures. Surreptitiously acquiring a PIN by watching it being keyed in can easily be prevented by providing visual shields on either side of the keypad. Attacks on the social level against Smart Card programmers are rendered pointless by making the procedures that are used public, and also by having third parties evaluate the program code that they produce. In this case, security depends only on secret keys, and the knowledge of the software developers is of no use to an attacker.

Attacks on Smart Cards on the physical level usually require technical equipment, since it is necessary to obtain physical access to the Smart Card microcontroller hardware in one way or another. The attacks can be either *static*, which means that no power is applied to the microcontroller, or *dynamic*, with the microcontroller operating. Static physical attacks impose no timing restrictions on the attacker, who can do his job at his own pace. With a dynamic attack, however, sufficiently fast data acquisition and measuring equipment must be available.

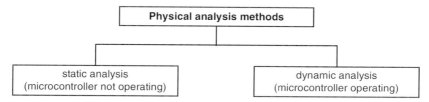

Figure 8.16 Classification of physical methods that can be used to analyze Smart Card microcontrollers

Up to now, most known successful attacks on Smart Cards have been on the logical level. These attacks arise from pure mental reflection or computations. This category includes classical cryptoanalysis, as well as attacks that exploit known faults in Smart Card operating systems and Trojan horses in the executable code of Smart Card applications.

Just as with the cryptoanalysis of cryptographic protocols, these attacks can be divided into passive and active types. In a passive attack, the attacker analyzes the ciphertext or cryptographic protocol without modifying it, and may for example make measurements on the semiconductor device. In an active attack, by contrast, the attacker tampers with the data transfer or the microcontroller.

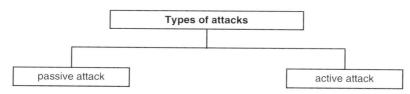

Figure 8.17 Classification of types of attacks on Smart Cards

The phases of the life cycle of a Smart Card as defined by the ISO 10202-1 standard[4] could be used with regard to the timing of possible attacks. However, this would result in verbose and long-winded descriptions, so for the sake of readability we have undertaken a simplification and classified the attacks into three intervals: (a) development, (b) production and (c) card usage.

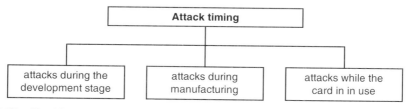

Figure 8.18 Classification of the timing of possible attacks

---

[4]   See also Chapter 10, 'The Smart Card Life Cycle'.

Attacks during development relate to system design, chip development, operating system development and the generation of applications. The term 'production' is used here to refer in general to all processes used to produce hardware. This covers the whole range from wafer production by the semiconductor manufacturer to card personalization and sending cards to their ultimate users. Card usage refers to the stage in which the Smart Cards are in the field, which means that they are being used by cardholders.

The consequences of an attack and the classification of attackers

In order to be able to realistically estimate the strengths and weaknesses of attacks on the security of a Smart Card, it is important that we first have at least a rough idea of the possible types of attackers. We can also use this information to help us devise defensive strategies and mechanisms.

As a rule, typical attackers have one of two basic motivations for their activities. The first of these is just plain greed, while the second is the desire for fame and status within a particular 'scene'. These two motivations have different consequences for the system operator. An attacker that seeks a financial reward for his activities can take the risk of becoming a 'card issuer' in his own right,[5] or he may attempt to blackmail the system operator. Both of these approaches can be combated using the usual judicial measures. If details of the attack become public, the reputation of the Smart Card system will be damaged. The worst damage to the reputation of the system operator occurs when a large number of cardholders lose money as a result of an attack.

The reputation of a Smart Card system is similarly impaired when an attack is prompted by a compulsion to perform scientific research rather than criminal tendencies. An attacker in this profession will consider his or her activities to be successful only if the results can be published in a suitable manner. The attacker is also under extreme pressure to publish these discoveries as quickly as possible, since in this trade, as is well known, only the first one past the post wears the laurels. The end result is that the system operator, with little or no warning, is faced with the publication of a detailed description of an attack on his system. Following this, the attack that has been made public is refined layer by layer by other interested parties, and also explained in terms that can be grasped by outsiders. The ultimate blow comes when programs that carry out the attack in a fully automated manner are published on the Internet.

A number of GSM network operators found themselves confronted with a similar series of events in the spring of 1998. However, here the attack on the COMP128 cryptographic algorithm had no large negative effects on normal network operation. From the point of view of the attacker, this manner of working also held a very large and very significant benefit. He was seen as the successful discoverer of a security leak, and thus as one of the 'good guys', and never had any serious concerns about any judicial consequences of his actions.

The quintessential conclusion that can be drawn from these scenarios is that it ultimately does not particularly matter to a system operator whether an attack comes from a 'good guy' or a 'bad guy'. With truly dangerous attacks, the financial damage and the damage to the reputation of the system are most often rather significant. In the worst case, the system must be shut down, all cards must be blocked and new cards that are immune to the attack must be issued. With large systems that have several million cards in use, such a process can take more than half a year.

---

5   This is the usual approach for producing self-reloading telephone cards.

The classification chart in Figure 8.19 shows the various types of attackers, based on experience and the previously described aspects. All four types can be equally dangerous to a Smart Card system, but they have different capabilities and options. The typical hacker has a moderate level of system knowledge, good creative ideas and usually a similar group of friends. He normally does not have an extensive amount of equipment, and his financial means are also limited. However, if he is competent and employs a suitable approach, he can certainly obtain access to a large amount of computing power, for example by means of an Internet campaign.

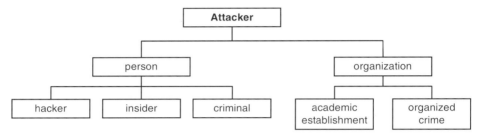

Figure 8.19   Classification of the possible types of attackers

All insiders form a special class of attackers, assuming in the first place that they have very good system knowledge. They may have access to hardware and software components, and the may know about weak points in the system. As long as only single individuals are involved, they are equivalent to hackers in terms of their resources and options. However, since insiders are neither anonymous nor especially numerous, it is usually possible to retrace their attacks.

The third group of individuals that can be regarded as potential attackers is criminals. They usually do not have a high level of technical knowledge, but they do have a lot of energy when it comes to achieving personal advantages (primarily financial) via their activities.

A potential source of attack that cannot be ignored in practice consists of academic institutions such as universities and technical institutes, with all their students and professors. They do not necessarily have special knowledge of specific Smart Card microcontrollers or applications, but they do have a large amount of generally useful knowledge. In addition, they have access to a large pool of qualified and inexpensive labor in the form of students and graduates, as well as adequate technical equipment in their laboratories. Many of these institutions also house a plentiful amount of computing power and highly motivated people with an experimental bent.

Organized criminal organizations naturally represent a completely different source of attacks on Smart Card systems. The amount of money that they have available allows them to acquire all the knowledge and tools necessary for a successful attack, either by purchasing it or by criminal means.

Classification of the attractiveness of an attack

In order to allow effective perimeter defense measures to be put in place, you should investigate the attractiveness of an attack for all of the candidate weaknesses of the system. This can be done in an objective mathematical manner by means of value analysis, in order to compute a prioritized list of probable attack targets. The scheme that is presented here is simplified, but it still allows us to make a relatively good estimate of the attractiveness of the different types of attack, and thus the probable lines of attack. Naturally, an attacker would normally choose an attack that requires the least effort and expense. The six criteria listed in Table 8.5 will consciously or unconsciously influence the attacker's behavior.

Table 8.5   Criteria for determining the effort and cost of an attack on the hardware or software of the security components, based on the prerequisites for the attack

Degree of attractiveness	Low	Medium	High
level of knowledge and skills required	high	medium	low
number of secrets required	many	moderate	few
amount of time required	much	moderate	little
acquisition of the necessary technical equipment (purchase or access)	difficult	moderate	easy
access to the components to be attacked	difficult	moderate	easy
value of the result (money or prestige)	low	medium	high

The lower the level of specific knowledge or skills required for an attack, the more attractive it is to an individual or an organization. An attack that does not require the knowledge of any secrets is similarly attractive. This is not inconsistent with Kerckhoff's principle, which says that security should depend only on the key and not on the cryptographic algorithm itself. This is because Kerckhoff's principle does not mean that you have to reveal everything you know in order to make a system secure. The presence of many secrets represents an enormous hindrance to the mounting of a successful attack.

Especially in the case of systematically searching for a key, the amount of time required plays an important role. The classic example of this would be breaking a cryptographic algorithm using a brute-force attack that takes an average of 10,000 years. No serious attack could be based on such an approach.

The attractiveness of an attack is equally dependent on the technical equipment that is needed for the attack. This need not necessarily refer only to the purchase of equipment, since it may be sufficient to be able to rent the equipment or somehow acquire access to the equipment. For example, a device that can generate and precisely position focussed ion beams costs several hundred thousand pounds, but such equipment can be rented by the day at research institutes, and some students can use this sort of equipment free of charge for their research work.

The availability of the components to be attacked also strongly influences the attractiveness of a particular type of attack. For example, you could attack a card-based electronic purse system either at home, by analyzing your own personal card and its card-specific keys, or at the system level by trying to analyze a security module with its system-wide master keys. The problem with the latter approach is that access to the security module is protected by multiple security measures.

This is by the way why Smart Cards for pay television are so strongly exposed to attack. The attacker can work undisturbed in his own living room, studying the communications and behavior of his Smart Card in order to try to duplicate them using a computer or a home-made electronic circuit. Nobody observes him or interferes with his work. However, if he were to attempt to do the same thing with a Smart Card terminal in a supermarket, the cashier would immediately forbid any further experiments and thus interrupt his work. A good review of the subject of security of electronic money with and without Smart Cards can be found in [BIS 96].

The final criterion, which is of decisive importance, is naturally the value of the reward for the attacker's troubles. His efforts must be paid back, either in a financial form or in the form of enhanced prestige. From this, it can be concluded that various field trials of electronic purses are only at risk of being attacked by hackers and academic groups. There are far too few locations where the cards can be used, and the businesses are mostly too simple (bakers, kiosks and the like), for any significant amount of money to be obtained from an attack.

8.2.2   Attacks and defense mechanisms during the development stage

Even in the development stage of microcontroller hardware and software for Smart Card operation systems, a wide variety of security measures are employed. Security, like quality, must be taking into account from the very beginning of a development project; it cannot be designed into a product as an afterthought.

With regard to attacks in the development stage, it can be generally said that access to the facilities in question is very difficult and the required level of expertise is very high. The attractiveness of an attack is thus correspondingly reduced. Nevertheless, the potential danger of a successful attack at this stage is significant, since there are very extensive possibilities for manipulating the hardware and software.

8.2.2.1   Smart Card microcontroller development

The development of the hardware for a Smart Card microcontroller takes many months. It involves only a few persons and takes place in access-controlled, supervised rooms within the facilities of a semiconductor manufacturer. The computer systems used for the design of the semiconductor are usually part of an independent network that is isolated from the rest of the world. This prevents any changes to the chip design to be made from outside, and also prevents outsiders from obtaining information about the internal design of the chip.

A very extensive amount of insider knowledge is needed to undertake manipulations to a chip design that would weaken its security, which means that this type of attack is probably very unlikely. In addition, nowadays the design and the protection mechanisms of almost all Smart Card chips are evaluated by independent testing agencies, so that an insider attack would not go undetected. However, it could be advantageous to an attacker to know the exact design criteria and the arrangement of the functional elements on the chip, since he would then know about the protective mechanisms and sensors present in the chip and the scrambling of busses and memories. This knowledge could later be useful to an attacker with regard to physical chip analysis.

Protection: design criteria
There are a number of basic criteria that apply to the definition of the functions of a Smart Card microcontroller. One of these is that the mechanisms for protection against static and dynamic attacks must actually work. Sensors and other protective elements are of little use if they can be too easily defeated or if they do not work in certain circumstances. An example is a sensor on the microcontroller chip that has such a large area that it can easily be destroyed by a needle, after which it no longer can fulfill its protective role.

A very important design criterion, which is different from the criteria used for standard chips, is that absolutely no undocumented mechanisms or functions must be present in the chip (*'that's not a bug, that's a feature'*). Such undocumented features are usually not completely tested, since only a few people know about them, and thus often have various errors and weak points. Since they are not documented, they could unintentionally be overlooked when the hardware is evaluated, and then possibly be used later for attacks. The use of such undocumented features is thus strictly prohibited, even though they can often be very helpful for the developer.

Protection: unique chip number
When the semiconductor hardware is being developed, all hardware security elements must be first defined and then converted into hardware for the resulting microcontroller. One such element, in addition to sensors and protective coverings, is WORM (*write once, read multiple*) memory, which is also referred to as an OTP (*one time programmable*) memory. When the semiconductor chips are manufactured, a unique chip number is written to this memory. This means that each chip is different and can be uniquely traced, and the Smart Cards can later be unambiguously identified within the system. In addition, chip numbers can also be used for the derivation of keys, and they allow the generation of 'blacklists' that can be used to take suspect cards out of circulation.

It should not be overlooked that although these numbers cannot be altered in the original chips, they naturally provide no protection against imitation chips using freely programmable microcontrollers. This means that security measures cannot be based on the presence of a particular chip number in the WORM memory of a particular chip. Such a unique number can only be used as the basis for true cryptographic security mechanisms. For example, a chip number can be used for the derivation of secret keys, which are in turn used in a challenge–response authentication process.

8.2.2.2   Smart Card operating system development
Software for Smart Cards is developed according to modern software development principles. Regardless of which development model is used (waterfall, spiral or whatever), certain general conditions must be observed.

The development computer always requires a separate, completely isolated network that does not allow any external access. The development tools, such as compilers and simulators, are software packages whose proper operation must be verified in dedicated tests. Sometimes two different compilers are even used, to be sure that the results are correct. Using software whose origins are not completely traceable is fundamentally prohibited, since such software would offer a possible means to manipulate development tools and consequently modify the programs to be generated.

Protection: development principles
Just as with hardware development, no undocumented features may be built into the software. For example, it would certainly be possible to convert the laborious black-box tests that are commonly used with Smart Cards into white-box tests by incorporating commands that could be used to read out arbitrary regions of memory. However, if such commands were inadvertently left in the operating system, it would be possible to use them to read the secret keys in real Smart Cards. In order to eliminate the possibility of such an attack, the creation of dump commands is prohibited as a matter of principle, even if their use would save valuable development time.

An additional principle is that programmers should never work alone on a project. This is already forbidden by considerations of software quality assurance, but the 'four eyes' principle must be observed for reasons of security as well. This effectively hinders attacks by insiders, since at least two developers must agree to work together on any attack. In addition, internal source code reviews are performed regularly, which assures the quality of the code and also supervises the development process.

Once the software development is finished, the entire source code and its functions are often inspected by an independent testing agency, as part of a software evaluation.[6] The main reason for performing this time-consuming and costly testing is to check for software errors, but it also has the effect of making it impossible for a developer to hide a Trojan horse (for example) in the operating system. In practice, such things can be found only by reviewing the entire code, since an experienced programmer is certainly able to finds means and techniques to hide a Trojan horse so that it cannot be found by a black-box test.

Protection: dividing knowledge
If several people work on a task, the result will be significantly more resistant to attack, due to the different opinions and experience of the people involved. The principle of dividing knowledge (shared secrets) is the opposite of the idea that 'everybody knows everything about everything'. In the development of security components, complete knowledge of the component should fundamentally never be vested in a single individual, since that person would then be a target for an attack. As in many military realms, knowledge is divided over several individuals in the development stage, so that although it is possible for experts to discuss particular subjects, there is never any single person who knows everything.

A similar situation exists with regard to completing of the Smart Card operating system, in which tables, program code and configuration data are loaded into the EEPROM. In addition to providing increased flexibility, this procedure also has a security aspect. This is because the card manufacturer, who receives the final, assembled ROM code for producing the fabrication masks, does not have complete knowledge of the operating system. The parts of the operating system that are located in the EEPROM are unknown to the chip manufacturer, so he cannot discover the complete security mechanisms and functions of the operating system by analyzing the ROM code.

8.2.3 Attacks and defense mechanisms during the production stage

Attacks during the production of chips or Smart Cards are typical insider attacks, since the production environments are closed. Access is strictly controlled, and every entry is logged. Nevertheless, security measures cannot be dispensed with in the production stage, since some technically very interesting and effective attacks can be carried out in this stage.

Protection: authentication during the finishing steps
Already at the wafer fabrication stage, the Smart Card microcontrollers are individualized using chip numbers and protected using transport codes. With recent operating systems, the transport code is chip-specific, and an authentication is a mandatory requirement for each access in the finishing process. Although this increases costs and the amount of time required to finish the chips, and naturally requires a security module for every machine, it considerably increases security.

---

6   See also Section 9.3, 'Evaluating and Testing Software'.

An obvious type of attack in the finishing process is to feed in dummy chips or dummy Smart Cards, which behave the same as genuine components but which for example include a 'memory dump' command. The earliest opportunity to exchange a dummy chip for a real chip is of course after the wafer has been divided into individual dice. This type of attack can be illustrated on the basis of a Smart Card for digital signatures.[7] The attacker swaps a dummy Smart Card for a genuine Smart Card at the initialization stage. This card is then initialized with genuine data and afterwards personalized. Since this Smart Card has all the functions of a real Smart Card, the process for generating the key for the asymmetric cryptographic algorithm will also be executed by the microcontroller. It obtains the necessary data for this from the initialization and personalization data. After this, the attacker must manage to recover possession of this card, and then he can read the secret signature key from the card using his special dump command. Since the associated public key has been signed by the trust center and is thus confirmed to be genuine, the attacker now knows everything necessary to produce as many duplicate cards as he wishes, all of which will be recognized as genuine.

This attack is unrealistic, since administrative measures are taken to prevent chips and Smart Cards from being taken into or removed from the finishing stations. In addition, mandatory authentication between the Smart Card and the security module of the finishing machine before every finishing step makes it difficult to swap chips or cards.[8]

8.2.4 *Attacks and defense mechanisms when the card is in use*

Access to the component to be attacked – the Smart Card – is usually much easier for the attacker after the Smart Card has been issued than in the previous phases of its life cycle. This is why the probability of attack is relatively high when the card is in use.

The idea of a self-destroying Smart Card microcontroller appears again and again in many publications, as a sort of panacea against all sorts of attackers. There are hardware security modules, such as those used for military applications, in which such mechanisms are sometimes employed, but such a defensive measure is not allowed in Smart Cards for a multitude of reasons. When it is without power, a Smart Card first of all has no way to recognize a potential attack, and there is also no possibility of any sort of active defense mechanism, since the Smart Card does not have any reserve source of energy. This is all aside from the fact that true self-destruction probably could not be imposed on cardholders, for purely legal reasons. Who would be responsible for the losses or damage that could occur under unfavorable circumstances, just because a Smart Card has incorrectly destroyed itself? In addition, true self-destruction is not at all necessary, since in almost all cases it is sufficient to erase the secret keys stored in the card.

There is yet another aspect to this subject, which relates to erasing keys or blocking Smart Cards. It is very difficult for a Smart Card to even recognize that it is being attacked. There is simply no sensor that can report 'Attack! Erase everything!' Too low a voltage or too high a clock rate could for example be a sign of an attack, but these conditions also occur in normal operation due to unfavorable ambient conditions. Contaminated contacts result in a high contact resistance and thus a low operating voltage. An excessive clock rate

---

[7] See also Section 13.6, 'Digital Signatures'.

[8] An extensive and detailed description of the usual cryptographic process for the initialization and personalization of Smart Cards is contained in Section 10.4, 'Phase 3 of the Life Cycle in Detail'.

can occur at a Smart Card terminal that is intended to be used with higher clock rates. Since recognizing an actual attack is so difficult, and usually not even possible, automatic mechanisms for blocking the card or erasing the keys are usually not employed.

In the following section, some attacks that can be considered to be 'classic' are described and explained. The descriptions of the attacks can be said to represent the 'state of the art'. They are intended primarily to provide people who are inexperienced in the subject of Smart Card security with a reasonably solid basic understanding, so that they will not in their ignorance reuse mechanisms that are already known to be vulnerable. The described defense mechanisms can be used to counter these attacks. These in turn can be avoided by somewhat modified attack scenarios, which results in the well-known cat and mouse game of measures and countermeasures for attacks and defenses.

The scenarios presented here are not an invitation to break the security of a Smart Card system, since they are without exception well known and also public. They do not represent any serious threat to the security of any modern, present-day Smart Card system, since they have long since been dealt with by suitable defense mechanisms. However, a few years ago, it is possible that a certain amount of success could have been achieved.

The attacks are divided into those that are directed against the chip hardware and those in which an attempt is made to break the Smart Card system at the logical level. The physical attacks and analysis methods can also be subdivided into static and dynamic attacks. In a static analysis, the chip is not operating, but it may be electrically powered. In a dynamic analysis, which is much more difficult to carry out, the chip operates with its full range of functions during the analysis.

8.2.4.1 Attacks on the physical level

Manipulations at the semiconductor level require a large amount of technical effort and expense. Depending on the attack scenario, the equipment required can include a microscope, a laser cutter, micromanipulators, focussed ion beams, chemical etching installations and very fast computers for the analysis, logging and evaluation of the electrical processes in the chip. This equipment and the knowledge of how to use it are available to only a few specialists and organizations, which strongly reduces the probability of an attack on the physical level. Nevertheless, a card or semiconductor manufacturer must assume that a potential attacker could employ all the equipment and facilities needed for such an attack, which means that suitable protection must be built into the hardware.

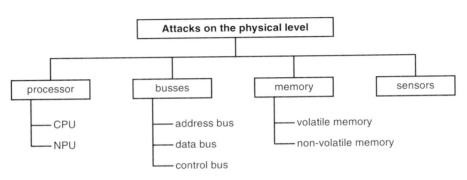

Figure 8.20 Classification of the points of attack at the physical level on a Smart Card microcontroller

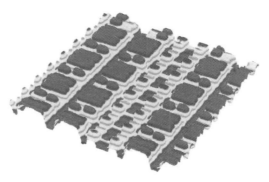

Figure 8.21   Graphic representation of the surface profile of a Smart Card microcontroller, as measured using an atomic force microscope. The maximum surface relief in this illustration is only 2.3 µm (Source: Giesecke & Devrient)

In order to conduct an attack on the physical level, a few preliminary steps are necessary. The thing that has to be done is to remove the module from the card, which can easily be done using a sharp knife. After this, the epoxy resin must be removed from the chip. Anderson and Kuhne [Anderson 96b] used fuming nitric acid for this with an infrared lamp as a heat source, followed by an acetone rinse to clean the chip. After this, the semiconductor chip is free and still fully operational. Many people think that the chip now lies unprotected before them and only has to be 'read out', but this is by no means so. An attacker still has to work through a manifold of security measures before he can gain access to the secrets.

The protective measures in the hardware can be divided into passive and active elements. The passive elements are based directly on the techniques used in semiconductor manufacturing. They include all processes and options that can be used to protect the memory region and the other functional parts of the microcontroller against various types of analysis.

There is a full spectrum of active elements available on a silicon chip to complement the passive possibilities offered by the semiconductor technology. Active protection means the integration of various types of sensors into the silicon crystal. These sensors are polled and evaluated by the Smart Card software as needed. This is naturally only possible when the chip with all its supply lines is fully activated. A chip without electrical power cannot measure any sensor signals, let alone evaluate them. Consequently, the line between useful protective elements and technical gadgetry is frequently particularly thin where sensors are concerned. A light-sensitive sensor that is supposed to prevent an optical analysis of the memory will not respond if the chip is located on the object carrier of an optical microscope without power or a clock signal. In addition, it is very easy to visually identify such a sensor on the chip surface and cover it with a drop of black paint, so that its protective function can easily be neutralized even when the chip is operating.

Long-term functional security is also important. For example, a temperature sensor that causes the Smart Card software to erase the entire EEPROM in response to a brief and non-damaging overheating of the chip, does not by any means contribute to increased functional security or security against an attack. For this reason, only a few sensors are normally employed in Smart Card microcontrollers.

In the following discussion, we explain the protective mechanisms of Smart Card microcontrollers that are the most important and the most often used in practice.

Static analysis of Smart Card microcontrollers

Protection: semiconductor technology
The dimensions of chip structures (lead widths, transistor sizes and so on) approach the limit of what is currently technically possible. The usual structural widths lie in the range of 1 to 0.35 μm, which is in itself no longer technically unusual. However, the density of the transistors on the surface of the silicon belongs to the highest level that can currently be achieved using standard lithographic manufacturing processes. These very fine structures alone make it nearly impossible to extract any information from the chip using analytic procedures, for which reason semiconductor technologies with 1-μm structure sizes are currently considered to be secure. This dimension is sure to be reduced in the future.

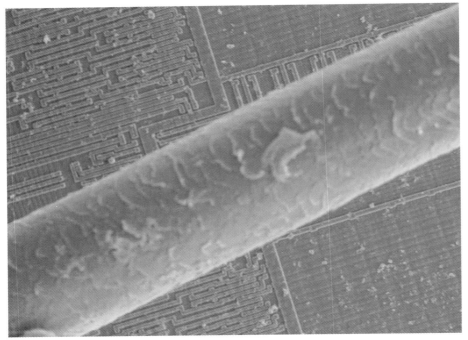

Figure 8.22   Photograph of a human hair in comparison with semiconductor structures of a Smart Card microcontroller, magnified 1000× (Source: Giesecke & Devrient)

Protection: chip design
'Standard cells' are frequently used for the design of semiconductor integrated circuits. These can contain the core elements of a processor or a particular type of memory. The advantage of using standard cells is it allows a semiconductor manufacturer to quickly produce a variety of different types of chips at a high quality level. This process, which has been developed for mass-produced components where security is not an issue, is not

allowed to be used for Smart Card microcontrollers. This is because the designs and functions of standard cells are known, and their use would thus provide a potential attacker with too much information and thus considerably simplify his task. The functional elements of Smart Card microcontrollers are developed especially for this application and are not used for any other purpose.

Protection: dummy structures
The use of dummy structures on the chip is a measure that is the subject of frequently controversial discussions among experts. Dummy structures are elements of the semiconductor that do not have any actual function, but instead are intended to confuse and mislead an attacker. The associated security is based purely on keeping the existence and locations of such structures secret. Dummy structures can also be monitored, so that any changes to them can be detected and can cause the chip to switch off. The main disadvantage of dummy structures is the additional room that they occupy on the chip.

Protection: busses on the chip
All internal busses of the chip, which connect the processor to the three different types of memory (ROM, EEPROM and RAM), are not brought to the outside of the chip. This means that it is not possible to directly make connections to these busses. It is thus not possible for an attacker to tap into the address, data or control bus of the microcontroller or to influence the bus signals in order to read out the memory contents. In addition, the busses on the chip are scrambled in a manner that is either static or chip-specific, so that the functions of the individual bus lines cannot be recognized from the outside.

Protection: memory design
The memory medium used for most programs is the ROM. The contents of ROM that is generally used in the industry can be read bit by bit using an optical microscope. It would not be particularly difficult to assemble these bits into bytes and then arrange these bytes to obtain the complete ROM code. In order to prevent exactly this type of analysis, the ROM is not located in the top level of the chip, which is the most easily accessible layer. It is instead located in the lower layers of the silicon. This impedes an optical analysis.

However, if the chip were to be glued to a carrier upside down and the rear surface were then ground off, it would be possible to read the contents of the ROM. For this reason, only ion-implanted ROM is used in Smart Card microcontrollers, since the contents of such a ROM cannot be seen using either visible or ultraviolet light. This also largely protects against selective etching, which is a process that can be used to attempt to etch the semiconductor to make the contents of the ROM visible again.

Protection: protective layers
The analysis of the electrical potentials on the chip while it is operating represents a threat. With a suitably high sampling resolution, this technique can be used to measure charge potentials (voltages) on very small regions of the crystal. With this information, it is possible to draw conclusions about the contents of the RAM while the chip is operating. This analysis can be very effectively prevented by placing current-carrying metallization layers on top of the memory cells in question. If these metallic layers are removed by chemical etching, the chip will no longer operate properly, since these layers are needed to distribute the supply voltages that the chip needs in order to function.

In addition, the chip can be fabricated with meandering current-carrying structures on top of the entire chip or on top of regions that need special protection, such as an underfrequency detector. The resistances of such structures can easily be monitored, or they can be incorporated into functional parts of the chip, such that the chip immediately stops working if they are damaged.

It is also conceivable to use opaque protective layers whose integrity is continuously monitored by phototransistors, which are easily implemented in semiconductor devices. As soon as such a layer is removed, this would immediately be detected and the chip could refuse to operate any further.

Attack and defense: reading out the volatile memory

As is well known, a RAM loses its data contents when its power supply is cut off. However, this does not occur if the memory cells are cooled to a temperature of –60 °C. Also, the content of the RAM is not necessarily lost if the stored data remain unchanged for a long time. The background of this effect is described in a paper by Peter Gutmann on the subject of the secure erasing of memory media [Gutmann 96]. Consequently, secret keys are not held in RAM any longer than is absolutely necessary, following which they are immediately erased or overwritten with other values. This minimizes the risk that traces of secret keys may be left in the RAM cells, and weakens attacks based on fixing the RAM contents by freezing or burning.

Reading out RAM cells is very difficult, since it requires detecting the switching states of the transistors involved. However, it is certainly possible to extract stored data from RAM cells using sophisticated electron microscopes and special contrast-enhancing processes. A prerequisite for this is the removal of the passivation layer and the metallization layers underneath the passivation layer, which protect the RAM against exactly this type of attack. Removing the metallization layers unavoidably causes the RAM cells to be destroyed, since part of their function in included in these layers.

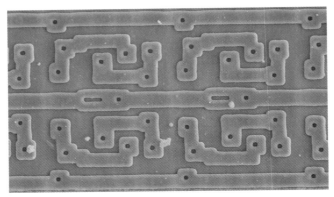

Figure 8.23 Photograph of several RAM cells magnified 3000×, without protection by means of additional metallization layers. The lower picture shows the electrical potentials of the same RAM cells, as measured using an electron beam tester with the chip in operation. The distribution of zeros and ones over the RAM can be clearly recognized. (Source: Giesecke & Devrient)

Protection: memory scrambling

Scrambling the memory on the microcontroller chip, which is similar to the well-established practice of scrambling the busses, is being used increasingly often. The security of this technique is based on the secrecy of the scrambling scheme for the memory cells. Memory scrambling is easily implemented and does not need much additional space on the chip. Without the relevant scrambling information, it is extremely difficult for an attacker to discover how the memory cells are actually addressed.

The EEPROM can also be scrambled using software. However, this requires complicated programming, and all write accesses must be protected by making them atomic processes, since otherwise the system would be very vulnerable to the sudden removal of the supply voltage. Software memory scrambling does however have the advantage that it can be made chip-specific and even dynamic, so that the memory contents can be redistributed within the memory in the course of a session.

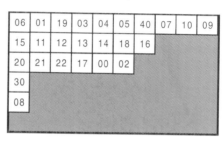

Figure 8.24   Comparison of a conventional semiconductor memory with a scrambled memory used in a Smart Card microcontroller

Attack and defense: encrypted PIN storage

This example of an attack at the physical level is actually a textbook case of an ineffective defensive mechanism. The basic idea of this defense is that if it is possible to read out the EEPROM of a Smart Card, an attacker should at least be prevented from learning the value of the PIN in this manner. There are naturally preventive measures that can be taken. The PIN need only be encrypted using a one-way function, and the result stored in the EEPROM as the reference value for a PIN comparison. A card-specific key could be used for the one-way function, so that the reference data for two identical PINs would be different in two different cards.

If it were now possible to read the reference data out of the EEPROM, it would appear to be impossible to derive the PIN from these data. However, a clever attacker would not even choose such an approach. If normal PINs consisting of four decimal digits are used, the number space of the PINS has a lower boundary of '0000' and an upper boundary of '9999'. This means that the number of possible PINs is exactly 10,000. If the attacker can read the entire memory of the Smart Card, he can also read the one-way function and its associated card-specific key. With this information, he sets about encrypting all possible PINs using the one-way function. After an average of 5000 attempts, he will have obtained a result that matches the reference value in the Smart Card, which means that he knows the

PIN. As can be seen, in this case using a one-way function to store the PIN does not provide any significant advantage.

Dynamic analyses of Smart Card microcontrollers

Protection: passivation layer monitoring
A passivation layer is placed on top of the microcontroller in the silicon at the conclusion of the fabrication process. This layer impedes oxidation (due to atmospheric oxygen) and other chemical processes at the surface of the chip. The passivation layer must always be removed before any sort of manipulation of the chip can be carried out. It should be borne in mind that although it is possible to chemically remove the passivation layer, the chip is then exposed to a large risk of oxidation, which can destroy it relatively quickly. A sensor circuit can employ resistance or capacitance measurements to determine whether the passivation layer is still present. If it is missing or damaged, this can either trigger an interrupt to the chip software or cause the complete hardware of the chip to be shut down, which reliably prevents any sort of dynamic analysis.

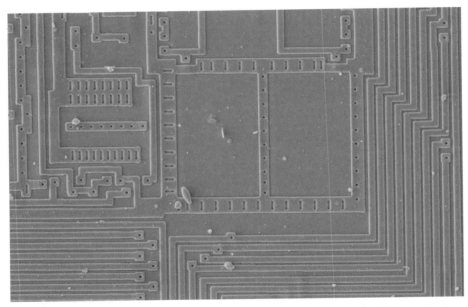

Figure 8.25 Photograph of a passivation layer monitor at 1000× magnification. The passivation layer detector consists essentially of the two rectangular semiconductor elements, and it uses capacitance measurements. (Source: Giesecke & Devrient)

Protection: voltage monitoring
A voltage monitor is present in every Smart Card microcontroller. It provides a well-defined shutdown of the IC in case the supply voltage exceeds its allowed lower or upper limits. This gives the software the assurance that it is not possible to operate the chip in marginal regions, in which the chip does not function properly. If such a voltage monitor were not present, it could for example happen that the program counter becomes unstable

when the chip is operated outside of its specified supply voltage range, which in turn could lead to uncontrolled jumps within the program or simply to processing errors. This faulty behavior could be used as the basis for determining secret keys by using the technique of differential fault analysis (DFA), which is described elsewhere in this book.

Another sensor that is partly based on the voltage detector is the power-on detector. This detector, which is also present in all chips, recognizes a power-on condition independently of the reset signal, and ensures that the chip is always placed in a defined initial state when power is first applied. The reasons for doing this are similar for those for voltage monitoring.

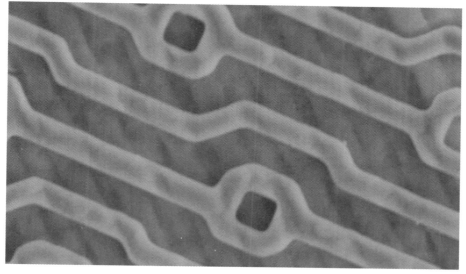

Figure 8.26 A modern passivation layer monitor at 8000× magnification. The track separation is 4 μm. This detector presumably uses resistance measurements. (Source: Giesecke & Devrient)

Protection: frequency monitoring

A Smart Card is always driven by an external clock, so its processing speed is completely determined outside the card. This means that, at least in theory, it is possible to run the microcontroller in single-step mode. This would provide outstanding opportunities for analyzing its operation, in particular with regard to measuring the operating current consumption (power analysis) and electrical potentials on the surface of the chip. In order to prevent such attacks, an underfrequency detector is built into the chip. This eliminates the possibility of reducing the clock rate to unallowed levels. The minimum clock rate given in most specifications is 1 MHz. However, the underfrequency detector has a wide tolerance range for technical reasons, so the chip usually stops working at around 600 kHz. This assures that the chip can will always work at the minimum specified clock rate of 1 MHz. Overfrequency detectors are also sometimes used, but nowadays the hardware is frequently constructed such that the chip cannot be used if the clock rate is too high.

In order to protect the microcontroller against the dangers of single-step operation, it is naturally necessary to secure the underfrequency detector with protective layers, so that any attempt to manipulate the detector will be noticed.

Protection: temperature monitoring
A temperature sensor is used in some types of chips, but the benefit of such a sensor is controversial. The chip is not damaged if the temperature briefly rises above the specified maximum value, and this does not in itself represent an attack. Shutting down the chip in this marginal situation, however, could result in an artificially increased failure rate, while it does not provide the operator of the Smart Card system with any additional security.

Protection: bus scrambling
In many Smart Card microcontrollers, the internal busses that drive the memory (and are not externally accessible) are scrambled. This means that the individual bus lines are not laid out next to each other in increasing or decreasing order, but are instead arranged randomly next to each other and 'swapped' several times, or even arranged in several layers on top of each other. This represents an additional hurdle for a potential attacker, who thus does not know which bus line is associated with which address bit or function.

Scrambling the bus lines was originally introduced only in a static version, with the same scrambling scheme used on every chip. This means that it would probably not be all that difficult for an attacker to discover the scrambling scheme over a moderate length of time, and thus be able to take it into account when tapping the busses.

The security provided by this technique can be improved by using chip-specific scrambling. This is naturally not achieved by using a different set of exposure masks for the busses of each chip, since this is currently either not technically possible or not affordable. Instead, scrambling is performed by randomizer circuits located just ahead of the memory. These can be driven by the chip serial number, for example. This technique is not difficult in terms of semiconductor technology, and it makes life considerably more difficult for someone who tries to tap the bus. Chip-specific scrambling that is different for each session would also be possible if a variable initial value is used for the randomizer.

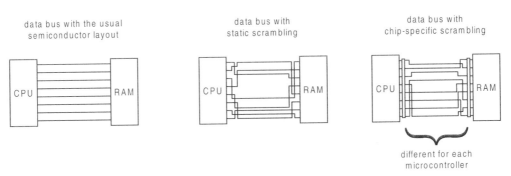

Figure 8.27 Bus scrambling in a Smart Card microcontroller illustrated using an 8-bit data bus between the CPU and the RAM. The data bus lines shown here represent information flows rather than electrical leads.

Protection: irreversible switching from the test mode to the user mode

All microcontrollers have a test mode that is used for verifying the chips during the fabrication process, and for executing internal test programs while the semiconductors are still in the wafer or after they have been packaged in modules by the manufacturer. The test mode allows types of access to the memory that are strictly forbidden when the chips are later in actual use. For technical reasons, however, it is an unavoidable requirement to be able to read data from the EEPROM in this mode.

The change from the test mode to the user mode must be irreversible. This can be realized by using a polysilicon fuse on the chip. In this case, a voltage is applied to a test point on the chip that is provided for this purpose, and this voltage causes the fuse to melt through. The chip is thus switched into the user mode using hardware. Normally, this cannot be reversed. However, a fuse is by its nature a relatively large structure on the surface of the chip. It is conceivable that the fuse could be mechanically bridged after the passivation layer has been partially removed where it covers the fuse. This would put the microcontroller back into the test mode, and the memory could be read out using the extended access options available in this mode. If the complete content of the memory is known, it is then easy to clone the Smart Card that has been read out.

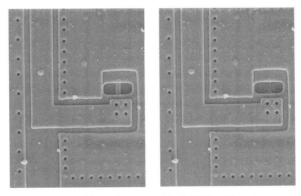

Figure 8.28 Photograph of a polysilicon fuse magnified 2000×. The picture on the left shows a fuse that is still intact, while that on the right shows a blown fuse. (Source: Giesecke & Devrient)

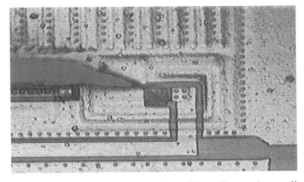

Figure 8.29 Photograph of a polysilicon fuse together with a microprobe needle, magnified 500×. A blown fuse could be bridged using a microprobe needle. (Source: Giesecke & Devrient)

In order to defend against this type of attack, most semiconductor manufacturers have adopted the practice of reserving a portion of the EEPROM for the switchover mechanism, in addition to using a fuse. If a certain unalterable value is located in this part of the memory, the chip has been irreversibly switched to the user mode. Even if the fuse is bridged over, the chip will not return to the test mode, since the additional logical switch in the EEPROM prevents this.

The security of the switchover from the test mode to the user mode can be increased even further by a very simple measure. The microcontroller chip is laid out on the wafer such that the test pads needed for making contact with the chip for performing the tests are simply sawed off when the wafer is divided into individual dice. In this case, neither a fuse nor any EEPROM cells are needed to switch between the modes, since the elements needed for the test mode are no longer present. In the same way, it would also be possible to replace the fuse that switches from the test mode to the user mode by a track that is irreversibly broken when the dice are sawed from the wafer. With present-day technology, it is not possible to make a connection to a sawn-through trace on the edge of a chip.

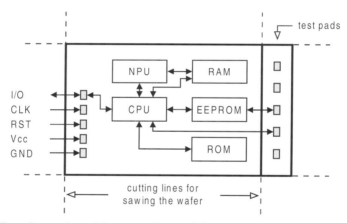

Figure 8.30   One of several possible ways to irreversibly remove the test pads used for testing the CPU and memory of a Smart Card microcontroller

Dynamic analysis and defense: tapping the memory busses of the microcontroller
Before the busses between the CPU and the memories of the microcontroller (ROM, EEPROM and RAM) can be tapped, the chip must be exposed and the passivation layer on the top surface of the chip must be removed. The passivation layer protects the chip against oxidation on the one hand, but it also protects the chip against attack, since its integrity is monitored by sensors. According to Anderson and Kuhn [Anderson 96b], it can be removed by etching with hydrofluoric acid. A laser cutter[9] can be used just as well to selectively cut openings in the passivation layer at the necessary locations.

After the passivation layer has been removed from the entire surface of the chip, or only from selected locations, it would be at least theoretically possible to make contact with the

---

[9]   A laser cutter is a device for drilling and cutting using a high-power laser beam. It has an accuracy of a fraction of a micron.

address, data and control busses for the memory using microprobe needles. If it is possible to make electrical connections to all the lines of these three busses, it is very easy to address the individual memory cells and to read any desired regions of the ROM and EEPROM. The chip does not have to be powered for this, and any desired type of connection jig can be used. The consequences of a successful attack using this method would be serious, since in principle it would make all the secret data in the non-volatile memory readable.

This method could be extended by making connections to the busses and then operating the chip in the normal manner. In this way, it would be possible to eavesdrop on the complete data traffic between the CPU and the memories, and this could be recorded using a sufficiently fast logic analyzer.

As already indicated, it is very difficult to make electrical contact with the individual traces on the chip. The number of connections needed for this attack is 16 for the address bus, 8 for the data bus and 1–4 for the control bus. In total, at least 25 simultaneous connections would have to be created between an external analysis computer and the traces on the chip. Even with modern micromanipulator technology, this is currently not possible, due to the very small dimensions of the semiconductor structures. However, it would be possible to use a focused ion beam generator, which is a tool that is commonly used in the semiconductor industry, to implant a sort of electrically conductive contact surface for each bus line. These then could be used as contact point for a microprobe needle. The effort required for this is however enormous.

Even if an attacker succeeded in making these connections, he would still have to determine how the busses have been scrambled before he could read the data successfully. This is because the individual bus traces are not arranged on the chip in an orderly fashion next to each other, but are instead arranged in an externally unrecognizable manner.

If markedly improved technology in the future should make it possible to make connections to the busses of current microcontrollers, this would probably not have any affect on security, since by that time the semiconductor structures will have become significantly finer than they presently are. In addition, micromechanical technology will probably always lag behind that of semiconductor technology, which is based on optical processes. This means that, even in the future, this sort of attack will probably not by suitable for significantly weakening the security of Smart Cards.

Dynamic analysis and defense: measuring the current consumption of the CPU
Already in 1995, in the first edition of this book, the following statement appeared at this point: 'The design of the processor is also critical with regard to security. A Smart Card processor must have nearly the same current consumption for all machine instructions. Otherwise, conclusions can be drawn regarding the instruction being processed, based on the current consumption. A certain amount of secret information can be deduced from these conclusions.' The fact that conclusions can be drawn about the instructions being executed by a processor, and even the data being processed, by analyzing the current consumption of the processor while the instructions are being executed, was thus already well known for years when Paul Kocher, Joshua Jaffe and Benjamin Jun published a paper on simple power analysis (SPA) and differential power analysis (DPA) in June of 1998 [Kocher 98].

The working principle of simple power analysis is relatively uncomplicated. The current consumption of a microcontroller is determined by measuring the voltage drop across a resistor connected in series with the power supply. The measurements are made with high time resolution using an analog-to-digital converter. With a high-performance processor, such as a Pentium or a PowerPC, no conclusions could be drawn about the instructions being executed, due to the complexity of the internal processes. The relatively simple designs of the 8051 and 6085 CPUs used for Smart Card microcontrollers, on the other hand, result in measurable and thus interpretable variations in the current consumption, depending on the instructions and data that are being processed.

Differential power analysis can reveal even finer differences in the current consumption of a microcontroller than can simple power analysis. With the DPA technique, the current consumption is first measured while the microcontroller is processing known data and then while it is processing unknown data. The measurements are repeated many times in order to eliminate the effects of noise by taking average values. The differences are calculated once the measurements have been completed, and conclusions regarding the unknown data are drawn from the results.

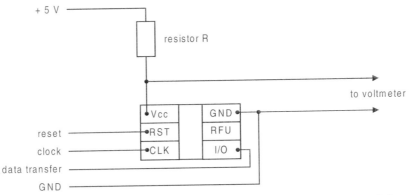

Figure 8.31   Circuit diagram of the connections to a Smart Card microcontroller needed to make simple current measurements using a series resistor

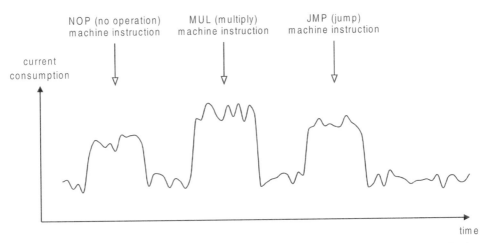

Figure 8.32   Simplified representation of variations in the current consumption of a Smart Card microcontroller with the execution of several different machine instructions. The current consumption may depend on both the data being processed and the machine instruction being executed.

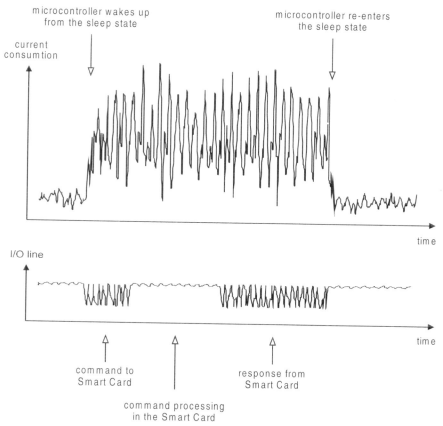

Figure 8.33 Simplified representation of the current consumption of a Smart Card microcontroller in the idle state and the current variations during operation. Based on the current drawn by the microcontroller, it is possible to recognize when the microcontroller awakes from the sleep state following the first falling edge on the I/O line, following which it exhibits a continuously varying power consumption that depends on the machine instruction being executed.

In the paper by Kocher *et al.*, 'high-order differential power analysis' (HO-DPA) is mentioned as a further extension of DPA. This involves measuring not only the current consumption of the microcontroller, but also other quantities that depend on the program being executed by the processor, such as the electromagnetic radiation of the chip. The measurement data collected in this manner, for both known and unknown data, can then be used in the same way as in the DPA technique to calculate differences that can be used to compute the secret data.

These three types of power analysis for Smart Card microcontrollers represent very serious forms of attack on hardware and software that has not been protected using suitable countermeasures. This is because the current consumption of some microcontrollers is definitely dependent on the machine instruction being executed as well as the data being processed by the instruction. In addition, the cost and complexity of the equipment needed for a successful attack using this method is relatively limited. However, there is a series of effective countermeasures, which are based on the one hand on suitably improved hardware and on the other hand on modified software.

The simplest hardware solution is to incorporate a fast-acting voltage regulator in the chip, which utilizes a series sense resistor to ensure that the current consumption of the microcontroller is independent of the machine instructions and data. Artificial current-noise generators on the chip are also an effective solution. A technically more complicated solution consists of modifying the semiconductor design of the processor such that it always has a constant power consumption.

There is presently an immense spectrum of possible software countermeasures. We can briefly describe some representative examples here. The simplest approach is to use exclusively machine instructions that have very similar current consumptions. In this case, machine instructions whose current consumption is significantly different from the average level are not allowed to be used in the assembler code. An additional solution is to have several different procedures for performing the same computations in cryptographic algorithms, and to randomly select one of these procedures each time it is actually needed. This makes it considerably more difficult for the observer to recognize a correlation between known and unknown machine instructions or processed data. In order to make it more difficult to obtain the data that must be acquired before a power analysis can be successfully carried out, all keys should be protected by irreversible retry counters. In addition, it is necessary to block free access to all commands (such as INTERNAL AUTHENTICATE) that can be used to pass any desired data through a cryptographic algorithm in the Smart Card. If commands of this sort absolutely must be used for some reason, the Smart Card must test the authenticity of the terminal before executing them. This restriction on the use of the available commands also makes it more difficult to collect reference data for a subsequent power analysis.

Analysis and defense: measuring the electromagnetic radiation of the CPU
It is at least theoretically possible to draw conclusions regarding the internal processes of the Smart Card microcontroller from measurements of its electromagnetic radiation, in the same manner as with differential power analysis. Magnetic fields with small dimensions and strengths can be measured using SQUIDs (superconducting quantum interference devices). However, this is technically enormously complicated, and the knowledge of the internal structure of the semiconductor device that is indispensable for this method is not generally available. In addition, ICs can be very effectively protected against this sort of attack by stacking several traces on top of each other, so that even if a magnetic field can be measured, it is not possible to determine which of the tracks is actually carrying the associated current.

Manipulation of the Smart Card microcontroller

Manipulation and defense: altering the memory contents of the Smart Card microcontroller
Directly reading the memory contents of a microcontroller is a possible attack scenario whose danger can be seen at first glance. A similar scenario that is almost as strong a form of attack is intentionally altering the data contents of a Smart Card microcontroller memory. This does not mean scattering random errors in the computation process of a cryptographic algorithm, which is used as the basis for differential fault analysis (DFA), but instead selectively changing the values of certain bits or bytes in the ROM or EEPROM.

Non-selective changes in all types of memory can be produced by (for example) exposing the module to X-rays or shining ultraviolet light on the exposed chip. EEPROM cells can be discharged by exposing them to ultraviolet light, which causes their contents to take on the value of the lowest-energy state. This process is exactly the same as erasing a conventional EPROM with an ultraviolet lamp. However, it cannot reasonably be used for an attack, since the attacker has no control over which EEPROM cells are switched.

The ultraviolet lamp can however be replaced by a collimated beam of light, or light from a laser, and this can be focussed to a fine point. This could certainly be used to alter the contents of individual memory cells. The advantage of using a laser is that it can supply enough power to also modify the contents of ROM cells. A focussed ion beam can also be used in a similar manner to change memory cells.

The changes that are possible can certainly be used for theoretically effective attacks. For example, the random number generator could be manipulated such that it no longer produces random numbers, but instead always supplies the same value. If this were possible, the authentication of the terminal by the Smart Card could be broken by a replay attack using a previously employed value.

Other types of attack could also be definitely undertaken if the contents of specific memory bits could be intentionally modified. For example, all S boxes of the DES algorithm could be intentionally altered to a uniform value of zero or one. This would mean that the DES would no longer be an encryption algorithm, but only a linear transformation [Anderson 96a].

If the exact location of the DES key in the EEPROM is known, and it is also possible to modify individual bits in the EEPROM (using focussed ultraviolet light, for example), it is naturally possible to utilize these conditions to mount an effective attack. The attack consists of setting an arbitrary bit of the key to 0 and then calling a command that employs the DES algorithm with the modified key. If the return code indicates a parity error in the key, the bit that has been modified was originally set to 1, while if no parity error is reported, the bit was already set to 0. The same procedure is then followed for the remaining 55 bits of the key, with the result that the secret key is known [Zieschang 98].

Many other types of attacks along the same lines are possible, such as selectively modifying program processes or altering pointer values. These attacks may look very simple and attractive on paper, but it would be very difficult to carry them out in actual practice. This is because the necessary conditions for a successful attack are not exactly easy to achieve, with the result that this type of attack remains an interesting but theoretical concept.

In order to alter bits selectively, an attacker must have detailed knowledge of the physical addresses of data and program code in the memory, and must also know the scrambling scheme used for the memory in question. In addition, all data and program routines that are important with regard to security are protected using checksums that are recalculated each time before using the data or routine. An attacker would therefore also have to selectively modify the checksum to match the modified data. You should also not overlook the fact that all protective layers covering the memory in question must be neutralized before any manipulation can take place. All of these considerations together reduce the attractiveness of this type of attack to almost nothing, even though it must be admitted that it sounds very attractive in theory.

8.2.4.2 Attacks on the logical level

The main prerequisite for attacks on the security of a Smart Card at the logical level is an understanding of the communications and information flow between the terminal and the Smart Card. It is more important to understand the software processes than to understand the processes that take place at the hardware level. In terms of information technology, the sample scenarios that are described here are located one level above attacks that primarily exploit the characteristics of the hardware.

Attack and defense: dummy Smart Cards

Probably the simplest imaginable type of attack is to use a Smart Card that has been custom programmed and includes additional logging and analysis functions. Up until a few years ago, this was almost unfeasible, since only a few companies could obtain Smart Cards or the microcontrollers used for making them. Nowadays, though, Smart Cards and configuration programs can be openly purchased from a number of firms. This naturally increases the options that are open to an attacker. Even without this, with some effort it is possible to assemble a working Smart Card using a plastic card, a standard microcontroller in an SMD package and a bit of dexterity. Such a card can at least be made to imitate the electrical interface of a real Smart Card and to behave the same way for data transfers. New possibilities are also offered by the Java technology for Smart Cards, which makes it easy to generate programs and load them into dummy cards.

With such a dummy card, it would be possible to log at least a part of the communications with a terminal and subsequently evaluate this information. After several attempts, it would probably be possible to execute part of the communications in exactly the same way as a genuine Smart Card.

Whether this can be put to advantage is doubtful, since all professionally designed applications have cryptographic protection for important activities. As long as the secret key is not known, the attack will not go any farther than the first authentication. Such an attack could be successful if the secret key is known, or if the complete application runs without any cryptographic protection. If such an application should exist, it is very doubtful that any advantages that could be gained by this type of attack would be significant enough to justify the effort needed to mount the attack.

Analysis: determining the command set of a Smart Card

The instruction classes and commands that are supported by a Smart Card are of course not often published, but it is very simple to determine what they are. This is more interesting with regard to completely determining the command set of a Smart Card than it is for an attack on the security of the Smart Card. It is however conceivable that an attack could be constructed on the basis of this information.

The structure of the procedure used to determine the command repertoire is illustrated in Figure 8.34. The first step is to generate a command APDU and send it to the Smart Card using a freely programmable terminal. The class byte in the APDU is changed for each APDU to cover the range from '00' to 'FF'. As soon as a return code other than 'invalid class' is received, the first valid class byte has been determined. There are usually two or three valid instruction classes, which can then be used to try out all possible instruction bytes in the next round. Command APDUs with various instruction bytes are sent to the Smart Card, and those that yield a return code other than 'unknown instruction' are noted. Given suitable software in the terminal, this method can determine which commands are supported by a particular Smart Card in one to two minutes. To a certain degree, some of the possible parameters of the commands so identified can also be determined in a similar manner.

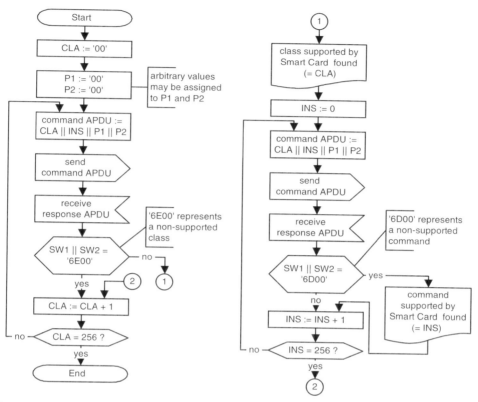

Figure 8.34 Basic procedure of an exhaustive search for all commands supported by a Smart Card operating system. This procedure will identify all supported commands only if command calling is not controlled by a state machine. Its works on the principle of systematically trying out all class byte (CLS) and instruction byte (INS) codes in turn, with no regard to any command contents (secure messaging, logical channels, V_{pp} control and so on).

The illustrated algorithm can be made considerably faster by using only class byte codes that are allowed by the ISO/IEC 7816-4 standard, and allowing the instruction byte to be an indexed variable. This strongly reduces the number space of the class byte by taking secure messaging and logical channels into account. A similar improvement can be made by using only even-valued instruction bytes, since as you know, odd-valued codes only contain the V_{pp} control information, which is no longer necessary.

The reason that this simple search algorithm for instruction classes, commands and parameters can be so effective is that practically all command interpreters in Smart Card operating systems evaluate received commands by starting with the class byte and working through the following bytes. The process is terminated as soon as the first invalid byte is recognized, and a suitable return code is generated and sent back to the terminal.

However, this procedure will only work if the Smart Card does not have a global state machine that monitors the command sequence. If such a state machine is present, it is at least possible to use this procedure to determine the command sequence in a step by step manner.

The advantage that an attacker can derive from this procedure may not appear to be that great, since the command set is normally not secret. However, it does at least provide a simple and fast method to determine all the commands. It can also be very useful in determining whether the producer of the operating system has incorporated any undocumented commands in the software.

Attack: tapping into the data transfer
A slightly modified Smart Card can be used to eavesdrop on the data transfer during a session and to manipulate the data if desired. The modifications consist of gluing an insulated dummy contact on top of the I/O contact surface, so that the original I/O interface is no longer connected to the I/O contact. The new (dummy) contact and the original I/O contact are connected to a fast computer. With suitable programming, this computer can delete or insert any desired data within the communications between the terminal and the Smart Card. If the computer is sufficiently fast, neither the terminal nor the card will detect any difference between normal and manipulated communications.

It is clear that the course of a session can be very strongly influenced using this method. Whether an attacker can derive any benefit from this method depends primarily on the application in the Smart Card. A well-known design principle says that eavesdropping on the communications, deleting data and inserting data must not be allowed to impair security. If this principle is not observed, an attacker can certainly obtain an advantage using this method. There are known cases of fraud using simulated memory cards.

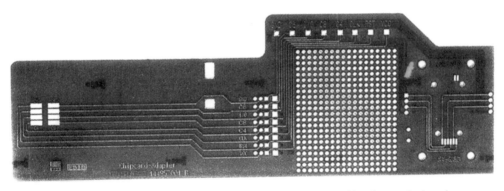

Figure 8.35 An adapter that can be used to extend a Smart Card outside of a terminal enclosure so that measurements can be made on the card. The eight contacts can be seen on the left, while an area for constructing electronic circuits can be seen on the right.

In order to provide protection against this type of attack, some terminals have shutters that cut off any wires attached to the Smart Card. Secure messaging can also be used very effectively here to allow any manipulation of the data during the data transfer to be reliably detected.

Many terminals may be used only under supervision, which makes it difficult to use manipulated cards with leads to an accompanying computer in such terminals. In summary, this type of attack can be regarded as theoretically very interesting and quite promising, but in practice as improbable and above all not too likely to be successful.

Attack and defense: cutting off the power supply

A type of attack that was successful with many Smart Cards until recently is to cut off power to the card at a certain time while a command is being executed. The background for this attack is that, with conventional programming, all write operations on EEPROM pages are carried one after the other. If the programmer has not been clever in arranging the order of the writing operations, cutting off the power at the right time can benefit the attacker.

This can be briefly illustrated using a very simplified example. In an electronic purse application, if the balance is increased before the log file is updated when a purse loading command is executed, at attacker would have a good chance of being to load a Smart Card for free. He would only have to switch off the power at the right time, or jerk the card out of the terminal with millisecond precision (!). The contents of the purse balance would then have been changed to the new value, but there would no log record for this transaction and no response to the command. With simple electronic purse systems in the past, such an attack was certainly a real possibility.

In order to determine the exact time to terminate processing, the attacker only has to use an electronic counter to count the number of clock pulses between the time when the command is sent, and then make a series of experiments with increasing clock counts until the proper time is determined. It hardly needs to be said that the entire procedure can be more or less automated using a computer.

	purse balance	purse balance file (in binary notation)
1. current purse balance	100 DM	°0110 0100°
2. deduct 10 DM	90 DM	
3. erase the EEPROM	255 DM	°1111 1111°
4. write the new purse balance	90 DM	°0101 1010°

Figure 8.36 Procedure for writing a new balance in an electronic purse. In this example, it is assumed that the erased state of the EEPROM represents a 1. This means that, due to the way the EEPROM works, it is necessary to erase the entire EEPROM page (which means setting all of its bits to 1) if only one bit in the page must be changed from 0 to 1. In this example, if the power for the Smart Card is cut off exactly after the EEPROM has been erased, which means after step 3, the purse balance would be set to its maximum value and the attacker would have created money. This can however be reliably prevented by using atomic processes.

Although though this type of attack sounds interesting and easy to copy, there are in practice several effective countermeasures. The simplest approach is to use a well-chosen sequence for the EEPROM write instructions. The EN 1546 standard for multi-sector electronic purses is well worth examining in this regard, since all the electronic purses described in this standard are explicitly protected against this sort of attack.

However, even a perfectly ordered sequence of writing operations cannot by itself achieve absolute protection. This can be made clear using another example. When the electronic purse of our previous example is being loaded, it may be necessary to erase the EEPROM before the writing process. If the erased state of the EEPROM corresponds to the maximum value of the purse balance, which by the way is usually the case, then it is only necessary to cut off the power at the right time to have the purse loaded to its maximum value. The suitable moment is when the erasing operation has just been completed and the writing operation has not yet been started.

Operating system designers know an effective countermeasure for this type of attack, which is the use of atomic processes as described in detail in Section 5.9. The characteristic of an atomic process is that it is indivisible, which means that it is executed either completely or not at all. This provides fully adequate protection against the type of attack just described. Even the optimally ordered EEPROM writing operations described in the EN 1546 standard require atomic processes in several locations to prevent this type of attack from being realized.

Attack and defense: current analysis during PIN comparisons
A technically very interesting attack on comparison features, such as PINs, can be carried out using a combination of physical measurements of a parameter and variation of logical values. This type of attack relates to all mechanisms in which data are sent to the Smart Card and compared in the card with corresponding values, with a retry counter being incremented according to the result of the comparison.

The attack works on the principle of measuring the current drawn by the card, for example by measuring the voltage drop across a resistor in the Vcc lead. If a suitable command containing the comparison data is sent to the card, it is possible to see from the current measurement whether the retry counter has been incremented, even before the return code has been received. The comparison value can be determined using this method if the return code is sent before the retry counter is written when the result of the comparison is positive. This can be done by sending all possible variations of the comparison value to the Smart Card, and cutting off power to the card before the retry counter has been incremented if the result is negative. A positive result can be clearly recognized by the associated return code, which is sent before the retry counter is written.

There are two basic ways to defend against this type of attack. The simplest defense consists of always incrementing the retry counter before making the comparison, and then reducing its value afterwards if necessary. Regardless of when the attacker cuts off the power to the card, it is not possible to gain any advantage, since the retry counter will have already been incremented. The second defense is more complicated, but it satisfies the same protective function. In this approach, the retry counter is incremented following a negative comparison and written to an unused EEPROM cell following a positive comparison. Both of these write accesses occur at the same time within the process, so the attacker can draw no conclusions with regard to the result of the comparison. He learns the result of the comparison only after receiving the return code, and at this point it is too late to prevent a write access to the retry counter by cutting off the power.

Attack and defense: timing analysis of PIN comparisons
Programmers always give a lot of attention to making programs execute as quickly as possible. Normally this is also an important consideration. However, the fact that the execution time of a process has been minimized can be used for an attack that definitely has a good chance of succeeding. If a PIN is sent to a Smart Card for comparison, the responsible comparison routine normally compares the PIN that it has received with the stored PIN value byte by byte. A programmer who is not security-conscious would program this routine such that the first difference between the two compared values would cause the

routine to immediately terminate and jump back to the calling program. This causes minute differences in the execution time for the comparison, which can definitely be measured using suitable equipment (such as a storage oscilloscope). This information could be used by an attacker to determine the otherwise secret PIN code in a relatively simple manner.

This type of attack was still effective with Smart Cards up to a few years ago. Nowadays, this is a known type of attack, and the comparison routines are constructed such that all digits of a PIN are always compared. This means that there is no time difference between positive and negative comparison results.

Protection: noise-free cryptographic algorithms
The security of a Smart Card application is based on the secret keys used for the cryptographic algorithm. In order to perform certain types of access or take certain actions with the card, the terminal must always first authenticate itself with the help of a secret key. It is understandable that the authentication of the terminal by the card is an interesting target for an attacker. The authentication of a Smart Card by the terminal is not interesting with regard to an attack on the card, since a Smart Card can be manipulated as desired using a (dummy) terminal.

The Smart Card authenticates the terminal by sending it a random number, which the terminal then encrypts and returns to the card. The Smart Card then performs the same encryption and compares the result with the value received from the terminal. If they match, the terminal has been authenticated and it receives a corresponding return code. If the authentication fails, the card sends a different return code. The starting point for the attacker is an analysis of the processing time between when the command is sent and when the response is returned by the Smart Card.

In the early 1990s, there were still some cryptographic algorithms in use whose execution times were significantly dependent on the key and plaintext involved. Based on the reduction in the key space that this represents, an attacker can use a brute force attack to search for the secret key. How long the search will take depends strongly on how noisy the algorithm is. The larger the time differences are, the smaller is the key space, and thus the easier and faster the search for the key. If the exact implementation of the algorithm in question on the target computer is known, this information can also be included as a reference for generating the timing tables. This sort of attack was made public under the name 'timing attack' in a publication by Paul Kocher in 1995 [Kocher 95], which is primarily concerned with time dependences of the RSA and DSS algorithms.

In principle, a timing analysis is very dangerous threat to the security of a Smart Card. However, since this type of attack has been known for a relatively long time, all present-day Smart Cards employ only noise-free cryptographic algorithms, which means that the time taken for encryption and decryption is independent of the input values. This blocks this sort of attack. However, the programmer has conflicting interests in this regard, since a noise-free algorithm usually requires more program code and is always slower than the noisy version. The reason for this is that a noise-free algorithm must be designed such that the path through the program is equally long for all plaintext data, ciphertext data and keys. This means that the longest necessary path is the reference value, and all other paths must be suitably modified to match this length.

To provide additional security, in some applications all authentication keys have their own retry counters, so that only a limited number of unsuccessful authentications can be carried out. Once the retry counter has reached its maximum value, the Smart Card blocks all further attempts at authentication.

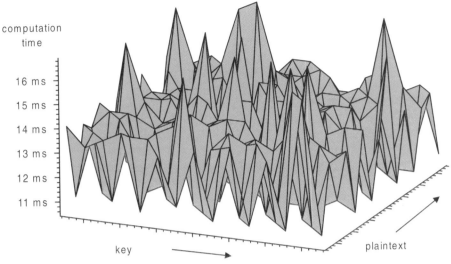

computation
time

16 ms

15 ms

14 ms

13 ms

12 ms

11 ms

key   plaintext

Figure 8.37   Example of the effects of ciphertext and plaintext data on a noisy cryptographic algorithm. The diagram shows a portion of the plaintext and ciphertext space. It was generated using an old implementation of the DES algorithm, with 100,000 iterations per measurement value.

Manipulation: differential fault analysis (DFA)

As is well known, the operation of electronic devices can be considerably impaired when they are exposed to electromagnetic disturbances. For instance, a mobile telephone can cause the processors of many types of small computer-controlled appliances to crash. The cause lies in the memory cells, whose contents can be altered by the high-frequency AC fields.

In 1996, Dan Boneh, Richard DeMillo and Richard Lipton published a study [Boneh 96] that describes a theoretical model for the determination of secret keys for asymmetric cryptographic algorithms by introducing scattered hardware faults. Since the three discoverers of this method worked at the Bell Communications Research (Bellcore) Laboratories at the time, this type of attack is often called the Bellcore attack.

Only two months later, Eli Biham and Adi Shamir published an extension of the Bellcore attack called *differential fault analysis* (DFA) [Biham 96], which also included symmetric cryptographic algorithms such as the DES. This meant that, at least in theory, many Smart Card applications were threatened with a new and serious form of attack.

The basic principle of both of these attacks is relatively simple. In the first step, an arbitrary plaintext is encrypted using the key to be broken, and the resulting ciphertext is saved. Following this, the operation of the card is disturbed while it is processing the cryptographic algorithm, for example by exposing it to ionizing radiation or high-frequency radiation, so that a single bit of the key in an arbitrary location is altered while the computation is being carried out. This yields a ciphertext that is incorrectly encrypted, due to the changed bit. This process is repeated many times, and all the results are saved for analysis. The rest of the process of determining the value of the key is purely mathematical. It is comprehensively described in the papers just mentioned.

The strength of this attack is primarily due to the fact that it is not even necessary to know the location of the altered bit in the secret key. Biham and Shamir state in their publication that with a single corrupted key bit, 200 ciphertext blocks are sufficient to compute the value of the secret DES key. If a triple DES (with a 168-bit key) is used in place of a simple DES, the number of required ciphertexts does not increase significantly. Even if more than one bit is altered, this attack remains effective; the only consequence is that more incorrectly encrypted ciphertexts are needed.

However, this type of attack is in practice not as simple as it sounds. If at all possible, only one bit should be altered, or at least only very few bits. However, if the entire microcontroller is simply bathed in microwave radiation, in most cases so many bits will be altered that the processor will generally crash without a chance. Consequently, attempts are made to induce the processor to make isolated computational errors by injecting specially prepared glitches[10] into the power or clock supplies. If the filter on the associated input leads cannot neutralize these glitches, they can produce the desired processing errors.

However, a Smart Card is not totally helpless in the face of either a Bellcore attack or DFA, as long as suitable preventative measures are employed. The simplest defense is to just compute the cryptographic algorithm twice and compare the two results. If they are identical, there has been no external attempt to alter any bits. This assumes that intentionally introduced scattered errors can never alter the same bit twice in a row. This is also a realistic assumption, since if it were ever to become possible to selectively alter specific bits in a Smart Card processor, then attacks that are much simpler and faster than DFA would be possible. The main disadvantage of double computation is the additional time that it requires, which can cause problems. This applies primarily to attacks on time-consuming asymmetric cryptographic processes such as RSA and DSS.

Another effective defensive measure against differential fault analysis can be achieved by always encrypting different plaintexts. The simplest solution is to prefix the plaintext to be encrypted with a random number. This means that the cryptographic algorithm always encrypts different data, which makes DFA no longer possible.

In summary, the Bellcore attack and differential fault analysis are unquestionably dangerous types of attack that can succeed with Smart Cards that do not have suitable protective measures. However, all Smart Card operating systems and applications were modified to protect them against these attacks shortly after they became known, so that neither the Bellcore attack nor DFA currently represents a serious threat.

Protective elements: Smart Card operating systems

Protective mechanisms in the hardware form the basis for the protective mechanisms in the operating system software. No potential weakness may be overlooked, since the three parts of the protective mechanisms – hardware, operating system and application – are linked in a logical AND relationship. This is comparable to a chain, in which the weakest link determines its breaking strength. If a particular mechanism fails in a Smart Card, the complete security collapses. In particular, the operating system is in turn the basis for the actual application, whose information and processes must be protected.

The following paragraphs deal specifically with protective measures against typical attacks, rather than general Smart Card security functions. However, most of these functions also contribute significantly to operational security and protection against attacks. For this reason, you are explicitly referred to the appropriate sections of Chapter 5.

---

[10]   A glitch is a very brief voltage dropout or voltage spike.

Protection: hardware and software tests following a reset

When the operating system is initialized, at least the most important parts of the hardware must be tested to see if they are in proper working order. A test of the RAM is for example indispensable, since all access conditions are stored in the RAM as long as the chip is operating, and the failure of a single bit could cause the entire security to collapse. The calculation and comparison of checksums for the most important parts of the ROM and EEPROM is equally necessary. The CPU is at least implicitly tested by sending the ATR, since the bulk of all possible machine instructions must be executed faultlessly for this to be possible. Explicit tests of the CPU, or any NPU that may be present, can usually be limited to sample testing, since completely testing all functions for flawless operation would take too much time and code.

If the operating system discovers a hardware or checksum error, there are two possible ways to proceed. The first option is for the software to immediately jump to an endless loop, which means that the ATR cannot be sent and no subsequent commands can be received. The main disadvantage of this is that the cause of this behavior cannot be recognized from the outside. You cannot tell, for example, whether the problem is a broken bonding wire, a fractured chip or a checksum error in the EEPROM. The second option, in which the Smart Card attempts to send a special ATR before switching itself off or entering an endless loop, is thus better. The error ATR at least gives the outside world an indication of what has happened inside the Smart Card. However, it must not be overlooked that just sending an error ATR requires a largely functional CPU, a few bytes of RAM and several hundred bytes of program code in the ROM.

Protection: layer separation in the operating system

Layer separation, with clearly defined transfer parameters between the individual layers, is a sign of a stable and robust Smart Card operating system. The consequences of design or programming errors that may be present in the operating system are minimized by a clean separation of the layers within the operating system. This naturally does not mean that such errors will not occur, but the effects of the errors will not be as extensive as in an operating system that is programmed in very compact and condensed code. Separation of the layers makes it difficult for any error that occurs in one layer to propagate to other layers.

Protection: controlling data transfers

Another very important element of security is to control the data transfer process in order to protect the memory against unauthorized accesses. The entire communication to and from the Smart Card takes place via an I/O interface that is controlled by the operating system. No other form of access is possible. This represents an effective form of memory protection in the Smart Card, since this ensures that the operating system always retains control of accesses to the memory regions.

The transmission protocol, which is controlled by the transport manager, must intercept all possible incorrect inputs. There must be no possibility of influencing the data transfer process by manipulating transfer blocks in order to cause data to be sent from the memory to the terminal without proper authorization.

Protection: checksums for important memory contents

The file structure, and in particular the file headers (or in other words the file descriptors), should be protected by means of checksums. This makes it possible for the operating system to at least detect any unintentional change to the data stored in memory. This requirement is especially important in light of the fact that the object-oriented access conditions for each file are stored in this part of the file.

All memory regions of the EEPROM that are vitally important for the Smart Card operating system must be protected using checksums (EDCs). Whenever such a region is accessed or the code it contains is called to be executed, the consistency of its contents must be verified before the access or code execution is allowed to proceed.

Protection: encapsulation of applications

Some operating systems encapsulate the individual DFs that contain the applications along with their files, so that individual applications are isolated from each other. This is however based on software protection alone, with no support from the hardware of the chip. The level of protection is thus not as high as it could be. Nonetheless, even this software approach to application encapsulation can be very advantageous in case of an error, since it makes it impossible for the file manager to exceed the boundaries of a DF without an explicit prior selection. The consequences of a memory error in a file are thereby at least limited to the associated DF.

If a memory management unit (MMU) is present in the hardware to provide hardware support to the operating system, different applications can be fully isolated from each other. In this case, even manipulated software within an application cannot obtain unauthorized access to the memory regions of other applications.

Protection: Camouflaging the activities of the operating system

Whenever data have to be written to the EEPROM, the charge pump in the chip must first be switched on. This increases the current consumption of the chip, which can easily be detected using simple measuring equipment. The design of the operating system must thus take into account the fact that it is always possible to determine externally when EEPROM write accesses occur. The software in the Smart Card must prevent an attacker from being able to take advantage of this.

It is therefore very important that it should not be possible to draw any useful conclusions about the processes and decisions in the machine program from measurements of the current drawn by the card. It would for example be fatal if were possible to use such measurements to reliably judge the outcome of a PIN comparison before the command execution has been completed and the return code sent. This information could very easily be used to analyze the value of the PIN.

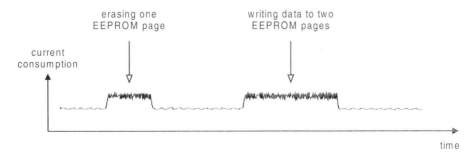

Figure 8.38   Approximate representation of the variation in the current consumption of a Smart Card when the charge pump is switched on

Protection: object-oriented access conditions

Early Smart Card applications were always based on two centrally managed access mechanisms. A disadvantage of centralized access controllers is that software or memory errors can affect the overall security of the Smart Card. Modern object-oriented file management systems, with access conditions stored in the individual files, have the advantage that only a single file is affected by a memory error, while the security of all other files remains intact. This is actually a fundamental property of all distributed systems. They are somewhat more difficult to program, but they provide significantly stronger security against attacks and errors due to their self-sufficiency.

Attack and defense: random number generators

The random numbers generated by the Smart Card are used in authentication to individualize a session, which means to make each session unique and different from all preceding and following sessions. The objective of this is to make it impossible to successfully replay data that has been obtained by tapping into a previous session. Another form of the attack would be to let the Smart Card generate so many random numbers that their sequence becomes predictable. Yet another possibility is to keep requesting random numbers from the Smart Card until the EEPROM memory of the random number generator no longer works properly, so that the same number is generated over and over again.

Any of these attacks could, if successful, bypass the authentication of the terminal by the Smart Card. Without exception, they work only with the first generation of Smart Cards. With modern operating systems, they will all fail. The cycle length of the random numbers that are generated is nowadays so great that the same random number never appears twice within the lifetime of an individual Smart Card. It is also no longer of any use to generate so many random numbers that problems start to occur with the EEPROM. If this happens, the generation of random numbers is simply blocked and any further authentication is thus prevented.

A high-quality random number generator must meet some additional requirements, such as producing non-predictable random numbers and having a long cycle length (the number of numbers that are generated before the generator repeats itself). In addition, all Smart Cards within a particular application must generate different random numbers. This may sound extremely obvious, but problems have repeatedly occurred in the past in this regard! This differing behavior is achieved by entering a starting value for the pseudorandom number generator when the Smart Card is initialized or personalized. This starting value is often called a *seed number*, in allusion to a biological seed that determines the growth of a plant. The design and evaluation criteria for pseudorandom number generators are extensively discussed in Section 4.9 ('Random Numbers'), along with methods to measure the quality of random numbers.

Protective elements of the Smart Card application

The protective mechanisms of the application are based on the corresponding mechanisms of the hardware and the operating system. The application is dependent on the extent to which these two lower levels fulfill their obligations with regard to protection, since it cannot correct for any errors in the hardware or the operating system. For example, if it is possible to read the contents of the EEPROM using an analysis procedure, even the most complicated and secure encryption processes are of no use at all, since the keys can be taken directly from the EEPROM by an attacker. An application must nevertheless be constructed such that the entire system is not compromised if an individual card is compromised.

Protection: simple mechanisms

In order to provide effectively protection against attacks, all mechanisms of an application should be designed to be as uncomplicated as possible, and they should always conform to the generally applicable principle of 'keep it simple'. To start with, this makes the implementation easier, and later on, it makes it easier to check the correct operation of the mechanisms and the effectiveness of protective mechanisms. It is extremely dangerous to assume that if something is only made complicated enough, it can thereby be protected against all possible attacks. As a rule, exactly the opposite is true. Complicated processes and mechanisms usually result in various things being forgotten or overlooked, which makes things that much easier for an attacker.

Fundamentally, the protective mechanisms provided by the operating system should always be utilized by the application. These have been tested for reliability, and the defense that they provide starts at a lower software level than that of the application. This is not intended to mean that an application does not need to have any protective mechanisms of its own, but the mechanisms that are already present in the operating system should always be used.

Protection: conservative access privileges

In addition to the principle of 'keep it simple', there is a second generally valid rule. This is that access privileges for the files and commands of a Smart Card should be granted as conservatively as possible. Access should be generally prohibited, and only allowed when it is absolutely necessary.

The advantages of this approach are that it makes it less likely that access to important data and commands will be granted unintentionally, and that it costs an attacker additional effort to obtain each piece of necessary information. This can considerably reduce the attractiveness of an attack, since it increases the overall amount of effort that is required.

Protection: state machines for command sequences

Attacking a Smart Card application is considerably more difficult if it is not possible to execute every command at any desired time an unlimited number of times. This can be realized by using a state machine that determines the allowed sequences of commands. If for example a mutual authentication of the terminal and the Smart Card is specified as the first required action, an attacker will have to overcome this protective barrier before he or she can execute any further commands.

Protection: double access security

The attacker's job is made considerably more difficult if the Smart Card files are protected not only by access conditions stored in the objects, but also by the use of a command state machine with predefined command sequences and allowable parameters. With this, the attacker cannot discover the specific features of the system by just trying each command or combination of commands in turn. If the command sequences are supervised by a state machine, only the commands defined in the application can be executed in the Smart Card. All other commands will be blocked in principle by the state machine. This considerably reduces the scope of the possibilities available to an attacker with regard to command manipulation.

Protection: various test levels

With bank notes, it has been standard practice for many years to support various test levels. This has to do with security features that can be checked by different groups of people or different types of machines, independent of each other. For example, many of the visual features, such as security threads and watermarks, can be checked by anyone on the street. For the checking the next level, you need an ultraviolet lamp in order to be able to recognize the fluorescent pigments in the paper. The features belonging to the next higher testing level are used by automated equipment to verify that the notes are genuine. A typical example is the infrared characteristics of the bank note. Yet other level of independent features is provided for tests performed by the central bank.

This entire system can easily be transferred to Smart Cards, with the logical consequence that not everybody or every piece of equipment can check all of the features. It is thus conceivable, for example, that a retail terminal for an electronic purse system could contain only some of the keys used for signature verification, rather than all of them. This does not weaken the system in a cryptographic sense, and it has the advantage that an attacker cannot compromise the entire system by learning the master key of a retail terminal. The only entity that knows all the keys in the system that are necessary for a complete transaction data set is the system operator, who thus can always recognize an attack by noting that the signatures are forged, and who can then introduce suitable countermeasures.

Protection: security features

Features incorporated into the microcontroller can offer additional security when a Smart Card is used. These are supplementary functional elements that are added to the microcontroller and that can be tested by the terminal in parallel with the software in the chip. Both analog and digital elements are used for this purpose. The security of these features is based on concealment, and the features are different for each application, which means that the chips are application-specific.

Protection: secure data transfers

There are certain risks involved in transferring data in potentially insecure surroundings. With relatively simple technical manipulations of the interface between the terminal and the card, it is possible to insert or delete almost any desired data within the normal data steam during a session. If this happens while sensitive data related to security are being transferred, an attacker could take advantage of this. In order to prevent this type of relatively simple and easily executed attack, a secure messaging procedure can be employed. However, complete encryption of data should be avoided as much as possible, with encryption only being used for the transfer of secret keys, for example. This is because the procedure requires the double encryption of the transferred data, which very sharply reduces the effective data transfer rate. An additional argument against complete data encryption has to do with data protection legislation. Almost all information that is written to the memory of a Smart Card is public. If this information is encrypted, nobody can then check what is actually written to or read from the card. In order to avoid any suspicions about the encrypted data, which in principle would be justified, the data should as much as possible remain unencrypted while it is being transferred.

Protection: error recovery functions

If a session is terminated for an undefined reason, or there are fundamental uncertainties regarding an earlier session, it is a major advantage if application-specific log files are present in the Smart Card. These files are maintained by the operating system, which updates them regularly during the session to reflect the current state of the application and any signatures or other data that may have been received from the terminal. The logged data are located in a cyclic file, in which the oldest record is always overwritten each time a new entry is made, so that the content of the oldest record is lost. If this file contains twenty records, for example, then information regarding the last twenty sessions can be stored for subsequent analysis of the course of each session. This information can be used to clear up a lot of uncertainties and to unambiguously clarify contested transactions and processes.

An additional argument for detailed log files in the Smart Card is the fact that they make certain error recovery functions possible. With a log file, it is possible to automatically restore the old state of the card (a *roll back*) in case the session is terminated in an undefined manner. This would otherwise require the exact procedure and sequence of events to be analyzed, possibly even with human help.

Protection: authentication

Unilateral authentication, which is well known due to its use with magnetic-stripe cards, ultimately amounts to nothing more than verification by the terminal that the card is genuine. A magnetic-stripe card, due to its passive nature, cannot verify the genuineness of a terminal. The introduction of Smart Cards has fundamentally changed this situation. Now the card can also test whether it has been inserted into a genuine terminal or is connected to a genuine background system. This has extensive consequences with regard to security, since it makes it possible for the card to also take active measures against unauthorized access attempts.

The possibilities that arise from the ability of the card and the terminal to perform a mutual authentication are substantial, and they are usually not exploited at anywhere near their full potential. A Smart Card should at minimum refuse to allow any further access attempts as long as the terminal cannot properly authenticate itself. This would make it impossible to undertake any sort of analysis of the Smart Card operating system in private, even if only to find out what commands are present.

Protection: on-line behavior

Terminals with integrated security modules can be used on their own to run applications with Smart Cards. Of course, periodic uploads and downloads to and from the background system are still necessary, but these usually occur only infrequently. With relatively large applications that have a large number of cards in circulation, however, it must at least be possible for the terminal to quickly make a connection to the background system if necessary, in order to provide direct end-to-end communication between a Smart Card and the background system. The importance of this increases with the size of the system and the advantages that an attacker could obtain by means of fraud. This is because a direct communications link to the Smart Card allows the background system to access the current database of the card, and if necessary to block the card. In addition, the keys in the background system are significantly more secure than those located in the many terminals in the field, even if the terminals have security modules. The background system can make very good statistical evaluations of the card data that it receives via sporadic end-to-end links to the Smart Cards.

All of these arguments are naturally particularly relevant to electronic purses based on Smart Cards. The 'urge' to go online can be triggered by random variables and timing windows that are stored in the Smart Card. An equally effective method is to use a counter in the card that demands an online connection with mutual authentication after a certain number of off-line transactions have taken place, or once a certain off-line transaction level has been achieved. At the end of the session, the background system can reset the counter or modify the values of the parameters that control the on-line behavior of the card.

Protection: black lists
The possibility of counterfeit Smart Cards being used in a system can never be completely excluded, no matter how well they may be protected against attacks. The Smart Card system must also incorporate effective mechanisms for protecting users by blocking stolen cards throughout the entire system. The procedures that are used for this purpose are strongly dependent on the application in question and the design of the system, but they can all be reduced to a few basic techniques.

In order to prevent forged or lost Smart Cards from being used, it is necessary to maintain lists in which either valid cards or invalid cards are noted on the basis of some unique feature. This is usually a number, such as the card number. From the viewpoint of impeccable system design, in which everything that is not explicitly permitted is implicitly prohibited, a list of valid cards would be the best. Such 'white lists' would however be far too large in large systems. This can be easily illustrated by noting that in a system with 10 million Smart Cards and an 8-byte card number, the white list would contain 80 MB of data.

This is why in practice black lists are used, in which only the blocked cards are noted. In the example just mentioned, the size of the list would be reduced to 800 kB if the number of blocked cards were one percent of the total. However, if it were necessary to block significantly more than one percent of the cards in the system due to attacks or loss, the size of the list would quickly become impractical even with this approach.

In order to further reduce number of data transfers and the amount of data that must be transferred between the entity that maintains the list and the entity that tests cards against the list, 'red lists' are sometimes also used. A red list identifies cards that are demonstrably forged, which should either be immediately confiscated or at least blocked for all further transactions. The number of entries in such a list lies in the two or three figures, even with large systems.

Smart Cards can be checked against these lists in real time with systems that work on-line. With systems that work partially or fully offline, updated black lists and red lists must be transferred to the terminals as often as possible. This should occur at least daily, since protective mechanisms that are based on blocking lists are otherwise no longer effective.

Attack and defense: computer viruses and Trojan horses
Until recently, computer viruses were entirely unknown with Smart Cards, since the downloading of program code was not technically feasible in the operational phase. Modern Smart Card operating systems, however, have mechanisms that allows program code to be downloaded to the Smart Card after it has been issued to the cardholder, and that also allows such programs to be executed. This means that, in principle, the necessary conditions for the existence of computer viruses in Smart Cards have been created. By definition, a computer virus is a program that can reproduce itself and thus spread to other computers. If such a program cannot reproduce itself, it is called a Trojan horse. Both types

of program have in common that under certain circumstances they can carry out unauthorized actions in the host computer. With a Smart Card, this could involve reading and sending out the values of secret keys.

In contrast to an ordinary PC, it is not a straightforward task to load a program into the memory of a Smart Card and then execute it. Security mechanisms have been provided to prevent the unauthorized running of programs. For example, it could be necessary to authenticate the terminal, depending on the application in question. In addition, the program code normally must be secured using at least a MAC or a digital signature. Some Smart Card operating systems also mutually isolate the memory regions of the individual applications by means of software or hardware, so that the applications in the Smart Card cannot influence each other. The result of these strong security measures is that quite likely neither computer viruses nor Trojan horses that are unintentionally downloaded when the card is already in use will be able to impair the functions or security of any applications.

Attack and defense: exhaustive key searches
One possible type of attack at the cryptographic level is an exhaustive search for a key. For this, the attacker needs a plaintext–ciphertext pair (or better yet, several pairs) and naturally the cryptographic algorithm that is used. He or she then encrypts the given plaintext with all possible keys in turn, until the given ciphertext is obtained. This key can then be tested with all other plaintext–ciphertext pairs on hand. If the correct encryption can be carried out in each case, then the key that has been identified is very likely the correct key. This procedure is basically suitable for all encryption algorithms, although it is not always the fastest method for determining the value of the secret key.

As early as 1993, Michael Wiener published plans for a special computer that was supposed to cost one million dollars and that could try out all DSS keys for a given plaintext–ciphertext pair within seven hours [Wiener 93]. This would make it possible to determine the value of a 56-bit DES key in 3.5 hours on average. A few years later, in 1997, the DES key for a plaintext–ciphertext pair made available by RSA Inc was determined in 97 days by systematic searching, using something more than 70,000 computers connected via the Internet [RSA 97]. The search rate during the final phase of this experiment amounted to around 0.7 percent of the DES key space every 24 hours.

Several different approaches are taken in practice to counter these attacks. The simplest measure, as is generally known, is to make the key space of the cryptographic algorithm so large that it is not possible to perform a systematic search within an acceptable length of time, even with very high computing power. This is why the DES algorithm is being replaced by the triple DES in more and more applications. The key space of the DES, when compared with currently available computing power, has simply become too small.

Another defense mechanism can be created very easily by constructing the application protocol such that pairs of plaintext and ciphertext do not occur. In most cases involving Smart Card applications, the data is not even encrypted, but simply provided with a MAC. Since the contents of multiple plaintext blocks are not reflected in the MAC in a straightforward manner, a brute force attack with a MAC is a great deal more arduous than the same type of attack on a plaintext–ciphertext pair.

If a random number is placed ahead of the plaintext in the Smart Card (*salted*) before the transferred data are either encrypted or used to calculate a MAC, the data to be encrypted are different every time the function is used. This means that the results are also different each time. This also makes in more difficult to perform an exhaustive search, since in many cases the random number does not have to be public. It could for example be a secret shared by the security module and the Smart Card. By the way, a random number that is prefixed to the data to be encrypted within the Smart Card also provides very good protection against attacks using differential fault analysis (DFA) and power analysis (PA), even if the random number is public.

It is also possible to make things a bit more difficult for the attacker by using dynamic keys (session keys), which change for each encryption operation. In this case, even if the attacker manages to determine the value of the key by some happy coincidence, this will not be of any use, since the key will be different for the following transaction.

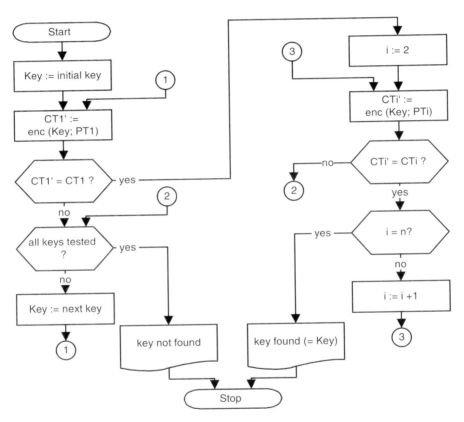

Figure 8.39   A basic procedure for performing an exhaustive search for a key, with a given cryptographic algorithm and several plaintext–ciphertext pairs, according to a proposal by James Massey [Massey 97]. The following abbreviations are used: CT = ciphertext; PT = plaintext; (Cti, Pti) = plaintext–ciphertext pair i; n = number of plaintext–ciphertext pairs.

9

Quality Assurance and Testing

Quality assurance, with its associated test procedures and methods, is particularly important in the case of Smart Cards. A Smart Card manufacturer must fabricate his products in very large numbers, with high quality and a low price. In contrast to other branches of the semiconductor industry, these products contain relatively complicated and sensitive microcontrollers together with software that mostly cannot be changed.

If we compare this situation with standard PC software, for example, the basic difference is obvious. In the latter case, it has become standard practice to replace the first issue of new software (usually identified by a '0' at the end of the version number) within a short time, ranging from a few weeks to at most one to two months, by revised and improved versions (with version numbers ending in 'a', 'b', 'c' and so on). This would be impossible with Smart Cards. The mask-programmed software is by nature fixed, and it is not possible to replace a large number of issued cards using any sort of recall campaign. Even with cards that are not used in the particularly sensitive area of financial transactions, such an action would cause lasting damage to the reputation of the card issuer, and the costs would be immense.

This is why quality assurance and testing are of fundamental importance in the production of Smart Cards. After the cards have been manufactured and delivered, it is simply impossible to 'stuff in' an improved version a short time later. This naturally means that a large amount of effort must be expended to manufacture a product that has as few errors as possible.

With regard to the various tests, a basic distinction must be made between qualification tests and production tests. Qualification tests are used to make a basic decision about whether the Smart Card in question can be used at all. These tests usually take place before the introduction of a new card body, chip, module or operating system. If the new or modified product meets the specified requirements, it is then qualified for production and can be manufactured in large numbers. After this, qualification tests are only performed infrequently on random samples.

A different sort of testing method is used for production tests. These tests are usually quick and straightforward, in order to satisfy the inescapable demand of mass production for short processing times and high throughput. They primarily involve only simple measurements of general mechanical and electrical parameters, together with sending suitable test commands to the Smart Card microcontroller.

Many test specifications for large Smart Card applications are primarily designed with an eye on interoperability between Smart Cards and terminals. A good example is the 'Subscriber Identity Module (SIM) Test Specification' in the GSM 11.17 specification, which occupies around 100 pages. It describes detailed tests for GSM Smart Cards that cover aspects ranging from the card body and general electrical parameters (including the supply voltage and current consumption) to data transmission protocols, commands and files. The GSM 11.17 tests are organized as follows:

- physical characteristics
- electrical signals and transmission protocols
- logical model
- security functions
- functions
- commands
- file contents

The organization of the individual tests in this specification is equally clear and practical. Each individual test consists of four parts. The first part contains a formal definition of the test and specifies its application. The second part lists the requirements to be satisfied, and the third part describes the objective of the test in detail. The last part specifies the actual test procedure.

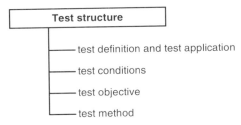

Figure 9.1   Basic organization of a GSM 11.17 test. This structure is so general that in principle, it can be used for all Smart Card tests.

9.1   Card Body Tests

There is presently only one international standard for testing cards with and without chips, which is the ISO/IEC 10373 standard. In Europe, there is also the EN 1292 standard, but this deals exclusively with Smart Cards and terminals, including their general electrical requirements. Standards relating to cards also often include individual tests and test procedures for checking the properties defined in the standard.

On the following pages, many of the usual tests and verifications for Smart Cards are briefly described in alphabetical order. The testing laboratories of card manufacturers usually have a repertoire of 120–150 different tests for cards.

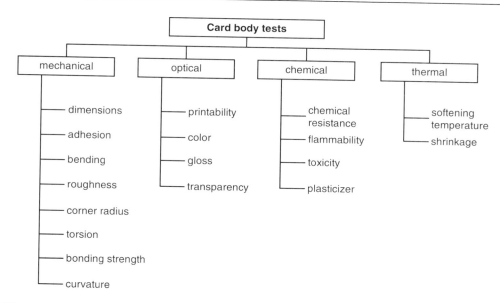

Figure 9.2   Classification of a selection of commonly used card body tests. A series of tests is necessary for each of the individual card elements (hologram, magnetic strip, chip and so on).

Standardized environmental conditions are a fundamental requirement for the test environment, which means that a temperature of 23° C ± 3° C and a relative humidity of 40 % to 60 % must be maintained in the test laboratory. The cards to be tested must be acclimatized to these conditions for at least 24 hours before the actual testing takes place.

Adhesion or blocking
(Basis: ISO 7810; test regulation: ISO/IEC 10373)
This test verifies whether the card's behavior changes when it is stored under certain environmental conditions. Five non-embossed cards are stacked together and uniformly subjected to a pressure of 2.5 kPa at 40° C with 90% relative humidity, for 48 hours. Following this, the cards are inspected for delamination, discoloration, surface changes and other visible changes.

Amplitude measurement
(Basis: ISO 7811-2; test regulation: ISO/IEC 10373)
This measurement verifies the signal amplitude and the resolution of the magnetic strip coding. A standardized read/write head that is passed along the magnetic strip at a precisely specified speed is used to make the measurement.

Bending stiffness
(Basis: ISO 7810; test regulation: ISO/IEC 10373)
In order to determine whether the card has the required bending stiffness, the left-hand side of the card is clamped to a depth of 3 mm, with the card facing downwards. The amount of bending is first measured with no load. A load of 0.7 N is then applied to the outer edge of the card, and the difference between the amount of bending under load the amount of bending with no load is calculated. The result indicates the stiffness of the card. The

bending stiffness test is often also performed at temperatures lower or higher than the usual testing temperature of 23° C.

Card dimensional stability and warpage with temperature and humidity
(Basis: ISO 7810; test regulation: ISO/IEC 10373)
Both the shape and the size of certain types of plastic change markedly in response to variations in atmospheric humidity. Consequently, the ability of the card to meet the standards must also be tested under these conditions. For this test, the card is placed flat on a surface and the temperature and humidity are varied. The testing conditions are –35° C, +50° C and +25° C at 5 % relative humidity and +25° C at 95 % relative humidity. The size and warping of the card are verified with respect to the standard values after it has been exposed to each of these conditions for 60 minutes.

Card dimensions
(Basis: ISO 7810; test regulation: ISO/IEC 10373)
This test measures the height, width and thickness of a non-embossed card. A force of 2.2 N is applied to the card, and its height and width are measured using a profile projector. For measuring the thickness, the card is divided into four equal rectangles, and the thickness of each rectangle is measured at the center using a micrometer at an applied force between 3.5 N and 5.9 N. The measured maximum and minimum values are both compared with the standard thickness.

Card warpage
(Basis: ISO 7810; test regulation: ISO/IEC 10373)
This test measures the amount of warpage of the card. The card is placed on a flat surface and the warpage is measured using a profile projector. This test is primarily intended to be used for cards that are stamped from a large sheet of base material.

Delamination
(Basis: ISO 7810; test regulation: ISO/IEC 10373)
The following test is only meaningful for multilayer cards, which, as you know, are assembled by laminating several layers of plastic. The top layer is separated from the core layer at some point using a sharp knife. Starting with this separation, the tester attempts to pull the two laminated foils apart. The necessary force is measured and compared to reference values.

Dynamic bending stress
(Basis: ISO 7816-1; test regulation: ISO/IEC 10373)
The dynamic bending test is illustrated in Figure 9.3. The card is flexed at a rate of 30 times per minute (0.5 Hz), either 2 cm across its length or 1 cm across its width (f). The card must remain undamaged after being flexed at least 250 times in each of the four possible directions (a total of 1000 bending cycles).

Dynamic torsion stress
(Basis: ISO 7816-1; test regulation: ISO/IEC 10373)
In the dynamic torsion test, the card is subjected to ±15° torsion about the longitudinal axis at a rate of 30 twists per minute (0.5 Hz). The standard requires 1000 torsion cycles without functional chip failure or visible mechanical damage to the card.

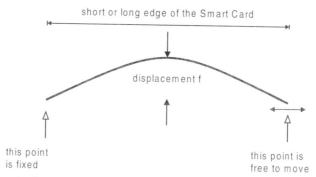

Figure 9.3   Schematic diagram of how the card is loaded for the dynamic bending test

Electrical resistance and impedance of contacts
(Basis: ISO 7816-1/2; test regulation: ISO/IEC 10373)
The electrical resistance of the contacts is an important criterion for reliable power supply and data transfer to the card's microcontroller. The resistance is measured using two test probes applied to two opposite corners of the smallest allowable contact surface rectangle, with a force of 0.5 N ± 0.1 N. The resistance between the two test probe contacts, which are gold plated and rounded to a radius of 0.4 mm, must be less than 0.5 Ω.

Electromagnetic fields
(Basis: ISO 7816-1; test regulation: ISO/IEC 10373)
In this test, the card is moved into a static electromagnetic field with a strength of 1000 Oe (79.6 H) at a maximum speed of 1 cm/s. The memory contents of the card must not change.

Embossing relief height of character
(Basis: ISO 7811-1; test regulation: ISO/IEC 10373)
In this test, the thickness of the card where it is embossed is measured using a micrometer, with an applied force between 3.5 N and 5.9 N.

Flammability
(Basis: ISO 7813; test regulation: ISO/IEC 10373)
The flammability of the card is measured by holding one edge at an angle of 45° in a specified Bunsen burner flame for 30 seconds (diameter 8.5 mm, height 25 mm).

Flux transition spacing variation
(Basis: ISO 7811-2; test regulation: ISO/IEC 10373)
This test determines whether the magnetic flux transitions that encode the individual bits in the magnetic strip are uniform and sufficiently strong. A read head is passed along the strip and the field variations are recorded. The measured results are compared to the value specified in ISO 7811-2.

Height and surface profile of the magnetic strip
(Basis: ISO 7811-2/4/5; test regulation: ISO/IEC 10373)
This test measures the height and uniformity of the surface of the magnetic strip. It generates a height profile using a special measuring device that is described in detail in the standard.

Light transmittance
(Basis: ISO 7810; test regulation: ISO/IEC 10373)
Some cards have an optical barcode on an embedded film. This test is suitable for determining the optical transparency of the covering layer and the rest of the card body. One side of the card is illuminated with a light source, and the light transmission is measured on the other side with a detector that is sensitive to 900-nm light.

Location of contacts
(Basis: ISO 7816-2; test regulation: ISO/IEC 10373)
This test is used to measure the locations of the contacts. The card is placed on a flat surface and subjected to a force of 2.2 N ± 0.2 N. Following this, the positions of the contacts relative to the edges of the card are measured using any desired method that has an accuracy of at least 0.5 mm.

Resistance to chemicals
(Basis: ISO 7810, ISO 7811-2; test regulation: ISO/IEC 10373)
The chemical resistance of the card body and the magnetic strip are investigated using these tests. Different cards are placed in the following precisely specified liquids at a temperature between 20°C and 25°C:

- 5% aqueous solution of sodium chloride
- 5% aqueous solution of acetic acid
- 5% aqueous solution of sodium carbonate
- 60% aqueous solution of ethyl alcohol
- 10% aqueous solution of sugar
- gasoline (according to ISO 1817)
- 50% aqueous solution of ethylene glycol

Each card is removed from the solution after one minute and either visually examined or tested using a magnetic strip reader.

Static electricity
(Basis: ISO 7816-1; test regulation: ISO/IEC 10373)
This test, which is only meaningful for Smart Cards, checks the chip's robustness with regard to electrostatic discharge (ESD). A 100-pF capacitor that has been charged to +1500 V and –1500 V in turn is discharged through a 1500-Ω current-limiting resistor into the chip's various contacts. There must be no damage to the functionality of the chip and no change to the contents of its memory due to the discharges.

Surface profile of contacts
(Basis: ISO 7816-1/2; test regulation: ISO/IEC 10373)
This test compares the surface profile of the individual contacts with the surface of the rest of the card. It is intended to ensure that the contacts lie in approximately the same plane as the overall card surface.

Surface roughness of the magnetic strip
(Basis: ISO 7811-2; test regulation: ISO/IEC 10373)
The surface roughness of the magnetic strip is measured using the same device as for the height and surface profile. However, it is used with a special probe tip that allows the

surface roughness of the magnetic strip to be determined. The reason that this test is important is that surface roughness is one of the major factors in the wear of the read/write heads in magnetic strip readers.

Ultraviolet light

(Basis: ISO 7816-1; test regulation: ISO/IEC 10373)

Since (E)EPROM memories lose their contents when they are exposed to ultraviolet light, there is a special test to determine whether the Smart Card is sensitive to ultraviolet light. The card is irradiated for 10–30 minutes by ultraviolet light with a wavelength of 254 nm and an energy density of 15 Ws/cm^2. The (E)EPROM data contents must not change as a result.

Vibrations

(Basis and test regulation: ISO/IEC 10373)

Since cards are often subjected to severe vibrations during transport and use (e.g. mobile phones in cars), an appropriate test is also necessary. It requires the card to be tested on a vibration table in each of the three axes with an amplitude not exceeding 1.5 mm over a frequency range from 10 Hz to 500 Hz. Both the chip functions and the memory contents must not be adversely affected.

Wear test for magnetic strip

(Basis: ISO 7811-2; test regulation: ISO/IEC 10373)

In order to establish how the magnetic strip responds to wear, test data are first written to the stripe. Then a dummy read/write head with a hardness of 110 HV to 130 HV and a radius of curvature of 10 mm is passed back and forth along the stripe 1000 times with an applied force of 1.5 N. Following this, the data are read again. The signal amplitude must lie within the limits specified in ISO 7811-2.

X-rays

(Basis: ISO 7816-1; test regulation: ISO/IEC 10373)

The contents of (E)EPROM memory cells can be changed by X-ray irradiation, just as with ultraviolet light. In order to test the X-ray resistance of the memory, the chip is irradiated by X-rays with an energy of 70 kV. The memory contents are then examined for any changes, or they are tested to see if they can still be written.

There are naturally many other things that can be tested, such as the number of insertion cycles, the wear resistance of the inks, the stability of the plasticizer and resistance to perspiration and saliva. Depending on where and how the card will be used, the appropriate tests must be selected and carried out.

9.2 Microcontroller Hardware Tests

One of the primary tasks of quality assurance, besides looking after the card body, is to ensure that the microcontroller is in good working order. This is the most important and most vulnerable component of a modern Smart Card.

The CPU and memory are subjected to a variety of tests, starting in the semiconductor fabrication stage. In order to be able to carry out these tests, every microcontroller has a test ROM, which contains various programs that allow external access to the CPU and memory.

In addition, there are sometimes special pads (contacts) that allow free access to the central buses of the processor. During production, needle probes are used to make contact with the appropriate pads on the chip so that the necessary test programs can be run. These pads are cut off when the chips are cut from the wafer, so they cannot later be used for attacks. With these chips, it is then no longer possible to access the internal bus.

Once the die has been packaged in a module, another test is naturally performed using the module contacts. This often consists of only executing an activation sequence and seeing whether an ATR can be received. If this is possible, it is assumed that the semiconductor has not suffered any serious damage while being packaged into the module and that all bonding wires are correctly connected. A similar ATR test is also made right after the module has been placed in the body of the card. This test checks whether the module has been damaged by being briefly heated during the embedding process.

The microcontroller is meticulously tested before the Smart Card is initialized. Test commands that are specifically allowed for this production step are used.[1] After successful completion of the test program, which lasts between 10 and 100 seconds, these commands are irreversibly blocked against further use. It is possible to perform these time-consuming tests out at this stage without reducing throughput by using a large number of initialization machines running in parallel, so that the duration of the test does not have a significant effect. The tests used here, to give some examples, check whether all EEPROM bytes can be written and again erased and whether the RAM is fully in working order. If the chip is scratched during the bonding process, this could prevent some EEPROM cells from being properly written or cause certain regions of the ROM to have incorrect contents.

Various final tests are performed after the card has been initialized and personalized, depending on the manufacturer. This is usually done using fully automatic, self-calibrating testers that can configure themselves by reading data relevant to the tests from the Smart Card, following which they carry out the tests accordingly.

In addition to the relatively simple and quickly executable tests undergone by all fabricated cards, there are also sample tests (batch tests) that are only carried out on selected individual cards. These cards, which are taken from ongoing production, can naturally also be subjected to destructive testing if necessary.

Qualification testing and continuous batch testing have also been addressed by the publication of the EN 1292 standard. It defines many different test procedures for microcontrollers. Typical sample and qualification tests for microcontrollers are:

- rise and fall times at the I/O contact (EN 1292)
- number of possible write/erase cycles in the EEPROM
- EEPROM data retention
- overfrequency and underfrequency detection for CLK
- overvoltage and undervoltage detection for Vcc
- I/O contact voltage (EN 1292)
- current consumption at the CLK input (EN 1292)
- current consumption at the Reset contact (EN 1292)
- current consumption at the Vcc input(EN 1292)
- current consumption at the Vpp input(EN 1292)

---

1 See also Section 7.13, 'Hardware Test Commands'.

Naturally, every card manufacturer also employs his own additional tests to cover special features of the embedded microprocessors in question. For example, there are special tests for the various sensors on the chip so that each of them can be suitably tested.

9.3   Evaluating and Testing Software

The testing of physical components, as represented by the bodies and modules of Smart Cards, can largely be carried out using standard conventional methods. Electrical characteristics can be determined equally well using automated test equipment. However, the situation with regard to the microcontroller software is somewhat different. Although testing software for errors has been steadily refined over the past 40 years, since the appearance of the first programs, and there are many known good procedures for producing programs with a low number of errors, it is still true that in everyday practice, software errors show up relatively frequently.

This is not a serious problem in most applications, since a revised version of the software can quickly be issued to eliminate the error. This is not so easily done with Smart Cards, since most of the software is located in the ROM of the microcontroller. A new version of the software necessitates a completely new production run by the semiconductor manufacturer, which takes around 8–12 weeks. If the Smart Cards have already been put into service, it is practically impossible to modify the existing software. It follows from these very strict general conditions that software for Smart Card microcontrollers must have an extremely low number of errors. The term 'error-free' would be even better, but given the present state of software generation, this remains a distant goal.

As is well known, the topic of software testing is extremely extensive. It is described in many books in all of its variations and approaches. We can only present a short sketch of this subject, which by now has become almost an independent branch of information technology. Consequently, in the following sections we discuss only certain special aspects of testing software for Smart Card microcontrollers. Glenford J. Myers' book [Myers 95] can be considered to be representative of the literature on this subject. We would also like to point out that military standards, in particular, contain many good and long-proven procedures for generating and testing software.

9.3.1   Evaluation

Due to their ability to store data securely, Smart Cards are primarily employed in security-sensitive areas. However, Smart Cards can be used to advantage not only for the secure storage of data, but also equally well for the secure execution of cryptographic algorithms.

The field of electronic financial transactions, in particular, is an expanding market for Smart Cards. Since enormous amounts of money will flow in a widely distributed system, the application provider or card issuer must have very great confidence in the semiconductor manufacturer, the producer of the operating system and the Smart Card personalizer. The application provider must be able to be absolutely certain that the software in the Smart Card performs the required financial transactions without any errors, and that the software is free of security leaks, not to mention trapdoors deliberately introduced into the software.

For example, suppose that a secret command could be sent to the Smart Card to read out the PIN and all secret keys. In the case of a GSM or Eurocheque card, the attacker would then be able to clone any number of cards and sell them in perfect working order.

However, the security requirements relate not only to the manufacturing of the Smart Cards, but equally well to the initialization and personalization of the cards, since the secret keys and PIN are loaded into the cards in these stages. The card issuer must place a high degree of trust in the card supplier with regard to security.

This applies as well to the fundamental security of the software in the Smart Card. Problems can arise even if a 'trap door' has not been intentionally included in the software to allow data to be spied out of the card. Faulty operation of the software could very well make it possible to read data from the card or write data to the card, using a command combination that is not used in normal procedures. Although the likelihood of such a coincidence is extremely low, it is nevertheless well known that it is impossible to guarantee that programs are free of errors in all circumstances, given the current state of software technology. It is certain that in the future, firms that make software for Smart Cards will no longer be able to deny all responsibility on the basis of such legalistic formulations.

There are only two ways in which the application provider can test the trustworthiness of a product. He can either test all possible variations the Smart Card software himself, or he can have it tested by a trustworthy party. The first option is frequently possible only to a limited degree, since the provider usually does not have all the necessary technical expertise and capabilities. The second option, which is assigning the tests to another party, is currently regarded by all concerned as an acceptable solution.

This same problem has existed for many years with software and systems developed for military use. It is thus not something that is new in the Smart Card world. In order to establish metrics for the trustworthiness of software products, which means to make it objectively measurable, the US National Computer Security Center (NCSC) in 1983 issued a catalog of criteria for evaluating the trustworthiness of information technology systems. NCSC was founded in 1981 by the American Department of Defense (DoD). The publication of 'Trusted Computer System Evaluation Criteria' (TCSEC) followed in 1985. This book had an orange binding, so it has come to be generally known as the 'Orange Book'. These criteria serve as guidelines to the NCSC for the certification of information technology systems.

The TCSEC has become an international model for practically all criteria catalogs in the information technology sector. In Europe, specifically European criteria have been defined, although they are based on the TCSEC. These were first published in 1990 as the 'Information Technique System Evaluation Criteria' (ITSEC), and a revised version was issued in 1991.

The Common Criteria (CC) were created in order to provide a uniform standard for testing the correctness of software. These could be said to represent the essential parts of the TCSEC and the ITSEC. The Common Criteria [Common Criteria 98] are also better organized for the evaluation of software than the TCSEC or the ITSEC. Although the first version of the Common Criteria was published as early as 1996, they have not yet supplanted the TCSEC or the ITSEC.[2] The Common Criteria have also been published as an international standard (ISO 15408). In contrast to the ITSEC, which has six levels, the Common Criteria have seven levels of trustworthiness. It is relatively easy to make the

---

2   The TCSEC, ITSEC and CC are available from many sites on the Internet.

transition from an evaluation based on the TCSEC or the ITSEC to one based on the Common Criteria, since all of these catalogs have many features in common. However, since in the Smart Card field in particular the ITSEC is still used as the essential basis for software evaluation, we will refer only to this catalog in the following discussion.

An evaluation process, regardless of the methodology used, has four characteristics. It must first be unbiased, which means that the evaluator must not have any preconceived ideas regarding the item to be evaluated or its producer. The second characteristic is that the evaluation process must be objective, and it must be structured such that the effects of personal opinions are minimized. The third characteristic is that the same result must be obtained if the evaluation process is repeated. The final characteristic is that the evaluation process must be reproducible, which means that a different tester or testing agency must be able to reach the same conclusions.

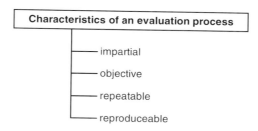

Figure 9.4 The four characteristics of an evaluation process

One of the most important points in any evaluation is defining the security targets for the target of evaluation (TOE). The target of the evaluation is the object to be tested, and the security targets describe the mechanisms to be tested. It is by the way possible to strongly simplify an evaluation by skillfully selecting the security targets, since elements that are critical with regard to security can thereby be excluded. This is just a trick that can be used to achieve a high evaluation level in the quickest and least expensive possible manner. Naturally, the actual security can only suffer as a result.

The ITSEC
Since the ITSEC is supposed to be valid for all possible information technology systems, and the document is only around 150 pages long, the security criteria must be described in a very abstract form. It is consequently very difficult to read and, like legislation, it occasionally requires outright interpretation.

The 1991 version of ITSEC assumes that there are three fundamental threats to any system, namely unauthorized access to data (breach of confidentiality), unauthorized alteration of data (breach of integrity) and unauthorized impairment of functionality (breach of availability). The security criteria are based on these three threats.

The threats can also be reduced outside the system by various measures. For example, an insecure computer system can be made considerably more secure if physical access to the system is controlled by security gates. Such aspects represent general conditions that also must be taken into consideration during an evaluation. The resulting residual threat can then be assessed using the ITSEC catalog of requirements.

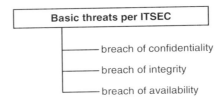

Figure 9.5   The three basic threat areas according to the ITSEC

The basic procedure for evaluating a system in terms of the ITSEC is to rate the mechanisms that it uses to maintain security with regard the to the tree defined basic threats. The ratings are made in a manner that is similar to grading a school assignment. The mechanisms used must be both correct and effective. The mechanisms are also judged with regard to basic functions that are relevant to security, which are also referred to as *generic categories.*

Table 9.1   Generic categories of mechanisms according to the ITSEC

Category
Identification and authentication
Access control
Evidence security
Log evaluation
Reprocessing
Genuineness
Reliability of service
Transfer security

Table 9.2 lists the requirements for the evaluation of software for a Smart Card. The amount of effort does not increase linearly for each successive level, but rather nearly quadratically. This means that it takes twice as much effort to go from level E2 to level E3 as it does to go from level E1 to level E2. The consequences of this are naturally most pronounced for evaluation levels E4 through E6. Complete evaluation of a medium-sized Smart Card operating system at the ITSEC E6 level can easily take several years and cost several million dollars.

In the evaluation, a fundamental distinction is made among informal, semi-formal and formal methods. An informal description of a function can perfectly well be compared with a textual description in this book. The Small-OS operating system presented in Section 5.11 is basically described semi-formally. A formal description, by contrast, can be logically tested and is produced using a formal notation (such as predicate logic).

In the ITSEC, the levels of security against threats are divided into several classes by function. Each of these classes is a collection of a particular combination of security functions, and it indicates the level of threat for which the system has a secure defense. There are six quality levels or evaluation levels (E1 through E6). E1 is the lowest level, while E6 is the highest level.

Table 9.2   Summary of the requirements for Smart Card software development as a function of the ITSEC quality level

Necessary information	Evaluation level					
	E1	E2	E3	E4	E5	E6
security targets	yes	yes	yes	yes	yes	yes
formal model of security policy	no	no	no	yes	yes	yes
descriptions of functions	informal	informal	informal	informal & semi-formal	informal & semi-formal	informal & formal
architectural design	informal	informal	informal	semi-formal	semi-formal	formal
detailed design	no	no	informal	semi-formal	semi-formal	semi-formal
implementation	nothing required	nothing required	nothing required	source code testing	source code testing	source code & object code testing
operation	yes	yes	yes	yes	yes	yes
level of rigor	-- state description --		---- presentation ----		----- explanation -----	

In all of the functional classes, the formal requirements imposed on the development process and the development environment are prescribed in a very abstract manner. This is also true of the information and specifications that the classes contain regarding operating documentation and the eventual operating environment. This information is presented in a form that can be applied to all possible technical variations of software development.

An ITSEC evaluation frequently includes an appendix that relates to the 'strength' of the mechanisms, which means their robustness against attacks. Three levels of strength are defined: low, medium and high. 'Low' characterizes protection against random, unintentional ingress into a secure environment. 'Medium' means that there is protection against attackers with limited resources. The highest level, 'high', means that there is protection against attackers who have very good technical knowledge and resources. Normally, the highest level of mechanism strength is used with ITSEC level E4 and higher. It should also be used as a matter of principle for Smart Cards, in the interests of system security.

The ZKA criteria

In addition to the international evaluation standards such as TCSEC, ITSEC and CC, there are also specific evaluation standards that have become established for large Smart Card applications. Two examples are the Visa criteria for security testing and the German ZKA criteria. The ZKA criteria are mandatory in Germany for Eurocheque cards with chips, and all Smart Card operating systems for this application are tested against these criteria by authorized evaluation bodies. The ZKA criteria are listed and explained in Table 9.3.

Table 9.3 The ZKA criteria for evaluating a Smart Card system (from Stefan Rother [Rother 98b])

Criterion	Explanation
component authentication	It must be possible to authenticate the security components of the system.
message integrity	Information relevant to security that is exchanged between the components must be protected against manipulation.
user authentication	Certain functions may be executed only after the user's PIN has been correctly entered in order to authenticate the user.
PIN and key secrecy	PINs and keys may never be transferred in plaintext outside of a secure area. To the extent that the PINs and keys are processed or stored in components of the operating system, these components must be protected against unauthorized reading or modification. The system must prevent carrying out an exhaustive PIN search.
logging	All events relevant to security must be logged in the components that are involved. The logs must be protected against manipulation.
key management	There must be mechanisms for the distribution, administration and exchange of keys. Only temporary keys may be used with symmetric cryptographic procedures, and keys must be kept separate according to their purposes.
hardware	All operations relevant to security must be protected against unauthorized accesses.
organizational precautions relating to manufacturing and personalization	Only the evaluated program code with the described parameter values may be used in actual operation.
process security	The correctness of the processes and data flows within the evaluated programs is ensured by a source code test.
other applications	Additional applications must not have any effects that endanger the security.
encryption procedures	The cryptographic algorithms that are used must represent the current state of the technology and must not be based on the secrecy of the procedure.
unambiguous representation	Each security component must be clearly identifiable in the overall system.
personnel requirement	Only suitable persons should be allowed access to the components and mechanisms that are relevant to security.

With testing based on the ZKA criteria, the prepared documents as well as the software and hardware are investigated. This is the main advantage of the Smart Card-specific ZKA criteria in comparison to the general TCSEC, ITSEC and CC criteria, which apply to all types of software.

In summary, the ZKA criteria are presently without doubt the best security for Smart Cards, since they meet the stringent demands of a large financial transaction system and are specially tailored to the particular interests of a system based on Smart Cards.

Table 9.4   Investigating the security of the specifications and source code of a Smart Card operating system based on the criteria (from Stefan Rother [Rother 98a])

Specifications and source code investigation		
1. operating system commands		7. memory and resource management
2. encryption and decryption		8. key administration
3. application commands		9. file attributes and file access privileges
4. key derivation		10. authentication mechanism
5. communications mechanisms		11. checksum algorithms
6. signature generation and signature testing		12. random number generator

Capsule summary

An actual Smart Card project takes place as follows. The card issuer first works with the operating system producer to define the operational requirements and the threats that must be taken into account. Following this, an agreement is reached regarding the necessary evaluation level. After this, the operating system is configured according to this evaluation level and the corresponding evaluation process, and the necessary documentation is provided. In the final step, the finished operating system with all its components can be evaluated by a suitable independent body, which is the evaluation agency, according to the previously agreed level.

An ITSEC evaluation has certain advantages for the card issuer and the application supplier, since they can be certain that the design and implementation of many security aspects are clearly defined as well as effective. This assurance is however coupled with some disadvantages in the dynamic Smart Card market. The development time is considerably increased by the need to conduct an evaluation, even in with the lowest functional classes. The additional effort needed to produce the necessary documentation also increases the development costs, which ultimately must be reflected in the price that the producer of the operating system charges for his product.

However, the major disadvantage is something completely different. Even with an evaluated system, it is not possible to guarantee that all processes and mechanisms function exactly as they are described in the relevant documentation. Evaluation does not mean that the party performing the evaluation completely tests the product, but only that it checks the documents, and possibly the source and object code, that it has received for the product in question. It is also equally important that the target hardware guarantees a level of security equivalent to that of the evaluation. It does not help at all to have error-free, secure software if it is possible to use the hardware as a 'back door' to bypass the software.

These four reservations with regard to evaluation must be kept in mind and carefully considered in each individual case.

9.3.2   Testing methods for software

Although it may sound obvious, there is one fact that should be clearly stated in advance. Testing a program does not mean systematically searching for errors, but rather carrying out a test procedure with the objective of finding out if there are errors in the program being tested. Testing thus has nothing to do with debugging, since tests are performed by testers and not by software developers.

As we have repeatedly emphasized, Smart Card microcontroller programs are not very large compared to other software. Nevertheless, they have their own specific features. This can be illustrated by considering some of the general specifications of a typical operating system.

In what is now an average chip, with 16 kB of ROM and 8 kB of EEPROM, the software requires around 20 kB of memory. This leaves 4 kB of EEPROM available for any applications. Presently, the software is usually programmed in assembler, and in our example it amounts to about 30,000 lines of program code. If this were printed at 60 lines to the page, it would fill 500 sheets of paper. The number of branches is around 2000, and even an experienced programmer needs around nine months to generate 20 kB of assembler code.

This numerical example clearly demonstrates that Smart Card operating system software is rather complex. On top of this, this software must be used almost exclusively in areas where security is a consideration. This means that the demand for a low level of errors cannot be met by simply using a few homemade tests during or following the software development. Instead, a suitable testing strategy is necessary.

With Smart Cards, as in other areas, the trend is moving away from assembly language programming to the use of high-level programming languages. The C language, which is relatively close to the hardware level, is being used increasingly often for Smart Cards. The object-oriented Java language will certainly also become just as important over the next few years. The use of high-level programming languages is necessary for large Smart Card operating systems with more than 30 kB of code, not only due to the reduced implementation time, but also because it reduces the number of errors.

This can be explained as follows. It can be assumed that the number of errors per line of source program code is nearly the same for almost all programming languages. Since the functional level of a high-level language is significantly higher than that of assembler, this means that the error density is lower for the same amount of executable code.

For example, if we assume the entirely realistic value of 1.5 errors for every 100 lines of program code, and in addition we assume that only half of all errors can be found with an acceptable amount of effort, then a tested program will still have 0.75 errors for every 100 lines of program code. With Java, the relationship between lines of source code and machine code (bytecode) is around 1 to 6. This means that the functionality of one line of Java code roughly corresponds to that of six lines of machine code. If we assume that all other conditions are the same and that the compilation process is largely error-free, the number of errors in a program can be reduced by a factor of six by using a more powerful language. Even if the actual value is lower than this in practice, this is still an exceptionally strong justification for using powerful programming languages.

9.3.2.1   Smart Card software testing

You must first consider the life cycle of Smart Card software before you can start to define a test strategy. You can use the waterfall model proposed by W. W. Royce, which has been known since 1970 and has been published in many forms. It is quite suitable for mask-programmed Smart Card operating systems. However, since this model is designed for very extensive software projects for PCs and mainframes, we will use a simplified version that is specially adapted for Smart Cards.

These five stages shown in Figure 9.6 are normally performed in sequence. It is however certainly possible for a problem encountered in a certain stage to make it necessary to go back one or more stages. This should be avoided as much as possible, since each iteration costs time and money.

In order to meet economic demands, such as time-to-market and a short software development time, it is often necessary to overlap the stages to a certain extent, instead of performing them in strict sequence. With this method, which is known as simultaneous engineering, parts of the software are split into individual modules as early as possible. These modules are then launched down the waterfall concurrently. Thus, it could be possible for Smart Card software containing only a data transmission protocol to already be at the system integration stage while the cryptographic algorithm for the same application is still being specified.

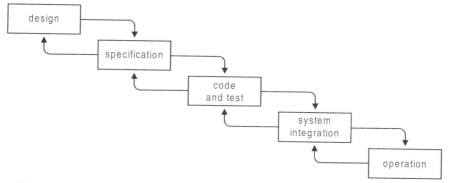

Figure 9.6   Simplified waterfall model for the production of Smart Card software. The individual steps can be regarded equally well as activities or as time intervals.

Design

The design stage includes establishing the basic definition of the objective and compiling the requirements in the form of a formal requirements document. This defines all the requirements that must be satisfied by the Smart Card software to be developed. The design stage also allows for the generation of proposed solutions in the form of preliminary designs. Put simply, this stage defines *what* the finished software has to do.

Specification

Following the design comes the next stage, which establishes *how* the software will do its job. For this purpose, it is necessary to generate precise specifications that are not subject to interpretation, and which completely define one of the possible solutions to the requirements produced in the design stage. Formally constructed specifications are best, since they allow the features, functions and processes of the software to be clearly and unequivocally defined. Specifications written in pseudocode, which can be tested for consistency and freedom from errors using computer programs, are well suited to this purpose. In a computer-aided software engineering (CASE) environment, such specifications can be used directly for generating the source code and test programs. They are sometimes referred to as 'executable specifications', which means specifications produced in a form that can be interpreted and further processed by computers.

Code and test
Once the specifications have been finalized and accepted, the flowcharts for assembler or C programming may be generated. This is followed by programming and the associated testing. The outcome of this stage is a fully programmed and tested Smart Card operating system.

System integration
Since Smart Cards can only work together with a larger system, the various system components must be merged in this stage. The result is the complete and error-free cooperation of all the parts, as well as the final documentation for the entire system.

Operation
This last stage in the software development process can only be used to modify any general parameters that are located in the issued cards. Large-scale software adaptations or modifications are no longer possible at this stage.

In the future, the ability to easily and quickly program Smart Cards using a high-level language (such as Java) will make it possible to also use 'evolutionary' procedure models in addition to the traditional waterfall model. In evolutionary models, the design, specification, and code and test stages, and in part even the system integration stage, are iterated several times with increasingly improved results. The objective is to quickly arrive at an optimum solution by expending minimal effort on the generation of specifications and working with fully functional prototypes.

It is a well known fact of long standing that in all kinds of projects, and particularly with software development, the cost of correcting an error increases as the project progresses. This fact should lead to the expenditure of an appropriate amount of time and effort in the first stages of the project, as represented by the waterfall model. If the design is incomplete or the specification is faulty, the cost of remedying the problem rises exponentially in the subsequent stages of the project.

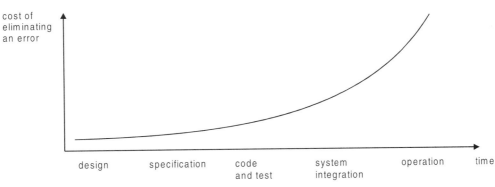

Figure 9.7   The cost of correcting an error as a function of the time when it is discovered

9.3.2.2   Test procedures and test strategies

Nowadays, is impossible to keep track of all the different methodologies and procedures that are available for testing software. However, only a few well-proven methods are necessary for testing Smart Card programs. This is one of the few benefits of the still unavoidable use of assembly language programming. We can draw on decades of experience and a large number of publications on the subject of testing software written in this language. Incidentally, software testing always means attempting to discover errors in the program, and not demonstrating that the program is correct.

All test procedures can be divided into static and dynamic types. In static procedures, the program code is analyzed and evaluated using various methods, either manually or automatically. The two most commonly used static testing methods can be briefly described and explained as follows.

Static program assessment using software tools   This consists of the analysis of various features of the program code using static techniques. The features that can be analyzed include the following:

- number of lines of code (LOC)
- number of lines of comments
- ratio of the amount of comments to the amount of program code
- structure of the program code
- number of functions
- nesting depth
- 'dead' code

Review   This consists of the formal analysis and evaluation of program modules by a team of assessors. This is sometimes referred to as a 'code walkthrough' or a 'code inspection'.

In contrast to static procedures, techniques for dynamic program analysis test the program while it is in operation, either manually or with the aid of computers. There are two fundamentally different approaches, plus a hybrid one.

Blackbox test   A blackbox test is based on the idea that the agent performing the test knows nothing about the internal processes, functions and mechanisms of the program to be tested. This means that all that can be done is to examine the input and output data with regard to their relationship to each other, as defined in the specifications.

Blackbox tests are the standard for Smart Card operating systems. They are also used for the security modules for terminal and computer systems. However, it is often incorrectly assumed that these tests can also discover Trojan horses or similar items that may be present, in addition to errors in the software. This assumption is used as an argument for doing without a relatively time-consuming and expensive program code analysis. While it may be possible to detect simple, unsophisticated trapdoors programmed into the system, or ones that have been inadvertently generated, an experienced programmer can easily create access possibilities that can never be detected by a blackbox test. This can be illustrated using a small example. It is not meant to serve as a model for a Trojan horse, since it is already known for a long time, but rather to enhance the awareness of the necessity of code inspections in security analyses.

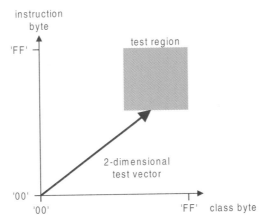

Figure 9.8   Example of a two-dimensional test vector for a blackbox test , which defines a specific test region by means of equivalence classes. With complex test cases, the test vectors can easily have ten or more dimensions. Due to the limited amount of time available for testing, the test region or regions must therefore be suitably restricted.

Almost all Smart Card operating systems contain a command for generating and issuing random numbers (GET CHALLENGE). It is conceivable that this command could be modified such that only the first 8-byte number that it issues is actually generated by the pseudorandom number generator. Each of the subsequent 'random' numbers would then be an 8-byte value from the EEPROM, XOR-ed with the first random number. A simple external program could then be used to read out the entire memory contents, including all the keys. This is by the way a very good example of applied steganography in Smart Cards.

With a blackbox test, there is no way to determine that a Trojan horse is concealed behind this command. Even a statistical analysis of the random numbers obtained would not detect any significant deviation from the usual pseudorandom numbers. The only way to recognize this manipulated program is to inspect the entire code of the operating system. This example illustrates only one of many possible options for altering a normal command in order to obtain the contents of the memory. Since only a few lines of program code are needed for this modification, the only effective way to combat such trapdoors is to completely review the source code.

Whitebox test   A whitebox test is often called a 'glassbox test', which clearly describes the concept. All internal data structures and processes are known and can be completely understood. The relevant program documentation is used to design and produce the tests, but the specification is always the sole authority. For decades, program flowcharts have been used to document programs in all dialects of assembly language, and they usually form the basis for evaluating the internal functions of the software in a whitebox test.

Since the exact program processes are known, it is natural for the tester to want to test all possible paths through the software. There are several ways to do this. One of them is *statement coverage*, in which every instruction in the program is executed at least once. This makes it very easy to discover if the program contains dead code, which is code that is never used, but it is too weak a test to be able to ensure that the desired functionality is present. A better method for this is *decision coverage*, which involves traversing all decisions in the program code at least once in each of their possible variations.

In order to be able to recognize these internal program processes during dynamic testing, it is necessary to use either sophisticated emulators for the Smart Card microcontrollers in question or to 'instrument' the program being tested. In instrumented program has special program code inserted just before every jump instruction, branch instruction and function call. This code collects location and parameter information when the program is run. An analysis program can be used to evaluate this information both statistically and graphically. Unfortunately, the additional program code alters the timing relationships of the program, and in the worst case, it can even cause the behavior of the program to change. This must be considered whenever this method is used.

	instruction coverage	1-2-3-4
		1-2-5-6
	decision coverage	1-2-3-4
		1-2-5-6
		1-2-5-3-4
		1-2-3-6
	decision coverage in all combinations	1-2-3-4
		1-2-5-6
		1-2-5-3-4
		1-2-5-3-6
		1-2-3-6

Figure 9.9   Example of the number of possible process paths in a program flowchart, for testing using statement coverage and decision coverage

An extension of the decision coverage criterion is to traverse all program decisions in all possible combinations, once for each combination. This covers all possible processing paths. However, the limited amount of time available for testing means that this is possible only with very small programs with a few hundred bytes of code. Even with programs on the order of 1000 bytes, testing all possible combinations cannot be carried out in a reasonable length of time.

Table 9.5 illustrates this in summary form on the basis of a typical Smart Card command interpreter. The function of this program module is to recognize a command located in the card's input buffer by means of the class and instruction bytes, and then to check the P1, P2, Lc and Le parameters. This size of the program code for this routine is around 200 bytes, and it contains 18 branches. The possible output values consist of five return codes and calls to 26 different command procedures.

Two other path coverage criteria are used in particular for testing Smart Card operating systems: *input coverage* and *output coverage*. The objective is to generate all possible input and output values. The output values are often restricted to the available return codes, since otherwise the number of variations would be too large.

Since the number of possible input values can also quickly reach a magnitude that makes testing impractical, due to the multitude of input values or the amount of time required, equivalence classes are usually employed. This reduces the large number of possible input values to a small number that can be tested in a reasonable amount of time. Equivalence classes are formed by selecting boundary cases on either side of the decision range, together with a value in the middle of the range.

Table 9.5 The number of possible test cases for various coverage methods in a whitebox test. A 200-byte Smart Card command interpreter is used for this example. An average processing time of 30 ms, including data transfer, was used to calculate the test duration.

Coverage method	Possible test cases	Test duration
one million random input values	1,000,000	$\approx$ 8 h
command coverage	10	$\approx$ 0.3 s
decision coverage	50	$\approx$ 1.5 s
decision coverage in all possible variations	50,000,000	$\approx$ 17 days
input coverage in all possible variations (5-byte header)	$\approx 1.1 \times 10^{11}$	$\approx$ 1000 years
input coverage with equivalent classes	15	$\approx$ 0.5 s
output coverage of the return codes	6	$\approx$ 0.2 s

For example, if the Smart Card command interpreter allows a range of 20 through 50 for the value of the P1 byte, the equivalence class would be formed using the values 19, 20, 50 and 51 for the boundary values and 35 (for example) as the midrange value. This set of values verifies the essential query conditions of the program. After this test, it could be assumed with a relatively high level of confidence that parameter range checking has been correctly implemented.

Particularly with assembly language programming, it is unfortunately necessary to take the properties of the target hardware into account when defining the equivalence classes. For instance, all arithmetic operations that can cause an overflow or underflow in the processor (due to the 8-bit or 16-bit architecture of the arithmetic unit) must be taken into account when forming the equivalence classes. Only then is it possible to be sure that underflows and overflows are correctly handled in the program.

Greybox test A greybox test represents a hybrid combination of the blackbox and whitebox tests. With such a test, some parts of the software are known, such as internal program processes. Greybox tests are primarily used in the integration phase with Smart Cards, since they allows errors in the interaction of the individual components to be very quickly and effectively detected and corrected. Naturally, appropriate test keys (which are public) are needed from the key management facility. Once this part of the integration tests is successfully concluded, the results can be checked using the real keys (*life keys*).

9.3.3 Dynamic tests for operating systems and applications

It is important to understand from the start that program testing can only demonstrate the presence of errors, not their absence. Assuming that only roughly half of the errors that are present are actually found, it can be safely concluded that the average program still contains a number of weak points.

In practice, an error rate of 0.7 per 100 lines of code, after testing, is often assumed for assembly-language programming. If we take the previously mentioned value of 30,000 lines of program code for a Smart Card operating system as representative, and we subtract two thirds of this number as comments, we can calculate that there are still around 70 errors in a fully tested and released operating system. Although most of these will never show up, in the right circumstances one error is sufficient to bypass all the security barriers of the system. It is very useful to always keep this fact in mind as a motivation for careful and well-considered testing.

Of course, there are natural limits to testing. Particularly in commercial projects, in contrast to research projects, the amount of time available and the maximum affordable cost are strongly limiting factors. In addition, testing becomes increasing more difficult and demanding as the number of errors present in the program decreases. The search for the last few errors must at some point come to an end, since the time and resources that can be expended on it are fundamentally limited.

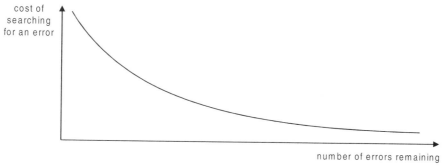

Figure 9.10   The cost of searching for errors as a function of the number of remaining errors

When a new version of the software is issued, you can in the first instance assume that it always contains fewer errors, since there has been an opportunity to analyze faults discovered during operation and eliminate them. Interestingly enough, this reduction in the number of errors does not continue indefinitely. Instead, the number of errors is usually seen to reach a minimum around the second version, following which it generally increases. This comes about simply because the necessary corrections are based on the original specifications and program code. After a certain time, which can vary greatly, it is likely that the correction of one error gives rise to one or more new errors. This leads to the curve shown in Figure 9.11. After a certain number of versions, it is thus significantly better to make a completely fresh start than to continue building on outdated concepts and repeatedly updated program code. Incidentally, this is true in almost all fields of technology.

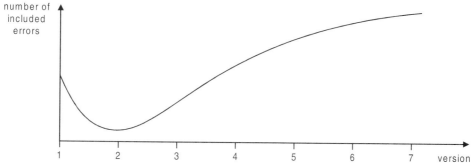

Figure 9.11   Empirical dependence of the number of errors in a program on the version number

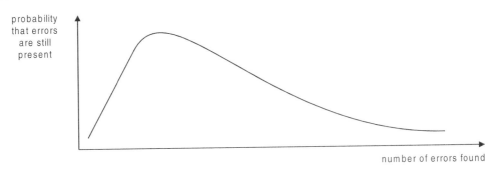

probability
that errors
are still
present

number of errors found

Figure 9.12   Empirically determined dependence of the number of residual errors in a program on the number of errors already found

In accordance with the IEEE 1008 standard ('Standard for Software Unit Testing'), three test levels can be distinguished for dynamic testing. The first of these is the *basic test* level, in which essentially the basic functions and successful execution of the individual commands are tested. The second level is the *capability test*, which encompasses boundary values and non-successful execution. The third level is the *behavior test*, in which the commands are tested in combination with each other.

Test procedures
There is a great difference between testing a new operating system and testing a new application. When a Smart Card operating system is tested, the entire program code must be tested for a wide variety of application cases. This requires a large number of different tests. With a new application, which as you know consists of only a DF and several EFs, the number of tests is reduced to match the amount of additional data and the identification and authentication procedure defined for the application.

If a new operating system must be tested, several test applications that are similar to some typical real applications are usually generated. Equivalence classes are then formed for 'typical' applications, in a manner of speaking. These form the basis for the individual tests that are subsequently performed.

The following approach to the testing of new Smart Card operating systems has become established in the course of several years in a wide variety of projects. Testing always starts with data transfer functions, since these form the basis for all further activities. Following this, all available commands are tested. If an application is involved, the next stage is file tests. If all these tests are completed successfully, testing the fixed processes can begin.

There are currently only a few international standards that regulate the construction and execution of tests for Smart Card operating systems and applications. A European standard (EN 1292) defines a few tests for the ATR and the T=1 transmission protocol. Relatively extensive tests for the operating system and application of GSM cards are defined in the GSM 11.17 specification.

In order to provide an overview, a selection of possible tests in a conventional sequence is presented below. This list does not pretend to be complete, and it is only meant serve as a detailed illustrative example. The purpose of the listed tests is to test the essential general parameters of a new operating system, including one or more applications.

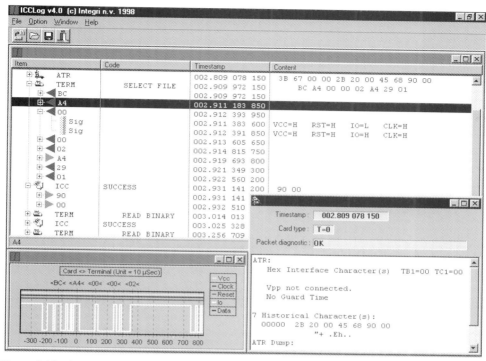

Figure 9.13 Screen display of a tool for verifying the communications between a terminal and a Smart Card on the physical and logical levels (Copyright © Integri [Integri])

Data transfer tests

- ATR (parity error detection, character repetition if T=0 is present, ATR structure and contents)
- PTS (PTS structure and contents)
- data transfer test at OSI layer 2 (start, data and stop bits, divider, convention)
- T=0 transmission protocol
  (parity error detection and character repetition, various routines)
- T=1 transmission protocol
  (CWT, BWT, BGT, resynch, error mechanisms, various routines)
- secure messaging

Tests for the available commands

- test all possible class bytes
- test all possible instruction bytes
- test all available commands with equivalence classes of the supported functions

Tests for available files

- test whether all files are present at the correct locations (MF, DF)
- test for correct file size
- test for correct file structure
- test for correct file attributes
- test for correct file contents
- test the defined access conditions (read, write, block, unblock etc.)

Tests for the available processes

- test the defined micro and macro state machines (for example, the command sequence)

As can readily be imagined, despite forming equivalence classes and employing a few other minimization procedures, a relatively large number of individual tests are required. We may assume that between 4000 and 8000 different tests must be defined to cover the essential test cases for a 20-kB Smart Card operating system. In this regard, a test that for example executes a loop that sends several hundred different values to the Smart Card is counted as a single test. It takes around one to two days to perform all these tests. The only way to manage such a large number of tests with a reasonable amount of effort and expense is to use a suitable database, which at the same time can also store the test results.

The 'tree and tabular combined notation' (TTCN), which is standardized in ISO/IEC 9646-3, is one of ways in which the tests can be formally described. Any desired test case can be described in a general and standardized form using this notation. An interpreter can then automatically generate the command APDUs for the card being tested, using this description. This allows largely automated test processes to be defined.

The structure of an idealized test tool is shown in Figure 9.14. The specification for the card's software, which is written completely in pseudocode, is located in an appropriate database. If the specification changes, the necessary modifications to the tests are made automatically. Another database contains all the tests, defined in a high-level language that can also be directly read by a computer. The two databases feed a test pattern generator, which generates the commands (i.e. TPDUs or APDUs as relevant) for the card being tested. A simulation of the real card, which is largely defined by the specification, is run in parallel. Since there are incompletely predictable processes in the real card (e.g. generating a random number), additional data must be sent to the simulated card. The real and simulated cards send their command responses to a comparator. If they are the same, the real card has provided the correct result, insofar as the simulation is the proper reference. All the data generated during a test run are stored in a protocol database, so that they can later be manually evaluated.

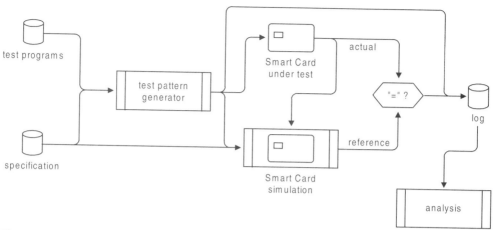

Figure 9.14   Conceptual design of a tool for testing Smart Card operating systems and applications

Figure 9.15   Screen display of an object-oriented, database-supported tool for testing Smart Card operating systems and applications. The definition of the data elements of a command APDU (INITIALIZE IEP for Load) is shown at the upper left. Below this is the associated reference simulation of the Smart Card. Part of the tree structure that defines the individual tests is displayed on the right. (Copyright © Integri [Integri])

10

The Smart Card Life Cycle

This chapter presents the life history of a Smart Card, from the origin of the semiconductor chip through the production of the card and finally to the recycling of the card materials. A section of this chapter is also dedicated to the life cycle of Smart Card applications, since this is very important with multi-application cards in particular, which are becoming more and more widely used. Separate chapters are dedicated to the most important stages of the life of a Smart Card, such as the basic aspects of Smart Card operating systems. These are explicitly referred to in the appropriate locations. Regarding the materials used for the card body, the various types of modules and other card components, we refer you directly to the appropriate sections of Chapter 3.[1]

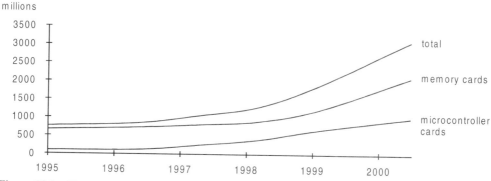

Figure 10.1 The international production of memory and microcontroller cards. The numbers are estimated values, since the various sources differ considerably. An average value has been used here.

A Smart Card basically consists of two completely different components. The first component is the card body with its printing, security features and possibly a magnetic strip. The second component, which is what makes the card body into a proper Smart Card, is the module incorporating the chip. This division of the card into two components applies equally well to memory cards and microcontroller cards.

The only aspect that can influence the construction of the card is the manner in which the data transfer takes place. Smart Cards with contacts employ electrical connections to the

---

[1] See Section 3.2, 'The Card Body'.

terminal by means of six or eight externally visible contact surfaces. Contactless Smart Cards contain coils of various sizes within the card body, which are connected to the chip module also embedded in the card body. This assembly naturally has a significant influence on the manufacturing of this type of card.

The manufacturing process is also influenced to a considerable degree by other elements of the card, such as the material used for the body of the card, the methods used for applying text to the card and the security features. However, regardless of all these options, there is one thing that has absolute priority: cost minimization. The fabrication of Smart Cards is a mass production process, in which the batch quantities start at around 10,000 pieces and can certainly reach the level of 10 million. A highly optimized production process is the most important prerequisite for the cost-effective manufacturing of high-quality card products.

10.1 The Five Phases of the Smart Card Life Cycle

In addition to the manufacturing process, the life cycle of a Smart Card depends on the application in which it is used. A GSM card, for example, has a considerably different career following its fabrication than a credit card containing a chip. However, cards that are used for different types of applications still have many common aspects.

The ISO 10202-1 standard attempts to define a card life cycle that is equally valid for all manufacturing methods as well as all the various applications. This standard is very strongly oriented towards financial transaction applications, and the information technology that is used in these applications, with less emphasis on the actual production of card bodies and chips. Nevertheless, it represents a quite successful attempt to provide a structured description of the life history of Smart Cards from the beginning to the end. This is why it is used here as the basis for the description of the Smart Card life cycle.

Table 10.1 Summary of the individual life cycle phases, according to the ISO 10202-1 standard

Life cycle phase	Typical activities
Phase 1: Production of the chip and the Smart Card	• designing the chip • generating the Smart Card operating system • fabricating the chips and modules • producing the card bodies • embedding the modules in the card bodies
Phase 2: Card preparation	• completing the Smart Card operating system
Phase 3: Application preparation	• initializing the application(s) • personalizing (individualizing) the application(s), both optically and electrically
Phase 4: Card utilization	• activating applications • de-activating applications
Phase 5: End of card utilization	• de-activating the applications • de-activating the card

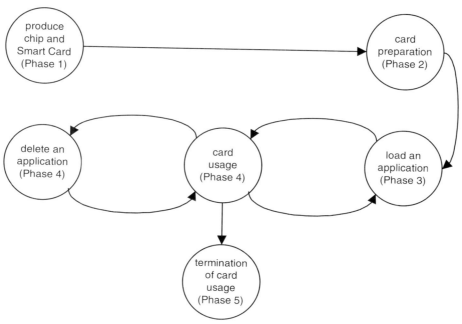

Figure 10.2   The life cycle of a Smart Card according to the phase model of the ISO 10202-1 standard. The downloading and deleting of applications, which is possible with multi-application Smart Cards, takes place in phases 3 and 4.

According to the ISO 10202-1 standard, the life of a card is divided into five phases, which are interconnected by precisely specified transitions. All stockpiles of cards that are made necessary by the production process, and all transportation paths between the various firms that perform the various steps of card production, must be secured either physically or cryptographically. This is necessary to preclude the manipulation or theft of partly finished products.

All production steps must naturally be accompanied by appropriate quality assurance procedures. Since Smart Cards are normally used for products in areas in which security is an issue, it is nowadays standard to guarantee the traceability of the production process in the sense of the ISO 9000 ff series of standards. At minimum, this means that all fabrication steps must be logged with batch and chip numbers. It must be possible to reconstruct the production steps undergone by each individual Smart Card, at any desired time after it has been manufactured. This makes it easier to analyze the cause of any production faults that may show up. Since each individual chip has a unique chip number, no two microcontrollers are identical following the semiconductor fabrication process, which makes it relatively easy to implement traceability on the basis of chip numbers. Fabrication traceability can be implemented either by storing the relevant information in a fabrication database or by writing all the information that is relevant to the fabrication of each chip in the chip itself. The ISO 10202-1 standard recommends storing the fabrication data in the chips, which has certain advantages compared with storing the data in a database. If the data are stored in the chips, it is possible to obtain the fabrication data for any chip without having to access a database, although this comes at the cost of precious space in the microcontroller EEPROM.

Table 10.2 Fabrication information that is typically stored in chips. This information can be written only once, after which it can only be read (the file or data object has the WORM attribute).

Phase of the Smart Card life cycle	Typical fabrication data
Phase 1: chip and Smart Card manufacturing	• identification of the chip manufacturer • identification of the fabrication facility • unique chip number • chip type • identification of the module packager • date and time of implanting the chip in the module
Phase 2: card preparation	• identification of the initializer • identification of the production machine • date and time of the initialization
Phase 3: application preparation	• identification of the personalizer • identification of the production machine • date and time of the personalization

10.2   Phase 1 of the Life Cycle in Detail

The first phase of the ISO 10202-1 standard life cycle can be subdivided into two parts. The first of these covers the generation of the Smart Card operating system and the semiconductor manufacturing process for the microcontroller, while the second part covers the entire technology for producing the card body.

10.2.1   Generating the operating system and producing the chip

Operating systems and other software for Smart Card microcontrollers are so complex that we have devoted a separate chapter to them, in which all aspects of the subject are explained.[2] However, we must not overlook the fact that a significant part of the basis for the security of the remainder of the card's life cycle is already established in the fabrication of the chip. No matter how high the quality of the operating system may be and how much cryptographic protection is used, they are of little use if all the secret data can be read from the chip thanks to an error in the design or development of the chip.

The fabrication of the semiconductor chips takes place in protected facilities with restricted access. Restricted access is relatively easy to achieve with cleanrooms, which can anyway only be entered via interlocked doorways. It is also important in terms of security, since it is the only way to guarantee that no ICs containing Trojan horses in their software can be smuggled into the system while the chips are being fabricated or separated into individual dice. This would otherwise be a very serious and relatively dangerous form of attack on the security of Smart Card applications.

---

[2]   See Chapter 5, 'Smart Card Operating Systems'.

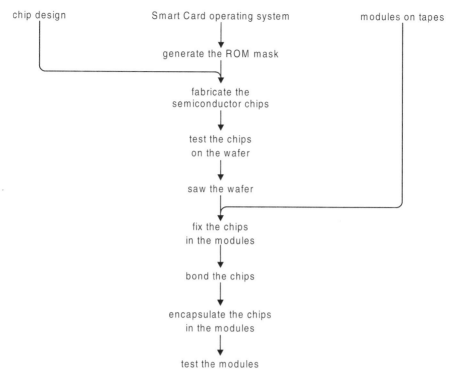

Figure 10.3   The production of chips and modules in the first phase of the Smart Card life cycle

Chip design

The geometric structure of a chip for a memory or microcontroller card should be as nearly as possible square, since this minimizes the risk of the chip being broken by the stresses that arise when the card is bent. Complete protection of the chip against bending stresses is in principle technically possible with an extremely stiff module package, but this is not desirable in practice. Such a stiff module would eventually cause the card body to crack, due to the alternating bending stresses to which the card is recurrently exposed.

The semiconductor elements that are used for the chip, such as the CPU or the numeric coprocessor, are normally standard elements[3] that have been technically modified for increased security. Semiconductor elements for the automotive industry are often used, since they must be designed to meet similarly severe environmental and reliability requirements. However, such elements must be modified as necessary to fully adapt them to the security demands imposed on Smart Card microcontrollers.

In the chip design process, the first step after establishing the functional specification is to generate a general chip architecture, with a block diagram of the circuit and a rough layout of the future microcontroller. Following this, the overall block diagram is refined step by step into logic blocks, gate functions, transistors and ultimately the geometric structures of the individual exposure masks. Each step is accompanied by a circuit simulation and extensive testing. This is a complex process, consisting of many individual

---

3   See also Section 3.4, 'Smart Card Microcontrollers'.

steps, and a fair amount of experience is necessary to arrive at the optimum arrangement of the elements of the chip. At the end of this process, sample chips are produced on a test fabrication line in a semiconductor manufacturing plant. These are the first reference devices, which are very precisely measured and exercised. A security assessment is often performed in parallel, although this assessment cannot be finished any earlier than the time when the first regular chips are produced.

The process of designing a chip can take several months to a year, until a fully operational chip is obtained that can meet all the necessary requirements for mass production. The fact that it takes so much time and effort to design a chip is the reason that the interval between successive generations of Smart Card microcontrollers is two to three years. Due to the high cost of making significant changes to an existing chip, the most substantial modifications mainly consist of 'shrinking' the chip to better utilize the wafer area, or making minor improvements or extensions to the hardware.

Smart Card operating systems

Software for Smart Card operating systems, and the applications that are based on the operating systems, must primarily be written using assembly language, due to the small memory capacities of the microcontrollers. Ideally, of course, the software would be written in a high-level language that is close to the hardware, such as C. However, the program code that even highly optimized compilers produce is 20–40% larger in volume than optimized assembler code. In addition, machine code produced by a compiler usually needs a lot of space in the RAM for passing parameters and storing the program stack, and RAM is always a scarce commodity in a Smart Card microcontroller. These are the main reasons why Smart Card software is still primarily developed in assembly language. Of course, this situation will change significantly in the future, and this will affect the costs of the entire software development process.

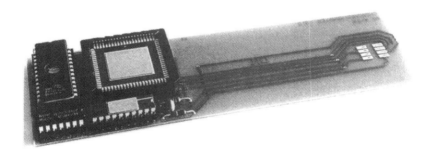

Figure 10.4 Mini emulator for a Smart Card, in which the mask-programmed ROM has been replaced by a removable EPROM in a DIL package. The large IC is a Smart Card microcontroller with all busses freely accessible (a 'bond-out' chip).

The tests for the software that is mainly located in the microcontroller ROM are very thorough and comprehensive, since it is almost impossible to correct any residual errors in this software after the chips have been manufactured.[4] The production of the chips always involves the generation of a ROM mask, which is essentially the software that will later be

---

[4]   See also Chapter 9, 'Quality Assurance and Testing'.

located in the ROM of the microcontroller and which cannot later be changed. If any software error is detected in the subsequent production stages, it can only be corrected by repeating all the preceding steps.

In order to make the best possible use of the available memory space in the microcontroller, the program code must be adapted to the specific type of chip that is used. Porting the software to another type of chip is thus only possible with additional effort and expense. Consequently, the time required to generate a complete ROM mask is around nine months. This can be significantly reduced if it is possible to use program code that is already on hand (in the form of software libraries). Once the development of the ROM mask has been completed, it can be formally handed over to a semiconductor manufacturer.

ROM masks and the fabrication of the semiconductor chips

Using the software that he receives in an EEPROM or on a diskette, the semiconductor manufacturer generates an exposure mask for the ROM of the microcontroller. This mask, which contains the program code, is referred to by the operating system designers as the 'ROM mask', or often simply as 'the mask'. It actually only one of approximately twenty photomasks that are necessary for fabricating the microcontroller. The microcontroller chips are built on suitably prepared high-purity silicon disks called *wafers*. The diameter of the wafers used for Smart Card microcontrollers is 6–8 inches (15.2–20.4 cm). With 0.8-μm technology, around 700 microcontrollers will fit on a 6-inch wafer.

Figure 10.5   A six-inch diameter wafer containing approximately 700 SLE44C80 microcontrollers. The milled straight edge ('primary flat') is used to identify the orientation of the wafer during manufacturing. (Source: Infineon)

In the semiconductor industry, the trend in the fabrication processes in the coming years is towards larger wafers and smaller structures. We can assume that in a few years, wafers with a diameter of 12 inches (30.6 cm) and 0.25-μm technology will become the established standard for the production of Smart Card microcontrollers. The cost of a fabrication plant at this level of technology is around 600 million dollars.

Only a few years ago, full-wafer masks were used, so that all 700 microcontrollers could be exposed at once. In this case, contact masks were normally used. As the dimensions of the chip structures became increasingly small, this was no longer possible, since the yield was not acceptable. With all new production methods, the set of photomasks represents only a single chip, rather than an entire wafer. These very delicate masks are made from plates of quartz glass, which is transparent to ultraviolet light. These plates carry the patterns for a chip, in the form of chrome-metal tracks. The track patterns are transferred to the glass plates by first coating the plates with a photosensitive layer and then using an electron-beam writer to expose the patterns. Following this, the sensitive layer is developed and the unexposed regions are removed by etching.

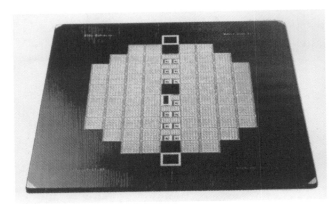

Figure 10.6   Quartz glass photomask for simultaneously exposing an entire wafer (Source: Philips)

The photomask is produced at a scale that is 5 or 10 times larger than the actual scale of the chip, which allows an image-enhancing reduction to be used when the wafer is exposed. The machines that are used to expose the wafers, which are called 'steppers' in the trade, are high-precision optical devices that can focus the image of the mask on the wafer with a precision of a fraction of a micrometer. They can also reposition the wafer with an equal level of precision. After the ultraviolet light exposure process for one chip has been completed, the wafer is moved by one step to the position of the next chip, and the exposure process is repeated at this position. The entire wafer is thus exposed one step at a time, until all the microcontrollers that it will contain have been exposed.

The entire wafer is coated with a light-sensitive lacquer called the 'photoresist'. Where the photoresist has been exposed to light, the lacquer is removed by etching, and the underlying wafer surface is then doped with impurity atoms. After the wafer has been cleaned several times and recoated with a new layer of photoresist, it is ready to be exposed using the next of the approximately 20 masks. Depending on the particular manufacturer and the fabrication process used, producing a finished wafer involves around 400 processing steps and takes six to twelve weeks. The actual processing time, however, is less than a week. The very long elapsed times in practice are primarily due to the queuing technique that is commonly used in the mass production of semiconductor devices.

In order to make the production process more economical, a group of several wafers (a *batch*) is always passed through the semiconductor fabrication machine each time. A typical batch consists of twelve wafers, and this is frequently the minimum quantity that is processed. It corresponds to around 10,000 chips for almost all semiconductor manufacturers. Processing a smaller quantity as a single batch would cause a significant amount of extra work in the production process, so that the fabrication costs would be just as high as for a full batch. However, some fabrication equipment allows 'shared' batches, in which different types of microcontroller chips with different ROM masks are produced on a single wafer. This allows smaller lots than the otherwise obligatory 10,000 pieces. However, by no means all types of microcontrollers can be produced in this manner, and not every fabrication machine can process shared batches.

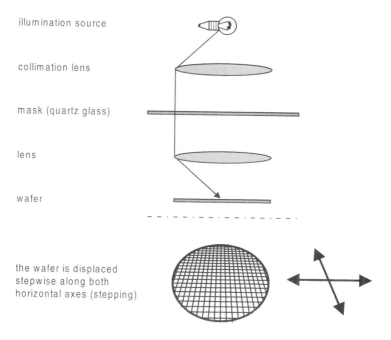

illumination source

collimation lens

mask (quartz glass)

lens

wafer

the wafer is displaced
stepwise along both
horizontal axes (stepping)

Figure 10.7   Exposing individual chips on a semiconductor wafer using a stepper.

The overall yield is around 80% with a well-tuned fabrication process. This means that only around 560 of the original 700 chips on the wafer can be used for the following production steps. However, with new production processes, the long-term yield can easily lie around 60%. This has a very negative effect on throughput and the profitability.

Chip tests on the wafer
In the next production step, the microcontrollers on the wafer are contacted using metal probes and individually tested. This requires making contact with each of the 700 microcontrollers on the wafer, either individually or in groups of up to eight, and then carrying out an electrical function test. Since there are not usually any supplementary contact groups for the microcontrollers, even at this production stage, only the five contacts that will later be used in the Smart Card can be used.

Figure 10.8   Cross sectional drawing of a stepper for the stepwise optical exposure of individual semiconductor chips (Source: ASM Lithography)

However, the functional elements are tested on the wafer significantly more intensively and extensively than later on, since the microcontrollers are still in the test mode at this stage. In the test mode, it is possible to read and write all the memories (RAM, ROM and EEPROM) without any restrictions. Any microcontrollers that fail this test are marked with a small colored dot. This allows the non-functional chips to be visually identified in the following steps, and they can be discarded after the wafer has been cut into individual dice.

Figure 10.9   Testing microcontrollers on the wafer. Contact probes are used to make connections to the IC, and each IC is individually tested (Source: Philips)

Figure 10.10 Section of a wafer with microcontroller ICs. The dots on top of some of the chips mark defective devices.

In addition, the ability to freely access the memory in the test mode is exploited to write chip-specific data to the EEPROM. This includes a serial number that must be used only once, so that it is unique for each chip. This individualizes each chip, and thereby each Smart Card. The benefit of this, in addition to certain security aspects, is that traceability as defined by the ISO 9000 ff standards is guaranteed by the unique chip number.

Sawing the wafer

After the devices have been tested on the wafer, the next step is to separate them. A thin self-adhesive foil is applied to the back of the silicon disk, so the individual chips will remain in position after the wafer has been cut. Special saws, with blades that are around 25 μm thick and spin at more than 30,000 rpm, are used to cut the wafer into pieces. Each piece of silicon crystal obtained by cutting the wafer holds a single microcontroller. These small pieces of silicon, with a maximum area of 25 mm^2, are called *dice* (*die* in the singular). Each die contains a microcontroller that will ultimately be incorporated into a Smart Card. Once the wafer has been separated into dice, the defective devices, which are marked with colored dots, are separated from the good devices and destroyed.

Figure 10.11 An individual die with a single match for size comparison. The die is approximately 0.12 mm thick.

Figure 10.12 Cutting a wafer into dice (Source: Philips)

Up to this point, it is not possible to tell whether the ROM software has been copied without any errors. For this reason, around ten dice are removed from the batch at this stage and mounted in ceramic DIL packages. The software producer receives these first sample devices, and now uses his test facilities to determine whether the software in the ROM functions correctly. The entire chip can also be tested. If an error in the software or hardware is detected at this point, the production process must be stopped, and the entire batch has only scrap value. Following this, after the error has been remedied, the production process must be started again from the beginning, with the generation of a new ROM mask by the semiconductor manufacturer. The lost time cannot be recovered, even with accelerated handling in the other production phases.

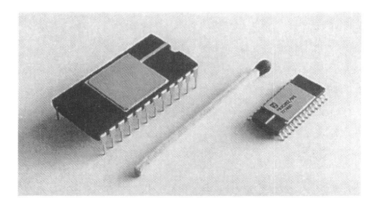

Figure 10.13 Examples of microcontrollers mounted in various types of ceramic packages for software testing

Mounting the chips in the modules

The next step in the production process, after the dice have been sawn from the wafer, is to mount them into modules. The modules increase the resilience of these very fragile bits of quartz crystal, and the electrical contacts located on their top surfaces will later be used to make the connections between the card and the terminal.

Chip modules are usually supplied on rolls of 35-mm plastic ribbon with perforated edges, which carry modules in side-by-side pairs. Depending on the size of the module, between 10,000 and 20,000 modules fit onto a single roll. The 35-mm plastic ribbon is simply referred to as 'tape' by insiders.

The width of the tape is incidentally the same as that of 35-mm film, which is commonly used in still and motion-picture cameras. The reason for using this format lies in the early days of Smart Card manufacturing. At that time, the 35-mm film format was chosen for the module carrier so that inexpensive transport and packaging techniques could be used, with a minimum of new development. This meant that existing commercial spools and winding equipment for films could be used for module tapes. Since changing to a different format was no longer feasible after this format had reached a certain level of general use, it is still employed today.

The bottom sides of the dice (the silicon base material) are permanently glued to the bottom sides of the modules. The dice can then be electrically connected to the contact surfaces of the modules in subsequent production steps.

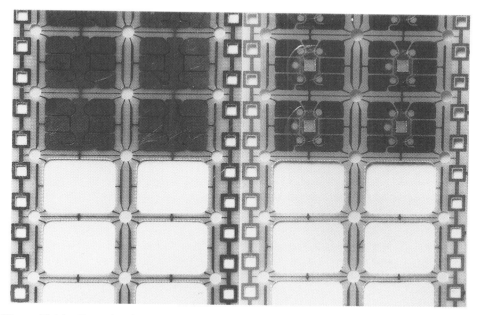

Figure 10.14   Example of a 35-mm plastic tape with attached pairs of modules. The front side of the tape picture is shown on the left, while the back side is shown on the right. The holes left by modules that have already been punched out of the tape can be seen in the lower parts of the pictures.

Figure 10.15   Example of the layouts of the front and rear sides of a module

Bonding the chips

After the dice have been glued to into the modules, the next step is to make the electrical connections to the bottom sides of the contact areas. This is done using very fine gold wire, which is welded to the aluminum contact pads on the die and the corresponding contact surfaces on the rear of the module. To prevent the bonding wires from being broken by temperature variations, each wire is formed into a loop. However, this must not be too large, since otherwise the bonding wire will not be fully covered in by the plastic material that is later poured over the chip. This would increase the risk of corrosion of the wire.

Encapsulating the chips

After the chip has been bonded, a black epoxy resin is poured over the chip and the rear surface of the module. This resin protects the fragile crystal against environmental influences such as humidity, twisting and bending. An opaque resin is used, because semiconductor devices are normally very sensitive to light and electromagnetic energy in the near-visible part of the spectrum.

After the chips have been encapsulated, the carrier tapes with the modules are wound onto large spools and packed into cardboard boxes. For small production runs, it is also possible to pack the modules individually in plastic containers. However, this is avoided when there are many devices, since it makes it difficult for the module implanter to use automated processing equipment.

If a new microcontroller is to be introduced into the market, or modified chip hardware has to be tested, the production process is often completed when the modules have been encapsulated. In this case, the modules are then passed through suitable tests and qualification stages. Only after these have been completed with no errors is it OK to start a new batch, with suitably modified software, for mass production. The situation is similar with a new operating system, in which case the production process also ends at this point. Following this comes the necessary qualification testing, which can take weeks or even months. If necessary, another pass through the correction loop with an improved version of the operating system may then take place.

Testing the modules

As a consequence of the production steps up to now – sawing the wafer, mounting and bonding the chips, and encapsulating the chips in the modules – between 3% and 7% of the dice will have become unusable. An additional test is therefore usually performed before the modules are packed and delivered. For this test, each module must be connected to the tester via the contact surfaces on the front side of the module. The first thing the tester does is to switch the microcontroller from the test mode to the user mode by blowing the polysilicon fuse and writing a special byte value to a particular address in the EEPROM. After this, it is no longer possible to externally access the memory for reading or writing without first satisfying defined security conditions.

The test computer next performs an ISO activation procedure and then tries to detect a valid ATR. If this is possible, it then tests the chip hardware using the commands included in the mask-programmed software. If all these tests are successful, the module has not been damaged by any of the previous production steps, and it can be built into a Smart Card.

10.2.2   Producing a card body without integrated coils

Card bodies for Smart Cards that do not have integrated coils can be mass produced using three different types of processes. These differ in terms of the durability of the card, the surface features and the allowed card elements. Many card manufacturers frequently offer only one type of process, rather than the full range of possibilities.

Lay persons often regard the manufacturing of card bodies as an uncomplicated, easily mastered technology that essentially only amounts to punching out a few pieces of plastic foil and gluing them together. However, this is by no means the case. The mass production of high-quality card bodies involves a multitude of complex manufacturing steps, and it demands excellent understanding of the necessary chemical processes for the plastic materials and the associated inks.

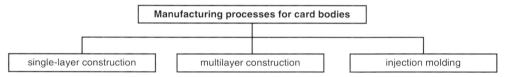

Figure 10.16   Classification of the basic manufacturing processes for plastic card bodies

The technically most complicated process is to construct the card body from several layers of plastic that are thermally bonded. This is called a *multilayer* construction, and the process of bonding the layers using heat and high pressure is called *lamination*. The thickness of the plastic foils used for the inner part of the card (the core foils) varies between 100 μm and 600 μm, while the thickness of the outer foils (overlay foils) ranges from 25 μm to 300 μm. A card body constructed in this manner allows a great degree of freedom in the form and layout of the card elements, is very stable and also allows security features to be inserted between the layers. This is, for example, what is done with the MM process for German Eurocheque cards.[5]

A single-layer construction for the card body, using a piece of 800-μm thick plastic sheet (*monofoil*), is a simplified version of multilayer construction. This process is less expensive, but the cards are less stable than multilayer cards. Above all, they allow a lot fewer choices for the design and layout of the card elements. For example, with this process it is not possible to have a transparent cover layer to protect the printed elements against scratching and rubbing.

---

5   See Section 3.1.2, 'Card components and security markings'.

The third process that can be used to produce a plastic card body is injection molding. This essentially results in a single-layer card body, with all of its advantages and disadvantages. There is however a small but significant difference. A thin printed foil (approximately 80 µm thick) can be placed in the mold, which allows injection-molded cards to be produced with printing right from the mold. This process, which is called in-mold labeling, has its limitations in comparison to offset printing or silk-screening, in terms of the layout and the inks that can be used. However, it has the advantage that it is not necessary to pass the cards through a single-card printing machine after they have been molded. An additional feature of the injection molding process is that the cavity for the chip module can formed in the molding process, so that it does not have to be milled out afterwards. There are also processes newly available in which the chip module is placed in the mold, so that it is already anchored to the card body when the card is molded. This technique also makes it unnecessary to perform many of the steps described below.

Figure 10.17   Overview of the commonly used types of card construction. The multilayer card is made up of outer cover layers (overlay foils) and internal layers (core foils).

Standard injection molding machines have a capacity of approximately 2000 card bodies per hour. Although the injection molding method is apparently inexpensive, it is nevertheless usually more expensive than stamping single-layer card bodies from a large sheet of plastic if very many cards are being made (more than one million). This is primarily due to the time-consuming handling of individual cards, which is one of the main cost factors.

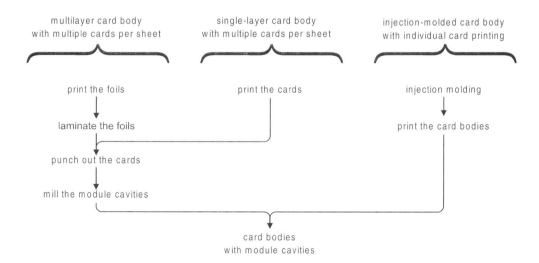

Figure 10.18   Producing card bodies for Smart Cards in Phase 1 of the Smart Card life cycle

Printing the foils

With regard to printing the card body, there is usually no difference between multilayer and single-layer cards. In the sheet printing process, large sheets of plastic are printed with multiple cards, which are then punched from the sheets. The sheets are normally big enough to allow 24 to 48 items (cards) to be printed on each sheet, which is fed one or more times through the individual inking stations of an offset or silk-screen printing press. The front and back sides of the card bodies must be printed separately.

A basic distinction must be made between offset printing and silk-screen printing, with regard to the printing processes used for cards. It is possible to print finer details on the card using offset printing than with silk-screen printing. In addition, the inks used for offset printing are hardened using ultraviolet light. This occurs immediately following printing, which has the advantage that the printed cards can be stacked right away. However, it is not possible to apply holograms or magnetic strips on top of UV-hardened coatings using the hot-stamp process,[6] since they are not thermoplastic. A similar consideration applies to internal foils for laminated card bodies, since the lamination process requires the foils to be thermoplastic. Consequently, the silk-screen printing process is used, in which the inks harden by the evaporation of a solvent and remain thermoplastic. Additional card elements can easily be added on top of surfaces printed in this manner.

In practice, the two printing processes are often used in combination. For example, large single-color areas and the background for the magnetic strip and hologram can be printed in silk-screen, while the fine details can be applied in a second step using offset printing. Silk-screen printing cannot achieve such a high level of detail.

In summary, offset printing is ideal for colored subjects with high resolutions and large piece quantities. However, if additional functional elements (such as holograms) must be permanently applied to the surface of the card, then either the entire card or at least the backgrounds for these elements must be printed with thermoplastic inks using the silk-screen process. Inks for offset printing cannot be used for this purpose, since they are hardened using ultraviolet light and are not thermoplastic.

A third process that is sometimes used is thermal transfer printing, in which a piece of colored foil is released by heating and transferred to a white card body. The colored foil bonds to the surface of the card. This process is frequently used as a purely black-and-white printing process to apply serial numbers to cards, but in principle, it can be used to reproduce all colors and shadings. However, it is slow and expensive, so it is primarily used for small quantities of cards in desktop personalization machines. It has a resolution of up to 300 dpi. An additional disadvantage of this process is that the applied colors are only stuck to the surface of the card, so that they can be scratched off.

The disadvantages of colors that only adhere to the surface of the card can be avoided by using thermal dye-sublimation printing. In this process, a print head heated to nearly $200°$ C presses the hot dye into the top plastic layer of the card body. The maximum penetration is 5 µm, which is sufficient to make the printing scratchproof. Thermal dye-sublimation printing otherwise has essentially the same characteristics as thermal transfer printing, including high costs, so it is also suitable only for relatively small quantities of cards. Even then, both processes can generally only be used to print on suitably prepared card surfaces, which limits their use. Nevertheless, they represent an entirely worthwhile complement to just-in- time printing for small and medium-sized quantities of cards.

---

[6]   This is a hot-gluing technique that requires a thermoplastic substrate.

Table 10.3 Summary of the most commonly used processes for printing cards

Characteristic	Offset printing	Silk-screen printing	Thermal transfer	Thermal dye sublimation
Sheet printing for mass production?	yes	yes	no	no
Resolution	very good	moderate	good	good
Surface coverage of the ink or dye	good	very good	satisfactory	satisfactory
Printed surface can be laminated?	no	yes	not usually	not usually
Scratch-resistant printing?	no	only with a laminated overlay foil	no	yes
Card-specific printing?	no	no	yes	yes
Cost	low	low	high	high

Laminating the foils

With a multilayer card body, after the foils have been printed, they are laminated at 100–150° C, and the required features are integrated at the same time. The printed foils are protected against scratching and rubbing by supplementary, laminated transparent cover foils, which are often also called overlay foils. Depending on the requirements of the customer, signature panels, magnetic strips and security features are also embedded into the cards or laminated on top of the cards at this stage.

It is certainly possible to use foils with different thicknesses for the front and back sides of the card. In general, though, it is better to use a symmetrical construction for mechanical reasons. This means that foils with the same thickness are used for the front and back of the card. This avoids possible problems with a 'bimetallic' effect that causes the card to curl.

Punching out the foils

Once the individual foils have been laminated, they must be brought to the desired card format. This is achieved using a punching process. The machines used for this can produce 4000–8000 cards per hour. The burr that can be seen or felt at the edges of some cards is an indication of a worn punch and die set.

Milling the module cavity

A necessary part of the process of producing a card body is milling a cavity to receive the module. There are also processes in which the foils are punched out in advance to produce an opening, into which the module is embedded when the card body is laminated. The cavity is also already present in the card body if it is produced by injection molding.

Since the back side of the module has a bump where the encapsulated die is located, a matching cavity must be formed in the card body by milling out an opening. A single-level cavity is most often disadvantageous with modern types of modules, so the cavity is usually milled with two or even three levels. This provides a larger contact area between the card body and the module, which improves the durability of the attachment of the module to the card body. In addition, it is mechanically significantly better if only the rim of the module is attached to the card body, with no physical contact between the die on the back side of the module and the card body. In a manner of speaking, the die 'floats' in the card body.

The first step in making the cavity is to mill an opening that is as large as the contact field of the module, and just as deep as the contact field is thick. Following this, an additional opening is milled in middle of the first opening, to make room for the encapsulated die. This yields a cavity with two steps.[7]

Figure 10.19   Example of a module cavity, which is a milled opening in the card body for implanting a module

The milling must be performed very precisely, since the thickness of the remaining card material at the deepest part of the cavity is only 0.15 mm. If the milling machine vibrates or rocks, the card body could be milled all the way through and thus be rendered useless. If the cavity is not deep enough, the module will stand proud of the surface of the card, which is only allowed within very narrow limits. This tricky production step is performed by fully automated machines, with the card bodies fed in from one bin and passed out to another bin. The throughput of one of these machines is around 1000 cards per hour.

Printing the card body

A second type of printing process, in addition to sheet printing, is printing the cards individually. In this process, the cards are printed after they have been separated. When the cards are printed individually, this always takes place before the cavity is milled. The throughput of machines for printing individual cards ranges up to 12,000 cards per hour. A variation of individual card printing is thermal transfer and thermal dye-sublimation printing in desktop personalization equipment. The throughput of such equipment is significantly lower, and lies around 300 cards per hour at most.

Applying card elements to the card body

Once the card bodies have been brought to the correct size, various elements such as holograms and magnetic strips are applied to them. The holograms, which are supplied in rolls, are permanently bonded to the card body using a thermal bonding technique (hot stamping or rolling).[8] After this, any attempt to remove this security feature will destroy it. Magnetic strips are applied to the card bodies using either a lamination process or a hot-stamping process.

---

7   See also Section 3.2.2, 'Chip modules'.
8   See also Section 3.1.2, 'Card components and security markings'.

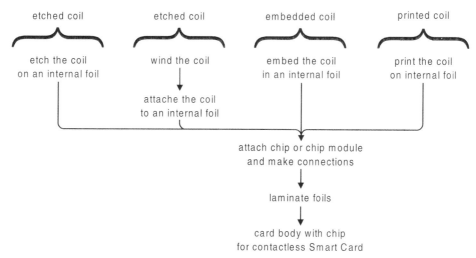

Figure 10.20   The production of contactless Smart Cards in Phase 1 of the Smart Card life cycle

10.2.3   Producing card bodies with integrated coils

Contactless Smart Cards need coils for transferring energy and data. At high frequencies, the coil can be made so small that it can be integrated into the chip module. With such a contactless card, the production process is thus nearly the same as for cards with contacts. The chip module with the coil is simply laminated between two or more plastic foils, or set into a cavity.

However, relatively low frequencies are used for most present-day contactless Smart Cards, which means the larger-diameter coils are necessary. These are usually rectangular with rounded corners, and measure approximately 75 mm by 45 mm. This means they are only slightly smaller than an ID-1 format card body. These coils normally have four turns, an inductance of around 4 µH and an electrical resistance of a few ohms. Printed coils are an exception; they have a resistance of around 300 ohms.

The production of special card bodies with integrated coils requires the standard production process to be adapted to meet the modified requirements. However, some general principles remain the same. For example, contactless Smart Cards are also mostly made in sets of 48 cards on large sheets, rather than individually, since this is significantly less expensive.

Connecting the chip to the coil

In the currently standard technique, the die that is to be connected to the coil is attached to a lead frame and electrically connected to two contact surfaces on the frame. These two contact surfaces are then used to make the electrical connections to the two ends of the coil. The reason for using a lead frame is that it allows more tolerance in the positioning of the electrical connections to the coil than if the coil is connected directly to the die. However, it is more expensive than connecting the coil directly to the die, although the latter process requires very precise positioning and sophisticated bonding techniques.

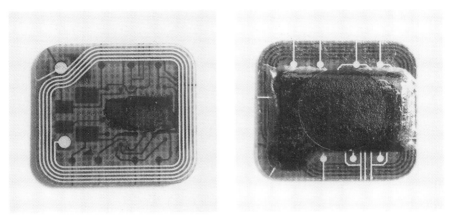

Figure 10.21   Front and rear views of a microcontroller chip module with an integrated coil

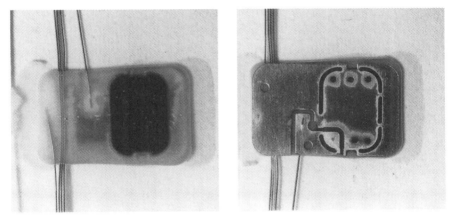

Figure 10.22   Front and rear views of a lead-frame module for contactless Smart Cards and a lead-frame module connected to a wound coil

If the coil is integrated into the card body, two different techniques are used to make the connections between the chip and the lead frame or coil. In the widely used wire bonding technique, the chip is connected using fine bonding wires. A significantly more elegant and less expensive solution is die bonding, in which the chip is pressed and glued against the lead frame or coil to make a direct electrical contact. This requires the chip to be flipped over (relative to the wire bonding technique), so this technique (and sometimes the chip itself) is called *flip-chip*. With both of these techniques, the chip is covered with a protective layer of plastic after the connections are made.

Etched coils
There are several ways to integrate a coil into a card body. In the first process that was developed for producing contactless Smart Cards, plastic foils coated with a 35-µm layer of copper are exposed and then etched to produce a coil in the copper layer. The etched track

is around 100 μm wide. After the etching, a chip is placed at the terminals of the coil and electrically connected to them. This assembly is then treated as an internal foil, to which overlay foils are laminated on the front and back sides. The Smart Card is then finished. This type of card is referred to as a contactless Smart Card with an etched coil.

Wound coils

The wound coil is a further development of the etched coil. Copper wire with a diameter of 150 μm is wound on a tapered coil form and then slid from the form onto a thin foil, to which it is attached by thermoplastic welding using heat and pressure. Following this, the chip or chip carrier is put in place and the foils are laminated. This technique is distinctly less expensive than etched coils, and it is more suitable for large quantities.

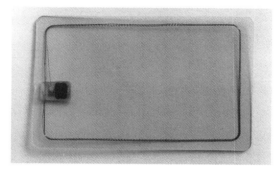

Figure 10.23   Example of a wound coil with a lead-frame module, in a card body that has been kept transparent for demonstration purposes

Embedded coils

Another method for producing wire coils is called the embedded coil technique. This works in a relatively simple manner. A coil of 150 μm-diameter copper wire is laid out directly on a plastic foil and simultaneously bonded to the foil using ultrasonic energy. A device called a 'sonotrode' is used to make the coil; it mechanically guides the wire and at the same time ultrasonically welds the wire to the plastic foil. After the coil has been made, the chip or chip carrier is connected to the coil and a cover foil is laminated on top.

Printed coils

Of all of the possible techniques, the printed coil technique is the most highly developed and the least expensive for mass production. The turns of the coil are printed on an internal foil using a silk-screen process with conductive ink. Silk screening must be used, since it allows enough ink to be applied to achieve an ink thickness of around 50 μm. This is necessary to prevent the electrical resistance of the coil from being too high. After the coil has been printed, the chip is connected by die bonding and encapsulated with a layer of epoxy resin. The final production step is laminating a protective cover foil on top of this assembly. The main advantage of this technique is its high throughput, due to the technical simplicity of the printing process. It is thus exceptionally well suited to the production of large numbers of cards. However, it takes a considerable amount of expertise to achieve the necessary level of quality in the printing and die bonding.

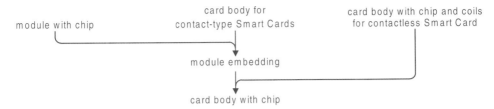

Figure 10.24   Implanting the module in the card body for a contact-type Smart Card, in Phase 1 of the Smart Card life cycle

10.2.4   Combining the card body and the chip

The final step in the production process is implanting the modules from the semiconductor manufacturer or module producer into the prefabricated card bodies from the card manufacturer. The mechanical aspects are the most important in this step. Nonetheless, a certain amount of specialized expertise is needed to durably anchor the modules in the cavities of the card bodies. This is not as simple as just pasting clippings in a scrapbook.

Implanting the module

Regardless of the method used to produce the card body and create a cavity for the module, the module must be embedded in the card body in the next step of the production process.

Normally, double-sided tape with hot-melt glue is used to attach the module to the card body. Only the supporting surface around the rim of the module is glued to the card body, while the encapsulated die in the middle of the module remains free. The module thus has a 'floating' attachment to the card body. To achieve this, the tape holding the adhesive must be pre-punched and then applied to the modules on the 35-mm carrier tapes so that only the edges of the modules are covered. After this, the individual modules are separated from the carrier tape and glued into the card bodies using the attached adhesive. The durability of the bond depends primarily on the proper combination of heat, pressure and time.

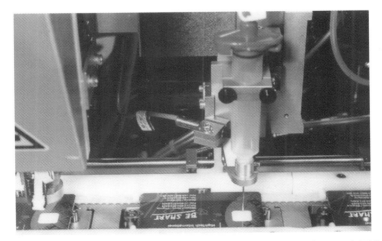

Figure 10.25   Placing the module in the cavity in an implanting machine (Source: Mühlbauer)

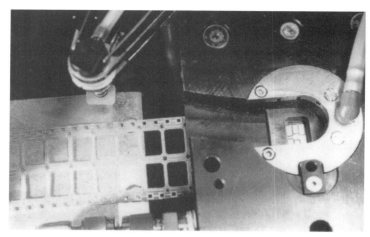

Figure 10.26   Stamping the modules from the tape in an implanting machine (Source: Mühlbauer)

The problem with this hot gluing process, which requires considerable expertise, is that the modules are briefly heated to around 180° C. This normally lasts approximately one second, but if it lasts too long the modules will be destroyed by being overheated. In any case, this brief heating artificially ages the chips, although this normally does not have any negative effect. The implanting machines used for card production can process around 2000 modules per hour, which amounts to one embedding operation every 1.8 seconds.

Other methods, such as using liquid cold-setting glues, are also used, but the hot-gluing method is still considered to be very reliable. The main problems with using liquid glues that are injected into the milled cavity are the lack of a clearly defined surface of adhesion and the tendency of the glue to harden over time.

Figure 10.27   Injecting liquid glue into the pre-milled module cavity in an implanting machine (Source: Mühlbauer)

Figure 10.28   Electrically testing the module after it has been glued into the card body in an implanting machine (Source: Mühlbauer)

Once the module has been implanted in the card body and all the non-personal features and printing have been applied to the card, the mechanical production of the Smart Card is complete.

10.3   Phase 2 of the Life Cycle in Detail

According to the ISO 10202-1 standard, Phase 2 of the Smart Card life cycle describes the loading of all data that are not card-specific, as well as the implanting of the chips in the prepared card bodies. Both Phase 2 and Phase 3 are frequently carried out by a single firm, although in such firms the two phases are normally completed separated, both organizationally and physically, for reasons of security.

A production planning and control (PPC) system is frequently used to coordinate these complicated production processes. The various processing machines draw their data from this system, and in parallel with this, they report current processing status to a central control room. This minimizes the time and costs involved in controlling the mass production of Smart Cards. An additional benefit of the PPC system is that networking the processing equipment makes the data needed for quality assurance and testing available for near-real-time evaluation.

Data transfer
The card issuer or application provider must provide the card personalizer with all the data that relate to his application. This includes for example the name of the application, the structure of the file tree, the necessary files and the file structures. This information is loaded into the cards when they are initialized. Furthermore, the personalizer also needs all customer and system-specific data, such as secret card-specific keys and the names and addresses of the cardholders. This information is transferred using diskettes, magnetic tapes or data telecommunications.

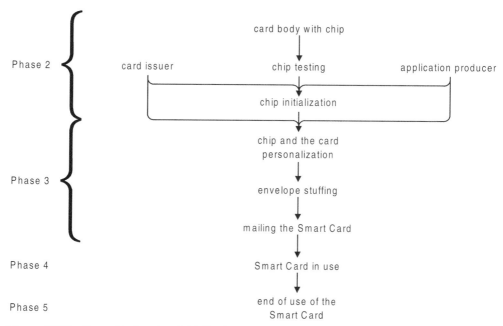

Figure 10.29   Smart Card testing, initialization, personalizing, use and end of use in Phases 2–5

The personalization data are almost always sensitive with regard to security, which is why the transport path and the transfer must be suitable protected. Consequently, the data are normally encrypted. The associated decryption key is naturally transported to the personalizer via a different route than the data. This means that the personalization data are worthless of they are lost, since it is not possible to decrypt them without the key.

However, there are many Smart Card applications in which no transfer of card-specific data takes place. The best known example is GSM cards, which are not manufactured for any specific card producer, but instead contain only individual data and keys. The necessary data sets are most often directly generated by the card producer and then sent back to the application operator, so that the latter knows which cards have been produced. The only sense in which a data transfer takes place is that the card producer receives first the data that are the same for all cards and second the initial and final values of the card-specific data. The data sets for each of the individual cards are then generated in security modules in the processing equipment.

Electrical tests

The first production step of this phase is an electrical test of the Smart Card. A basic test is made by carrying out the ISO Smart Card activation sequence, to which the card must respond with a valid ATR. If the ATR can be received and it meets expectations, then it is certain that at least of core of the microcontroller is operational. Following this come special tests for the hardware components, such as the ROM, EEPROM and RAM. Special machines that can process several cards in parallel are used to achieve high throughput with these tests, some of which can take up to several seconds. Typically, machines with carrousel construction and a throughput of up to 3500 cards per hour are used.

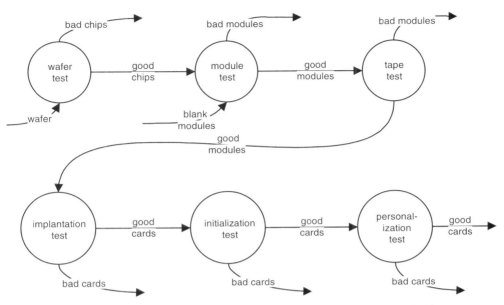

Figure 10.30   Material flow diagram showing the typical electrical tests for Smart Card microcontrollers that are performed during production

wafer test:	wafer test performed by the semiconductor manufacturer
module test:	test performed by the module implanter after bonding the chip
tape test:	module test on the 35-mm tape
implantation test:	Smart Card test after the module has been implanted
initialization test:	Smart Card test after the card has been initialized
personalization test:	Smart Card test after the card has been personalized

Figure 10.31   Contact head of an incoming inspection machine for modules on 35-mm tape, which can process 16 cards in parallel

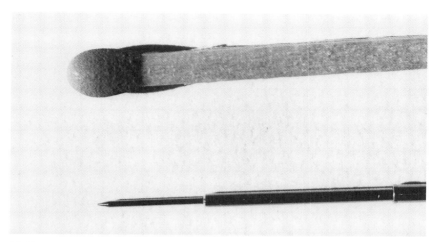

Figure 10.32   Detailed view of a contact needle for the contact unit of a high-performance Smart Card test station. Although the shape of the contact probe needle does not conform with the ISO/IEC 7816-2 standard, this type of needle is often used in production facilities, due to its reliability.

The preferred way to test the operation of the EEPROM is to write a 'checkerboard' pattern, such as 'AA' (°1010 1010°) or '55' (°0101 0101°), to the individual bytes. However, since with large EEPROMs in particular this would take very long, a trick is sometimes used to shorten the test. Instead of using the specified EEPROM writing time, which might for example be 3.5 ms per page, only one tenth of this time is used (350 µs in this example). The data will be retained in the EEPROM for only a few minutes with such a short writing time, but this does not cause any problems in this case, since testing the checkerboard pattern in the memory is already completed a few seconds after the data have been written. The advantage of this dynamic form of EEPROM programming is a significantly reduced testing time with the same level of quality. The same technique is sometimes used when the transmit and receive buffers of the I/O manager are located in the EEPROM instead of the RAM. The reduced writing time in this case produces a marked increase in the effective data transfer rate.

There is another interesting trick that is used for the electrical tests. In order to reduce the amount of time required to load the data in the subsequent production steps, a final test pattern (such as '00') is written to the entire EEPROM, but in this case using the normal writing time. Since the value that is already stored in the memory is known in the subsequent completion, initialization and personalization processes, only the data that are different from this value have to be actually written to the EEPROM. A similar technique can be used to set the contents of the EEPROM to a value that makes it unnecessary for subsequent writing operations to first erase the page to be written. Both of these tricks distinctly reduce the times required to carry out all subsequent production steps in which data are written to the EEPROM.

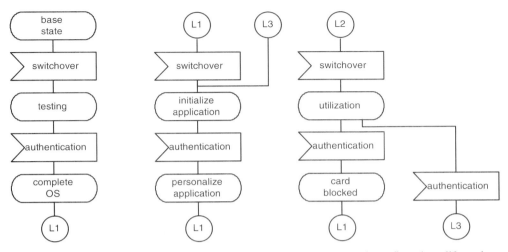

Figure 10.33   Smart Card operating system state machine for implementing a five-phase life cycle

Completion

Most operating systems are only partially located in the mask-programmed ROM of the Smart Card. The link tables, and part of the program code, are loaded into the EEPROM of the Smart Card only after an authentication with a secret key. The process of loading the EEPROM part of the operating system is called *completing* the operating system.

This approach allows small adaptations to be made to the ROM program code for error rectification or special application cases, without requiring a new ROM mask. The Smart Card operating system is not fully present in the Smart Card until the EEPROM data have been written to the card. After this, it is possible to execute all application commands, such as SELECT and READ RECORD.

Card completion, in which the data are the same for all cards for a particular application, is performed using machines that operate on several cards in parallel, just as with the incoming inspection of the cards.

Initialization

The completion of the card provides it with the software that is necessary for the next production step, which consists of loading all the data that belong to an application and are the same for all Smart Cards in that application. These are first the application data that do not vary from card to card, and second all other non-personal data that are also the same for every Smart Card. This step is called *initialization*.

Seen from the file level, initialization consists of creating all necessary files (MF, DFs and EFs) and filling them as much as is possible with the application data. With modern operating systems, this is done using the CREATE, UPDATE BINARY and UPDATE RECORD commands. This processing step is the last one in which all Smart Cards are treated the same. Consequently, initialization can also be carried out using fast machines that operate on several cards in parallel. The card-specific application data and the personal data are not loaded into the Smart Card until the following step, which is *personalization*.

Figure 10.34   An initialization machine that can process 40 cards in parallel (Source: Mühlbauer)

The reason for making a distinction in the production process between general, global data and specific, personal data has to do with minimizing production costs. Personalization machines that can write specific data to each individual Smart Card under the required security conditions are technically complex and have a throughput of around 700 cards per hour. They are usually also equipped with relatively slow labeling units for the card bodies. This results in high unit costs for loading the data into the cards. Consequently, an attempt is always made to load all global data, which does not differ from card to card, into the cards using simpler and faster initialization machines, which can process around 3500 cards per hour.

The bottleneck for both initialization and personalization is transferring the data to the card and writing it to the EEPROM. The time required for write accesses to the EEPROM cannot presently be reduced, due to technical limitations. However, the time required to transfer the initialization and personalization data can be strongly reduced by increasing the clock rate and reducing the divider value. For example, many initialization and personalization machines work with data transfer rates of up to 115 kbit/s instead of the 9600 bit/s, which is the normal operational value for Smart Cards. In some cases, this can reduce the initialization or personalization time by a factor of up to two.

The following numerical example clearly illustrates that even very small improvements in processing times can be worthwhile in the mass production of Smart Cards. We assume that one million cards are to be initialized with 4 kB (4096 bytes) of data each, and that there are two initialization machines available that work two shifts per day (16 hours per day). The initialization is carried out using 40 commands, and employs the T=1 transmission protocol with 12 data bits for each byte of data transferred. In addition, the EEPROM write cycle time is 3.5 ms for a 4-byte page, and a prior erase operation is not necessary. The transport time for the initialization machines, which are not equipped with terminals that work in parallel, is 1 second per Smart Card, and any preparation time (for loading or emptying bins, for example) is not taken into account. The resulting cycle time is thus the sum of the EEPROM writing time, the data transfer time and the transport time.

Using the formulas given in Section 14.2 ('Formulas for Estimating Processing Times'), with a data transfer rate of 9600 bit/s, we obtain a processing time of 90.7 days. If the data transfer rate is increased to 38.4 kbit/s, the time required to process one million cards drops to 52.5 days. A data transfer rate of 115 kbit/s would be ideal, since with this rate we could finish the card production more than 46 days earlier than at 9600 bit/s.

From this example, it is clear that in particular when a large amount of data has to be stored in the Smart Cards, it is worthwhile to invest time and effort in optimizing the processing. The described increases in the data transfer rate depend only on the Smart Card operating system, and do not require any special chip hardware, such as would be necessary for writing the data to the EEPROM faster. Consequently, it is possible to reduce the initialization time with all suitably prepared Smart Cards.

Table 10.4 Processing time for the initialization of Smart Cards with different data transfer rates. The basic assumptions and general conditions are explained in the text.

Data transfer rate	9600 bit/s	38,400 bit/s	115,200 bit/s
EEPROM writing time	3.584 ms	3.584 ms	3.584 ms
Data transfer time	5.870 ms	1.468 ms	0.489 ms
Resulting period	10.454 ms	6.051 ms	5.073 ms
Resulting processing time	90.7 days	52.5 days	44.0 days

10.4   Phase 3 of the Life Cycle in Detail

Phase 3 primarily covers the part of the life cycle consisting of the visual and electrical personalization of the Smart Card. As with Phase 2, this phase normally occurs in a highly automated production environment that is designed for processing large numbers of cards.

Personalization (individualization)

The next step in producing a Smart Card that is ready to be issued to the user is *personalization*, which is sometimes called 'individualization'. The term 'personalization', in the wider sense, means that all data that are assigned to a particular person or a particular card are entered into the Smart Card. These could for example be a name and address, but they could also be card-related keys. The only thing that matters is that the data are specific to a particular card.

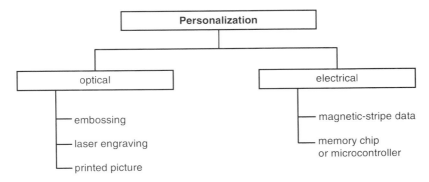

Figure 10.35   Classification of the elements of a Smart Card that can be personalized

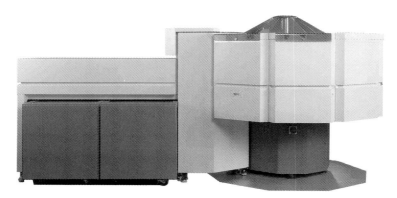

Figure 10.36   View of a modular personalization system (Luchs 5000), which includes an integrated laser labeling assembly (Source: Giesecke & Devrient)

A basic distinction is made between optical and electrical personalization. The embossing characters, as well as text or pictures applied to the card using laser engraving, constitute the optical part of personalization. The electrical part consists of loading personal data into the microcontroller and writing data to the magnetic strip. The processing time for optical personalization depends very strongly on the particular features, and cannot be generally specified. Electrical personalization usually takes between 5 and 20 seconds, depending on the amount of data.

Names and similar card-specific, character-based information is embossed onto the card by a machine in which metal letter punches are hammered against the rear side of the card at great speed and with considerable force. Since this is a relatively simple procedure, but one which is very loud and produces a lot of vibration, the machines in question are usually physically separated from the rest of the processing equipment. Laser engraving equipment, with which regions just below the overlay foil of the card body can be blackened using a laser beam, is very often employed instead of mechanical embossing. This technique is also useful if it is necessary to have a black-and-white picture on the card body.

The data for the chip are written to the memory in the same way as for initialization. However, to the extent that this involves secret keys, cryptographically protected data transfers[9] are often used to prevent an attacker from deriving any benefit from tapping the data line. For cards that are used for financial transactions, an even more complex method is sometimes used. This involves using a special security module in the personalization machine to re-encrypt the encrypted personalization data received from the card issuer and then load it directly into the Smart Card. The advantage of this method is that the personalizer does not know the secret data in the card and also has no possibility of spying it out by tapping the data lines.

The technical trend in Smart Card personalization is heading more and more in the direction of a process that is cryptographically fully secured. This means that in principle the work can be carried out by inexpensive service firms in non-secured facilities. Nowadays, there are also processes in which the personalizer receives the card-specific data recorded on a CD-ROM. In this case, the production data set with its associated card-specific key is inseparably linked to the unique chip number of the microcontroller. Among

---

[9]   See also Section 6.6, 'Secure Data Transfers'.

other things, this makes it impossible for the personalizer to produce duplicates of Smart Cards, unless he can somehow manipulate the operating system. However, this method has the disadvantage that some of the delivered data sets cannot be used if any of the chips are faulty, since the defective chips are no longer available. If this method is used, the personalizer must always report back to the party that generated the data to let them know which chips have actually been processed. With the personalization procedures that are presently commonly used, this is not necessary, since it is easy to reproduce a faulty card. Incidentally, this is also why the personalization facilities of card producers are always secured areas.

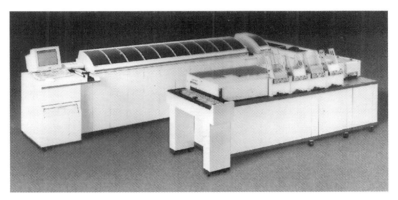

Figure 10.37   View of a modern personalization system (Data Card 9000), which includes integrated postal processing (Source: Data Card)

Unfortunately, the cryptographic procedures and security measures that are used in the realm of personalization are largely secret, so it is not possible for us to describe any particular application. However, Figure 10.38 shows an example of an initialization process followed by a personalization process, seen from a cryptographic perspective. Both of these production steps must take place in separate rooms with separate personnel, in order for the cryptographic protection to be effective.

The illustrated procedure works as follows. In the initialization, a card-specific key (KD) is derived in a security module, using the unique chip number and a master key (KM). This key is sent as plaintext to the card, where it is stored. Naturally, a lot of other data must be written to the Smart Card during the initialization, but the generation and storage of the card-specific key (KD) is the only cryptographically relevant step.

Following this, the card is personalized. This can take place either immediately following the initialization, but it may also take place several weeks later. The important factor is that the personalization must be completely separate from the initialization, to prevent a KD that has been illicitly acquired during the initialization from being used during the personalization to decrypt the card-specific data.

In the personalization process, the personalization data that have been encrypted with a common key are decrypted for each individual card by the security module. This is necessary because the generator of the personalization data does not know the individual chip numbers, which are independently generated by the semiconductor manufacturer. The security module then computes the card-specific key (KD) from the card number that it receives from the Smart Card and the master key (KM). Now the security module and the Smart Card have a shared secret in the form of the KD. This is used to encrypt the personalization data, which are then transferred in encrypted form to the Smart Card, where they are decrypted and written to the appropriate locations in the EEPROM. This process provides complete cryptographic protection of the personalization procedure. It protects the data to be used for personalization against being spied out, as long as the key (KD) that is written to the card during the initialization remains secret.

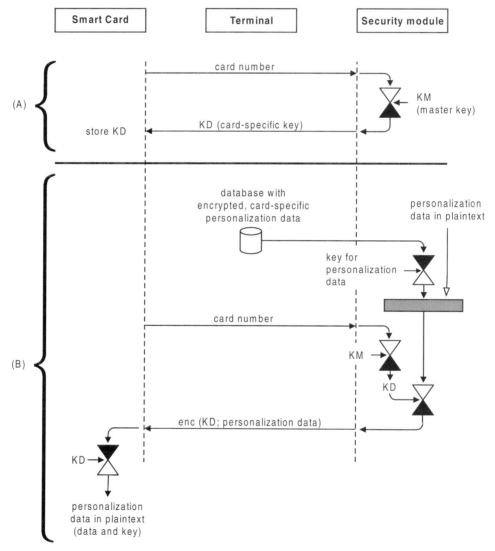

Figure 10.38   Schematic representation of a possible procedure for (a) initialization and (b) personalization with cryptographic protection for transferring data and keys. Only the cryptographically relevant processes are shown.

At the conclusion of the personalization process, the personalization machine runs several quality control tests on the finished Smart Card by. In the latest machines, for example, each card is scanned by a camera, and the optical personalization is evaluated by a computer and checked against a production database. In case of an error, the card is ejected into a faulty-card bin and a new copy of the card is automatically produced. Normally, the personalization data in the microcontroller are also checked. However, this is technically difficult to do, since read access is no longer allowed to many of the files. Consequently, special security modules for these tests are frequently present in the personalization machines. These modules contain secret master keys with which the personalized keys in the Smart Cards can be tested for correctness, possibly via an authentication.

Another approach is to provide command strings and corresponding response strings for each individual card to the personalizer. The personalizer then sends these commands in the correct sequence to the Smart Card and compares the responses received from the card with the accompanying the commands. If they do not match, the Smart Card is not behaving as expected, and a personalization error must have occurred. With this method, it is not necessary to have a special security module for the tests in the personalization machine.

Once the Smart Cards are personalized, it is normally not possible to reverse any of the processing, which means that an incorrectly personalized Smart Card is worthless. The electrical personalization is the more prone to error, and any errors that occur in this process with a large batch of cards would result in great financial damage and the loss of a large amount of time. There are thus a few Smart Card operating systems that allow the complete personalization to be deleted following a suitable authentication. With regard to the operating system, the Smart Card behaves afterwards the same as after the semiconductor fabrication step or the completion step. This capability is sometimes used for test cards, since it allows the software in the card to be modified without having to scrap it every time. Such mechanisms in the Smart Card operating system are occasionally also enabled for regular cards, so that a card can be depersonalized if necessary.

As a rule, the process of personalizing Smart Cards is only undertaken for typically ten thousand or more cards. In many applications, however, it must be possible to reproduce individual customer-specific Smart Cards. For example, it must be possible to replace a defective or lost Eurocheque Smart Card within a few days, since otherwise the cardholder will no longer be able to obtain any money from cash dispensers. There is an increasing demand for this sort of just-in-time personalization equipment in connection with a rising degree of customer friendliness. It is usually installed alongside the mass-production personalization equipment, receives the card data via data telecommunications and uses Smart Cards that have already been initialized and held as partly-finished products. With this sort of card production, it can be guaranteed that the end user (the cardholder) can receive a new card within 24 hours if necessary. This sort of equipment, which is designed for fast turnaround, is naturally not suitable for the mass production of Smart Cards.

Packing and shipping

The final processing step in the production of Smart Cards is packing and mailing the cards. This is not necessary with some types of cards, such as pre-paid phone cards, which are frequently shipped *en masse* to the card issuer. However, for more sophisticated and expensive cards, it is standard for the cardholder to receive a personalized letter with his or her new card. With some applications, such as credit cards, the cardholder also receives a letter with the PIN. For reasons of security, this is sent separately and a few days later than the card. The area in which all of these activities take place is often called the 'lettershop'.

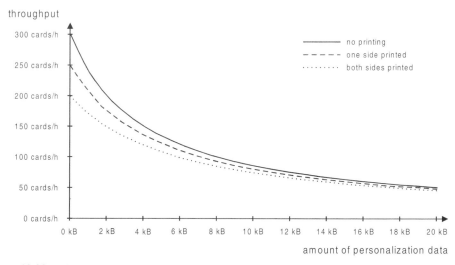

Figure 10.39   Throughput diagram for electrical personalization with single-sided and double-sided card printing, using a desktop personalization machine

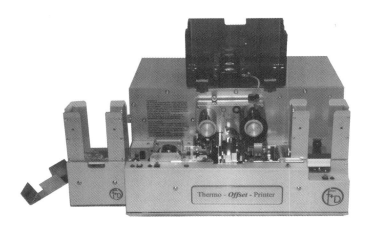

Figure 10.40   Example of a desktop personalization machine for electrical personalization and double-sided color printing with a resolution of 300 dpi. The input stack of cards is located on the right-hand side, while the stacks of good and rejected cards are located on the left. (Source: F + D)

The PIN letter envelope is made with a carbon-paper coating on the inside. This allows a slip of paper inside the envelope to be printed from the outside using a dot-matrix needle printer. It is also constructed such that an unauthorized person cannot read the printed PIN code without visibly damaging the envelope. These measures ensure that it is not possible for someone to spy out PIN codes without being noticed, even while the PIN letters are being generated. High-performance printing systems for PIN letters can print up to 34,000 documents per hour.

Figure 10.41   View of an installation for attaching cards to their associated letters. These are then stuffed into envelopes along with any necessary attachments. This machine can prepare and stuff up to 7000 envelopes per hour. (Source: Böwe Systec)

For mailing the cards, the personal information (such as the cardholder's name and address) are either read from the card or retrieved from the production database, depending on the card type. This information is printed on a 'card carrier', which is a pre-printed letter, using a high-throughput laser printer. The letter may have two punched slots to hold the corners of the card. Alternatively, a strip of easily-removable adhesive material is often used to attach the card to the letter. Following this, the card carrier is folded and inserted into an envelope. After the envelope has been franked, the Smart Card with the personalized letter is ready to be mailed to the cardholder. High-performance envelope stuffing machines have a throughput of around 7000 letters per hour.

The final quality control step is to automatically weigh the finished letters containing the cards. The weight of the card, which is around 6 grams, is easily enough for reliable verification that each envelope actually contains a card.

In order to minimize postage costs, it is common to presort the letters by postal code before handing them over to the post office. This optimization is most easily realized by producing the cards in the order that is necessary to satisfy the postal sorting criteria (such as a regional code followed by a local code).

The production steps and phases that have been described thus far represent a mass production process, which is standard for cards such as GSM cards and credit cards with chips. Other applications or card issuers may have other basic requirements with regard to card production. For example, some GSM Smart Cards are personalized on site in the shop and then handed directly to the customer. The customer naturally receives a good impression of the competence and capability of the shop if he or she can receive a personalized card immediately after signing up and paying. However, this depends very strongly on marketing policy and the security requirements of the card issuer. In contrast to this example, the production of card bodies and modules is basically independent of the ultimate card issuer or his marketing considerations, and thus largely the same for all applications.

Table 10.5   Summary of the relative cost factors for two types of Smart Cards, containing microcontrollers with different memory capacities

Component or production step	Smart Card with: ≈ 6 kB ROM ≈ 1 kB EEPROM ≈ 128 bytes RAM	Smart Card with: ≈ 16 kB ROM ≈ 8 kB EEPROM ≈ 256 bytes RAM
Die	50.0 %	65.0 %
Module	25.0 %	15.0 %
Card body	12.5 %	10.0 %
Initialization and personalization	12.5 %	10.0 %

10.5   Phase 4 of the Life Cycle in Detail

Phase 4 of the life cycle of a Smart Card is well known to a normal card user from daily experience with his or her own cards. New applications can be downloaded or activated, and applications already present in the card can be de-activated if necessary. Since the majority of this book addresses this phase, it is not necessary to deal with it any further here.

10.6   Phase 5 of the Life Cycle in Detail

Phase 5 of the life cycle of a Smart Card according to the ISO 10202-1 standard defines all measures relating to the termination of the use of the card. Specifically, these are the de-activation of the application(s) in the Smart Card, followed by the de-activation of the Smart Card itself. Both of these processes, however, are purely theoretical with most Smart Cards. In practice, cards are either thrown into the trash or carefully labeled and filed away by collectors for some indeterminate length of time. Only rarely are cards returned to the card issuer.

However, there are commands for the de-activation of individual applications and the complete Smart Card. The ISO/IEC 7816-9 commands DELETE FILE, DEACTIVATE FILE, TERMINATE DF and TERMINATE CARD USAGE are explicitly intended to be used to herald the final stage of the application life cycle.[10] Alternatively, the EN 726-3 LOCK command can be used to permanently block files, in order to make an individual application or the entire Smart Card unavailable for further use.

These commands are essential for managing individual applications in multi-application cards, but they are rarely used with present-day Smart Cards, which mostly incorporate more or less only one application. The easiest way to end the life of a Smart Card is to simply cut it into pieces with a pair of scissors. Anyone can do this, and some card issuers recommend this method for 'terminating' Smart Cards.

Nevertheless, in some cases, it would certainly be justified to return the Smart Cards to their issuer. They still contain secret keys that in part are still valid. If a potential attacker could come into possession of several hundred or even a thousand cards, he would have a

---

[10]   See also Section 7.8, 'File Management Commands'.

significantly larger pool of data for analyzing the hardware and software of the Smart Card than if he had only a few cards. The statistical investigation of a large number of cards will always be more informative than the investigation of individual cards.

With this consideration in mind, as well as well-known environmental considerations, some card issuers therefore collect expired cards when they issue new cards. In addition, collection bins for used-up telephone cards are often placed next to card phones. Effective recycling of cards is only possible after the cards have first been collected.

Recycling

We must honestly admit that the recycling of Smart Cards is not very well developed. For one thing, there are presently simply not enough cards collected for a proper recycling process, and the amount of material to be recycled is anyhow not all that large. In 1997, approximately 40,000 metric tons of plastic were used in the whole world for the production of Smart Cards. Even under the fully idealistic assumption that an equal mass of cards can be separated out and fed back into a recycling process, this amounts to a negligibly small amount in comparison to the total yearly world production of plastics, which is around 13 million metric tons.

This will however change with the steadily increasing use of cards. Recycling of Smart Cards in particular is a problematic subject. The card body, which is laminated from several layers of various types of plastic, is a highly heterogeneous material. In addition, the cards are printed with several different kinds of ink and contain holograms, signature panels and magnetic strips, all of which add to the number of different materials in the mix. Large amounts of homogeneous materials can only be accumulated during card production, such as the scrap material from punching card bodies out of single-layer plastic sheets. It is relatively easy to reuse these materials, and may card manufacturers already do so.

Table 10.6   Summary of the major components of Smart Cards, in terms of weight

Component	Material	Weight
card body	various plastics (e.g. PVC, PC, ABS)	4.400 g
inks on the card body	resins and pigments	very low
magnetic strip	iron oxide and similar materials, ink and adhesive	very low
hologram	aluminum and adhesive	very low
microcontroller (10 mm^2)	silicon with various doping elements	0.009 g
bonding wires	gold or aluminum	very low
encapsulation blob for the microcontroller	epoxy resin	0.010 g
adhesive to hold the module in the card body	epoxy resin	very low
module with six contacts	epoxy resin, glass fibers, nickel, aluminum, gold	0.170 g
module with eight contacts	epoxy resin, glass fibers, nickel, aluminum, gold	0.180 g

With Smart Cards that are no longer used, on the other hand, it is currently practically impossible to separate the cards into homogeneous sorts of material. The presently proposed recycling method is to punch the modules out of the cards and then shred the rest of the card bodies. The plastic shreddings can be used to produce low-quality plastic items (garden ornaments are a typical example of this type of recycling). The modules can also be finely ground, and the metals that they contain can be recovered using electrolytic processes. These processes are however presently not used anywhere on a large scale. In addition, it is not entirely clear that this sort of complex recycling truly protects the environment better than simple incineration or burial.

In the case of contactless Smart Cards with coils of copper wire or conductive ink embedded in the card body, it is *de facto* impossible to separate the material of the card into individual sorts of plastic.

Particularly in the case of multilayer cards, the only practical approach is high-temperature incineration, which some people rather arrogantly refer to as 'energy recycling'. If the temperature is sufficiently high, relatively few harmful materials are released. It remains to be seen whether this solution will be considered to be acceptable in the long term. After all, even though a single Smart Card weighs only 6 grams, the net weight of one million such cards is 6 metric tons.

11

Smart Card Terminals

The only connection between a Smart Card and the outside world is the serial interface. There is no other way in which data can be exchanged. An additional device that makes the electrical connection to the card is thus needed. In this book, such a device is always referred to as a *terminal*. However, other terms are used, such as interface device (IFD), chip-accepting device (CAD), chip-card reader (CCR), Smart Card reader[1] and Smart Card adapter. The basic functions, which are to supply power to the card and to establish a data-carrying connection, are the same for all of these devices.

Any terminal that consists of more than just a contact unit, a voltage converter and a clock generator always has its own processor and associated memory. This usually has an 8-bit or 16-bit architecture. In simple equipment, the processor can be part of a microcontroller, but it is often a component of a single-board computer. Terminals are usually programmed by their manufacturers using the C language alone, although a few terminals are programmed in C++.

The difficulties associated with allowing third parties to program terminals are the same as with executable program code in Smart Cards, which means that the same sorts of solutions are employed. The Europay Open Terminal Architecture (OTA), with a Forth interpreter, was one of the first attempts at a solution in 1996, and Java for terminals is already on the horizon. The EMV specification also explicitly includes the concept of downloadable program code.

Terminals do not have hard disk drives of their own, which means that they must store their programs and data in battery-backed RAM, EEPROM or flash EEPROM. The amount of memory available for this is usually on the order of a megabyte.

In contrast to Smart Cards, which all have very similar technical constructions, terminals are built in many different ways. A fundamental distinction can be made between portable and fixed terminals. Portable terminals are battery-powered, while fixed terminals are preferably powered from the mains network or via the data interface. Terminals can also be classified according to their user interface. Portable devices in particular may have displays and simple keypads to allow the most important functions to be used on site.

Fixed terminals also often have displays and keypads, but they have permanent links to higher-level computer systems as well. If the terminal does not have a man–machine interface (i.e. no display and no keypad), there must be a direct connection to a computer to provide the link between the Smart Card and the user.

---

[1] The term 'Smart Card reader' should not be understood to mean that data can only be read from the card using this device. Write accesses are naturally also possible.

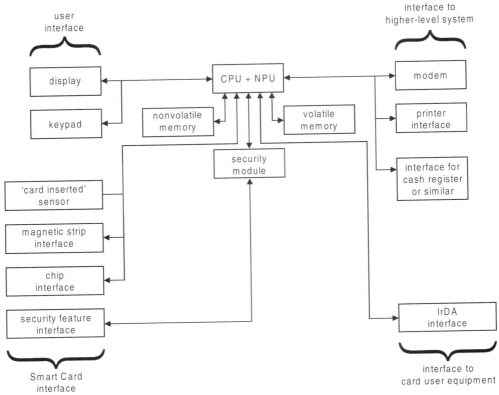

Figure 11.1   Typical architecture of a Smart Card terminal, with a display, keypad, magnetic-stripe
reader and security module. Such terminals are often used at point-of-sale locations to allow
payments to be made with a wide variety of cards (credit cards, debit cards and electronic purses).
A keypad that is specially protected against tampering (a PIN pad) can be used if necessary. This
diagram shows the basic energy and data flows, and is not a schematic diagram.

There are also a small number of terminals that have infrared interfaces. These usually
support the international IrDA (Infrared Data Association) standard, and are used for direct
communication with a personal digital assistant (PDA) or a mobile telephone. The
advantage of this is that the user can assume that his or her own device is trustworthy and
thus does not have to enter a PIN into an unfamiliar terminal.

The division into portable and fixed terminals leads to a further distinguishing feature,
which is how the terminal is used. An *on-line* terminal has an uninterrupted connection to a
remote computer during operation, and this computer takes over part of the control
function. A typical example is a terminal used for physical access control, which is
completely controlled by a background system to which it is permanently connected.

The opposite type of terminal is an *off-line* terminal. Such a terminal works completely
independently with respect to any higher-level system. However, although there are very
many on-line terminals, there are practically no 'pure' off-line terminals. All off-line
terminals occasionally exchange data with a background system, if only to request a new
blacklist or update the terminal software.

Figure 11.2   Example of a typical Smart Card terminal, which can be connected to a computer via a serial interface (Giesecke & Devrient model CCR2)

Figure 11.3   Example of a portable Smart Card terminal for financial transactions using credit cards, debit cards and electronic purses (Giesecke & Devrient model ZVT 900). This terminal has an integrated security module and a built-in printer, and can be used off-line.

In typical applications within a building, the physical link between the terminal and the remote computer is either an electrical cable or a fiber-optic cable. However, the link can also be formed by a telephone connection to the nearest computer center, as is the case with point-of-sale payment terminals. This can be either a dial-up link or a permanent link, depending on the application. Since permanent links (leased lines) are expensive, the tendency is more and more to use the telephone line only as necessary, in order to reduce operating costs. This means that the terminal must be equipped with a dial-up modem.

Smart Card terminals in the form of PC cards (formerly called PCMCIA cards) do not readily fit into this classification scheme. They can be used both on-line and off-line, and with both desktop and portable computers. In principle, such terminals are just simple and usually inexpensive hardware interfaces between a Smart Card and a computer. The only prerequisite for using a PC-card terminal is a PC card slot, which must be either a type I slot (3.3 mm high) or type II slot (5 mm high), depending on the manufacturer. Some PC-card Smart Card terminals contain expansion memory for the Smart Card and coprocessor

ICs for mass data encryption and decryption, in addition to the Smart Card interface. These terminals, which are only a few millimeters thick, are certainly the most versatile of all. They open application areas for Smart Cards that in some cases are totally new. With such terminals, it is now possible for Smart Cards to work together with standard PCs and standard software without additional cables, power supplies and external hardware. The spectrum of possible applications is very wide. It includes access protection for specific PC functions, software copy protection and e-mail transfers protected by digital signatures.

Figure 11.4   View of a typical Smart Card terminal in PC-card format (Gemplus model GPR400)

There are also 'diskette terminals' available. These provide a simple means to exchange data between a Smart Card and a PC. Such a terminal has the form of a 3.5-inch diskette and contains a very thin contact unit, power-control electronics, a battery and a coil for transferring data to and from the read/write head of the diskette drive. There is enough room in the 3.3-mm thick diskette terminal to insert a Smart Card. On the PC side, all that is needed is a suitable software driver to handle the data exchange. This is a good way to integrate Smart Cards into existing systems in an uncomplicated and economical manner.

Many years of R&D activity separate the earliest two-chip Smart Cards and modern-day Smart Cards, with their very powerful microcontrollers. Terminals have undergone a similar technical evolution over the same period. The first terminals often had very primitive mechanical and electrical constructions, partly due to inexperience. The consequence of this was that Smart Card microcontrollers were frequently damaged and thus failed prematurely. Nowadays, most terminal manufacturers have overcome these 'teething troubles', and a development stage has been reached in which external design features play a bigger role in the buyer's choice of terminal than technical features and specifications, which are generally nearly the same for all terminals and all manufacturers.

In functional terms, a Smart Card terminal consists of two parts: a contact unit for the card and a terminal computer. A card reader, into which the Smart Card is inserted so that it can be electrically contacted, has essentially only a mechanical function. The terminal computer is needed to drive the terminal electrically, to manage the user interface and to create a link to a higher-level system. In the simplest case it can be a single microcontroller, while in technically more sophisticated solutions, it is a single-board computer.

Figure 11.5   Photograph of a pocket-sized terminal for memory cards and processor cards that includes its own keypad. The loudspeaker mounted on the rear of the terminal can be used to automatically dial telephone numbers using tone dialing, under the control of the Smart Card. (Source: Giesecke & Devrient)

11.1   Mechanical Characteristics

When a Smart Card is inserted into a terminal, at least two things must happen in a mechanical sense. First, the card's contacts must be electrically connected to the terminal computer. This is handled by the contact unit. Second, the terminal must detect the fact that a card has been inserted. This can be handled by a microswitch or a light barrier, for example. One drawback of the latter is that its reliability can be affected by dirt or cards with transparent bodies. A mechanical switch is always the most effective solution.

Terminals differ very greatly in terms of the contact units and contacts that are used. The GSM 11.11 specification imposes certain limits on the insertion force and the shape of the contacts, and almost terminals use these values. According to this specification, the contact elements should be rounded rather than pointed. The radius of curvature of the contact tips should be at least 0.8 mm. This largely prevents scratching of the contact surfaces of the card. In addition, the force required to insert the card into the contact unit is significantly lower if the contacts have rounded leading edges than if they are pointed.

According to the GSM specification, the maximum force exerted by a single contact must not exceed 0.5 N under any circumstances (the EMV specification allows 0.6 N). This is intended to protect the chip located beneath the contact surfaces, since this piece of silicon crystal could break under greater stress.

Although the location of the contacts on the card is internationally standardized by ISO and should thus be the same everywhere, a French national standard (AFNOR) has the chip nearer the top edge of the card. Consequently, there are terminals that have two contact heads. This allows both ISO and AFNOR contact locations to be supported. This technically complicated solution is of interest in systems in which Smart Cards with ISO and AFNOR contact positions are used together. This is only a transitional situation, since ISO specifies that the AFNOR location should no longer be used. Several French banking applications, for example, employ terminals with dual contact heads. This allows both the old AFNOR cards and the newer ISO cards to be used during the transition period.

Problems can occur with the connections between the terminal and the Smart Card, especially with portable terminals and terminals installed in vehicles. Such terminals, in particular those in vehicles, are often subjected to high accelerations, which can cause the contacts to briefly separate from the card's contact surfaces. You can imagine that a vehicle travelling over cobblestones at a certain speed can cause the spring-loaded contacts to oscillate at their resonant frequency. If the card is electrically activated at the time, it is simply impossible to predict what will happen.

In the extreme case, when all contacts simultaneously lift free and then reconnect with the card, the card would probably execute a power-on sequence and then send an ATR. However, in this situation it is certain that the electrical activation sequence will not comply with the ISO standard, which means that the chip may eventually fail if this occurs repeatedly. In any case, the brief power interruption will naturally result in the loss of all states that have been achieved in the card during the open session. Depending on the application, this means that the PIN may have to be entered again or the user re-authenticated.

If only one contact lifts free, the consequences depend strongly on which contact it is. If it is the I/O contact, the communications link will only be temporarily disturbed. This can be remedied by error recovery mechanisms. If a different contact lifts free, the card will be reset. In this case, the communications link must be re-established from the very beginning.

In order to prevent the contacts from lifting free due to acceleration forces, the contact force can be increased, but the upper limit is still 0.5 N per contact. There is no simple, satisfactory technical solution to this problem, but the probability of contact separation can be minimized by sensible placement of the terminal. For example, the terminal can be mounted so that the contacts are perpendicular to the main axis of acceleration.

In any case, the terminal software must be able to independently re-establish communications if the contacts have briefly lifted free of the card. The millions of GSM sets in daily use demonstrate that Smart Cards can be used in portable equipment without any problems.

The service life of the contacts and the technical construction of the terminals vary immensely. The service life is also strongly affected by environmental conditions, such as temperature, humidity and the like. An MTBF (mean time between failures) of 150,000 insertion cycles, however, is considered to be a normal value for a terminal.

Contact units with wiping contacts

The technically simplest terminals, which are thus also the least expensive, have only wiping contacts in the form of leaf or disc springs. No other mechanical contact components are present in these simple terminals. However, with such a simple spring-based unit, the contact surfaces and part of the card are always dragged across the contacts when the card is inserted and withdrawn, which produced scratch marks. These are undesirable, for both aesthetic and technical reasons.

Repeated scratching of the gold-plated contact surfaces of the card gradually wears away the protective gold layer, and the exposed metal underneath this plating will then oxidize. This adversely affects the electrical connection. The user may have to insert and remove the card several times in order to rub off the oxide layer, so that a satisfactory electrical connection can be made.

Mechanically driven contact units

The next higher class of terminals does not have fixed sliding contacts, but instead a mechanism that presses the contact unit against the contact surfaces of the card when the card is inserted in the terminal. A lever arrangement converts the force used to insert the card into a force perpendicular to the contact surfaces.

An optimally designed mechanism also produces a very small amount of movement of the contact unit along the length of the card while the contacts are being applied to the card. This ensures reliable electrical contact with the card. The sliding motion rubs away any light soiling on the contact surfaces. The contact pins themselves are also individually spring-loaded, in order to ensure a well-defined contact pressure for each contact surface.

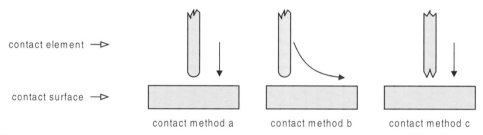

Figure 11.6   Methods for making electrical contact with Smart Cards. Method (a) with rounded contact pins is unreliable, since soiling of the contact surface will adversely affect the reliability of the electrical contact. Methods (b) and (c) represent good solutions for the two types of contact pins illustrated. The sharp-edged contact pins shown in (c) slightly penetrate the contact surface, which can be seen under a microscope as small surface nicks.

Electrically driven contact units

The technically most complex solution, which is also the best mechanical solution, is a terminal with an electrically driven contact unit. Here a set of parallel contact pins is driven by a motor or electromagnet to make perpendicular contact with the card from above, with a slight sideways movement. Due to the complexity of this electromechanical construction, the terminal is relatively large. However, this type of terminal is well suitable for use in professional applications, in which many millions of contact cycles must be made without maintenance. It is therefore typically used in automated teller machines (ATMs) and personalization machines employed in Smart Card manufacturing.

Card ejection

The Smart Card is normally inserted manually, which means without any assistance from the terminal. Only ATMs have self-feeding card readers, which use a conveyor mechanism to lead the card to the contact unit within the machine. This is not the case with ordinary terminals, which differ only in the manner in which the card is ejected. Simple terminals do not automatically eject the card, which means that the card must be manually removed from the reader. Two different techniques are used for this, called 'push–push' and 'push–pull'. With a push–push contact unit, the card is inserted by hand as usual, and it must be pressed again and then pulled to remove it. This is not ergonomically desirable, since this sequence of motions is unnatural. The result is that people often forcibly pull the card out of the terminal. This causes the contact pins to be scraped over the contact surfaces and the body of the card, since the contacts have not been released by the mechanism. Push–pull contact units better match normal motion sequences, since the card is simply pushed into the terminal to insert it and pulled out of the terminal to remove it.

Terminals that automatically eject the card have a spring that is tensioned by inserting the card. This can be released by the terminal computer via an electromagnet. This causes the card to be partially extended from the terminal, rather than fully ejected, so that the user can grasp it and pull it out completely.

Card-ejecting readers have one major advantage relative to other types. The ejection of the card very clearly signals the end of the session to the user, and at the same time, it reminds the user not to forget the card in the terminal. This reminder is often emphasized by an audible beep. This practical benefit is the main reason for using card-ejecting readers.

Cash dispensers, in particular, are usually able to retain Smart Cards if necessary. Since they routinely have self-feeding card readers, it is naturally technically feasible to route the card to a special retention bin in the machine if necessary, rather than to the exit slot. From a technical viewpoint, retaining cards presents no major problems, as long as the terminal is large enough to hold the extra mechanism and the retention bin. In certain circumstances, however, there can be legal problems if the card user is also the legal owner of the card.

Card removal

The reliability of a system based on Smart Cards can suffer severely if users can withdraw the card from the terminal at any time during a session. For one thing, this causes the card to be disconnected from the power supply without following the prescribed deactivation sequence. For another, it could interrupt EEPROM read or write processes, causing the contents of a file to be fully undefined. This could cause the card to fail completely. For these reasons, it is advantageous to use terminals with card-ejecting readers that are designed such that it is impossible to manually pull the card out of the terminal. A hidden mechanical emergency ejector can be provided to remove a Smart Card from the terminal in case of a power failure. Normally, however, the terminal can determine when to return the card to the user, and the user cannot interfere with the processes that are executed.

11.2   Electrical Characteristics

With the exception of the contact unit, a terminal consists mainly of electronic components. These are used to provide the interfaces to the user and the background system, and to electrically drive the contacts. The terminal's electromechanical parts and the Smart Card itself must be supplied with electrical signals. The only information that is directly provided by the contact unit is whether a card has been inserted. The only signal that is sent directly to the contact unit is the activation of the automatic card ejector, if such a device is present.

The card interface consists of the five contacts for the ground, supply voltage, clock, reset and data transfer signals. Once the electrical contacts have been made, it is very important with regard to the service life of the card that the activation sequence specified by ISO/IEC 7816-3 is followed exactly. Otherwise, the chip may be electrically overloaded, which increases the failure rate. It is also important to observe the proper deactivation sequence, since otherwise the same problems will occur.

In this regard, there is an important consideration with simple terminals that allow the user to remove the card manually. Whenever the contact unit detects that the card is being withdrawn, the terminal's electronic circuitry must immediately execute a deactivation sequence. This is the only way to prevent the contacts from sliding across the contact field of the card while they are possibly still energized, which would produce results that have little in common with a standard deactivation sequence. However, the consequences of such an unallowed card withdrawal can be even more serious, since shorts may occur between the leads if the contacts are worn or slightly bent. The mild sparking due to the discharge of capacitors in such a situation will damage both the contact elements and the contact surfaces of the card.

With regard to electric circuitry, almost all terminal manufacturers have realized by now that short-circuit protection is indispensable. If this point is neglected, a single Smart Card with shorted contact surfaces can cause the electrical demise of very many terminals. Incidentally, shorted cards crop up regularly, partly due to vandalism and partly due to technical defects.

Short-circuit protection should extend to the point that every contact can be connected to any other contact or group of contacts without any repercussions. Ideally, the circuitry that drives the Smart Card should be fully electrically isolated from the remaining circuitry of the terminal. This is standard practice in public cardphones in Germany, since it also largely protects the equipment against externally applied voltages as well as shorts.

The voltage needed for writing and erasing EEPROM pages is generated by the microcontroller via a charge pump on the chip. This can draw currents of up to 100 mA for intervals of a few nanoseconds. The same effect, in a reduced form, is produced by the transistor switching processes in the CMOS integrated circuits. Even very fast regulator circuits in the power supply cannot handle these short spikes, with the consequence that the supply voltage for the card collapses due to the high current load, and the EEPROM write or erase cycle fails. In extreme cases, the voltage dropout can be so severe that the processor lands outside of its stable operating area and a system crash occurs.

The solution is to connect a capacitor as close as possible to the contacts. A ceramic capacitor of about 100 nF is suitable, as it can release its charge very quickly. The leads to the Smart Card must be as short as possible, so that the lead resistance and inductance do not significantly affect the ability of the circuit to meet the increased current demand within the necessary interval. A brief increase in the current demand can be met by drawing charge from the capacitor until the voltage regulator can respond to the change. This is a simple and economical way to avoid power supply problems.

For electronic financial transactions in particular, it is nowadays standard to equip the terminal with a real-time clock. This is required for reasons of traceability and user protection. According to the EMV specification for credit card terminals, the clock may not be off by more than 1 minute per month. This is not technically difficult, since suitably accurate clock components are available as single-chip solutions. In addition, the clock can be adjusted every time the terminal makes an on-line connection to the higher-level system. Radio time signal receivers in terminals have so far not achieved any practical importance, since signal reception is too strongly affected by the screening effects of the site where the terminal is installed. Standard time code signals usually cannot be received inside a reinforced concrete structure, for example.

11.3   Security Techniques

Terminals may contain a very large variety of security mechanisms. The spectrum ranges from mechanically protected enclosures to security modules and sensors for the various card features. In pure on-line terminals, whose only function is to convert the electrical signals that pass between the background computer system and the Smart Card, there is normally no need for additional built-in security features. In such cases, security is handled entirely by the computer that controls the terminal.

However, as soon as data must be input to the terminal or the terminal must operate independently of the higher-level system, it is necessary to incorporate suitable mechanisms to provide additional system security. The possibilities are almost unlimited, but they depend very strongly on the Smart Card in question and its security features.

With a typical Smart Card, whose body is very basic and only serves as a carrier for the microcontroller, there are usually no security features on the card body. There is thus no need for the terminal to check such features. In contrast, Smart Cards for financial transactions are usually hybrid cards, which means that they have a magnetic stripe in addition to a chip, in order to maintain compatibility with older systems. However, hybrid cards also possess the usual features that enable the terminal to check their genuineness independently of the chip. Corresponding sensors must therefore be present in the terminal.

Terminals that work off-line, either completely or occasionally, must contain master keys for the cryptographic algorithms that are used, since card-specific keys cannot be derived without these keys. These master keys are very sensitive with regard to security, since the entire security of the system is based on them. In order to guarantee their security and confidentiality at all times, they are not stored in the normal electronic circuitry of the terminal, but in a separate security module within the terminal that has special mechanical and electrical protection.

This security module can for example be a single-board computer encapsulated in epoxy resin, which can exchange data with the actual terminal computer only via an interface. The secret master keys are never allowed to leave the security module, but are used only internally to perform computations. In a typical application example, the security module receives an individual card number or chip number from the Smart Card via the terminal computer, and it uses this number to derive a card-specific key. This key is then used within the security module to compute a signature or perform authentication.

Modern versions of this module, which is normally the size of a matchbox, contain extensive sensor systems for detecting attacks. They are also largely self-contained electrically, so they can actively resist attacks, even when their power supply is cut off. If an attack is detected, the usual defense is to erase all keys, so that an attacker is left with only a circuit board circuit encased in epoxy resin inside a metal case, with no data worth analyzing.

Due to the high cost of good security modules, the trend in recent years is to use Smart Cards instead. Although this leads to certain restrictions in terms of memory size, sensors and self-reliance, the level of security is generally adequate, even for electronic payment applications. Cards in the IC-000 format (plug-in) are used to limit the physical size.

Since security modules in Smart Card format are not permanently built into the terminal, but can be exchanged, they are eminently suitable for extending the terminal hardware. The following example illustrates this. Static unilateral RSA authentication will become increasingly important in the next few years, partly because it is called for in the international EMV specification for chip-carrying credit cards. Since RSA authentication is so computationally demanding that it cannot be performed by normal terminal processors within an acceptable length of time, permanent built-in security modules present a problem. However, if the terminal uses plug-in format Smart Cards as security modules, it is easy to exchange the security module. Relatively expensive Smart Cards containing supplementary arithmetic coprocessors can be used for the security modules, and the terminal can execute an RSA computation at high speed once its software has been suitably modified.

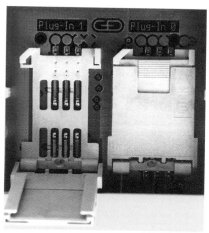

Figure 11.7   Example of two contact units for plug-in format security modules, located next to each other in a Smart Card terminal

In the future, a variety of card issuers will bring chip-carrying debit and credit cards on the market. These cards will all use different keys and key derivation and authentication procedures. Furthermore, certainly not all card issuers will be willing to reveal secret data and procedures to manufacturers of security modules. In all probability, the approach that will be taken is for a card issuer or group of card issuers to issue a common 'terminal card' that can carry out all the processes relevant to the security of their collective systems, and that can execute these processes within the terminal. This card will be addressed using one of the two standardized transmission protocols (T=0 or T=1), and it will largely behave just like a standard Smart Card. The only difference will be that the terminal card will contain functions related to secret master keys, key derivation procedures and the collection of security-related data (such as sales balances). The terminal will only look after the user interface and uploading or downloading data to or from the background system. All technical security aspects will be handled by the card. This means that the terminal must be able to work with several terminal cards, rather than only one. A specific card will be automatically selected according to the card issuer and the selected function. The demand for several independent terminal cards has been taken into account in the latest terminals. Some of them have up to four contact units for plug-in cards. They can thus serve terminal cards from several different card issuers in parallel, without mutual interference.

In addition to the mechanical protection of the terminal by a robust housing that can only be opened using special tools, and the inclusion of a security module in the terminal, a commonly-used security measure is mechanical protection against the unauthorized tapping of the data transfer to the Smart Card. This consists of a sort of guillotine arrangement that cuts through any wires that may run from the card to the exterior of the card reader after the card has been inserted. The purpose of this device, which is called a 'shutter', is to prevent tapping or manipulation of the messages sent between the card and the terminal. It can be actuated either electrically or simply by inserting the card. If the wires cannot be cut due to their thickness or composition, the shutter cannot close completely. This is detected by the terminal electronics, and no power is applied to the card, so no communication takes place.

Communication between the terminal and the Smart Card must fundamentally be designed such that tapping or tampering cannot impair the security of the system. Shutters should thus not actually be necessary. Nevertheless, security can be improved somewhat if a would-be attacker's life is made more difficult. There is a big difference between being able to easily tap into the data exchange and having to first overcome a few hurdles. On the other hand, a shutter makes the terminal bigger and more expensive, and there are very few models that still close precisely after several thousand operating cycles. The system design should therefore not rely entirely on this sort of mechanical protection.

11.4   Linking a Terminal with PC/SC

In order to use Smart Card in a PC environment, it is necessary to have a terminal connected to the PC and to have support from the PC software. The difficulty here is naturally that in the past, each type of terminal required its own software driver to be installed in the PC. Each driver in turn had its own software interfaces, so that in practice it was not possible to generate terminal-independent software.

In Germany, the development of a specification for a software link between terminals and PCs was started at a relatively early date. This resulted in the *Multifunktionales Kartenterminal* (MKT) specification, which was published in 1994 in several versions. It is primarily oriented towards the interests of the health care sector, but it is presently used as a foundation for many other types of terminals within Germany.

The subject of terminal-independent integration of Smart Cards in PC programs is certainly also very important outside of the German market. Consequently, in May 1996, work was started on the generation of an international standard for the link between cards and PCs. The following companies assisted in this work: Bull, Hewlett-Packard, Microsoft, Schlumberger, Siemens Nixdorf, Gemplus, IBM, Sun, Toshiba and Verifone.

Version 1.0 of the 'Interoperability Specification for ICCs and Personal Computer Systems' was published in December 1997. It consists of eight parts, as described in Table 10.1. The working group was known as PC/SC (for personal computer/Smart Card), and this acronym is also used as an abbreviated name for the specification. It can be obtained via the Internet from the WWW server of the specification group [PC/SC].

At least in principle, the PC/SC is platform-independent, since it works on all Windows-based PCs, and these make up the majority of personal computers. It allows Smart Cards to be integrated into any desired applications and is largely independent of the programming language used, since it supports widely used languages such as C, C++, Java and BASIC. The only prerequisites are that a suitable driver must be available for the terminal to be used, and the Smart Card must be PC/SC-compatible. However, this compatibility requirement is reasonably non-critical, since the scope has been kept relatively broad.

Table 10.1   Summary of the eight parts of the PC/SC specification

PC/SC Specification	Contents
Part 1: Introduction and Architecture Overview	This is the basis for all other parts of the specification. It identifies the relevant standards, summarizes the system architecture and the hardware and software components, and lists definitions and acronyms.
Part 2: Interface Requirements for Compatible IC Cards and Readers	This defines the physical characteristics of contact-type Smart Cards. It specifies basic electrical properties, such as the operating power source and the reset behavior, and defines the data elements, structures and allowed processes of the ATR and PTS. There is a summary of the basic aspects of data transfers at the physical level, and the T=0 and T=1 protocols are both described.
Part 3: Requirements for PC-Connected Interface Devices	The requirements imposed on the terminal and the supported terminal features (display, keypad and so on).
Part 4: IFD Design Considerations and Reference Design Information	Information for designing terminals, with reference to PS/2 keyboard interfaces and USB interfaces.
Part 5: ICC Resource Manager Definition	Detailed descriptions of the technical aspects of the ICC Resource Manager, including the associated classes.
Part 6: ICC Service Provider Interface Definition	Detailed descriptions of the technical software aspects of the ICC Service Provider and Crypto Service Provider, including the associated classes.
Part 7: Application Domain and Developer Design Considerations	Description of the utilization of the PC/SC specification from the application perspective.
Part 8: Recommendations for ICC Security and Privacy Devices	Compilation and definition of recommended functions and mechanisms that should be supported by a PC/SC Smart Card. This includes the file system (MF, DF and EF), associated file access conditions, necessary system files in the Smart Card (for keys, PINS and so on), commands, return codes and cryptographic algorithms.

The easiest way to gain an overall view of the PS/SC specification is to view it in terms of the defined hardware and software components. The following seven components are described in terms of their functions and mutual relationships:

- ICC-aware application,
- ICC service provider,
- crypto service provider,
- ICC resource manager,
- IFD handler,
- IFD,
- ICC.

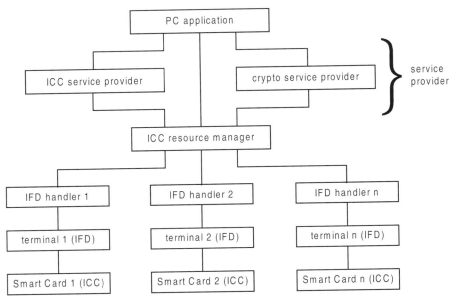

Figure 11.8   Summary of the software architecture of the PC/SC specification for linking Smart Cards to PC operating systems

In the following material, each of these components is briefly explained in terms of its tasks and functions, in the order listed.

ICC-aware application
This is an application that runs on a PC and that would like to use the functions and data of one or more Smart Cards. It can also be an application that runs under a multiuser operating system with multitasking and multithreading.

Service provider
The function of the service provider is to encapsulate the individual functions of a Smart Card, independent of whatever operating system is used in the Smart Card. For example, it is possible to select a file via the API of the service provider without knowing the Smart Card command used for this, or having any idea of the coding for this command.

The service provider component is split into the *ICC service provider* and the *crypto service provider*. This is done to avoid problems with export restrictions that many countries have with regard to cryptographic algorithms. The separate crypto service provider, which handles all functions that can cause export problems, can be omitted. In this case, the PC/SC interface can be used for all functions except cryptographic functions.

The service provider does not have to be a single piece of software. It can also consist of multiple software components linked by a network. In this way, it is for example possible to locate the crypto service provider on a cryptographically secured or high-performance computer that is isolated from the remainder of the PC/SC components.

ICC resource manager

The ICC resource manager is the most important component of the PC/SC architecture. It manages all resources that are necessary to integrate Smart Cards into the operating system. It must fulfill three important assignments.

First, it is responsible for recognizing connected terminals and Smart Cards. It must also notice when a Smart Card has been inserted or removed from a terminal, and respond to such events by providing suitable messages.

The second assignment is to manage the allocation of terminals to one or more applications. A terminal resource can be exclusively assigned to a particular application. However, if several applications access the same terminal simultaneously, this terminal must be identified and managed by the ICC resource manager as a shared resource.

The third assignment is to prepare transaction primitives. A transaction primitive is formed by binding the commands belonging to a particular function into a group. This ensures that these commands will be executed in an uninterrupted sequence. Otherwise, it would be possible for two uncoordinated applications to concurrently access a Smart Card, each with its own sequence of commands. The problems that this would cause can most easily be illustrated by considering the following example. In a Smart Card, only one file can be selected at a time. If two different applications attempt to select different files at the same time with SELECT FILE commands, and then read data from the Smart Card using read commands (such as READ BINARY), it is completely undefined which file will actually be read. This depends only on the order in which the commands arrive at the Smart Card. A much more complicated situation, but no less tricky, arises if complex processes (such as paying using an electronic purse) must be carried out between several applications and a single Smart Card. The ICC resource manager ensures that command sequences that belong together cannot be split up or interrupted by other commands, and so ensures that the individual processes are executed one after the other.

IFD handler

The IFD handler is a sort of driver that is specific to each individual terminal. Its task is to link the terminal located at a defined interface to the PC, and to reflect the individual characteristics of the terminal onto the PC/SC interface. The IFD handler represents, in a manner of speaking, a data channel from the PC to a particular terminal.

IFD (interface device)

The IFD component of the PC/SC specification is a terminal connected to the PC via an interface. The interface is arbitrary, so the terminal can for example be connected to the computer via an RS232 interface, a USB (universal serial bus) interface or a PC card interface. The terminal must meet the ISO/IEC 7816-1/2/3 standards, which among other things means that it must support both of the asynchronous data transmission protocols (T=0 and T=1). Optionally, it may support synchronous transmission protocols (2-wire, 3-wire and I^{2}C bus) for memory cards, as specified by the ISO/IEC 7816-10 standard. In the terminal, in addition to a display, the PC/SC specification supports a numeric keyboard, a fingerprint scanner and other biometric sensors for user identification.

ICC (integrated chip card)

The PC/SC specification requires microprocessor cards that are compatible with the ISO/IEC 7816-1/2/3 standards to be supported. If the terminal allows it, memory cards that comply with the ISO/IEC 7816-10 standard may also be included.

12

Smart Cards in Payment Systems

The original primary application of Smart Cards with microcontrollers was user identification in the field of telecommunications. In recent years, however, Smart Cards have established themselves in another market sector, namely electronic payment systems. Due to the large number of cards in use, the market potential of this sector is enormous. This is underlined by the fact that more than one hundred million credit cards have been issued throughout the world.[1] The future applications of electronic purses include replacing conventional means of payment (banknotes and coins), shopping via global networks and pay-per-view television.

Smart Cards are by nature particularly suitable for financial transaction applications. They can easily and securely store data, and their convenient size and robustness make them easy for everyone to use. Since Smart Cards can also actively execute complicated computations without being influenced by external factors, it is possible to develop totally new approaches to performing financial transactions. This is very clearly illustrated by electronic purses in the form of Smart Cards, which are possible only in this medium.

Electronic payment systems and electronic purses benefit everyone involved. For banks and merchants, they reduce the costs associated with handling cash. Offline electronic purses largely eliminate the costs of data telecommunication for the transactions. The risk of robbery and vandalism is reduced, since electronic systems contain no cash to be stolen. For merchants, the fact that transactions are processed more quickly is also a persuasive argument, since it means that cash flows can be optimized. Vending machines and ticket dispensers can be made simpler and cheaper, since assemblies to check coins and banknotes are not needed. Electronic money can be transferred via any desired telecommunication channel, so it is not necessary to regularly collect money from the machines.

Customers also benefit from the new payment methods, although to a minimal degree. It is not necessary to always have change on hand, and it is possible to pay quickly at a vending machine or ticket dispenser.

Ultimately, however, it is potential users of a system who determine the success or failure of a payment system. If the benefits for them are too marginal, they will not use the system and will choose other means of payment. An electronic purse, after all, is just a new means of payment that complements rather than replaces other existing means of payment,

---

[1] As of autumn 1998.

such as credit cards and cash. It is unlikely that these options will be entirely supplanted by electronic purses in the form of Smart Cards, nor do we need to fear that means of payment that have provided reliable service for many years will disappear entirely.

12.1   Payment Transactions with Cards

The simplest approach to card-based payment transactions is to use magnetic-stripe cards to store information that is used for online authorization. After the card has been checked against the black list and solvency has been verified, the funds can be transferred directly from the cardholder's bank account to that of the merchant. With Smart Cards, the scenario is slightly different, but in principle it remains the same. The Smart Card is logically linked to a bank account, and after unilateral or mutual authentication of the background system and the card, a previously entered amount is transferred. Naturally, PIN verification is also performed in the Smart Card or background system during the transaction.

Both of these scenarios are based on a background system that makes all the decisions. They do not by any means fully exploit the capabilities of Smart Cards. However, there are other means and methods of making payments that can be implemented by exploiting these capabilities. Some of these are described below.

12.1.1   Electronic payments with Smart Cards

There are three fundamental models for electronic payments using Smart Cards: (a) credit cards, in which payment is made after a service is rendered (*pay later*), (b) debit cards, in which payment is made when the service is rendered (*pay now*) and (c) electronic purses, in which payment is made before the service is rendered (*pay before*). We describe these three models below, as well as a variation on them.

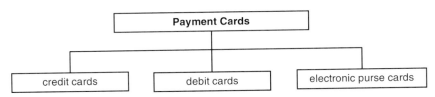

Figure 12.1   Classification of payment cards

Credit cards

The original idea of using a plastic card to pay for goods or services comes from credit cards. The principle is simple: you pay using the card, and the corresponding amount is later debited from your account. The cost of this process is borne by the merchant, who usually pays a fee that depends on the amount of the transaction. The fee is usually around two to five percent of the purchase price.

Up to now, most credit cards have not included chips. The disadvantage of such cards is that they have a relatively low level of protection against forgery. Consequently, card issuers experience significant losses due to fraudulent cards, since the merchant is guaranteed payment. Evidently, up to now, these losses have been lower than the cost of introducing cards with chips. However, credit cards will probably be supplemented with chips in the not too distant future, in order to reduce the steadily increasing cost of fraud.

Debit cards

The country in which debit cards are most widely used is Germany. A debit card, which may be a magnetic stripe card or a Smart Card, allows the amount of the payment to be transferred to the account of the merchant or service provider as a direct part of the payment process. With both debit cards and credit cards, the actual payment process is normally authorized by a credit check via a background system. There is usually a threshold level above which this must occur, so it is not always necessary to make a connection to the background system for small purchases. The threshold level is on the order of 400 DM.

Electronic purses

With an electronic purse, 'electronic money' is loaded into the card before any payment is made. This can be done in exchange for cash or using a cash-free process. When a purchase is actually made, the balance in the card is reduced by the amount of the payment, and at the same time the balance of the electronic purse of the second party (who is usually the merchant) is increased by the corresponding amount. The merchant can later submit the electronic money received in this manner to the operator of the electronic purse system, and receive the corresponding amount in real money. The user of an electronic purse thus exchanges real money for an electronic form of money that is loaded in his or her Smart Card. When a purchase is made, the cardholder exchanges this electronic money for goods or services.

This system has three significant disadvantages for the user. The first is that when the card is loaded, the user receives electronic money in exchange for real money. Financially, the user gives the operator of the purse system an interest-free loan, since it could take a few weeks to actually spend the electronic money, while the real money immediately becomes the property of the system operator. The amount of interest may be small for a single user, but in total it represents a substantial source of supplementary income for the purse system operator. In many field trials conducted up to now, it has been found that in industrialized countries the average amount in an individual electronic purse is around 150 DM. The total average amount of money in an electronic purse system is called the 'float'. Assuming that 10 million cards are in use and the interest rate is 5%, the total annual interest on the float amounts to 150 million DM. There are no additional expenses to offset this amount. The interest lost by an individual cardholder in this example is only 7.50 DM, which he or she will not regard as a major disadvantage. In addition to this interest income from the float, the purse system operator receives additional income in the form of unspent electronic money, due to cards that end up in collections and non-returned defective cards.

A second disadvantage is that a real problem arises if the purse operator goes bankrupt. This is because the card user has exchanged real money, whose value is guaranteed by the state within certain limits, for electronic money in a Smart Card. If the purse operator goes bankrupt, the electronic money could suddenly become worthless, and the user would have lost his or her money. Consequently, efforts are now being made in some countries to restrict the operation of electronic purse systems to banks and similar institutions. At minimum, the lodging of a security deposit with a government agency is required, so that the amount of money loaded in the Smart Cards is covered in the event that the system operator goes bankrupt.

Now comes the third distinct disadvantage for the user. What can the holder of an electronic purse do if it no longer works? If the card is anonymous, not even the purse system operator can determine the amount of money that was last stored in the card. The purse holder will also find it practically impossible to provide convincing proof of how much money was still in the card. If the chip is destroyed, the electronic money is thus irrevocably lost. Unfortunately, a Smart Card is much less robust than banknotes or coins, for understandable reasons.

In practice, the current solution to this problem amounts to a compromise. Since the last amount that was loaded into the card online is known, as well as the purse balance at the time of this transaction, the approximate amount in the purse can be calculated. This amount is then paid to the client. However, if the same client repeatedly makes claims due to faulty Smart Cards, the system operator will limit his goodwill and the client will not receive anything more, since there is a risk that the card failures are fraudulent.

Open and closed system architectures

A distinction must be made between open and closed architectures for electronic payment systems. An open system is fundamentally available to multiple application providers, and it can be used for general payment transactions between various bodies. In contrast, a closed system can be used only for payments to a single system operator.

The technical aspects of this can be briefly illustrated using a telephone card with a memory chip as an example. With memory cards, all that happens when a payment is made is that a counter is irreversibly decremented. It is not necessary for the terminal to keep an exact account of the number of units that have been deducted, but only to ensure that the counter in the card is always properly decremented whenever the service is used (that is, whenever a call is made using the card). The terminal is thus a sort of machine for destroying units of electronic money. Of course, in practice a balance is kept for each terminal, but the amounts that are deducted are only booked in the internal accounts of the purse system operator. Fraud associated with the accounting for the deducted amounts between the terminal owner and the purse system operator is impossible in principle, since both of them are part of the same organization (in this example, the telephone company).

In an open system, the terminal owner and purse system operator can be completely different firms. The purse system operator must therefore be able to verify that the accounts for the terminal turnover amounts are correct and not manipulated. This must be taken into consideration from the very beginning in the system design, since otherwise account settlement between the terminal owner and purse operator will be very difficult or impossible. In the previous example with a memory card, the system concept makes it impossible for the terminal operator to convincingly guarantee the purse system operator that the amounts submitted are correct. All the terminal operator can do is to present a certain number of units to be settled on account. He cannot present any forgery-proof signatures for the amounts paid, as he could with a genuine electronic purse system.

System structure and terminal integration

The system structure of an electronic payment system using Smart Cards can be either centralized or decentralized. With payment systems in particular, system security is the most important issue. There is thus frequently a tendency to use centralized systems, since this gives the system operator complete control of the system.

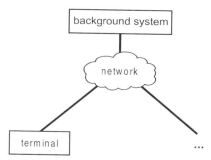

Figure 12.2   The basic structure of a centralized system for electronic payment transactions. All of the illustrated connections are permanent.

In concrete terms, a centralized system means an online system in which every payment transaction is executed directly via an online link to the background system. If a communications link cannot be established, no payment is possible. Nevertheless, a centrally operated system has certain advantages. For example, incoming transactions can be directly compared with the current black list in real time. Key exchanges can be carried out directly by the background system without any delays. The software in the terminals and the general parameters in the cards can be updated directly and with little additional effort, since a direct link to the background system must be established for each transaction.

However, these advantages are countered by several major disadvantages. In many countries, telecommunication charges are so high that it is not reasonable for merchants to have permanent links to background systems or to dial up a background system for each transaction. In some areas, the telephone network is not sufficiently reliable to allow an online link to the higher-level computer to be established at any desired time. Due to their active nature, Smart Cards are excellent for decentralized systems, since they contain part of the system security 'on site'. This is also their main advantage compared with passive magnetic stripe cards, which cannot compel the system to perform specific processes.

In particular, the use of electronic purses with automated equipment, such as vending machines and ticket dispensers, compels the use of a decentralized system. This is because electronic purses can operate completely independently for weeks or months, and they do not provide any means to establish contact with an existing communications system. A decentralized system is therefore often preferred. In addition, a decentralized system has significantly better behavior with regard to failures. If the background system fails in a centralized system, all payment transactions are completely blocked. In a decentralized system, by contrast, the consequences of a temporary failure usually do not even reach as far as the merchants' terminals.

Decentralized systems also have certain disadvantages, mainly in the area of system management. The reason is that online connections can only be established at certain intervals, and as a rule only by the terminals. However, it is essential for system security that the terminals always work with the current black list. This is one of the reasons why many systems require each terminal to establish an online connection to the background system at least once a day. This is used to transfer the accumulated transaction data to the background system, and in response various types of administration data are transferred to the terminal. Some examples of administration data are new terminal software, new key sets, the current black list and data to be loaded into customers' cards.

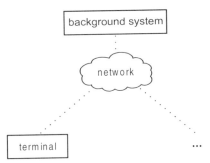

Figure 12.3   The basic structure of a decentralized system for electronic payment transactions. All of the illustrated connections can be established as needed.

In practice, a compromise is often made between pure decentralized and centralized system structures, in order to combine the advantages offered by both types without having to also accept their disadvantages. This consists of allowing both the terminal and the Smart Card to force an online connection under certain conditions. If this connection cannot be established, the payment transaction does not take place.

Some typical conditions are: (a) online authorization is required for payments above a certain amount, which can usually be set by the system operator individually for each Smart Card; (b) the number of offline transactions and the amount of time since the last online transaction can be used to decide whether to go online; (c) a random number generator can be used to force a certain percentage of all transactions to take place online. Some systems also have a special button on the terminal that forces an online transaction. This can be pressed by the sales staff if they suspect that the customer is using a manipulated card.

Table 12.1   Typical actions and conditions that trigger an online connection between a Smart Card and the background system

Action or condition	Usual value
transaction type (e.g. cash payout)	---
POS conditions (e.g. PIN pad present)	---
parameter-driven random selection	10 %
manual request at the merchant terminal	1 %
first use of the Smart Card	---
number of offline transactions performed since the last online transaction	10
accumulated offline amount since the last online transaction	1000 DM
time since the last online transaction	7 days
exceeding an adjustable threshold value for the payment amount	400 DM

All of these criteria ensure that, on average, every card makes a direct connection to the background system within a defined and statistically computable time interval. The system operator thus recovers the direct control that was originally lost by using a decentralized system. Terminals and automated machines that have only a small turnover can be excluded from these online constraints, since even in the case of fraud only small losses can occur. This eliminates the need to have a connection to a communications network, since the data exchange can be performed manually by service personnel.

12.1.2   Electronic money

Electronic money must have certain properties if it is to be used with the same flexibility as normal money. If these are wholly or partly absent, the options available to users of electronic money are necessarily more or less limited. The essential properties that are necessary to minimize the difference between electronic money and real money are described below.

Processable
An important, although in principle trivial, property of electronic money is that it can be completely and automatically processed by machines. This is the only way in which large systems can be operated economically.

Transferable
Electronic money must not be bound to a particular medium, such as Smart Cards. It must be possible to transfer electronic money using any desired medium, such as a network or computer.

Divisible
Electronic money must be divisible, so that arbitrary sums can be paid without having to have recourse to using normal money. This property in turn is analogous to normal money, which, although not arbitrarily divisible, is available in a sufficient number of different denominations that normal purchases can be made using a small number of coins and banknotes.

Decentralized
Centrally structured payment systems can be easily monitored by the purse system operator, and the possibility of fraud is very limited. The best example of such a system is the online authorization of credit card purchases. However, centralized systems suffer from many disadvantages. They are expensive, susceptible to technical disturbances, inflexible, and difficult to modify or extend. Decentralized systems minimize these disadvantages. This can be seen very clearly in the private sphere, in which money changes hands without any involvement by a central body. Electronic money should also have this property, since it otherwise cannot compete with normal money. As applied to payment transaction systems, this means in concrete terms that it must be possible to make offline payments and also to make payments directly from one purse to another purse. The property of allowing direct payments between purses (purse-to-purse transactions) is sometimes called transferability.

Monitorable
Despite the demand for anonymity, electronic money must allow the purse system operator to monitor the system. This is the only way in which manipulations and security gaps can be detected and remedied. It is exactly analogous to the situation with normal money, in which every citizen is obliged to immediately report the appearance of forged money to the appropriate authorities. In the case of electronic money, the purse system operator is the

only one who can monitor the consistency of payment traffic, and is thus responsible for guarding against fraud and forgery.

Secure

A fundamental property of electronic money must naturally be security against forgery. Any system will collapse with a short time if it is possible to forge or duplicate money, in whatever form, or manipulate payment flows. This is another reason why cryptographic functions are used so extensively in the field of electronic payment transactions, since this is the only way to achieve the required level of security.

Anonymous

Anonymity means that it is impossible for anyone to make any connections between payments and individual persons. This value of this depends very much on the perspective of the observer. From a technical perspective, the purse issuer desires a system with as little anonymity as possible, so that he can monitor the system in the best possible manner. The possibility of fraud is very limited in non-anonymous systems, since anyone who performs a fraud can quickly be identified. Government agencies, such as the police and tax authorities, have similar interests. Non-anonymous electronic money would provide them with considerably more scope for monitoring financial transactions than they have enjoyed up to now with normal money.

The position of the user of the purse is diametrically opposed to this. He considers the current payment methods using normal money as a very good state of affairs, and regards complete anonymity and non-traceability of payment flows as the optimum solution.

The property of anonymity in particular is frequently compromised by purse system operators in the interest of system security. For example, payments are anonymous in most systems, but loading the electronic purse is not. This allows relatively good system monitoring to be carried out, simply and with a minimum of effort.

At first sight, some of the above properties appear to be directly contradictory. For example, complete anonymity and perfect system monitoring are in many cases mutually exclusive. However, this area is only starting to develop, and there are already systems being planned in which these two properties can definitely be realized simultaneously.

There are two properties of real money that are not mentioned in the above list, although they are highly significant. The first is that real money is legal tender that must be accepted by all residents of a particular country. In almost all countries, vendors of goods or services are obliged to accept the legal currency of that country as a means of payment. The second property relates to the stability of the currency. Except for a few countries with high rates of inflation, the legal currency in circulation has a stable value. If this is not the case, the consequence is that people resort to barter or the use of foreign currencies.

12.1.3   Basic system architecture options

Electronic payment systems based on Smart Cards can be constructed in a wide variety of ways. For economic reasons, they are often based on existing systems, most of which are based on magnetic-stripe cards. However, there is no single basic model that applies to all payment systems, since the requirements vary too widely. We can therefore only describe the basic principles of such systems, in terms of their essential components.

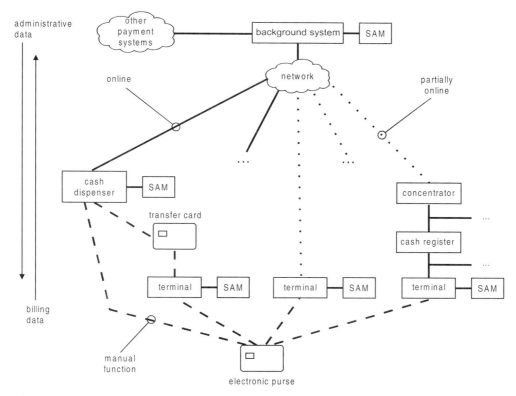

Figure 12.4 Example architecture of an electronic purse payment system (SAM = security module)

Large Smart Card payment systems basically consist of four different components. These are the background system, the network, the terminals and the cards.

Background system

The background system consists of two parts: *clearing* and *management*. The clearing subsystem maintains the accounts for all the banks, merchants and cardholders that participate in the system, and it books all incoming transaction data. It also looks after the system monitoring tasks. A simple example of such a task is maintaining a running balance to check whether the total of the amounts submitted to the clearing system exceeds the total amount of money in the electronic purses. If it does, this means that an attacker has loaded money into the Smart Cards without the knowledge of the background system.

The management part of the background system controls all administrative processes, such as distributing new black lists, switching to new key versions, sending software updates to the terminals and so on. This subsystem also generates data sets for personalizing Smart Cards.

The background system has complete control of the electronic payment system, regardless of the system structure. Even with systems that work completely offline, the background system establishes the global system parameters and monitors the security and operation of the system.

Network

The network links the background system to the terminals. The links can be line-oriented (e.g. ISDN) or packet-oriented (e.g. X.25). As a rule, the network is totally transparent to the messages, which are passed unmodified from the sender to the receiver.

Terminals

The various types of terminals can be classified as either loading terminals or payment terminals, according to their functions with respect to payment transactions. They can also be classified as automated terminals or attended terminals. The classic example of an automated terminal is a cash dispenser (ATM). In electronic purse systems, automated terminals are mostly used only for loading cards. It would naturally also be conceivable to allow an electronic purse to be emptied using such a terminal, with the balance paid out in cash. Attended terminals are typically located at supermarket checkouts and in retail shops. They are always used to pay for goods. In some systems, terminals in banks can also be used to load Smart Cards in exchange for a cash payment.

Smart Cards

Smart Cards are the most widely distributed component of the system. They may be used as electronic purses, but they can also be used as security modules in various types of terminals. Another area of application is transporting data between various system components. Cards for this purpose, which are called transfer cards, are used to manually transfer transaction data between a terminal that works completely offline and one that works online (such as a cash dispenser).

The example system shown in Figure 12.4 further clarifies the system components and their logical links. The background system is connected to the various components via a transparent network. This can be the background system of a different operator, or a component of the system itself.

Electronic purses are primarily loaded using cash dispensers. These often work online, although they can also work offline for a limited time if there is a network failure. They therefore contain their own security modules, in which all the relevant keys for normal operation and key derivation processes are stored.

There are also electronic purse payment systems that work completely offline. Two examples are parking meters and taxi meters with terminals. A transfer card can be used to transport the transaction data from the security module to a cash dispenser, from which they reach the background system via the network. In the other direction, the terminal receives current administration data, such as black lists and software modifications.

The second type of payment terminal is connected to the network via an online link that is set up as necessary. This type of terminal normally works offline, but it periodically connects to the background system in order to exchange any available billing and administrative data.

The third type of payment terminal has no direct connection to the network. For example, it could be connected to a supermarket cash register that in turn is connected to a concentrator located in the store. This concentrator, which is normally a PC acting as a server, might connect to the background system once a day via the network. The necessary data exchange takes place during this connection.

The 'Quick' electronic purse system in Austria and the 'Geldkarte' system in Germany are similar to the example system just described, and the Visa Cash electronic purse system largely corresponds to the described system. For large applications, it is quite common to employ a distributed system architecture, consisting of several different background systems that work in parallel. This architecture allows a number of different purse systems, with more than one system operator, to be operated with mutual compatibility.

12.2 Prepaid Memory Cards

With electronic payment systems in particular, we must not neglect memory cards. They are produced in very large numbers in the form of prepaid electronic purses, and they are used in many applications. This certainly will not change over the next few years. Although memory cards will slowly but steadily be replaced by microcontroller Smart Cards, their strength lies in their unbeatably low price. The prepaid telephone cards that are widely used in many countries,[2] which are simply discarded once they are used up, are a typical example of electronic purse cards containing memory chips.

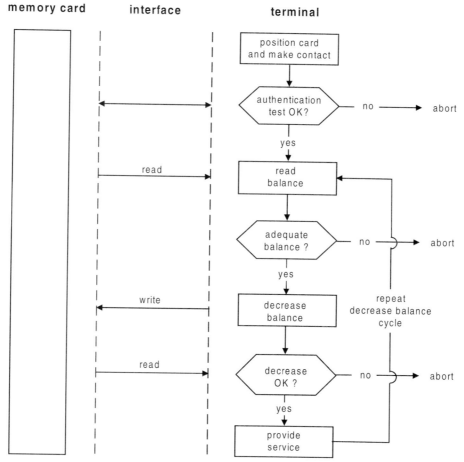

Figure 12.5 The devaluation cycle of a prepaid memory card, as seen from the terminal

---

[2] See also Section 13.1, 'Public Cardphones in Germany'.

Memory cards[3] contain only some control logic and an irreversible down counter to allow them to perform their functions. Newer versions also allow unilateral authentication of the card by the terminal. For this purpose, the control logic is supplemented by a simple encryption function, whose task is to encrypt a random number received from the terminal using a secret key stored in the card, and then return the result to the terminal. This is the only way the terminal can be sure that the memory card being used is genuine.

In addition to the fact that it is difficult to establish the authenticity of memory cards, they have a disadvantage that limits their use as a general electronic payment medium. Since such cards can easily be manipulated, it is difficult to construct a payment system that can support various independent service providers (such as kiosks and taxis). This is because it is not possible to have protected communications between the terminal and the card. Consequently, the correct allocation of payments, and thus overall system monitoring, can in principle only be performed using indirect methods.

Memory cards enjoy advantages with respect to microcontroller cards in certain very limited application areas, due to their low cost, but they are not particularly suitable for open applications in payment systems. Future payment systems will certainly operate primarily with microcontroller Smart Cards, since these are significantly more flexible.

12.3   Electronic Purses

The idea of implementing an electronic purse in a Smart Card goes back to the early days of Smart Card technology. However, only in recent years has this concept been translated into an actual design and entered the implementation stage.

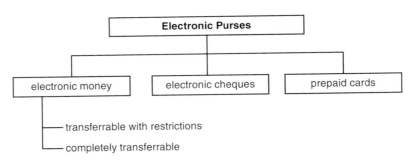

Figure 12.6   Classification of electronic purse systems based on Smart Cards. Electronic money is not tied to a card, and it can be denominated as desired. Electronic cheques have fixed denominations. Electronic purses contain electronic money that is tied to the card and can be freely denominated.

If you take a normal purse containing coins and banknotes as a reference, you can easily see what characteristics electronic money has in the eye of the user. You must put money in a purse before you can use it, so it cannot be used like a debit or credit card, where you pay only at the time the goods are received or even later. Instead, it is used like a telephone card, which is also paid in advance. The payment process must be quick and simple, since otherwise the level of user acceptance is low. Furthermore, all payments from a purse are anonymous, which means that it is not possible to retrace who bought what at which time.

---

3   Chip structure for memory cards; see also Section 2.3.1, 'Memory cards'.

The most annoying characteristic of a purse becomes apparent if it is lost. The money it contains is irretrievably gone, but this does not have to be true with an electronic purse.

The greatest advantage of a purse, or rather the money it contains, is that it is accepted everywhere within a given country. It is precisely this factor that is missing in most existing electronic purse systems. With a telephone card, all you can do is make telephone calls, and nothing else. This is typical of a closed application. An ideal electronic purse, on the other hand, can be used in more than one sector, and thus allows its user to make payments in many different shops using a single purse.

12.3.1   The CEN EN 1546 standard

The European Commission decided in 1990 to have the *Comité Européen de Normalisation* (CEN) produce a European standard for a multi-sector electronic purse system. Work on the standard started in 1991. Up to 1998, the individual project teams had spent around eight man-years working on developing this standard. Since a number of independent experts have participated in this effort, it very unlikely that there are security gaps in this standard. It is thus already quasi-evaluated. The essential parts are presently no longer subject to change, and they will be published as a European standard after the final vote takes place.

The CEN EN 1546 standard is a public standard, and the processes of its individual functions are described in great detail. It is therefore very suitable for illustrating payment and loading processes in an electronic purse system. With many existing systems, it is not possible to illustrate these processes in as much detail, since the relevant commands, processes and internal functions are confidential. This standard is thus very useful for visualizing the fundamental external and internal processes of an electronic purse system.

The typical application areas are clearly indicated by the first systems based on this standard. The Danish system operator Danmønt has introduced a purse compliant with this standard into their existing system. In Austria, the Eurocheque card that is issued throughout the country includes an electronic purse (called 'Quick') that is based on the EN 1546 standard, in addition to other applications. However, the largest international application of this standard is the Visa Cash system. This is one of several electronic purses offered by Visa.

The CEN EN 1546 standard for electronic purse systems is titled 'Inter-sector Electronic Purse' and consists of four parts. The first part, 'Concepts and structures', describes the overall system. This basic document defines and explains, in abstract form, all the logical components and their interconnections.

The second part uses these basic concepts to describe the security architecture for both the overall system and its individual components. It covers not only the mechanisms for maintaining security, but also possible attacks and the necessary countermeasures.

Part three, 'Data elements and interchanges', contains the descriptions and definitions of the data elements needed for the electronic purse system. It also covers the commands and their associated responses related to the Smart Card and the security modules that are used.

The final part describes the state machines and the states of the devices used. It uses a symbolic representation similar to the well-known flowchart diagram. This formal notation, which is known as SDL notation, is derived from the CCITT Z.100 recommendation.[4]

The standard, which encompasses around 300 pages, thus contains a complete description of an electronic purse system, including the Smart Cards, the terminals with their security

---

[4]   See also Section 4.2, 'SDL Notation'.

modules and the background and clearing systems. Its objective is the standardization of large electronic purse systems with very many Smart Cards and wide geographical distribution.

The advantage of a general standard for electronic purse systems is primarily that it allows individual, independently operated systems to be mutually compatible. As with GSM, this gives the user the option of being able to use his or her card in the future to make payments using the systems of other purse providers. This is an essential prerequisite for the success of this sort of payment system.

However, we must make a small comment at this point. The EN 1546 standard provides a large amount of freedom with regard to actual implementations, and it regards itself to be more of a framework than a detailed specification of individual bits and bytes. It is thus perfectly possible for two different systems to be completely compliant with this standard and still be mutually incompatible, for example because they use different cryptographic algorithms.

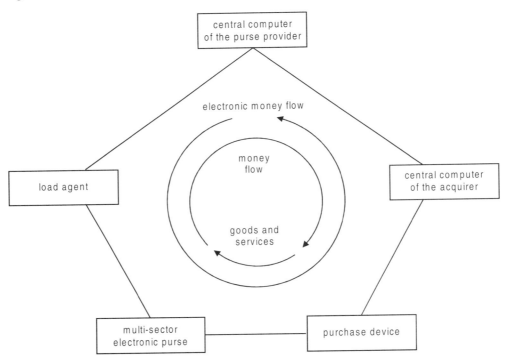

Figure 12.7   Basic structure of an inter-sector electronic purse system and the associated payment flows, according to EN 1546

The *purse provider* has the overall responsibility for the system and is also the system manager. It is comparable to a GSM network operator. The term *purse holder* is defined in the standard to refer to the electronic purse user. This is the person who makes payments using the electronic purse application in the card, and who receives goods or services in return.

In addition, there are three other participants that perform tasks in the system. The *service provider* offers goods or services that are accepted by the user and paid for using an electronic purse. The *acquirer* is responsible for setting up and managing the data links between the purse issuer and the service providers. He may also consolidate the individual transactions arriving from the payment facilities, so that the purse provider only receives collective certificates. The *load agent* is the counterpart to the service provider, since he can re-load the electronic purse in exchange for a payment.

These five participants need not all be real persons or companies; they may also be virtual. However, real technical components are allocated to each of them, distributed according to their security. The secure components prevent any external manipulation of the data that are processed or stored within them. Such manipulation is at least theoretically possible with the non-secure components. However, the system as a whole is designed such that the manipulation of any of the components identified as non-secure in Figure 12.8 will not affect the overall security of the system

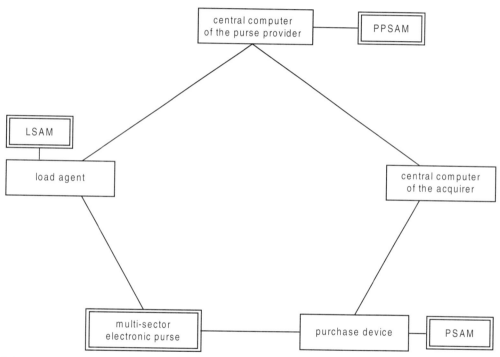

Figure 12.8   Components and connections of electronic purse systems according to EN 1546. The components with a single outline are not secure, while those with a double outline are secure.

The acronym 'IEP' stands for 'inter-sector electronic purse' and refers to the use of an inter-sector electronic purse in a Smart Card. A payment device is used to pay for received goods or services. It is a terminal with keypad and display, and it must also have a security module. The term *secure application module* (SAM) is used in the standard to refer to all types of security module. A SAM contains all secret keys necessary for actions between the IEP and the central computer of the purse provider. The keys naturally never leave the security module, but are used only inside the SAM by the cryptographic algorithms.[5] In many cases, therefore, the links between the system's secure components are all direct. The non-secure components are used only to transparently relay the security-related data.

Smart Cards have made it possible for the first time to convert the idea of an electronic purse system into reality. They are thus the main subject of the system description in the CEN EN 1546 standard. All the associated files, commands, states and processes are defined and described in this standard. In order to define the entire system, there are similar parts that cover the security module and the other components.

However, since this is a standard rather than a specification, the purse provider is naturally given a great degree of freedom and many options. A variety of functions can be used to construct the purse. For example, a simple system that only allows loading and paying with the purse can be easily implemented. This can be further built up with functions for canceling payments, modifying purse parameters and making currency conversions. The exact selection of the many complex options is largely left to the purse provider, who must choose the options that best meet his particular needs.

The most important aspects of an inter-sector electronic purse system with regard to the IEP, which is the Smart Card, are described below.

EN 1546 data elements

Labels for all data elements were introduced to allow the data used in the entire system for the electronic purse application to be referred to unambiguously. Data flows and data processing can be represented, simply and unambiguously, in a mathematically correct notation using these very short reference names. The standard also contains a simple data dictionary, which describes the corresponding data contents and associated formats for the standardized data elements.

Table 12.2 (part 1)   Summary of the most important standard data elements of EN 1546

Data element	Description
ALG_{IEP}	cryptographic algorithm used by an IEP
AM_{IEP}	authentication mode required by an IEP
AP_{IEP}	application profile of an IEP
BAL_{IEP}, BAL_{PSAM}, BAL_{PPSAM}	balance of an IEP, PSAM or PPSAM
$BALmax_{IEP}$	maximum balance of an IEP
CC_{IEP}, CC_{PSAM}, CC_{PPSAM}	completion code from an IEP, PSAM or PPSAM
CT	collection status
$CURR_{IEP}$, $CURR_{LDA}$, $CURR_{PDA}$	actual currency for an IEP, LDA or PDA
$DACT_{IEP}$	activation date of an IEP
DD	discretionary data
$DDEA_{IEP}$	deactivation data of an IEP

---

5   A possible key management system for an EN 1546 electronic purse system is described in Section 4.7.6, 'Key management example'.

Table 12.2 (part 2)   Summary of the most important standard data elements of EN 1546

Data element	Description
$DEXP_{IEP}$	expiry date of an IEP
ID_{IEP}, ID_{PSAM}, ID_{PPSAM}	identifier for an IEP, PSAM or PPSAM
IEP	inter-sector electronic purse
IK_{IEP}, IK_{PSAM}, IK_{PPSAM}	key information for an IEP, PSAM or PPSAM
LDA	load device application
LSAM	load SAM
M_{LDA}, M_{PDA}	transaction amount for load or purchase
$MTOT_{IEP}$, $MTOT_{PSAM}$	total transaction amount for a purchase
NC	number of collection
NI	number of individual transactions
NT_{IEP}, NT_{LSAM}, NT_{PSAM}	transaction number for an IEP, PSAM or PPSAM
PDA	purchase device application
PP_{IEP}, PP_{PSAM}, PP_{PPSAM}	purse provider identifier for an IEP, PSAM or PPSAM
PPSAM	purse provider SAM
PSAM	purchase SAM
R	random number
S_1	IEP signature
S_2	PSAM or PPSAM signature
S_3	IEP signature
S_4	PSAM signature
SAM	secure application module
TM	total amount
TRT	transaction type and status

Files

The complete electronic purse application is contained in a dedicated DF in the Smart Card. All the files necessary for proper operation are contained in this DF. In addition, information relating to the card, the chip, other applications and the like is stored in several files directly under the MF.

The data elements needed to operate the purse are located in six EF files within a directory for the purse. Table 12.3 summarizes the files and the data elements contained in them. The EF_{IEP} file specifies the general parameters of the purse, which form the basis for all transactions that take place. The EF_{IK} file contains specific information for every available key. EF_{BAL} contains the amount that is currently in the purse and available to the user. The three log files are used exclusively to record all transactions, separated by function. Only with these files is it possible to cancel a payment or handle errors. There are separate log files for loading, paying, modifying the purse parameters and making currency conversions. All log files have a cyclic structure, in order to record the latest transactions.

Table 12.3   The files and data elements needed for an electronic purse, in accordance with EN 1546. The log files for exchange rate calculations and parameter changes are not shown.

File	Function	Data elements	Descriptions
EF_{IEP}	Fixed data and parameters for the purse	PP_{IEP}	IEP purse provider identifier
		ID_{IEP}	IEP identifier
		$DEXP_{IEP}$	IEP expiry date
		$DACT_{IEP}$	IEP activation date
		$DDEA_{IEP}$	IEP deactivation date
		AM_{IEP}	IEP authentication mode
		AP_{IEP}	IEP application profile
		DD	user-specific data
EF_{IK}	Information relating to all keys	ALG_{IEP}	IEP cryptographic algorithm
		IK_{IEP}	IEP key information
		DD	user-specific data
EF_{BAL}	Purse balance	BAL_{IEP}	EP balance
		$CURR_{IEP}$	IEP currency
		$BALmax_{IEP}$	IEP maximum balance
		DD	user-specific data
EF_{TFIELD}	Transaction field	NT_{IEP}	EP transaction number
EF_{LLOG}	Log file for loading	TRT	transaction type and state
		NT_{IEP}	IEP transaction number
		BAL_{IEP}	IEP balance (new balance)
		M_{LDA}	transaction amount for loading
		$CURR_{LDA}$	currency for loading
		ID_{PPSAM}	purse provider identifier for PPSAM
		CC_{IEP}	IEP completion code
		DD	user-specific data
EF_{PLOG}	Log file for payment	TRT	transaction type and state
		NT_{IEP}	IEP transaction number
		BAL_{IEP}	IEP balance (new balance)
		M_{PDA}	transaction amount for payment
		$CURR_{PDA}$	currency for payment
		ID_{PSAM}	purse provider identifier for PSAM
		NT_{PSAM}	PSAM transaction number
		CC_{IEP}	IEP completion code
		DD	user-specific data

Commands[6]

The files form the foundation of the purse, and the commands are built on this foundation. Eight different commands are necessary, of which three belong to the ISO/IEC 7816-4 standard: SELECT FILE, READ BINARY and READ RECORD. These are only used to select the electronic purse application using its AID, and subsequently to read various data from the purse files as necessary.

The other five commands were developed specifically for use with electronic purses. They are always used in pairs for individual transactions, since in principle they function as a sort of mutual authentication. During the authentication process, the data needed for the transaction are also exchanged. The commands and responses are naturally structured such that tampering at the interface between the card and terminal can be immediately detected. Any tampering causes immediate abortion of the transaction and logging of the event.

All purse commands directly access data elements in the purse files for both reading and writing. The files are automatically selected by the operating system prior to these accesses. For example, if basic purse data are needed during the execution of a command, the operating system selects the EF_{IEP} file and the required data element is made available to the command. All transactions, and the most important data, are recorded in the relevant log files as part of the command–response cycle.

EN1546 defines the commands listed in Table 12.4 and specifies their internal functions in detail.

Table 12.4   Specific commands for electronic purses as defined in EN 1546

Command	Function
INITIALIZE IEP	initialization for subsequent purse command
LOAD IEP	load the purse, cancel a previous payment and error recovery
DEBIT IEP	pay with the purse; confirm payment
CONVERT IEP CURRENCY	convert currency
UPDATE IEP PARAMETER	modify general purse parameters

The standard does not provide commands for verifying or changing PINs, since these functions are not needed for the proper operation of the purse. However, additional commands for PIN verification and management can be included in the purse application as necessary, without causing any interference or problems with the existing purse commands.

States

As may already be apparent from the command summary, each transaction consists of an introductory initialization command and a further command that closes the transaction. In order to fix the sequence of commands, the state diagrams define the necessary states and state transitions in the application. This naturally requires the card to contain a state machine. Depending on its current state, the card will accept or reject various commands.

---

6   For a detailed explanation of the commands, see Section 7.10, 'Commands for Electronic Purses'.

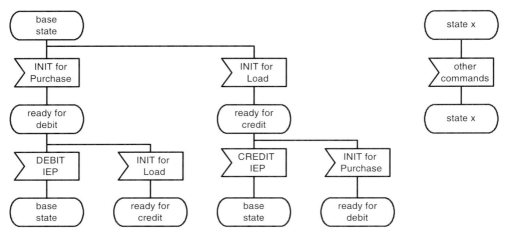

Figure 12.9   Simplified state diagram for loading and paying with an EN 1546 electronic purse

Cryptographic algorithms[7]

The entire security of the system is based on a cryptographic algorithm. The messages exchanged between the components all have an appended signature, so that manipulation can be detected. This is the only protection for messages, which are always exchanged in plaintext.[8] The message exchange is set up such that any desired type of cryptographic algorithm can be used to generate the signature. The symmetric DES algorithm is currently most often used, but the standard also allows asymmetric algorithms such as RSA or DSS. This algorithm independence is a great advantage, since it considerably extends the useful life and flexibility of the standard.

Processes

The standard does not just specify files, commands and states; it also describes and explains the related processes. These are specified in detail in terms of their data elements, using a pseudolanguage similar to BASIC. This was necessary because the security of some processes depends strongly on the internal command processing sequence. During a transaction, for example, the relevant log file must always be updated before the response is sent to the terminal. The processes and transactions for all components are also precisely specified. Descriptions are provided for the following electronic purse processes for Smart Cards:

- loading,
- paying,
- canceling a payment,
- correcting an error,
- converting currencies,
- changing the purse parameters.

---

7   See Section 4.6, 'Cryptology'.
8   With the DES algorithm, the message is followed by a 4-byte MAC as a signature.

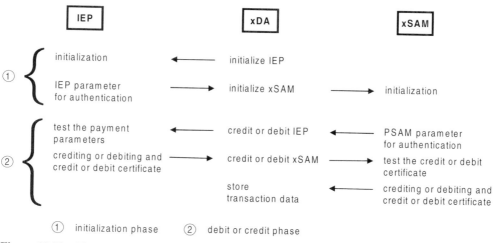

Figure 12.10   The basic process of an EN 1546 electronic purse transaction (phases 1 and 2)

In addition, the commands for reading files can of course also be used for monitoring purposes. However, the actions involved may vary, depending on the purse provider and the objective of the monitoring.

There is an essential principle that determines the structure of all specified processes. This is that it must never be possible to create electronic money by manipulating the data transfers between the components or interrupting transactions. The worst consequence of such actions is limited to the destruction of electronic money. This eliminates certain types of attack in advance. Incidentally, this principle is incorporated in the designs of nearly all electronic purse systems.

Each process is basically divided into three phases. In the first phase, complete initialization of the participating components takes place. The second phase provides the actual execution of the function involved. The third phase, which is optional, is used to confirm the previous actions. Successful execution of the first two phases amounts to a unilateral (or optionally, mutual) authentication of the two components.

In all processes, the actual purse functions (payment, loading, etc.) are interleaved with unilateral or mutual authentication. This minimizes the time taken by the transactions and increase security, since it significantly reduces the required number of commands. This is similar to the situation with the standard ISO commands INTERNAL AUTHENTICATE and EXTERNAL AUTHENTICATE, which can be replaced by the non-standard command MUTUAL AUTHENTICATE without affecting functionality or security.

Loading procedure
Before an electronic purse can be used to make payments, it must first be loaded using the procedure shown in Figure 12.11. This procedure illustrates the option in which the electronic purse is loaded directly at a terminal by an online background system with its associated security module (PPSAM). The standard provides for other options, such as loading a purse via a security module in a terminal (LSAM). However, the procedure described here is commonly used in current systems, since it gives the system operator complete control over loading.

Table 12.5 Abbreviations for functions and processes used to illustrate payment transactions according to the EN 1546 standard (see Figures 12.11 and 12.13)

Abbreviation	Meaning
Ax	Unique label for an action
Cx	Unique label for a command
Parameters (...)	Request to a participant, with the indicated data elements
Response (...)	Response to previous request, with the indicated data elements
Rx	Unique label for a response
Sign (...)	Generate a signature for the indicated data elements
Verify (...)	Verify the given data elements or function
Write (...)	Write the given data elements to a file

In this example, an electronic purse (IEP) is loaded by the background system via a terminal (LDA) using the security module (PPSAM). The user first inserts the card in the terminal, which executes a reset. In the ATR, the IEP sends the terminal various general parameters for the subsequent communication process. After this, the terminal selects the electronic purse DF in the card. Once this has been done successfully, the user inserts the amount of money to be loaded into the terminal, using an acceptable currency. The amount and the currency used are sent to the PPSAM with the first purse command. The PPSAM checks the indicated currency and the amount that is still allowed to be loaded. In response, it returns three data elements to the terminal.

The terminal adds the load amount (M_{LDA}) and the associated currency ($CURR_{LDA}$) to the data received from the PPSAM and sends all of this information to the IEP using the command INITIALIZE IEP for Load. The card then checks, among other things, whether the purse balance after the load amount is added would exceed the maximum allowed amount in the purse ($BALmax_{IEP}$). If it would not, the IEP increments a transaction counter (NT_{IEP}), computes the session key ($KSES_{IEP}$) and calculates a signature (S_1). These are returned to the terminal, together with a few other data elements.

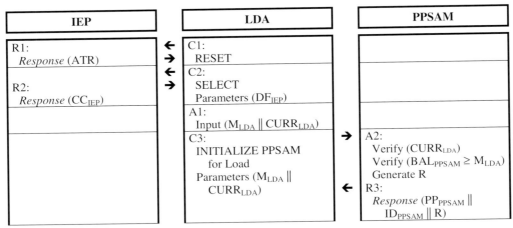

Figure 12.11 (part 1) Procedure for online loading an electronic purse (IEP) via a terminal (LDA) using a security module (PPSAM)

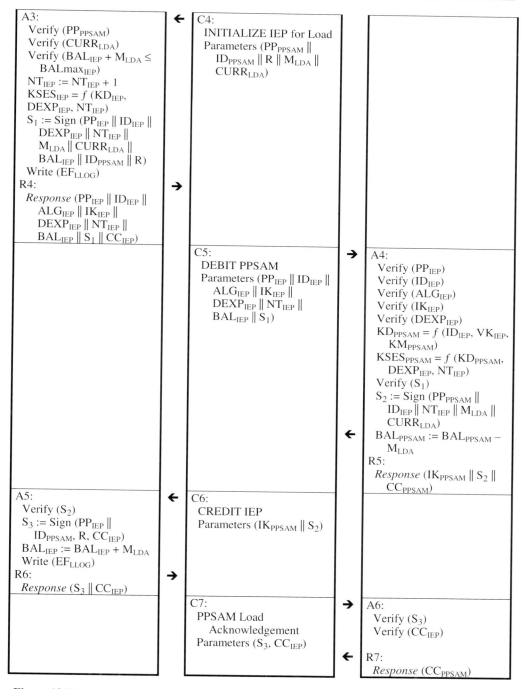

A3:
  Verify (PP$_{PPSAM}$)
  Verify (CURR$_{LDA}$)
  Verify (BAL$_{IEP}$ + M$_{LDA}$ ≤
    BALmax$_{IEP}$)
  NT$_{IEP}$:= NT$_{IEP}$ + 1
  KSES$_{IEP}$ = f (KD$_{IEP}$,
    DEXP$_{IEP}$, NT$_{IEP}$)
  S$_1$:= Sign (PP$_{IEP}$ || ID$_{IEP}$ ||
    DEXP$_{IEP}$ || NT$_{IEP}$ ||
    M$_{LDA}$ || CURR$_{LDA}$ ||
    BAL$_{IEP}$ || ID$_{PPSAM}$ || R)
  Write (EF$_{LLOG}$)
R4:
  Response (PP$_{IEP}$ || ID$_{IEP}$ ||
    ALG$_{IEP}$ || IK$_{IEP}$ ||
    DEXP$_{IEP}$ || NT$_{IEP}$ ||
    BAL$_{IEP}$ || S$_1$ || CC$_{IEP}$)

C4:
  INITIALIZE IEP for Load
  Parameters (PP$_{PPSAM}$ ||
    ID$_{PPSAM}$ || R || M$_{LDA}$ ||
    CURR$_{LDA}$)

C5:
  DEBIT PPSAM
  Parameters (PP$_{IEP}$ || ID$_{IEP}$ ||
    ALG$_{IEP}$ || IK$_{IEP}$ ||
    DEXP$_{IEP}$ || NT$_{IEP}$ ||
    BAL$_{IEP}$ || S$_1$)

A4:
  Verify (PP$_{IEP}$)
  Verify (ID$_{IEP}$)
  Verify (ALG$_{IEP}$)
  Verify (IK$_{IEP}$)
  Verify (DEXP$_{IEP}$)
  KD$_{PPSAM}$ = f (ID$_{IEP}$, VK$_{IEP}$,
    KM$_{PPSAM}$)
  KSES$_{PPSAM}$ = f (KD$_{PPSAM}$,
    DEXP$_{IEP}$, NT$_{IEP}$)
  Verify (S$_1$)
  S$_2$:= Sign (PP$_{PPSAM}$ ||
    ID$_{IEP}$ || NT$_{IEP}$ || M$_{LDA}$ ||
    CURR$_{LDA}$)
  BAL$_{PPSAM}$:= BAL$_{PPSAM}$ −
    M$_{LDA}$
R5:
  Response (IK$_{PPSAM}$ || S$_2$ ||
    CC$_{PPSAM}$)

A5:
  Verify (S$_2$)
  S$_3$:= Sign (PP$_{IEP}$ ||
    ID$_{PPSAM}$, R, CC$_{IEP}$)
  BAL$_{IEP}$:= BAL$_{IEP}$ + M$_{LDA}$
  Write (EF$_{LLOG}$)
R6:
  Response (S$_3$ || CC$_{IEP}$)

C6:
  CREDIT IEP
  Parameters (IK$_{PPSAM}$ || S$_2$)

C7:
  PPSAM Load
    Acknowledgement
  Parameters (S$_3$, CC$_{IEP}$)

A6:
  Verify (S$_3$)
  Verify (CC$_{IEP}$)
R7:
  Response (CC$_{PPSAM}$)

Figure 12.11 (part 2)   Procedure for online loading an electronic purse (IEP) via a terminal (LDA) using a security module (PPSAM)

All that the terminal does after this command is to relay the received data elements to the PPSAM. Here they are checked against the permitted range of values, and both a card-specific key (KD_{PPSAM}) and a session key ($KSES_{PPSAM}$) are generated. If the subsequent verification of signature S_1 is successful, the card has been authenticated, since it must know the secret key for computing S_1. The PPSAM then generates signature S_2 and sends it to the terminal together with the key information (IK_{PPSAM}). The terminal again only relays the data to the card, this time using the command LOAD IEP.

The card now verifies signature S_2. If this is successful, the PPSAM has also been authenticated by the IEP. The balance in the purse (BAL_{IEP}) is then increased. The IEP next generates a third signature (S_3), which is sent to the terminal to confirm that the balance has been successfully increased. The final command, which transfers this signature to the PPSAM, concludes the whole loading procedure.

The procedure just described is one of many possible options. It is often used in practice, since it is very common to perform purse loading transactions online. EN 1546 also contains options for loading via a special loading security module (LSAM). Such decentralized modules can be built into special loading terminals, such as cash dispensers.

Payment procedure

The following example, which is illustrated in Figure 12.13, demonstrates the payment procedure in terms of the relevant components: the electronic purse (IEP), the terminal (PDA) and the security module in the terminal (PSAM).

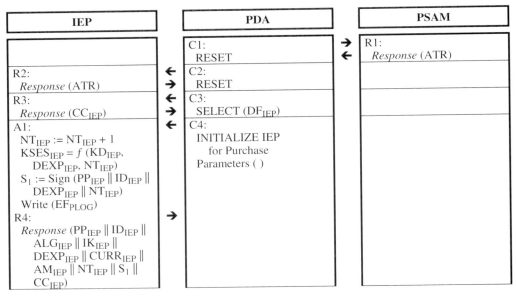

Figure 12.12 (part 1)   Transaction procedure for paying using an electronic purse (IEP) with a terminal (PDA) and security module (PSAM) in accordance with EN 1546

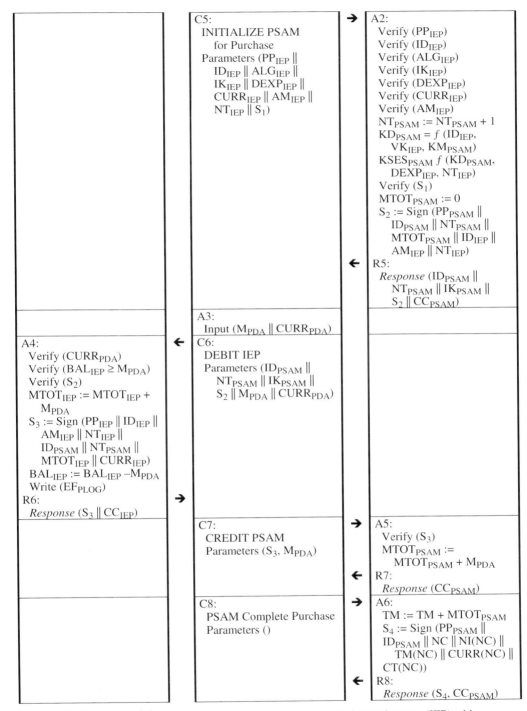

Figure 12.12 (part 2)   Transaction procedure for paying using an electronic purse (IEP) with a terminal (PDA) and security module (PSAM) in accordance with EN 1546

After the card has been inserted in the terminal, the terminal executes a reset in order to demand an ATR from both the PSAM and the IEP. If either ATR does not match the expected value, the terminal aborts the payment procedure. If the ATRs match their expected values, the terminal selects the purse DF in the IEP. If the file selection cannot be executed, the procedure is also aborted. For reasons of clarity, however, both this and general error handling are not portrayed.

After selecting the purse DF in the IEP card, the terminal sends the initialization command INITIALIZE IEP for Purchase. The IEP card receives this command, increments the transaction counter and then generates a signature (S_1) for various data elements. It then sends the data and the signature to the terminal.

The terminal next sends the initialization command INITIALIZE PSAM for Purchase to the PSAM. This command simply relays the data received from the card to the PSAM. The PSAM checks these data, which means that the expiry date ($DEXP_{IEP}$), currency ($CURR_{IEP}$), cryptographic algorithm used (ALG_{IEP}) and the other data received are compared with values stored in the PSAM. If all the comparisons are successful, the transaction counter (NT_{IEP}) is incremented. If any comparison fails (e.g. the expiry date of the IEP has been reached), further command processing is immediately aborted and an appropriate return code is sent to the terminal (PDA).

The PSAM next generates a derived key using the data sent by the IEP and generates a session key, and then checks signature S_1. If the signature is correct, it follows that all the transferred data are authentic, and the IEP is at the same time authenticated by the PSAM. In other words, the PSAM knows that the card containing the electronic purse is genuine.

In the following step, the PSAM also produces a signature (S_2), which is sent to the terminal together with a few additional data elements. The amount to be paid (M_{PDA}) and the associated currency ($CURR_{PDA}$) are now entered at the terminal. The terminal then sends the entered amount (M_{PDA}) and the data previously received from the PSAM to the card, using the command DEBIT IEP. The card now checks whether there is enough money in the purse to make the payment. If there is, it then verifies signature S_2. If the signature is correct, the data has not been tampered with during the transfer, and the PSAM has also been authenticated by the IEP. This is because only a genuine PSAM can have the secret key necessary to generate the signature S_2. The appropriate amount is subtracted from the purse balance, the log file is updated and a third signature (S_3) is generated to confirm the debit transaction just performed.

The signature S_3 and the debited amount are sent via the terminal to the PSAM, which verifies S_3. If this signature is correct, the amount debited in the IEP is added to an internal data element ($MTOT_{PSAM}$). An additional command, PSAM Complete Purchase, updates the PSAM balance by adding $MTOT_{PSAM}$ to the purse balance (TM). Finally, the PSAM receives a signature (S_4) to confirm that the payment was successfully made.

The procedure just described is a very simple example of the various payment procedures described in EN 1546. Other possibilities exist, including a special fast debiting procedure for card phones, as well as one that can produce a receipt at the end of the transaction.

The files, commands and processes described above for the card are also specified for all other important system components. This applies above all to the security module, since system security relies solely on this module. Statistical methods may be employed to monitor the overall operation of the system, which in the case of large applications may consist of tens of thousands of terminals and several hundred thousand Smart Cards. Generating a complete account for every individual card would conflict with demand for anonymity, and anyhow would require far too much computation. However, as experiments have shown, the security of the overall system can be verified on an ongoing basis at an acceptable cost by using sample checks to monitor the money flows.

The European EN 1546 standard laid the cornerstone for a multi-sector Smart Card electronic purse system with Smart Cards. The standard includes almost all processes and functions in use at the time that were considered worthwhile. There is only one function so far that has not yet been described, although it is very important for card users. This is the purse-to-purse transaction, which means transferring electronic money directly from one purse to another. There is presently no description of this type of transfer in the EN 1546 standard. However, various bodies are attempting to have this payment transaction included in the standard.

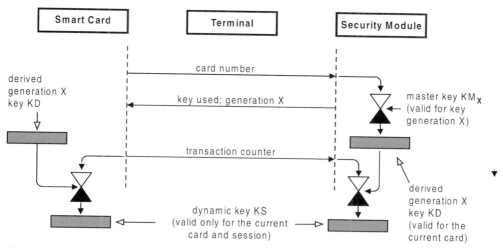

Figure 12.13   A possible key derivation process for an EN 1546 electronic purse system. The key depends on the combination of a card-specific key from a certain generation that is passed to the card and a session-specific transaction counter. The key that is so generated may be used for making payments or debiting monetary units.

12.3.2   The Mondex system

There are presently several large payment systems in the world that use Smart Cards as a central component. Only very few of these systems are based on an electronic purse, in which monetary units are stored directly in the card and not in a background system. Of these, there is only one system that can claim to permit electronic payments that correspond to those using normal money. This is the Mondex system.

The idea behind this concept, which is currently unique, was born in 1990. After five years of development, the first field trial was carried out in July 1995 in the southern English city of Swindon. A wide variety of shops were included in this trial, including newsstands, snack bars, supermarkets and travel agents, as well as filling stations and telephones. The maximum amount in the purse was set at £500 for the trial, but this value can in principle be set to any desired level. Following this trial, which was widely reported in the press, there have been additional field trials in many different regions, but up to now the system has not been introduced in any country as a national system.

Mondex is a consortium of three firms: British Telecom, National Westminster Bank and Midland Bank. Its purpose is to create a means of payment that can be used like cash but does not have the disadvantages of cash. The results of this technical development are intended to be franchised to banks and other firms. Mondex is currently one of the few electronic purse systems offered as a complete system, from the cards to the background system. This is also why there are relatively high expectations for this technology.

The Smart Card operating system used for Mondex is not limited to electronic purse systems. It is a multifunctional, general-purpose system that can be used for multiple applications in a single Smart Card. This operating system is called 'Multos', and it is marketed internationally by Maosco [Maosco], primarily in the card-based payment systems sector. A special feature of Multos is that it supports downloading software to cards in the field. This software is written using a language similar to C called MEL (Multos executable language), which is processed by an interpreter in the Smart Card.

The system

Since the Mondex purse is designed to behave in the same way as real money, purse-to-purse transactions are naturally possible. This allows cardholders to make payments among themselves, without the intervention or knowledge of a bank or similar organization. The system is completely open and anonymous, and as many participants as desired can be involved. Figure 12.14 shows the system participants and the possible flows of funds.

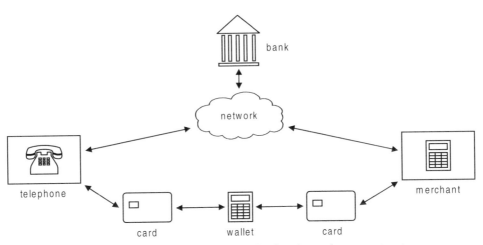

Figure 12.14   Money flows and participants in the Mondex electronic payment system

The electronic purse is located in the chip of a conventional ID-1 card with contacts. A matchbox-sized key ring attachment with a display can be used to view the balance in the purse. If the card is inserted in this mini-terminal, the current purse balance and the last ten transactions can be viewed. A 'wallet' is needed to transfer electronic money to the purse of another cardholder. This device, which resembles a pocket calculator, has a small keypad and display. It also has a built-in security module and a terminal for the electronic purse. To perform a purse-to-purse transaction, the user inserts the first Smart Card into the wallet and then enters the amount to be transferred. This amount is transferred from the electronic purse to the wallet's security module. The second card is then inserted into the wallet, and the amount is transferred to it from the security module. This completes the transaction.

Another device in this payment system is a telephone with a built-in terminal. It allows money to be transferred over the telephone line during a call. A typical application is ordering goods from a mail-order catalogue. In this case, it is possible to pay at the time that the order is placed. This technique can naturally also be used to load the purse via the phone, or to conduct a transaction between two cardholders. If the card is loaded from a bank account, a four-figure PIN must of course be entered for security reasons, in order to protect the account holder against unauthorized withdrawals.

Figure 12.15   A 'wallet' terminal for purse-to-purse transactions and parameter changes, and a 'balance reader' that displays the most recent transactions and the current balance. Both are part of the Mondex electronic purse system. (Source: Mondex)

Each electronic purse can accept up to five different currencies. As soon as the balance for a particular currency reaches zero, it is possible to load a different currency in the card. The purse can be blocked with a simple command and unblocked by entering a four-digit PIN, to prevent unauthorized use.

The merchant terminals contain security modules that use the same type of Smart Card as that of the customer. It would thus also be possible to pay for other goods in turn using the merchant's security module. Interestingly enough, this could make the theft of such a card worthwhile, as it could then be used just like a normal purse. However, this problem has been recognized and preventative measures have been taken. The merchant cards can be configured to allow only the receipt of electronic money. Debiting the card is only possible during an online transaction with the merchant's bank. As you can see, electronic money is not necessarily immune to theft. It all depends on whether it can be used by a thief. To the extent that the merchant terminal has online access to the bank (possibly via a dial-up link), it can be set up to automatically transfer money from the merchant card to the merchant's bank account whenever a particular balance is reached.

Security mechanisms and the payment procedure

All specifications relating to the payment procedure and the security model of the Mondex system are confidential. This makes it very difficult to obtain exact technical details of the system and its individual components. We can therefore provide only a broad technical summary that illustrates some of the mechanisms and procedures used in the system.

The microcontroller currently used is a Hitachi H8/3102. For mass production, a processor specially developed for Mondex is planned, with a numerical coprocessor and a suitable amount of memory, since the application requires around 5 kB in the EEPROM. It is likely that a symmetric cryptographic algorithm, such as DES, is currently used. Considering that a special processor with a numerical coprocessor will be used in the future, it can be assumed that this will be changed to an asymmetric algorithm for increased security. The RSA algorithm could be used, for example. In principle, though, the system is independent of the cryptographic algorithm used. It does not rely on special properties of a particular algorithm, but only uses (digital) signatures to protect data transfers. In this regard, it differs little from the EN 1546 multi-sector European electronic purse systems.[9]

Since the Mondex system is operated in a completely decentralized fashion, there must be a special procedure for switching key versions and algorithms. Each issued card contains at least two totally different cryptographic algorithms with several associated keys. If it is necessary to switch to another key version, or even to use a different algorithm, an appropriate parameter is set in all Smart Cards that make an online links to the background system. These cards can in turn set the same parameter in all cards with which they execute payment transactions. This snowball effect produces a system-wide switch to the new general parameters within a very short time, due to the exponential increase in rate of data propagation. This would happen even if the background system only modified the parameter in a single card. This is a very effective, fast and simple method of changing global data in a decentralized payment system.

Naturally, it must also be possible to isolate particular cards in the system. This can be done in three different ways. First, suspect cards identified by black lists can be recognized and retained by the machine into which the card is inserted. This is however usually only possible with cash dispensers, since only they have the technical resources to retain cards. Second, the black lists are loaded into all the terminals, and these are able to block cards so that they can no longer be used for transactions. Third, all issued electronic purse cards only allow a certain number of transactions to take place, after which they are automatically blocked. The block can be removed by an online query after the card has been checked against the black list, so the card does not have to be replaced. This ensures that a card with an electronic purse cannot be used indefinitely without any control by the background system.

A typical payment transaction between two Smart Cards in the Mondex system is divided into two phases, which are shown graphically in Figure 12.16. In the first phase, a registration of the current transaction takes place, during which all the data needed for the subsequent money transfer are exchanged. In the second phase, Smart Card 2 sends the desired amount to Smart Card 1. The complete data set is signed digitally, so it cannot be manipulated during the transfer. After receiving the data, Smart Card 1 checks the signature

---

[9]   See also Section 12.3.1, 'The CEN EN 1546 standard'.

to verify both the authenticity of Smart Card 2 and the authenticity of the data transfer. If all checks are satisfactorily concluded, the desired amount is debited from Smart Card 1 and sent to Smart Card 2, together with a digital signature. Smart Card 2 checks this signature to eliminate the possibility that the data have been manipulated, and at the same time to authenticate Smart Card 1. If all checks are satisfactorily concluded, the amount is credited to the purse. Following this, Smart Card 2 generated a confirmation that the amount was properly credited, adds a digital signature and sends these to Smart Card 1. The transaction is completed when this confirmation of payment has been received and successfully been checked.

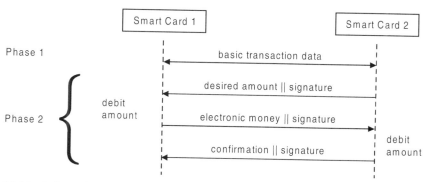

Figure 12.16   Information flow for a transaction between two Smart Cards in the Mondex system

Both cards contain log files, and they have suitable processes to allow a transaction to be correctly resumed from the appropriate point if it is interrupted. These error recovery processes are very important, since otherwise electronic money could be destroyed if a transaction were interrupted. Each of the participating cards has three separate log files for storing transaction-related data. The first is the transaction log, which stores various data relating the last ten successful payment transactions. The second is the pending log, which contains all the data accumulated during a transaction that are needed if error recovery should become necessary. The third is the exception log, which stores all transactions that were not completed successfully. If all the records in this file have been written, the Smart Card is automatically blocked. The cardholder must then unblock it via an online transaction, during which the log file entries are loaded into the background system and analyzed, following which the these entries are deleted.

Summary

The Mondex system is currently the only completely open electronic payment system using an electronic purse. It permits all transactions that are possible with normal cash. In addition to this, it also allows payment transactions via various telecommunications media, such as the telephone system. If the card containing the purse is lost, the money held in it is naturally also lost, just as with a real purse containing cash. However, this makes the system completely anonymous, which is sure to bolster user acceptance. The Mondex system is, to a certain extent, a simulation of a real money circuit.

Since it is in principle impossible to demand a fee for each individual transaction with such a system, a question that naturally arises quite quickly is how the system operator can generate any revenue. The investments needed to set up and operate the system are not exactly trivial. In the Swindon field trial, each electronic purse user was charged a relatively low fee of £1.50 per month. The merchants naturally also paid fees. Although it would be possible to charge clearing fees for the merchants' turnovers, the completely open nature of the system naturally leaves the merchants free to use their accumulated electronic money to make purchases from each other. The system operator could also generate revenue by offering various services to the cardholders and merchants. A significant advantage of Mondex for the system operator is that the clearing costs are nearly zero, since clearing in the original sense is no longer necessary. Particularly with very low-value payments (micropayments), the clearing costs in many systems can be very large in comparison with the actual turnover. In some electronic purse systems, for example, the complete clearing costs, including transaction recording via shadow accounts, lies in the region of 5 Pfennig per transaction. This means that the clearing cost for a pack of chewing gum purchased from a vending machine (for 20 Pfennig) is no less than 25% of the purchase price. This is completely unacceptable for a merchant. In the Mondex system, however, these costs would not arise.

In the coming years, the Mondex system will influence the market for electronic payment systems in many ways. At the international level, several large banks are considering the introduction of such a payment system. The possibility of making card-to-card transactions, which is viewed by many national banks as a security risk, can be disabled in the latest version of the system. This has strongly increased its level of acceptance. We can hardly wait to see what will develop.

12.4   The Eurocheque System in Germany

Germany is different from other countries with regard to card-based payment systems, in that traditionally debit cards (Eurocheque cards) have been used much more than credit cards. This type of card can be used in many places to make payments, after the user has entered a 4-digit PIN. The amount to be paid is immediately deducted from an account that is associated with the card. The merchant must pay a fixed fee for each payment transaction, but this is not all that high. Credit cards are not particularly popular in Germany, because the fee is a percentage of the amount paid. There is also a variation in which the customer consents to have the purchase price transferred from his account to that of the merchant via a debit note (this is called *POS ohne Zahlungsgarantie*, or POZ). In this case, the Eurocheque card serves only as a reference to the customer's bank account, and an online liquidity check is made to this account. In this case, however, the merchant does not receive a payment guarantee, as he would with a credit card transaction or a normal Eurocheque or Geldkarte transaction.

Since Eurocheque transactions usually have to be authorized online by a background system, the merchant also must pay the costs of the individual data transfers or the rent on a leased line. Since this is only worthwhile for the merchant if there is a large turnover and many purchases are made using Eurocheque cards, improvements to this system have been sought for some time. The acceptance of Eurocheque cards among merchants would increase dramatically if the high telecommunications costs could be eliminated. This means that a system that can work offline is needed.

In 1993, the *Zentraler Kreditausschuß* (ZKA), which is a working group of the leading associations of the German banking industry, issued a call for tenders for the design of a multifunctional chipcard (MFC)[10] that would be suitable for electronic payment systems. Several firms made suitable submissions. One of them was selected and awarded the rights to the design by the ZKA. Due to changes in the general technical requirements, this was then extensively revised. As a result of these revisions, there are now a total of five interface specifications for Eurocheque cards with chips, each of which covers a particular subject area. Unfortunately, these specifications are confidential, so it is not possible for us to publish any detailed information about them. However, we can present brief summaries of the subject areas, which are:

- data structures and commands,
- the electronic cash system,
- electronic purses,
- key management,
- personalization.

These five documents describe a payment card whose functionality corresponds to that of the current Eurocheque card and which also contains an electronic purse. It is also possible to load any desired additional applications in the card after it has been issued.

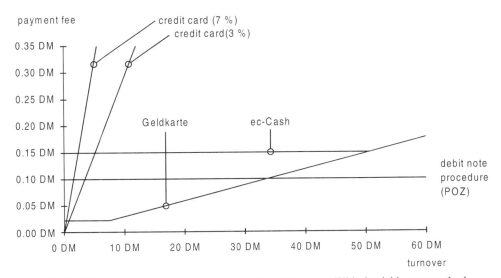

Figure 12.17   Merchant fees for electronic payments in Germany. With the debit-note method (POZ), 10 Pf are charged per transaction, regardless of the amount. Here the customer signs a debit note to indicate his agreement to pay, and the merchant's bank can then debit the payment amount from the purchaser's account. When a purchase is made with a Geldkarte, the merchant must pay at least 0.3 percent of the purchase amount, with a minimum fee of 2 Pf. If the customer uses ec-Cash, the merchant must pay at least 0.3 percent of his sales turnover, but the minimum fee is 15 Pf. With credit cards, the merchant must pay between 3% and 7% of the sales turnover, depending on the contract, but there can also be a contractual minimum amount.

---

10   The term 'multifunctional chip card' was coined at this time and in this context.

Prior to the nationwide introduction of the new card, a large-scale field trial was conducted in a region around Ravensburg and Weingarten (near Lake Constance) with a population of 250,000 people. The trial involved around 100,000 Eurocheque cards and approximately 500 terminals. Following this, the wide-scale introduction of the card throughout Germany started in the fall of 1996. Up to now, around 40 million Eurocheque cards with chips have been issued in Germany, all of which must be replaced every three years. At the end of 1998, 250,000 terminals had been installed in Germany, and 300 million transactions take place each year with these terminals. The combined turnover of magnetic-stripe and chip-based transactions with ec-Cash and Geldkarte amounts to 38 billion Deutsche Marks per year. The 41,000 cash dispensers in Germany are also gradually being upgraded with Smart Card terminals. One of the largest Smart Card payment systems in the world has thus emerged from this field trial.

User functions

The new German Eurocheque card with a chip has several different user functions. The card can be used to make online or offline payments at a suitable terminal after entering a PIN code. The amount due is then deducted from the associated account by the bank that issued the card. This application is always referred to in this chapter as 'ec-Cash', although there are also other designations for it.

Naturally, it is also possible to use a Eurocheque card with a chip to obtain cash from the cash dispensers of various banks, but in terms of functionality this actually belongs in the realm of ec-Cash. It goes without saying that the Smart Card also supports a variety of lobby machine functions, such as printing out an account statement. In addition, the card also contains a prepaid electronic purse called 'Geldkarte', which can be used to make payments without entering a PIN code. This purse is available in anonymous and non-anonymous versions. It can be repeatedly reloaded using suitable loading terminals (located at bank counters and cash dispensers), either against a cash payment or via an ec-Cash transaction.

One of the basically new functions of the Eurocheque card is the possibility of downloading additional applications, with all their associated files and commands, after the card has been issued to the cardholder. It remains to be seen whether this capability will be used in the future, and if so when, since suitable authorization from the bank that issued the card as well as a security certification are naturally required before an application can be downloaded.

The overall system in brief

As is usual with large payment systems, there is a central clearing system for the new German Eurocheque Smart Cards. Several computer centers (4–6 are planned) make the settlements among the accounts of the merchants, cardholders and participating financial institutions. Figure 12.18 shows an overview of the clearing process for the electronic purse application in the non-anonymous version. By the way, the clearing body in Germany is called the *Börsenevidenzzentrale* (BEZ).

The entire payment traffic is based on two accounts: the purse settlement account, which always reflects the current balance of the electronic purse in the card, and the card account, which for example could be the current account of the cardholder. The purse settlement account is a 'shadow account' that is maintained in parallel to the electronic purse. If an amount of money is loaded into the electronic purse, a corresponding booking is made to the purse settlement account at the same time, since this process always has to be carried out online. Depending on whether the payment process takes place online or offline, the amount of the payment is booked against the purse settlement account either at the same time or at a later time. The main advantage of this account-based system is that the purse balance can be reconstructed after a certain amount of time if the electronic purse is lost or becomes unusable. Of course, it is not possible to ensure that payments are anonymous with this approach. However, there is an anonymous version of the electronic purse with a shadow account, but with no connection to a customer account.

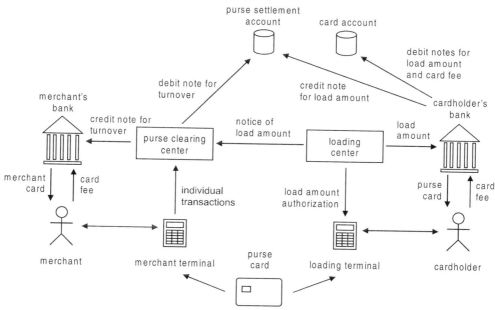

Figure 12.18   Basic structure of the payment system for the German Geldkarte, which is an account-linked electronic purse system and thus not anonymous

Eurocheque Smart Cards

There are several different types of cards in the German Eurocheque system. However, the Smart Cards for normal bank customers can be divided into several subgroups, as follows:

- Eurocheque cards (ec-Cash and Geldkarte linked to an account) (non-anonymous),
- Geldkarte linked to an account (non-anonymous),
- Geldkarte not linked to an account (anonymous).

There are two DFs in a Eurocheque card, one of which holds the ec-Cash application and the other the Geldkarte[11] application. The DF for ec-Cash is missing from cards that contain an electronic purse. The account link, which determines whether the card is anonymous, is created only by certain data elements.

---

[11]   The term 'Geldkarte' is ambiguous, so that it unfortunately must always be understood in context. It can refer to either an electronic purse as an application in a Smart Card, or the Smart Card itself.

There is a merchant card in the ID-000 format (plug-in) for use in merchant terminals. It contains all the commands and files needed to carry out financial transactions. This card can be seen as the security module of the terminal. One of the unusual and technically interesting features of this system is that it has both real and virtual merchant cards. A real merchant card is a normal Smart Card in the plug-in format. A virtual merchant card is only the software simulation of a real merchant card, which runs in the protected environment of a security module of a merchant terminal.

This solution was originally a compromise to allow terminals without sockets for plug-in cards to also be used in the new system. In the meantime, this solution has turned out to have some very positive technical features. For instance, a virtual merchant card can easily be replaced via remote maintenance, since it consists only of software. In addition, its service life is significantly longer than that of a real card, since it is not subject to the detrimental effects of a limited number of EEPROM write cycles. Finally, a good hardware security module is at least as secure as a Smart Card, since its security mechanisms are always active, thanks to its built-in power buffer.

The overall information-technological design, as well as the security module of the card, is strongly based on the European EN 726-3 standard. Elements of the ISO/IEC 7816 set of standards have also been incorporated. Finally, the EMV specification for credit cards also made important contributions to the specification for the Eurocheque card. However, there are some deviations from these standards, which could cause problems in the future with regard to interoperability with other systems. For example, secure messaging is application-specific and is not based on any standard. The mechanism for assigning EFs to specific applications is also completely independent of any existing standard.

In terms of the general technical parameters that have been made public in the specification, the card is based on many previously existing standards. Its dimensions naturally match those of the ID-1 format, and thus are the same as the present Eurocheque card. In addition, it is built as a hybrid card with both a chip and a magnetic stripe, which means that there will not be any compatibility problems during the transition from currently standard terminals with magnetic-stripe readers to new terminals with Smart Card contact units. The card uses the T=0 transmission protocol as defined by EMV-1 Version 1, rather than that of ISO/IEC 7816-3 Amd 1. Since these two definitions are not entirely identical, there can be mismatch problems when certain protocol mechanisms of one or the other standards are used. The Triple-DES algorithm is used for the cryptographic processes. One of the interesting security aspects is that the complete software as well as the hardware of the Smart Card must be have security certifications based on the ZKA criteria catalog.[12]

The file management system supports several levels of DFs, as well as file selection using Short FIDs, FIDs and AIDs. Only the linear fixed and cyclic data structures are supported, and these can be implicitly selected in the appropriate commands. The maximum size of either of the record-oriented file structures is 254 records of 255 bytes each. Naturally, no presently existing application actually fully exploits the maximum size.

The file management system uses a special mechanism to assign EFs to particular applications. This function, which is implemented by means of two non-standardized commands, allows EFs to be assigned to an application outside of the boundaries of a DF. This makes it possible within a particular DF to use a Short FID to select an EF that is located in a different DF. This means that a particular EF can be assigned to several different DFs. This corresponds in principle to the alias mechanism used in many PC

---

12   See also Section 9.3.1, 'Evaluation'.

operating systems. The objective is to make EFs containing general information available to several applications, across application boundaries, without using complicated selection procedures. An EFs that is assigned to several applications in this manner can be selected using a Short FID or a FID, and then be read or written after the necessary security state has been attained. All files have specific access conditions that make reading and writing depend on previously attained states (such as via a PIN entry). With this object-oriented system, it is also possible make file accesses depend on secure data transfers. This means that there are file attributes that can compel the use of secure messaging for any access.

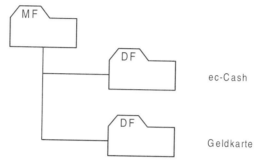

Figure 12.19   Basic file tree of a German Eurocheque Smart Card containing both the ec-Cash and Geldkarte applications

The available commands can be divided into three classes. The first class consists of commands that are compliant with ISO/IEC 7816-4, but which have a reduced range of functions compared to the standard. The second class consists of administration commands that are specific to Eurocheque, which are used for management purposes in the card. They can be used to create new files, delete existing files and enter new commands into the card. The third class is extension commands, which are used to achieve the functionality of the ec-Cash and Geldkarte applications. The administration and extension commands are purely specific to Eurocheque cards, and in principle they have no connection to any international standards.

As you can see from this brief description, the Eurocheque card has a relatively large range of functions. This unavoidably results in a large memory demand. Consequently, the target hardware that is presently used is primarily microcontrollers with 16 kB of ROM and 8 kB of EEPROM. A pure electronic purse card without the ec-Cash functions needs only around 4 kB of EEPROM. Altogether, these memory demands mean that the largest currently available chips must be used to hold the extensive amount of program code as well as the application data for ec-Cash and Geldkarte (around 1.5 kB apiece). The following types of microcontrollers are used (among others): H8/3102, MC 68HC05 SC28, ST 16SF48 and SLE 44C80S.

Supplementary applications
The operating system of the Eurocheque card has commands and mechanisms for downloading executable program code. However, this must be different for each type of microcontroller and operating system, since only address-dependent machine code can be

downloaded. The amount of logistical effort that this requirement creates for downloading new commands is the main reason why this mechanism is presently not used.

However, it is not absolutely necessary to load programs in the Smart Card for a supplementary application. In most cases, it is sufficient to have files available that have suitable access privileges. The Eurocheque card specification includes commands for creating files. The amount of administrative effort that is needed to implement an additional application in each individual card via a clearing center is however very high, which why this mechanism is also very seldom used.

Instead, several files are stored in the Eurocheque cards when they are personalized, to provide room for new applications at some time after the cards have been issued. The savings bank association has given the name 'Space Manager' to this technique for managing files for supplementary applications. The EFs in question have linear fixed structures, contain several records according to the amount of available memory, and have a uniform record size of 55 bytes. This curious and fairly large record size is a compromise between the minimum amount of memory needed to store an electronic driver's license and the amount of free memory that was available in the first implementation. Normal Eurocheque cards thus contain one EF with a single 55-byte record and a second EF with two 55-byte records. Account-linked Geldkarten have three additional EFs containing one, four and ten 55-byte records, respectively. One or more records of a file are thus reserved for a supplementary application. In the German Eurocheque world, such records are often referred to as 'drawers'.

Since the Space Manager files are only located in the cards of the savings bank association, and the Eurocheque cards of other banks are usually not compatible with this scheme, a common solution for all banks is being developed under the auspices of the ZKA. The new solution is called 'Empty Space Manager' (*Leerraumverwalter*), and it strongly resembles the Space Manager.

Summary

The German Eurocheque card system is presently one of the largest and most complex payment systems using Smart Cards. This applies not only to the payment procedures, but also to the logistics of the infrastructure for manufacturing the chips, personalizing the cards and issuing the cards. After all, every three years, approximately 40 million cards must find their way into the customers' hands within less than three months.

The security analyses of the microcontrollers and software for the operating system and applications have also set certain standards, since the acceptance criteria are severe and they must be constantly adapted to new circumstances (such as DFA, DPA and so on). The technically intricate compatibility tests are also interesting. They must ensure that software produced by more than twenty different masks, with ten different types of microcontrollers, works with more than thirty different types of terminals without any problems.

The question of whether this large system can expand on the European level, or even come to predominate, depends on many different conditions. Field trials in France, Luxembourg and Poland are being planned. An additional perspective for this system is shown by several pilot experiments, in which the Geldkarte is being used as a payment medium in the German part of the Internet. All that the customer needs for this is a simple, inexpensive terminal connected to a home PC, along with the associated software. The merchant counterpart is a security module or a special terminal that is connected to the customer's PC via the Internet.

The fact that more than 40 million Smart Cards are in use in the German Eurocheque system has an effect on all payment system projects based on Smart Cards. The knowledge gained from the German experience with multi-application Smart Cards will lead to a considerable amount of additional development in both Smart Card operating systems and the associated microcontroller hardware.

12.5   Credit Cards with Chips

Specifications and standards for using Smart Cards in a wide variety of application areas have been available for many years. However, traditionally there has been a strong concentration on telecommunications applications, which means telephone and GSM cards. This situation has changed markedly in the last two to three years. The European electronic purse standard (EN 1546) is a good example of this development. However, the most important example in the area of payment systems is the EMV specification, which is named after its three initiators: Europay, Mastercard and Visa. It describes in detail all aspects of credit cards containing microcontroller chips. A corresponding specification for matching terminals is also now available.

In the autumn of 1993, the three internationally active credit card companies Europay, Mastercard and Visa started work on a specification titled 'IC Card Specifications for Payment Systems'. Version 1 was published relatively soon afterwards, in October 1994. In mid-1995, a revised specification (Version 2) was completed. The final version of the specification, called 'EMV '96', was released at the end of June 1996. It is backward compatible with Version 2. After ambiguities had been cleared up and small errors corrected, a completely revised and compatible version of the EMV specification was published in the summer of 1998. It is also available on the Internet [Europay, Mastercard, Visa], and it can be recommended without reservation as a worthwhile subject of study.

Several factors motivated these three credit card issuers to prepare a specification for credit cards with chips within such a short time. First, existing credit cards with magnetic strips can be very easily forged. Nowadays, the only real obstacle is the hologram, which is largely secure against forgery. All other card features can be copied relatively easily. The second important factor is the additional functions that a microcontroller card can offer. Electronic purses, bonus points and telephone functions are only some of the possibilities.

The specification for credit cards with chips is divided into three parts. The first part draws heavily on the ISO/IEC 7816-1/2/3 set of standards. It describes the electromechanical characteristics, logical interface and transmission protocols. According to this part, the Smart Card has an ID-1 format[13] with ISO contact locations. It must have a supply voltage of 5 V ± 0.5 V, a maximum current consumption of 50 mA and a clock rate of 1–5 MHz, among other things. The contact force must not exceed 0.6 N per contact. Data transmission at the physical level is essentially identical to ISO/IEC 7816-3. This applies to the time interval for individual bits,[14] the ATR[15] and the two transmission protocols[16] (T=0 and T=1). The specification of the APDU[17] is identical to that in the ISO/IEC 7816-4 standard.

---

13   See also Section 3.1.1, 'Card formats'.
14   See also Section 6.1, 'The Physical Transfer Layer'.
15   See also Section 62, 'Answer to Reset'.
16   See also Section 6.4, 'Data Transmission Protocols'.
17   See also Section 6.5, 'Message Structure: APDUs'.

The second part of the specification defines the data elements and commands. Its relies heavily on the ISO/IEC 7816-4 standard, with additional commands specific to EMV. The third part, which is called 'Transaction Processing', describes the data and processes needed for payment transactions. An additional general specification deals with the terminals to be used and establishes their most important parameters, including the keypad layout, data elements and user prompts. The appendix contains a further summary that defines the possible terminal types. The entire specification is kept very general and only addresses basic parameters for terminals.

Since typical credit cards are mass-produced articles, the manufacturing costs must naturally not be too high. However, since these depend predominantly on the embedded microcontroller, the credit card application was designed from the start to use a minimum amount of memory. A conventional EMV application without supplementary functions thus fits into a processor with 6 kB of ROM, 1 kB of EEPROM and 128 bytes of RAM. Even in the Smart Card field, these values represent the lower end of the range of available chips, but they do make for an inexpensive product.

The EMV specifications are to a certain degree basic documents that specify the minimum requirements for the various card issuers. The current version (Version 2) leaves a number of issues open. For example, risk management of the terminal and the Smart Card during transactions is not precisely specified. Consequently, the EMV application is only described here in outline, since many details are either not fully specified or are certain to be changed.

Files and data elements
The specification for credit cards with chips only states that a tree structure[18] must be used for the files. The application itself is located in its own DF, which is selected via an AID (application identifier) and which contains all the data elements for the credit card application. These are stored in EFs whose structures may be either linear fixed or linear variable. FIDs (file identifiers) for EFs are limited to no more than five bits, so that all EFs must be selected implicitly by Short FIDs. The main difference between this and similar specifications is that no particular EFs and FIDs are specified. This is anyhow not necessary for the functions used in the payment process, since all the necessary data can be processed using existing commands. There is only one file directly below the MF (EF$_{DIR}$ as per ISO/IEC 7816-5), which contains all the information relating to the applications present in the card.

All the data elements in the terminal–card system are specified using unambiguous templates and tags. They can be addressed within the application using the specified commands, without their precise locations in the file tree or a particular file having to be known. This makes it possible to leave the definition of the file structure to the card issuer, since it does not affect the execution of payment transactions.

Commands
Technically speaking, only three commands are necessary for the actual execution of payment functions. However, additional commands are certainly needed for personalization, management, special functions and supplementary applications. These fall outside the scope of the EMV specification. The following commands must be available,

---

18   See also Section 5.6, 'Smart Card Files'.

according to the requirements of the EMV specification, with all return codes being analogous to ISO/IEC 7816-4:[19]

- EXTERNAL AUTHENTICATE (as a subset of ISO/IEC 7816-4)
- GENERATE APPLICATION CRYPTOGRAM (specific to EMV)
- GET DATA (specific to EMV)
- GET PROCESSING OPTIONS (specific to EMV)
- READ RECORD (as a subset of ISO/IEC 7816-4)
- SELECT (as a subset of ISO/IEC 7816-4)
- VERIFY (as a subset of ISO/IEC 7816-4)

Cryptography

The cryptographic mechanisms used in an application are by nature highly dependent on the associated general conditions. This can be seen especially well in the EMV application. A basic premise of the system design is that the terminals do not have security modules. This makes it impossible to use symmetric cryptographic algorithms, since the keys could not be kept secret. Security modules are not used, because they would considerably increase the cost of the terminal, and international key management for terminals that work partly offline is very complicated and expensive.

In order to make Smart Card authentication by the terminal possible under these conditions, an asymmetric cryptographic algorithm must be used. EMV uses unilateral static authentication with card-specific keys.[20] This does not allow the card to check the authenticity of the terminal, but this is not essential in EMV application, since debiting is not performed in the card. The card only generates a transaction certificate for the terminal. This transaction certificate is not anonymous with regard to the cardholder, and it can be submitted to the relevant card issuer only by an authorized (known) merchant. This largely eliminates most possible forms of fraud, since a valid transaction certificate can be exchanged for money only by an authorized merchant known to the card issuer.

In principle, any desired cryptographic algorithm can be used, since both the associated data elements and the algorithm itself are unambiguously identified by TLV-coded data structures. Version 2 of the EMV specification allows either the SHA-1 (as per FIPS 180-1) or the ANSI X9.30-2 hash function to be used.[21] The cryptographic algorithm used is not DSS, as might be expected from the use of SHA-1, but rather the RSA algorithm[22] (ANSI X9.31-1). The length of the key may be between 512 and 1024 bits, and it is indicated by an appropriate coding (signature identifier). Small numbers, such as 3 or $2^{16}+1$, are recommended for the public key to minimize the computation time.

If the Smart Card has established an online link with the background system, it is possible to protect the data transmission by secure messaging[23] as per ISO/IEC 7816-4. A symmetrical cryptographic algorithm is used with this end-to-end communication, namely DES. In this case, it is possible to do so without compromising system security, since both the background system and the card can securely store the secret key.

---

[19]   See also Section 7.10, 'Commands for Credit and Debit Cards'.
[20]   See also Section 4.10.3, 'Static asymmetric authentication'.
[21]   See also Section 4.8, 'Hash Functions'.
[22]   See also Section 4.6.2, 'Asymmetric cryptographic algorithms'.
[23]   See also Section 6.6, 'Secure Data Transfers'.

System structure and transaction processes

The system structure for payment systems is traditionally very centralized. There are usually several background systems, which are either individually or collectively responsible for a certain region (such as Germany). The computer centersfor the background systems, which are equipped with high-performance computers, are interconnected by the independent network of the card issuer. This network supports data exchanges for clearing and increases operational reliability, since if one center fails, its work can be taken over by other centers. Individual terminals are connected to the background system via the public telephone system and data networks, such as ISDN and X.25. An acquirer, who routes and bundles the transaction data, may be located between the terminals and the background systems. However, this depends strongly on the individual card issuer and country. At the terminal level, there are two different options: data may be exchanged directly with the acquirer, or a concentrator belonging to a merchant or chain of shops may be used. Both of these options are possible, and both are used in practice.

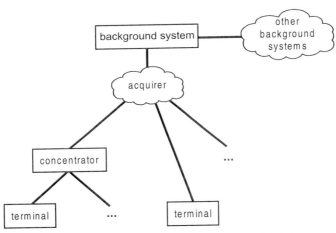

Figure 12.20   Basic structure of a payment system for credit cards with chips

The process of making a payment with a 'smart' credit card is not much different from making a payment with the old type of credit card. The customer presents his or her card to the cashier, and it is inserted into a Smart Card terminal. If a terminal is not available, the payment can be made in the conventional manner using the normal magnetic stripe or embossed characters, which are still present on the card. However, even if a terminal is present, it is still possible to verify the identity of the card user by means of a signature. In this case, the associated transaction receipt has a code that indicates that the card user has been identified in this manner. The other option is for the card user to enter a four-digit PIN. This can be checked online by the background system, or offline by the Smart Card. If a PIN is used for identification, the transaction certificate that is generated will indicate which the type of PIN check was performed. This process is shown in summary form in Figure 12.21.

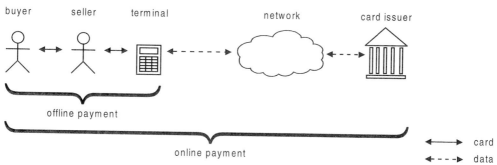

Figure 12.21   Basic infrastructure for a payment transaction with a Smart Card, according to the EMV specification

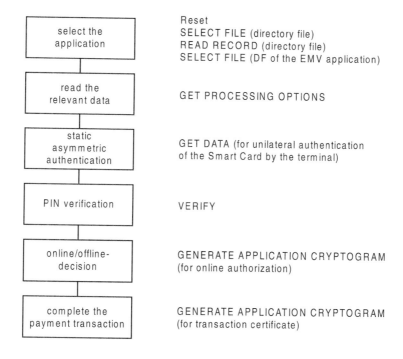

Figure 12.22   A very simplified portrayal of the payment process according to the EMV specification, as seen by the terminal

At the detail level, the individual functions and the process of a successful payment transaction with a Smart Card are naturally somewhat more complicated. An example is shown in Figure 12.22, which illustrates the fundamental mechanisms. Both the card and the terminal determine the exact course of the transaction, depending on various transaction data, such as the amount involved. For instance, if the amount to be paid exceeds a certain limit, the card requests an online authorization of the payment by the background system. The terminal must then set up a link to the background system and have payment approved.

Only after the background system has approved the request does the card generate a valid transaction certificate, which the merchant can submit for clearing. The purpose of online authorization is to minimize the financial risk to the card issuer. Since a 'smart' credit card must regularly report to the background system, based on a number of conditions, the maximum loss in the event that a card is stolen or tampered with can be held within precisely calculable limits.

Future developments

In terms of its structure and contents, the EMV specification allows the card issuer a lot of room for individual initiative, which means that further developments and variations are possible. This flexibility is certain to be utilized to the full in the future by various firms, and the first extensions are already in sight. For example, each of the three credit card issuers involved in drawing up the EMV specification is working on an electronic purse that can be put into a Smart Card as a debit application.[24] Particularly with vending machines and small purchases, a prepaid electronic purse definitely has advantages, since the fee that otherwise must be paid for a credit card payment is not applicable.

The increasing commercial use of worldwide networks such as the Internet will in the near future require secure, internationally available and widely accepted means of payment. Credit cards with chips and a few additional functions would be eminently suitable for this, since they have already achieved international acceptance and widespread use, independent of any particular country or currency. Work is presently being carried out in this area. The difference between this and the payment process described above is not great, since the Internet can take the place of the acquirer. In principle, all that is necessary is that a Smart Card terminal is available to the customer.

Although the EMV specification still has some inadequacies and also leaves a few points open, it has achieved nearly the same importance in the Smart Card industry as the GSM 11.11 specification. There are practically no new Smart Card operating systems on the market that do not claim to be 'EMV-compatible'. Due to the number of credit cards with chips that can be expected to be in circulation in the future, this specification is likely to represent the minimum standard for all future applications.

---

24   See also Section 12.3.1, 'TheCEN EN 1546 standard'.

13

Sample Applications

In this chapter, we present some sample Smart Card applications and describe them in broad terms. They are all based on the previous chapters, which deal with all relevant aspects of Smart Card technology, and they illustrate the extent and complexity of large Smart Card applications.

Another objective of this chapter is to present systems using Smart Cards, in which the cards are only one of several components. In such systems, the functionality, user friendliness and (in most cases) nearly all of the system security depend on the Smart Cards that are used. However, these systems should always be viewed as a whole, since they work satisfactorily only when all their components work together harmoniously.

13.1 Public Cardphones in Germany

Starting in the summer of 1989, the German state telephone company (Telekom) began introducing public cardphones in the entire country. Prior to this, several field tests were carried out using systems from various manufacturers. These started as early as 1983 in four different regions in Germany with different characteristics (conurbations, cities and rural areas). Several different types of cards were also tested in these trials, including magnetic-strip cards, hologram cards and cards made from different types of material (paper and plastic).

The conclusion drawn from these initial field tests was that plastic memory cards were the most suitable for use with cardphones. The decisive factors were the achievable level of security and upward compatibility, as compared with other types of cards. In December of 1986 and July of 1987, large-scale field trials of a system based on memory cards were started in sixteen major cities. These were successfully concluded in May of 1989. In the spring of 1999, there were more than 90,000 public cardphones in Germany, and more than 300 million prepaid phone cards have been sold since the system was introduced. Internationally, by the way, the numbers are 8.3 million cardphones and more than 1100 million cards.

The cardphones installed by Telekom in Germany can in principle be operated using two different types of cards. The first type is the phone card, which is produced and used in very large numbers. In commercial terms, this is a prepaid card, which the user purchases before using it. In technical terms, it is a memory card with an irreversible counter, a security feature and synchronous data transmission.

The other type of card is the telephone charge card, which is not as widely distributed. It resembles a credit card, since the user pays for the services (telephone calls) only after they have been received, by means of a monthly settlement against his or her bank account. In technical terms, a charge card is a Smart Card with a microcontroller. The data transfer to the cardphone takes place using the block-oriented, asynchronous T=14 transmission protocol. Since this type of card is not particularly important, we will not discuss it any further.

The entire cardphone system is constructed as a distributed (decentralized) system with several successive layers of computers. The lower levels can work completely autonomously for several days if a system breakdown occurs, without affecting normal telephone operations. The two parts of the system that the normal user sees, which are the phone cards and the cardphones, are described here.

In order to survive as long as possible under the severe operational conditions to which it is exposed, the cardphone has a sturdy metallic enclosure with openings for the controls and indicators, such as the keypad and display. The terminal is completely electrically isolated from the rest of the electronic assemblies. It is also short-circuit-proof, in order to protect it against vandalism and other attempts to disturb its operation by shorting its contacts.

The cardphone is controlled by a powerful microcontroller. The control software can be remotely updated, which means that it is not necessary for a service technician to update the software on location by replacing EPROMs. The control processor of the cardphone can also directly exchange information with the higher-level computer systems while a conversation is taking place, using a data-over-voice (DOV) modem. This is primarily necessary for telephone charge cards, since the accumulated charges are immediately sent to the background system for billing.

The main control processor of the cardphone also looks after communications with the phone card or charge card. With a charge card, it communicates using the T=14 protocol, which is specified by Telekom and is used only in Germany. Otherwise, it uses the synchronous protocol employed by the memory card in question.

The built-in terminal can supply the memory card with an external programming voltage that is adjustable between 5 V and 25.5 V in 255 steps. However, since practically all new memory cards have an internal voltage converter that generates the programming voltage from the normal supply voltage, this is actually no longer technically necessary. It is only present for compatibility with older generations of phone cards, whose contents are still good.

If a charge card is used (a 'real' Smart Card), the terminal can choose a clock frequency ranging from 1.2 to 9.8 MHz. With synchronous cards, the clock frequency must be reduced to around 20 kHz in order to obtain usable communications.

The terminals have massively constructed shutters for protection against 'dummy' cards. The shutter has an impact cutter that slices through any wires or cables that pass through the card slot. This prevents tapping or manipulating the communications between the card and the terminal.

The chips that are used for phone cards have both ROM (which can be mask programmed by the semiconductor manufacturer) and EEPROM storage. These hold card-specific data and the counter for the card balance.

A charge pump in the chip generates the programming voltage for the EEPROM, so it is no longer necessary to apply an external voltage for this. This means that the chip needs only the normal operating voltage. For protection against copies, the chip has a hardware security feature whose function is secret. In the future, memory chips that allow a unilateral authentication of the card by the outside world will also be used.

Modern phone card chips have only six contacts. This is because only five contacts are actually necessary for the full functionality of the memory chip, and all other contacts are unused. If eight contacts are present, production costs are higher, since the module is bigger and thus more expensive. In addition, it takes longer to mill the cavity for the module in the card body, which reduces the throughput of the production equipment and increases the marginal cost. Consequently, practically all new phone cards have only six contacts. The six or eight leads from the chip to the contacts are assigned as shown in Figure 13.1.

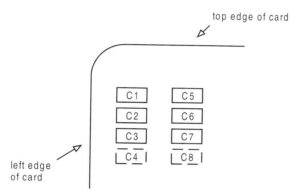

Figure 13.1   Contact assignments for a phone card. The contacts that are not listed are not used.

C1 supply voltage (5 V ± 5%)	C2 control input
C3 clock input ($\approx$20 kHz)	C5 earth (ground)
C7 data transfer	

Prepaid Telekom phone cards contain the data described below in the chip memory.

Serial number   Each phone card has a 7-digit serial number. This allows individual cards to be blocked if there is suspicion of fraud. The serial number is stored in a part of the EEPROM that is blocked against overwriting or erasing.

Date of manufacture   When the chip is manufactured, the year and date of its fabrication are permanently written to the EEPROM. This date refers only to the embedded chip, and not to the card itself.

Manufacturer code   After the chip has been embedded in the card body, the card manufacturer writes a specific code to the memory of the card. This identifies the manufacturer, and it cannot later be altered.

Initial value   The initial value of the card is also stored in every phone card, in addition to the current value.

Remaining balance   This is the only data element in the card that is functionally necessary for a prepaid phone card. The cardphone deducts the individual billing units one after the other from a five-digit irreversible octal counter in the EEPROM. The counter thus consists of five bytes of eight bits each. This gives a maximum count capacity of 8^5 (32,768). When the card is completed, the counter is set to the desired initially value of the card, which means that its count is equal the value of the card in units of Pfennigs. After this, the counter can be decremented toward zero whenever the card is used. Once the counter has reached zero, the card has been fully used up.

Transaction process

If you insert a phone card in a cardphone, the first thing that happens is that the shutter closes the card slot, and at the same time the terminal's contacts are applied to the contact surfaces of the card. After this, the cardphone executes an ISO activation sequence and thus activates the card. Following this, the terminal sets the address pointer in the card to zero and reads the first 16 bits. If this is successful, the phone card is in good working order. The 16 bits are interpreted by the terminal as a sort of ATR that contains various operating parameters for the card. The terminal next checks these parameters to see whether the inserted card is one of the allowed card types that it can correctly use with all functions. The card number, which the terminal has also read from the card, is at the same time sent via the DOV modem to the next higher level of the system. Here it is compared with the black list of blocked cards. If the black list contains no entry for this card, the card may be used, and the cardphone receives a corresponding response.

The address pointer is now set to point to the region of the fee counter, and the current value of the card is read from the EEPROM and shown on the display. After the user has entered the telephone number and a connection has been successfully made, the cardphone is regularly advised via the DOV modem when it is time to deduct charge units. Each time this occurs, the cardphone decrements the counter in the memory card and immediately reads out its new value. This allows it to check whether the amount was correctly deducted. If this is not the case, the connection is immediately broken.[1] If the cardphone calculates that the remaining balance of the card will be used up within the next 20 seconds, it generates a warning tone, and the card can exchanged for a new one without interrupting the conversation.

13.2   Contactless Memory Cards for Air Travel

The system described in this section is different in several ways from the usual applications for Smart Cards. The differences relate to the fundamental system architecture and the manner in which the card transfers data and receives power. The system is the new ticket-free flight system operated by the German airline Lufthansa. It is based on contactless memory cards, which do not have to be inserted into a terminal.[2] The basic system concept also differs from that of all other Smart Card applications described here, in that the card is only used to identify the user, while all application data are held in the background system.

For some years now, Lufthansa has been issuing frequent-flyer cards and cards for its bonus system. These cards have embossing as well as magnetic strips for automatic processing. In addition, certain types of cards can also be provided with a chip to allow them to be used with German public cardphones and to add a credit card function. The idea was to extend the capabilities of this existing family of cards with supplementary applications relating to boarding and ticketing, while retaining reverse compatibility. A

1   Devaluation cycle for a memory card; see also Section 12.2, 'Prepaid Memory Cards'.
2   See also Section 2.3.3, 'Contactless Smart Cards'.

second objective to make as few modifications as possible to existing systems. An equally important consideration was that the new cards should be easy for customers to use.

In the full-up version, the result is a multi-application card with a magnetic strip, embossing, a hologram, a memory chip with contacts and a memory chip without contacts. A large number of different applications can be served by this card, without any compatibility problems with earlier types of cards.

Before this system was introduced for Lufthansa's entire operations, a pilot project was conducted on the Frankfurt–Berlin route. The new cards were issued to 600 customers of all different types, and many thousands of flights were made on this route between May and December of 1995. To allow the customers to use the cards, suitable automated service kiosks were installed at both airports. Each of these was a PC-based system with a touch-sensitive color screen, a printer, and a transceiver for the contactless Smart Cards.

Applications in the card

A wide variety of applications is available to the customer in the new card. Here we consider the fully equipped version of the card. There are of course simpler versions as well, such as a card with only the contactless memory chip functions. With this chip, the traveler can check himself or herself in at the service kiosk. This naturally leads to higher throughput, with shorter waiting times and/or faster processing. Lufthansa's bonus system is also integrated into the contactless chip, so that it is not necessary to use any other card or enter additional information.

Using suitable automated equipment located in the air terminal, the traveler can check in and receive a printed 'boarding information' form. This contains all the essential information regarding the booked flight, as is found on a conventional ticket. If necessary, the traveler can also make a seat selection at the same time. Flights can also be booked by phone, using the number embossed on the card. In the future, Lufthansa plans to manage all of this without paper tickets, since all relevant information can be retrieved from the background system via the card. Naturally, this does not rule out making flight reservations by phone or fax, which will continue to be possible.

In addition to these functions, the new card can also carry the elements needed for a credit cart, namely the appropriate data on the magnetic strip and a hologram. It is also possible to embed a contact-type chip in the ISO location. This is currently used for a phone card chip.

The overall system

In Smart Card terms, all the applications based on the contactless chip are constructed very simply. The card is used only for identification, and it employs a memory chip that supports dynamic authentication by the background system.[3] After authentication, all the information held by the card is read. Currently, this is the customer number, customer name and customer profile. With this information, the background system can match the card to a particular person, after which the booking data, bonus system points and all other applications are made available. The card is thus used only as a kind of key, while the data and mechanisms belonging to the various applications are always held in the background system. This has considerable advantages in this case, since the background system and all its required databases, programs and interfaces have been available for a long time.

---

3   See also Section 4.10.1, 'Symmetric unilateral authentication'.

Another advantage of this system is how it deals with lost or defective cards. This is a problematic issue that up to now has always been neglected in other systems with multi-application cards. If applications are loaded into the card after personalization, when the cardholder receives a new card, he or she must contact all the individual application providers in order to have them reload their applications into the card. Thanks to its centralized system architecture, the Lufthansa Smart Card system does not have this problem. If a card is lost or becomes defective, the customer receives a new one and the old card is blocked system-wide by means of its number. This also does not require a large amount of logistical effort, since all airports served by Lufthansa have access to the necessary data via the well-established Lufthansa network.

The contactless card
The new Smart Card has the internationally standard ID-1 format. The memory chip embedded in the card body uses inductive coupling, with a single coil for both power and data transfers. With this technique, the terminal can both read and write data at a distance of up to 10 cm.[4] If the card is in an ordinary wallet, it is even possible to carry out the data exchange with the terminal without removing the card from the wallet. It is only necessary to hold the wallet next to the terminal. Since the chip and the coil are both located inside the card, the graphic layout need not be affected by these components. Furthermore, contactless technology eliminates the problem of worn contacts, since there simply aren't any contacts.

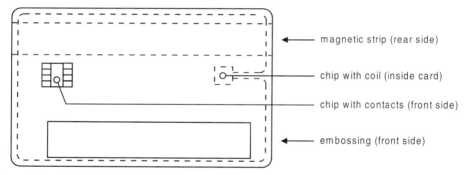

magnetic strip (rear side)

chip with coil (inside card)

chip with contacts (front side)

embossing (front side)

Figure 13.2   Physical appearance of the new Lufthansa Smart Card in the full-up version

The typical transaction time between the terminal and the card in this system is around 100 ms, and a clock frequency of 13.56 MHz is used. The memory chip used (SLE 44R35) contains 1 kB of EEPROM, and it can be unilaterally dynamically authenticated by the outside world. Since presently only 48 bytes of data are stored in the memory (for the customer number, name and profile), it would be possible to include new application areas in the future – although in this system, new applications naturally do not require additional memory in the Smart Card.

Future prospects
The field trial on the Frankfurt–Berlin route was very successful. Lufthansa planned to start issuing production cards in March of 1996, first to frequent fliers and later to all interested

---

4   See also Section 3.6, 'Contactless Cards'.

customers. To this end, suitable equipment was quickly added to Lufthansa's customer processing areas in all German airports, so that the new contactless card could be used everywhere. Going beyond this, it was planned in principle to install this new technology in all airports served by Lufthansa. The benefit for the customer is that it takes less time to check in before the flight.

Although this Smart Card application has an atypical system structure, it does offer a somewhat different and interesting perspective on the Smart Card world. Since all application data are held in the background system, all issues related to data protection are also lodged with this system. This means that data protection legislation, independent of the country, has no influence on the data content of the card. It also avoids the well-known problem of limited memory space for multi-application cards, since the question of which provider may write which data to the card does not exist. Applications in the card are also completely isolated from each other, and the system design prevents applications from interfering with each other. Finally, we may note that information that is valuable to the system operator is always stored securely in the background system, rather than in cards where it may go astray.

This new system is also a model of seamless progression from one card technology to the next generation. This 'gentle' migration from one level of technology to the next one is a very important factor, since prior investments are not lost and it is not necessary to build up a new system from scratch. Of course, the price of this benefit is a relatively new and expensive card technology. However, it is quite conceivable that this technology will become very influential in this sector, since it makes possible faster and more efficient customer processing than do existing tickets with magnetic strips.

13.3   Health Insurance Cards

In Germany, all members of a public health insurance plan were issued a health insurance card (*Krankenversichertenkarte* or KVK) by the end of 1994. Since 1996, the private health insurance plans have begun to issue their own Smart Cards, which are compatible with the public KVKs. These cards, of which more than 72 million have been issued, have thus penetrated German society more than any other type of card, including phone cards.

Originally, it was only planned to introduce a magnetic-strip card, but it was decided to use Smart Cards instead, due to considerations of possible future developments. The least costly solution in this case was to use memory cards, since at that time microcontroller cards were much more expensive. Nonetheless, the system is designed such that the memory cards can be replaced by 'real' Smart Cards in further stages of development over the next few years. In this way, a nationwide system is in the process of being developed. It can form the basis for a future health-care Smart Card, if this becomes necessary.

The health insurance member's card has two basic functions for the insured person. Its first function is to identify the person to the doctor who treats that person. It thus replaces the paper health insurance certificate, which is no longer necessary. Its second function is to act as a machine-readable data carrier for the computer in the doctor's clinic. The terminal is normally connected to a PC in the clinic, which also drives the terminal. The card can be read in the terminal, and the resulting data can be further processed automatically. If the doctor manages his or her practice using conventional methods (that is, without a computer), the terminal can directly transfer the data from the card to a printer, where they can be transferred to a form.

There are three different places where the data in the health insurance member's card can be accessed. The first is the doctor's clinic, where data can only be read from the card. It is not intended that data can be written in this situation, and the terminal software would anyhow prevent this. The second location is within the health insurance organization itself, where again only read access to the data is allowed. Here the insured person can read and check the personal information that is stored in the card. The third location is also within the health insurance organization, where special terminals allow data to be written to the card. This can for example be necessary if the insured person moves to a new address. However, many insurers have decided to simply issue a new card to the insured person in this case, with instructions to return the old card for destruction. This is logistically simpler and thus less costly.

In the initial phase of the KVK project, a wide variety of information about the patient was open to discussion. Some people wanted to include all possible information in the card, ranging from the blood type to allergies, so that it would be a sort of emergency care card. However, after all objections based on data protection were dealt with, only the personal information listed in Table 13.1 was left to be stored in the card's chip.

Table 13.1   Data elements and TLV coding of the German health insurance card, according to the technical specifications for German health insurance cards (1993)

Data element	Length	Tag
Birth date of the insured (DDMMYYYY)	8 bytes	'88'
Checksum for the entire template (XOR)	1 byte	'8E'
First name of the insured	1–28 bytes	'85'
Insurance card number (VKNR)	5 bytes	'8F'
Legal system (east/west)	1 byte	'90'
Name extensions of the insured	1–15 bytes	'86'
Name of the insurer	2–28 bytes	'80'
Number of the insured	6–12 bytes	'82'
Number of the insurer	7 bytes	'81'
City or town	2–22 bytes	'8C'
Postal code	4–7 bytes	'8B'
State	1–3 bytes	'8A'
Status of the insured	4 bytes	'83'
Street name and house number	2–28 bytes	'89'
Surname of the insured	2–28 bytes	'87'
Template for data regarding the insured	70–212 bytes	'60'
Title of the insured	2–15 bytes	'84'
Expiry date of the card (MMYY)	4 bytes	'8D'

The requirement that the information in the card be generally known was also one of the prerequisites for the approval of the overall system. No secret information, and no information not known to the insured person, may be present in the card. It must also not be possible to write additional data to the card at a later date without authorization. To fundamentally prevent the writing of data to the card, neither the doctors nor the insurance organizations received terminals that could write to the cards. Only a few administrative terminals located in the insurance organizations can write data to the cards. However, no special authentication key is needed for this, so data could easily be written using any suitably equipped terminal.

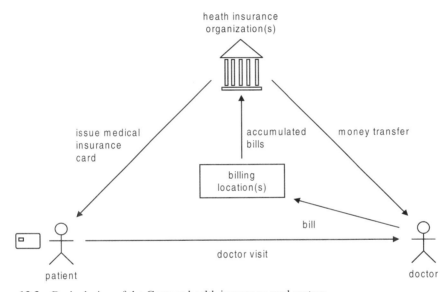

Figure 13.3   Basic design of the German health insurance card system

Externally, a health insurance card behaves as though it contains only a single transparent file (EF) that can be read as desired using length and offset parameters. Specific administrative terminals can also write to the memory, but for reasons of data protection, this is an exception.

When the structure of the data elements was specified, it was very important to allow for future extensions or modifications without causing any compatibility problems. Consequently, all personal data in the health insurance card are structured using the ANS.1 data description language. They are contained in the card's memory in TLV structures. This allows additional data objects to be added in the future, or the codes used for existing data objects to be modified. The tags that are to be used are prescribed by a specification, so that the data elements of all health insurance cards are structured in the same manner.

The health insurance card is not a microprocessor card. It is a memory card, with hardware similar to what has been used for years in phone cards. The EEPROM that is used must have a capacity of at least 256 bytes. This is equal to the amount of all necessary data held in the health insurance card. If the EEPROM is exactly this large, the necessary data just fit and it is physically impossible to write any additional data to the card in violation of data protection regulations.

Table 13.2 Sample data set of a KVK for a privately insured person, with all data elements decoded. The listed data correspond to a non-existent person, and the XOR checksum is intentionally incorrect. The following abbreviations are used:

T:	tag	
L1:	tag within the length code, indicating that the following byte (L2) is the length code	

L, L2:	length of the following data	
NU:	not used	
V1:	identifier of the personalizer	

	x0	x1	x2	x3	x4	x5	x6	x7	x8	x9	xA	xB	xC	xD	xE	xF
0y	'A2'	'13'	'10'	'91'	'46'	'0B'	'81'	'15'	'44'	'45'	'47'	'2B'	'44'	'32'	'88'	'F1'
				ATR and manufacturing data												
1y	'9F'	'61'	'0B'	'4F'	'06'	'D2'	'76'	'00'	'00'	'01'	'01'	'53'	'01'	'07'	'60'	'81'
	T	L		T	L			AID				T	L	V1	T	L1
2y	'88'	'80'	'18'	'56'	'65'	'72'	'65'	'69'	'6E'	'74'	'65'	'20'	'4B'	'56'	'20'	'20'
	L2	T	L	"V"	"e"	"r"	"e"	"i"	"n"	"t"	"e"	" "	"K"	"V"	" "	" "
3y	'20'	'20'	'20'	'20'	'34'	'30'	'33'	'33'	'35'	'33'	'31'	'81'	'07'	'30'	'30'	'30'
	" "	" "	" "	" "	"4"	"0"	"3"	"3"	"5"	"3"	"1"	T	L	"0"	"0"	"0"
4y	'30'	'30'	'30'	'30'	'8F'	'05'	'30'	'30'	'30'	'30'	'30'	'82'	'07'	'38'	'31'	'36'
	"0"	"0"	"0"	"0"	T	L	"0"	"0"	"0"	"0"	"0"	T	L	"8"	"1"	"6"
5y	'34'	'35'	'30'	'31'	'83'	'04'	'30'	'30'	'30'	'32'	'90'	'01'	'31'	'85'	'09'	'41'
	"4"	"5"	"0"	"1"	T	L	"0"	"0"	"0"	"2"	T	L	"1"	T	L	"A"
6y	'6C'	'65'	'78'	'61'	'6E'	'64'	'65'	'72'	'87'	'05'	'52'	'61'	'6E'	'6B'	'6C'	'88'
	"l"	"e"	"x"	"a"	"n"	"d"	"e"	"r"	T	L	"R"	"a"	"n"	"k"	"l"	T
7y	'08'	'30'	'31'	'30'	'31'	'31'	'39'	'36'	'35'	'89'	'15'	'50'	'72'	'69'	'6E'	'7A'
	L	"0"	"1"	"0"	"1"	"1"	"9"	"6"	"5"	T	L	"P"	"r"	"i"	"n"	"z"
8y	'72'	'65'	'67'	'65'	'6E'	'74'	'65'	'6E'	'73'	'74'	'72'	'2E'	'20'	'32'	'30'	'30'
	"r"	"e"	"g"	"e"	"n"	"t"	"e"	"n"	"s"	"t"	"r"	"."	" "	"2"	"0"	"0"
9y	'8B'	'05'	'38'	'30'	'30'	'30'	'30'	'8C'	'07'	'4D'	'7D'	'6E'	'63'	'68'	'65'	'6E'
	T	L	"8"	"0"	"0"	"0"	"0"	T	L	"M"	"ü"	"n"	"c"	"h"	"e"	"n"
Ay	'8D'	'04'	'30'	'31'	'30'	'37'	'8E'	'01'	'42'	'C0'	'54'	'20'	'20'	'20'	'20'	'20'
	T	L	"0"	"1"	"0"	"7"	T	L	XOR	T	L					
By	'20'	'20'	'20'	'20'	'20'	'20'	'20'	'20'	'20'	'20'	'20'	'20'	'20'	'20'	'20'	'20'
				unused memory region (filled with space characters)												
Cy	'20'	'20'	'20'	'20'	'20'	'20'	'20'	'20'	'20'	'20'	'20'	'20'	'20'	'20'	'20'	'20'
Dy	'20'	'20'	'20'	'20'	'20'	'20'	'20'	'20'	'20'	'20'	'20'	'20'	'20'	'20'	'20'	'20'

Ey	'20'	'20'	'20'	'20'	'20'	'20'	'20'	'20'	'20'	'20'	'20'	'20'	'20'	'20'	'20'	'20'
Fy	'20'	'20'	'20'	'20'	'20'	'20'	'20'	'20'	'20'	'20'	'20'	'20'	'20'	'20'	'20'	'00'
																NU

The clock-synchronous data transmission protocol depends on the specific type of chip that is used. Each terminal must therefore be able to fully execute all possible protocols. The card body can be manufactured using injection molding or a multilayer technique. The working life of the health insurance card is specified to be six years. After this time, the insured person automatically receives a new card. This means that around 15 million new cards must be issued each year.

All terminals that are connected to a computer are controlled by the computer, using the T=1 transmission protocol according to ISO/IEC 7816-3 Amd 1. There is one restriction in this regard, which is that data chaining may not be used in the protocol. Doing so would not add any functionality to the application, but would only increase amount of memory needed in the terminal. By the way, this is a typical example of the fact that real applications often use only the necessary parts of the standards, and it is rare for all of the specified functions of a standard to actually be implemented.

There are only three possible commands that the terminal can execute. The first command is a reset to the health insurance card, followed by the reception or reading of the ATR. This command is always used at the start of a session to activate the health insurance card. The second command is READ BINARY with the ISO coding. The terminal can use this command to read selected parts of the data or the entire data. The third command is WRITE BINARY, also in accordance with ISO/IEC 7816-4, but this command is only available in administrative terminals. It is blocked in all other terminals. The health insurance cards of the private insurers have write protection that is implemented using PINs that are known only to the insurance organization. The cards for the public health insurance plans can be freely written, if the necessary command is known.

If there is a direct link to a computer, all that the terminal essentially does is to convert between the T=1 protocol and the hardware-dependent synchronous protocol of the health insurance card. Nevertheless, it is quite evident that changing over to microprocessor cards would not be difficult or expensive. With microprocessor cards, the terminal would only have to the transparently relay the command received from the clinic computer. The response from the card could also be transparently passed back to the control computer, without any processing by the terminal.

13.4   Electronic Toll Systems

In some countries, it is common to demand a toll for using certain roads. In contrast to blanket fees that are paid for stickers, tolls are usage-dependent. In other words, the amount paid depends on how often the road is used and the type of vehicle. Up to now, tolls have usually been paid in cash at tollbooths. A few isolated electronic systems, using various types of cards, have been used over the past few years in toll (road-pricing) systems. However, all systems up to now have the disadvantage that they significantly impede the flow of traffic, since the vehicles must either stop or slow to a walking pace. The toll stations also require a lot of space.

In order to overcome these problems, the German traffic ministry decided in 1993 to start a large-scale field trial of automatic toll collection systems. The road selected was the A555 *Autobahn* between Cologne and Bonn. Various systems offered by ten different firms were tested between May 1994 and June 1995.

Several basic requirements were used to evaluate the systems that were tested. First, traffic should be able to flow normally and unimpaired past the toll collection points. However, in Germany 'normal traffic flow' can mean anything up to 250 km/h. Therefore, it should not be possible to evade toll collection by driving very fast. Furthermore, toll booths or collection baskets were not allowed be used for paying tolls, since the equipment of the desired system had to be suitable for mounting on flyovers and sign bridges. In addition, the systems used had to support both single-lane and multi-lane traffic. It is technically undesirable to separate the traffic into single lanes, since this strongly restricts traffic flow.

A supplementary requirement, which was not originally foreseen, arose during the course of the project. This was complete anonymity with respect to toll collection, and it increasingly came to be regarded by the general public as the critical factor for the entire trial. It should not be possible to generate vehicle movement profiles or monitor the routes traveled by specific vehicles. All proposed systems had a payment option with which vehicle anonymity was maintained as long as the toll was paid. As soon as a toll collection failed, the vehicle was photographed. The registered owner could then be tracked down and sent a suitable penalty notice. Naturally, this did not actually occur during the trial, since only 'play money' was used in the trial.

Almost all the proposed automated toll collection systems used a Smart Card to hold the electronic currency. That is why we address this subject here, since it may well become important to the Smart Card industry in the future.

All systems have a device mounted in the vehicle that is known as either the OBU (on-board unit) or IVU (in-vehicle unit), as well as additional equipment as necessary. This device is powered by the car battery. Inside the passenger compartment, each system has a Smart Card terminal, a display and a simple keypad. There is also a link to the outside world, which may be unidirectional or bi-directional, depending on the system. This is usually a microwave link in the 5.795–5.805 GHz frequency band, as recommended by CEPT for this application. Some systems used radio signals in the 400–500 MHz range or infrared light. The disadvantage of using infrared light is naturally that transmission is strongly affected by weather conditions, such as heavy snow or fog. In the systems that participated in the trial, the stated mass-production price of the OBU was between 100 and 300 DM, depending on the configuration.

The control stations were installed along the motorway on flyovers and sign bridges as necessary. No modifications to the roadway construction were necessary. The Smart Cards used did not contain sophisticated electronic purses, but only very simple and fast debiting commands. This was in part due to the fact that no real money was used, and in part due to the short time available for the transaction. In some systems, the optimization process went so far as to reduce the card's ATR to four bytes, in order to leave sufficient time for the actual debiting. This is understandable in light of the requirement for unrestricted traffic flow. At 250 km/h, a vehicle covers 70 m per second. The control station has a communications range of approximately 5 m in the 5.8-GHz frequency band. This results in a dwell time of around 70 ms, during which the vehicle is within range of the station. The card must execute the following processes within this interval:

- card reset and ATR transmission ($\approx$ 10 ms),
- DES encryption to authenticate the card ($\approx$ 12 ms),
- EEPROM write access to store the new balance ($\approx 2 \times 3.5$ ms),
- data transmission from and to the card ($\approx$ 30 ms).

Additional time is needed for data transmission between the OBU and the electronic toll station. You can calculate that the time available is very tight. The Smart Cards are all driven at the maximum allowed clock frequency (usually 5 MHz). The data transfer rate between the OBU and the toll station, at 1 Mbit/s, is substantially higher than that between the OBU and the card. It thus has little influence on the total transfer time.

Here we describe three representative systems, chosen from the ten systems that were involved in the field trial. Since some of the systems used were nearly identical, except for the fine technical details, we have not identified any of the systems by name. They can be considered to be representative for this sort of application.

System 1

The first system has a classic infrastructure design. The necessary equipment is installed on two closely spaced flyovers or sign bridges. The system is naturally designed so that toll collection is not affected if the vehicle changes lanes between the two stations. There are two different payment modes. In the postpaid mode, the accumulated toll fees are collected from a bank account, as with a credit card. This makes it very difficult to preserve vehicle anonymity. In the prepaid mode, prepaid Smart Cards are used, and the appropriate amount is deducted from the balance in the card each time a toll is collected.

Technically, the system works as follows. At the first station, the OBU and Smart Card are activated. Following this, the toll is levied and a general assessment of the vehicle is made in order to classify it as a car or lorry. Since the amount of the toll depends on the vehicle class, it is necessary to check the class that is noted in the card. This is done by measuring the vehicle's height profile when it passes under the station. This is sufficient for reliable classification. When the vehicle is within the communications range of the immediately following second station, another link to the card is set up via the OBU. The second electronic toll station checks whether the toll payment initiated by the first station has been successfully executed. If this is not the case, the vehicle is photographed. Nowadays, a vehicle can be uniquely identified from its numberplate using a completely automatic process. The registered owner of the vehicle can then be sent an appropriate fine.

The system can be further extended as desired with various other functions. For example, the card can be checked against a black list when a payment is made, in order to eliminate the use of stolen cards. Another possibility that was considered was to photograph all passing vehicles in certain special cases, such as after bank robbery with escape by car, in order to obtain data regarding the escape route.

The advantage of this system is that the OBUs have a simple construction and consequently a low cost. However, this means that the electronic toll stations must be correspondingly more complicated and expensive.

System 2

The second system is not based on a roadway infrastructure. It uses the principle of virtual toll stations, which exist only as databases in the vehicle OBUs. These contain all the coordinates of available toll stations, as well as the corresponding charges for road sections. As soon as a vehicle enters a defined stretch of road, the relevant fee is deducted from the Smart Card. Naturally, this process is far less time-critical that of System 1.

The OBU knows the vehicle's coordinates at any point in time by means of an attached GPS module. The GPS, or 'global positioning system', is a worldwide system for determining locations, developed for military purposes in the USA under Department of Defense contracts. This started in 1973, and it was operationally complete in 1993. The GPS consists of 24 satellites that transmit encoded radio signals from an altitude of 20,000 km. Each signal contains the transmission time, the satellite position and a satellite identifier. A suitable receiver, which nowadays can be as small as a pack of cigarettes, can receiver the satellite signals in the 1.6-GHz band and use them to determine its geographic position. The accuracy for civilian applications is around 10–20 m. This can be reduced to a few meters using an improved technique with differential measurements (differential GPS), which relies on an additional signal broadcast by a terrestrial transmitter. Presently, this signal can be received only in Central Europe. For military applications, positions can be located with an accuracy of around one meter throughout the world.

Of course, the second toll system also requires monitoring, but this is done by means of random samples, similar to current speed controls. The main advantages of this system are that it is not necessary to install any equipment along the roadway and that the tolls are levied at virtual locations. However, control is more difficult, and Smart Cards cannot be used in the postpaid mode, since no information is exchange between the OBU and the outside world. On the other hand, this provides an advantage in terms of anonymity.

System 3

The third system is the most ambitious technical solution for automatic vehicle toll collection. The OBU is equipped with a GPS receiver and a GSM mobile phone. The position of the vehicle is determined by the GPS unit, and the OBU uses the mobile phone to obtain billing data. As with the other two systems, the OBU contains a terminal with a Smart Card, which is involved in billing the toll fees.

As in System 2, the toll stations are only virtual locations along the roads, specified in terms of geographic coordinates. In this system, the motorway operator can also modify the tariffs at will by means of bi-directional communications with the OBU. The tolls can thus vary depending on the time, place, vehicle class and ambient conditions. The accumulated tolls are calculated by the OBU in the vehicle.

In order to maintain a certain level of anonymity, the toll fees deducted from the Smart Card in the OBU are not immediately forwarded to the background system. Instead, they are accumulated in the card until a certain level is reached. The accumulated amount is then transmitted by mobile phone for checking and billing. Thanks to the bi-directional data exchange, this system can naturally support both debit (prepaid) cards and credit (postpaid) cards. However, anonymity cannot be maintained if the tolls are paid after the fact. With prepaid Smart Cards, which are used like electronic purses, the fact that only accumulated tolls are paid means that the motorway operator cannot generate vehicle movement profiles. However, it is of course possible for the GSM network operator to continuously determine the location of a moving vehicle via its permanently activated mobile phone.

In this system, as in the others, continuous monitoring of the moving traffic is necessary to check for drivers who try to avoid paying tolls. This is done using automatic cameras mounted on bridges and mobile checkpoints. The associated monitoring computers can read the vehicle registration number from the OBU or Smart Card via the mobile phone, and compare it with the numberplate of the vehicle. If these do not match, or if no link to the OBU can be established, the car is photographed and a penalty process is initiated.

The main advantage of this system is that no specific infrastructure has to be set up, apart from that needed for monitoring. However, this means that the OBU is certainly more costly than in the other two systems. Two other arguments in favor of this system are that it supports two different payment methods (prepaid and postpaid), and that data can be transmitted to the vehicle.

13.5   The GSM Network

The GSM network (the *D-Netz* in Germany) was originally a European standard for mobile telephony. Since more and more countries and telecommunication firms have embraced this standard, it has now been extended throughout the world. The original meaning of GSM, 'Groupe Spécial Mobile', was changed in the process to 'Global System for Mobile Telecommunications'. The first parts of this network became operational in Europe in July of 1991. In 1999, 293 mobile phone networks based on the GSM standard were operating in 120 countries, with more than 120 million subscribers.[5] The common legal basis for all GSM operators is the 'Memorandum of Understanding' (MoU), which was first signed in 1987 by all European network operators.

The specification of the GSM network was initiated in 1982 under the auspices of the CEPT (*Conférence Européenne des Postes et Télécommunications*). This work was later continued by the ETSI (European Telecommunications Standards Institute). The specification of the GSM Smart Card, or SIM (subscriber identity module), was started in January 1998 by the Subscriber Identification Module Expert Group (SIMEG). In 1994, the Special Mobile Group 9 (SMG9) was formed from the SIMEG and took over its tasks and authorities. This group is now responsible for the SIM specification.

On the occasion of the tenth anniversary of the SIM standard, the SMG9 published the slogan shown in Figure 13.4. This indicates rather clearly the size and importance that the Global System for Mobile Communications has currently achieved, and how much pride there is in one of the essential elements of the system–the subscriber identity module.

```
                     Billions of Calls
                   Millions of Subscribers
          Thousands of Different Types of Telephones
                   Hundreds of Countries
                  Dozens of Manufacturers ...

                    ... AND ONLY ONE CARD
```

Figure 13.4   Slogan published by the SMG9 on the tenth anniversary of the GSM 11.11 standard

---

5   The 100 millionth subscriber joined in July 1998. [GSM] provides a good overview of the current number of users and all network operators.

The specification is divided into two parts. The first, GSM 02.17 (ETS 300509), covers general functional characteristics. The second part, GSM 11.11, describes the interface and logical structures and is around 100 pages long. As part of a restructuring, the GSM specifications were given a new numbering scheme, so the GSM 11.11 standard is now designated prETS 300608.

Both specifications have their own specialized vocabulary. The terms used are defined very precisely in the specification, and they are only valid in the GSM context. Extensions to the specification have appeared since 1994, and these are collectively referred to as Phase 2. They contain additional data elements and mechanisms, so that fixed dialing numbers (FDMs) and accumulated call meters (ACMs) are supported. Currently, work is underway on the specifications for Phase 2+, which is currently designated GSM 11.11 Version 7.1.0. An extension to GSM 11.11, called GSM 11.12, describes GSM cards with 3-volt technology.

In addition to these standards, there is also GSM 11.14 Version 5.6.0, which is titled 'SIM Application Toolkits'. It offers network operators the possibility of downloading their own phone control applications 'over the air' (OTA) into the Smart Cards. The GSM 11.14 standard precisely specifies, for example, how to drive the display, how to scan the keypad, how to send short messages (short message service, or SMS) and other functions related to such applications.

The GSM network is a cellular digital mobile telephone network operating in the 900-MHz band. The term 'cellular' means that the coverage area is subdivided into circular cells with a diameter of 1–40 km. The size of the cell depends on the number of calls and the terrain. In order to completely cover a country the size of Germany ($\approx 360,000$ km^2), approximately 3000 overlapping cells are needed. The commonly used cells have the following approximate diameters:

- large cells: 10–40 km
- macrocells: 1–10 km
- microcells: 0.1–1 km
- picocells: 50–100 m.

The original GSM specifications have been extended, and now also cover an 1800 MHz system known as DCS1800 (digital cellular system) or PCS1900 (personal communication network).[6] Due to the higher frequency and reduced transmitter power, the cell diameter in the DCS1800 system does not exceed 20 km. It is therefore is used mainly in conurbations, rather than in rural areas with low subscriber densities. The only significant difference between GSM and DCS1800 lies in the transmitter and receiver components on either side of the air interface.

In the satellite-based Inmarsat mobile telephone network, which has been in use for several years, modified GSM cards are used for user identification. Another extension of the SIM, which is distinguished by having little supplementary data and a special cryptographic algorithm, is the Smart Card for the international Iridium mobile telephone network [Iridium]. In its final configuration, this system is planned consist of 66 satellites orbiting at an altitude of 780 km. These are equivalent to GSM base stations. The Iridium system uses a frequency of around 1616 MHz.

---

6   DCS1800 is called *E-Netz* in Germany.

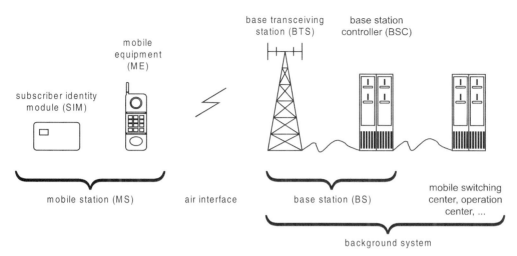

Figure 13.5   Basic structure of a mobile station in the GSM network

Network structure

A base station (BS) is located at the center of each cell. Its first job is to establish contact with the mobile phone across the air interface, and its second job is to feed calls into the regular telephone network. Data are transferred across the air interface in encrypted form at 13 kbit/s, using a lossy compression process with technically sophisticated error correction mechanisms (frequency hopping, convolutional coding and interleaving).[7]

A mobile phone is referred to in GSM as an MS (mobile station). It consists of two physically and logically distinct parts, the ME (mobile equipment) and the SIM (subscriber identity module). The ME is the radio and encryption part of the mobile station, while 'SIM' is another name for a GSM Smart Card. These two parts together make up a functional mobile phone.

From an information-technological perspective, the GSM background system is made up of two parts: the switching system and the base station system. The job of the base station system is to exchange information with the mobile stations via many base transceiver stations using radio communications (the 'air interface'). This is where the actual telephone conversations are handled, in addition to some administration data. A group of several base transceiver stations is in turn managed by a base station controller. If a mobile station leaves the range of the current transmitter, the base station controller also takes care of handover to another base station system.

A group of several base station systems is managed by a mobile services switching center. This also looks after forwarding calls to other networks, such as the national public telephone system. In order for a subscriber to be reached by a caller, the mobile services switching center must always know the subscriber's location or the identity of the base transceiver station within whose range the mobile station is currently located. Information relating to the current location of the mobile station, as well as other data regarding the called mobile station, is located in the home location register of the switching system.

---

7   See also [Eberspächer 97].

Another database, called the visitor location register, holds the locations of all mobile stations that are currently within range of the switching system in question. GSM subscribers of other network operators can also be temporarily listed in this database.

At the highest level in the hierarchy stands the authentication center, which has sole authority over the keys and algorithms needed to authenticate mobile stations (SIMs). Whenever a subscriber is authenticated at the start of a conversation or during a conversation, this is performed by the authentication center.

This is the general structure of a GSM system with a single operator. In reality, of course, several other components and databases are necessary, such as those related to call billing.

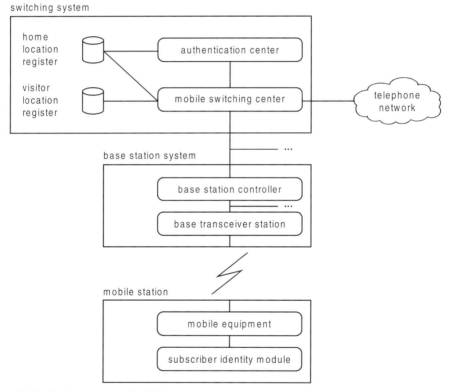

Figure 13.6   Basic structure of a GSM network with a single network operator

In the GSM, the SIM can have two different card formats. The ID-1 format is used with mobile phones in which the SIM is exchanged relatively frequently. A SIM in the ID-000 (plug-in) format is used in mobile phones that must be physically very small, in which the SIM is only rarely exchanged. The only difference between the two types of cards is the size of the card, since they otherwise have identical logical and physical characteristics.

The job of the SIM in the GSM network is to ensure that only authorized persons can access the network, so that a properly functioning billing system can be set up. In order to do this, it must be able store data, guard accesses to the stored data and execute a cryptographic algorithm under secure conditions.

In order to understand these functions, you must realize that the SIM, and thus the mobile equipment, is authenticated by the relevant network operator. This is a unilateral authentication of the SIM by the background system. The data transferred between the mobile station and base station across the air interface are encrypted, to prevent unauthorized persons from eavesdropping. Text information that is received from the network in the form of short messages can be stored in the SIM and read out as necessary.

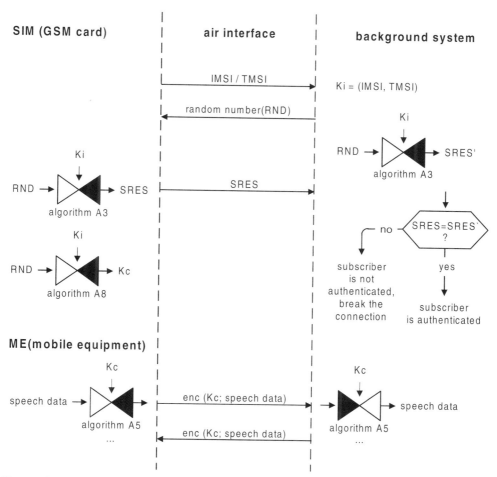

Figure 13.7   The cryptographic functions of the SIM in the GSM network. They are used to authenticate the SIM and to encrypt data passing between the mobile station and the base station.

Each SIM is identified by a number that is unique within the entire GSM system. This number, which is at most eight bytes long, is called the IMSI (international mobile subscriber identity). The subscriber can be identified by this number within all GSM networks, everywhere in the world. In order to keep the subscriber's identity as confidential as possible, the network uses a TMSI (temporary mobile subscriber identity) instead of the IMSI whenever possible. A TMSI is valid only within part of the relevant GSM network.

The card-specific keys for authenticating and encrypting data at the air interface can be derived from the IMSI. However, encryption is not carried out in the SIM (the Smart Card), since the Smart Card does not have sufficient computational and data transfer capacity for real-time encryption of voice data. Instead, the SIM computes a temporary derived key for transmission encryption, and passes this key to the mobile equipment. The latter contains a powerful encryption unit that can encrypt and decrypt voice data in real time.

When a subscriber wants to conduct a conversation, the mobile station sets up a link to the base station with the best reception, and relays the IMSI or TMSI from the SIM to this base station. If the subscriber's IMSI is registered there, the mobile equipment receives a random number via the air interface, which it passes on to the SIM. The SIM uses this number as a plaintext block for encryption with the A3 algorithm, whose key is specific to the card and the subscriber and is derived from the IMSI. The result is a ciphertext block that is transferred from the mobile equipment to the base station via the air interface.

The background system, which is permanently linked to the base station, derives the card-specific key from the IMSI and then performs the same computation as the SIM. After the SIM's ciphertext block has arrived at the background system, it only has to be compared with the block computed by the background system to determine if the subscriber is authentic and thus authorized to place the call.

Both the SIM and the background system use the A8 algorithm with the random number and the card-specific key to compute a temporary key, which is used for data encryption over the air interface. The encryption takes place in the mobile equipment, since the Smart Card is too slow for the real-time encryption and decryption of voice data. The computed key is therefore passed from the SIM to the mobile equipment, which performs data encryption and decryption using the A5 algorithm.

If the mobile equipment is in located the region of a foreign network, it is possible to make calls via the base stations of that network. This is known as 'roaming'. However, the home network does not provide the secret key (Ki) or the equally secret cryptographic algorithm to the foreign network for authenticating the SIM. Instead, it provides the combination of RND, SRES and Kc. These are adequate for authenticating the SIM and setting up an encrypted link to the mobile equipment.

The GSM network is currently the largest international application for Smart Cards. It was the first application ever in which Smart Cards had to meet the requirements of several different national and international system operators. This global application has established a standard for the Smart Card industry that covers both the commands and the properties of the card itself. The sheer pressure of this de facto standardization forces all standards organizations to maintain their standards compatible with GSM, at least in all essential aspects, in order to achieve international acceptance. This is why we describe GSM here in such detail.

The subscriber identity module (SIM)
The SIM has a hierarchical file system, with an MF and two DFs. The DFs contain the EFs with the application data. The EFs may have transparent, linear fixed or cyclic structures. The GSM 11.11 specification defines 22 Smart Card commands, which are identified by the class byte 'A0'.[8]

---

8   For a listing of the commands, see Section 6.5.1, 'Structure of the command APDU'.

There is one special feature regarding entering the 4-digit PIN, which incidentally is called the 'chip holder verification' (CHV). With a special command and the correct PIN, the used can disable further PIN interrogations by the card, so that it is no longer necessary to enter the PIN before each call. The user must accept the risk that a lost card can be used illegally to make telephone calls until it has been blocked by the network operator. The user can re-enable PIN checking using a separate command.

The communications between the mobile equipment and the SIM use the T=0 data transmission protocol with standard parameters. However, the data transmission convention can be freely selected by the card via the ATR. A PTS is allowed for, but it is currently not used, since the T=0 transmission protocol is specified as compulsory.

The GSM 11.11 specification defines 48 different EFs for the application data. These are collected in two DFs. The file identifiers (FIDs) are unusual in that the first byte of the DF is always '7F'. The first byte of the FID of an EF directly under the MF must be '2F', and for an EF under a DF it must be '6F'. In addition to the specified files, each network operator can store his own files in the SIM, for maintenance and administrative purposes.

Table 13.3   Simplified file tree of a GSM card, showing only the most important files

File type	FID	Structure and size	Description
MF	'3F00'	—	root directory
EF$_{ICCID}$	'2FE2'	transparent, 10 bytes	Smart Card identification number
DF$_{TELECOM}$	'7F10'	—	telecom DF
EF$_{ADN}$	'6F3A'	linear fixed, x bytes	abbreviated dialing numbers
EF$_{FDN}$	'6F3B'	linear fixed, x bytes	fixed dialing numbers
EF$_{LND}$	'6F44'	cyclic, x bytes	last number dialed
EF$_{SMS}$	'6F3C'	linear fixed, 176 bytes	short messages
EF$_{SMSS}$	'6F43'	linear fixed, x bytes	short message service status
EF$_{SMSP}$	'6F42'	linear fixed, x bytes	short message service parameters
DF$_{GSM}$	'7F20'	—	GSM DF
EF$_{LP}$	'6F05'	transparent, x bytes	preferred language
EF$_{KC}$	'6F20'	transparent, 9 bytes	key Kc
EF$_{SPN}$	'6F46'	transparent, 17 bytes	service provider name
EF$_{PUCT}$	'6F41'	transparent, 5 bytes	price per unit and currency table
EF$_{SST}$	'6F38'	transparent, 4 bytes	SIM service table
EF$_{IMSI}$	'6F07'	transparent, 9 bytes	IMSI
EF$_{LOCI}$	'6F7E'	transparent, 11 bytes	TMSI + location information
EF$_{PHASE}$	'6FAE'	transparent	GSM phase information

The Smart Card's identification number, which is unique throughout the system, is held in a transparent EF located immediately below the MF. The available directories are a DF for GSM data and a DF for telecommunication data.

The GSM DF contains, for example, an EF (EF$_{LP}$) that stores the user's preferred language for displaying information on the mobile equipment. Another EF (EF$_{IMSI}$) holds the IMSI used for user identification and billing. Yet another file (EF$_{TMSI}$) stores the current TMSI, together with supplementary location data. Since this file must be rewritten each time the base station changes and each time a call is made, it is specially protected by the Smart Card operating system. The normal lifetime of the EEPROM pages, which is usually no more than 10,000 write/erase cycles, would not be adequate, since this information must be written far more often during the life of the SIM. The file EF_{PHASE} identifies the phase of the GSM 11.11 that specifies the SIM. Currently, this is usually '2'.

The second DF in the SIM is the telecom DF. It contains an EF with abbreviated dialing numbers (EF_{ADN}), so that the user always has his or her most important numbers on hand in the SIM. Another EF (EF_{FDN}) contains fixed dialing numbers. If a certain mechanism in the GSM application has been activated, only the fixed numbers can be called, and all other numbers are blocked. The next three files (EF_{SMS}, EF_{SMSS} and EF_{SMSP}) contain any short messages and various related status data. Short messages can be received via the air interface and then read out from the SIM at any desired time. Finally, the EF_{LND} file stores the last number dialed.

It was originally intended to replace GSM Smart Cards every two years, in order to avoid failures due to the limited number of write/erase cycles in the EEPROM. However, since few problems have so far come to light in this respect, most application providers replace cards only if they actually fail. This provides considerable savings to the provider, who only has to deal with the logistics of replacing defective cards. In addition, a much smaller number of cards have to be replaced, since most cards last significantly longer than two years. This reduces procurement costs, since the Smart Cards only have to be replaced when they no longer work.

The SIM Application Toolkit and supplementary applications

The original GSM system specification viewed the GSM card as only a medium for identifying the user (via a PIN) and an authentication token for billing security, independent of the mobile phone. In the course of time, however, the desire to use the GSM card for additional functions has come to the fore. For example, a mobile phone is also an adequate medium for checking the balance in your bank account or for receiving interesting and possibly important messages, such as football scores. However, it was not technically possible to realize such 'value-added services' (VAS) with the existing functionality of the GSM card. In response to this, the GSM 1.14 specification was generated. The first version of this specification, which is titled 'SIM Application Toolkit', was published in 1996. The current version is 5.6.0.

The SIM Application Toolkit is a building block system that enables any desired application to be implemented in a GSM card. GSM 11.14 specifies in detail such matters as driving the display, scanning the keypad, sending short messages and additional functions that must be used with a supplementary application. All supplementary applications are based on EF files with linear fixed data structures. The data contents of the records in question are linked by pointers, and they contain the instructions for controlling the mobile equipment.

Only a few new commands are necessary for the SIM Application Toolkit. The actual functionality is obtained via data in individual files and records, rather than by something like downloaded program code. However, the use of downloaded code as an improved solution is a medium-term objective. The methods that are currently under discussion range from downloaded native program code to interpreted languages such as Java.[9]

---

[9] See also Section 5.10.2, 'Java Card'.

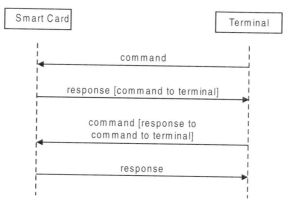

Figure 13.8   The extended protocol process between the mobile equipment and the SIM (GSM card), as used with the SIM Application Toolkit according to the GSM 11.14 specification. The data section of the response that the card sends to a (periodic) command from the mobile equipment contains a command to the terminal. The terminal executes this command and returns its response to the card in the data section of a command.

Since only the mobile equipment can act as the master in the communications between the mobile equipment (the mobile phone) and the SIM, it is necessary to create a way for the SIM to take on the master role if necessary. The solution to this problem is relatively uncomplicated. The mobile equipment periodically sends a polling command to the SIM (typically every 30 seconds). If necessary, the SIM can send its own command to the mobile equipment, as part of the response to the polling command.

The master–slave relationship between the mobile equipment and the SIM in effectively reversed in this manner. This makes it possible for the card to scan the keypad, present its on information on the display or activate the phone beeper, all on its own initiative. The short message mechanism can also be used to exchange data between the card and the background system. This can be used to regularly poll a news server, for example, with the result being displayed as an e-mail or short message.

Over-the-air (OTA) communications
It is frequently necessary to establish a direct link to the GSM card immediately after it has been issued. Especially for setting up supplementary applications that are implemented using the SIM Application Toolkit after the card has been issued, this sort of communication is essential. Consequently, mechanisms have been created that can be used to set up an end-to-end communication link between the background system and a GSM card via the air interface. This over-the-air communication can be used to modify data in the SIM and to create or delete files, among other things.

Since these requirements were not covered by the original GSM specification, and it is by no means easy to make changes to a system of this size, a trick is used for end-to-end communication with the GSM card. The short messages that are supported by the system are used as containers for messages to and from the SIM. This means that it is only necessary to make a few adaptations in the background system and the newly issued card. All intermediate parts of the system can remain unaltered.

OTA communication works as follows. Suppose that the background system wants to send a command (such as CREATE FILE) to a particular GSM card. It generates a short message to the card in question and embeds this command in the message, protected by various cryptographic techniques (MAC protection and encryption). As soon as a mobile phone with the target GSM card logs in to the system, the short message is delivered via the service channel. It is not necessary to set up a voice link for this purpose. The mobile phone automatically passes the received message to the GSM card for storage in the EF_{SMS} file. However, the GSM card recognizes the message as a command from its structure, and processes it exactly like a normal command. In response, the card can generate a short message that is returned to the background system in the normal manner.

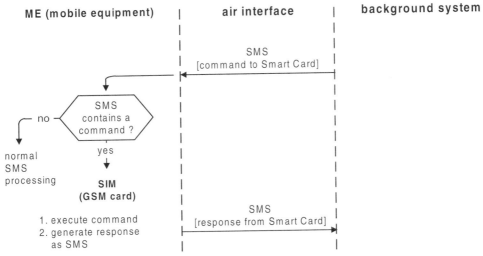

Figure 13.9 The procedure for exchanging data between the background system and a GSM card using short messages, commonly referred to as over-the-air communications

This artifice makes it possible to set up a bi-directional end-to-end link between the background system and the GSM card that is transparent with regard to all intermediate system elements. The GSM card can thus be addressed as though it were located in a terminal connected to a PC. This communication channel can be used to modify existing data, for example. The usual application for OTA is to modify the abbreviated calling numbers for service functions, which are stored in the GSM card. For example, if a new number is assigned to directory services, the number stored in the card can be automatically updated. However, it is possible to use this method for significantly more complex tasks. For example, OTA can be used to create new files in the GSM card or to download executable program code for extensive supplementary applications. The possibilities that OTA offers the network operator are nearly unlimited.

Future developments
The GSM application represents an international breakthrough for Smart Cards, and it is effectively *the* standard for cards and operating systems. The GSM commands and mechanisms may appear outdated in comparison with the latest developments in the Smart Card sector, but GSM was and still is the pioneer for large, international Smart Card applications. After all, all subsequent applications could (and still can) profit from the

experience and problems of the GSM application. In many respects, the GSM 11.11 and 11.14 specifications form the basis for all new Smart Card applications, which are certain to be more sophisticated than GSM.

The trend in the development of mobile phones is to include more and more functions of a personal digital assistant (PDA), in addition to the telephone function. Since it is relatively difficult to externally manipulate the software in a mobile phone, it can also be considered to be a 'trusted device'. The consequences of this can already be seen in the form of certain service functions and mobile phones with hardware extensions. For example, some mobile phones have a bi-directional infrared interface that complies with the IrDA specification, and 'dual-slot' mobile phones are also available.

These developments make it technically possible to make payments using a mobile phone and an electronic purse, together with a suitable POS system. If it is necessary to enter a PIN, this can in the future be done using the relatively secure mobile phone keypad instead of an unfamiliar terminal. The necessary data are exchanged via the infrared interface, and it is not even necessary to set up a telephone link to the nearest base station. It is also conceivable that an electronic purse could simply be reloaded online via the second contact unit of the mobile phone. Naturally, credit cards and debit cards can also be used to make purchases, and possibly even to pay the monthly bill for the mobile phone. The range of possibilities is very large, and we can presently only suggest its extent.

Even an established system such as GSM must be further developed in order to meet new demands and satisfy additional customer wishes. This is currently taking place in small steps. It has led to the changes and extensions of Phase 2 and Phase 2+, as well as the OTA functions. However, at some point it will become necessary to make a large stride in the development, in order to bring all of the extensions, modifications and special cases into a new self-contained system. The preliminary work on this system, which is called the Universal Mobile Telecommunication System (UMTS), is already underway, and the first large field trials are expected to take place after 2001. The subscriber identity module will be renamed the 'user identity module' (UIM) or 'universal subscriber identity module' (USIM). This will be used for mutual authentication with the background system with the aid of asymmetric cryptographic algorithms, which in all probability will be elliptic curves.

13.6   Digital Signatures

The are two fundamental prerequisites for the use of legally binding digital signatures. The first is a microcontroller Smart Card with a powerful numeric coprocessor, and the second is an unambiguous general legal framework. The Smart Card is used for the secure storage of the secret signature key and the generation of the digital signature. The legislation establishes the binding conditions that apply to all participants in this Smart Card application.

Digital signatures can of course also be used as a closed solution that is completely independent of any legislative background. However, in this case suitable contractual arrangements between the participating parties are necessary to make the digital signature binding on the person who produces the signature. In the business-to-business sector, for example, the use of digital signatures has been standard practice for several years. However, if the objective is an open system that works with parties that are not known to each other, suitable legal conditions are required. This legislative framework allows digital signatures to be made equivalent to normal signatures, and makes it possible to bring suit in court with regard to a document so signed, just as with a handwritten signature.

The applications for legally recognized digital signatures, which are thus equivalent to handwritten signatures, are unlimited. In the simplest case, they can be used to sign electronic letters and orders. A digital signature can also be used in an electronic payment system to sign a deposit slip for a debit note, for example, or to sign a remittance order in a home banking system. The most important use for digital signatures will certainly be with regard to contracts of all sorts.

Standardized general conditions

In order for digital signatures to be utilized interoperably in an open commercial environment, they must meet the relevant security requirements of various standards. As far as signature cards are concerned, the applicable standards are the international ISO/IEC 7816-4 standard for general Smart Card commands and 7816-8 for signature commands. For the card bodies, electrical properties, and data transfers, the relevant standards of the ISO/IEC 7816 series must be taken into account. The authentication process between the Smart Card and the rest of the world is covered by ISO/IEC 9796-2. Basic mechanisms and methods for digital signatures are handled by ISO/IEC 14888. In addition, the structure and coding of the certificate normally comply with the X.509 standard.

A digital signature card can be used to generate a signature that is equivalent to a real signature. This means that large monetary or material values can be involved, depending on circumstances and the document bearing the signature. The system elements relating to security must therefore be evaluated in the context of a digital signature application. The proven means for this is the ITSEC criteria catalog.

The general legal conditions in Germany

In order for digital signatures to be recognized as signatures that are generally valid, binding and subject to suit, there is no alternative to arranging for the necessary general legal conditions. The state of Utah in the USA was the first to introduce legislation in this area, in 1995. This legislation has served as a model for legislation in many other countries, including German signature legislation.

On June 13, 1997, the *Gesetz zur Regelung der Rahmenbedingungen für Informations- und Kommunikationsdienste* (IuKDG) was passed in the German federal legislature (*Bundestag*). This came into force on August 1, 1997. Article 3 of this legislation is the digital signature law, which is called the *Signaturgesetz* (SigG) for short. This article is divided into 16 sections. The first paragraph of the first section states the objective of the signature law. It reads, 'The objective of the legislation is to establish general conditions for digital signatures, under which they can be considered to be secure and which allow forgeries of digital signatures or falsifications of the signed data to be reliably detected.' This statement clearly indicates that no particular technical solution is specified, but only general conditions with regard to the use of digital signatures. As a natural consequence, the legislative requirements are described at a relatively abstract level.

The essential requirements of the German signature law are the following:

- The operation of a certification authority (that is, a trust center) is subject to a permit from the responsible public authority.
- Any person who requests a certificate must be reliably identified by the certification authority.

- The signature key certificate may bear a pseudonym of the key holder in place of the actual name.
- The actual name behind a pseudonym must be revealed in response to a formal request from certain public authorities.
- The certification authority must make the certificate for the public signature key available in a manner that allows it to be investigated online.
- A certificate can be blocked under certain conditions, such as if this is requested by the holder of the signature key.
- The technical components for the generation and verification of digital signatures must clearly and unambiguously represent the data to which the digital signature refers.
- Digital signatures from other countries can also be recognized in Germany by means of suitable agreements.

According to the signature law, a signature key certificate, which is usually made in compliance with the X.509 standard, must contain the following information:

- the name or pseudonym of the signature key holder,
- the public signature key that is assigned to the signature key holder,
- the designations of the hash algorithm and signature algorithm used for the certificate,
- the serial number of the certificate,
- the initial and final validity dates for the certificate,
- the name of the certification agency,
- any specifications that limit the use of the signature key to specific applications.

The regulation and description of digital signature applications cannot be handled by legislation alone. A hierarchy of requirements is necessary for this, with the legal regulation–the signature law–at the top of the hierarchy. This is followed by the associated implementation rules, which in Germany are called the signature regulation (*Signaturverordnung* or SigV). On the next lower level, which already describes the requirements for digital signatures in technically very concrete terms, we find the measures catalogs. In Germany, these are issued by the regulation agencies for telecommunication and postal services. These three levels are obligatory for all applications of digital signatures in Germany, to the extent that the digital signatures are to be equivalent to regular signatures.

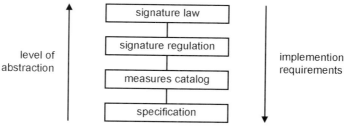

Figure 13.10   The hierarchic structure of the requirements for a digital signature application in Germany. The system of requirements is formed in a top-down manner, which means that the lower-level documents are based on those that are higher up in the list.

The level below the measures is the technical specifications. These define concrete solutions in unambiguous and non-interpretable form. The specifications depend on the application operator in question, and they must be aligned to the three higher levels of the hierarchy. However, if an application operator does not need the digital signature to have a legally binding character, he can naturally set up his system in whatever manner best meets his objectives. This is referred to as a legally non-compliant solution, as opposed to a legally compliant solution.

The German digital signature regulation (SigV), which was passed by the federal government on October 8, 1997 and came into force on November 1, 1997, contains the implementation rules for the signature law (SigG). For each of the sixteen sections of the signature law, it contains detailed descriptions that are worded significantly less abstractly than the law. The following statements are contained in the SigV, among others:

- A person who requests a certificate must be identified by means of a personal identity document, passport or in some other suitable manner.
- The data medium (usually a Smart Card with a signature key) must be held in personal safekeeping, and the associated identification data (such as a PIN) must be kept secret.
- When a digital signature is verified, the validity of the signature key certificate at the time that the signature was generated must also be verified, and a check must be made to see if there are any restrictions on the use of the signature.
- A certificate cannot be valid for more than five years.
- The blocking of a certificate may not be reversed.
- A measures catalog with suitable security measures must be generated.
- The certification authority (trust center) must be audited by the responsible public authority.
- The signature keys must be generated in a manner such that that the probability of two keys being the same verges on zero.
- Signature keys may not be duplicated.
- All security modifications to technical components must be recognizable to the user in question.
- The signature key may be used only after its holder has been identified by means of possession and knowledge.
- The signature key itself may not be revealed when it is used.
- A list of suitable algorithms for generating signature keys, computing hash values of signed data and generating and verifying digital signatures will be published in the Federal Gazette (*Bundesanzeiger*). The time interval during which these may be used will also be indicated.
- If signed data are to remain valid for longer than the time allowed for the approved algorithms, the data must be signed anew, with a time stamp, before the expiry date of the algorithm.
- The essential components of a legally compliant signature application must be tested against the ITSEC criteria using the mechanism strength 'high'. The testing level depends on the individual component and how it is used. Table 13.4 lists the components and the associated test levels.

Table 13.4   Testing levels for various components related to digital signatures

ITSEC test level E2 (high)	ITSEC test level E4 (high)
▪ components for acquiring and testing identification data, and for displaying the data to be signed	▪ components for generating keys
▪ components for displaying signed data and testing certificates	▪ components for storing and using signature keys (usually Smart Cards)
▪ components for generating time stamps	▪ components for generating and testing digital signatures that are commercially offered to third parties for their own use

The measures catalogs for the signature law and the signature regulation specify the technical requirements for this system. They are distinctly more concrete than the signature law. The technical specifications are based in turn on the measures catalogs. They specify the technical implementation in detail. In Germany, the most important specification for digital signatures in the DIN specification entitled *Spezifikation der Schnittstelle zu Chipkarten mit digitaler Signatur-Anwendung/Funktion nach SigG und SigV* ('Specification of the Interface to Smart Cards with Digital Signature Applications and Functions according to the SigG and the SigV').

The following subjects related to signature cards are dealt with in the DIN specification:

- answer to reset (ATR) and protocol type selection (PTS),
- data transmission protocols (T=0 and T=1),
- files, data objects and data formats,
- authentication of the cardholder,
- computation and verification of signatures,
- logging the signature generation process,
- procedures for generating and verifying signatures,
- cooperation with terminals and the associated command sequences.

System structure

The two essential components of a system for digital signatures are the trust center and the signature card. These two entities thus command most of our attention, since all the processes related to security take place in them. In this regard, in a system for digital signatures, we can make a fundamental distinction between issuing the card and the processes of signing and verification of signatures.

Issuing a digital signature card

The future user of a digital signature card must first register himself or herself at a registration authority. The person to be registered must appear in person and present a recognized proof of identification, such as a personal identification card. At this point, the future user of the digital signature card can elect to have a pseudonym appear in place of his or her actual name. The registration authority transfers the information it has received and verified to the trust center, which initiates the generation of a key and the personalization of a signature card. These activities can take place in the trust center, but they can also be carried out by a third party or directly within the registration authority. In the process, the public key of the new Smart Card is signed by the certification service of the trust center, which makes it into a recognized genuine public key. The signed public key is then entered into the public key directory of the trust center as a valid key, which makes it available for use by every system subscriber. After all this is finished, the new subscriber receives his or her digital signature card along with a PIN letter. This completes the issuing of the card, which can now be used.

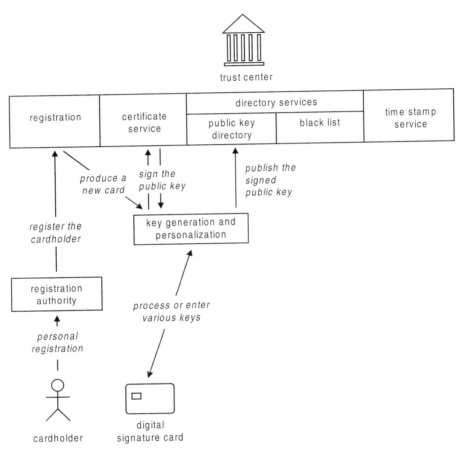

Figure 13.11   The basic procedure for issuing a digital signature card

Signing and verifying documents
When an electronic document is signed, it is compressed to a hash value and then signed using the digital signature card with the secret key. This process can initiated only after the cardholder (that is, the signer) has unambiguously been identified, which currently means after the cardholder's PIN code has been entered. In the future, a test based on a biometric feature (such as a fingerprint) could be used for identification. The system elements that are used for signing the data are the signature card, a terminal, and normally a PC.

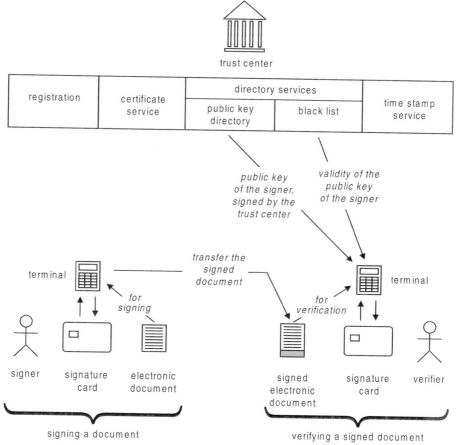

Figure 13.12   Basic procedures for signing and verifying a document with a digital signature

The digitally signed document can now be sent in any desired manner. In order to test the document, it is necessary to again compute the hash value. If necessary, the public key of the signer can be retrieved from the public key directory of the trust center. With this key, it is now possible to verify whether the digital signature of the electronic document is genuine. In order to be sure that the secret key of the signer has not been compromised, the revocation list (black list for keys) of the trust center must be consulted. The signature verifier can thereby be assured that the secret key of the signer has not been blocked.

If the verifier needs to have a confirmation that the revocation list has been consulted, in a form that can be verified by a third party, he can obtain a suitable confirmation with reference to the verification data via the time stamp service of the trust center.

The system structure and the illustrated processes for issuing a digital signature card and signing and verifying a document with a digital signature are examples, which may differ from actual implementations. However, they represent a realistic if simplified overview of a typical digital signature system.

The trust center (TC)

The trust center is the most important element in a digital signature system, aside from the signature card. It has six different tasks: (1) registering new users (registry), (2) generating keys and personalizing new digital signature cards, (3) providing certification services, (4) providing directory services for public keys, (5) providing directory services for black lists (revocation lists) and (6) providing a time stamp service.

The registry service collects personal information and verifies the identity of a new user of a digital signature card. A registration agency normally acts as an intermediary between the trust center and a new user, so that the user does not have to personally appear at the trust center. It is only necessary to generate a key and personalize the signature card when a new card is issued. This does not necessarily have to take place within the trust center; it can for example be done by the card manufacturer. The key can be generated either inside the card or outside the card. The certification service of a trust center signs public keys for digital signature cards using a secret key belonging to the trust center, so that they can be recognized as being genuine. A trust center can naturally have more than one secret key for generating certificates.

The directory service for public keys contains the public keys for the digital signature cards, signed by the trust center. The directory service for the revocation list contains a list of all blocked keys, which belong to digital signature cards that for example have been lost or compromised. According the German signature law, these directory services must be accessible via generally available public networks.

The time stamp service need not necessarily be integrated into a trust center, but it is certainly available in most trust center implementations. This service is used to attach the current date and time to electronic information that presented to the trust center. The received information, together with the date and time, are then signed using the secret key of the time stamp service. This allows the supplier of the information to prove to a third party that the information signed by the time stamp service must have been available no later than the indicated time.

The digital signature card

A secure hardware environment is necessary for storing and using private signature keys. Smart Cards represent an ideal solution to this requirement, since they are physically small, inexpensive and offer a high level of protection against external reading or modification of the stored data. It can be asserted without reservation that digital signatures only became possible with the advent of Smart Cards.

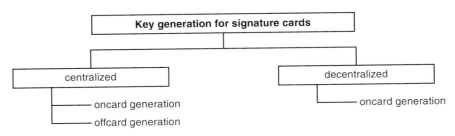

Figure 13.13   The possible options for generating keys for signature cards

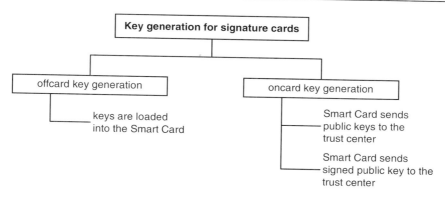

Figure 13.14   The three basic methods for key generation and key distribution between the signature card and the trust center

The cryptographic algorithms that are used for this purpose call for a microcontroller with a supplementary numeric processing unit (NPU), so that the signatures can be generated and tested in an acceptable length of time. In theory, a signed document could also be tested outside of the signature card, since all the necessary information is also available there. Whether this will become significant in practice remains to be seen. In terms of security, however, it is better to perform the test inside the card.

Public and private keys for digital signature cards can be generated in either a centralized or a decentralized manner. In any case, the keys must be generated in a cryptographically secured environment, such as inside a security module or Smart Card. It is also important that the public key is then signed using the secret key of the trust center. This is necessary to ensure that the private and public keys of the signature card are recognized as genuine by the trust center. Without this recognition, it is not possible for a third party to verify that the digital signature of an electronic document actually comes from a genuine subscriber of a digital signature system

If key generation is centralized, all key pairs are generated in one place and signed using the secret key of the trust center, immediately after they have been generated. Both the generation of the keys and the signing by the trust center can be performed inside the card (*oncard*) or outside the card (*offcard*). Decentralized key generation can only be performed oncard, since the card is the only cryptographically secured environment in this context. The advantages and disadvantages of the two methods are listed in Table 13.5. In practice, centralized key generation will probably become predominant, since it is very secure and also fits with the conventional manufacturing process for Smart Cards.

In Germany, the most important document for the signature card interface is the previously mentioned DIN specification. It extensively describes the necessary commands, files and processes for a digital signature card.

The signature application is located in a DF directly under the MF. The DIN specification provides nine EFs for this application, which contain all the necessary data for the application. All data elements used are TLV encoded. The EFs and their contents are summarized in Table 13.6.

Table 13.5   Comparison of the three methods that can be used for generating keys. The method listed in the middle column is normally used in large real-life applications

	Offcard key generation	Oncard key generation, with public keys passed to the trust center for signing	Oncard key generation, with signed public keys passed to the trust center
Pro	• fast personalization • reproduction of identical cards is possible	• secret signature key never leaves the Smart Card	• secret signature key never leaves the Smart Card
Neutral	• key escrow is technically possible and presents no problems	• key escrow is difficult	• key escrow is difficult
Con	• the secret key is generated offcard	• the public key must be transported to the trust center via a secure route • reproduction of identical cards is not possible	• the secret key of the trust center is in the Smart Card • reproduction of identical cards is not possible • personalization is time consuming

Table 13.6   Simplified file tree of a signature card according to the German signature law, showing the most important files

File type	FID	Structure and size	Description
MF	'3F00'	---	root directory
EF$_{GDO}$	'2F02'	transparent, x bytes	serial number of the card and name of the card owner
DF$_{SigG}$	not defined	---	digital signature application DF (AID = 'D2 76 00 00 66 01')
EF$_{KEY}$	---	undefined, x bytes	PINs; public and secret keys
EF$_{SSD}$	'1F00'	transparent, x bytes	security service description (SSD)
EF$_{C.ICC.AUT}$	'C000'	transparent, x bytes	authentication certificate of the signature card
EF$_{C.CA.AUT}$	'C008'	transparent, x bytes	authentication certificate of the certification authority
EF$_{DM}$	'D000'	transparent, 8 bytes	display message (DM) that allows the user to recognize that the card is genuine
EF$_{C.CH.DS}$	'C100'	transparent, x bytes	signature certificate of the card owner
EF$_{C.CA.DS}$	'C108'	transparent, x bytes	signature certificate of the certification authority
EF$_{PK.CA.DS}$	'B000'	transparent, x bytes	public key of the certification authority
EF$_{PROT}$	'A000'	cyclic, x bytes	signature generation log

The following commands are used for signature cards:

- APPEND RECORD
- EXTERNAL AUTHENTICATE
- GET CHALLENGE
- INTERNAL AUTHENTICATE
- MANAGE SECURITY ENVIRONMENT
- PERFORM SECURITY OPERATION
- READ BINARY
- READ RECORD
- SELECT FILE
- UPDATE BINARY
- MANAGE CHANNEL (optional)
- GET RESPONSE (optional for T=0)

The hash functions and signature algorithms that are approved for German signature cards are SHA-1, RIPEMD-160, RSA, DSA and ECC. Since it would take too long to compute the hash value of a large electronic document inside the Smart Card, the hash value can be computed in a PC and then passed to the card for signing. Alternatively, the hash value can be computed up to the final block in the PC, following which the final block can be computed from the corresponding part of the document in the Smart Card and then signed. It is also possible to completely compute the hash value in the Smart Card, but this is frequently not practical, due to the well-known performance limitations of the card.

Summary and prospects

In Germany, the signature law has created the necessary legal framework to allow digital signatures to be used reliably in everyday life. Their main use will probably not be for private purchase contracts for houses or wills, but instead for daily activities such as sending e-mail and ordering or paying for goods via public networks, which by definition are not secure.

It can safely be assumed that, in addition to the legally compliant solutions, there will also be many that are not legally compliant, since the implementation of a non-compliant solution is easier and much less costly. The requirements imposed on the security of the components used are set very high in Germany, and thus do not favor inexpensive and quick solutions. However, this level of security is necessary if digital signatures are to be made equivalent to handwritten signatures. If it were possible to successfully manipulate digital signatures, the whole system would lose its credibility, and in the worst case it would have to be closed down.

Another important consideration in this context is the behavior of multinational organizations, which are not likely to pay too much attention to the laws of any individual country. This means that it is certainly possible for a solution that is not legally compliant to become established as an international quasi-standard.

In spite of these risks, the digital signature solution that has arisen in Germany has international influence, both on legislation and on the technical implementation of similar projects. If the potential of this system is utilized, Germany will find itself with a technically superior and exemplary solution to the implementation of digital signature systems.

14

Application Design

The first section of this chapter contains general notes and technical data related to the use of Smart Cards. This information, which has been distilled from many applications, can be directly used for designing Smart Card applications. It thus represents a brief summary of the current state of the technology.

The type of card body used is not critical at the design stage, since almost any type of module can be incorporated into any type of card body. However, problems often arise due to the limited memory capacity of the microcontroller used, so we devote a considerable amount of attention to this subject.

Much of the technical data depend strongly on the actual hardware used, but in spite of this, it is still possible to make estimates for a projected application that are adequate for practical purposes. This is in part because the physical and electrical properties of all available Smart Card microcontrollers are nearly the same, within certain limits. The main differences are actually only in the memory capacities.

In the second section of this chapter, we describe the working principles of a number of tools that allow even complex applications to be created simply and quickly, without any programming. In the remaining sections, several possible applications are illustrated by means of examples. These are built up systematically, so that the reasons for using the various mechanisms and routines can easily be grasped.

The notes and examples presented in this chapter have been intentionally laid out in a general and flexible manner, so that they can be used as a sort of design template to give form to any desired application. Additional system design information that is specifically related to security issues is located in Section 8.2.4.2 ('Attacks on the logical level').

14.1 General Notes and Characteristic Data

14.1.1 Microcontrollers

Production

If a new application requires a special Smart Card operating system to be programmed, this will take a great deal of time. As rule of thumb, it takes around nine months to design, program and test a completely new operating system. If library routines and existing program modules can be used, a single programmer will still need about four to six months for the job. Carrying out the activities in parallel is possible only to a limited extent, due to the very complex nature of assembly language programming. As a result, generating a new

Smart Card operating system takes a minimum of two to three months with current software technology, even if no expense is spared and as many people as possible are assigned to the job.

Once the fabrication of the microcontroller is started in the plant of a semiconductor manufacturer, it takes from eight to ten weeks until finished dice are available in moderate quantities (several tens of thousands). Incorporating the dice in the modules then takes another one to two weeks. The turnaround time for producing the microcontrollers is thus around 12 weeks.

Life expectancy

A Smart Card's life expectancy primarily depends on the stiffness of the card body, the corrosion resistance of the module contact surfaces and the number of possible EEPROM write/erase cycles.

The life expectancy of the card body strongly depends on the application area. With typical GSM cards, which are permanently located in a mobile phone and are never removed, there is practically no wear on the card body. At the other extreme, such as with an employee identification card that is used in the canteen as well as for access control, the card body may crack after two to three years. Carrying the card in a wallet held in a hip pocket increases the likelihood of card body failure.

The second limitation on the life expectancy of the card is the maximum number of insertion cycles. The card contacts have hard gold plating to increase their wear resistance, and can survive about 20,000 contact cycles. After this, the contact surfaces will have become so scratched that dirt and grease will adhere to them and hinder reliable electrical contacts. In addition, the gold plating will be largely worn off, which leads to oxidation. This also adversely affects the electrical conductivity of the contact surfaces. Naturally, the life expectancy of the contacts (and thus the card) also depends very much on the type of contact unit used and the applied contact force. If the contact unit has an optimal design, the life expectancy of the card may be increased by a factor of two to four.

Figure 14.1   Photograph of a module after 50,000 insertion cycles in a testing machine. Very high-quality contacts were used in the contact unit of the testing machine. In practice, the contact surfaces may be abraded considerably faster, depending on the type of contact unit used.

The greatest limitation on the life expectancy of a Smart Card comes from the limited number of write/erase cycles in the EEPROM.[1] Most semiconductor manufacturers usually guarantee between 10,000 and 100,000 cycles per EEPROM page. However, in practice, the EEPROM will not fail until after 300,000 to 600,000 cycles when used at room temperature with small supply voltage deviations.

The failure of an EEPROM page is gradual rather than abrupt. Two indications of incipient failure are (a) that the EEPROM cannot be set to the desired value on the first write attempt and (b) that data written to the EEPROM are no longer present in memory after a few hours. If a memory page continues to be used under these conditions, it will not be possible to store any data at all in the EEPROM page after a few thousand additional cycles, since the contents will be lost at once. However, only one page at a time is affected (the page size is usually 4 or 32 bytes). Other memory pages remain unaffected by the failure of a particular page. This fact can be utilized in the context of the memory organization for error recovery strategies.

Data bits are stored in EEPROM as electrical charges held in tiny capacitors. Like all such devices, they have leakage currents that cause a loss of charge over time, which results in the loss of the stored information. This effect is accelerated at high temperatures. The data retention time of the EEPROM is thus not unlimited, and the value that is guaranteed by the manufacturer is only ten years. The problem with specifying the data retention time is that it cannot be measured directly, since this requires waiting ten years. The discharge time can only be calculated by determining the value of the current leakage. Since this is almost impossible to measure, the maximum data retention time is determined indirectly, by extrapolating the results of tests made with different values of certain ambient conditions, such as temperature and programming voltage. The data retention time also depends on the number of write/erase cycles. The specified value of ten years applies to maximum-stress conditions, so that normally a much longer lifetime can be expected. In terms of application design, you can assume that data will be retained for ten years, but if possible you should avoid actually reaching this value.

A blanket statement regarding the life expectancy of a Smart Card can only be made if all the general conditions are taken into consideration. A practical guideline is three to five years, for a normal application that does not make any special demands on the card. In some areas, such as GSM, many firms are switching over to replacing cards only as necessary (when they actually fail), which is economically advantageous.

Data transfer times

The speed at which commands can be processed with a Smart Card depends primarily on the data transmission rate of the interface between the terminal and the Smart Card, in addition to the internal instruction execution rate.[2]

Since an asynchronous serial interface is used, each byte transferred must be extended to 12 bits, since a start bit, a parity bit and two stop bits must be transmitted in addition to the 8 bits of useful data. This means that you can assume a working value of 1.25 ms per byte at a data transmission rate of 9600 bit/s. If you add a margin of 20% for the obligatory transmission protocol overhead, you arrive at a value of 1.5 ms per byte. This is sufficiently accurate for making practical estimates of the durations of data transfers.

---

[1] See also Section 3.4.2, 'Memory types'.
[2] See also Section 6.4, 'Data Transmission Protocols'.

Table 14.1   Typical data transmission rates for Smart Cards

Function	Data rate with a 3.5-MHz clock	Data rate with a 4.9-MHz clock
Data transmission, divider value = 372	9600 bit/s	13,212 bit/s
Data transmission, divider value = 512	6975 bit/s	9600 bit/s
Data transmission, divider value = 372, T=0 or T=1	≈ 7680 bit/s	≈ 10,570 bit/s
Data transmission, divider value = 372, T=1 and secure messaging (Authentic mode)	≈ 3800 bit/s	≈ 5200 bit/s
Data transmission, divider value = 372, T=1 and secure messaging (Combined mode)	≈ 1900 bit/s	≈ 2600 bit/s

Algorithm execution speeds

Since Smart Cards are often used as protected computers for processing algorithms, we list a few typical values for processing times in Table 14.2. These are average values, which can certainly vary from one implementation to the next. The listed values are for clock rates of 3.5 and 4.9 MHz.

All of the listed values, except for the time required to write data to the EEPROM, are directly proportional to the applied clock frequency. The throughput values can be increased by a factor of 1.4 if the clock frequency is raised from 3.5 MHz to 4.9 MHz, which is possible with every currently available microcontroller. The time required to write or erase EEPROM data is independent of the applied clock rate, and cannot be reduced.

Table 14.2   Typical throughput values for Smart Card algorithms programmed in assembler or C

Function	Actual time		Throughput	
	3.5 MHz	4.9 MHz	3.5 MHz	4.9 MHz
XOR computation	1 μs/byte	0.7 μs/byte	1 MB/s	1.4 MB/s
CRC computation	0.2 ms/byte	0.14 ms/byte	5 kB/s	7.1 kB/s
Write or erase EEPROM data (page size 4 bytes)	3.5 ms (4 bytes)	3.5 ms (4 bytes)	1142 byte/s	1142 byte/s
Generate a random number (8 bytes)	26 ms	18.5 ms	—	—
DES computation (8-byte block, ECB/CBC mode)	17 ms (8 bytes)	12 ms (8 bytes)	470 byte/s	660 byte/s
DES computation of a MAC (8-byte block)	17 ms (8 bytes)	12 ms (8 bytes)	470 byte/s	660 byte/s
IDEA computation (8-byte block, ECB/CBC mode)	12.3 ms (8 bytes)	8.8 ms (8 bytes)	650 byte/s	910 byte/s
RSA computation without CRT (512-bit key with NPU)	308 ms (64 bytes)	220 ms (64 bytes)	208 byte/s	290 byte/s
RSA computation without CRT (768-bit key with NPU)	910 ms (96 bytes)	650 ms (96 bytes)	105 byte/s	148 byte/s
RSA computation without CRT (1024-bit key with NPU)	2 s (128 bytes)	1.4 s (128 bytes)	64 byte/s	91 byte/s

14.1.2   Applications

In addition to the specific factors relating to the microcontroller, other aspects need to be considered during the design and development stages of a Smart Card application. With regard to application design, these primarily relate to key management, the management of application data and the basic principles of data exchange.

Key management

In all applications that use cryptographic algorithms, security is based on the secrecy of the associated key. If a secret key becomes known for any reason, then all the security mechanisms based on it are worthless. In principle, this eventuality cannot be ruled out with complete certainty, so suitable precautions must be taken.[3] The most trivial and most expensive option is to replace all the cards, but this is economically out of the question in a large application. In practice, therefore, the following measures are used to minimize the consequences of a key becoming known:

- Fundamentally, only card-specific keys should be located in Smart Cards. Each card then has its own keys, which can only be used to clone the one card if they should become known. The associated master key must remain secret under all conditions.
- In order to provide protection in case a master key becomes compromised, several generations of card-specific keys can be stored in the card. It is then possible to switch over to a new key generation, if this becomes necessary.
- Keys can be separated according to their intended use. In practice, this means for example that an authentication key may not be used for data encryption. This helps to minimize the consequences of a key becoming known.

These three principles should always be observed in every large application, as they guarantee the application operator a significantly more secure system. They may even save him a great deal of money that would otherwise be spent on replacing all the cards. However, each of these recommendations increases the number of keys in the system. This can quickly lead to memory problems, and it means that a more-or-less elaborate key management system will be required. In practice, therefore, it is necessary to consider very carefully whether all three of these measures must be used, or whether it is possible to accept certain compromises.

Data management

The principles that are standard in the information technology industry should be adhered to with regard to storing data in Smart Cards. For an application, this means that numerical classification schemes should be avoided as far as possible, since even small modifications or extensions often cause such schemes to collapse. Numerical identification schemes, on the other hand, are frequently too abstract, so in practice mixed schemes predominate.

Telephone number systems provide a good example of a mixed numbering scheme. The first part of the number (the exchange number) is classificatory. If you know this number, you can say with certainty in which area the telephone is located. The second part of the number (the subscriber number) is purely identificatory, since at least in small towns, it provides no information about the location of the subscriber. The two parts together are a typical mixed number, which can be extended in a fairly straightforward manner.

---

[3] See also Section 4.7, 'Key Management'.

ASN.1-coded data objects[4] are very suitable for flexibly handling data objects that have several different versions. Based on current practice, a good estimate of the resulting coding overhead is around 25% of the data volume. This applies only to small data sets, but these are common in the Smart Card domain.

One basic remark relating to the memory capacities of Smart Cards can be made here. Applications with more than 2 kB of useful data are rare, since the available amount of memory is usually very limited.

When planning a new application, you should make at least an approximate calculation of the amount of memory that will have to be reserved in the Smart Card. You should include not only the useful data, but also the necessary amount of administrative data. In recent operating systems, which have object-oriented file management systems and allow several applications to be present in one Smart Card at the same time, the proportion of administrative data is relatively high. The size of the header for each file usually ranges between 16 and 32 bytes. However, the amount of administrative data per file is fixed, which means that it does not depend on the amount of useful data in the file. It makes no difference to the file header whether there is only 1 byte or 200 bytes of useful data in the file. Consequently, you should avoid creating a separate file for each data element, since otherwise too much memory will be wasted on administration.

Data exchange
You must keep two main things in mind with regard to data exchanges between the terminal and the Smart Card. The first thing is that the card's serial interface is very slow in comparison with the performance of modern computer systems. Although the normally used rate of 9600 bit/s makes the data transfer very robust with regard to interference, it also means that data exchange takes a great deal of time. It is therefore a good idea to limit the data exchanged between the terminal and the card to essential items. The terminal should not make a second request during a session for any data that it has already received. Naturally, this does not apply to applications in which time has a very low priority. However, if people are involved in the process, minimizing the volume of data exchanged via the interface must always be a priority.

With regard to data exchange, the following points are also relevant. If you use secure messaging,[5] all data can in principle be protected very well at the terminal/card interface. An attack at this interface is thus nearly impossible. However, secure messaging sharply reduces the effective transfer rate. Consequently, secure messaging should only be used for data exchange when it is unavoidable for reasons of system security.

With the exception of secret keys, it is almost always sufficient to just use a MAC to protect the useful data (Authentic mode). This does not reduce the transfer rate as severely as the additional encryption required by the Combined mode. Also, in the Authentic mode, the data transmission is transparent at the interface and can easily be checked externally. This may be necessary in connection with data protection legislation, since it makes it possible to see what data passes across the interface, at any time and without knowledge of the secret keys.

---

4 See also Section 4.1, 'Data Structures'.
5 See also Section 6.6, 'Secure Data Transfers'.

14.1.3   System considerations

Security

Many applications use the DES as the cryptographic algorithm. There are many reasons for this, but in most cases the main reason is probably that the DES is very well known, so the application developer does not have to venture into unknown territory. However, there are many other good data encryption and decryption algorithms besides the DES. The fact that the DES is well known also means that it is subjected to the most attacks, one of which may sometimes be successful. In the case of large Smart Card projects, therefore, you should consider using some other cryptographic algorithm instead of the DES. This will minimize the consequences to your system of a successful attack on the DES. The current GSM cryptographic algorithms provide an excellent illustration of this philosophy. They were specially developed for this application and are totally independent of the DES.

User interface

User acceptance is decisive for a successful Smart Card project. Although this primarily relates to the man–machine interface of the terminal, the interface between the terminal and the card also has an effect. Experience shows that user acceptance suffers when transactions take longer than one second. If this happens, the user often thinks that there is a technical problem and attempts to pull the card out of the terminal, which results in an uncontrolled abortion of the session. In order to avoid this, all processes between the terminal and the Smart Card should be optimized so that they take less than one second to execute.

Above all, in this regard you should not overlook the fact that if the level of user acceptance is inadequate, the system will experience a significantly higher level of technical problems due to unforeseeable interventions (such as prematurely withdrawing the card from the terminal).

Design

When designing a data processing and management system using Smart Cards, you should always apply Kerckhoff's principle. This principle states that the security of a system should rely solely on its secret keys. It is very difficult to test a system whose security is based on the confidentiality of the data, since the tests can only be carried out by a very limited number of people. Since this group of people has usually been involved in the design of the system, testing on this basis cannot be very effective. The alternative of making all the internal data available to the testers has its own disadvantages, since this could give a potential attacker free access to the data. A compromise approach is usually taken in practice, in which the fundamental design of the system is open but some special things are kept secret.

If, completely counter to Kerckhoff's principle, certain information within a Smart Card project must remain confidential, then it is logical to not have security of this information rest on a single person. It is better to divide the knowledge between several people, so that each of them is familiar with part of the system, but none of them knows the entire system.

With regard to designing systems based on Smart Cards, you should always keep the overall system in view, rather than concentrating only on the card and its immediate surroundings. This is the most commonly seen mistake in the course of a Smart Card project. Since the Smart Card is an active device, you are always dealing with a distributed system, whose individual elements must act autonomously and in which pure master–slave relationships do not prevail. After the entire system has been integrated, all of its elements must work smoothly with each other, rather than just the card with the terminal. You must therefore acquire an overall perspective of the project from the very start.

14.2   Formulas for Estimating Processing Times

In the design of Smart Card applications, or in the design of new Smart Card commands, it is frequently necessary to estimate processing times. Even with experience, it is relatively difficult to obtain sufficiently accurate numbers using estimates alone, and the estimation process is rather imprecise. In this section, we have collected a set of basic formulas for calculating processing times relating to Smart Card operating systems and data transfers via contact interfaces. If these formulas are properly used, they will give results that are acceptably accurate. However, they should by no means be regarded as foolproof, and their accuracy has certain limitations. They are intended to be used for numerical estimates with an error tolerance on the order of ten percent. We can thus immediately recommend that you add an appropriate safety margin in critical situations.

   The names of the variables in the following formulas are intended to be self-explanatory, which means that relatively long names are necessary. Unless otherwise indicated, all fixed times are based on a standard clock frequency of 3.5712 MHz. The stated times can however be adjusted for other clock frequencies using the proportionality factor **PF** from Formulas (14.9) and (14.10). In all cases, it is assumed that the commands in question are executed without any errors, and that no errors occur in any EEPROM operations or data transfers. With regard to data transfers, it is further assumed that the processing of a command starts immediately after the last bit has been received, which means that the shortest possible character waiting time (CWT) has been selected.

Command processing
The following three formulas form the basis for all timing calculations. They divide the total processing time into two parts: the data transfer time and the processing time inside the Smart Card:

$$t_{total} = t_{data\_transfer} + t_{SC\_internal} \tag{14.1}$$

$$t_{data\_transfer} = t_{data\_transfer\_command} + t_{data\_transfer\_response} \tag{14.2}$$

$$t_{SC\_internal} = t_{command\_interprerter} + t_{command\_processing} \tag{14.3}$$

   The time that the command interpreter needs for its activities depends only on the frequency of the applied clock signal, which is here taken to be 3.5712 MHz:

$$t_{command\_interpreter} \approx 1.5 \text{ ms} \tag{14.4}$$

$$t_{command\_processing} = t_{EEPROM} + t_{cryptoalgorithm} + t_{command\_code} \tag{14.5}$$

   The variables in Formulas (14.1) through (14.5) are defined and described in Table 14.1.

   The exact processing time of a command can only be determined by means of a detailed analysis of the program routines at the machine-code level. Program branches and loops would lead to very complex formulas, which would be unusable in practice. Consequently, we have simply divided the commands into three groups according to their complexity. Simple commands, such as SELECT FILE or READ BINARY, need the least time. Moderately complex commands, such as INTERNAL AUTHENTICATE, need somewhat more time for their internal processes. Highly complex commands, such as DEBIT IEP, need even more time. In this regard, you should keep in mind that these blanket values do not cover actions such as calling a cryptographic algorithm or performing an EEPROM write operation, but only the necessary internal computations and queries.

$$t_{command\_code} \approx 5 \text{ ms (for simple commands)} \tag{14.6}$$

$$t_{command\_code} \approx 12 \text{ ms (for moderately complex commands)} \tag{14.7}$$

$$t_{command\_code} \approx 20 \text{ ms (for highly complex commands)} \tag{14.8}$$

Table 14.3   Definitions and descriptions of the variables for Formulas (14.1) through (14.8)

Variable	Unit	Description
$t_{data\_transfer}$	s	data transfer time for a command and its associated response
$t_{data\_transfer\_response}$	s	data transfer time for the response to a command
$t_{data\_transfer\_command}$	s	data transfer time for a command
t_{EEPROM}	s	time required to write data to the EEPROM
t_{total}	s	time required to receive a command, process it and send the response
$t_{SC\_internal}$	s	time required to process a command inside the Smart Card
$t_{command\_processing}$	s	time required to execute a command
$t_{command\_code}$	s	time required to execute a special part of a command (such as a cryptographic algorithm)
$t_{command\_interpreter}$	s	time required to analyze a command and call the associated program code
$t_{cryptoalgorithm}$	s	time required to execute a cryptographic algorithm

Proportionality factor for predefined functions
If the predefined time values that are given for certain functions assume that a particular clock frequency is used, the proportionality factor **PF** can be used as necessary to convert them to values that correspond to the clock frequency that is actually used. The variables in Formulas (14.9) and (14.10) are defined and described in Table 14.4.

$$PF = \frac{f_{reference}}{f_{actual}} \tag{14.9}$$

$$t_{actual} = t_{reference} \cdot PF \tag{14.10}$$

Table 14.4 Definitions and descriptions of the variables for Formulas (14.9) and (14.10)

Variable	Unit	Description
f_{actual}	MHz	actual clock frequency
$f_{reference}$	MHz	reference frequency for which the time value is specified
t_{actual}	s	actual duration of an action
$t_{reference}$	s	duration specified for a particular clock frequency
PF	---	proportionality factor for functions whose processing time depends on the clock frequency

EEPROM operations

Before data can be written to the EEPROM, the content of the affected part of the EEPROM may first have to be erased, depending on content of the data to be written. With some Smart Card microcontrollers, the page size for erasing can be different from the page size for writing. This is taken into account in the following formulas.

To determine whether an EEPROM page must first be erased, you need to know the current content of the page and the content of the new data. For conservative estimates, however, you should always assume that the page to be written must first be erased.

One small comment is appropriate with regard to the durations of EEPROM write and erase operations. The microcontrollers that are currently commonly used in Smart Cards do not have internal time bases, so the only timing reference for the operating system is the externally applied clock signal. If the microcontroller has a maximum specified clock frequency of 5 MHz, for example, then all EEPROM write routines will be designed for this frequency. This means that if the actual clock frequency is less than the maximum value, the EEPROM write time will be proportionally longer. For precise calculations, this should be taken into account by use of the proportionality factor. However, this all depends on the maximum clock frequency, which varies according to the microcontroller type and is also a parameter of the Smart Card operating system. Consequently, this effect is not taken into account here. In the future, it will anyhow not be such a significant factor, since the latest microcontrollers have internal time bases and can thus carry out EEPROM operations with fixed timing, independent of the frequency of the applied clock. The variables in Formulas (14.11) through (14.13) are defined and described in Table 14.5.

$$t_{EEPROM} = t_{EEPROM\_erase} + t_{EEPROM\_write} \tag{14.11}$$

$$t_{EEPROM\_erase} = \frac{t_{page\_erase}}{PS_{erase}} \cdot n \tag{14.12}$$

$$t_{EEPROM\_write} = \frac{t_{page\_write}}{PS_{write}} \cdot n \tag{14.13}$$

Table 14.5 Definitions and descriptions of the variables for Formulas (14.11) through (14.13) for EEPROM operations

Variable	Unit	Example	Description
n	byte	20 bytes	number of bytes to be written to the EEPROM; must be rounded up to the relevant page size
PS_{erase}	byte	32 bytes	page size for erasing
PS_{write}	byte	4 bytes	page size for writing
t_{EEPROM}	s	21 ms	time required to write n bytes to the EEPROM, including prior erasing if necessary
t_{EEPROM_erase}	s	3.5 ms	time required to erase n bytes in the EEPROM
t_{EEPROM_write}	s	17.5 ms	time required to write n bytes in the EEPROM
t_{page_erase}	s/byte	3.5 ms (4 bytes)	time required to erase one EEPROM page
t_{page_write}	s/byte	3.5 ms (4 bytes)	time required to write one EEPROM page

Data transfer

The time required to transfer the command and the subsequent response depends primarily on the amount of data to be transferred. The structures of the basic transport protocol data units (TPDUs) and application protocol data units (APDUs) are described in detail in Sections 6.1.2 ('The T=0 transmission protocol') and 6.1.3 ('The T=1 transmission protocol'). The variables in Formulas (14.14) through (14.17) are defined and described in Tables 14.6.

$$t_{command} = t_{byte\_transfer} \cdot n_{command\_data} \tag{14.14}$$

$$t_{response} = t_{byte\_transfer} \cdot n_{response\_data} \tag{14.15}$$

$$n_{command\_data} = n_{level\_2} + n_{header} + n_{body} \tag{14.16}$$

$$n_{response\_data} = n_{level\_2} + n_{trailer} + n_{body} \tag{14.17}$$

The total time required to transfer a single byte depends on the clock rate, the bit-rate adjustment factor and the clock-rate conversion factor. The variables in Formula (14.18) are defined and described in Table 14.7.

$$t_{byte\_transfer} = \frac{1}{D} \cdot \frac{F}{f} \cdot n \tag{14.18}$$

Table 14.6   Definitions and descriptions of the variables for Formulas (14.14) through (14.17)

Variable	Unit	Typical value	Description
$n_{response\_data}$	byte	---	number of transferred data bytes
n_{body}	byte	---	number of bytes in the command body or associated response body. If the command includes a data part, it contains a 1-byte or 2-byte length parameter for the command body or response body.
n_{header}	byte	4	number of bytes in the command header For the T=1 protocol, these are the CLS, INS, P1 and P2 bytes.
$n_{command\_data}$	byte	---	number of transferred bytes
$n_{level\ 2}$	byte	4	number of bytes for transmission layer 2 For the T=1 protocol, theses are the NAD, PCB, LEN and EDC bytes
$n_{trailer}$	byte	2	number of bytes needed for the trailer of a response. These are the SW1 and SW2 bytes.
$t_{response}$	s	---	time required to transfer the response to a command
$t_{byte\_transfer}$	s	---	time required to transfer one byte
$t_{command}$	s	---	time required to transfer a command

Table 14.7   Definitions and descriptions of the variables for Formula (14.18)

Variable	Unit	Typical value	Description
D	bit·byte/MHz·ms	1 bit·byte/MHz·ms	bit-rate adjustment factor
f	MHz	3.5712 MHz	clock frequency
F	1	372	clock-rate conversion factor
n	bit	12 bits	number of bits per byte (1 start bit, 8 data bits, 1 parity bit, 2 stop bits)
$t_{byte\_transfer}$	ms	1.25 ms	transfer time for one byte

Worked-through example: Smart Card READ BINARY command

Here we present a sample calculation of the estimated processing time for a Smart Card command with a simple structure. We have chosen the READ BINARY command for this example. As general conditions, we have selected the T=1 transmission protocol with a divider value of 372, and a clock frequency of 5 MHz.

The time required for the data transfer can be calculated using Formula (14.18):

$$t_{\text{byte\_transfer}} = \frac{1}{D} \cdot \frac{F}{f} \cdot n = \frac{1}{1 \dfrac{\text{bit} \cdot \text{byte}}{\text{MHz} \cdot \text{ms}}} \cdot \frac{372}{5\text{MHz}} \cdot 12 \text{ bits} \approx 0.89 \text{ ms/byte}$$

For a transfer using the T=1 protocol, four bytes are needed for level 2 in addition to the data for level 7. The command header of READ BINARY has a length of four bytes (CLA, INS, P1, and P2), and the associated body contains one byte (Le). The response consists of the data that have been read, with a length of (Le), and a trailer consisting of the status bytes SW1 and SW2. With this information, we can use Formulas (14.16) and (14.17) to express the amount of data to be transferred as a function of the value of *Le*:

$$n_{\text{command\_data}} = n_{\text{level\_2}} + n_{\text{header}} + n_{\text{body}}$$

$$n_{\text{command\_data}} = (4 + 4 + 1) \text{ bytes} = 9 \text{ bytes}$$

$$n_{\text{response\_data}} = n_{\text{level\_2}} + n_{\text{trailer}} + n_{\text{body}}$$

$$n_{\text{response\_data}} = (4 + Le + 2) \text{ bytes} = (Le + 6) \text{ bytes}$$

From this, we can determine the times required to transfer the command and the response using Formulas (14.14) and (14.15), respectively:

$$t_{\text{command}} = t_{\text{byte\_transfer}} \cdot n_{\text{command\_data}} = (0.89 \text{ ms/byte}) \cdot (9 \text{ bytes}) = 8.01 \text{ ms}$$

$$t_{\text{response}} = t_{\text{byte\_transfer}} \cdot n_{\text{response\_data}}$$

$$t_{\text{response}} = (0.89 \text{ ms/byte}) \cdot (Le + 6 \text{ bytes}) = 0.89 \, (Le + 6 \text{ bytes}) \text{ ms/byte}$$

The command is a simple command, so an execution time of 5 ms at a clock frequency of 3.5712 MHz has been assumed. We can convert this to correspond to the actual clock frequency of 5 MHz using Formulas (14.9) and (14.10):

$$PF = \frac{f_{\text{reference}}}{f_{\text{actual}}} = \frac{3.5712 \text{ MHz}}{5 \text{ MHz}} = 0.714$$

$$t_{\text{actual}} = t_{\text{reference}} \cdot PF = (5 \text{ ms}) \cdot 0.714 = 3.57 \text{ ms}$$

READ BINARY does not require any data to be written to the EEPROM, nor is it necessary to call a cryptographic algorithm. The time required for the command processing inside the Smart Card can thus be calculated as follows:

$$t_{\text{command\_processing}} = t_{\text{EEPROM}} + t_{\text{cryptoalgorithm}} + t_{\text{command\_code}}$$

$$t_{\text{command\_processing}} = 0 \text{ ms} + 0 \text{ ms} + 3.57 \text{ ms} = 3.57 \text{ ms}$$

Under the additional assumption that the command interpreter needs around 1.5 ms to do its job with a 3.5712-MHz clock, we can now calculate the internal execution time of the command:

$$t_{actual} = t_{reference} \cdot PF = (1.5 \text{ ms}) \cdot 0.714 \approx 1 \text{ ms}$$

$$t_{SC\_internal} = t_{command\_interpreter} + t_{command\_processing} = 1 \text{ ms} + 3.5 \text{ ms} = 4.5 \text{ ms}$$

All of the values determined thus far can now be inserted into Formula (14.2), to yield an expression for the total processing time for the READ BINARY command as a function of the amount of data read:

$$t_{data\_transfer} = 8.01 \text{ ms} + (Le + 6 \text{ bytes}) \text{ ms/byte}$$

$$t_{total} = t_{data\_transfer} + t_{SC\_internal} = 8.01 \text{ ms} + (Le + 6 \text{ bytes}) \text{ ms/byte} + 4.5 \text{ ms}$$

$$t_{total} = (12.51 + 0.89 (Le + 6) \text{ byte}^{-1}) \text{ ms} = (17.85 + 0.89 Le \text{ byte}^{-1}) \text{ ms}$$

The result of these calculations is a reasonably good order-of-magnitude match to the empirically determined formula for READ BINARY at a clock rate of 3.5712 MHz (Formula (14.21) in the following section). For example, suppose that 5 kB of data must be written to the EEPROM to initialize the card. These are transferred using the T=1 protocol and a divider value of 372, with a clock rate of 3.5712 MHz. The calculated time for transferring a single byte, from Formula (14.18), is:

$$t_{byte\_transfer} = \frac{1}{D} \cdot \frac{F}{f} \cdot n = \frac{1}{1 \dfrac{\text{bit} \cdot \text{byte}}{\text{MHz} \cdot \text{ms}}} \cdot \frac{372}{3.5712 \text{ MHz}} \cdot 12 \text{ bits} \approx 1.25 \text{ ms/byte}$$

Assuming that the initialization command has a 4-byte header (CLA, INS, P1 and P2), the length parameter is one byte (Lc), 100 bytes of useful data are transferred per command and the response consists only of SW1 and SW2, we can calculate the number of bytes of data that are transferred for the command and response:

$$n_{command\_data} = n_{level\_2} + n_{header} + n_{body}$$

$$n_{command\_data} = (4 + 4 + 1 + 100) \text{ bytes} = 109 \text{ bytes}$$

$$n_{response\_data} = n_{level\_2} + n_{trailer} + n_{body} = (4 + 0 + 2) \text{ bytes} = 6 \text{ bytes}$$

From this, we can determine the duration of the transfer for the command and response:

$$t_{command} = t_{byte\_transfer} \cdot n_{command\_data} = (1.25 \text{ ms/byte}) \cdot (109 \text{ bytes}) = 136.25 \text{ ms}$$

$$t_{response} = t_{byte\_transfer} \cdot n_{response\_data} = (1.25 \text{ ms/byte}) \cdot (6 \text{ bytes}) = 7.5 \text{ ms}$$

One hundred bytes of data must be written to the EEPROM. Under the additional assumption that one EEPROM page is 4 bytes and the write time is 3.5 ms per page, we can determine the time required to write the data to the EEPROM for each command:

$$t_{EEPROM\_write} = \frac{t_{page\_write}}{PS_{write}} \cdot n = \frac{3.5\ ms}{4\ bytes} \cdot 100 = 3.5\ ms \cdot 25 = 87.5\ ms$$

We also assume that the EEPROM does not have to be erased before the write operation, since this has already taken place during the testing of the microcontroller:

$$t_{EEPROM} = t_{EEPROM\_erase} + t_{EEPROM\_write} = 0\ ms + 87.5\ ms = 87.5\ ms$$

It is not necessary to call a cryptographic algorithm for initialization, and the command that is used has a simple internal structure. An execution time of around 5 ms can therefore be assumed for the command code:

$$t_{command\_processing} = t_{EEPROM} + t_{cryptoalgorithm} + t_{command\_code}$$

$$t_{command\_processing} = 87.5\ ms + 0\ ms + 5\ ms = 92.5\ ms$$

Now we have to insert the values calculated using Formulas (14.3) and (14.2), and this completes our calculation of the time required for an initialization command with 100 bytes of data. The command interpreter needs 1.5 ms on top of this:

$$t_{SC\_internal} = t_{command\_interpreter} + t_{command\_processing} = 92.5\ ms + 1.5\ ms = 94\ ms$$

$$t_{data\_transfer} = t_{data\_transfer\_command} + t_{data\_transfer\_response}$$

$$t_{data\_transfer} = 136.5\ ms + 7.5\ ms = 144\ ms$$

$$t_{total} = t_{data\_transfer} + t_{SC\_internal} = 144\ ms + 94\ ms = 238\ ms$$

We have thus calculated that it takes 238 ms to transfer 100 bytes of data to the Smart Card, write it to the EEPROM and return a response to the terminal to confirm successful processing of the command.

However, according to our initial assumptions, a total of 5 kB of data (5120 bytes) must be written to the memory. To simplify the calculations, we assume that this takes approximately 52 times as long:

$$t_{initialization} = t_{total} \cdot 52 = 12.4\ s \approx 13\ s$$

According to our calculations, the initialization process should take 12.4 seconds. If we include a small safety margin, we can assume that the initialization will not take longer than 13 seconds. However, any transmission errors or EEPROM errors that might occur during initialization are not taken into account in our calculations. Such errors must be regarded as a sort of 'force majeure' that can only be dealt with statistically.

14.3   Processing Times of Typical Smart Card Commands

The formulas in this section are based on timing measurements of actual Smart Card operating systems. Linear equations have been fitted to these measurements. The basis for all the formulas is a Smart Card operating system with the T=1 transmission protocol, a clock rate conversion factor of 372 and a bit rate adjustment factor (D) of 1. The microcontroller is clocked at 3.5712 MHz, and the write/erase cycle time of the EEPROM is 3.5 ms for a 4-byte page. It is also assumed that no error occurs during data transfers or any necessary EEPROM write/erase operations. The formulas are valid for all values of n between 1 and 254.

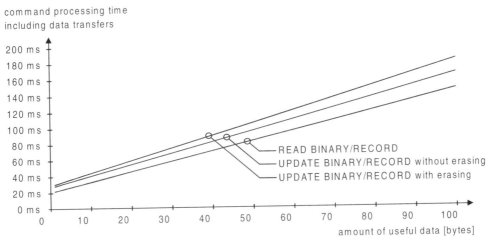

Figure 14.2   Processing times for Smart Card commands as a function of the amount of useful data, including the transfer times of the command and the response. This chart is based on the T=1 transmission protocol, a clock-rate conversion factor of 372, a clock frequency of 3.5712 MHz and an EEPROM write/erase cycle time of 3.5 ms for a 4-byte page.

Formulas (14.19) and (14.20) are intended to be used to calculate times for the SELECT FILE command, with the option of file selection using a 2-byte FID or an n-byte DF name:

$$t_{\text{total SELECT FILE with FID}} \approx 23 \text{ ms} \tag{14.19}$$

$$t_{\text{total SELECT FILE with DF name}} \approx (20.75 + 1.26 \cdot n) \text{ ms} \tag{14.20}$$

Formulas (14.21) through (14.24) can be used to estimate the times required by read commands for files with transparent and record-oriented structures. The variable n is the number of bytes to be read. These formulas can also be used if READ BINARY or READ RECORD is used with implicit file selection, since the time difference due to processing the implicit file selection is negligible:

$$t_{\text{total READ BINARY}} \approx (20.77 + 1.26 \cdot n) \text{ ms} \tag{14.21}$$

$$t_{\text{SC\_internal READ BINARY}} \approx (2.02 + 0.01 \cdot n) \text{ ms} \tag{14.22}$$

$$t_{\text{total READ RECORD}} \approx (20.70 + 1.26 \cdot n) \text{ ms} \tag{14.23}$$

$$t_{\text{SC\_internal READ RECORD}} \approx (1.95 + 0.01 \cdot n) \text{ ms} \tag{14.24}$$

Formulas (14.25) through (14.3)2 can be used to estimate the processing times for UPDATE BINARY and UPDATE RECORD commands, with and without implicit file selection. The command duration depends primarily on whether it is necessary to erase the relevant EEPROM page prior to the write operation. The formulas are therefore divided into two sets according to this condition. With the formulas that include erasing prior to writing, it is always assumed that all pages must be erased. The number of bytes to be written is denoted by n.

$$t_{\text{total UPDATE BINARY without erasing}} \approx (25.55 + 1.39 \cdot n) \text{ ms} \tag{14.25}$$

$$t_{\text{SC\_internal UPDATE BINARY without erasing}} \approx (6.8 + 0.14 \cdot n) \text{ ms} \tag{14.26}$$

$$t_{\text{total UPDATE BINARY with erasing}} \approx (27.26 + 1.54 \cdot n) \text{ ms} \tag{14.27}$$

$$t_{\text{SC\_internal UPDATE BINARY with erasing}} \approx (8.51 + 0.29 \cdot n) \text{ ms} \tag{14.28}$$

$$t_{\text{total UPDATE RECORD without erasing}} \approx (25.35 + 1.38 \cdot n) \text{ ms} \tag{14.29}$$

$$t_{\text{SC\_internal UPDATE RECORD without erasing}} \approx (6.7 + 0.14 \cdot n) \text{ ms} \tag{14.30}$$

$$t_{\text{total UPDATE RECORD with erasing}} \approx (27.13 + 1.54 \cdot n) \text{ ms} \tag{14.31}$$

$$t_{\text{SC\_internal UPDATE RECORD with erasing}} \approx (8.4 + 0.28 \cdot n) \text{ ms} \tag{14.32}$$

14.4   Typical Command Execution Times

The tables in this section are based on the average times for the successful execution of Smart Card commands. Measurements were made on various types of Smart Cards with various operating systems. The listed values are averages; actual values can differ significantly from the listed values in individual cases, depending on the operating system. All measurements were made with the T=1 transmission protocol, a clock rate conversion factor of 372, a clock frequency of 3.5712 MHz, an EEPROM write/erase cycle time of 3.5 ms for a 4-byte page and the DES algorithm with a processing time of 17 ms/8 bytes.

The execution times for the commands marked with a '⊗' show a strong dependence on manner in which the command is implemented and the extent of the supported functions. Consequently, times for these commands can vary widely.

Table 14.8 Average execution times for the READ BINARY command, as measured with several different Smart Card operating systems. The time behavior of this command is similar to that of READ RECORD. The values listed in parentheses are the amount of data read. The measurement conditions are described in the text.

Command	Execution time without data transfer	Execution time with data transfer
READ BINARY (1 byte)	2.02 ms	22.02 ms
READ BINARY (2 bytes)	2.03 ms	23.28 ms
READ BINARY (3 bytes)	2.04 ms	24.54 ms
READ BINARY (4 bytes)	2.04 ms	25.79 ms
READ BINARY (5 bytes)	2.05 ms	27.05 ms
READ BINARY (10 bytes)	2.12 ms	33.37 ms
READ BINARY (20 bytes)	2.23 ms	45.98 ms
READ BINARY (50 bytes)	2.54 ms	83.79 ms
READ BINARY (100 bytes)	2.98 ms	146.73 ms

Table 14.9 Average execution times for the UPDATE BINARY command without prior erasing of the affected EEPROM pages, as measured with several different Smart Card operating systems. The time behavior of this command is similar to that of UPDATE RECORD. The values listed in parentheses are the amount of data read. The page size is 4 bytes and the read/erase time is 3.5 ms. All other measurement conditions are described in the text.

Command	Execution time without data transfer	Execution time with data transfer
UPDATE BINARY, no erase (1 byte)	6.95 ms	26.95 ms
UPDATE BINARY, no erase (2 bytes)	7.01 ms	28.26 ms
UPDATE BINARY, no erase (3 bytes)	7.03 ms	29.53 ms
UPDATE BINARY, no erase (4 bytes)	7.11 ms	30.86 ms
UPDATE BINARY, no erase (5 bytes)	7.12 ms	32.12 ms
UPDATE BINARY, no erase (10 bytes)	7.33 ms	38.58 ms
UPDATE BINARY, no erase (20 bytes)	12.33 ms	56.08 ms
UPDATE BINARY, no erase (50 bytes)	18.16 ms	99.41 ms
UPDATE BINARY, no erase (100 bytes)	18.81 ms	162.56 ms

Table 14.10   Average execution times for the UPDATE BINARY command with prior erasing of the affected EEPROM pages, as measured with several different Smart Card operating systems. The time behavior of this command is similar to that of UPDATE RECORD. The values listed in parentheses are the amount of data read. The page size is 4 bytes and the read/erase time is 3.5 ms. All other measurement conditions are described in the text.

Command	Execution time without data transfer	Execution time with data transfer
UPDATE BINARY, with erase (1 byte)	9.42 ms	29.42 ms
UPDATE BINARY, with erase (2 bytes)	9.51 ms	30,76 ms
UPDATE BINARY, with erase (3 bytes)	9.52 ms	32.02 ms
UPDATE BINARY, with erase (4 bytes)	9.48 ms	33.23 ms
UPDATE BINARY, with erase (5 bytes)	9.62 ms	34.62 ms
UPDATE BINARY, with erase (10 bytes)	9.85 ms	41.10 ms
UPDATE BINARY, with erase (20 bytes)	17.41 ms	61.16 ms
UPDATE BINARY, with erase (50 bytes)	25.87 ms	107.12 ms
UPDATE BINARY, with erase (100 bytes)	35.34 ms	179.09 ms

Table 14.11   Average execution times of typical commands, as measured with several different Smart Card operating systems. The measurement conditions are described in the text.

Command	Execution time without data transfer	Execution time with data transfer
ASK RANDOM (8-byte random number)	26 ms	55 ms
CREDIT ⊗	175 ms	222 ms
DEBIT ⊗	235 ms	270 ms
EXTERNAL AUTHENTICATE	22 ms	51 ms
GET CARD DATA (8 bytes)	4 ms	33 ms
INITIALIZE IEP for Load ⊗	89 ms	173 ms
INITIALIZE IEP for Purchase ⊗	135 ms	201 ms
INTERNAL AUTHENTICATE	26 ms	65 ms
INVALIDATE	15 ms	34 ms
MUTUAL AUTHENTICATE	95 ms	163 ms
REHABILITATE	15 ms	33 ms
SEEK	3 ms	22 ms
SELECT FILE (with an 8-byte AID)	3 ms	32 ms
SELECT FILE (with a 2-byte FID)	3 ms	24 ms
VERIFY (8-byte PIN) ⊗	27 ms	56 ms

14.5   Application Development Tools

There are presently several PC-based programs for designing Smart Card applications. They allow complete applications to be developed quickly and easily, without knowledge of the internal aspects of the Smart Card operating system being used.

With these tools, the first task is usually to construct a file tree to hold the various applications (i.e. DFs) and their associated EFs. It is naturally necessary to specify the file structures and relevant access conditions for the EFs. If the Smart Card operating system has a state machine for commands, its parameters can also be defined using the graphical user interface of the application generation program. Some application development programs can also run a consistency check on the state machine. Since the application needs various data and keys in its EFs, a link to a database can be set up after the files have been defined. The contents of the EFs in the individual cards are thereby linked to the data sets in the database.

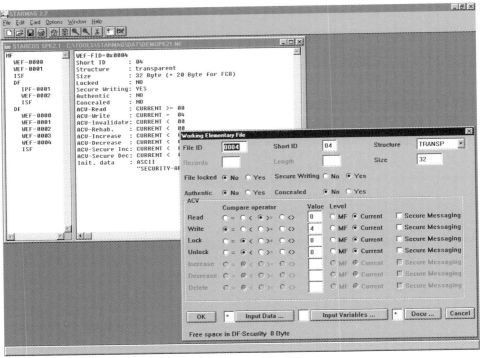

Figure 14.3   Screen display of a PC-based software tool for generating Smart Card applications (Source: Giesecke & Devrient)

Once the complete application has been defined using the application generator, various general parameters of the Smart Card operating system can be set, such as the transmission protocol and the divider value. The application can now be experimentally loaded into one or more Smart Cards with appropriate memory capacity. After several test cards have been produced, they can be tried out in a terminal. If it turns out that modifications are necessary, it is possible to delete the application in the Smart Card and then load a revised version.

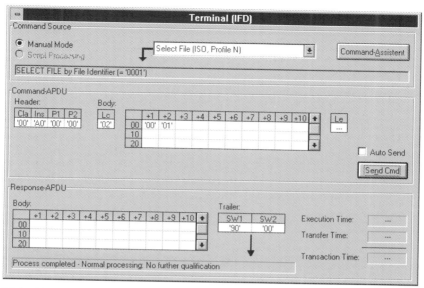

Figure 14.4   Terminal window of the 'Smart Card Simulator' simulation program [Rankl, Hanser]

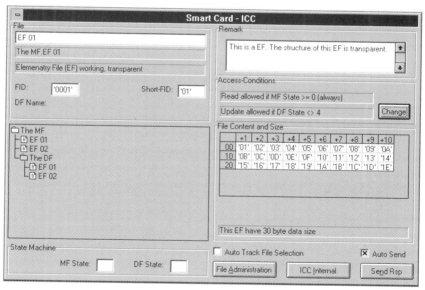

Figure 14.5   Smart Card window of the 'Smart Card Simulator' simulation program [Rankl, Hanser]

If all the tests have been concluded satisfactorily and a relatively large number of cards are needed, they can be produced in a standard card production facility. The application data generated by the PC program (files, commands, states, and so on) form the basis for the completion, initialization and personalization of the Smart Cards. The turnaround time for producing finished cards thus remains very low, despite a high degree of flexibility.

Smart Card simulators are also available, in addition to these application development tools. A Smart Card simulator behaves in exactly the same way as a normal Smart Card, but it is only a dummy Smart Card linked to a PC interface by a cable. Suitable software in the PC simulates the card in real time. Naturally, applications can also be generated and tested in the PC, as described above. However, a PC is always needed to perform the simulation, and this can frequently cause problems, due to the size of the necessary equipment.

14.6   The Course of a Smart Card Project

The course of a Smart Card project is shown in Figure 14.6. Currently, the card manufacturer also develops the associated microcontroller software. This means that both the first phase (A) and the last phase (E) of the project are carried out by the same firm. Phases B and C are performed by a semiconductor manufacturer. The dice can be built into the modules either directly by the semiconductor manufacturer or by the card manufacturer. All of phase E, which includes initialization, personalization and so on, is always performed by the card manufacturer. The card manufacturer also usually manages the Smart Card project, while the other firms are more or less subcontractors.

This brief description says very little about the time required for the individual production processes. However, this should not be underestimated, since several different firms work together to produce Smart Cards, and the elapsed time for some processes can be many weeks. Typical times for the completion time of the most important production steps are shown in Figure 14.6. This example is based on the following assumptions:

- 50,000 cards are to be produced;
- a new operating system must be generated, based largely on existing libraries;
- all firms involved have mid-range production capacities;
- each production process can start immediately after the necessary parts are received.

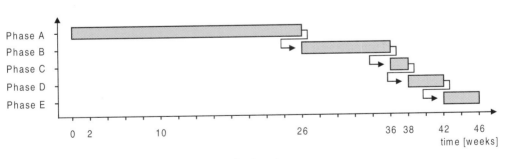

Figure 14.6   Gantt chart for a typical Smart Card project

Phase A: mask generation	6 months
Phase B: semiconductor fabrication	10 weeks
Phase C: module production	2 weeks
Phase D: card body production	4 weeks
Phase E: initialization and personalization	4 weeks

14.7   Design Examples for Smart Card Applications

The following examples illustrate typical Smart Card applications. These are mid-range data processing applications that do not require extensive system design. Their typical users would be medium-sized firms. The background system employed may be a PC in a secure environment, which means that both the installation costs and operational costs are at the low end of the scale.

These simple examples demonstrate very effectively how a typical Smart Card application is constructed. We explain the construction step by step, and demonstrate how to gradually build up to the finished application in the Smart Card. In each case, we pay only cursory attention to the terminal and the background system, but these aspects of the system can be deduced from the available information.

All of these examples are based on the principle of distributing information among many individual, separate systems. This contrasts with commonly used centralized mainframe solutions, in which all the information is located in one place. If such an approach is translated into a Smart Card application, it means that the card is only a sort of proof of identification, while all information is stored in an omniscient background system. If an application built this way must be extended, a regularly observed consequence is that the all-powerful background system must also undergo an expensive and time-consuming upgrade.

Here we have attempted to take a different approach. The background system is only responsible for the management and consistency of the overall system. All other information is held locally in the cards. A global database is of course necessary for system administration, so that lost or faulty cards can be reproduced from data on hand. However, this database should not be necessary for the normal operation of the system.

A distributed Smart Card system may be thought of as a large tree with many branches, which like all trees draws its energy from photosynthesis in its many leaves. Energy production is distributed among the leaves and takes place in many places simultaneously. Figuratively, an effective and well-designed Smart Card application behaves in the same way. The information is stored in decentralized cards, and is thus protected against every form of attack. The large mass of information also does not burden the background system, which only has to deal with the centralized management tasks. The actual system processes are decentralized, just like the process of photosynthesis in the tree's leaves, and take place in parallel in localized terminals and Smart Cards. This makes it very simple to extend the overall system by adding more terminals and cards, without needing to worry about any major impact on the background computer system.

The opposite approach to the system just sketched out is to concentrate all of the necessary activities in a central background system. In our analogy, this would amount to moving photosynthesis from the tree's leaves to its trunk, which would result in a huge trunk. The overall system would then not only be very large and expensive, it would also be extremely vulnerable to disturbances in the background system. However, this should be avoided as much as possible.

In their ignorance of the characteristics of Smart Cards, many would-be system operators make the mistake of designing the whole system top-down. When it comes to the part of the system that is most critical in terms of security, namely the terminals and the Smart Cards, they then stipulate that these components should be made secure in some more or less undefined manner.

By contrast, the proposed designs described here represent a bottom-up approach, in which the system and its required characteristics are defined by starting with the lowest-level object (the Smart Card) and working upwards. The risk of security gaps can be very effectively minimized with this approach, since the system is built up securely from the smallest unit all the way to the top.

14.7.1   Electronic purse system for arcade games

Situation and objectives
This Smart Card application is intended to provide functions that make it possible to pay small amounts into arcade game machines. Smart Cards should replace the coins that are conventionally used with such machines, in order to reduce operating costs.

There are two types of terminals. A loading terminal has a coin slot with a coin tester and can load electronic 'currency' into Smart Cards. A debiting terminal, which operates largely autonomously, debits this currency from the Smart Card.

Requirements
The entire system should be anonymous, while still allowing all payment streams to be monitored. If there is a suspicion of fraud, it must be possible to identify individual cards and selectively block them.

Since this is a payment system application, and considering that the equipment used is completely automatic and operates without human supervision, the design should aim for a medium level of security.

Proposed solution
The solution is based on a simple closed purse system, which has been specially designed for the purposes of this application. It could of course easily be used for similar applications by making small modifications to the files and processes. We have avoided using an electronic purse system that is compliant with the CEN EN 1546 standard, as such a solution would be more expensive for the application provider than the proposed solution. In addition, our objective is to demonstrate the principles of a simple closed purse system.

In a central location, there is an automated machine that can accept both coins and banknotes and then load the equivalent amount of electronic currency into a Smart Card. Neither a PIN nor any other user input is necessary, as the electronic purse is anonymous. The only record keeping that is performed for the amounts loaded into the cards is based on individual card numbers, which are unique within the system.

A PC looks after the management of all data and the payment flows. The PC also contains a database that holds general information for all issued cards. A daily or weekly balance calculation can be used to check that the money flows in the system remain closed.

The loaded monetary units are debited from the electronic purse in the Smart Card at a payment (debit) terminal. A display on this terminal shows the user the amount debited. To keep the cost of the terminals as low as possible, secure messaging is used to protect data transfers, instead of a shutter. Each terminal has a security module to store the secret key and keep track of the amounts paid, sorted by card number. At regular intervals, the data so obtained are transferred by wire or a special transfer card to the administration PC, which looks after evaluating the data. The file tree of the proposed solution is shown in Table 14.12. It needs around 100 bytes in EEPROM, depending on the Smart Card operating system.

Table 14.12   File tree of the 'arcade games' sample application

File	FID	Structure	Description
MF	'3F00'	—	Smart Card root directory
DF	—	—	directory for the 'arcade games' application
EF 1	'0001'	transparent	date of issue and card number
EF 2	'0002'	cyclic	amount
EF 3	'0003'	linear fixed	key 1 key 2 key 3 key 4

Table 14.13   Keys required for the 'arcade games' sample application

Key	Usage	Function	State transition
key 1	MUTUAL AUTHENTICATION	mutual authentication of the terminal and the Smart Card – paying with the purse – blocking the purse	$x \rightarrow 1$
key 2	MUTUAL AUTHENTICATION	mutual authentication of the terminal and the Smart Card – unblocking the purse – card management	$x \rightarrow 2$
key 3	MUTUAL AUTHENTICATION	mutual authentication of the terminal and the Smart Card – loading the purse	$x \rightarrow 3$
key 4	secure messaging	protecting the data transfer – authentic mode	—

The regular data exchanges between the debiting terminals and the administration computer can be used to maintain a black list of blocked cards in the terminals. If a terminal determines that an inserted card is on the black list, it blocks the EF2 file containing the electronic money. After this, the card can no longer be used for payments. The user must have the card unblocked at the administration terminal, and at the same time, a check can be made to see why the card appears on the black list.

The keys that are needed for this application are listed in Table 14.13. We have done without derived and card-specific keys in the interest of a simple overall system.

The proposed solution is very suitable for paying for services received from automatic equipment. Human supervision is unnecessary. However, it is not essential to use a special machine to automatically load 'money' into the cards. Cards could also be loaded manually at a service counter, in exchange for a cash payment equal to the amount to be loaded. With minor modifications, the system outlined here can also be used in a launderette or canteen.

Figure 14.7 illustrates the basic procedure for loading electronic currency into the electronic purse, and Figure 14.8 similarly illustrates the procedure for making a payment.

Table 14.14   Access conditions for the 'arcade games' application
(≥ 0: always, < 0: never, SM: secure messaging)

File	Read	Write	Block	Unblock	Increase amount	Decrease amount
EF 1	≥ 0	$= 2$	< 0	< 0	< 0	< 0
EF 2	$\geq 0 \wedge$ SM	< 0	$= 1$	$= 2$	$= 3$	$= 1$
EF 3	< 0	< 0	$= 1$	$= 2$	< 0	< 0

Smart Card		Terminal	User
	↔	SELECT FILE (DF)	*Output*
	↔	SELECT FILE (EF 1)	"insert money"
	↔	READ BINARY	
	↔	SELECT FILE (EF 2)	
	↔	READ BINARY	
	↔	ASK RANDOM	
	↔	MUTUAL AUTHENTICATION	
		enable secure messaging	
	←	INCREASE	
Payment [... ‖ Return code]	→	IF (Return code = OK) THEN command successfully executed ELSE abort	*Output* "xx DM loaded in the Smart Card"

Figure 14.7   Basic command sequence for loading electronic monetary units into the purse in the 'arcade games' application

Smart Card		Terminal	User
	↔	SELECT FILE (DF)	
	↔	SELECT FILE (EF 1)	
	↔	READ BINARY	
	↔	SELECT FILE (EF 2)	
	↔	READ BINARY	
	↔	ASK RANDOM	
	↔	MUTUAL AUTHENTICATION	
		enable secure messaging	
	←	DECREASE	
Response [... ‖ Return code]	→	IF (Return code = OK) THEN command successfully executed ELSE abort	*Output* "xx DM debited from the Smart Card"

Figure 14.8   Basic command sequence for making a payment from the purse in the 'arcade games' application

14.7.2   Access control system

Situation and objectives

The objective of this application is to create a Smart Card-based, graduated access control system for a number of rooms and computer systems. This means that certain doors and computers will be fitted with terminals that will allow people to pass through the associated door or use the associated computer, after communication with a Smart Card. It is important to be able to define various security levels, for which access can be limited to specific groups of users. Access will be granted after successful authentication and identification of the user. The necessary proof will be possession of a genuine card and knowledge of its associated PIN. If both of these criteria are satisfied, access will be granted. The terminals must be able to maintain simple black lists, so that 'lost' cards cannot be used for access and such cards can be permanently blocked if necessary.

Requirements

In order to maximize the user acceptance of the solution, the time required for any communication process between the terminal and the Smart Card, together with the subsequent granting of access, must not significantly exceed one second. A longer interval will sooner or later significantly impede user acceptance of the system and encourage users to use various tricks to circumvent security measures, such as propping open doors. Users must be able to select their own PIN codes, in order to avoid users writing PIN codes on the cards.

The system should be designed for a moderately low level of security, as it is fairly unlikely to be subjected to elaborate attacks or analysis. The acquisition and operating costs of the system must not exceed those of a good conventional key–lock system, since the latter would otherwise represent a more economical alternative. The system and Smart Card must be designed to allow time card and canteen billing functions to be incorporated into the system at some later date.

Proposed solution

Simple terminals with ten-digit numeric keypads will be firmly attached to the appropriate doors and computers. These terminals can work autonomously, and they are fitted with economical and exchangeable security modules (such as Smart Cards in the plug-in format). They can grant authorized persons access to the associated doors or computers. Any terminals at a critical or sensitive location can if necessary independently establish a link to the PC that serves as the central system computer, via a two-wire cable. This simple structure satisfies the requirement for low operating costs.

The central computer has a simple multitasking operating system, so that it can execute several tasks in parallel. It is connected to a supplementary terminal that is responsible for the management of the entire system.

The Smart Cards used in this system must have an operating system that can manage several applications, and that can create files (DFs and EFs) in the Smart Card file tree after the card has been issued, so that additional applications can be loaded as necessary after the cards have been issued. The amount of EEPROM required by the current application and the projected future applications will not exceed 1 kB. The cards will be purchased from a card manufacturer complete with the operating system, and then configured appropriately using the management computer.

All cards will initially have a standard and easily remembered PIN. The simplest option in this case is '0000'. This eliminates the cost of generating PINs and preparing PIN letters. Upon receipt of the card, each user must change the standard PIN to some other number using the management terminal, since all terminals will reject a PIN entry of '0000'.

If the user forgets the PIN or the retry counter reaches its limit, the system terminal can be used to enter a new PIN and reset the retry counter, following an authentication.

Since the system is required to have an intermediate level of security, and system management should not be too costly, a severely limited key management scheme is appropriate. Neither derived keys nor multiple generations of keys are used in this system. The keys only need to be separated by function, which leads to the arrangement shown in Table 14.15. The file tree that must be present in the Smart Card is described in Table 14.16, and the file access conditions are listed in Table 14.17.

The file EF2, which contains the authorization data for access to rooms and computers, has a record-oriented structure. All records all have the same length (linear fixed). Each record has an entry that indicates whether the card user is allowed to enter a particular room. It is also possible to define security levels, so that it is not necessary to explicitly list each room. Access can then be globally restricted to certain areas.

Table 14.15   The keys needed for the 'access control' application

Key	Application	Function	State transition
key 1	MUTUAL AUTHENTICATION	application management – creating new files – writing to files – unblocking an application	$x \rightarrow 1$
key 2	MUTUAL AUTHENTICATION	mutual authentication of the terminal and the Smart Card – access authorization – blocking an application	$x \rightarrow 2$
PIN	VERIFY CHV	user identification	$2 \rightarrow 3$

Table 14.16   The file tree for the 'access control' application

File	FID	Structure	Description and contents
MF	'3F00'	—	Smart Card root directory
DF	—	—	directory for the 'access control' application
EF 1	'0001'	transparent	last name, first name, department
EF 2	'0002'	linear fixed	authorization level
EF 3	'0003'	linear fixed	key 1; key 2; PIN

Table 14.17   File access conditions for the 'access control' application
(≥ 0: always, <0: never)

File	Read	Write	Block	Unblock	Create DFs and EFs
DF	—	—	—	—	= 1
EF 1	≥ 0	= 1	<0	= 1	—
EF 2	= 3	= 1	= 2	= 1	—
EF 3	<0	<0	= 2	= 1	—

Since experience shows that it is frequently necessary to modify and restructure access control systems, a TLV structure should be used for the record contents. This allows extensions and modifications to be carried out in a technically elegant manner.

Only standard commands that are provided by commercially available ISO-compliant or ETSI-compliant operating systems are used for the Smart Cards. This means that nothing has to be programmed in the Smart Cards, which considerably reduces acquisition costs. The following commands are needed:

ASK RANDOM	READ BINARY
CHANGE CHV	REHABILITATE
CREATE	UNBLOCK CHV
INVALIDATE	SELECT FILE
MUTUAL AUTHENTICATION	VERIFY CHV
	WRITE BINARY

Figure 14.9 (overleaf) shows the typical command sequence for an access control session. If necessary, the terminal can use READ BINARY directly following the ATR to read the user's name from the file and compare it to a black list. If the user's name is on the black list, further use of the card for access control can be prohibited by blocking all EFs with the INVALIDATE command. If necessary, the application can be reactivated at the management terminal with REACTIVATE, following mutual authentication.

If the system operator decides to also use the Smart Cards for canteen billing, a new application with its own DF and EFs must be generated. There are two ways to do this. Either all employees must bring their cards to the management terminal, or the necessary files can be automatically downloaded during access checks. The second approach is certainly less expensive and more user-friendly, since it does not require any extra administrative effort.

The required control of access to computer systems is completely analogous to the process just described. The only difference is that instead of a door release mechanism being activated, a signal is sent to the computer to tell it to grant access to the user.

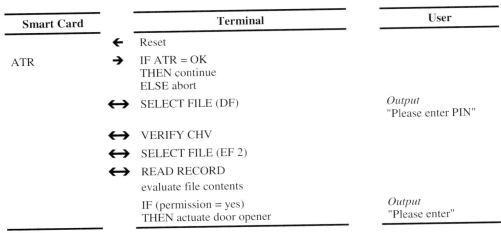

Smart Card	Terminal	User
	← Reset	
ATR	→ IF ATR = OK THEN continue ELSE abort	
	↔ SELECT FILE (DF)	*Output* "Please enter PIN"
	↔ VERIFY CHV	
	↔ SELECT FILE (EF 2)	
	↔ READ RECORD evaluate file contents	
	IF (permission = yes) THEN actuate door opener	*Output* "Please enter"

Figure 14.9 Access control command sequence for the 'access control' application

14.7.3 *Testing the genuineness of a terminal*

Situation and objectives
There are situations in which it should be possible for a card user to test whether a terminal is genuine. One example is a terminal in a supermarket, in which the user must enter a PIN after inserting the card. A counterfeit terminal could be used to spy out secret PIN codes.[6] If the card is subsequently stolen, the thief who has learned the PIN can use the card to make purchases or obtain money from a cash dispenser.

In the summer of 1997, a counterfeit cash dispenser at the Marienplatz in Munich was used in this manner to illicitly collect magnetic-stripe data and associated PIN codes. If Smart Cards are used, good protection against this type of attack can be provided with a suitable application design.

Requirements
It is necessary to design a component of a Smart Card application that allows the card user to recognize a counterfeit Smart Card terminal (but not a manipulated terminal). The user must not need any additional technical aids or equipment to check the terminal.

Proposed solution
The proposed solution involves storing a password, known only to the card user, in a file in the Smart Card. This file can be read by the terminal only after it has successfully authenticated itself with respect to the Smart Card via a secret key.

After this authentication process, the terminal is allowed to read the password from the file and show it on its display. As soon as the card user sees the password and verifies that it is correct, he or she can assume that the terminal is genuine, since only the user knows the password. Only after he or she has verified the password will the user enter the PIN that makes the rest of the transaction possible.

---

[6] See also Section 8.1.1, 'Testing a secret number'.

The procedure just described is recommended in the DIN specification for German signature cards, for example, in order to allow card users to determine whether public signature terminals are genuine.[7]

An important limitation of the solution must be mentioned. This is that it allows a counterfeit terminal to be recognized, but not a manipulated terminal. If it were possible to modify the terminal software without losing the secret key in the process, it would be possible for a manipulated terminal to correctly authenticate itself with respect to the Smart Card and then display the password. This limitation should taken into account in any application in which this technique is used. However, this is basically not a critical issue, since a terminal that can be manipulated on this scale will allow significantly more extensive forms of attack than just spying out PIN codes.

The proposed solution, which is presented in the form of specific files and access conditions in Tables 14.18 and 14.19, is not a complete Smart Card application. Instead, it is a sort of design template that can be woven into any desired application. Consequently, the FIDs and state transitions, as well as the procedure illustrated in Figure 14.10 (overleaf), can be modified as necessary to use different values or command sequences. This example is primarily intended to convey the basic idea of how the genuineness of a terminal can be tested, rather than to serve as a concrete application.

Table 14.18   Keys needed for testing the genuineness of a terminal

Key	Usage	Function	State transition
key 1	EXTERNAL AUTHENTICATION	authentication of the terminal by the Smart Card	$x \rightarrow 1$
PIN	VERIFY CHV	identification of the user	$1 \rightarrow 2$

Table 14.19   File tree and access conditions for testing the genuineness of a terminal (≥ 0: always, < 0: never)

File	FIC	Structure	Read	Write	File contents
EF 1	'0001'	transparent	$= 1$	$= 2$	password
EF 2	'0002'	linear fixed	< 0	< 0	key 1, PIN

---

[7] See also Section 13.6, 'Digital Signatures'.

Smart Card		Terminal	User
	←	Reset	
ATR	→	IF ATR=OK THEN continue ELSE abort	
	↔	EXTERNAL AUTHENTICATE (with key 1)	
	↔	IF (authentication is successful) THEN continue ELSE abort	
	↔	SELECT FILE (EF 1)	
	↔	READ BINARY	*Output the contents of the password file*
			IF (password correct) THEN terminal is genuine
			ELSE abort
	↔	VERIFY CHV	*Output* "Please enter PIN"

Figure 14.10   Command sequence for verifying the genuineness of a terminal

15

Appendix

15.1   Glossary

1-μm, 0.8-μm, … technology

In the manufacturing of semiconductor chips, the performance of the technology used is traditionally expressed in terms of the dimension of the smallest possible transistor structure on the semiconductor material. This is usually the width of the gate oxide strip. Currently, the smallest possible structure width is around 0.18–0.25 μm. It is naturally always possible to make structures on the chip that are bigger than the minimum dimension.

μP card

This is another term for a microprocessor card (*qv*).

3DES

See Triple DES.

Acquirer

An entity that sets up and manages the data links and the data exchanges between the operator of a payment system and the individual service providers. The acquirer can combine the individual transactions that it receives, so that the operator payment system receives only collective certificates.

Administrative data

Data that are only used for the management of useful data and that have no other particular significance in an application.

AFNOR

The *Association Française de Normalisation* is a French standards organization based in Paris.

AID (application identifier)

The AID identifies an application in a Smart Card. It is defined in the ISO/IEC 7816-5 standard. A part of the AID can be registered nationally or internationally, in which case it is reserved for the registered application and is unique in the entire world. The AID consists of two data elements: the RID (registered identifier) and the PIX (proprietary identifier). See also Section 5.6.1, 'File types'.

ANSI

The American National Standards Institute is an American standards organization based in New York.

Application

An application consists of the data, commands, procedures, states, mechanisms and algorithms located within a Smart Card to enable it to be used within the context of a particular system. An application and its associated data are usually located in a dedicated DF directly below the MF.

Application operator

An entity that runs an application on Smart Cards. Generally the same as the application provider.

APDU

An APDU (application protocol data unit) is a software data container that is used to package data so that they can be exchanged between a Smart Card and a terminal. An APDU in converted into a TPDU (transport protocol data unit) by the transmission protocol, and then sent via the serial interface to the Smart Card or the terminal. APDUs can be divided into command APDUs and response APDUs. See also Section 6,5. 'Message Structure: APDUs'.

API

An API (application programming interface) is a meticulously specified software interface within a program that provides third parties with standardized access to the functions of the program.

Applet

An applet is a program written in the Java programming language, which is executed in the virtual machine of a computer. For reasons of security, the functionality of an applet is restricted to a previously defined program environment. In the Smart Card world, applets are sometimes called 'cardlets', and they usually correspond to Smart Card applications.

ASN.1

'Abstract syntax notation 1' is a data description language. It allows data to be clearly defined and presented independent of any particular computer system. ASN.1 is defined by ISO/IEC 8824 and ISO/IEC 8825.

Assembler

An 'assembler' is a program that translates assembly-language program code into machine language that can be executed by a processor. After the assembly process, it is usually necessary to link the resulting code using a linker program. However, the term 'assembler' is also often used as a short form of 'assembly language program code'.

ATR

An ATR (Answer to Reset) is a sequence of bytes sent by a Smart Card in response to a (hardware) reset. It includes various parameters relating to the transmission protocol for the Smart Card, among other things. See also Section 6.2, 'Answer to Reset (ATR)'.

Authentication

The process of proving the genuineness of an entity (such as a Smart Card) by means of a cryptographic procedure. Put simply, authentication amounts to using a fixed procedure to determine whether someone is actually the person he or she claims to be.

Authenticity

A property possessed by an entity or message that is genuine and unaltered.

Authorization

Testing whether a particular action is allowed to be carried out is known as 'authorization'. It amounts to giving someone the authority to do something. For example, when a credit card transaction is authorized by the credit card issuer, the card data are checked to see if the data are correct, the amount of the purchase is less than the allowed limit and so on. The payment is then allowed if all checks are satisfactory. An authorization can be achieved by means of the authentication of the entity in question (such as a Smart Card). Put simply, an authorization amounts to giving someone permission to perform some particular action.

Auto-eject reader

A terminal that can automatically eject an inserted card in response to an electrical or mechanical signal.

Background system

Any sort of computer system that looks after the processing and management of data above the level of the terminals.

Bellcore attack

See Differential fault analysis.

Big-endian

See Endianness.

Binary-compatible program code

See Native code.

Black list

A list in a database that marks all cards that are no longer allowed to be used in a particular application.

Blackbox test

A blackbox test is based on the assumption that the entity performing the test has no knowledge of the internal processes, functions and mechanisms of the software to be tested.

Boot loader

A boot loader is a (usually small) program that is used to load other, larger programs, for example via a serial interface. A boot loader is usually used to load the actual program code in a new chip or a new piece of electronic equipment. In many cases, the boot loading process can be carried out only once.

Brute force attack

An attack on a cryptographic system based on computing all possible values of a key.

BSI

The German *Bundesamt for Sicherheit in der Informationstechnik* (BSI) was founded in 1990. It is the successor of the German *Zentralstelle für das Chiffrierwissen*. The BSI advises agencies and establishes general conditions for cryptographic applications in Germany. It also offers additional services, including the certification of the security features of computerized data systems.

Buffering

A typical type of attack on magnetic-strip cards. It consists of reading the data from the magnetic strip and writing it back again after it has been modified using the terminal (for example, with an altered value for the retry counter).

Bug fix

Additional program code that remedies a known software error (for example, a 'work around').

Bytecode

Bytecode is the name given to the code that a Java compiler produces from the source code. It is a sort of intermediate code that is executed by the Java Virtual Machine (JVM). Bytecode is standardized by the Sun Corporation.

CAP file

A CAP file (card application file) is the data exchange format used between the Java Offcard Virtual Machine and the Java Oncard Virtual Machine.

Cardlet

See Applet.

Classfile

A classfile stores a Java program that has been compiled (translated into bytecode) along with supplementary information. It is executed by the Java Virtual Machine after it has been loaded.

Card accepter

An entity with which cards for a particular application can be used. A typical example is a merchant that accepts credit cards for making payments.

Card body

A plastic card that forms an intermediate product in the manufacturing of Smart Cards. It is further processed in subsequent production steps and receives additional functional elements, such as the embedded chip.

Card issuer

An entity that is responsible for issuing cards. With single-application cards, the card issuer is usually also the application provider, but this need not necessarily be the case.

Card manufacturer

An entity that produces card bodies and embeds modules in them.

Card owner

The card owner is the natural or legal person that has legal control over the card, and who can do as he wishes with the card. With credit and debit cards, the card owner is very often the bank that issues the card, and the customers who use the cards are only cardholders.

Cardholder

The cardholder is the person who has the actual right to possess and use the card. The cardholder need not necessarily be the (legal) card owner.

Card possessor

The person who has possession of a card.

Card reader

A device with relatively simple electrical and mechanical construction that can receive Smart Cards and make electrical contact with them. In contrast to a terminal, a card reader does not have a display or a keypad. In spite of the name, card readers usually can also be used to write data to the card.

Card user

The person using a card, who perforce possesses the card, but who need not necessarily be the legal owner of the card.

Cavity

The opening in the card body, usually produced by milling, that receives the implanted chip module.

CCITT

The *Comité Consultatif International Télégraphique et Téléphonique* was an international committee for telephone and telegraph services based in Geneva. It has now taken on more responsibilities and has been renamed as the ITU.

CCS

A CCS is a cryptographic checksum for data that can be used to detect manipulation of the data during storage. If data are protected by a CCS while being transferred, the term MAC (message authentication code) is used.

CEN

The European standards organization *Comité Européen de Normalisation*, located in Brussels, is composed of all national European standards organizations and is the official institution of the European Union for European standardization.

CEPT

The *Conférence Européenne des Postes et Télécommunications* is a European standards organization for the national telecommunications companies.

Certificate

A certificate is a signed public key issued by a trustworthy body that allows the key to be recognized as authentic. The most widely used and best known specification of the structure and coding of certificates is the X.509 standard.

Certification authority (CA)

A certification authority is entity that certifies public keys for digital signatures, which means that it guarantees that they are genuine. It does this by signing the user's public key using its own private key, and if necessary it makes the signed public keys available in a directory. The CA can itself generate the necessary key pairs (private and public).

Chip card

A general term for a card (usually plastic) that contains one or more semiconductor chips. It can be either a memory card or a microprocessor card. In English-speaking countries, the term 'Smart Card' is used instead of 'chip card'.

Clearing

In electronic payment systems, the process of settling the account between the entity that accepts an electronic payment (usually a merchant) and the associated bank.

Clearing system

A computer-based background system that carries out centralized account settlements in an application for electronic payments.

Clock-rate conversion factor

The clock-rate conversion factor (CRCF) defines the length of one bit during a data transfer, in terms of the number of clock cycles per bit interval. The short form 'divider' is commonly used as an equivalent term.

Cloning

Attacking a Smart Card system by making a complete copy of the ROM and EEPROM of a microcontroller.

Closed application

A Smart Card application that is only available to the application operator and cannot be used generally.

Closed purse

An implementation of a closed application for an electronic purse. A closed purse can be used only within the limits of what is allowed by the application operator, and not for general payment transactions.

Cold reset

See Reset.

Combicard

See Dual-interface card.

Command APDU

A command APDU is a command sent from a terminal to a Smart Card. It consists of a command header and an optional command body. The command header in turn consists of a class byte, an instruction byte, and two parameter bytes P1 and P2. The command APDU is described in detail in Section 6.5.1, 'Structure of the Command APDU'.

Common Criteria

The Common Criteria are a criteria catalog for the evaluation and certification of information technology systems. In the future, they should replace national and international criteria catalogs, such as TCSEC and ITSEC. Version 1.0 of the Common Criteria [*www.commoncriteria.org*] was published in 1996 by the NIST. Since then, they have been internationally standardized as ISO 15408. The currently valid revision is Version 2.0 of 1998.

Compiler

A compiler is a program that translates a program written in a language such as BASIC or C into machine language that can be executed by a processor. After the compilation process, it is normally necessary to link the code using a linker program.

Completion

Completing the operating system by loading the portion that is located in the EEPROM. This allows the operating system to be modified and adapted after the chips have been manufactured, without requiring the production of a new ROM mask. Identical data are written to each Smart Card during completion, so it is in principle a sort of initialization.

Core foil

See Internal foil.

COS

The designation 'COS' (card operating system) for a Smart Card operating system has taken root throughout the world, and it is frequently seen in product names such as STARCOS, MPCOS and the like. See also Chapter 5, 'Smart Card Operating Systems'.

CRC

A cyclic redundancy check (CRC) is a simple and widely used form of error detection code (EDC) for the protection of data. The CRC must be determined based on an initial value and a divider polynomial before it can be used.

Contactless Smart Card

Card with which energy and data are transferred without contact via electromagnetic fields.

Credit card

A card, with or without a chip, that indicates that the holder has been extended credit, and with which payment takes place some time after the goods or services have been received ('buy now, pay later'). Embossed credit cards are typical examples.

Cryptographic algorithm

A cryptographic algorithm is a computational rule with at least one secret parameter (the key) that can be used to encrypt or decrypt data. There are symmetric cryptographic algorithms (such as the DES algorithm) that use the same key for encryption and decryption, and asymmetric cryptographic algorithms (such as the RSA algorithm) that use a public key for encryption and a private (secret) key for decryption.

Debit card

A card, with or without a chip, that indicates that the holder has been extended credit, but with which payment takes place when the goods or services are received ('pay now').

Debugging

Debugging means searching for and eliminating errors, with the objective of detecting and correcting as many errors in a software program as possible. Debugging is normally carried out by software developers, and is not the same thing as testing.

Delamination

Delamination is the undesired separation of foils that have been attached to each other (laminated) using heat and pressure. Delamination of a card can, for example, be caused by printing overly large areas between the core and overlay foils with non-thermoplastic ink, such as typically used in offset printing.

Deterministic

A process or procedure is said to be deterministic if it always produces the same results for a given set of initial conditions. The opposite of this is 'probabilistic'.

DF name

The DF name, like the FID, is a characteristic of a DF. It has length of 1–16 bytes. It is used for selecting the DF, and it can contain a registered AID (application identifier) that is 5–16 bytes long and makes the DF internationally unique. See also Section 15.4, 'Registration Authorities for RIDs'.

DF

A DF (dedicated file) is a directory in the file system of a Smart Card. The root directory (MF) is a type of DF.

Die, dice

A die (plural dice) is a small flat piece of crystalline silicon on which a single semiconductor integrated circuit is fabricated (such as a microcontroller).

Differential fault analysis (DFA)

The principle of differential fault analysis was published in 1996 by Dan Boneh, Richard A. DeMillo and Richard J. Lipton, who were all employees of Bellcore [Boneh 96]. The procedure is based on intentionally introducing scattered errors into a cryptographic computation in order to determine the secret key. In the original procedure, only public-key algorithms were named, but within a few months this method of attack was very quickly further developed [Anderson 96a], with the result that presently all cryptographic algorithms can in principle be attacked in this manner if they do not have special protective measures.

Differential cryptoanalysis

Differential cryptoanalysis is a method for computing the value of a secret key by using plaintext–ciphertext pairs with certain differences but the same key. The manner in which these differences propagate with additional DES rounds is analyzed to determine the key.

Digital signature

A digital signature is used to establish the authenticity of an electronic message or document. Digital signatures are usually based on asymmetric cryptographic algorithms, such as the RSA algorithm. The legal validity of digital signatures is regulated by law in many countries (as by the *Signaturgesetz* in Germany, for example). Digital signatures are sometimes referred to as 'electronic signatures'.

Digital fingerprint

This term is often used to refer to the hash value of a message (e.g. generated using SHA-1).

Divider

A commonly used short form for 'clock-rate conversion factor' (*qv*).

Download

Transferring data from a higher-level system (background or host system) to a lower-level system (e.g. terminal). The opposite of this is 'upload'.

DRAM

Dynamic random access memory (DRAM) is a type of RAM that has a dynamic structure that requires both a constant power supply and periodic refreshing in order to preserve its contents. DRAM cells are effectively capacitors. DRAM takes up less space on the chip than SRAM and is thus less expensive, but SRAM has a shorter access time.

Dual-interface card

The term 'dual-interface card' refers to a Smart Card that has both contactless and contact-type interfaces for transferring data to and from the card. The term 'combicard' is also used.

Dual-slot cellphone

This refers to a mobile phone that has a second externally accessible contact unit for an ID-1 Smart Card, in addition to the user card (e.g. SIM). With a dual-slot cellphone, it is for example possible to use existing electronic purse Smart Cards with the phone to make payments via the cellular phone network.

Duplicate

This means transferring genuine data to a second card with the objective of producing one or more identical (cloned) cards. 'Duplicate' usually means the same thing as 'clone'.

ECC

An error correction code (ECC) is a checksum for data. With an ECC, errors in the data can be detected with a certain probability, and sometimes also fully corrected. The acronym ECC also stands for 'elliptic curve cryptosystem'.

e-commerce

This is an abbreviation of 'electronic commerce', which refers to all forms of service and trade that utilize public networks (primarily the Internet), as well as the payment traffic based on them.

EDC

An error detection code (EDC) is a checksum for data. With an EDC, errors in the data can be detected with a certain probability. Typical examples of EDCs are the XOR and CRC checksums used in various data transmission protocols.

EEPROM

EEPROM (electrically erasable programmable read-only memory) is a type of non-volatile memory that is used in Smart Cards. An EEPROM is divided into memory pages and its contents can be altered or erased, but there is a physically determined upper limit to the number of write or erase accesses.

EF

EFs (elementary files) represent the actual data storage elements of the file tree of a Smart Card. An EF can have either the property 'working' (for use by the terminal) or the property 'internal' (for use by the Smart Card operating system). An EF has an internal structure (transparent, linear fixed, linear variable, cyclic, ...).

Electronic purse (e-purse)

An electronic purse is a card with a chip that must be loaded with an amount of money before it can be used for making payments ('pay before'). Some typical examples are the German *Geldkarte*, the Visa Cash card and the Mondex card. Electronic purses may also allow purse-to-purse transactions.

Embossing

Part of the physical personalization process, in which raised characters are impressed into the plastic card body so that they stand proud of the surface of the card.

EMV (EMV specification)

General specifications for financial transaction cards with chips and their associated terminals, prepared by Europay, Mastercard and Visa. These specifications have achieved the status of international industry standards for credit and debit cards and electronic purses. In the financial transaction area, they are thus the counterpart to the GSM11.11 telecommunications standard.

Endianness

The term 'endianness' indicates the sequence of the bytes within a byte string. 'Big-endian' means that the most significant byte stands at the beginning of the byte string, which consequently means that the least significant byte stands at the end of the string. 'Little-endian' refers to the opposite order, which means that the least significant byte comes first and the most significant bit comes last.

EPROM

An EPROM (erasable programmable read-only memory) is a type of non-volatile memory that is now very rarely used in Smart Cards. It can only be erased by ultraviolet light, so it can be used only for WORM storage (write once, read multiple) in Smart Cards.

ETS

European Telecommunication Standard (ETS) refers to standards issued by ETSI. They are primarily concerned with European telecommunications.

ETSI

The European Telecommunications Standards Institute, based in Sophia Antipolis, France, is the standardization body of the European telecommunication companies. It looks after defining standards in the area of European telecommunications. The most important ETSI standard in the area of Smart Cards is the series of GSM standards (GSM 11.11 ff).

etu (elementary time unit)

An elementary time unit is the duration of one bit in a data transfer to a Smart Card. The absolute duration of an etu is not fixed; instead, it is defined in terms of the frequency of the clock signal applied to the card and the clock-rate conversion factor ('divider').

Fault tree analysis

Fault tree analysis refers to a testing method in which every program execution path in the program code is traversed in order to search for possible faults.

FID

The file identifier (FID) is a two-byte characteristic of a file. Each MF, DF and EF has a FID. The FID of the MF is always '3F00'. See also Section 5.6, 'Smart Card Files'.

FIPS

The term 'Federal Information Processing Standard' (FIPS) refers to American standards that are issued by the NIST.

Floor limit

A floor limit relates to the level of a purchase for which authorization by a third party is required. No authorization is necessary below this limit, but authorization must always be obtained for purchases above the limit, since otherwise payment may not be possible and is not guaranteed.

Garbage collection

Garbage collection is a function that collects memory that is no longer used by an application and makes it available as free memory. In the past, garbage collection was implemented as an interrupt to normal program execution. In modern computer systems, garbage collection is a low-priority thread that constantly searches the memory for regions that are no longer in use and returns them to the free memory pool.

Greybox test

A greybox test is a mixed form that combines elements of blackbox and whitebox tests, in which the entity performing the test knows some of the internal processes, functions and mechanisms of the software to be tested.

GSM

The 'Global System for Mobile Communications' is a specification for an international, terrestrial mobile telephone system. Although originally intended for use in a few countries of central Europe, it has developed into an international standard for mobile telephones. The designated successor of GSM will be UMTS (Universal Mobile Telecommunications System).

Hard mask

The term 'hard mask' means that the complete program code is largely contained in the ROM. This saves space in comparison to a soft mask, since ROM cells are significantly smaller than EEPROM cells. However, it has the disadvantage that an actual exposure mask must be generated for the semiconductor manufacturing process. The turnaround time is thereby increased considerably in comparison with a soft mask. Hard masks are normally used with large numbers of chips for Smart Cards that have largely identical functionality. The opposite of a hard mask is a soft mask, with which essential functions are in EEPROM.

Hash function

A hash function is a procedure for compressing data by means of a one-way function, so that it is not possible to recompute the original data. A hash function produces a fixed-length result for any arbitrary input value, and it is designed so that any change to the input data has a very high probability of effecting the computed hash value (output value). SHA-

1 is a typical representative of hash algorithms. The result of a hash function is a hash value, which is often also referred to as a digital fingerprint.

Hologram

A photographic exposure made using a holographic process is called a hologram. It is a three-dimensional image of the photographed object. The object in the photograph can thus be seen from different angles, depending on the viewing angle of the observer. The holograms that are normally used with Smart Cards are embossed holograms, which produce reasonably satisfactory three-dimensional images.

Hot list

A list in a database that notes all Smart Cards that probably have been manipulated and must not be accepted under any circumstances.

Hybrid card

Refers to a card with two different card technologies. Typical examples are cards with both magnetic strips and chips, or Smart Cards with optical storage layers on their surfaces.

ID-l card

Standard Smart Card format (length ≈ 85.6 mm, width ≈ 54 mm, thickness ≈ 0.76 mm).

Identification

Procedure for proving the genuineness of a device or a person by comparing a presented password to a stored reference password.

IEC

The International Electrotechnical Commission [IEC] was founded in 1906 and is based in Geneva. Its job is to establish international standards for electrical and electronics technology.

Initializer

A body that performs initialization.

Initialization

Loading the fixed, person-independent data of an application into the EEPROM. A synonym for this is 'pre-personalization'.

Intelligent memory card

A memory card with additional logic circuitry for extra security functions that monitor memory accesses.

Internal foil

An internal foil is one of the foils that is located inside the stack of foils that are laminated together to make a card. It is therefore sometimes also called a 'core foil'. Normally, the internal foil is laminated between two cover foils, and these three foils together form the card. The internal foil often carries security features or electrical components, such as the coils for contactless Smart Cards.

Interpreter

An interpreter is a program that carries out 'run-time' translation of the instructions of a program language such as BASIC or Java into machine language instructions that can be executed by a processor. Each translated instruction is executed immediately upon being translated. An interpreted program always runs more slowly than compiled program code, due to the fact that the translation takes place while the program is running. However, interpreters allow a significantly higher degree of hardware-independent programming than do compilers.

ISO

The International Organisation for Standardisation [ISO] was founded in 1947 and is based in Geneva. Its task is to support the establishment of international standards, in order to enable the free exchange of goods and services. The first ISO standard was published in 1951; it deals with temperatures with regard to length measurements.

ITSEC

The Information Technique System Evaluation Criteria (ITSEC), published in 1991, are a catalog of criteria for the evaluation and certification of the security of information technology systems in Europe. The further development of the ITSEC and its combination with various national criteria resulted in the Common Criteria.

ITU

The International Telecommunications Union (ITU) is an international organization for the coordination, standardization and development of global telephone services, based in Geneva. It is the successor to the CCITT [ITU].

Java

Java is a hardware-independent, object-oriented programming language developed by the Sun Corporation. It is widely used on the Internet. Java source code is translated into a standardized bytecode by a compiler, and the bytecode is then normally interpreted by a 'virtual machine' built on the target hardware (Intel, Motorola, ...) and operating system platform (Windows, MacOS, Unix, ...). There are already processors (such as PicoJava) that can directly execute Java bytecode.

Key fault presentation counter

See Retry counter.

Key management

This refers to all administrative functions relating to the generation, distribution, storage, updating, destruction and addressing of cryptographic keys.

Kinegram

A kinegram shows different images when viewed at different angles. It can show an apparently 'moving' image that changes in jerks, or it can show two completely different images, depending on the viewing angle. Kinegrams are similar to holograms, which show three-dimensional images, but not identical to them.

Lamination

The process of gluing together thin sheets of material using heat and pressure.

Laser engraving

A process for blackening special plastic layers by heating them with a laser beam. This is also colloquially referred to as 'lasing'.

Linker

The job of a linker is to convert the symbolic memory addresses of compiled or assembled program code into absolute or relative memory addresses.

Little-endian

See Endianness.

Load agent

The load agent is the body that carries out the loading of electronic money into an electronic purse. It is in a manner of speaking the counterpart to the service provider, which can remove money from the purse in exchange for goods or services.

MAC

A MAC (message authentication code) is a cryptographic checksum for data that allows manipulation of the data during the transfer process to be recognized. If a MAC is used to protect stored data, the term CCS (cryptographic checksum) is used instead.

Magnetic card

A commonly used but technically incorrect short form of 'magnetic-strip card'.

Magnetic-strip card

A card with a magnetic strip on which data may be read and subsequently read. The magnetic strip normally holds three data tracks with differing data recording densities. Tracks 1 and 2 are only read after the card has been issued, while data may also be written to track 3 while the card is in normal use. The magnetic material in the strip may have either a high-coercivity characteristic or a low-coercivity characteristic.

Memory card

A card with a chip that has a simple logic circuit with additional memory that can be read and/or written. Memory cards can also have supplementary security logic blocks, which for example can make authentication possible.

MF

The MF, or master file, of a Smart Card file system is a special type of file. It is the root directory of the file tree and is automatically selected when the Smart Card is reset.

Microprocessor card

A microprocessor card is a card with a microcontroller chip, which contains a CPU, volatile memory (RAM) and non-volatile memory (ROM, EEPROM and the like). A microprocessor card can also contain a numerical coprocessor (NPU) so that it can quickly execute public-key cryptographic algorithms. Cards of this type are sometimes referred to as 'cryptocards' or 'cryptocontroller cards'.

Module

The component that carries and supports a die, with a set of contact elements arranged on its surface, is called a module.

Module manufacturer

An entity that attaches dice to blank modules and produces the electrical connections between the die and the module contacts.

Mono-application Smart Card

A Smart Card that contains only one application.

Monofunctional Smart Card

A monofunctional Smart Card is a processor chip card whose operating system supports only one particular application, and which may even be optimized for this application. Management functions for applications, such as creating and deleting files, are supported either in a very limited form or not at all by such cards.

MoU

The Memorandum of Understanding (MoU) is the common legal basis for all GSM network operators.

Multi-application Smart Card

A multi-application Smart Card is one that contains more than one application, such as a bank card that also has a telephone function.

Multifunctional Smart Card

The term 'multifunctional Smart Card' normally refers to a processor chip card that supports more than one application and that has suitable management functions for loading and deleting applications and files. However, this term is used in such an inflated sense that nowadays there is hardly any Smart Card operating system that cannot be upgraded to be 'multifunctional'. Some humorists like to assert that every card is fundamentally multifunctional, since they all can at least be used to scrape the ice off a frosted windshield.

Multitasking

A computer system that supports multitasking allows several programs to be run quasi-simultaneously. Each of the programs that can be run in parallel is usually located in a separate address space that is protected against access by other programs. The programs that run in parallel can exchange data with each other only by means of special mechanisms. Multitasking is not the same as multithreading, in which a single program performs several different tasks quasi-simultaneously. A computer system may support both multitasking and multithreading.

Multithreading

A computer system that supports multitasking allows a single program to perform several different tasks quasi-simultaneously. The individual threads of a program normally utilize a common address space. Multithreading is not the same as multitasking, in which several different programs run in parallel, each with its own separate address space. A computer system may support both multitasking and multithreading.

Native code

Native code means a program whose instructions are in the particular machine language of the processor that executes the program.

NBS

Before 1988, the NIST was known as the National Bureau of Standards (NBS).

NCSC

The American National Computer Security Center (NCSC) is a subagency of the NSA. It is responsible for testing security products, and it publishes criteria for secure computer systems, including the TCSEC.

Negative file

See Black list.

Negative result

The case in which a logical decision leads to a bad or undesired result.

Nibble

The four most significant or least significant bits of a byte.

NIST

The American National Institute of Standards and Technology (NIST) is a division of the US Department of Commerce and is responsible for the national standardization of information technology. It was known as the NBS until 1988. The NIST publishes the FIPS standards.

Noiseless

A property of a cryptographic algorithm that always takes the same amount of time to encrypt or decrypt data, regardless of the key, plaintext and ciphertext involved. If a cryptographic algorithm is not noiseless, the size of the key space can be markedly reduced by analyzing the processing time behavior of the algorithm. This allows the key to be determined significantly faster than with a brute-force attack.

Non-repudiation

The non-repudiability of a message refers to a cryptographic procedure that ensures that the recipient of a message cannot refuse to acknowledge (repudiate) the contents of the message. The sender of the message can thereby prove that the person to whom the message was sent actually received the message. Non-repudiation is thus equivalent to a registered letter with return receipt in ordinary postal systems.

Non-volatile memory

A type of memory (such as ROM or EEPROM) that retains its contents even without power.

NSA

The American National Security Agency (NSA) is the official US government institution for communication security. It reports directly to the Department of Defense, and one of its functions is to monitor foreign communications and decode them. The development of new cryptographic algorithms and the restriction of the use of existing algorithms also fall under the authority of this agency.

Numbering

Numbering is the process of embossing or printing numbers on a Smart Card. It is typically used in the production of anonymous prepaid phone cards in order to give each card a visible and unique identification number.

One-way function

A one-way function is a mathematical function that is easily computed but whose inverse function requires a large amount of computational effort.

Open application

An application in a Smart Card that is available to various service providers (such as merchants and vendors of services) without requiring a mutual legal relationship.

Open purse

The implementation of an open application for an electronic purse. It can be used for general payment transactions with various service providers.

Operating system producer

An entity that programs and tests an operating system.

Optical memory card

Card in which information is 'burnt' into a reflective surface layer (similar to a CD).

Padding

Padding means extending a data string with filler data in order to bring the string to a particular length. The length of the resulting string most often must be an integral product of a certain block size (such as 8 bytes) so that it can be further processed, for example by a cryptographic algorithm.

Page-oriented

Refers to memory structures in which a number of bytes are organized into a 'page', which can be written or erased only as a group. In Smart Card microcontrollers, only the EEPROM is page-oriented. The usual size of a memory page is 4 or 32 bytes. However, nowadays there are microcontrollers that have a page orientation that can vary within certain limits, such as 1–32 bytes, instead of a fixed orientation.

Passivation

A passivation layer is a protective layer on top of a semiconductor chip that protects it against oxidation and other chemical processes. The passivation layer must be removed before the semiconductor can be manipulated.

Patch

In software development terms, a patch is a short program that extends or modifies the behavior of an existing program. It is often written in machine code. Patches are usually used to make quick and uncomplicated corrections to program errors.

Patent

A patent is a document that gives an inventor the right to the exclusive exploitation of his or her invention for a limited amount of time and in one or more countries. The maximum term of a patent is usually 20 years.

Pay before

The expression 'pay before' refers to the money flow for cards that are used for payment transactions. The 'real' money flows out of the card owner's account before the goods or services are actually purchased. Typical representatives of 'pay before' cards are electronic purse cards, which the user must load with electronic money prior to making a purchase.

Pay later

The expression 'pay later' refers to the money flow for cards that are used for payment transactions. The 'real' money flows out of the card owner's account only some time after the goods or services are actually purchased. Typical representatives of 'pay later' cards are credit cards, with which it may take up to several weeks after the purchase before the money is transferred from the account of the payer to the account of the merchant.

Pay now

The expression 'pay later' refers to the money flow for cards that are used for payment transactions. The 'real' money flows out of the card owner's account at the same time as the goods or services are purchased. Typical representatives of 'pay now' cards are debit cards, such as the Eurocheque card, which allow the money to be transferred from the account of the payer to the account of the merchant at the time that the purchase occurs.

Personalizer

An entity that carries out personalization.

Personalization

The process in which a card is assigned to a person. This can take place by means of physical personalization (e.g. embossing or laser engraving) as well as by means of electronic personalization (loading personal data in the memory of the Smart Card).

PIN pad

In its original sense, a PIN pad is a terminal data-entry keypad that has special mechanical and cryptographic protection. In general usage, the entire terminal is often referred to as a PIN pad.

PKCS#1...11

The Public Key Cryptographic Standards (PKCS) are computation rules published by RSA Inc that relate to the use of asymmetric cryptographic algorithms, such as the RSA algorithm.

Plug-in

A Smart Card with a very small format, which is primarily used in GSM phones (length ≈ 25 mm, width ≈ 15 mm, thickness ≈ 0.76 mm).

Polling

Polling is the regular querying or sampling of an input channel under software control in order to detect the content of an incoming message. Depending on the repetition rate of the queries, polling can require a considerable amount of processing power. Normally, interrupt-driven sampling that is supported by the processor hardware is preferred to pure software polling.

POS

POS (point of sale) refers to any location at which a particular item or service is sold.

Positive result

The case in which a logical decision yields a good or intended result.

Pre-personalization

This is another name for initialization (*qv*).

Processor

The most important subassembly of a microcontroller. It executes the machine instructions in the order defined by the program and performs memory accesses. The term CPU (central processing unit) is often used as a synonym for 'processor'.

Processor card

This is a short form of 'microprocessor card' (*qv*).

Purse holder

A person who possesses a Smart Card containing an electronic purse.

Purse provider

The organization that is responsible for the overall functionality and security of an electronic purse system. This is usually the issuer of the electronic money for the cards. The purse provider normally also guarantees the redemption of the electronic money.

Purse-to-purse transaction

Transfer of electronic monetary units from one electronic purse directly to another, without the intervention of a third, higher-level system. Normally, this functionality means that the purse system must operate anonymously and that the electronic purses must use a single common key for this function.

RAM (random-access memory)

A type of volatile memory that is used in Smart Cards as working storage. RAM loses its contents in the absence of power. SRAM and DRAM are types of RAM with special technical properties.

Record

A record or set of data is a certain quantity of data that is similar to a string.

Red list

See Hot list.

Reset

A reset means that the computer (in this context, a Smart Card) is restored to a clearly defined initial state. A 'cold reset' or 'power-on reset' means that the reset is initiated by switching the power off and then on again. A 'warm reset' is initiated by a signal on the reset lead of the Smart Card, without affecting the supply voltage.

Response APDU

The Smart Card sends a response APDU as its answer to a command APDU received from a terminal. The response APDU consists of optional response data and a mandatory 2-byte portion containing the status words SW1 and SW2. It is described in detail in Section 6.5, 'Message Structure: APDUs'.

Retry counter

A counter that accumulates negative results and determines whether a particular secret (PIN or key) may continue to be used. If the retry counter reaches its maximum value, the secret is blocked and can no longer be used. The retry counter is normally reset to zero when the operation is completed successfully (positive result).

ROM (read-only memory)

A type of non-volatile memory that is used in Smart Cards. It is mainly used to store programs and static data, since the contents of a ROM cannot be altered.

ROM mask

An exposure mask used to produce the ROM in the semiconductor fabrication process. This expression is also used to refer to the data contents of the ROM in a Smart Card microcontroller.

Sandbox

See Virtual machine.

Scrambling

An intentionally confusing layout of the address, data and control busses on a microcontroller chip, so that it is not possible to recognize the functions of individual bus lines without inside information. 'Static' scrambling means that the busses of a given series of chips are all scrambled in the same way. 'Dynamic' scrambling means that the busses are scrambled differently for each different chip.

Secure messaging

A method for protecting data transferred via an interface against manipulation (by means of a MAC, i.e. 'authentic mode') or eavesdropping (by means of encryption, i.e. 'combined mode').

Security module

An assembly that is secured both mechanically and computationally and is used for the storage of secret data. It is also called a secure application module (SAM) or a hardware security module (HSM).

Service provider

A service provider in a Smart Card system is an entity that offers services that are utilized and paid for by a card user. In the case of payment system with electronic purses, a service provider is the entity that receives money from the purse system owner in exchange for his electronic money.

Session

The interval between when the activation and deactivation sequences of a Smart Card, during which both the complete data exchange and the necessary computational mechanisms take place.

SET

The Secure Electronic Transaction (SET) standard is a protocol for payment transactions for carrying out secure credit card payments via the Internet, as defined by Visa and Mastercard. It does not necessarily require the payer to have a Smart Card, since it can be implemented fully in software on a PC. An extension of the SET, called C-SET (Chip-SET), is up to now only relevant inside France and is not yet internationally standardized.

Short FID

A Sort FID is a 5-bit identifier for an EF that can have a value from 1 through 31. It is used within a write or read command (such as READ BINARY) for the implicit selection of an EF.

Shutter

A shutter is a mechanical assembly in a terminal that cuts off any wires leading from the card. This is intended to prevent manipulation of the communications. If it is not possible to cut through any such wires, the inserted circuit card will not be electrically activated.

Signaturgesetz (SigG)

This refers to the German signature law (*Signaturgesetz*), or in full Article 3 of the *Gesetz zur Regelung der Rahmenbedingungen for Informations- und Kommunikationsdienste (Informations- und Kommuinkationsdienste-Gesetz – IuKDG)* of 13 June 1997. This legislation prescribes the general conditions for the use of digital signatures in Germany.

Signaturverordnung (SigV)

The German *Signaturverordnung* (signature regulation) of 8 October 1997 translates the general conditions prescribed by the *Signaturgesetz* into concrete terms, to the extent necessary to allow lists of specific measures to be generated as recommendations for the practical use of digital signatures. For example, the *Signaturverordnung* describes the necessary procedure for the generation of signature keys and identification data, as well as the necessary security concepts and the necessary testing stages for the signature components according to the ITSEC.

SIM

Subscriber Identity Module (SIM) is another name for a GSM-specific Smart Card. This can be the same size as a standard credit card (ID-1 format), or it can be a small plug-in card in the ID-000 format. The SIM carries the secret authentication data for the network operator and also contains user-specific data, such as the telephone number of the mobile phone (see also Section 13.5, 'The GXM Network'). The successor to the SIM in the UMTS is the USIM (*qv*).

SIMEG

The Subscriber Identification Module Expert Group (SIMEG) was a group of experts working within the framework of ETSI who defined the specifications for the interface between Smart Cards and mobile telephones (GSM 11.11). The name 'SIMEG' was replaced in 1994 by 'SMG9'.

Smart Card

Strictly speaking, the expression 'Smart Card' is an alternative name for a microprocessor card, in that it refers to a chip card that is 'smart'. Memory cards thus do not properly fall into the category of Smart Cards. However, the expression 'Smart Card' is generally used in English-speaking countries to refer to all types of chip cards.

Smartcard

The term 'Smartcard' is a registered trademark of the Canadian company Groupmark [Groupmark].

SMG9

The Special Mobile Group 9 (SMG9) is a group of experts working within the framework of ETSI who define the specifications of the interface between Smart Cards and mobile telephones (GSM 11.11). It is composed of representatives of card manufacturers, mobile phone manufacturers and network operators. The SMG9 was formerly called the SIMEG.

Soft mask

The term 'soft mask' means that part of the program code based on the Smart Card operating system in ROM is located in the EEPROM. Programs in the EEPROM can be easily modified by overwriting, which means that they are 'soft'. The term 'mask' in this case is actually not correct, since it is not necessary to produce a semiconductor exposure mask for a program stored in EEPROM. Soft masks are typically used for small numbers of cards (e.g. for field trials) in rapid prototyping. The opposite of a soft mask is a hard mask, with which the essential functions are in the ROM.

Specification

In this book, the work *specification* is used to refer to any document that resembles a standard (*qv*) but which is generated or issued by (for example) a company or an industrial group, rather than by a national or international standardization authority. The terms 'specification' and 'standard' are sometimes used interchangeably, but this is not actually correct.

SRAM (static random-access memory)

A static RAM needs only a constant supply of power to retain its contents; it does not need to be periodically refreshed. The access time of a SRAM is less than that of a DRAM, but SRAM takes up more space on the chip and is thus more expensive.

Stack

A stack is a data structure in which the most recently entered data object is the first to be retrieved (last in, first out). Probably the best known stack is the program stack, onto which return addresses are placed when subroutines are called.

Standard

A standard is a document that contains technical descriptions and/or precise criteria that are used as rules and/or definitions of characteristics and features, in order to ensure that materials, products , processes and services can be used for their intended purposes. In this book, the term *standard* is always used in connection with a national or international standardization authority (such as ISO, CEN, ANSI or ETSI). A standard is not the same as a specification (*qv*).

State machine

A part of a program that determines the course of a process by means of a predefined state diagram, which consists of specific states and state transitions.

Steganography

The objective of steganography is to conceal messages within other messages such that they no longer can be recognized by a naïve observer (man or machine). For example, a text could be encoded and hidden in an image file in such a way that it only marginally modifies the image, so that the changes to the image are practically unobservable.

Super Smart Card

The term 'Super Smart Card' refers to a Smart Card with integrated complex card elements such as a display and keypad.

TCSEC

The Information Technique System Evaluation Criteria (TCSEC) were published in 1985 by the NCSC. They are a catalog of criteria for the evaluation and certification of the security of information technology systems in the United States. The national TCSEC were followed by the internationally valid Common Criteria.

TDES

See Triple-DES.

Terminal

A terminal is the complement to a Smart Card. It is the device that allows the Smart Card to receive electrical power and exchange data. Some terminals have displays and keypads.

Testing

Testing means checking whether an already debugged program is in good working order. The primary objective of testing is not to look for errors in the program, but rather to check out the expected functions. Testing is thus not the same thing as debugging.

Thread

See Multithreading.

TLV format

A data format conforming to ASN.1, which uses a prefix label (*tag*) and a length code (*length*) to uniquely describe a particular data object (*value*). The TLV format also allows chained data objects.

TPDU

See APDU.

Transaction number

A transaction number (TAN) , in contrast to a PIN, is valid for only one transaction, which means that it can be used only once. Normally, the user receives several TANs printed on a slip of paper (as four-digit numbers, for example), and these must be used exactly in the prescribed order for the individual transactions or series of sessions.

Transfer card

A transfer card is a Smart Card that is used as a transport medium to carry data between two entities. It contains a large data memory for this purpose, and it normally contains keys for authenticating whether the data to be transferred are allowed to be read or written by the entity in question.

Transmission protocol

In the Smart Card world, the term 'transmission protocol' refers to the mechanisms used to send and receive data between a terminal and a Smart Card. A transmission protocol describes in detail the OSI protocol layers that are used

Transport protocol

An alternate name for transmission protocol (qv).

Trap door

A trap door is a mechanism in software or an algorithm that is intentionally included in order to allow security functions or protective mechanisms to be circumvented.

Triple-DES

The Triple-DES algorithm, which is also referred to as TDES and 3 DES, is a modified DES encryption. It consists of calling the DES algorithm three times in succession, with alternating encryption and decryption. If the same key is used for all three DES calls, the Triple-DES encryption corresponds to a normal DES encryption. However, if two or three different keys are used, the Triple-DES encryption is significantly stronger than a single DES encryption. See also Section 4.6.1, 'Symmetric cryptographic algorithms'.

Trojan horse

Historically, this refers to the wooden horse that enabled Odysseus to gain entry into the heavily fortified city of Troy. In modern usage, it refers to a program that performs a specific 'foreground' function but also can execute additional functions unbeknownst to the user. A Trojan horse is introduced purposely into a computer system or host program. In contrast to a virus, it cannot reproduce itself.

UIM (user identification module)

An outdated term for USIM (qv).

Unicode

Unicode is a further refinement of the well known ASCII code for characters. In contrast to the 7-bit ASCII code, Unicode employs 16 bits for coding. This allows the characters of the most widely used languages of the world to be supported. The first 256 Unicode characters are identical to the ISO 8859-1 ASCII characters. The WWW site of the Unicode consortium is [Unicode].

Upload

'Upload' means to transfer data from a lower-level system (such as a terminal) to a higher-level system (such as a background or host system). It is the opposite of 'download'.

Useful data

Data that are directly needed by an application.

User

A person who uses a Smart Card. This need not necessarily be the cardholder.

USIM

Universal Subscriber Identity Module (USIM) is another name for a UMTS-specific Smart Card. This can be the same size as a standard credit card (ID-1 format), or it can be a small plug-in card in the ID-000 format. The USIM carries the secret authentication data for the network operator and also contains user-specific data, such as the telephone number of the mobile phone.

Virgin card

A virgin card is one that has not been implanted with a chip and has not yet been optically or electronically personalized. It is essentially a printed, undistinguished card body, such as is made in the mass production of cards.

Virtual machine (VM)

A virtual machine is a microprocessor that is simulated in software. Among other things, it has its own opcodes for machine instructions and a (also simulated) address space. This makes it possible to generate software that is independent of the particular features of specific hardware. For example, the virtual address space of a VM can be many times larger than the actual address space that provided by the hardware. The term 'sandbox' is often used in the Java milieu to refer to the closed environment of the VM.

Visa Easy Entry (VEE)

Visa Easy Entry is a method for simple migration from magnetic-strip credit cards to credit cards with microcontroller chips. It is accomplished by storing the name of the cardholder and all the data from the magnetic strip in an EF under a DF that is reserved for Visa. When a payment is made using the credit card, the terminal reads the data that are needed for the transaction from the chip instead of from the magnetic strip. The advantage of this approach is that only the POS terminals have to be upgraded to include a Smart Card contact unit, while the entire background system can be used as before without any modifications.

Volatile memory

A type of memory (e.g. RAM) that retains its contents only as long as power is applied.

Warm reset

See Reset.

White list

A database list that notes all Smart Cards that are allowed to be used in a particular application.

Whitebox test

A whitebox test, which is often also called a 'glassbox' test, is one in which it is assumed that the entity performing the test has complete knowledge of all internal processes and data of the software to be tested.

Work-around

In software development, a work-around is something that circumvents a known problem by 'programming around' it. A work-around does not eliminate the actual error, but only eliminates its negative effects on the rest of the program. Work-arounds are typically made in the EEPROM portion of mask-programmed Smart Card operating systems, since the program code in the ROM cannot be changed after the chip has been manufactured.

WWW (W3)

The World Wide Web (WWW) is a part of the international Internet. It is best known for its ability to link any desired documents by means of hyperlinks and the integration of multimedia objects in documents.

X.509

The X.509 standard defines the structure and coding of certificates. It is internationally the most commonly used standard for certificate structures.

ZKA

The *Zentraler Kreditausschuß* (ZKA) in Germany is a committee that coordinates the electronic payment procedures of the German banks. The ZKA is composed of the following banking associations: the *Deutsche Sparkassen- und Giroverband* (DSGV), the *Bundesverband der Deutschen Volks- und Raiffeisenbanken* (BVR), the *Bundesverband deutscher Banken* (BdB) and the *Verbund öffentlicher Banken* (VÖB). The chairman of the ZKA is chosen from each of the four member associations in yearly rotation.

15.2   Literature

The following publications are sorted first by the last name of the author and then in ascending order of the date of publication. Publications that appeared in newsgroups or discussion forums on the Internet have 'Internet' as the source.

[Anderson 92]	Ross J. Anderson: *Automatic Teller Machines*, Internet, December 1992
[Anderson 96a]	Ross J. Anderson, Markus G. Kuhn: *Improved Differential Fault Analysis*, Internet, November 1996
[Anderson 96b]	Ross J. Anderson, Markus G. Kuhn: *Tamper Resistance – a Cautionary Note*, USENIX Workshop, November 1996
[Bellare 95a]	Mihir Bellare, Juan Garay, Ralf Hause, Amir Herzberg, Hugo Krawczyk, Michael Steiner, Gene Tsudik, Michael Waidner: *iKP – A Family of Secure Electronic Payment Protocols*, Internet, 1995
[Bellare 95b]	Mihir Bellare, Philip Rogaway: *Optimal Asymmetric Encryption – How to encrypt with RSA*, Internet, 1995
[Bellare 96]	Mihir Bellare, Philip Rogaway: *The Exact Security of Digital Signatures – How to Sign with RSA and Rabin*, Internet, 1996
[Beutelsbacher 93]	Albrecht Beutelsbacher: *Kryptologie*, 3rd edition, Vieweg Verlag, Braunschweig 1993

[Beutelsbacher 96]   Albrecht Beutelsbacher, Jörg Schwenk, Klaus-Dieter
                            Wolfenstetter: *Moderne Verfahren der Kryptografie*,
                            Vieweg Verlag, Braunschweig, 1996

[Biham 91]   Eli Biham, Adi Shamir: *Differential Cryptoanalysis of
                            DES-like Cryptosystems*, Journal of Cryptology, Vol. 4,
                            No. 1, 1991

[Biham 93]   Eli Biham, Adi Shamir: *Differential Cryptoanalysis of the
                            Data Encryption Standard*, Springer-Verlag, New York,
                            1993

[Biham 96]   Eli Biham, Adi Shamir: A New Cryptoanalytic Attack on
                            DES, Internet, 1996

[BIS 96]   Bank for International Settlements: *Security of Electronic
                            Money – Report by the Committee on Payment and
                            Settlement Systems and the Group of Computer Experts of
                            the Central Banks of the Group of Ten Countries*,
                            Basel, August 1996

[Boneh 96]   Dan Boneh, Richard A. DeMillo, Richard J. Lipton: *On
                            the Importance of Checking Computations*, Internet, 1996

[Bronstein 96]   I. N. Bronstein, K. A. Semendjajew:
                            Taschenbuch der Mathematik, 7th edition,
                            B. G. Teubner Verlagsgesellschaft, Leipzig, 1997

[Buchmann 96]   Johannes Buchmann: *Faktorisierung großer Zahlen*,
                            Spektrum der Wissenschaft, September 1996

[Common Criteria 98]   Common Criteria, Version 2.0, NIST,
                            Gaitersburg, October 1998

[Dhem 96]   J. F. Dhem, D. Veithen, J.-J. Quisquater: *SCALPS:
                            Smart Card Applied to Limited Payment Systems*,
                            UCL Crypto Group Technical Report Series,
                            Université Catholique de Louvain, 1996

[Dictionary of Computing 91] *Dictionary of Computing*, Oxford University Press,
                            Oxford, 1991

[Diffie 76]   Whitfield Diffie, Martin E. Hellman: *New Directions
                            in Cryptography*, Internet, 1976

[Eberspächer 97]   Jörg Eberspächer, Hans-Jörg Vögel: *GSM – Global
                            System for Mobile Communication*,
                            B.G. Teubner Verlag, Stuttgart, 1997

[EG 91]   Europäische Gemeinschaften – Kommission: *Kriterien
                            für die Bewertung der Sicherheit von Systemen der
                            Informationstechnik (ITSEC)*, Version 1.2, June 1991

[Fenton 96]   Norman E. Fenton, Shari Lawrence Pfleeger: *Software Metrics*, Thomson Computer Press, London, 1996

[Finkenzeller 98]   Klaus Finkenzeller: *RFID-Handbuch*, Carl Hanser Verlag, München/Wien, 1998

[Franz 98]   Michael Franz: *Java – Anmerkungen eines Wirth-Schülers*, Informatik Spektrum, Springer-Verlag, Berlin, 1998

[Freeman 97]   Adam Freemann, Darrel Ince: *Active Java – Object Oriented Programming for the World Wide Web*, Addison-Wesley, Reading, MA, 1997

[Fumy 94]   Walter Fumy, Hans Peter Ries: *Kryptographie*, 2nd edition, R. Oldenbourg Verlag, München/Wien, 1994

[Gentz 97]   Wolfgang Gentz: *Die elektronische Geldbörse in Deutschland*, Diplomarbeit an der Fachhochschule München, München, 1997

[Glade 95]   Albert Glade, Helmut Reimer, Bruno Struif: *Digitale Signatur*, Vieweg Verlag, Braunschweig, 1995

[Gosling 95]   James Gosling, Henry McGilton: *The Java Language Environment – A White Paper*, Sun Microsystems, USA, 1995

[Grün 96]   Herbert Grün: *Card Manufacturing Materials and Environmental Responsibility*, Presentation at CardTech/SecurTech, Atlanta, GA, May 1996

[GSM 95]   Proceedings of the Seminar for Lation America Decision Makers by GSM MoU Association and ECTEL: *Personal Communication Services based on the GSM Standard*, Buenos Aires 1995

[Gutmann 96]   Peter Gutmann: *Secure Deletion of Data from Magnetic and Solid-State Memory*, USENIX Konferenz, San Jose, CA, 1996

[Gutmann 98a]   Peter Gutmann: *Software Generation of Practically Strong Random Numbers*, Internet, 1998

[Gutmann 98b]   Peter Gutmann: *X.509 Style Guide*, Internet, 1998

[IC Protection 97]   *Common Criteria for IT Security Evaluation Protection Profile – Smartcard Integrated Circuit Protection Profile*, Internet, 1997

[Isselhorst 97]   Hartmut Isselhorst: *Betreiberorientierte Sichrheitsanforderungen für Chipkarten-Anwendungen*, Card-Forum, Lüneburg, 1997

[Kaliski 93]   Burton S. Kaliski Jr.: *A Layman's Guide to a Subset of ASN.1, BER and DER*,
                        RSA Laboratories Technical Note, Internet, 1993

[Kaliski 96]   Burton S. Kaliski Jr.: *Timing Attacks on Cryptosystems*,
                        RSA Laboratories, Redwood City, CA, 1996

[Karten 97]   Zeitschrift cards: *Zur Sicherheit der ec-Karte PIN: Das Urteil des OLG Hamm*,
                        Fritz Knapp Verlag, Frankfurt, August 1997

[Knuth 97]   Donald Ervin Knuth: *The Art of Computer Programming, Volume 2:Seminumerical Algorithms*, 3rd edition,
                        Addison-Wesley/Longman, Reading, MA, 1997

[Kocher 95]   Paul C. Kocher: *Timing Attacks on Implementations of Diffie-Hellmann, RSA, DSS, and Other Systems*, Internet, 1995

[Kocher 98]   Paul C. Kocher, Joshua Jaffe, Benjamin Jun: *Introduction to Differential Power Analysis and Related Attacks*,
                        Internet, 1998

[Lender 96]   Friedwart Lender: *Production, Personalisation and Mailing of Smart Cards – A Survey*, Smart Card Technologies and Applications Workshop, Berlin, November 1996

[Lindholm 97]   Tim Lindholm, Frank Yellin: *Java – Die Spezifikation der virtuellen Maschine*, Addison-Wesley, Reading, MA, 1997

[Massey 88]   James L. Massey: *An Introduction to Contemporary Cryptology*, Proceedings of the IEEE, Vol. 76, No. 5, May 1988, pp 533–549

[Massey 97]   James L. Massey: *Cryptography, Fundamentals and Applications*, 1997

[Meister 95]   Giesela Meister, Eric Johnson: *Schlüsselmanagement und Sicherheitsprotokolle gemäß ISO/SC 27 – Standards in Smart Card-Umgebungen,* in: Albert Glade, Helmut Reimer, Bruno Struif: *Digitale Signatur,*
                        Vieweg Verlag, Braunschweig, 1995

[Menezes 93]   Alfred J.Menezes:
                        Elliptic Curve Public Key Cryptosystems,
                        Kluwer Academic Publishing, Boston, MA, 1993

[Menezes 97]   Alfred J. Menezes, Paul C. van Oorschot, Scott A. Vanstone: *Handbook of Applied Cryptography*,
                        CRC Press, Boca Raton, FL, 1997

[Merkle 81]	Ralph C. Merkle, Martin E. Hellman: *On the Security of Multiple Encryption*, Internet, 1981
[Meyer 82]	Carl H. Meyer, Stephen M. Matyas: *Cryptography*, John Wiley & Sons, New York, 1982
[Meyer 96]	Carsten Meyer: *Nur Peanuts – Der Risikofaktor Magnetkarte*, c't, July 1996
[Müller-Maguhn 97 a]	Andy Müller-Maguhn: *„Sicherheit" von EC-Karten*, Die Datenschleuder, Ausgabe 53, 1997
[Müller-Maguhn 97 b]	Andy Müller-Maguhn: *EC-Karten Unsicherheit*, Die Datenschleuder, Ausgabe 59, 1997
[Myers 95]	Glenford J. Myers: *The Art of Software Testing*, 5th edition, John Wiley & Sons, New York, 1995
[Nebelung 96]	Brigitte Nebelung: *Das Geldbörsen-Konzept der ec-Karte mit Chip*, debis Systemhaus, Bonn, 1996
[Odlyzko 95]	Andrew. M. Odlyzko: *TheFuture of Integer Factorization*, AT&T Bell Laboratories, 1995
[Otto 82]	Siegfried Otto: *Echt oder falsch? Die maschinelle Echtheitserkennung*, Betriebswirtschaftliche Blätter, Heft 2, February 1982
[Peyret 97]	Patrice Peyret: *Which Smart Card technologies will you need to ride the Information Highway safely?*, Gemplus, 1997
[Pfaffenberger 97]	Bryan Pfaffenberger: *Dictionary of Computer Terms*, Simon & Schuster/Macmillan, New York, 1997
[Piller 96]	Ernst Piller: *Die „ideale" Geldbörse für Europa*, Card-Forum, Lüneburg, 1996
[Pomerance 84]	C. Pomerance: *The Quadratic Sieve Factoring Algorithm*, Advances in Cryptology – Eurocrypt 84
[Press 92]	William H. Press, Saul A. Teukolsky, William T. Vetterling, Brian P. Flannery: *Numerical Recipes in C – The Art of Scientific Computing*, 2nd edition, Cambridge University Press, Cambridge, 1992
[Rivest 78]	Ronald L. Rivest, Adi Shamir, Leonard Adleman: *Method for Obtaining Digital Signatures and Public-Key Cryptosystems*, Internet, 1976
[Robertson 96]	James Robertson, Suzanne Robertson: *Vollständige Systemanalyse*, Carl Hanser Verlag, München, Wien 1996

[Rother 98 a]   Stefan Rother: *Prüfung von Chipkarten-Sicherheit,*
                         Card-Forum, Lüneburg 1998

[Rother 98 b]   Stefan Rother: *Prüfung von Chipkarten-Sicherheit,*
                         in *Tagungsband Chipkarten,* Vieweg Verlag,
                         Braunschweig 1998

[RSA 97]   RSA Data Security Inc.: *DES Crack Fact Sheet,*
                         Internet, 1997

[Schief 87]   Rudolf Schief: *Einführung in die Mikroprozessoren und
                         Mikrocomputer,* 10th edition,
                         Attempto Verlag, Tübingen, 1987

[Schindler 97]   Werner Schindler: *Wie sicher ist die PIN?,* speech
                         presented at the conference „Kreditkartenkriminalität",
                         Heppenheim, October 1997

[Schlumberger 97]   Schlumberger: *Cyberflex – Programmers Guide,* Version
                         6d, April 1997

[Schneier 96]   Bruce Schneier: *Applied Cryptography,* 2nd edition,
                         John Wiley & Sons, New York, 1996

[Sedgewick 97]   Robert Sedgewick: *Algorithmen,* 3rd edition, Addison-
                         Wesley, Bonn/München/Reading, MA, 1997

[SigG 97]   Gesetz zur Regelung der Rahmenbedingungen for
                         Informations- und Kommunikationsdienste (Informations-
                         und Kommunikations-Gesetz – IuKDG), 13. June 1997,
                         Artikel 3 – Gesetz zur digitalen Signatur
                         (Signaturgesetz – SigG)

[Silverman 97]   Robert D. Silverman:
                         Fast Generation of Random, Strong RSA Primes,
                         RSA Laboratories Crypto Byte, Internet, 1997

[Simmons 92]   Gustavus J. Simmons (Editor): *Contemporary
                         Cryptology,* IEEE Press, New York, 1992

[Simmons 93]   Gustavus J. Simmons: *The Subliminal Channels in the
                         U.S. Digital Signature Algorithm,* Proceedings of
                         Symposium on State and Progress of Research in
                         Cryptography, Rome, 1993

[Sommerville 90]   Ian Sommerville: *Software Engineering,*
                         Addison-Wesley, Wokingham, 1990

[Stix 96]   Gary Stix: *Herausforderung „Komma eins",*
                         Spektrum der Wissenschaft, February 1996

[Stocker 98]   Thomas Stocker: *Java for Smart Cards* in Tagungsband
                         Smart Cards, Vieweg Verlag, Braunschweig, 1998

[Tannenbaum 95]   Andrew S. Tannenbaum: *Moderne Betriebssysteme*,
                         2nd edition, Carl Hanser Verlag, München/Wien ,1995

[Tietze 93]   Ulrich Tietze, Christoph Schenk:
                         Halbleiter-Schaltungstechnik, 10th edition,
                         Springer-Verlag, Berlin, 1993

[Vedder 97]   Klaus Vedder, Franz Weikmann: *Smart Cards –
                        Requirements, Properties and Applications*, ESAT-
                        COSIC course, Katholische Universität Leuven, 1997

[Weikmann 92]   Franz Weikmann: *SmartCard-Chips – Technik und
                        weitere Perspektiven*, Der GMD-Spiegel 1'92,
                        Gesellschaft for Mathematik und Datenverarbeitung,
                        Sankt Augustin, 1992

[Weikmann 98]   Franz Weikmann, Klaus Vedder: *Smart Cards
                        Requirements, Properties and Applications*,
                        in: Tagungsband Smart Cards,
                        Vieweg Verlag, Braunschweig, 1998

[Wiener 93]   Michael J. Wiener: *Efficient DES Key Search*,
                        Crypto 93, Santa Barbara, CA, 1993

[Yellin 96]   Frank Yellin: *Low Level Security in Java*, Internet, 1996

[Zieschang 98]   Thilo Zieschang: *Differentielle Fehleranalyse und
                        Sicherheit von Chipkarten*, Internet, 1998

15.3   List of Standards with Comments

This section contains an extensively commented directory of standards and specifications that are relevant to cards with and without chips. This directory primarily takes into account international standards, rather than local, country-specific standards. It lists standards produced by official standards organizations (such ANSI, CEN, ETSI and ISO) as well as quasi-standards that are relevant to Smart Cards, such as the EMV specification and the Internet RFCs.

In addition to the commented directory, a summary of possibly helpful compilations, summaries and sources for standards related to specific subjects is contained in Table 15.1. Industry standards in particular are often available free of charge via the WWW, which is unfortunately usually not true of official standards.

All standards and specifications are listed here in the order of the names of the issuing institutions and their numerical designations, ignoring prefixes (such as 'pr') and status indications (such as 'DIS'). The date listed is the date at which the currently valid version first appeared. The most important standards for Smart Cards are marked with a '♦'.

Table 15.1   Summary of directories and sources of international stnadards and specifications that are useful in the field of Smart Cards

Subject	Source
General standards overviews	The following Web addresses are useful starting points for further research:
	[ANSI] [ETSI] [ISO]
	[CEN] [IEC] [ITU]
	[DIN] [IEEE] [SEIS]
Cryptography	Extensive summaries of standards for cryptography are located in [Menezes 97] and [Schneier 96].
	The following Web addresses are useful starting points for further research:
	[Ascom] [Kocher] [Tatu]
	[Certicom] [Kryptocom] [UCL]
	[Counterpane] [MIT] [Uni Siegen]
	[Entrust] [R3]
	[JYA] [RSA]
RFC	The Web address [RFC] is a useful starting points for further research.
GSM	[Eberspächer 97] contains an extensive summary of standards that are specific to GSM. The [GSM] site is a useful starting point for further research.

ANSI X 9.9: 1986	Financial Institution Message Authentication
NSI X 9.17: 1985	Financial Institution Key Management
ANSI X 9.19: 1986	Financial Institution Retail Message Authentication
ANSI X 9.30	Public Key cryptography using irreversible algorithms for the financial services industry
– 1: 1993	Part 1: The Digital Signature Algorithm (DSA)
– 2: 1993	Part 2: The Secure Hash Algorithm (SHA-1)
ANSI X 9.31 Part 1: 1993	Public Key cryptography using irreversible algorithms for the financial services industry – Part 1: The RSA Signature Algorithm
ANSI X 3.92: 1981	Data Encryption Algorithm *Describes the DES algorithm.*
ANSI X 3.106: 1983	American National Standard for Information Systems – Data Encryption Algorithm – Modes of Operation

IEEE 828: 1990

Standard for Software Configuration Management
Plans

IEEE 1363: 1998

Standard for RSA, Diffie-Hellman and Related
Public-Key Cryptography

♦ *This is a very extensive and comprehensive
standard, which covers almost all aspects of
asymmetric cryptographic algorithms, including
key generation, using digital signatures, key
exchange and encryption.*

ANSI / IEEE 829: 1991

Standard for Software Test Documentation

*Describes the methods and necessary
documentation for testing software.*

ANSI / IEEE 1008: 1987

Standard for Software Unit Testing

Describes basic methods for testing software.

ANSI / IEEE 1012: 1992

Software Verification and Validation Plans

*Specifies the necessary test activities and test
plans for software development. This standard is
based on the waterfall model for software
development.*

CCITT Z.100: 1994

Specification and Description Language (SDL)

DIN 44 300 – 1 ...9: 1988

Informationsverarbeitung – Begriffe

Defines many data processing terms.

EMV

Integrated Circuit Card [Application/Terminal]
Specification for Payment Systems

♦ *This is the most important set of standards for
Smart Cards used in payment systems. It is a joint
publication of Europay, Mastercard and Visa.
There are three parts, which deal with Smart
Cards, payment applications and the associated
terminals.*

EMV `96 Version 3.1.1: 1998

Integrated Circuit Card Specification for Payment
Systems

♦ *Specifies the mechanical and electrical
properties of Smart Cards and terminals. Contains
definitions of the activation and deactivation
sequences, data transfer at the electrical level, the
ATR and its associated data elements. In addition,
it specifies the T=0 and T=1 transmission
protocols, the structure of the APDU, logical
channels, secure messaging, card commands,*

	return codes and data elements, along with the associated TLV coding.
EMV `96 Version 3.1.1: 1998	Integrated Circuit Card Application Specification for Payment Systems
	Specifies transaction processes and security functions for Smart Cards used in payment systems. Also specifies the processes and working methods of authorization, user verification by PIN entry and terminal-based risk management.
EMV `96 Version 3.1.1: 1998	Integrated Circuit Card Terminal Specification for Payment Systems
	Specifies mandatory and optional requirements for terminals that supports EMV-compliant Smart Cards. Addresses possible configurations, functional and security-related requirements, possible and allowed user messages (including the character set used) and the acquirer interface. Also outlines the architecture of the terminal software and defines the basic features of a model interpreter for executable program code for terminals. The appendix contains advice on the technical design of the terminal and examples of terminals in point-of-sale equipment, cash dispensers and vending machines.
EMV `96 Errata Version 1.0: 1998	ICC Specifications for Payment Systems
	Clarifies ambiguities and corrects known errors in the three parts of the EMV specification.
EN 726-1: 1994	Identification card systems – Telecommunications integrated circuit(s) card and terminals – Part 1: System overview
EN 726-2: 1995	Identification card systems – Telecommunications integrated circuit(s) card and terminals – Part 2: Security framework
EN 726-3: 1994	Identification card systems – Telecommunications integrated circuit(s) card and terminals – Part 3: Application independent card requirements
	♦ *Defines file structures, commands, return codes and files for generally usable functions, and basic Smart Card mechanisms for telecommunications applications. This is the ETSI counterpart to ISO/IEC 7816-4, and it provides the framework for GSM 11.1l.*

EN 726-4: 1994	Identification card systems – Telecommunications integrated circuit(s) card and terminals – Part 4: Application independent card related terminal requirements
EN 726-5: 1995	Identification card systems – Telecommunications integrated circuit(s) card and terminals – Part 5: Payment methods *Defines various payment methods along with the associated data structures, data elements and processes for Smart Cards. These payment methods are designed to be used for telecommunications applications.*
EN 726-6: 1995	Identification card systems – Telecommunications integrated circuit(s) card and terminals – Part 6: Telecommunication features
EN 726-7: 1996	Identification card systems – Telecommunications integrated circuit(s) card and terminals – Part 7: Security Module
EN 1038: 1995	Identification card systems – Telecommunication applications – Integrated circuit(s) card payphone *Defines the fundamentals for using Smart Cards with public cardphones. This standard contains primarily references to previous standards, and it specifies the various places in the system where a security module can be effectively used to authenticate a phone card.*
prEN 1105: January 1995	Identification card systems – General concepts applying to systems using IC cards in inter-sector environments – Rules for inter-application consistency *Defines the basic demands placed on a Smart Card in order to ensure that it can be used for multiple applications. It primarily contains references to prior standards, as well as various rules for Smart Cards and terminals.*
prEN 1292: 1995	Additional Test Methods for IC Cards and Interface Devices *Defines tests for the general electrical parameters of Smart Cards and terminals and the basic data transfer between Smart Cards and terminals. This standard is an extension to ISO/IEC 10 373.*

EN 1332	Identification card systems – Man–Machine Interface
- 1 pr: 1995	Part 1: Design principles and symbols for the user interface
- 2 pr: 1995	Part 2: Definition of a Tactile Identifier for ID-1 cards
	Defines a tangible recess in ID-1 cards to allow the orientation of the card to be recognized.
- 3 pr: 1995	Part 3: Keypads
- 4 pr: 1995	Part 4: Coding of user requirements for people with special needs
prEN 1545-1: 1995	Identification card systems – Surface transport applications – Part 1: General
prEN 1545-2: 1995	Identification card systems – Surface transport applications – Part 2: Transport payment
prEN 1545-3: 1995	Identification card systems – Surface transport applications – Part 3: Tachograph
prEN 1545-4: 1995	Identification card systems – Surface transport applications – Part 4: Vehicle and driver licencing
prEN 1545-5: 1995	Identification card systems – Surface transport applications – Part 5: Freight
EN 1546	Identification card systems – Inter-sector electronic purse
	♦ *The most internationally most important standard for electronic purses, which forms the foundation for most purse systems. This set of standards has been kept relatively general, and thus includes many options, but it is a very good and complete description of an electronic purse.*
- 1 prEN: 1997	Part 1: Definition, concepts and structures
	Defines concepts relating to the entire set of standards, and describes the basic concepts and structures of multi-sector electronic purse systems.
- 2 prEN: 1996	Part 2: Security architecture
	Describes the notation used for security mechanisms, the security architecture and the associated processes and mechanisms, for multi-sector electronic purse systems.

- 3 prEN: 1996	Part 3: Data elements and interchanges *Describes the data elements, files, commands and return codes used between all components of a multi-sector electronic purse system.*
- 4 prEN: 1997	Part 4: Data objects *Describes the TLV mechanism for reading arbitrary data elements from files, and also provides a detailed presentation of the components and states of a state machine for a multi-sector electronic purse system. Also includes a list of tags for all data elements used.*
EN 13 343	Identification card systems – Telecommunications IC cards and terminals – Test methods and conformance testing for EN 726-3
- 1 prEN: 1998	Part 1: Implementation Conformance Statement (ICS) pro-forma specification
- 2 prEN: 1998	Part 2: Test Suite Structure and Test Purposes (TSS & TP)
- 3 prEN: 1998	Part 3: Abstract Test Suite (ATS) and Implementation extra Information for Testing (IXIT) pro-forma specification
EN 13 344	Identification card systems – Telecommunications IC cards and terminals – Test methods and conformance testing for EN 726-4
- 1 prEN: 1998	Part 1: Implementation Conformance Statement (ICS) pro-forma specification
- 2 prEN: 1998	Part 2: Test Suite Structure (TSS) and Test Purposes (TP)
- 3 prEN: 1998	Part 3: Abstract Test Suite (ATS) and Implementation extra Information for Testing (IXIT) pro-forma specification
EN 13 345	Identification card systems – Telecommunications IC cards and terminals – Test methods and conformance testing for EN 726-7
- 1 prEN: 1998	Part 1: Implementation Conformance Statement (ICS) pro-forma specification
- 2 prEN: 1998	Part 2: Test Suite Structure and Test Purposes (TSS & TP)
- 3 prEN: 1998	Part 3: Abstract Test Suite (ATS) and Implementation extra Information for Testing (IXIT) pro-forma specification

PC/SC: Revision 1: December 1997	Interoperability Specification for ICCs and Personal Computer Systems
	This extensive, detailed specification forms the basis for linking Smart Cards and terminals to the resource management system of the Microsoft 16-bit and 32-bit operating systems.
- 1	Part 1: Introduction and Architecture Overview
- 2	Part 2: Interface Requirements for Compatible IC Cards and Readers
- 3	Part 3: Requirements for PC-Connected Interface Devices
- 4	Part 4: IFD Design Considerations and Reference Design Information
- 5	Part 5: ICC Resource Manager Definition
- 6	Part 6: ICC Service Provider Interface Definition
- 7	Part 7: Application Domain and Developer Design Considerations
- 8	Part 8: Recommendations for ICC Security and Privacy Devices
ENV 1257	Identification card systems – Rules for Personal Identification Number handling in intersector environments
	Illustrates and explains security aspects related to using PINs, from the transfer of the PIN to the cardholder (PIN letter) to entering the PIN using a keypad (PIN pad).
- 1 prENV: 1997	Part 1: PIN presentation
- 2 prENV: 1997	Part 2: PIN protection
- 3 prENV: 1997	Part 3: PIN verification
prETS 300 331: 1993	Radio Equipment and Systems; Digital European Cordless Telecommunications; Common Interface; DECT Authentication Module
	Describes Smart Cards used in the DECT system. Encompasses all associated commands, files, access conditions and authentication processes. Also defines the dimensions of the Mini-ID and Plug-in card formats. This standard is heavily based on the GSM11.11 specification.
prETS 300 641	see GSM 11.12 (identical to this standard)
prETS 300 977	see GSM 11.12 (identical to this standard)

FIPS 46: 1977	Data Encryption Standard (DES) ♦ *This standard describes the DES algorithm.*
FIPS 74: 1981	Guidelines for Implementing and Using the NBS Encryption Standard
FIPS 81: 1980	DES Modes of Operation
FIPS 180-1: 1994	Secure Hash Standard (SHA-1) ♦ *This standard describes the SHA-1 hash function.*
FIPS 186: 1994	Digital Signature Standard (DSS) ♦ *This standard describes the DSS algorithm and the SHA hash function, which has however been replaced by the FIPS 180-1 (SHA-1) function.*
FIPS 140-1: 1994	Security Requirements for Cryptographic Modules
GSM 1.04 (ETR 100)	Digital cellular telecommunications systems (Phase 2); Abbreviations and acronyms
GSM 11.11 Version 7.1.0: 1998	Digital cellular telecommunications system (Phase 2+) – Specification of the Subscriber Identity Module – Mobile Equipment (SIM–ME) interface ♦ *Complete specification of the interface between a Smart Card and a GSM mobile phone. Includes definitions of the formats of the ID-1 and plug-in cards, as well as the general mechanical parameters of the card and the contacts. Specifies general electrical values and the ATR and PTS. Defines data structures, security mechanisms, commands and return codes. Defines all data elements and files necessary for GSM, as well as the processes corresponding to the individual functions. The GSM 11.11 specification is published by ETSI as prETS 300.977.*
GSM 11.12 Version 4.3.1: 1998	Digital cellular telecommunications system (Phase 2); Specification of the 3 Volt Subscriber Identity Module – Mobile Equipment (SIM–ME) interface *Specification for 3-V SIMs, including a compatibility list for SIMs programmed according to previous specifications. This standard only includes differences and extensions for the GSM 11.11 specifications with regard to 3-V SIMs.*

GSM 11.14 Version 5.6.0: 1997

Digital cellular telecommunications system (Phase 2 +); Specification of the SIM Application Toolkit for the Subscriber Identity Module – Mobile Equipment (SIM–ME) Interface

Defines and extensively describes the SIM Application Toolkit for GSM cards. This toolkit allows a network operator to use the 'over-the-air' (OTA) interface to load an application that can control the mobile phone. For example it specifies in detail the prescribed procedures for driving the display, scanning the keypad, sending short messages and performing other functions in connection with a related application.

GSM 11.17 V 0.5.1: 1997

Digital cellular telecommunications systems (Phase 2); Subscriber Identity Module (SIM) test specification

Specifies the test environment, test equipment, test hierarchy and individual test cases for testing GSM cards. The described tests are without exception aimed at the electrical and data processing aspects of GSM cards. To meet these objectives, tests for the supply of electrical power, data transfers, file management, commands and typical processes used in the GSM application are specified in detail. This standard is a very good and extensive illustration of how GSM tests should currently be constructed and executed.

ISO 639: 1997

Codes for the representation of names of languages

ISO 3166: 1993

Codes for the representation of names of countries

ISO 3309: 1984

Information processing systems – Data communication – High-level data link control procedures – Frame structure

Defines the message structure for data transfers and the message protection using an error detection code (CRC).

ISO/IEC 4217: 1995

Codes for the representation of currencies and funds

ISO 4909: 1987

Bank cards – Magnetic stripe data contents for track 3.

ISO/IEC 7501   Identification cards – Machine readable travel documents

     - 1: 1997   Part 1: Machine readable passport

     - 2: 1997   Part 2: Machine readable visas

     - 3: 1997   Part 3: Official travel documents

ISO 7810: 1995   Identification cards – Physical characteristics

Describes the most important physical characteristics of cards without chips, and defines the card dimensions for ID-1, ID-2 and ID-3 formats.

ISO 7811   Identification cards – Recording technique

This set of standards is an important reference for the mechanical aspects of cards. It specifies the configuration of the essential card elements.

     - 1: 1995   Part 1: Embossing

An exact definition of the ten numeric characters and the basic method used to emboss cards.

     - 2: 1995   Part 2: Magnetic stripe

Defines the size and position of the magnetic strip on a card. In addition, specifies the physical characteristics of the magnetic material and the coding of the characters on the magnetic strip.

     - 3: 1995   Part 3: Location of embossed characters on ID-1 cards

Defines the possible locations for embossing on ID-1 cards.

     - 4: 1995   Part 4: Location of read-only magnetic tracks – Tracks 1 and 2

Defines the positions of the read-only tracks (tracks 1 and 2) on an ID-1card.

     - 5: 1995   Part 5: Location of read–write magnetic track – Track 3

Defines the position of the read–write track (track 3) on an ID-1card.

     - 6: 1996   Part 6: Magnetic stripe – High-coercivity

ISO 7812   Identification cards

     - 1: 1993   Part 1: Numbering system

Specifies a numbering scheme for the manufacturer of ID cards.

- 2: 1993	Part 2: Application and registration procedures *Specifies the registration authority and includes a form for registering an application. Also contains a checksum algorithm modeled on the Luhn algorithm (modulo 10 checksum).*
ISO 7813: 1995	Identification cards – Financial transaction cards *Defines the basic physical properties, size and embossing of ISO 7810 ID-1 cards to be used for financial transactions. Also defines the data contents of tracks 1 and 2 of the magnetic strip.*
ISO/IEC 7816	Identification cards – Integrated circuit(s) cards with contacts ♦ *The most important set of ISO standards for microcontroller Smart Cards. The first three standards focus primarily on the cards and chip hardware. The remaining standards specify almost all mechanisms and properties of Smart Card operating systems and associated data processing.*
- 1: 1998	Part 1: Physical characteristics *Defines the physical characteristics of a card with a contact-type chip, and the associated tests for this type of card.*
- 2: 1996	Part 2: Dimensions and location of the contacts *Defines the size and position of the contact elements of a Smart Card, as well as the arrangement of the chip, the magnetic strip and the embossing. Also describes the method for measuring the positions of the contacts on the Smart Card.*
- 3: 1997	Part 3: Electronic signals and transmission protocols ♦ *The most important ISO standard for the general electrical parameters of a microcontroller Smart Card. It specifies all basic electrical characteristics (such as the 3-V and 5-V supply voltages), stopping the clock and reset behavior for cold and warm resets. It also defines the data elements, structure and possible process sequences of the ATR and PTS. A large part of this standard deals with basic aspects of data transfer at the physical level (such as the divider value) and the definition of the two transmission protocols (T=0 and T=1), and it includes extensive examples of communications processes.*

- 4: 1995	Part 4: Inter-industry commands for interchange
	◆ *The most important application-level ISO standard for Smart Cards. It defines the file organization, file structures, security architecture, TPDUs. APDUs, secure messaging, return codes and logical channels. The majority of this standard is taken up by an extensive description of commands for Smart Cards. The fundamental Smart Card mechanisms for inter-industry applications are also described.*
- 4 Amd. 1: 1997	Part 4 – Amendment 1: Use of secure messaging
- 5: 1994	Part 5: Numbering system and registration procedure for application identifiers
	Defines numbering schemes for the unique identification of national and international applications in Smart Cards. Also defines the exact data structure of the AID and explains the registration procedure for applications.
- 5 Amd. 1: 1996	Part 5 – Amendment 1: Registration of identifiers
- 6: 1996	Identification cards – Integrated circuit(s) cards with contacts – Part 6: Inter-industry data elements
	Defines the data objects (DOs) and associated TLV tags for inter-industry applications. Also explains related TLV structures and procedures for reading data objects from Smart Cards.
- 6 Amd. 1 DIS: 1997	Part 6 – Amendment 1: IC Manufacturer registration
- 7 DIS: 1997	Part 7: Interindustry commands for Structured Card Query Language (SCQL)
	Defines supplementary Smart Card commands that form an extension to ISO/IEC 7816-4. Defines the basic principles of a database derived from SQL, and specifies the commands for the associated SCQL accesses to Smart Cards.
- 8 CD: 1998	Part 8: Security related interindustry commands
	This part of the standards set is fully dedicated to the functions and commands relating to security. As an extension to ISO/IEC 7816-4, it defines additional mechanisms for secure messaging. It also describes many commands for cryptographic functions, such as digital signatures, hash computations, MAC computations and the encryption and decryption of data.

- 9 CD: 1998	Part 9: Enhanced inter-industry commands
	This standard is divided into three parts. The first part describes the life cycle of a Smart Card application at the file level in terms of states. The second part describes access control objects (ACOs) that can be used to regulate file accesses. The third part defines search commands for searching file contents and administration commands for creating and deleting files, which are needed for managing applications.
- 10 CD: 1998	Part 10: Electronic signals answer to reset for synchronous cards
	For memory cards, this is the counterpart to Part 3 of this set of standards. It specifies the essential electrical characteristics of memory cards, and it defines the data objects and the construction and possible sequences of the ATR for synchronous cards.
- 11 WD: 1998	Part 11: Card structure and enhanced functions for multi-application use
	Defines the stages of the life cycle of a Smart Card, with the associated commands for the activation and de-activation of applications. Also specifies the working principles and commands for loading and executing program code in Smart Cards.
ISO 8372: 1992	Modes of Operation for a 64-Bit Block Cipher Algorithm
	♦ *Defines four modes for encryption algorithms with 64-bit blocks (such as DES): Electronic Codebook (ECB), Cipher Block Chaining (CBC), Output Feedback (OFB) und Cipher Feedback (CFB). The block encryption modes described in ANSI X 3.106 and FIPS 81 form a subset of this standard.*
ISO 8583: 1993	Financial transaction card originated messages – Interchange message specifications
	Standard for data transfers between a terminal and its host system. In Germany, the communication between debit card terminals and the background system is based on this standard.

ISO 8730: 1990	Banking – Requirements for message authentication *Foundations of data security relating to data transfers and the generation and testing of MACs. The appendix contains many numerical examples, as well as a description of a DES pseudorandom number generator.*
ISO 8731	Banking – Approved algorithms for message authentication
– 1: 1987	Part 1: DEA *A very short standard in which the DEA is indicated to be suitable for MAC computation. Also contains a brief description of parity calculation for DES keys.*
– 2: 1992	Part 2: Message authenticator algorithm *Defines a fast algorithm for MAC computation in banking applications. The appendix contains numerical examples as well as an exact description of the algorithm.*
ISO 8732: 1988	Banking – Key management *Extensive standard that deals with the fundamentals and procedures for key management between two or more participating bodies, with the aid of symmetric cryptographic algorithms.*
ISO/IEC 8824: 1990	Information technology – Open Systems Interconnection – Specification of Abstract Syntax Notation One (ASN.1) *Defines the basic coding rules of ASN.*
ISO/IEC 8825: 1990	Information technology – Open Systems Interconnection – Specification of Basic Encoding Rules for Abstract Syntax Notation One (ASN.1) *Defines the data description language ASN.1.*
ISO/IEC 9075: 1992	Information technology – Database languages – SQL2 *Defines the database query language Structured Query Language (SQL), which is a superset of the Smart Card-based database query language SCQL.*
ISO/IEC 9126: 1991	Information technology – Software product evaluation – Quality characteristics and guidelines for their use

ISO 9564   Banking – Personal Identification Number
                                            management and security

   - 1: 1991   Part 1: PIN protection principles and techniques

                                            *Fundamentals of PIN selection, PIN management
                                            and PIN protection for general banking
                                            applications. The appendices define general
                                            requirements for PIN entry devices, among other
                                            things, as well as recommendations for the layout
                                            of suitable keypads and advice regarding the
                                            erasing of sensitive data on various media, such
                                            as magnetic tape, paper and semiconductor
                                            memories.*

   - 2: 1991   Part 2: Approved algorithm(s) for PIN
                                            encipherment

                                            *A very short standard that defines the DES as an
                                            algorithm for PIN encryption.*

ISO/IEC 9646-3: 1992   Information technology – Open Systems
                                            Interconnection – Conformance testing
                                            methodology and framework – Part 3: The Tree
                                            and Tabular Combined Notation (TTCN)

                                            *An extensive standard that describes a general
                                            high-level language for specifying tests. TTCN is
                                            used in a few isolated cases in the Smart Card
                                            environment.*

ISO/IEC 9796   Information technology – Security techniques –
                                            Digital signature scheme giving message recovery

                                            *Defines processes for generating and verifying
                                            digital signatures with message recovery. The
                                            appendix contains several numerical examples of
                                            key generation, signature generation and
                                            signature verification.*

   - 1: 1999   Part 1: Mechanisms using redundancy
   - 2: 1997   Part 2: Mechanisms using a hash function

ISO/IEC 9797: 1994   Information technology – Security techniques –
                                            Data integrity mechanism using a cryptographic
                                            check function employing a block cipher
                                            algorithm

                                            *Defines the computation of a message
                                            authentication code (MAC) using a block-oriented
                                            encryption algorithm. This standard is a
                                            generalization of the MAC computation of
                                            ISO 8731, ANSI X9.9 and ANSI X9.19.*

ISO/IEC 9798	Information technology – Security techniques – Entity authentication
	♦ *This set of standards describes in detail various cryptographic techniques for the authentication of one, two or three participating entities. It is the most important reference on the subject of authentication.*
- 1: 1991	Part 1: General model
	Specifies the terminology and notation used in the other parts of this set of standards.
- 2: 1994	Part 2: Mechanisms using symmetric encipherment algorithms
	Specifies authentication processes that are based on symmetric cryptographic algorithms.
- 3: 1993	Part 3: Entity authentication using a public key algorithm
	Specifies authentication processes that are based on asymmetric cryptographic algorithms.
- 4: 1995	Part 4: Mechanisms using a cryptographic check function
	Specifies authentication processes that are based on cryptographic check functions.
- 5: 1997	Part 5: Mechanisms using zero-knowledge techniques
	Specifies authentication processes that are based on zero-knowledge processes.
ISO 9807: 1991	Banking and related financial services – Requirements for message authentication (retail)
ISO/IEC 9979: 1995	Information technology – Security techniques – Procedures for the registration of cryptographic algorithms
ISO 9992	Financial Transaction Cards – Messages between the Integrated Circuit Card and the Card Accepting Device
- 1: 1990	Part 1: Concepts and structures
- 2: 1998	Part 2: Functions, messages (commands and responses), data elements and structures
	Defines commands, processes, and data elements for Smart Cards used in payment systems. Contains the definitions of tags used in payment systems and many cross-references to other standards in the ISO/IEC 7816 series.

- 4 DIS: 1993	Part 4: Common data for interchange
- 5 CD: 1991	Part 5: Organization of data elements
ISO/IEC 10 116: 1995	Information technology – Security techniques – Modes of operation for an *n*-bit block cipher algorithm
	Describes the four standard modes (ECB, CBC, CFB, OFD) for using a block-oriented encryption algorithm. An appendix contains detailed comments regarding the use of each of the four modes, and another appendix contains corresponding numerical examples.
ISO/IEC 10 118	Information technology – Security techniques – Hash functions
	General fundamentals of hash functions, as well as the associated padding methods.
- 1: 1994	Part 1: General
- 2: 1999	Part 2: Hash functions using an *n*-bit block cipher algorithm
	Defines hash functions that are based on block-oriented encryption algorithms. Algorithms with single-length and double-length keys are described. The appendix contains a matching numerical example for each type of key, based on the DES algorithm.
- 3: 1996	Part 3: Dedicated hash functions
- 4: 1996	Part 4: Hash functions using modular arithmetic
ISO/IEC 10 170	Information technology – Security techniques – Key Management
- 1 CD: 1995	Part 1: Key Management Mechanisms Using Asymmetric Techniques
- 2 CD: 1995	Part 2: Key Management Mechanisms Using Symmetric Techniques
ISO 10 202	Financial Transaction Cards – Security architecture of financial transaction systems using Integrated Circuit Cards
- 1: 1991	Part 1: Card life cycle
- 2: 1996	Part 2: Transaction process
- 3 DIS: 1995	Part 3: Cryptographic key relationship
- 4: 1996	Part 4: Secure application modules
- 5: 1998	Part 5: Use of algorithms

- 6: 1994	Part 6: Card holder verification
- 7: 1998	Part 7: Key Management
	Defines general mechanisms for key management and key derivation. Both symmetrical and asymmetrical processes are described.
- 8: 1998	Part 8: General principles and overview
ISO/IEC 10 373: 1993	Identification cards – Test methods
	♦ *Fundamental standard for card testing. Contains precise descriptions of test methods for card bodies and card bodies with implanted chips. The individual tests are described in detail, with many explanatory illustrations. This standard has been supplanted by the following set of standards.*
ISO/IEC 10 373	Identification cards – Test methods
- 1 DIS: 1998	Part 1: General
- 2 DIS: 1998	Part 2: Magnetic stripe technologies
- 3 CD: 1998	Part 3: Integrated circuit cards
	Specifies the test environment, test methods and test procedures for electrical tests of contact-type Smart Cards. Also specifies detailed procedures for checking contact locations, testing the power supply, testing ATR and PTS data transfers and testing data transmission protocols.
- 4 CD: 1998	Part 4: Contactless integrated circuit cards
- 5 DIS: 1998	Part 5: Optical memory cards
- 6 WD: 1998	Part 6: Proximity cards
- 7 WD: 1998	Part 7: Vicinity cards
ISO/IEC 10 536	Identification cards – Contactless integrated circuit(s) cards
- 1: 1992	Part 1: Physical characteristics
	Defines the physical characteristics of contactless Smart Cards and the associated test methods.
- 2: 1995	Part 2: Dimension and location of coupling areas
	Specifies the dimensions and location of the coupling surfaces for contactless cards, and the operation of card terminals having card slots or surface interfaces.
- 3: 1996	Part 3: Electronic signals and reset procedures
	Defines the electrical signals of the inductive and capacitive elements used to couple the Smart Card to the terminal.

- 4 CD: 1997	Part 4: Answer to reset and transmission protocols
	Specifies the data transfer on the physical level, as well as the structure and date elements of the ATR and PTS, for contactless Smart Cards. Defines the T-2 data transmission protocol, with many sample scenarios of protocol sequences.
ISO 11 568	Banking – Key management
- 1: 1994	Part 1: Introduction to key management
- 2: 1994	Part 2: Key management techniques for symmetric ciphers
- 3: 1994	Part 3: Key life cycle for symmetric ciphers
- 4: 1998	Part 4: Key management techniques for public key cryptosystems
- 5: 1998	Part 5: Key life for public key cryptosystems
- 6: 1998	Part 6: Key management schemes
ISO/IEC 11 693: 1994	Identification cards – Optical memory cards
ISO/IEC 11 694	Identification cards – Optical memory cards and devices – Linear recording method
- 1: 1994	Part 1: Physical characteristics
- 2: 1995	Part 2: Dimensions and location of the accessible optical area
- 2 DAM 1: 1997	Part 2, Amendment 1: Optical card layout
- 3: 1995	Part 3: Optical properties and characteristics
- 4: 1996	Part 4: Logical data structures
ISO/IEC 11 770	Information technology – Security techniques – Key management
- 1: 1996	Part 1: Key management framework
- 2: 1995	Part 2: Mechanisms using symmetric techniques
- 3: 1995	Part 3: Key management mechanisms using asymmetric cryptographic techniques
ISO 13 491	Banking – Secure cryptographic devices
- 1: DIS	Part 1: Concepts, requirements and evaluation methods
- 2: DIS	Part 2: Audit check lists for devices used in magnetic stripe card systems
ISO/IEC 13 888	Information technology – Security techniques – Non-repudiation
- 1: 1997	Part 1: General

- 2: 1998	Part 2: Mechanisms using symmetric techniques
- 3: 1997	Part 3: Mechanisms using asymmetric techniques
ISO/IEC 14 443	Remote-coupling communication cards
- 1 CD: 1998	Part 1: Physical characteristics
- 2 CD: 1998	Part 2: Radio frequency interface
- 3 WD: 1998	Part 3: Initialisation and anticollision
- 4 WD: 1998	Part 4: Transmission protocols
ISO/IEC 14 888	Information technology – Security techniques – Digital signature with appendix *Specifies basic mechanisms and procedures for a digital signature with appendix. This standard is independent of any particular asymmetric cryptographic algorithm.*
- 1 CD: 1997	Part 1: General
- 2 CD: 1997	Part 2: Identity-based mechanisms
- 3 CD: 1997	Part 3: Certificate-based mechanisms
ISO/IEC WD 10 460: 1995	Identification cards – Integrated circuit(s) cards with contacts – Integrated circuit with voltages lower than 3 volts
ISO/IEC 15 693	Contactless integrated circuit(s) cards – Vicinity cards
- 1 CD: 1998	Part 1: Physical characteristics
- 2 WD: 1998	Part 2: Air interface and initialisation
- 3 WD: 1999	Part 3: Protocols
- 4 WD: 1999	Part 4: Registration of Application issuers
ISO/IEC 15 946	Information technology – Security techniques – Cryptographic techniques based on elliptic curves
- 1 WD:1997	Part 1: General
- 2 WD: 1998	Part 2: Digital signatures
- 3 WD: 1998	Part 3: Key establishment
ITU X.509: 1997	Information Technology – Open Systems Interconnection – The Directory: Authentication Framework ♦ *Specifies the structure and coding of certificates. It is internationally the most commonly used basis for certificate structures, and it is the same as ISO/IEC 9594-8.*

Java Card 2.1: 1998	♦ *This industrial standard is the basis for Java in Smart Cards. It was generated by the Java Card Forum and published by the Sun corporation.*
Application Programming Interface	*This part of the standard identifies the limitations of Java in Smart Cards as compared with the full implementation of Java.*
Language Subset and Virtual Machine Specification	*This very short part of the standard specifies the framework of Java Card and shows the design concept of Smart Card applets, the way that Java works with APDUs and the structure of the Java file system in Smart Cards.*
Programming Concepts	*This very short part of the standard specifies the framework of Java Card and shows the design concept of Smart Card applets, the way that Java works with APDUs and the structure of the Java file system in Smart Cards.*
PKCS #1 V 1.5: 1993	RSA Encryption Standard ♦ *Describes how the RSA algorithm performs encryption and decryption*
PKCS #3 V 1.4: 1993	Diffie–Hellman Key-Agreement Standard *Describes how a key exchange procedure between two bodies works using the Diffie–Hellman process.*
PKCS #11 V 2.0 Draft: 1997	Cryptographic Token Interface Standard ♦ *This is the international* de facto *standard for an API for calling cryptographic functions. The API is called 'Cryptoki' (cryptographic token interface), and it includes functions such as RC2, RC4, RC5, MD5, SHA-1, DES, Triple-DES, IDEA, RSA, DSA, MAC computation and key generation for a wide variety of cryptographic algorithms.*
RFC 1750: 1994	Randomness Recommendations for Security *Illustrates the operating principles of various types of random number generators. Includes recommendations for techniques for designing high-quality pseudorandom number generators for PCs, based on these operating principles.*
RFC 1115: 1992	The MD2 Message-Digest Algorithm
RFC 1186: 1992	The MD4 Message-Digest Algorithm
RFC 1321: 1992	The MD5 Message-Digest Algorithm

15.4   Registration Authorities for RIDs

The form for registering a RID is located in the appendix of the ISO/IEC 7816-5 standard. An application for an international RID is normally made via the relevant national authority, and there is a fee. The addresses of national registration authorities, as well as the registration procedures for RIDs, can usually be requested from national standardization authorities.

Table 15.2   Registration authorities for ISO/IEC 7816-5 RIDs

Region	Organization
International	TeleDanmark KTAS attn: ISO/IEC 7816-5 Registration Authority Teglholmsgade 1 1790 Copenhagen V Denmark
Germany	RID German National Registration Authority c/o GMD Bruno Struif Rheinstraße 75 D - 64 295 Darmstadt Germany

15.5   Events

Table 15.3 lists trade fairs, congresses and conventions that have Smart Cards or a related subject as at least one of their major themes. The indicated places and times are typical for the past several years, but they can change in the future, depending on the event organizer. As you can see, a traveler with an interest in the subject can visit an event in a different country every month of the year.

Table 15.3 (part 1)   Selected annual events relating Smart Cards and cryptography

Event name	Place	Time
Asia Crypt	Asia	Autumn
Card Tech / Secure Tech [CTST]	USA (Arlington, Virginia)	September
Card Tech West [CTST]	USA (San Jose, California)	December
Cards Africa	South Africa (Johannesburg)	November
Cards Asia	Singapore	February
Cards Australia	Australia (Melbourne)	August
Cards Latin America	Chile (Santiago de Chile)	July
Cartes	France (Paris)	October
CeBit	Germany (Hanover)	March
Crypto	USA (Santa Barbara, California)	Summer
Euro Crypt	Europe	Spring

Table 15.3 (part 2)   Selected annual events relating Smart Cards and cryptography

Event name	Place	Time
GSM World Congress	somewhere in the world	February
OmniCard	Germany (Berlin)	January
Smart Card	United Kingdom (London)	February

15.6   World Wide Web Addresses

The following list of World Wide Web addresses does not claim to be complete. It should be seen as a cross-section of the various companies and institutions that are active in the field of Smart Cards. The listed addresses are thus entirely suitable for use as starting points for further research. Thanks to the hypertext structure of HTML documents, many of the listed sites contain links to other interesting documents and World Wide Web locations. Large collections of links are explicitly marked as 'link farms'.

When using this list, you should bear in mind that the Internet is very dynamic, so the addresses can very quickly become outdated. This is also why we do not list specific documents, but have limited the listings to subdirectories. Even these are frequently changed when a Web server is reorganized, so in case of doubt we recommend that you use the address up to the organization or country code (*.com, *.de and so on). After this, you can manually select currently valid directories on the Web server via the home page.

The classification of the Internet addresses or firms is based on the main areas of activity. However, many of the listed firms are active in several of the indicated areas; this is normally shown explicitly. To the extent that it is reasonable to do so, the country in which the firm or organization is located is also noted.

As a rule, you can find the postal address of a firm and the telephone number of a contact person on the home page of the firm. Consequently, postal addresses are not included in the list. If you have a specific need for particular information, we generally advise you to use appropriate search terms (keywords) and a powerful search engine to comb through the World Wide Web. This at least will ensure that you are working with a current cross-section of information.

Table 15.4 (part 1)   Summary of the descriptive categories used in the list of Web addresses

Category	Explanation
development systems	software development systems for Smart Card operating systems
attacks	attacks on Smart Cards, Smart Card terminals, security modules etc.
operating systems	operating systems for Smart Cards
biometrics	user authentication by means of biometric features
e-money	electronic payments and electronic money, with and without Smart Cards
research	private or government research institute

Table 15.4 (part 2)   Summary of the descriptive categories used in the list of Web addresses

Category	Explanation
semiconductor manufacturer	manufacturer of semiconductors for Smart Cards, memory chips and/or microcontrollers
card issuer	issuer of cards and/or Smart Cards
card manufacturer	manufacturer of cards with or without chips
cryptography	cryptography relating to Smart Cards
link farm	collection of links to other Internet sites
machines for card manufacturing	machines and equipment for manufacturing cards
standards, standardization	standards relating to Smart Cards and cryptography
patents	patents relating to Smart Cards
security technology	security technology relating to Smart Cards
software	PC software for Smart Cards, Smart Card simulations
terminals	manufacturer of terminals for cards with or without chips
university	university or technical institute
events	trade fairs, conferences and congresses relating to Smart Cards
publisher	journals and books about Smart Cards

[a la Card]   **a la Card**
                        publisher in the Smart Card field
                        http://www.alacard.de/

[ACS]   **Advanced Card Systems Ltd., Hong Kong**
                        Smart Cards, terminals
                        http://www.acs.com.hk/

[Aladdin]   **Aladdin Knowledge Systems Ltd., USA**
                        terminals, software, Smart Cards
                        http://www.aks.com/
                        http://www.aladdin.de/

[Aliro]   **Aliro Ltd., Israel**
                        security technology
                        http://www.aliro.com/

[AM]   **American Magnetics, USA**
                        terminals
                        http://www.magstripe.com/

[Amazing]   **Amazing Controls Inc., USA**
                        Smart Cards, terminals
                        http://www.amazingcontrols.com/

[AmEx]   **American Express, USA**
                        card issuer
                        http://www.americanexpress.com/

[Anderson]   **Ross Anderson's Home Page, Great Britain**
                    information about attacks on Smart Cards
                    http://www.cl.cam.ac.uk/users/rja14/

[ANSI]   **ANSI, USA**
                    standards
                    http://www.ansi.org/

[Ascom]   **Ascom AG, Switzerland**
                    cryptography; IDEA; security technology
                    http://www.ascom.ch/

[Ashling]   **Ashling Microsystems Ltd., Ireland**
                    development systems
                    http://www.ashling.com/

[ASM]   **ASM Lithography, The Netherlands**
                    machines for semiconductor manufacturing
                    http://www.asml.com/

[Aspects]   **Aspects software Ltd., Scotland**
                    software and hardware for testing Smart Cards und terminals
                    http://www.aspects-sw.com/

[Atmel]   **Atmel, USA**
                    Smart Card microcontrollers
                    http://www.atmel.com/

[AU Systems]   **AU-System Ego AB, Sweden**
                    security modules; Smart Cards; personalization equipment
                    http://www.ego.ausys.com/

[BdB]   **Bundesverband deutscher Banken (BdB), Germany**
                    http://www.bdb.de/

[Bellcore]   **Bellcore, USA**
                    security technology; 'Bellcore attack'
                    http://www.bellcore.com/

[Brokat]   **Brokat GmbH, Germany**
                    electronic commerce, home banking
                    http://www.brokat.com/
                    http://www.brokat.de/

[BSI]   **Bundesamt for Sicherheit in der Informationstechnik (BSI), Germany**
                    articles on the subject of security
                    http://www.bsi.bund.de/

[Bull]   **Bull CP8 Smart Card and Terminals, France**
                    Smart Cards; terminals
                    http://www.cp8.bull.net/

[Cardshow]   **The Smart Card Cyber Show, France**
                    http://www.cardshow.com/

[CCC]   **Chaos Computer Club e.V., Germany**
                    attacks on Smart Cards; cryptographic algorithms
                    http://www.ccc.de/

[CEN]   **CEN**
                    standards
                    http://www.cenorm.be/

[Certicom]   **Certicom Corp., Canada**
                      cryptography, ECC
                      http://www.certicom.ca/

[Computational]   **Computational Logic Inc., USA**
                      Java
                      http://www.cli.com/

[Counterpane]   **Counterpane, USA**
                      cryptography
                      http://www.counterpane.com/

[CTST]   **CardTech/SecurTech Conference, USA**
                      events relating to Smart Cards
                      http://www.ctst.com/

[Cybercash]   **Cybercash Inc., USA**
                      electronic money, Cybercash
                      http://www.cybercash.com/

[Dai Nippon]   **Dai Nippon Printing Co. Ltd., Japan**
                      Smart Card manufacturer
                      http://www.dnp.co.jp/

[Dallas Semi]   **Dallas Semiconductor, USA**
                      semiconductor manufacturer; security processors
                      http://www.dalsemi.com/
                      http://www.ibutton.com/

[Datacard]   **Datacard, USA**
                      production machines for Smart Cards
                      http://www.datacard.com/

[De La Rue]   **De La Rue Card Systems, Great Britain**
                      Smart Card manufacturer
                      http://www.delarue.com/

[DI]   **Digital Instruments, Germany**
                      scanning probe microscopes
                      http://www.digmbh.de/

[Diebold]   **Diebold Inc., USA**
                      system integrater; cash dispensers
                      http://www.diebold.com/

[Digicash]   **Digicash, The Netherlands**
                      electronic money; e-Cash
                      http://www.digicash.com/

[DIN]   **Deutsches Institut for standardization e.V. (DIN), Germany**
                      standards
                      http://www.din.de/

[Diners]   **Diners Club, USA**
                      card issuer
                      http://www.diners-club.com/

[DPA]   **Deutsches Patentamt, Germany**
                      patents
                      http://www.deutsches-patentamt.de/

[Drexler]   **Drexler Technology Corp., USA**
                       cards with optically writeable and readable regions
                       http://www.lasercard.com/

[ECC]   **The Error Correcting Codes (ECC) Home Page, Japan**
                       link farm for error detection and correction codes
                       http://imailab.iis.u-tokyo.ac.jp/~robert/codes.html/

[e-Commerce]   **US Department of Trade and Commerce, USA**
                       e-commerce
                       http://www.ecommerce.gov/

[Entrust]   **Entrust, Canada**
                       cryptography
                       http://www.entrust.com/

[Eracom]   **Eracom Pty. Ltd., Australia**
                       cryptography
                       http://www.eracom.com.au/

[ETHZ]   **Eidgenössische Technische Hochschule Zürich**
                       http://www.inf.ethz.ch/

[ETSI]   **ETSI**
                       standards
                       http://www.etsi.fr/

[Europay]   **Europay International, Belgium**
                       card issuer; EMV specification; SET specification; electronic commerce
                       http://www.europay.com/

[F + D]   **F + D Feinwerk- und Druktechnik GmBH, Germany**
                       desktop personalization machines
                       http://www.fuddruk.de/

[Fischer]   **Fischer International Systems Corporation, USA**
                       terminals
                       http://www.fisc.com/

[GD]   **Giesecke & Devrient GmbH, Germany**
                       Smart Cards; operating systems; terminals
                       http://www.gdm.de/
                       http://www.gdasia.com/

[Gemplus]   **Gemplus S.C.A., France**
                       Smart Cards, operating systems, terminals
                       http://www.gemplus.com/

[GIS]   **General Information Systems Ltd., Great Britain**
                       electronic payments; terminals
                       http://www.gis.co.uk/

[GMD]   **Gesellschaft für Mathematik und Datenverarbeitung (GMD), Germany**
                       operating systems; digital signatures
                       http://www.gmd.de/

[GPT]   **GPT Card Technology, Great Britain**
                       Smart Cards
                       http://www.gpt.co.uk/

[Groupmark]	**Groupmark Ltd., Canada** Smart Card manufacturer *http://www.groupmark.com/*
[GSM]	**GSM MoU Association** link farm relating to GSM *http://www.gsmworld.com/*
[Gutmann]	**Peter Gutmann's Security and Encryption Links** *http://www.cs.auckland.ac.nz/~pgut001/links.html/*
[GZS]	**Gesellschaft for Zahlungssysteme (GZS) GmbH, Germany** electronic payments; clearing *http://www.gzs.de/*
[Hanser]	**Carl Hanser Verlag GmbH, Germany** Handbuch der Chipkarten, The Smart Card Simulator *http://www.hanser.de/*
[Hekuma]	**Hekuma, Germany** injection-molded cards *http://www.emedia.com/tech/hekuma/hekuma.html/*
[Hitachi]	**Hitachi Ltd. Japan** Smart Card microcontrollers, terminals *http://www.hitachi.com/* *http://www.hitachi.co.jp/*
[Hypercom]	**Hypercom Corp., USA** terminals *http://www.hypercom.com/*
[Hyperion]	**Hyperion, Great Britain** consultant *http://www.hyperion.co.uk/*
[IA]	**Integrity Arts, USA** Java for Smart Cards *http://www.integrityarts.com/*
[IATA]	**International Air Transport Association (IATA)** *http://www.iata.org/*
[IBI]	**Institut für Bankinformatik an der Universität Regensburg (IBI), Germany** *http://www.ibi.de/*
[IBM]	**IBM** Smart Card manufacturer; terminals *http://www.ibm.zurich.ch/* *http://www.chipcard.ibm.com/* *http://www.research.ibm.com/* *http://www.zurich.ibm.ch/Technology/Security/publications/1995/rap.html/*
[IEC]	**IEC** standards *http://www.iec.ch/*
[IEEE]	**IEEE** standards *http://www.ieee.org/*

[Infineon]	**Infineon AG, Germany**
	semiconductor manufacturer
	http://www.infineon.com

[Ingenico]	**Ingenico, France**
	terminals
	http://www.ingenico.com/

[Innovatron]	**Groupe Innovatron, France**
	patents; consultants
	http://www.innovatron.com/

[Innovonics]	**Innovonics Inc., USA**
	payments with Smart Cards
	http://www.innovonics.com/

[Integri]	**Integri, Belgium**
	testing Smart Card operating systems
	http://www.integri.com/

[Intel]	**Intel**
	semiconductor manufacturer
	http://www.intel.com/

[Intellect]	**Intellect, Australia**
	terminals
	http://www.intellect.com.au/

[InterNIC]	**Network Information Server for RdW**
	http://www.internic.net/

[Iridium]	**Iridium, USA**
	information about the Iridium mobile phone network
	http://www.iridium.com/

[ISO]	**ISO**
	standards
	http://www.iso.ch/

[ITU]	**ITU**
	standards
	http://www.itu.ch/

[Java Lobby]	**Java Lobby, USA**
	Java
	http://www.javalobby.org/

[Javasoft]	**Javasoft, USA**
	Java, Java for Smart Cards
	http://www.javasoft.com/

[JCB]	**JCB, USA**
	card issuer
	http://www.jcb.com/

[JCF]	**Java Card Forum, USA**
	Java, specifications for Java on Smart Cards
	http://www.javacardforum.org/

[JOS]	**Java Operating System, USA**
	Java operating system
	http://www.jos.org/

[JTC1]   **ISO, Joint Technical Committee One**
                    international standardization
                    http://www.jtc1.org/

[JYA]   **JYA, USA**
                    cryptography
                    http://www.jya.com/

[Kaba]   **Kaba Holding AG, Germany**
                    contactless Smart Cards; Legic
                    http://www.kaba.com/

[Keil]   **Keil GmbH, Germany**
                    development systems
                    http://www.keil.com/

[Kocher]   **Paul Kochers Cryptography Page**
                    http://www.cryptography.com/

[Krone]   **Krone GmbH, Germany**
                    terminals
                    http://www.krone-gmbh.de/

[Kryptocom]   **Kryptokom GmbH, Germany**
                    cryptography
                    http://www.kryptokom.com/

[Litronic]   **Litronic Inc., USA**
                    security systems
                    http://www.litronic.com/

[Logika]   **Logika Comp Spa, Italy**
                    personalization systems
                    http://www.logika.it/

[MagTek]   **MagTek Inc., USA**
                    terminals
                    http://www.magtek.com/

[Maosco]   **Maosco Ltd., Great Britain**
                    Smart Card operating system
                    http://www.multos.com/

[Mark Twain]   **Mark Twain Bank, USA**
                    electronic money (e-Cash)
                    http://www.marktwain.com/

[Mastercard]   **MasterCard International, USA**
                    card issuer; EMV specification; SET specification; electronic commerce
                    http://www.mastercard.com/

[Meinen Ziegel]   **Meinen, Ziegel & Co., Germany**
                    card manufacturing machines
                    http://www.royonix.com/

[Micromedia]   **Micromedia AG, Germany**
                    biometrics (dynamic signature)
                    http://www.micromedia.de/

[Microsoft]   **Microsoft, USA**
                    Crypto-API, PC/SC
                    http://www.microsoft.com/

[Mikron]   **Mikron AG, Germany**
                      contactless Smart Cards
                      http://www.mikron.de/

[Millicent]   **Digital Equipment Corp., USA**
                      electronic money (Millicent)
                      http://www.millicent.digital.com/

[MIT]   **Massachusetts Institute of Technology (MIT), USA**
                      cryptography
                      http://www.mit.edu/

[Mondex]   **Mondex International Ltd., Great Britain**
                      electronic purse, Smart Card operating system
                      http://www.mondex.com/

[Motorola SPS]   **Motorola SPS Inc., USA**
                      Smart Card manufacturer
                      http://www.mot-sps.com/

[Motorola]   **Motorola, USA**
                      semiconductor manufacturer
                      http://design-net.com/csic/SMARTCRD/smartcrd.htm/

[Mühlbauer]   **Mühlbauer GmbH, Germany**
                      card manufacturing machines
                      http://www.muehlbauer.de/

[NC]   **Network Computer, Inc., USA**
                      Open-Card Framework
                      http://www.nc.com/

[Netscape]   **Netscape**
                      standards (SSL)
                      http://www.netscape.com/

[NICDE]   **Network Information Server for DE**
                      http://www.nic.de/

[NIST]   **National Institute of Standards and Technology (NIST), USA**
                      standards (FIPS); specifications (DES, SHA-1, DSS); Common Criteria for
                      information technology security
                      http://www.csrc.ncsl.nist.gov/
                      http://www.nist.gov/

[NSA]   **National Security Agency (NSA), USA**
                      http://www.nsa.gov/

[Oberthur]   **Oberthur Smart Cards, USA**
                      Smart Cards
                      http://www.oberthurkirk.com/

[Oki]   **Oki, Japan**
                      microcontrollers; terminals
                      http://www.oki.com/
                      http://www.oki.co.jp/

[Orga]   **Orga GmbH, Germany**
                      Smart Card manufacturer; operating systems; terminals
                      http://www.orga.com/

[OTP]	**Open Trading Protocol** payments in public networks *http://www.otp.org/*
[PC/SC]	**PC/SC Working Group, USA** PC/SC specification *http://www.smartcardsys.com/*
[PGP]	**Pretty Good Privacy** *http://www.pgp.com/*
[Philips]	**Philips, Germany** semiconductor manufacturer *http://www.philips.com/* *http://www.semiconductors.philips.com/*
[Protechno]	**Protechno Card GmbH, Germany** desktop personalization machines *http://www.protechno-card.com/*
[Proton]	**Proton** electronic purse (Proton) *http://www.protonworld.com/*
[R3]	**R3 Security Engineering, Switzerland** cryptography *http://www.r3.ch/*
[Racom]	**Racom Systems Inc., USA** card manufacturer, semiconductor manufacturer *http://www.racom.com/*
[Rankl]	**Home page of Wolfgang Rankl** Given the very dynamic nature of the Internet, it may at some time be necessary to relocate my home page. A reference to the current location of my home page can always be found at the home page of Carl Hanser Verlag [Hanser]. *http://www.wrankl.de*
[RFC]	**RFC Server** Internet standards; RFC *http://www.yahoo.com/* (one of many)
[RSA]	**RSA Inc., USA** cryptography *http://www.rsa.com/*
[SA]	**Strategic Analysis Inc., USA** consultants *http://www.sainc.com/*
[SCARD]	**Smart Card Developer Association, USA** attacks, software *http://www.scard.org/*
[Schlumberger]	**Schlumberger Ltd., France** Smart Card manufacturer; terminals; operating systems *http://www.slb.com/*
[SE Transaction]	**Secure Electronic Transaction LLC, USA** SET homepage *http://www.setco.org/*

[Secude]   **Secude GmbH, Germany**
                      security technology
                      http://www.secude.com/

[Security   **Security Dynamics Technologies Inc., USA**
Dynamics]   security technology
                      http://www.securid.com/

[SEIS]   **SEIS Swedish Secured Electronic Information in Society Specification, Sweden**
                      standards
                      http://www.seis.se/

[Semper]   **SEMPER**
                      e-commerce
                      http://www.semper.org/

[SEPT]   **SEPT, France**
                      standards
                      http://www.sept.fr/

[Smart Card   **The Smart Card Club**
Club]   *http://www.smartcardclub.co.uk/*

[Smart Card   **The Smart Card Forum**
Forum]   *http://www.smartcardforum.org/*

[STM]   **ST Microelectronics, France**
                      semiconductor manufacturer
                      http://www.st.com/

[Sun]   **Sun, USA**
                      Java, Java electronic commerce Framework (JECF)
                      http://www.java.sun.com/

[Tasking]   **Tasking Inc., USA**
                      development systems
                      http://www.tasking.com/

[Tatu]   **Tatu Ylönen's Security and Encryption Links**
                      link farm related to cryptography
                      http://www.cs.hut.fi/crypto/

[TC]   **TC Trustcenter GmbH, Germany**
                      trust center
                      http://www.trustcenter.de/

[Techno Data]   **Techno Data, Germany**
                      magnetic-strip cards, Smart Cards
                      http://www.technodata-ibk.com/

[Telesec]   **Telesec GmbH, Germany**
                      Smart Card operating systems
                      http://www.telekom.de/telesec

[Teletrust]   **Teletrust, Germany**
                      http://www.teletrust.de/

[Tenth   **Tenth Mountain Systems, Inc., USA**
Mountain]   SET compliance testing
                      http://www.tenthmtn.com/

[TI]	**Texas Instruments Inc., USA** semiconductor manufacturer *http://www.ti.com/*
[TNO]	**TNO (Netherlands Organization for Applied Research), The Netherlands** hardware testing of microcontrollers *http://www.tno.nl/*
[TR]	**Report on Smart Cards** *http://www.tr.com/*
[Ubiq]	**UbiQ Inc., USA** personalization *http://www.ubiqinc.com/*
[UCL]	**UCL – Microelectronics Laboratory, Belgium** cryptography *http://www.dice.ucl.ac.be/*
[Uni Siegen]	**Universität Siegen, Germany** cryptography *http://www.uni-siegen.de/*
[Unicode]	**Unicode Konsortium** *http://www.unicode.org/*
[Utimaco]	**Utimaco AG, Germany** terminals *http://www.utimaco.com/*
[UTM]	**University of Tennessee at Martin, USA** prime numbers http://www.utm.edu/research/primes/
[Verifone]	**Verifone Inc., USA** terminals *http://www.verifone.com/*
[Visa]	**Visa International, USA** card issuer; EMV specification; SET specification; electronic commerce *http://www.visa.com/* *http://www.visa.de/*
[W3C]	**World Wide Web Consortium (W3C)** Internet standardization *http://www.w3.org/*
[Wiley]	**John Wiley & Sons, Inc., Great Britain** The Smart Card Handbook, The Smart Card Simulator *http://www.wiley.co.uk/*
[Zeitcontrol]	**Zeitcontrol Cardsystems GmbH, Germany** Smart Cards *http://www.zeitcontrol.de/*
[ZKA]	**Zentraler Kreditausschuß (ZKA), Germany** *http://www.zka.de/*

15.7   Characteristic Values and Tables

15.7.1   ATR interval

Table 15.5   Time interval within which the ATR must be sent following a reset

Clock rate	Minimum time (400 clocks)	Maximum time (40,000 clocks)
1.0000 MHz	0.400 ms	40.000 ms
2.0000 MHz	0.200 ms	20.000 ms
3.0000 MHz	0.133 ms	13.333 ms
3.5712 MHz	0.112 ms	11.201 ms
4.0000 MHz	0.100 ms	10.000 ms
4.9152 MHz	0.081 ms	8.138 ms
5.0000 MHz	0.080 ms	8.000 ms
6.0000 MHz	0.067 ms	6.667 ms
7.0000 MHz	0.057 ms	5.714 ms
8.0000 MHz	0.050 ms	5.000 ms
9.0000 MHz	0.044 ms	4.444 ms
10.0000 MHz	0.040 ms	4.000 ms

15.7.2   ATR data element conversion tables

The following tables are based on the definition of the ATR data elements CWT and BWT in the ISO/IEC 7816-3 standard. The indicated times are for a clock rate of 3.5712 MHz, with three different values for the clock rate conversion factor (divider) F.

Table 15.6 (part 1)   Conversion table for CWT (all times based on a 3.5712-MHz clock)

CWI	CWT	CWT (F = 93)	CWT (F = 186)	CWT (F = 372)
0	12 etu	0.313 ms	0.625 ms	1.250 ms
1	13 etu	0.339 ms	0.677 ms	1.354 ms
2	15 etu	0.391 ms	0.781 ms	1.563 ms
3	19 etu	0.495 ms	0.990 ms	1.979 ms
4	27 etu	0.703 ms	1.406 ms	2.813 ms
5	43 etu	1.120 ms	2.240 ms	4.479 ms
6	75 etu	1.953 ms	3.906 ms	7.813 ms
7	139 etu	3.620 ms	7.240 ms	14.479 ms
8	267 etu	6.953 ms	13.906 ms	27.813 ms
9	523 etu	13.620 ms	27.240 ms	54.479 ms
10	1,035 etu	26.953 ms	53.906 ms	107.813 ms
11	2,059 etu	53.620 ms	107.240 ms	214.479 ms

Table 15.6 (part 2)   Conversion table for CWT

CWI	CWT	CWT (F = 93)	CWT (F = 186)	CWT (F = 372)
12	4,107 etu	106.953 ms	213.906 ms	427.813 ms
13	8,203 etu	213.620 ms	427.240 ms	854.479 ms
14	16,395 etu	426.953 ms	853.906 ms	1,707.813 ms
15	32,779 etu	853.620 ms	1,707.240 ms	3,414.479 ms

Table 15.7   Conversion table for BWT (all times are based on a 3.5712 MHz clock)

BWI	BWT	BWT (F = 93)	BWT (F = 186)	BWT (F = 372)
0	1,011 etu	26.328 ms	52.656 ms	105.313 ms
1	2,011 etu	52.370 ms	104.740 ms	209.479 ms
2	4,011 etu	104.453 ms	208.906 ms	417.813 ms
3	8,011 etu	208.620 ms	417.240 ms	834.479 ms
4	16,011 etu	416.953 ms	833.906 ms	1,667.813 ms
5	32,011 etu	833.620 ms	1,667.240 ms	3,334.479 ms
6	64,011 etu	1,666.953 ms	3,333.906 ms	6,667.813 ms
7	128,011 etu	3,333.620 ms	6,667.240 ms	13,334.479 ms
8	256,011 etu	6,666.953 ms	13,333.906 ms	26,667.813 ms
9	512,011 etu	13,333.620 ms	26,667.240 ms	53,334.479 ms

15.7.3 Determining the data transmission rate

Table 15.8 (part 1)   Common data transmission rates for various clock rates, using standard values for the divider F and the transmission adjustment factor D

Clock rate	F=372, D=1/64 ⇒ 8	F=372, D=1/32 ⇒ 16	F=372, D=1/16 ⇒ 32	F=372, D=1/8 ⇒ 64
1.0000 MHz	125,000 bit/s	62,500 bit/s	31,250 bit/s	15,625 bit/s
2.0000 MHz	250,000 bit/s	125,000 bit/s	62,500 bit/s	31,250 bit/s
3.0000 MHz	375,000 bit/s	187,500 bit/s	93,750 bit/s	46,875 bit/s
3.5712 MHz	446,400 bit/s	223,200 bit/s	111,600 bit/s	55,800 bit/s
4.0000 MHz	500,000 bit/s	250,000 bit/s	125,000 bit/s	62,500 bit/s
4.9152 MHz	614,400 bit/s	307,200 bit/s	153,600 bit/s	76,800 bit/s
5.0000 MHz	625,000 bit/s	312,500 bit/s	156,250 bit/s	78,125 bit/s
6.0000 MHz	750,000 bit/s	375,000 bit/s	187,500 bit/s	93,750 bit/s
7.0000 MHz	875,000 bit/s	437,500 bit/s	218,750 bit/s	109,375 bit/s
8.0000 MHz	1,000,000 bit/s	500.000 bit/s	250,000 bit/s	125,000 bit/s
9.0000 MHz	1,125,000 bit/s	562,500 bit/s	281,250 bit/s	140,625 bit/s
10.0000 MHz	1,250,000 bit/s	625,000 bit/s	312,500 bit/s	156,250 bit/s

Table 15.8 (part 2)   Common data transmission rates for various clock rates

Clock rate	F=372, D=1/4 ⇒ 93	F=372, D=1/2 ⇒ 186	F=372, D=1 ⇒ 372	F=512, D=1 ⇒ 512
1.0000 MHz	10,753 bit/s	5,376 bit/s	2,688 bit/s	1,953 bit/s
2.0000 MHz	21,505 bit/s	10,753 bit/s	5,376 bit/s	3,906 bit/s
3.0000 MHz	32,258 bit/s	16,129 bit/s	8,065 bit/s	5,859 bit/s
3.5712 MHz	38,400 bit/s	19,200 bit/s	9,600 bit/s	6,975 bit/s
4.0000 MHz	43,011 bit/s	21,505 bit/s	10,753 bit/s	7,813 bit/s
4.9152 MHz	52,852 bit/s	26,426 bit/s	13,213 bit/s	9,600 bit/s
5.0000 MHz	53,763 bit/s	26,882 bit/s	13,441 bit/s	9,766 bit/s
6.0000 MHz	64,516 bit/s	32,258 bit/s	16,129 bit/s	11,719 bit/s
7.0000 MHz	75,269 bit/s	37,634 bit/s	18,817 bit/s	13,672 bit/s
8.0000 MHz	86,022 bit/s	43,011 bit/s	21,505 bit/s	15,625 bit/s
9.0000 MHz	96,774 bit/s	48,387 bit/s	24,194 bit/s	17,578 bit/s
10.0000 MHz	107,527 bit/s	53,763 bit/s	26,882 bit/s	19,531 bit/s

15.7.4   Sampling times

Table 15.9 is based on data transfers that comply to the ISO/IEC 7816-3 standard. The indicated times have been calculated for a clock rate of 3.5712 MHz.

Table 15.9 (part 1)   Data transmission sampling times with a divider value of 372

	Start	Lower limit	Midrange	Upper limit	End
Start bit	0 clocks 0.000 µs	112 clocks 31.250 µs	186 clocks 52.083 µs	260 clocks 72.917 µs	372 clocks 104.167 µs
Data bit 1/8	372 clocks 104.167 µs	484 clocks 135.417 µs	558 clocks 156.250 µs	632 clocks 177.083 µs	744 clocks 208.333 µs
Data bit 2/7	744 clocks 208.333 µs	856 clocks 239.583 µs	930 clocks 260.417 µs	1,004 clocks 281.250 µs	1,116 clocks 312.500 µs
Data bit 3/6	1,116 clocks 312.500 µs	1,228 clocks 343.750 µs	1,302 clocks 364.583 µs	1,376 clocks 385.417 µs	1,488 clocks 416.667 µs
Data bit 4/5	1,488 clocks 416.667 µs	1,600 clocks 447.917 µs	1,674 clocks 468.750 µs	1,748 clocks 489.583 µs	1,860 clocks 520.833 µs
Data bit 5/4	1,860 clocks 520.833 µs	1,972 clocks 552.083 µs	2,046 clocks 572.917 µs	2,120 clocks 593.750 µs	2,232 clocks 625.000 µs
Data bit 6/3	2,232 clocks 625.000 µs	2,344 clocks 656.250 µs	2,418 clocks 677.083 µs	2,492 clocks 697.917 µs	2,604 clocks 729.167 µs
Data bit 7/2	2,604 clocks 729.167 µs	2,716 clocks 760.417 µs	2,790 clocks 781.250 µs	2,864 clocks 802.083 µs	2,976 clocks 833.333 µs

Table 15.9 (part 2)   Data transmission sampling times with a divider value of 372

	Start	Lower limit	Midrange	Upper limit	End
Data bit 8/1	2,976 clocks	3,088 clocks	3,162 clocks	3,236 clocks	3,348 clocks
	833.333 µs	864.583 µs	885.417 µs	906.250 µs	937.500 µs
Parity bit	3,348 clocks	3,460 clocks	3,534 clocks	3,608 clocks	3,720 clocks
	937.500 µs	968.750 µs	989.583 µs	1,010.417 µs	1,041.667 µs
Guard time/ Stop bit 1	3,720 clocks	3,832 clocks	3,906 clocks	3,980 clocks	4,092 clocks
	1,041.667 µs	1,072.917 µs	1,093.750 µs	1,114.583 µs	1,145.833 µs
Guard time/ Stop bit 2	4,092 clocks	4,204 clocks	4,278 clocks	4,352 clocks	4,464 clocks
	1,145.833 µs	1,177.083 µs	1,197.917 µs	1,218.750 µs	1,250.000 µs

15.7.5   The most important Smart Card commands

Table 15.10 (part 1)   Summary of standardized Smart Card commands

Command	Function	INS	Standard
APPEND RECORD	Insert a new record in a file with a linear fixed structure. If the file has a cyclic structure, the highest numbered record is replaced.	'E2'	ISO/IEC 7816-4
APPLICATION BLOCK	Reversibly block an application.	'1E'	EMV `96
APPLICATION UNBLOCK	Unblock an application.	'18'	EMV `96
ASK RANDOM	Request a random number from the Smart Card.	'84'	EN 726-3
CHANGE CHV	Change the PIN.	'24'	GSM 11.11
			EN 726-3
CHANGE REFERENCE DATA	Change the PIN.	'24'	ISO/IEC 7816-8
CLOSE APPLICATION	Reset all attained access condition levels.	'AC'	EN 726-3
CONVERT IEP CURRENCY	Convert currency.	'56'	EN 1546-3
CREATE FILE	Create a new file.	'E0'	ISO/IEC 7816-9
			EN 726-3
CREATE RECORD	Create a new record in a record-oriented file.	'E2'	EN 726-3
CREDIT IEP	Load the purse (IEP).	'52'	EN 1546-3
CREDIT PSAM	Pay the IEP against the PSAM.	'72'	EN 1546-3
DEACTIVATE FILE	Reversibly block a file.	'04'	ISO/IEC 7816-9

Table 15.10 (part 2)   Summary of standardized Smart Card commands

Command	Function	INS	Standard
DEBIT IEP	Pay from the purse.	'54'	EN 1546-3
DECREASE	Reduce the value of a counter in a file.	'30'	EN 726-3
DECREASE STAMPED	Reduce the value of a counter in a file that is protected using a cryptographic checksum.	'34'	EN 726-3
DELETE FILE	Delete a file.	'E4'	ISO/IEC 7816-9 EN 726-3
DISABLE CHV	Disable the PIN request.	'26'	GSM 11.11 EN 726-3
DISABLE VERIFICATION REQUIREMENT	Disable the PIN request.	'26'	ISO/IEC 7816-8
ENABLE CHV	Enable the PIN request.	'28'	GSM 11.11 EN 726-3
ENABLE VERIFICATION REQUIREMENT	Enable the PIN request.	'28'	ISO/IEC 7816-8
ENVELOPE PUT / ENVELOPE	Embed another command in a command to provide cryptographic protection.	'C2'	EN 726-3 ISO/IEC 7816-4
ERASE BINARY	Set the contents of a file with a transparent structure to the erased state.	'0E'	ISO/IEC 7816-4
EXECUTE	Execute a file.	'AE'	EN 726-3
EXTEND	Extend a file.	'D4'	EN 726-3
EXTERNAL AUTHENTICATE	Authenticate the outside world with respect to the Smart Card.	'82'	EN 726-3 ISO/IEC 7816-4
GENERATE AUTHORISATION CRYPTOGRAM	Generate a signature for a payment transaction.	'AE'	EMV-2
GENERATE PUBLIC KEY PAIR	Generate a key pair for an asymmetric cryptographic algorithm.	'46'	ISO/IEC 7816-8
GET CHALLENGE	Request a random number from the Smart Card.	'84'	ISO/IEC 7816-4
GET DATA	Read TLV-coded data objects.	'CA'	ISO/IEC 7816-4
GET PREVIOUS IEP SIGNATURE	Repeat the computation and output of the last received IEP signature.	'5A'	EN 1546-3
GET PREVIOUS PSAM SIGNATURE	Repeat the computation and output of the last received PSAM signature.	'86'	EN 1546-3

Table 15.10 (part 3)   Summary of standardized Smart Card commands

Command	Function	INS	Standard
GET RESPONSE	Request data from the Smart Card (used with the T=0 transmission protocol).	'C0'	GSM 11.11 EN 726-3 ISO/IEC 7816-4
GIVE RANDOM	Send a random number to the Smart Card.	'86'	EN 726-3
INCREASE	Increase the value of a counter in a file.	'32'	GSM 11.11 EN 726-3
INCREASE STAMPED	Increase the value of a counter in a file that is protected using a cryptographic checksum.	'36'	EN 726-3
INITIALIZE IEP	Initialize IEP for a subsequent purse command.	'50'	EN 1546-3
INITIALIZE PSAM	Initialize PSAM for a subsequent purse command.	'70'	EN 1546-3
INITIALIZE PSAM for Offline Collection	Initialize PSAM for offline booking of the amount.	'7C'	EN 1546-3
INITIALIZE PSAM for Online Collection	Initialize PSAM for online booking of the amount.	'76'	EN 1546-3
INITIALIZE PSAM for Update	Initialize PSAM for changing the parameters.	'80'	EN 1546-3
INTERNAL AUTHENTICATE	Authenticate the Smart Card with respect to the outside world.	'88'	EN 726-3 ISO/IEC 7816-4
INVALIDATE	Reversibly block a file.	'04'	GSM 11.11 EN 726-3
ISSUER AUTHENTICATE	Verify a signature of the card issuer.	'82'	EMV-2
LOAD KEY FILE	Load keys in files using cryptographic protection.	'D8'	EN 726-3
LOCK	Permanently block a file.	'76'	EN 726-3
MANAGE ATTRIBUTES	Change the access conditions for a file.	tbd	ISO/IEC 7816-9
MANAGE CHANNEL	Control the logical channels of a Smart Card.	'70'	ISO/IEC 7816-4
MANAGE SECURITY ENVIRONMENT	Change the parameters for using cryptographic algorithms in the Smart Card.	'22'	ISO/IEC 7816-8
MUTUAL AUTHENTICATE	Mutually authenticate the Smart Card and the terminal.	'82'	ISO/IEC 7816-8
PERFORM SCQL OPERATION	Execute an SCQL instruction.	'10'	ISO/IEC 7816-7
PERFORM SECURITY OPERATION	Execute a cryptographic algorithm in the Smart Card.	'2A'	ISO/IEC 7816-8
PERFORM TRANSACTION OPERATION	Execute an SCQL transaction instruction.	'12'	ISO/IEC 7816-7

Table 15.10 (part 4)   Summary of standardized Smart Card commands

Command	Function	INS	Standard
PERFORM USER OPERATION	Manage users in the context of SCQL.	'14'	ISO/IEC 7816-7
PSAM COLLECT	Execute PSAM online booking of an amount.	'78'	EN 1546-3
PSAM COLLECT Acknowledgement	End PSAM online booking of an amount.	'7A'	EN 1546-3
PSAM COMPLETE	End paying the IEP against the PSAM.	'74'	EN 1546-3
PSAM VERIFY COLLECTION	End PSAM offline booking of an amount.	'7E'	EN 1546-3
PUT DATA	Write TLV-coded data objects.	'DA'	ISO/IEC 7816-4
REACTIVATE FILE	Unblock a file.	'44'	ISO/IEC 7816-9
READ BINARY	Read from a file with a transparent structure.	'B0'	GSM 11.11 EN 726-3 ISO/IEC 7816-4
READ BINARY STAMPED	Read from a file with a transparent structure that is protected using a cryptographic checksum.	'B4'	EN 726-3
READ RECORD / READ RECORD(S)	Read from a file with a linear fixed, linear variable or cyclic structure.	'B2'	GSM 11.11 EN 726-3 ISO/IEC 7816-4
READ RECORD STAMPED	Read from a file with a linear fixed, linear variable or cyclic structure that is protected using a cryptographic checksum.	'B6'	EN 726-3
REHABILITATE	Unblock a file.	'44'	GSM 11.11 EN 726-3
RESET RETRY COUNTER	Reset a retry counter.	'2C'	ISO/IEC 7816-8
RUN GSM ALGORITHM	Execute a GSM-specific cryptographic algorithm.	'88'	GSM 11.11
SEARCH BINARY	Search for a text in a file with a transparent structure.	'A0'	ISO/IEC 7816-9
SEARCH RECORD	Search for a text in file with a linear fixed, linear variable or cyclic structure.	'A2'	ISO/IEC 7816-9
SEEK	Search for a text in file with a linear fixed, linear variable or cyclic structure.	'A2'	GSM 11.11 EN 726-3
SELECT / SELECT (FILE)	Select a file.	'A4'	GSM 11.11 EN 726-3 ISO/IEC 7816-4
SLEEP	Place the Smart Card in a power-saving mode (command no longer used).	'FA'	GSM 11.11

Table 15.10 (part 5)   Summary of standardized Smart Card commands

Command	Function	INS	Standard
STATUS	Read various data from the currently selected file.	'F2'	GSM 11.11 EN 726-3
TERMINATE CARD USAGE	Permanently block a Smart Card.	tbd	ISO/IEC 7816-9
TERMINATE DF	Permanently block a DF.	tbd	ISO/IEC 7816-9
UNBLOCK CHV	Reset a PIN retry counter that has reached its maximum value.	'2C'	GSM 11.11 EN 726-3
UPDATE BINARY	Write to a file with a transparent structure.	'D6'	GSM 11.11 EN 726-3 ISO/IEC 7816-4
UPDATE IEP PARAMETER	Change the general parameters of a purse.	'58'	EN 1546-3
UPDATE PSAM Parameter (offline)	Execute the offline modification of the parameters in the PSAM.	'84'	EN 1546-3
UPDATE PSAM Parameter (online)	Execute the online modification of the parameters in the PSAM.	'82'	EN 1546-3
UPDATE RECORD	Write to a file with a linear fixed, linear variable or cyclic structure.	'DC'	GSM 11.11 EN 726-3 ISO/IEC 7816-4
VERIFY	Verify the received data (such as a PIN).	'20'	ISO/IEC 7816-4 EMV-2
VERIFY CHV	Verify the PIN.	'20'	GSM 11.11 EN 726-3
WRITE BINARY	Write to a file with a transparent structure using a logical AND/OR process.	'D0'	EN 726-3 ISO/IEC 7816-4
WRITE RECORD	Write to a file with a linear fixed, linear variable or cyclic structure using a logical AND/OR process.	'D2'	EN 726-3 ISO/IEC 7816-4

15.7.6   Summary of utilized instruction bytes

Tables 15.11 through 15.13 identify the INS codes that are used with various class bytes. The odd-numbered codes in the shaded columns may not be used to encode commands, since the T=0 transfer protocol uses these codes to control the programming voltage.[1]

The following symbols are used in these tables:

⊛	EMV-2	✳	GSM 11.11	☆	ISO/IEC 7816-8
✛	EN 726-3	✪	ISO/IEC 7816-4	✻	ISO/IEC 7816-9
✢	EN 1546-3	✭	ISO/IEC 7816-7		

---

1   See also Section 6.4.2, 'The T=0 transfer protocol'.

Table 15.11   INS byte codes used with a class byte (CLA) of '00'.

	x0	x1	x2	x3	x4	x5	x6	x7	x8	x9	xA	xB	xC	xD	xE	xF
0y															✪	
1y	★		★		★											
2y	✪		☆		☆		☆		☆		☆		☆			
3y																
4y	❋				❋		☆									
5y																
6y																
7y	✪															
8y			✪		✪				✪							
9y																
Ay	❋		❋		✪											
By	✪		✪													
Cy	✪		✪								✪					
Dy	✪		✪				✪				✪			✪		
Ey	❋		✪		❋											
Fy																

Table 15.12   INS byte codes used with a class byte (CLA) of '80'.

	x0	x1	x2	x3	x4	x5	x6	x7	x8	x9	xA	xB	xC	xD	xE	xF
0y																
1y																
2y																
3y																
4y																
5y	✤		✤		✤		✤				✤					
6y																
7y	✤		✤		✤		✤		✤		✤		✤		✤	
8y	✤		✤ ⊗		✤		✤									
9y																
Ay															⊗	
By																
Cy																
Dy																
Ey																
Fy																

Table 15.13   INS byte codes used with a class byte (CLA) of 'A0'.

	x0	x1	x2	x3	x4	x5	x6	x7	x8	x9	xA	xB	xC	xD	xE	xF
0y					✛ ✳											
1y																
2y	✛				✛		✛		✛				✛			
3y	✛		✛ ✳		✛		✛									
4y					✛ ✳											
5y																
6y																
7y							✛									
8y			✛		✛		✛		✛ ✳							
9y																
Ay			✛ ✳		✛ ✳											
By	✛ ✳		✛ ✳		✛		✛									
Cy	✛ ✳		✛													
Dy	✛		✛		✛		✛ ✳		✛				✛ ✳			
Ey	✛		✛		✛											
Fy			✛ ✳									✳				

15.7.7   Smart Card command encoding

Tables 15.14 through 15.19 show the most important codes for some sample Smart Card commands. For the sake of clarity, it is assumed that neither secure messaging nor addressing by logical channels is used. Refer to the ISO/IEC 7816-4 standard for the complete coding of these and other Smart Card commands.[2]

Table 15.14   Codes for the Case 1 and Case 3 command SELECT FILE, with the major options

Data element	Code	Remark	
CLA	'00'	Class byte reserved for ISO/IEC 7816 commands	
INS	'A4'	Instruction byte for SELECT FILE, which is the command for selecting a file (MF, DF or EF)	
P1	X	$X = $ '00' $\wedge Z =$ Read $\wedge W = $ Read	Select the MF
		$X = $ '00'	Select a file using its FID
		$X = $ '04'	Select a file using its DF name
P2	'00'	---	
Lc	Z	$X = $ '00', $Z = 2$	Select a file using its FID
		$X = $ '04', $Z = 1 \dots 16$	Select a file using its DF name
DATA	W	$X = $ '00', $W = $ FID	Select a file using its FID
		$X = $ '04', $W = $ DF name	Select a file using its DF name

---

2   See also Section 6.5.1, 'Structure of the command APDU'.

Table 15.15   Codes for the Case 2 command READ BINARY according to ISO/IEC 7816-4, with the most important options

Data element	Code	Remark		
CLA	'00'	Class byte reserved for ISO/IEC 7816 commands.		
INS	'B0'	Instruction byte for READ BINARY, which is the command for reading data from a file with a transparent structure.		
P1	...	P1.b8 = 0	Read data from the currently selected file with an offset; offset = (P1.b7 ... P1.b1 ‖ P2).
		P1.b8 = 1	Read data following implicit file selection using a Short FID with an offset; Short FID = (P1.b5 ... P1.b1), offset = P2.
P2	...	P1.b8 = 0	Part of the offset the data to be read.
		P1.b8 = 1	The offset to the data to be read.
Le	Z	Z = 0:	Read all data until the end of the file.
		Z > 0:	Z is the number of bytes to be read.

Table 15.16   Codes for the Case 3 command UPDATE BINARY according to ISO/IEC 7816-4, with the most important options

Data element	Code	Remark		
CLA	'00'	Class byte reserved for ISO/IEC 7816 commands.		
INS	'D6'	Instruction byte for UPDATE BINARY, which is the command for writing data to a file with a transparent structure.		
P1	...	P1.b8 = 0	Write data to the currently selected file with an offset; offset = (P1.b7 ... P1.b1 ‖ P2).
		P1.b8 = 1	Write data following implicit file selection using a Short FID with an offset; Short FID = (P1.b5 ... P1.b1), offset = P2.
P2	...	P1.b8 = 0	Part of the offset to data to be written.
		P1.b8 = 1	The offset to the data to be written.
Lc	Z	Z is the number of bytes to be written.		
DATA	...	The bytes to be written.		

Table 15.17 (part 1)   Codes for the Case 2 command READ RECORD according to ISO/IEC 7816-4, with the most important options

Data element	Code	Remark	
CLA	'00'	Class byte reserved for ISO/IEC 7816 commands.	
INS	'B2'	Instruction byte for READ RECORD, which is the command for reading data from a file with a record-oriented structure.	
P1	...	P1.b3 = °0°	P1 contains a record number.
		P1.b3 = °1°	P1 contains a record identifier.

Table 15.17 (part 2) Codes for the Case 2 command READ RECORD

P2	...	P2.b8 ... P2.b4 = °00000°	Read data from the currently selected file.
		P2.b8 ... P2.b4 ≠ °00000°	Read data after implicit file selection with a Short FID; Short FID = (P1.b8 ... P1.b4)
		P2.b3 ... P2.b1 = °000°	Read the first record with the record identifier passed via P1.
		P2.b3 ... P2.b1 = °001°	Read the last record with the record identifier passed via P1.
		P2.b3 ... P2.b1 = °010°	Read the next record with the record identifier passed via P1.
		P2.b3 ... P2.b1 = °011°	Read the previous record with the record identifier passed via P1.
		P2.b3 ... P2.b1 = °100° ∧ P1 = 0	Read the current record.
		P2.b3 ... P2.b1 = °100° ∧ P1 ≠ 0	Read the record with the record identifier passed via P1.
		P2.b3 ... P2.b1 = °101°	Read all records from the record number passed via P1 until the end of the file.
		P2.b3 ... P2.b1 = °110°	Read all records from the end of the file until the record number passed via P1.
Le	Z	Z = 0:	Read all bytes until the end of the record.
		Z > 0:	Z is the length of the record.

Table 15.18 Codes for the Case 3 command UPDATE RECORD according to ISO/IEC 7816-4, with the most important options

Data element	Code	Remark	
CLA	'00'	Class byte reserved for ISO/IEC 7816 commands.	
INS	'DC'	Instruction byte for UPDATE RECORD, which is command for writing data to a file with a record-oriented structure.	
P1	X	P1 = 0	Write the current record.
		P1 ≠ 0	Write the record whose record number is passed via P1.
P2	...	P2.b8 ... P2.b4 = °00000°	Write data in the currently selected file.
		P2.b8 ... P2.b4 ≠ °00000°	Write data following implicit file selection using the Short FID; Short FID = (P1.b8 ... P1.b4).
		P2.b3 ... P2.b1 = °000°	Write the first record.
		P2.b3 ... P2.b1 = °001°	Write the last record.
		P2.b3 ... P2.b1 = °010°	Write the next record.
		P2.b3 ... P2.b1 = °011°	Write the previous record.
		P2.b3 ... P2.b1 = °100°	Write the record whose record number is passed via P1.
Lc	...	Length of the record to be written.	
DATA	...	The record to be written.	

Table 15.19   Codes for the Case 3 command VERIFY according to ISO/IEC 7816-4, with the most important options

Data element	Code	Remark	
CLA	'00'	Class byte reserved for ISO/IEC 7816 commands.	
INS	'20'	Instruction byte for VERIFY, which is the command for comparing passed-in data to reference data (typically a PIN).	
P1	'00'	---	
P2	...	P2 = '00'	No explicit information is transferred.
		P2.b8 = °1°	Reference data that is valid for the entire Smart Card (global reference data) is used.
		P2.b8 = °1°	Reference data that is valid for one or more applications (specific reference data) is used.
		P2.b5 ... P2.b1	Reference data identification number.
Lc	...	Length of the transferred comparison value.	
DATA	...	The transferred comparison value (usually a PIN)	

15.7.8   Smart Card return codes

The return codes described in Table 15.20 are classified according to the scheme used in the ISO/IEC 7816-4 standard.[3] The following status codes are used:

NP: process completed, normal processing   EE: process aborted, execution error
WP: process completed, warning processing   CE: process aborted, checking error

Table 15.20 (part 1)   Selected standardized Smart Card return codes according to ISO/IEC 7816-4

Return code	Status	Meaning	Standard
'61xx'	NP	Only with T=0: 'xx' bytes of data can still be requested using a GET RESPONSE command.	ISO/IEC 7816-4
'6281'	WP	The returned data may be faulty.	ISO/IEC 7816-4
'6282'	WP	Fewer bytes than specified by the Le parameter could be read, since the end of the file was encountered first.	ISO/IEC 7816-4
'6283'	WP	The selected file is reversibly blocked.	ISO/IEC 7816-4
'6284'	WP	The file control information (FCI) is not formatted.	ISO/IEC 7816-4
'63Cx'	WP	The counter has reached the value 'x' $(0 \leq x \leq 15)$ (the exact significance of this depends on the command).	ISO/IEC 7816-4
'6581'	EE	Memory error (e.g. with a write operation).	ISO/IEC 7816-4

---

3   See also Section 6.5.2, 'Structure of the response APDU'.

Table 15.20 (part 2)   Selected standardized Smart Card return codes

Return code	Status	Meaning	Standard
'6700'	CE	Length incorrect.	GSM 11.11 EN 726-3 ISO/IEC 7816-4
'6800'	CE	Functions in the class byte .not supported (general).	ISO/IEC 7816-4
'6881'	CE	Logical channels not supported.	ISO/IEC 7816-4
'6882'	CE	Secure messaging not supported.	ISO/IEC 7816-4
'6900'	CE	Command not allowed (general)	ISO/IEC 7816-4
'6981'	CE	Command incompatible with file structure.	ISO/IEC 7816-4
'6982'	CE	Security state not satisfied.	ISO/IEC 7816-4
'6983'	CE	Authentication method blocked.	ISO/IEC 7816-4
'6985'	CE	Use conditions not satisfied.	ISO/IEC 7816-4
'6986'	CE	Command not allowed (no EF selected).	ISO/IEC 7816-4
'6A00'	CE	Incorrect P1 or P2 parameters (general).	ISO/IEC 7816-4
'6A80'	CE	Parameters in the data portion are incorrect.	ISO/IEC 7816-4
'6A81'	CE	Function not supported.	ISO/IEC 7816-4
'6A82'	CE	File not found.	ISO/IEC 7816-4
'6A83'	CE	Record not found.	ISO/IEC 7816-4
'6A84'	CE	Insufficient memory.	ISO/IEC 7816-4
'6A87'	CE	Lc inconsistent with P1 or P2.	ISO/IEC 7816-4
'6A88'	CE	Data referenced by command not found.	ISO/IEC 7816-4
'6B00'	CE	Parameter 1 or 2 incorrect.	GSM 11.11 EN 726-3 ISO/IEC 7816-4
'6D00'	CE	Command (instruction) not supported.	GSM 11.11 EN 726-3 ISO/IEC 7816-4
'6E00'	CE	Class not supported.	GSM 11.11 EN 726-3 ISO/IEC 7816-4
'6F00'	CE	Command aborted – more exact diagnosis not possible (e.g. operating system error).	GSM 11.11 EN 726-3 ISO/IEC 7816-4

Table 15.20 (part 3)   Selected standardized Smart Card return codes

Return code	Status	Meaning	Standard
'9000'	NP	Command successfully executed.	GSM 11.11 EN 726-3 ISO/IEC 7816-4
'920x'	NP	Writing to EEPROM successful after 'x' attempts.	GSM 11.11 EN 726-3
'9210'	CE	Insufficient memory.	GSM 11.11 EN 726-3
'9240'	EE	Writing to EEPROM not successful.	GSM 11.11 EN 726-3
'9400'	CE	No EF selected.	GSM 11.11 EN 726-3
'9402'	CE	Address boundaries exceeded.	GSM 11.11 EN 726-3
'9404'	CE	FID not found, record not found, comparison pattern not found.	GSM 11.11 EN 726-3
'9408	CE	Selected file type does not fit with command.	GSM 11.11 EN 726-3
'9802'	CE	No PIN defined.	GSM 11.11 EN 726-3
'9804'	CE	Access conditions not satisfied, authentication failed.	GSM 11.11 EN 726-3
'9835'	CE	ASK RANDOM or GIVE RANDOM not executed.	GSM 11.11 EN 726-3
'9840'	CE	PIN verification not successful.	GSM 11.11 EN 726-3
'9850'	CE	INCREASE or DECREASE could not be executed because a limit has been reached.	GSM 11.11 EN 726-3
'9Fxx'	NP	Command successfully executed, and 'xx' bytes of data are available to be sent.	GSM 11.11 EN 726-3

15.7.9   Selected chips for memory cards

Table 15.21 lists a selection of various types of memory chips for Smart Cards. Similar components are also available from other manufacturers.

You should bear in mind that this information is for reference only and is subject to change, due to technical progress and market forces. Before deciding to use a particluar type of chip, you should always consult the manufacturer.

Table 15.21 (part 1)   Summary of selected memory chips for Smart Cards

Manufacturer and type	Memory capacity		Additional information	
Philips	ROM:	---	Vcc:	5 V
PCB 2032	PROM:	---	Icc:	3 mA
	EEPROM:	256 bytes	W/E cycles:	10,000
			W/E time:	5 ms
			HW:	---
Philips	ROM:	16 bits	Vcc:	5 V
PCB 7960	PROM:	48 bits	Icc:	3 mA
	EEPROM:	40 bits	W/E cycles:	10,000
			W/E time:	5 ms
			HW:	104-bit counter
ST Microelectronics	ROM:	---	Vcc:	3–5.5 V
ST14C02C	PROM:	---	Icc:	2 mA
	EEPROM:	2 kbits	W/E cycles:	1,000,000
			W/E time:	10 ms
			HW:	I^2C bus interface
ST Microelectronics	ROM:	---	Vcc:	3–5.5 V
ST14C04C	PROM:	---	Icc:	2 mA
	EEPROM:	4 kbits	W/E cycles:	1,000,000
			W/E time:	10 ms
			HW:	I^2C bus interface
ST Microelectronics	ROM:	---	Vcc:	2.7–5.5 V
ST15E32F	PROM:	---	Icc:	2 mA
	EEPROM:	32 kbits	W/E cycles:	100,000
			W/E time:	10 ms
			HW:	I^2C bus interface
ST Microelectronics	ROM:	---	Vcc:	5 V
ST1305	PROM:	---	Icc:	2 mA
	EEPROM:	192 bits	W/E cycles:	1,000,000
			W/E time:	5 ms
			HW:	counter for 262,144 units

Table 15.21 (part 2)   Summary of selected memory chips for Smart Cards

Manufacturer and type	Memory capacity		Additional information	
ST Microelectronics ST1333	ROM:	16 bits	Vcc:	5 V
	PROM:	---	Icc:	2 mA
	EEPROM:	272 bits	W/E cycles:	1,000,000
			W/E time:	5 ms
			HW:	counter for 32,767 units; unilateral authentication
Infineon SLE 4404	ROM:	16 bits	Vcc:	5 V
	PROM:	144 bits	Icc:	1.5 mA
	EEPROM:	256 bits	W/E cycles:	100,000
			W/E time:	5 ms
			HW:	---
Infineon SLE 4406	ROM:	16 bits	Vcc:	5 V
	PROM:	56 bits	Icc:	1.5 mA
	EEPROM:	32 bits	W/E cycles:	10,000
			W/E time:	5 ms
			HW:	counter for $\approx$ 20,000 units
Infineon SLE 44R35	ROM:	---	Vcc:	5 V
	PROM:	---	Icc:	3 mA
	EEPROM:	1 kB	W/E cycles:	100,000
			W/E time:	2 ms
			HW:	PIN logic for additional write protection; unilateral authentication; for contactless memory cards
Infineon SLE 5536	ROM:	16 bits	Vcc:	5 V
	PROM:	185 bits	Icc:	2.5 mA
	EEPROM:	36 bits	W/E cycles:	100,000
			W/E time:	3 ms
			HW:	counter for $\approx$ 20,000 units, unilateral authentication
Infineon SLE 4442	ROM:	---	Vcc:	5 V
	PROM:	32 bits	Icc:	3 mA
	EEPROM:	256 bytes	W/E cycles:	100,000
			W/E time:	2.5 ms
			HW:	---

15.7.10   Selected microcontrollers for Smart Cards

Table 15.22 lists a selection of various types of microcontrollers for Smart Cards. Similar components are also available from various other manufacturers.

You should bear in mind that this information is for reference only and is subject to change, due to technical progress and market forces. Before deciding to use a particluar type of chip, you should always consult the manufacturer.

The following abbreviations are used in the table:

Vcc:	Supply voltage range
Clock:	Clock frequency range
Icc:	Current consumption of the chip at the indicated frequency (first value: normal operation; second value: low-power mode with clock; third value: low-power mode with no clock)
Size:	Size of the die
Structure:	minimum structure width on the chip
Page:	Size of one EEPROM page (for write accesses)
W/E cycles:	guaranteed number of write/erase cycles per EEPROM page
W/E time:	Cycle time for writing or erasing one EEPROM page
HW:	additional hardware on the chip
#	information not available

Table 15.22 (part 1)   Summary of selected microcontrollers for Smart Cards

Manufacturer and type	CPU	Memory capacity		Additional information	
Atmel AT90 SC1618A	8051	Flash:	16 kB	Vcc:	2.7–5.5 V
		EEPROM:	8 kB	Clock:	1–10 MHz
		RAM:	512 bytes	Icc:	#; 200 µA (sleep)
				Size:	#
				Structure:	0.35 µm
				EEPROM	
				Page:	1–64 bytes
				W/E cycles:	100,000
				W/E time:	3.5 ms
				Flash EEPROM	
				Page:	64 bytes
				W/E cycles:	1 000
				W/E time:	1.5 ms
				HW:	RISC CPU; random number generator; two 16-bit timers

Table 15.22 (part 2)   Summary of selected microcontrollers for Smart Cards

Manufacturer and type	CPU	Memory capacity		Additional information	

Atmel AT90 SC1616C	AVR	Flash:
		EEPROM:
		RAM:
			16 kB
			16 kB
			1 kB
				Vcc:
				Clock:
				Icc:
				Size:
				Structure:
				EEPROM
				  Page:
				  W/E cycles:
				  W/E time:
				Flash EEPROM
				  Page:
				  W/E cycles:
				  W/E time:
				HW:
					2.7–5.5 V
					1–5 MHz
					#; 200 µA (sleep)
					#
					0.35 µm
					
					1–64 bytes
					100,000
					3.5 ms
					
					64 bytes
					1,000
					1.5 ms
					RISC CPU; NPU; random number generator; two 16-bit timers

Atmel AT90 SC3232C	AVR	Flash:
		EEPROM:
		RAM:
			32 kB
			32 kB
			1 kB
				Vcc:
				Clock:
				Icc:
				Size:
				Structure:
				EEPROM
				  Page:
				  W/E cycles:
				  W/E time:
				Flash EEPROM
				  Page:
				  W/E cycles:
				  W/E time:
				HW:
					2.7–5.5 V
					1–5 MHz
					#; 200 µA (sleep)
					#
					0.35 µm
					
					1–64 bytes
					100,000
					3.5 ms
					
					64 bytes
					1000
					1.5 ms
					RISC CPU; NPU; random number generator; two 16-bit timers

Hitachi H8/3153	H8	ROM: EEPROM: RAM:	32 kB 16 kB 1024 bytes	Vcc: Clock: Icc: Size: Structure: Page: W/E cycles: W/E time: HW:	2.7–5.5 V 1–10 MHz or 1–5 MHz # # 0.8 µm 64 bytes 100,000 5 ms RISC-like CPU; 2 I/O

Table 15.22 (part 3)   Summary of selected microcontrollers for Smart Cards

Manufacturer and type	CPU	Memory capacity		Additional information	
Hitachi H8/3112	H8	ROM:	24 kB	Vcc:	2.7–5.5 V
		EEPROM:	8 kB	Clock:	1–10 MHz or 1–5 MHz
		RAM:	1056 bytes	Icc:	#
				Size:	#
				Structure:	0.8 µm
				Page:	1–32 bytes
				W/E cycles:	100,000
				W/E time:	5 ms
				HW:	RISC-like CPU; 2 I/O; NPU
Hitachi H8/3114	H8	ROM:	32 kB	Vcc:	2.7–5.5 V
		EEPROM:	16 kB	Clock:	1–10 MHz or 1–5 MHz
		RAM:	2 kB	Icc:	#
				Size:	#
				Structure:	0.5 µm
				Page:	1–32 bytes
				W/E cycles:	10,000
				W/E time:	max.15 ms
				HW:	RISC-like CPU; two I/O; NPU
Motorola MC 68HC05 SC24	6805	ROM:	3 kB	Vcc:	5.0 V ± 10 %
		EEPROM:	1 kB	Clock:	1–5 MHz
		RAM:	128 bytes	Icc:	2.5 mA at 5 MHz; 200 µA
				Size:	10.4 mm^2 (2.8 mm × 3.7 mm)
				Structure:	1.2 µm
				Page:	4 bytes
				W/E cycles:	10,000
				W/E time:	10 ms
				HW:	five I/O
Motorola MC 68HC05 SC27	6805	ROM:	16 kB	Vcc:	3–5.5 V
		EEPROM:	3 kB	Clock:	1–5 MHz
		RAM:	240 bytes	Icc:	5 mA/5 MHz, 500 µA, 50 µA
				Size:	21 mm^2 (4.2 mm × 5.0 mm)
				Structure:	1.2 µm
				Page:	4 bytes
				W/E cycles:	10,000
				W/E time:	10 ms
				HW:	five I/O

Table 15.22 (part 4)   Summary of selected microcontrollers for Smart Cards

Manufacturer and type	CPU	Memory capacity		Additional information	
Motorola MC 68HC05 SC27	6805	ROM:	16 kB	Vcc:	3–5.5 V
		EEPROM:	3 kB	Clock:	1–5 MHz
		RAM:	240 bytes	Icc:	5 mA/5 MHz, 500 µA, 50 µA
				Size:	21 mm^2 (4.2 mm × 5.0 mm)
				Structure:	1.2 µm
				Page:	4 bytes
				W/E cycles:	10,000
				W/E time:	10 ms
				HW:	five I/O
Motorola MC 68HC05 SC28	6805	ROM:	12.5 kB	Vcc:	3–5.5 V
		EEPROM:	8 kB	Clock:	1–5 MHz
		RAM:	240 bytes	Icc:	5 mA at 5 MHz; 50 µA
				Size:	26 mm^2 (4.9 mm × 5.3 mm)
				Structure:	1.2 µm
				Page:	4 bytes
				W/E cycles:	10,000
				W/E time:	10 ms
				HW:	five I/O
Motorola MC 68HC05 SC29	6805	ROM:	12.8 kB	Vcc:	5.0 V ± 10 %
		EEPROM:	4 kB	Clock:	1–5 MHz
		RAM:	512 bytes	Icc:	#
				Size:	26 mm^2
				Structure:	#
				Page:	4 bytes
				W/E cycles:	100,000
				W/E time:	10 ms
				HW:	NPU; PLL (only for NPU); five I/O
Philips P83W86x	8051	ROM: 20; 20; 20; 28 kB EEPROM: 2; 4; 8; 16 kB (= x) RAM: 256; 256; 384; 512 bytes		Vcc:	2.7–5.5 V
				Clock:	1–8 MHz
				Icc:	#
				Size:	#
				Structure:	#
				Page:	1–32 bytes
				W/E cycles:	100,000
				W/E time:	4.5 ms
				HW:	two I/O; two 16-bit timers; random number generator; internal timing for EEPROM writing

Table 15.22 (part 5)   Summary of selected microcontrollers for Smart Cards

Manufacturer and type	CPU	Memory capacity	Additional information	
Philips P83W85x	8051	ROM: 20; 20; 28 kB EEPROM: 4; 8; 16 kB (= x) RAM: 800; 800; 1024 bytes	Vcc: Clock: Icc: Size: Structure: Page: W/E cycles: W/E time: HW:	2.7–5.5 V 1–8 MHz # # # 1–32 bytes 100,000 4.5 ms two I/O; two 16-bit timers; random number generator; internal timing for EEPROM writing; NPU
Philips SmartXA	8051	ROM: 32 kB EEPROM: 8; 16; 32 kB (= x) RAM: 1568; 1568; 2048 bytes	Vcc: Clock: Icc: Size: Structure: Page: W/E cycles: W/E time: HW:	2.7–5.5 V 1–8 MHz # # # 1–32 bytes 100,000 4.5 ms two I/O; two 16-bit timers; random number generator; internal timing for EEPROM writing; NPU
ST Microelectronics ST16SF4x	6805	ROM: 16 kB EEPROM: 1.2; 2; 4; 8; 16 kB (x = 1, 2, 4, 8, F) RAM: 384 bytes	Vcc: Clock: Icc: Size: Structure: Page: W/E cycles: W/E time: HW:	2.7–5.5 V 1–5 MHz 10 mA at 5 MHz; # # 1 μm 1–32 bytes 100,000 2.5 ms two I/O; simple MMU
ST Microelectronics ST 16CF54B	6805	ROM: 16 kB EEPROM: 4 kB RAM: 512 bytes	Vcc: Clock: Icc: Size: Structure: Page: W/E cycles: W/E time: HW:	4.5–5.5 V 5 MHz 10 mA at 5 MHz; # # 1 μm 1–32 bytes 100,000 2.5 ms NPU; two I/O; simple MMU; PLL for NPU

Table 15.22 (part 6)   Summary of selected microcontrollers for Smart Cards

Manufacturer and type	CPU	Memory capacity		Additional information	
Infineon SLE 44C10S	8051	ROM:	7 kB	Vcc:	2.7–5.5 V
		EEPROM:	1 kB	Clock:	1–7.5 MHz
		RAM:	256 bytes	Icc:	10 mA at 5 MHz; 100 µA
				Size:	6.1 mm^2 (2.62 mm × 2.31 mm)
				Structure:	0.8 µm
				Page:	1–16 bytes
				W/E cycles:	100,000
				W/E time:	3.6 ms
				HW:	---
Infineon SLE 44C42S	8051	ROM:	15 kB	Vcc:	2.7–5.5 V
		EEPROM:	4 kB	Clock:	1–7.5 MHz
		RAM:	256 bytes	Icc:	8 mA at 5 MHz; 100 µA
				Size:	8.1 mm^2 (2.66 mm × 3.06 mm)
				Structure:	0.8 µm
				Page:	1–16 bytes
				W/E cycles:	100,000
				W/E time:	3.6 ms
				HW:	---
Infineon SLE 44C80S	8051	ROM:	15 kB	Vcc:	2.7–5.5 V
		EEPROM:	8 kB	Clock:	1–7.5 MHz
		RAM:	256 bytes	Icc:	8 mA at 5 MHz; 100 µA
				Size:	10 mm^2 (2.66 mm × 3.76 mm)
				Structure:	0.8 µm
				Page:	1–32 bytes
				W/E cycles:	100,000
				W/E time:	3.6 ms
				HW:	---
Infineon SLE 44C160S	8051	ROM:	15 kB	Vcc:	2.7–5.5 V
		EEPROM:	16 kB	Clock:	1–7.5 MHz
		RAM:	256 bytes	Icc:	8 mA at 5 MHz; 100 µA
				Size:	10 mm^2 (2.66 mm × 3.76 mm)
				Structure:	0.8 µm
				Page:	1–32 bytes
				W/E cycles:	500,000
				W/E time:	3.6 ms
				HW:	---

Table 15.22 (part 7)   Summary of selected microcontrollers for Smart Cards

Manufacturer and type	CPU	Memory capacity		Additional information	
Infineon SLE 44CR80S	8051	ROM: EEPROM: RAM:	15 kB 8 kB 706 bytes	Vcc: Clock: Icc: Size: Structure: Page: W/E cycles: W/E time: HW:	2.7–5.5 V 1–7.5 MHz 10 mA at 5 MHz; 100 µA 10 mm^2 (2.66 mm × 3.76 mm) 0.8 µm 1–32 bytes 100,000 3.6 ms NPU
Infineon SLE 66C160S	8051 derivative	ROM: EEPROM: RAM:	32 kB 16 kB 1280 bytes	Vcc: Clock: Icc: Size: Structure: Page: W/E cycles: W/E time: HW:	2.7–5.5 V 1–7.5 MHz 5 mA at 5 MHz, 100 µA 21 mm^2 0.6 µm 1–64 bytes 500,000 3.6 ms random number generator; timer; CRC processing unit
Infineon SLE 66CX160S	8051 derivative	ROM: EEPROM: RAM:	32 kB 16 kB 1980 bytes	Vcc: Clock: Icc: Size: Structure: Page: W/E cycles: W/E time: HW:	2.7–5.5 V 1–7.5 MHz 10 mA at 5 MHz; 100 µA 21 mm^2 0.6 µm 1–64 bytes 500,000 3.6 ms random number generator; timer; CRC processing unit; NPU

Index